STRING THEORY, QUANTUM GRAVITY AND THE UNIFICATION OF THE FUNDAMENTAL INTERACTIONS

STRING THEORY, QUANTUM GRAVITY AND THE UNIFICATION OF THE FUNDAMENTAL INTERACTIONS

Edited by

Massimo Bianchi
Francesco Fucito
Enzo Marinari
Augusto Sagnotti
Dipartimento di Fisica
Università di Roma "Tor Vergata"

World Scientific
Singapore • New Jersey • London • Hong Kong

Published by

World Scientific Publishing Co. Pte. Ltd.
5 Toh Tuck Link, Singapore 596224
USA office: 27 Warren Street, Suite 401-402, Hackensack, NJ 07601
UK office: 57 Shelton Street, Covent Garden, London WC2H 9HE

British Library Cataloguing-in-Publication Data
A catalogue record for this book is available from the British Library.

STRING THEORY, QUANTUM GRAVITY AND THE UNIFICATION OF THE FUNDAMENTAL INTERACTIONS

ISBN-13 978-981-02-1161-5
ISBN-10 981-02-1161-9

Foreword

Within the E.C. Contract SC1-0934-C, the University of Rome "Tor Vergata" organized a Meeting for all the Institutions in this European Collaboration. A limited number of experts from Institutions outside the Collaboration also took part in the Meeting, which was held in Rome, at the "Accademia dei Lincei", on September 21–26, 1992.

During the past decade, the emergence of a widespread interest in String Theory brought a number of experts from different areas of Theoretical Physics into close contact with this fascinating issue. The convergence of ideas and techniques from different fields, often stimulated by the wide applicability of many of the results, led to a rapid progress in several areas not directly related to Particle Physics. On the other hand, more traditional applications of String Theory to hadronic physics did not lie at the center of attention for a while. Presently, after a number of exciting developments in Two-Dimensional Conformal Field Theory, Random Surfaces and Quantum Gravity, one is facing a progressive dispersion of the traditional lines of research in this area. The diverse nature of the contributions will be apparent to the reader of this volume. Indeed, the lectures deal with low-energy manifestations of String Theory, with some mathematical offsprings of Two-Dimensional Conformal Field Theory, with non-critical models, with Liouville theory, with the Random-Surface approach, as well as with attempts to clarify black hole physics and non-perturbative gauge theories using string-inspired models.

It is a pleasure to thank the European Community Science Committee for making this Meeting possible. We are also grateful to the University of Rome "Tor Vergata", to the University of Rome "La Sapienza" and to I.N.F.N. for providing partial support, as well as to the "Accademia Nazionale dei Lincei" for hosting the Meeting. Finally, we would like to thank Prof. Paul Windey for proposing this fruitful Collaboration, Prof. Gianni Jona-Lasinio for much help and advice, and our Scientific Secretaries Liù Catena and Rita D'Ottavi for their warm and efficient cooperation.

On behalf of all the participants, we are happy to dedicate this volume to Prof. François Englert, on the occasion of his sixtieth birthday. Prof. Paul Windey, Head of our Collaboration and a former student of Prof. Englert, kindly wrote the brief scientific profile that follows.

Roma, June 1992

Massimo BIANCHI
Francesco FUCITO
Enzo MARINARI
Augusto SAGNOTTI

François Englert
Courtesy of Arie Mandelbaum

A BRIEF PROFILE OF FRANÇOIS ENGLERT

Clearly this is not the place to review in detail the many major contributions that François has given to Theoretical Physics. After all, we decided not to devote a long session to the history of his accomplishments, for we thought he might be more pleased with an offering of current research papers than with recollections of past things, whatever their importance to modern thought. Still, there are few salient facts that we would like to mention à propos François. What characterizes him the best is probably his unfaltering fascination for what used to be called "natural philosophy". Rather insensitive to fads and fashions, François has always focused on those few hard problems where any advance could provide a glimmer of light on the inner workings of nature. Very few active physicists can compare with him as far as breadth of interest and depth of insight go. François has made significant contributions to areas ranging from the theory of ferromagnetism to superconductivity, from Particle Physics to Quantum Gravity. His best known contribution is certainly his discovery with Robert Brout of the mechanism of dynamical symmetry breaking which is still at the heart of modern Particle Physics. No less important perhaps, although less recognized, is the first model of an inflationary universe proposed much before it became the accepted dogma in cosmology.

All of us who have had the great joy and privilege of working with François, or just the pleasure of meeting him briefly during conferences, know that if there is a lot to be learned by reading his papers, there is also much to be learned from his unique style of doing Physics. In particular, younger physicists should keep in mind that people like François value so much new concepts and pictures that they are never really afraid to err in the pursuit of lofty ideas.

During this meeting we had the pleasure to see François as enthusiastic and imaginative as ever, and we hope that by offering him these Proceedings as an after-conference surprise we did not spoil any of his walks in the wonderful city of Rome by forcing him to attend more talks than he might have wished.

HAPPY BIRTHDAY FRANÇOIS

PUBLICATIONS OF F. ENGLERT

1) F. Englert and C. Jauquet, Théorie du dipôle replié, Revue E.F. III, 157 (1955).

2) F. Englert, Application de la théorie des groupes au calcul du couplage spin-orbite dans les cristaux, Académie royale de Belgique, Bulletin de la Classe des Sciences, 5e série, **43**, 273 (1957).

3) F. Englert, Interaction entre un petit et un grand système, Il Nuovo Cimento **10**, 560 (1958).

4) P. Aigrain and F. Englert, "Les semiconducteurs", Monographie Dunod, 203 pp - Dunod, Paris (1958) - Version espagnole, Eudeba, Buenos Aires (1966).

5) F. Englert, Comportement d'un petit système quantique dans un milieu faiblement dissipatif, Journal of Physics and Chemistry of Solids **11**, 78 (1959).

6) F. Englert, Renormalisation de masse d'un électron dans un corps noir, Académie royale de Belgique, Bulletin de la Classe des Sciences, 5e série, **45**, 782 (1959).

7) R. Brout and F. Englert, Linked Cluster Expansion in Quantum Statistics, Physical Review **120**, 1519 (1960).

8) F. Englert and R. Brout, Dielectric Formulation of Quantum Statistics, Physical Review **120**, 1085 (1960).

9) F. Englert, Effective Hamiltonian Formalism in the Exciton Problem and in Superconductivity, Proceedings of the International Conference on Semiconductor Physics (Prague), 34 (1960).

10) F. Englert, Theory of a Heisenberg Ferromagnet in the Random Phase Approximation, Physical Review Letters **5**, 102 (1960).

11) F. Englert and M. Antonoff, Band Theory of Ferromagnetism, Physica **30**, 429 (1964).

12) F. Englert, Linked Cluster Expansion in the Statistical Theory of Ferromagnetism, Physical Review **129**, 567 (1963).

13) R. Stinchombe, G. Horwitz, F. Englert and R. Brout, Thermodynamic Behavior of the Heisenberg Ferromagnet, Physical Review **130**, 155 (1963).

14) G. Horwitz, F. Englert and R. Brout, Zero Temperature Properties of Many Fermion Systems, Physical Review **130**, 404 (1963).

15) J. De Coen, F. Englert and R. Brout, Linked Cluster Expansion in N-Boson Problem, Physica **30**, 1293 (1964).

16) F. Englert, Fonctions de réponse et leurs relations avec les phénomènes d'équilibre, publ. in "Etudes des phénomènes irréversibles, Morgins 1963, 207-250", Bureau - Lausanne.

17) F. Englert and R. Brout, Broken Symmetry and the Mass of Gauge Vector Mesons, Physical Review Letters **13**, 321 (1964).

18) F. Englert and R. Brout, A Possible Argument which Eliminates the Hitherto Unrealized Representations of SU_3, Physical Review Letters **12**, 682 (1964).

19) F. Englert and R. Brout, Remark on the Misuse of the Active Interpretation of General Relativity, Physics Letters **16**, 250 (1965).

20) F. Englert, R. Brout and M.F. Thiry, Vector Mesons in the Presence of Broken Symmetry, Il Nuovo Cimento **43**, 244 (1966).

21) R. Brout and F. Englert, Gravitational Ward Identity and the Principle of Equivalence, Physical Review **141**, 1231 (1966).

22) C. De Dominicis and F. Englert, Potential Correlation Function Duality in Statistical Mechanics, Journal of Mathematical Physics **8**, 2143 (1967).

23) F. Englert and C. De Dominicis, Lagrangian Field Theory in Terms of Green's Functions, Il Nuovo Cimento **53A**, 1007 (1968).

24) F. Englert and C. De Dominicis, Self Generating Interactions, Il Nuovo Cimento **53A**, 1021 (1968).

25) F. Englert, R. Brout and H. Stern, Field Theoretic Formulation of Bootstrap Theory, Il Nuovo Cimento **58A**, 601 (1968).

26) F. Englert and R. Brout, Current Algebra and Partial Symmetry, Il Nuovo Cimento **55A**, 543 (1968).

27) R. Brout and F. Englert, Asymptotic Form Factors and Scattering Amplitudes, Physics Letters **27B**, 647 (1968).

28) F. Englert, R. Brout, P. Nicoletopoulos and C. Truffin, The Optical Limit of Relativistic Scattering, Il Nuovo Cimento **64A**, 561 (1969).

29) R. Brout, F. Englert and C. Truffin, Chiral Symmetry and Linear Trajectories, Physics Letters **29B**, 590 (1969).

30) F. Englert, R. Brout and C. Truffin, Meson Mass Spectrum, Physics Letters **29B**, 686 (1969).

31) F. Englert, R. Brout and H. Stern, Mass Quantization Conditions from Infinite Superconvergence and Chiral Symmetry, Il Nuovo Cimento **66A**, 845.

32) F. Englert, Self Consistent Approach to the Fine Structure Constant, Il Nuovo Cimento **16A**, 557 (1973).

33) F. Englert, J.-M. Frère and P. Nicoletopoulos, Vertex Bootstrap for the Fine Structure Constant, Il Nuovo Cimento **19A**, 395 (1974).

34) R. Brout, F. Englert and C. Truffin, $\beta\omega\pi$ Decay and SU_3 Breaking, Physical Review **D9**, 2694 (1974).

35) F. Englert, J.-M. Frère and P. Nicoletopoulos, Low Energy Eigenvalue Condition from Dynamically Broken Gauge Symmetry, Physics Letters **52B**, 433 (1974).

36) F. Englert and R. Brout, Dynamical Theory of Weak and Electromagnetic Interactions, Physics Letters **49B**, 77 (1974).

37) F. Englert, J.-M. Frère and P. Nicoletopoulos, Infrared Versus Ultraviolet Driven Dynamical Mass Generation in Ladder Approximation, Physics Letters **59B**, 346 (1975).

38) F. Englert, J.-M. Frère and P. Nicoletopoulos, Dynamical Symmetry Breakdown in Pure Yang Mills Theory, Nuclear Physics **B95**, 269 (1975).

39) F. Englert, Dynamical Symmetry Breakdown, published in "Weak and Electromagnetic Interactions at High Energies, Cargèse 1975, Part A 265-190", Plenum Press, New York and London (1976).

40) F. Englert, E. Gunzig, C. Truffin and P. Windey, Conforma Invariant General Relativity with Dynamical Symmetry Breakdown, Physics Letters **57B**, 73 (1975).

41) F. Englert, R. Gastmans and C. Truffin, Conformal Invariance in Quantum Gravity, Nuclear Physics **B117**, 407 (1976).

42) F. Englert, Mechanisms of Dynamical Symmetry Breaking, International Center for Theoretical Physics reports (Trieste) IC/76 (1976).

43) F. Englert and P. Windey, Quantization Conditions for 't Hooft Monopoles, Physical Review D, 2728 (1976).

44) R. Brout, E. Englert and W. Fischler, Magnetic Confinement in Non Abelian Gauge Theory, Physical Review Letters **36**, 649 (1976).

45) R. Brout, F. Englert and J.-M. Frère, On the Origin of the Pion in Confinement Schemes, Nuclear Physics **B134**, 327 (1978).

46) F. Englert and P. Windey, Electric Confinement and Magnetic Superconductors, Nuclear Physics **B135**, 529 (1978).

47) R. Brout, F. Englert and E. Gunzig, The Creation of the Universe as a Quantum Phenomenon, Annals of Physics **115**, 78 (1978).

48) F. Englert, Electric and Magnetic Confinement Schemes, published in "Hadronstructure and Lepton-Hadron Interactions, Cargèse 1977, 503-560", Plenum Press, New York and London (1979).

49) F. Englert and P. Windey, Dynamical and Topological Considerations on Quark Confinement, Physics Reports **49**, 173 (1979).

50) R. Brout, F. Englert and E. Gunzig, The Causal Universe (First Prize for 1978 awarded by Gravity Research Foundation), General Relativity and Gravitation Journal **10**, 1 (1979).

51) R. Brout, F. Englert and P. Spindel, Cosmological Origin of the Grand-Unification Mass Scale, Physical Review Letters **43**, 417 (1979).

52) R. Brout, F. Englert, J.-M. Frère, E. Gunzig, P. Nardone, C. Truffin and P. Spindel, Cosmogenesis and the Origin of the Fundamental Length Scale, Nuclear Physics **B170**, 228 (1980).

53) F. Englert, The Creation of the Universe as the Breakdown of a Grand Unified Symmetry, published in "Physical Cosmology, Les Houches, session XXXII, 1979, 515-532", North Holland Publishing Company, Amsterdam - New York - Oxford (1980).

54) R. Brout and F. Englert, TCP Conservation in Cosmology, Nuclear Physics **B180**, 181 (1981).

55) J.-M. Blairon, R. Brout, F. Englert and J. Greensite, Chiral Symmetry Breaking in the Action Formulation of Lattice Gauge Theory, Nuclear Physics **B180**, 439 (1981).

56) F. Englert and J.-M. Frère, Time-dependent Perturbation Theory with Permanent Effects, Nuclear Physics **B180**, 468 (1981).

57) A. Casher and F. Englert, The Quantum Era, Physics Letters **104B**, 117 (1981).

58) F. Englert, Quantum Field Theory and Cosmology, published in "Theory of Fundamental Interaction, Cargèse 1981", Plenum Press, New York and London (1982).

59) F. Englert, Spontaneous Compactification of Dimensional Supergravity, CERN Preprint TH 3394 (1982), Physics Letters **119B**, 339 (1982).

60) B. Biran, B. De Wit, F. Englert and H. Nicolaï, Gauged N=8 Supergravity and its Breaking from Spontaneous Compactification, CERN preprint TH 3489 (1982), Physics Letters **124B**, 45 (1983).

61) F. Englert, From Quantum Cosmology to Quantum Gravity published in "Unification of the Fundamental Particle Interactions II", edited by J. Ellis and S. Ferrara, Plenum Press, New York and London (1983).

62) F. Englert, M. Rooman and P. Spindel, Supersymmetry Breaking by Torsion and Ricci-Flat Squashed Seven-Spheres, Physics Letters **127B**, 47 (1983).

63) F. Englert, M. Rooman and P. Spindel, Symmetries in Eleven-Dimensional Supergravity Compactified on a Parallelized Seven-Spheres, Physics Letters **130B**, 50 (1983).

64) B. Biran, A. Casher, F. Englert, M. Rooman and P. Spindel, The Fluctuating Seven-Sphere in Eleven-Dimensional Supergravity, Physics Letters **134B**, 179 (1984).

65) A. Casher, F. Englert, H. Nicolaï and M. Rooman, The Mass Spectrum of Supergravity on the Round Seven-Sphere, Nuclear Physics **B243**, 173 (1984).

66) F. Englert, The Quest for Unification, published in "Recent Developments in Quantum Field Theory", eds. J. Ambjorn, B.J. Durhuus and J.L. Petersen, Elsevier Sc. Pub. North Holland, 39 (1985).

67) A. Casher, F. Englert, H. Nicolaï and A. Taormina, Consistent Superstrings as Solutions of the d=26 Bosonic String Theory, CERN preprint TH 4220/85, Physics Letters **162B**, 121 (1985).

68) F. Englert and A. Neveu, Non-Abelian Compactification of the Interacting Bosonic String, CERN preprint TH 4168/85, Physics Letters **163B**, 349 (1985).

69) F. Englert, Metric Space-Time from Field Propagation on Fractal Structures, CERN preprint TH 4091/85, From SU(3) to gravity, ed. E. Gotsman and G. Tauber, 35 (1985).

70) F. Englert, H. Nicolaï and A. Schellekens, Superstrings from 26 Dimensions, Nuclear Physics **B264**, 514 (1986).

71) F. Englert, Hidden Superstrings, Festschrift in honor of Y. Nambu to be published (1986).

72) F. Englert, J.-M. Frère and M. Rooman, Metric Space-Time as Fixed Point of the Renormalization Group Equations on Fractals Structures, Nuclear Physics **B280**, 147 (1987).

73) F.A. Bais, F. Englert, A. Taormina and P. Zizzi, Torus Compactification for Non Simply Laced Groups, Nuclear Physics **B279**, 529 (1987).

74) F. Englert, From Disorder to Space-Time Geometry, Foundations of Physics **17**, 621 (1987).

75) R. Brout and F. Englert, Cosmologie Quantique, dans "La Nouvelle Encyclopédie Diderot, Vol. : Aux Confins de l'Univers", ed. J. Schneider, Fayard, Fondation Diderot, Paris (1987).

76) F. Englert, From Classical to Quantum Cosmology in "XVII G.I.F.T. International Seminar on Cosmology and Particle Physics", eds. E. Alvarez, C.A. Dominguez, L.E. Ibanes and M. Quiros (World Scientific, Singapore, 1987).

77) Y. Aharonov, F. Englert and J. Orloff, Macroscopic Fundamental Strings in Cosmology, Physics Letters **B199**, 366 (1987).

78) F. Englert, A. Sevrin, W. Troost, A. Van Proeyen and P. Spindel, Loop Algebras and Superalgebras based on S^7, J. Math. Phys. **29**, 281 (1988).

79) F. Englert, J. Orloff and T. Piran, Fundamental Strings and Large Scale Structure Formation, Physics Letters **B212**, 423 (1988).

80) F. Englert, String Thermodynamics and Cosmology, Proceedings of the Boulder NATO Workshop on Superstrings, Plenum Press (1988).

81) F. Englert, Quantum Physics without Time, Physics Letters **B228**, 111 (1989).

82) F. Englert and J. Orloff, Universality of the Closed String Phase Transition, Nuclear Physics**B334** ,472 (1990)

83) F. Englert, From Quantum Correlations to Time and Entropy, published in "The Gardener of Eden, edited by P. Nicoletopoulos and J. Orloff, in Physicalia Magazine, Vol. 12 special issue", in honour of R. Brout's 6Oth birthday (1990).

84) A. Casher and F. Englert, Entropy Generation in Quantum Gravity, Class. Quantum Gravity **9** , 2231 (1992).

85) A. Casher and F. Englert, Black Hole Tunneling Entropy and the Spectrum of Gravity (U.L.B.-TH 8/92; TAUP 2017/92).

Contents

GRAVITATIONAL SCATTERING AT PLANCKIAN ENERGIES: THE EIKONAL AND BEYOND

DANIELE AMATI
International School for Advanced Studies
and INFN - Sezione di Trieste
Trieste, 34100, Italy

There has been, recently, a revival of interest in very high energy scattering in gravity. To be more precise, by the scattering of two massless (gravitons) - or small mass - particles at c. m. energies E for which $GE^2 \geq 1$, where G is the Newton constant, and for fixed transfer momentum t , i.e. $Gt \ll 1$ (small angle scattering).

This problem has been tackled since 1987 and different approaches have been proposed[1-5]. They all coincide at the leading level, i.e. all powers of GE^2. In terms of the impact parameter b , the leading amplitude has the eikonal form

$$S_0 = e^{2i\delta_0(b,s)} \qquad (s \equiv 4E^2)$$
$$\delta_0(b,s) = -Gs \log b/\lambda \qquad (in \quad 4 \quad dimensions) \tag{1}$$

showing the typical infrared singular Coulomb phase for $D = 4$.

This result has been obtained, with four different methods, by:

1) 't Hooft[1] by the quantum scattering of a graviton in the external metric generated by the other;
2) Muzinich and Soldate[2] with an eikonal approach;
3) Ciafaloni, Veneziano and myself[3] (ACV) studying the scattering of gravitons in superstring theory;
4) Verlinde and Verlinde[4] with a gauge choice for the metric that leads to a topological like regime in high energy Einstein quantum gravity.

The leading amplitude in terms of momentum transfer, or scattering angle, is then given by

$$\frac{1}{s}A^{eik}(s,q) = \int_0^\infty d^2b e^{i\vec{q}\cdot\vec{b}} e^{-2iGs\log b} = \frac{8\pi Gs}{q^2}\left(\frac{4}{q^2}\right)^{-iGs}\frac{\Gamma(1-iGs)}{\Gamma(1+iGs)} \tag{2}$$
$$(\vec{q}^2 = -t = s, sen^2\theta/2).$$

The ratio of two Γ functions in (2) represents an irrelevant t independent phase for real s but shows poles (first noticed by 't Hooft[1]) for positive integer values of iGs. Analogously, the partial wave amplitude

$$\frac{i}{8\pi}A_J^{eik} = e^{2i\delta_J^0} = \left(\frac{s}{4}\right)^{iGs}\frac{\Gamma(J+1-iGs)}{\Gamma(J+1+iGs)} \tag{3}$$

shows poles at $J=n$ for $iGs=n(n=1,2,\ldots)$.

Do these poles have a physical meaning? This is hard to believe due to the fact that they are originated by the singularity of the integrand in (2) at $b=0$. Indeed any smoothening - as for instance[5] the replacement of $\log b$ by $\frac{1}{2}\log(b^2+b_0^2)$– would eliminate them. On the other hand, the leading expression of Eq. (1) is unwarranted at small b due to non-leading effects that may become relevant.

Anyhow, interesting physics that goes beyond the semiclassical expression of Eq. (1) has to imply the evaluation of non-leading contributions. The only approach where this evaluation has been attempted is the one of Ref. (3) based on superstrings. The others are not suited to do it: it is obvious for the first two and I will show later why it is true for the fourth one.

Let me review briefly the different origins of non-leading effects. In superstring theory there is a scale (the string length λ_s) that is related to the Planck lenght λ_p by $\lambda_p = g\lambda_s$ where g is the string coupling constant (the one appearing in string-loop expansions). In the perturbative approach g is assumed to be smaller than unit , i.e. $\lambda_p < \lambda_s$.

Beyond the leading level, the analysis shows of course typical string effects, as excitations of string states, which give contributions of order $\frac{\lambda_s}{b}$ to $A(s,b)$. Other non-leading contributions depend on how large is $R \equiv G\sqrt{s}$ (the Schwarzild radius of the whole process). Indeed, in the asymptotic regime considered,

$$\frac{R}{\lambda_s} = \frac{\sqrt{G}}{\lambda_s}\sqrt{Gs} = \frac{\lambda_p}{\lambda_s}\sqrt{Gs} = g\sqrt{Gs} \tag{4}$$

can be smoller or greater than 1 depending on how the two conditions of that regime $g<1$, $Gs>1$ are realized. For $R<\lambda_s$, we prooved that as long as string effects are negligible i.e. $b \ll \lambda_s$ (or, better, $b \ll \lambda_s\sqrt{\log s}$) no other important corrections are generated so that the leading contribution (1) is justified. It gives rise to the classical Einstein deflection

$$\theta = \frac{G\sqrt{s}}{b} = \frac{R}{b} \tag{5}$$

which is therefore justified up to a maximum angle

$$\theta_M = \frac{R}{\lambda_s\sqrt{\log s}} < 1. \tag{6}$$

For smaller b no classical trejectory may be defined or, said differently, for diminishing θ, below θ_M , the values of b explored increase. This implies a new relation between explored distance and momenta

$$x \sim \frac{\hbar}{q} + q\alpha' \log s \tag{7}$$

showing a minimum explorable distance

$$x_{min} = \lambda_s\sqrt{\log s}. \tag{8}$$

This is reasonable if we think that in string theory there is no probe which is smaller that the string length.

The situation is different if $R > \lambda_s$ (actually $R > \lambda_s\sqrt{\log s}$). Then new corrections appear of order $\left(\frac{R^2}{b^2}\right)^n$. They change the classical Einstein deflection by increasing it

$$sen\frac{1}{2}\theta_{cl} = \frac{R}{b}(1+\left(\frac{R}{b}\right)^2+\ldots). \tag{9}$$

They look classical and may appear even for $b \gg \lambda_s$. If the leading eikonal may be described by diagrams

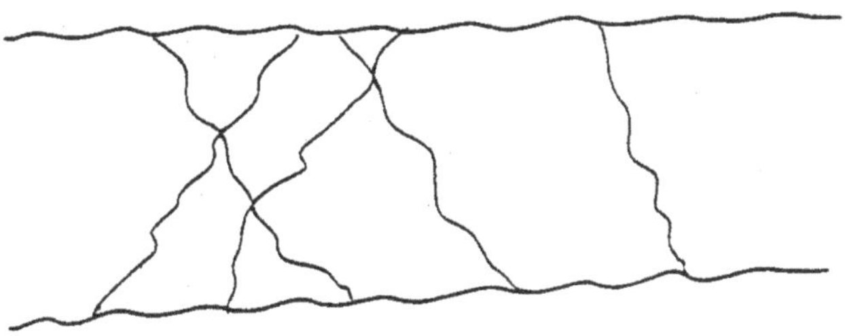

Fig. I Graviton-exchange diagrams providing the leading eikonal approximation

the classical non-leading corrections in question come from diagrams

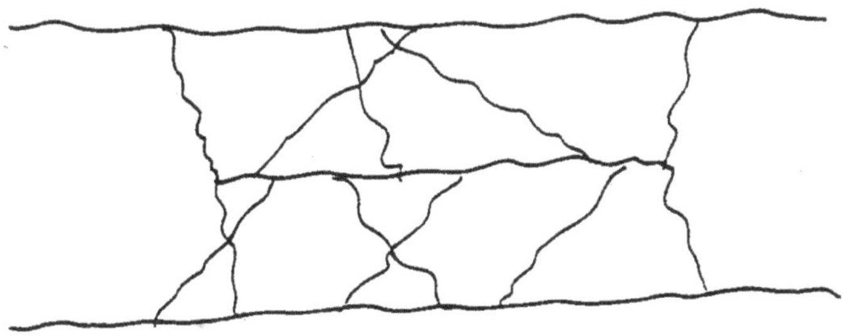

Fig. II A typical diagram giving a quantum correction to the eikonal

These non-leading contributions to the scattering amplitude - hard to extract due to the presence of i.r. singularities - have the form[6]

$$A(b,E) = e^{2i(\delta_0+\delta_1+\delta_2+\ldots)} \tag{10}$$

where, up to two loop computation

$$\delta_1 = \frac{6}{\pi}\frac{G^2 s}{b^2}\log s \qquad (purely \quad real) \tag{11}$$

$$\delta_2 = 2\frac{G^3 s}{b^2}(1+\frac{2i}{\pi}(1+\log\frac{b}{\lambda})\log s).$$

The imaginary part of δ_2 is due to the production of gravitons with both physical polarizations. The i.r. scale λ reflects the soft graviton emission that allows a Bloch-Nordsieck interpretation giving rise to a finite $Re\delta_2$ and thus a finite contribution to the Einstein deflection. For this, it is essential that the i.r. singular Coulomb phase is indeed not modified in a b dependent way as explicitly checked at the 2 loop level.

Higher loops provide $\left(\frac{R}{b}\right)^{2n}$ terms that, if resumed, may build a singularity in $\frac{R}{b}$. This could represent novel dynamical features - as black hole formation - generated by a consistent quantum gravity theory. But higher loops are difficult to compute specially if plagued by i.r. singularities that are hard to deal in gravity due to the non linear character of the theory.

The (Verlinde)[2] attempt[4] seemed a welcomed simplification trhough the identification of the relevant modes in the asymptotic regime: though a conveninet gauge choice the asymptotic dynamics would greatly simplify. In a gauge in which the metric is block diagonal in the 2 longitudinal and 2 transverse modes (i.e. $g_{LT}=0$) the longitudinal and transverse modes would disentangle at high energies allowing a topological approach for the longitudinal over and a pertubative approach for the transverse. Unfortunately this approach does not allow the computation of non-leading corrections. Indeed δ_2 of Eq. (11) involves the creation of the physical graviton with LT polarization that implies an LT fluctuation around a flat metric. That cannot be reconduced to a block diagonal form by any diffeomorphism. Moreover, the proportional production of the other physical graviton, with $LL-TT$ polarization, shows the difficulty to disentangle the longitudinal and transverse dynamics.

Nevertheless the idea to separate the longitudinal and transverse modes allowing however a coupling among them may be fruitful. In this sense, a phenomenological action for high energy (QCD and gravity) scattering has been proposed by Lipatov[7]. As in Ref. (4), the trasverse modes propagate longitudinal and the longitudinal ones (the only coupled with $\sqrt{GE}$ to the external scattering particles) transevrsally. They also interact with coupling $\sqrt{G}$ in the different trilinear combinations.

The leading eikonal is generated by the action evaluated at its classical solution taking into account the sources (external particles) but omitting trilinear interactions. A perturbative treatment of the equation of motion and the insertion of the correspondence solutions into the action should give the development of the eikonal as in Eq. (10). We are presently investigating if the i.r. problem can be dealt by a correct treatment of boundary terms arising from integration by parts needed to find the equation of motion. If the effective action reproduces the perturbative expansion of the amplitude - at least up to the 2 loop level - we may gain confidence in its reliability. A non perturbative solution of the equations of motion would then determine a classical action that could resum, directly at the level of a generalized eikonal, the classical correction at all orders thus showing the onset of singularities or gravitational instabilities.

References

1. G.'t Hooft, *Phys. Lett.* **198B** (1987) p.61.
2. I.Muzinich and M.Soldate,*Phys. Rev.* **D37** (1988) p.353.
3. D.Amati, M.Ciafaloni and G.Veneziano, *Phys. Lett.* **197B** (1987) p.81; *Int. J.Mod. Phys.* **A3** (1988) p.1615.

4. E.Verlinde and H.Verlinde, "Scattering at Planckian energies", *Princeton preprint* **PUTP-1279** (1991); see also R.Kallosh, "Geometry of scattering at Planckian energies", *Stanford preprint* **SU-ITP 903** (1991).
5. D.Amati, M.Ciafaloni and G.Veneziano, *Phys. Lett.* **B289** (1992) p.87.
6. D.Amati, M.Ciafaloni and G.Veneziano, *Nucl. Phys.* **B347** (1990) p.550.
7. L.N.Lipatov, *Nucl. Phys.* **B365** (1991) p. 614.

STRING LOOP CORRECTIONS TO GAUGE AND YUKAWA COUPLINGS

I. ANTONIADIS
Centre de Physique Théorique, Ecole Polytechnique
91128 Palaiseau, France

and

T.R. TAYLOR
Department of Physics, Northeastern University
Boston, MA 02115, U.S.A.

ABSTRACT

We report on the recent progress in computing the effective supergravity action from superstring scattering amplitudes beyond the tree approximation. We discuss the moduli-dependent string loop corrections to gauge, gravitational and Yukawa couplings.

1. Introduction

The basic property of string theory, which makes it so attractive from the point of view of particle physics, is that the physical couplings and masses are in principle calculable. They are determined by the vacuum expectation values (VEVs) of massless scalar fields, like the dilaton and moduli. The latter are scalar fields with flat potential, whose VEVs correspond to compactification parameters and determine the size and shape of the internal space. Any serious attempt to compute the low-energy parameters from string theory must address two basic questions: 1) How do masses and couplings depend on the VEVs of dilaton, moduli, Higgs scalars *etc.*? and 2) What fixes these VEVs? In the past several years, there has been some steady progress towards answering the second question, although the general perception is that the problem of scalar VEVs still escapes a satisfactory solution. On the other hand, there has been a lot of progress towards answering the first question, which we will generically call the problem of moduli dependence of physical parameters.

A very efficient method of studying the moduli dependence of low-energy parameters, developed in the last couple of years, relies on the computation of the effective supergravity action describing the physics of massless string excitations. The moduli dependence of the effective action can be determined by evaluating the appropriate superstring scattering amplitudes.[1–5] In this process, supergravitational interactions are determined directly from superstring theory. The moduli-dependent loop corrections obtained in this way give rise to the so-called threshold corrections to superstring unification parameters,[6] and determine the boundary conditions of gauge and Yukawa couplings at the string unification scale M_{SU}. They are also relevant for non-perturbative phenomena in string theory, which could generate dynamically potential for moduli providing a mechanism that fixes their VEVs. They may also be relevant for the mechanism of supersymmetry breaking, either via gaugino condensation,[7] or with a large compactification radius.[8]

The massless spectrum of any four-dimensional heterotic superstring model contains the supergravity multiplet, gauge multiplets, and a large number of chiral multiplets. In addition, there is a dilaton which belongs to a very distinct supersymmetry multiplet, together with the two-index antisymmetric tensor – the Kalb-Ramond field. The dilaton VEV plays the role of the string loop expansion parameter. Since the Kalb-Ramond field is equivalent to a pseudoscalar axion, one usually represents the dilaton and its supersymmetric partners by one chiral multiplet S. The most general $N = 1$ supergravity action, describing local interactions involving up to two derivatives, is characterized by three functions of chiral superfields:[9] the real Kähler potential K which determines the kinetic terms, the analytic superpotential W related to the Yukawa couplings, and the analytic gauge function f associated with the gauge couplings.

At the tree level, the general stucture of the f-function and the Kähler potential is common to all compactifications:

$$f^{(0)} = kS\ , \qquad K^{(0)} = -\ln(S+\bar{S}) + G^{(0)}(Z,\overline{Z})\,, \tag{1}$$

where k is the integer level of the Kač-Moody algebra that generates the gauge group, and Z denote chiral superfields other than S. Thus, the coupling constant g_a of a gauge group factor G_a is given at the tree-level by $g_a^2 = g^2/k$, where g is the four-dimensional string coupling. The function $G^{(0)}(Z,\overline{Z})$ as well as the superpotential $W(Z)$ depend on the details of compactification and can be determined by using a number of different methods. At low energies, the most relevant interactions among matter and gauge fields correspond to effective operators of dimension four. Consequently, one may expand $G^{(0)}$ and W up to quadratic and cubic order in matter fields, respectively, with moduli-dependent coefficients which one is interested to compute. These are known in many examples, in particular for the case of (2,2) compactifications which exhibit the property of special geometry implying that both functions are given in terms of a single prepotential.[1,10]

2. Infrared divergences and threshold corrections

The tree-level effective action describes interactions of massless string excitations at energies below the string scale. These include contact interactions due to the propagation of heavy particles in one-(massive)-particle reducible diagrams. The masses of heavy particles depend on the moduli, *e.g.* the radii of compactified dimensions, therefore the induced massless particle interactions are also moduli-dependent. In order to compute the loop corrections to the effective action, one should integrate all diagrams involving heavy particles propagating inside loops. In string theory, it is very difficult to separate heavy from massless particles in higher genus diagrams, therefore the computation of the effective action becomes more subtle than in field theory. In fact integration over massless particles gives rise to infrared divergences for on-shell amplitudes, associated with the running of low energy couplings.[6,11] In the analogous field-theoretical computations, such logarithmic divergences are usually regulated by going off-shell, to momentum $\mu^2 \neq 0$. It is very important to realize that in string theory, as well as in quantum field theory, the momentum-dependence of coupling constants is a purely infrared effect, and therefore the corresponding β-function coefficients of the $\mu^2 \to 0$ divergence depend on the massless particle content only.

Consider for instance the one-loop case. A generic on-shell amplitude $\mathcal{A}$ corresponding to some physical coupling of the low-energy theory is written as an integral over the complex Teichmüller parameter $\tau = \tau_1 + i\tau_2$ of the world-sheet torus inside its fundamental domain $\Gamma \equiv \{|\tau_1| \leq \frac{1}{2},\ |\tau| \geq 1\}$:

$$\mathcal{A} = \int_\Gamma \frac{d^2\tau}{\tau_2} A(\tau, \bar{\tau}). \tag{2}$$

The presence of massless particles propagating in the loop implies that the integrand A goes to a constant A_0 as $\tau_2 \to \infty$, and the the integral over the Teichmüller parameter diverges in the infrared. When the logarithmic divergence is regularized and compared to the field-theoretical $\overline{DR}$ scheme, it is converted to $\ln(M_{SU}^2/\mu^2)$, where μ is the infrared cutoff and $M_{SU} \simeq 5 \times g \times 10^{17}$ GeV is the string unification scale.[6] The remaining finite part of the integral yields the moduli-dependent string threshold corrections:

$$\mathcal{A} = A_0 \ln \frac{M_{SU}^2}{\mu^2} + \int_\Gamma \frac{d^2\tau}{\tau_2} [A(\tau, \bar{\tau}) - A_0]. \tag{3}$$

When computing the moduli dependence of threshold corrections we use two main ingredients: 1) $N = 1$ space-time supersymmetry to relate physical couplings to amplitudes involving pseudoscalars which receive contributions only from odd spin structures, and thus are much easier to compute. 2) The invariance of the superstring and its low energy effective theory under the discrete duality group. This generalizes the simple $R \to 1/R$ duality, where R is the radius of a circle on which one internal dimension is compactified to the case of more general compactifications. For instance, in the case of a plane one can define two complex parameters, each of them generating an $SL(2, Z)$ group of duality transformations $T \to 1/T$ and $T \to T + i$.

3. Gauge couplings

The moduli dependence of threshold corrections to gauge couplings can be determined in a way that circumvents the problem of infrared divergences. It is based on the observation that the β-function which is the coefficient of the infrared divergence is moduli independent and drops when one considers derivatives with respect to the moduli. Consider for instance the 3-point scattering amplitude involving one modulus T and two gauge bosons. As a consequence of supersymmetry, the CP-even and the CP-odd part of this amplitude are proportional,[2] implying:

$$\partial_T g^{-2}(T,\overline{T}) = i\Theta_T \ , \tag{4}$$

where Θ_T is the axionic coupling of the pseudoscalar component of T to gauge bosons. If Θ_T is the derivative with respect to T of a Θ-angle which is a function of moduli, Eq. (4) implies that gauge couplings and the corresponding Θ-angles are obtained as the real and the imaginary part, respectively, of analytic gauge functions f, consistently with the general form of $N = 1$ supergravity.[9] However, due to the integration over massless particles in the physical amplitude, Θ_T may not in general be integrable and a useful relation to examine is the integrability condition:

$$\partial_T \partial_{\overline{T}} g^{-2}(T,\overline{T}) = \frac{i}{2}(\partial_{\overline{T}}\Theta_T - \partial_T\Theta_{\overline{T}}) \stackrel{?}{=} 0 \ . \tag{5}$$

Explicit one loop calculation in orbifolds shows that the integrability condition Eq. (5) is indeed violated and the full moduli dependence of threshold corrections to gauge couplings can be computed.[2] At higher loops, this dependence satisfies a non-renormalization theorem implying that the one-loop result is exact.[3] For general compactifications, a general formula can be derived by integrating the one-loop expression of the axionic coupling Θ_T in Eq. (4):[4,5]

$$\frac{1}{g^2} = -\frac{i}{32\pi^2}\int_\Gamma \frac{d^2\tau}{\tau_2}\bar{\eta}^{-2}\mathrm{Tr}_R F(-1)^F(Q^2 - \frac{k}{4\pi\tau_2}), \tag{6}$$

where η is the Dedekind eta function, Q is the gauge group generator, and the trace is over the Ramond sector of the internal $N = 2$ superconformal theory with $U(1)$-charge operator F. Note that the integral in Eq. (6) is infrared divergent. The coefficient of the logarithmic divergence $\frac{d\tau_2}{\tau_2}$ is:

$$b = \frac{1}{32\pi^2}(-3\mathrm{Tr}Q_V^2 + \mathrm{Tr}Q_M^2), \tag{7}$$

where the two terms are the contribution of gauginos and matter fermions with $U(1)$-charges $F = \pm 3/2$ and $F = \pm 1/2$, respectively. The expression of b in Eq. (7) coincides with the field-theoretical one-loop β-function of gauge couplings in $N = 1$ supersymmetric Yang-Mills theory. It is worth noting that the same quantity $\mathrm{Tr}F(-1)^F$ was studied in the massive case, where $F = F_L - F_R$, as a new kind of "index" for $N = 2$ theories which depends only on chiral deformations.[12]

4. Gravitational couplings

The above results for gauge couplings are extended for the case of gravitational couplings.[4] In this case, there is no moduli dependent correction to the Planck mass, at least up to the one loop order, and the role of gauge couplings and Θ-angles is played by the coefficients of the Gauss-Bonnet combination and the $R\tilde{R}$ term, respectively. Eq. (6) is then valid for the gravitational coupling after replacing the gauge group generator Q^2 with

$$Q^2_{\mathrm{grav}} \equiv -\frac{1}{i\pi}\partial_{\bar{\tau}} \ln(\bar{\eta}^2). \tag{8}$$

The infrared divergence can now be identify with the four-dimensional one-loop trace anomaly:

$$b_{\mathrm{grav}} = \frac{1}{32\pi^2}[\frac{1}{6}(-3N_V + N_S) - \frac{11}{3}(-3 + N_{3/2})], \tag{9}$$

where N_S, N_V and $N_{3/2}$ denote the number of chiral, vector and spin-3/2 (massless) supermultiplets, while the first term in the second bracket accounts for the contribution of graviton and dilaton supermultiplets. This expression agrees with the field-theoretical calculations of trace anomaly coefficients, where the antisymmetric tensor contribution is *different* from that of a scalar.[13]

5. Integrability condition and duality anomalies

Going back to the integrability condition (5), using the expression of Eq. (6) one finds:[4,5]

$$\begin{aligned}\partial_T\partial_{\overline{T}} g^{-2}(T,\overline{T}) &= \frac{-i}{2(2\pi)^4}\int_\Gamma d^2\tau \partial_\tau \int d^2\zeta\, \bar{\eta}^{-2} \langle Q^2 \Psi_T(\zeta)\overline{\Psi}_{\overline{T}}(0)\rangle_{\mathrm{odd}} \\ &\quad + \frac{k}{8(2\pi)^5}\int_\Gamma \frac{d^2\tau}{\tau_2{}^2}\int d^2\zeta\, \bar{\eta}^{-2}\, \langle \Psi_T(\zeta)\overline{\Psi}_{\overline{T}}(0)\rangle_{\mathrm{odd}}\,, \end{aligned} \tag{10}$$

where Ψ_T ($\overline{\Psi}_{\overline{T}}$) are the corresponding chiral (anti-chiral) primary fields of the underlying $N = 2$ internal superconformal theory with dimension ($\frac{1}{2}$,1). Eq. (10) gives the non-harmonicity of gauge couplings and contains two parts:

1. The group-dependent part proportional to Q^2 is a total derivative in τ and its contribution to the integral comes only from the boundary of the moduli space, namely the degeneration limit $\tau_2 \to \infty$. It is therefore determined only from massless particles. In fact, all massless chiral fermions couple off-shell with the pseudoscalars[9] and generate one-loop anomalous couplings of pseudoscalars to gauge bosons or gravitons which are non-local. A detailed analysis of this term[4] was shown to reproduce the field theory computation of the anomalous graphs.[14,15]

2. The universal part (*i.e.* the term proportional to k) can be identified with the one-loop correction $G^{(1)}_{T\overline{T}}$ to the Kähler metric. The corresponding one-loop contribution to the Kähler potential is:

$$K^{(1)} = -\ln[1 - \frac{2}{(S+\overline{S})}G^{(1)}(Z,\overline{Z})]\,. \tag{11}$$

$G^{(1)}(T,\overline{T})$ gives rise to one-loop kinetic terms that mix the moduli T with the dilaton S which then couples universally to gauge bosons and gravitons at the tree-level. This is usually called the Green-Schwarz term because it can be interpreted as the compactification of the ten-dimensional term involved in the Green-Schwarz anomaly cancellation mechanism.[14]

The non-harmonicity of gauge couplings is also related to duality anomalies.[14,15] In fact the first term in Eq. (10), associated with one-loop anomalous graphs, generates anomalies in the duality transformations at the effective field theory level. These are partially cancelled by the contribution of the massive string modes which produce *local* one-loop corrections to the gauge kinetic functions $f^{(1)}(T)$, obtained after integrating Eq. (10) to Eq. (6). There remains a gauge group-independent anomaly cancelled by the second term of Eq. (10), which plays a similar role as the Green-Schwarz term in the cancellation of ten-dimensional gauge anomalies.

6. One loop Kähler metric and Yukawa couplings

As a consequence of the supersymmetric non-renormalization theorems in the background field method, the superpotential does not receive any loop corrections. However, the physical Yukawa couplings defined by the fermion-scalar-fermion scattering amplitudes may receive loop corrections which arise from the wave function renormalization factors, *i.e.* from the corrections to the Kähler metric. The computation described in the previous Sections 3-5 remains valid if one replaces the modulus T with any matter field which is singlet with respect to the gauge group under consideration. Thus, the one-loop correction to the Kähler metric is given by the second term of Eq. (10) and corresponds to the variation of a Kähler potential given by the second term of Eq. (6):

$$G^{(1)} = \frac{i}{16(2\pi)^3}\int_\Gamma \frac{d^2\tau}{\tau_2^2}\bar{\eta}^{-2}\mathrm{Tr}_R F(-1)^F \, . \tag{12}$$

This can be checked independently by an alternative computation of the one-loop three-point amplitude involving two complex scalars and the antisymmetric tensor field.[5] The result is:

$$G^{(1)}_{i\bar{\jmath}} = \frac{1}{8(2\pi)^5}\int_\Gamma \frac{d^2\tau}{{\tau_2}^2}\int d^2\zeta\, \bar{\eta}(\bar{\tau})^{-2}\, \langle \Psi_i(\zeta)\overline{\Psi}_{\bar{\jmath}}(0)\rangle_{\mathrm{odd}}\, , \tag{13}$$

in agreement with the field-theoretical expression from Eq. (10).

In the case of matter metric, the integral over the Teichmüller parameter τ in Eq. (13) is infrared divergent. As in the case of gauge couplings, these divergences are due to massless particles propagating in the loop. The coefficients of divergent terms correspond to the one-loop anomalous dimensions:[5]

$$\gamma_i = -\frac{1}{64\pi^2}\{-\frac{4}{k}C_2(R_i) + \frac{1}{g^2}\sum_{j,k}|\lambda_{ijk}|^2\}, \tag{14}$$

where $C_2(R_i)$ is the quadratic Casimir of the representation R_i to which the i-th field belongs, and λ_{ijk} are the physical Yukawa couplings. The comparison with the field-theoretical anomalous dimensions shows that the string computation implicitly uses a gauge in which the superpotential remains unrenormalized. Again as in the case of gauge couplings, the momentum-dependence of the physical Yukawa couplings in string theory turns out to be determined by the corresponding field-theoretical β-functions.[16] The remaining finite part of $G^{(1)}_{i\bar{j}}$ gives the string threshold corrections to wave function factors. These corrections determine the boundary conditions for the physical Yukawa couplings λ_{ijk} at the unification scale:

$$\lambda_{ijk}(M_{SU}) = \lambda_{ijk}^{\text{tree}}\,[1 + g^2\,(Y_i + Y_j + Y_k)]^{-1/2}, \tag{15}$$

where Y_i is defined as the finite part of $G^{(1)}_{i\bar{\imath}}/G^{(0)}_{i\bar{\imath}}$.

7. Explicit examples in orbifold models

As an example we will consider the symmetric orbifold models. These are a particular case of $(2,2)$ compactifications, which possess $N=2$ world-sheet supersymmetry in both left and right moving sectors. The gauge group is $E_8 \otimes E_6 \otimes H_2$, with all Kač-Moody levels $k=1$ and with H_2 a model dependent factor of rank-2. The relevant matter fields transform as 27 or $\overline{27}$ under E_6 and they are in one-to-one correspondence with the moduli : 27's are related to $(1,1)$ moduli and $\overline{27}$'s to $(1,2)$ moduli. Furthermore, the moduli metric is block-diagonal with respect to these two types of moduli. An interesting consequence of the right-moving $N=2$ tree-level Ward-identities is the property of special geometry, which relates the tree-level moduli metric to the Yukawa couplings.[1,10]

Here, we mainly focus our attention on the case of the three untwisted moduli T_j, $j=1,2,3$, describing the size of the three internal compactified planes, which are in one-to-one correspondence with three untwisted families A_j. In this case, the superpotential and the tree-level Kähler potential are:

$$W = w_{123}A_1A_2A_3\,, \qquad G^{(0)} = -\sum_{j=1}^{3}\ln(T_j + \overline{T}_{\bar{\jmath}} - A_j\bar{A}_{\bar{\jmath}})\,, \tag{16}$$

where w_{123} are numerical constants. The tree-level physical Yukawa couplings are then $\lambda_{123}^{\text{tree}} = \frac{g}{\sqrt{2}}w_{123}$.

At the one-loop level, one has to sum over all sectors of boundary conditions of the orbifold group. The untwisted sector, which preserves $N=4$ space-time supersymmetry, gives vanishing contribution to all β-functions and threshold corrections. The twisted sector preserving only $N=1$ supersymmetry leads to moduli-independent corrections, since the masses of the corresponding string states do not depend on the geometry of the compactified space. Therefore, the gauge and untwisted Yukawa couplings receive moduli-dependent corrections only from sectors that preserve $N=2$ space-time supersymmetry, which appear when one of the three internal planes remains invariant under the boundary conditions. The variation $\partial_T\partial_{\overline{T}}g_a^{-2}$ can be easily evaluated from Eq. (10) with the result:[2,3]

$$\partial_T\partial_{\overline{T}}g_a^{-2} = \tilde{b}_a G^{(0)}_{T\overline{T}}\ , \qquad G^{(0)}_{T\overline{T}} = \frac{1}{(T+\overline{T})^2}\ , \tag{17}$$

where the coefficient $\tilde{b}_a$ is proportional to the β-function of the gauge group factor G_a in the corresponding $N = 2$ theory. It can be writen as a sum of two terms,

$$\tilde{b}_a = \tilde{b}_a^{\rm FT} + \tilde{b}^{\rm GS}\ , \tag{18}$$

where $\tilde{b}_a^{\rm FT}$ is the field-theoretical group-dependent contribution and $\tilde{b}^{\rm GS}$ is the universal Green-Schwarz term, in correspondence with the two terms of Eq. (10). Explicit evaluation[5] of the second integral shows that $\tilde{b}^{\rm GS}$ does indeed subtract the contribution of the $N = 1$ sector from the field-theoretical coefficient, so that $\tilde{b}_a$ is given entirely by $N = 2$ sectors. The same equation (17) is also valid for the gravitational couplings with a coefficient $\tilde{b}_{\rm grav}$ proportional to the trace anomaly of the corresponding $N = 2$ theory.[4]

After integrating the differential equation (17) using the $SL(2,Z)$ duality symmetry one finds:

$$\begin{aligned}\frac{1}{g_a^2(M_{SU})} &= \frac{1}{g^2} - \tilde{b}_a \ln[|\eta(iT)|^4(T+\overline{T})] + c_a \\ &+ \frac{B\overline{B}}{T+\overline{T}}[\frac{1}{2}C_2(G_a) + T(27)(1+\frac{\chi}{6} - \frac{2}{g^2}\sum_{k,l}|\lambda_{Bkl}|^2)] + \dots\ ,\end{aligned} \tag{19}$$

where c_a are moduli-independent constants and we also included the contribution of the blowing-up moduli B at the lowest order.[4] The latter deform orbifolds to smooth Calabi-Yau manifolds and are in one-to-one correspondence with twisted families. χ is the number of generations given by the Euler number, and $T(27)$ is the Dynkin index of the matter representation which equals 3 and 0 for E_6 and E_8, respectively.

The one-loop Kähler metric and threshold corrections to Yukawa couplings can be explicitly computed from Eqs. (13) and (15). For the untwisted moduli one obtains $G^{(1)}_{T\overline{T}} = \tilde{b}^{\rm GS} G^{(0)}_{T\overline{T}}$, where $\tilde{b}^{\rm GS}$ was defined in Eq. (18). A consequence of this multiplicative finite renormalization of the moduli metric is that special geometry is in general violated beyond the tree approximation. This is expected since from field-theoretical point of view special geometry is a consequence of $N = 2$ space-time supersymmetry and $\tilde{b}^{\rm GS}$ is non vanishing only in $N = 1$ sectors.[5]

The wave function threshold corrections of an untwisted field A_j (27 or $\overline{27}$) associated with the j-th plane depends only on its moduli T_j, and *not* on the moduli of other planes:[5]

$$Y_j = 2\tilde{\gamma}_j \ln[|\eta(iT_j)|^4(T_j + \overline{T}_j)] + y_i\ , \tag{20}$$

where y_j is a moduli-independent constant and the coefficient $\tilde{\gamma}_j$ is the anomalous dimension of the A_j-field in the corresponding $N = 2$ supersymmetric theory. Since this field belongs to an $N = 2$ vector supermultiplet, one has $\tilde{\gamma}_j = -\tilde{b}_j/2$, where $\tilde{b}_j/2$ is the corresponding beta function coefficient of any gauge subgroup that transforms A_j non-trivially in the embedding $N = 2$ theory. The threshold corrections to Yukawa couplings can be computed by substituting Eq. (20) into Eq. (15). The boundary conditions for the one-loop untwisted Yukawa couplings are then given by:

$$\lambda_{123}(M_{SU}) = \frac{g_{E_6}(M_{SU})}{\sqrt{2}} w_{123} \left[1 + g^2_{E_6}(M_{SU}) y_{123}\right]^{-1/2}, \tag{21}$$

where y_{123} are moduli-independent constants, and $g_{E_6}(M_{SU})$ is the one-loop E_6 gauge coupling at the unification scale, *cf.* Eq. (19):

$$\frac{1}{g^2_{E_6}(M_{SU})} = \frac{1}{g^2} - \sum_{j=1}^{3} \tilde{b}^j_{E_6} \ln[|\eta(iT_j)|^4 (T_j + \overline{T}_j)] + c_{E_6} \,. \tag{22}$$

As a result, the boundary relation between the untwisted Yukawa couplings and the E_6 gauge coupling at the unification scale does not receive any moduli-dependent corrections at the one-loop level.

8. Conclusions

Superstring theory provides a unique example of a perfectly consistent unification of gauge and gravitational interactions, within the framework of local supersymmetry. The efforts described here have been motivated by the desire to understand the low-energy limit of such a consistent theory. As a result, we have now a very good understanding of the low-energy physics of supergravity theory that describes not only the classical limit of string theory, but also some interesting string loop effects, in particular the threshold corrections. From the three functions which determine the effective $N = 1$ supergravity, the two analytic ones are subject to non-renormalization theorems. The superpotential is given entirely at the tree-level, while the gauge kinetic function at the one loop. On the other hand, the one-loop correction to the real Kähler potential is closely related to the new "index" of $N = 2$ superconformal theories. The infrared divergences were identified with the field-theoretical β-functions of gauge and Yukawa couplings, while the finite part define the moduli-dependent threshold corrections which are calculable in any given model. Here, we presented explicit examples in the case of orbifolds. In addition to possible applications for low-energy physics, this "bottom–up" approach to superstring theory may well provide some new insights into the Planck-scale physics.

Acknowledgements

This work was supported in part by the Northeastern University Research and Scholarship Development Fund, in part by the National Science Foundation under grant PHY-91-07809, in part by the EEC contracts SC1-0394-C, SC1-915053 and SC1-CT92-0792, and in part by a CNRS-NSF collaborative grant.

References

1. L.J. Dixon, V.S. Kaplunovsky and J. Louis, *Nucl. Phys.* **B329** (1990) 27.
2. L.J. Dixon, V.S. Kaplunovsky and J. Louis, *Nucl. Phys.* **B355** (1991) 649.
3. I. Antoniadis, K.S. Narain and T.R. Taylor, *Phys. Lett.* **B267** (1991) 37.
4. I. Antoniadis, E. Gava and K.S. Narain, *Phys. Lett.* **B283** (1992) 209; *Nucl. Phys.* **B393** (1992) 93.
5. I. Antoniadis, E. Gava and K.S. Narain and T.R. Taylor, preprint NUB-3057 (1992).
6. V.S. Kaplunovsky, *Nucl. Phys.* **B307** (1988) 145 and Erratum, preprint Stanford-ITP-838 (1992).
7. J.-P. Derendinger, L.E. Ibáñez and H.P. Nilles, *Phys. Lett.* **B155** (1985) 65; M. Dine, R. Rohm, N. Seiberg and E. Witten, *Phys. Lett.* **B156** (1985) 55; S. Ferrara, N. Magnoli, T.R. Taylor and G. Veneziano, *Phys. Lett.* **B245** (1990) 409.
8. I. Antoniadis, *Phys. Lett.* **B246** (1990) 377; I. Antoniadis, C. Muñoz and M. Quirós, preprint CPTH-A206.1192 (1992).
9. E. Cremmer, S. Ferrara, L. Girardello and A. Van Proeyen, *Nucl. Phys.* **B212** (1983) 413.
10. S. Ferrara, Proceedings of *STRINGS 1992*, Roma, September 1992 (World Scientific P. C.); P. Candelas, Proceedings of 26th Erice Workshop *From Superstrings to Supergravity*, December 1992 (World Scientific P. C.); and references therein.
11. T.R. Taylor and G. Veneziano, *Phys. Lett.* **B212** (1988) 147.
12. S. Cecotti, P. Fendley, K. Intriligator and C. Vafa, preprint HUTP-92/A021 (1992); S. Cecotti and C. Vafa, preprint HUTP-92/A044 (1992).
13. M.J. Duff, Proceedings of *Supergravity 1981*, Trieste 1981, p. 197.
14. J.-P. Derendinger, S. Ferrara, C. Kounnas and F. Zwirner, *Nucl. Phys.* **B372** (1992) 145; G.L. Cardoso and B.A. Ovrut, *Nucl. Phys.* **B369** (1992) 351; M.K. Gaillard and T.R. Taylor, *Nucl. Phys.* **B381** (1992) 577.
15. J. Louis, in *Proceedings of the Second International Symposium on Particles, Strings and Cosmology*, P. Nath and S. Reucroft eds. (World Scientific, 1992), p. 751; S. Ferrara, C. Kounnas, D. Lüst and F. Zwirner, *Nucl. Phys.* **B365** (1991) 431; L. Ibáñez and D. Lüst, *Nucl. Phys.* **B382** (1992) 305.
16. R. Barbieri, S. Ferrara, L. Maiani, F. Palumbo and C.A. Savoy, *Phys. Lett.* **B115** (1982) 212.

SCHWINGER EFFECT IN STRING THEORY

C.BACHAS
Centre de Physique Théorique, Ecole Polytechnique
91128 Palaiseau, France

ABSTRACT

I discuss the quantum instability of an electric field in a theory of open strings.

One of the few available hints on the interplay between quantum theory and gravity is the Hawking emission of a black hole [1]. This can be described as pair creation in the background gravitational field. Of the two produced particles one stays trapped inside the horizon, while the other escapes to infinity where it emerges with a thermal spectrum at an effective temperature $T_H = (8\pi GM)^{-1}$. In deriving his famous result Hawking assumed a Schwarzild form for the metric, and minimally-coupled non-self-interacting quantum fields. In string theory we expect two kinds of corrections to this result: *(a)* Quantum or string-loop effects, which are for sure important for small black holes of mass $M \simeq M_{Planck} \simeq 1/\sqrt{\alpha'}g_{string}$, and *(b)* σ-model or α' corrections, which when $g_{string} \ll 1$ can become important much earlier, for $M \simeq M_{Planck}/g_{string} \simeq 1/\sqrt{\alpha'}g^2_{string}$. Assuming the spectrum stays thermal, these latter corrections will presumably modify the expression for the temperature [2] by replacing $GM \to GM \times f(GM/\sqrt{\alpha'})$, where $f(y) \simeq 1 + o(\frac{1}{y^2})$ for large y. Knowledge of the function f, which parametrizes the effects of the non-minimal coupling of strings to gravity, could be a crucial first step before addressing the paradoxes related to black-hole evaporation.

A similar but simpler problem arises when one considers the quantum instability of strong electromagnetic fields. Schwinger's field-theoretic calculation [3], for minimally-coupled non-self-interacting particles, gives the following result for the rate of pair production per unit volume and time:

$$w = \frac{2J+1}{8\pi^3} \sum_{k=1}^{\infty} (-)^{(2J+1)(k+1)} \left(\frac{eE}{k}\right)^2 \exp\left(-\frac{k\pi m^2}{|eE|}\right). \tag{1}$$

Here E is the constant external electric field, e and m are the charge and mass of the produced particles, and $J = 0$ or $\frac{1}{2}$ is their spin. This rate, obtained as the imaginary part of the one-loop vacuum-energy density, vanishes non-perturbatively fast with the external field, showing that pair-production is a quantum-tunneling phenomenon. The potential barrier has, roughly-speaking, a height of order $2m$ and a width of order m/eE.

This calculation can be also done exactly, if point-particles are replaced by open strings [4]. The key observation [5] is that the effect of a constant electric field $F^{01} = E$ is equivalent to an orbifold-like twist of the light-like coordinates $X^{\pm} = \frac{1}{\sqrt{2}}(X^0 \pm X^1)$, by an imaginary angle $\pm i\epsilon$, where

$$\epsilon = \frac{1}{\pi}[arcth(\pi e_1 E) + arcth(\pi e_2 E)] \ . \tag{2}$$

Here e_1 and e_2 are the two charges at the string endpoints, and we have set $2\alpha' = 1$. Taking into account the above twist, and the anomalous commutation relation of the center-of-mass coordinates $[x^+, x^-] = \frac{-i}{(e_1+e_2)E}$, one arrives [4] at the following expression for the annulus contribution to the vacuum energy density of oriented bosonic strings :

$$\mathcal{F} = \int_0^\infty \frac{dt}{t} \ (2\pi^2 t)^{-\frac{D}{2}} \ Z(t) \ f_A(t, e_1 E, e_2 E) \ , \tag{3}$$

where $Z(t) = \sum_{states} e^{-\frac{\pi t}{2} m_S^2}$ is the usual partition function of an open string, D is the number of non-compact space-time dimensions, and the entire field dependence is contained in the correction factor

$$f_A = \frac{\pi(e_1 + e_2)Ete^{-\pi t\epsilon^2/2}}{2sin(\pi t\epsilon/2)} \prod_{n=1}^{\infty} \frac{(1 - e^{-\pi t n})^2}{(1 - e^{-\pi t(n+i\epsilon)})(1 - e^{-\pi t(n-i\epsilon)})} \ . \tag{4}$$

This factor goes to one in the zero-field limit, and has evenly-spaced simple poles at $t = \frac{2k}{|\epsilon|}$ $(k = 1, 2, ..)$ on the integration axis. These poles give in turn rise to an imaginary part of $\mathcal{F}$, and hence to a string pair-production rate

$$w_{bos} = \frac{1}{2(2\pi)^{D-1}} \sum_{states \ S} \frac{(e_1 + e_2)E}{\epsilon} \sum_{k=1}^{\infty} (-)^{k+1} \Big(\frac{|\epsilon|}{k}\Big)^{D/2} \exp\Big(-\frac{\pi k}{|\epsilon|}(m_S^2 + \epsilon^2)\Big) \ . \tag{5}$$

The above result stays valid for unoriented strings [4], for which the projection onto physical states is achieved by adding to the annulus the contribution of the Möbius-strip [6]. For the fermionic string, on the other hand, a straightforward extension of the calculation [4] gives

$$w_{sustr} = \frac{1}{2(2\pi)^{D-1}} \sum_{states \ S} \frac{(e_1 + e_2)E}{\epsilon} \sum_{k=1}^{\infty} (-)^{(k+1)(a_S+1)} \Big(\frac{|\epsilon|}{k}\Big)^{D/2} \exp\Big(-\frac{\pi k m_S^2}{|\epsilon|}\Big), \tag{6}$$

where $a_S = 0$ or 1 according to whether the state S belongs to the Neveu-Schwarz or the Ramond sector.

The string-pair-production rate, eqs.(5,6), reduces to Schwinger's result, eq.(1), in the weak-field limit, as has been noticed before by Burgess [7]. The non-minimal electromagnetic coupling of strings manifests, however, itself for stronger E, and is essentially summarized by the replacement:

$$(e_1 + e_2)E \to \epsilon \simeq (e_1 + e_2)E \times [1 + o\big((E\alpha')^2\big)] \ . \tag{7}$$

The non-linear function ϵ, as well as the string-pair-production rate, diverge at the limiting field value $E_{lim} = 1/2\pi\alpha' max(e_i)$. The existence of a limiting field has been noticed also at the classical level [8,5,7], but the above result clearly establishes the responsible dissipation mechanism. Analogous results for strong gravitational, or non-abelian fields would be extremely interesting. The latter could in particular drastically modify our current semi-phenomenological [9] understanding of fragmentation in Q.C.D.

Before concluding let me note that while the string-structure seems to enhance the effects of an electric field, it moderates those of a magnetic field. A magnetic field, $F^{12} = B$, twists the complex space-coordinate $X = \frac{1}{\sqrt{2}}(X^1 + iX^2)$ by a real phase [5]

$$\tilde{\epsilon} = \frac{1}{\pi}[arctan(\pi e_1 B) + arctan(\pi e_2 B)] \ , \tag{8}$$

which determines the spacing between Landau levels. For sufficiently massive string states this spacing reads $\tilde{\epsilon}/2m$. It reduces to the usual cyclotron frequency for weak B, but levels off towards a limiting value $1/4\alpha' m$ as the field becomes stronger. This behaviour can be understood as due to a maximum allowed centripetal acceleration. Notice that the degeneracy of any given Landau level is the same as for a point particle, since the commutation relation of the center-of-mass coordinates is not modified [5]. Finally let me point out that the annulus amplitude in the presence of a B-field has no poles on the integration axis . Thus, as in point-particle field theory, a magnetic field cannot dissipate by spin-polarizing the vacuum.

This research was supported in part by EEC contracts SC1-0394-C and SC1-CT92-0792.

References

1. S.W. Hawking, Nature **248** (1974) 30 ; Comm.Math.Phys. **43** (1975) 199.
2. C.G.Callan, R.C.Myers and M.J.Perry, *Nucl. Phys.* **B311** (1988/89) 673.
3. J.Schwinger, Phys.Rev. **82** (1951) 664.
4. C.Bachas and M.Porrati, Phys.Lett. **296B** (1992) 77.
5. A.Abouelsaood, C.G.Callan, C.R.Nappi and S.A.Yost, *Nucl. Phys.* **B280** [FS18] (1987) 599.
6. See for instance A.Sagnotti, in *Non-Perturbative Quantum Field Theory*, G.'t Hooft et al eds., (Plenum, 1988), p. 521.
7. C.P.Burgess, Nucl.Phys. **B294** (1987) 427.
8. E.S.Fradkin and A.A.Tseytlin, Phys.Lett. **163B** (1985) 123; V.V.Nesterenko, Int.J.Mod.Phys. **A4** (1989) 2627.
9. B.Andersson, G.Gustafson, G.Ingelman and T.Sjöstrand, Phys.Rep. **97** (1983) 33.

Quantum Models of Black Hole Evaporation

F.Belgiorno, A.S.Cattaneo
Dipartimento di Fisica, Universitá di Milano and INFN sez. di Milano
20133 Milano, Italy

F.Fucito
Dipartimento di Fisica, Universitá di Roma II "Tor Vergata" and INFN sez. di Roma II
00173 Roma, Italy

and

M.Martellini
Dipartimento di Fisica, Universitá di Roma I "La Sapienza"
00185 Roma, Italy

ABSTRACT

The discovery of black-hole evaporation represented in many respects a revolutionary event in scientific world; as such, in giving answers to open questions, it gave rise to new problems part of which are still not resolved.
Here we want to make a brief review of such problems and examine some possible solutions.

1. BH Evaporation: Open Questions

The great problem of bh-evaporation is well exemplified by the following question: which is the black-hole final state?
A complete answer to this question contains the solution to the other great enigma: is there a loss of information in bh-evaporation?
That is, is it possible to recover the information contained in the initial state of bh-evolution? In dimension $D = 4$ these question have no answer. Hawking's work[11] is based on two distinct approximations:

a)the semiclassical approximation in which the bh background geometry is classical,

b)the emitted radiation has no backreaction on the geometry.

It is well known that both these approximations fail at $m_{bh} \sim M_{Planck}$; at such energy scales it is necessary to make a full quantum treatement of the gravitational field.

The so called "Information Loss Problem" consists in what follows: it is reasonable to think that we can associate to black-holes a great entropy content, because we are not able to measure and specify its internal microstates; our only physical possibility to speak about a bh is to assign to it a macrostatus defined generally by three measurable quantities: its mass, its angular momentum, its charge. As soon as the bh forms, because of the event horizon, all the informations about the initial state and microscopic structure of the collapsing star are lost. This is why we say that the "bh has no hair".

The bh entropy is proportional to the area of its horizon. During the evaporation the horizon shrinks causing the bh entropy to decrease. Can this cause a loss of information?

This same problem can also be reformulated looking for the existence of a unitary S-matrix[12].

2. Bidimensional Models

Given the difficulties of the four dimensional case much work has been recently concentrated on the study of bidimensional quantum models that could give some insight, at least qualitatively, about the physical case.

Our starting point is the action containing the coupling of gravity to a dilaton field

$$S_{DG} = \frac{1}{2\pi}\int d^2x\sqrt{-g}e^{-2\Phi}(R + 4(\nabla\Phi)^2 + 4\lambda^2). \tag{2.1}$$

As it is well known it is the coupling of the dilaton field Φ to the Einstein metric that makes the theory non trivial. The main advantages of such models are:

i) the theory is perturbatively renormalizable so that, at least in principle, we can indagate its quantum aspects;

ii) we can think about S_{DG} as deriving from an off-critical string theory in a low energy approximation;

iii)after some manipulations the dilaton-gravity can be mapped into some soluble model

The theory, using light-cone coordinates and in the conformal gauge $g_{+-} = -\frac{e^{2\rho}}{2}$ (ρ is the Liouville mode) looks like:

$$S_{DG} = \frac{1}{\pi}\int d^2\sigma[e^{-2\Phi}(2\partial_+\partial_-\rho - 4\partial_+\Phi\partial_-\Phi + \lambda^2 e^{2\rho})] \tag{2.2}$$

3. The CGHS Model

The purpose of this model[6] is to provide a description of:

i)the creation of a bh by means of matter shock waves

ii)the backreaction problem

This is achieved by summing to the action Eq.(2.1) a matter term:

$$S_M = \frac{-1}{4\pi} \int d^2 x g^{\frac{1}{2}} \sum_{i=1}^{N} (\nabla f_i)^2, \tag{3.1}$$

where the f_i's are conformal massless matter fields minimally coupled to gravity whose energy tensor in light cone coordinates is proportional to $\delta(x^+ - x^0)$.

One way to keep the backreaction into account is to include at the Lagrangian level the integrated trace anomaly which is local in our gauge: the asymptotically flat coordinates trace anomaly contribution is:

$$S_{anomaly} = -\frac{N}{12\pi} \int d^2\sigma \partial_+ \rho \partial_- \rho. \tag{3.2}$$

In the large N limit, we can perform a 1/N expansion to find that the anomaly term is of the same order of magnitude of the classical contribution. Furthermore the quantum effects of gravity and of the dilaton can be neglected in this approximation.

An analysis of the model reveals two different regions of the Penrose diagram. The first is a weak coupling region where $Exp(-2\Phi) \gg \frac{N}{12}$. The second is a strong coupling region where $Exp(-2\Phi) \ll \frac{N}{12}$. These two regions are separated by a critical time-like line $Exp(-2\Phi_{cr}) = \frac{N}{12}$.

The problem[5,13,14] is that $\Phi \sim \Phi_{cr}$ leads to a naked curvature singularity or, in other words, the semiclassical approximation breaks down. Moreover we find that the Hawking temperature is mass independent contrary to our physical expectations.

4. The Bilal-Callan-De Alwis Model

One possible way out of these problems[4,7,8,9] is to build two-dimensional models derived from low energy string theories using the many possible vacua that lead to vanishing beta functions. The recipe to do this is:

i) to replace the $\frac{N}{12}$ coefficient in Eq.(3.2) with a generic trace anomaly coefficient κ to be determined by imposing the vanishing of the total central charge;

ii) to modify the potential term by replacing it with conformal vertices of weight (1,1) which in the classical limit $\Phi \to -\infty$ reproduce the cosmological term.

The models thus obtained are Liouville-like. They are still ill-defined at the quantum level as their kinetic energy goes to zero at a certain critical value of the dilaton field for which also the curvature is singular. A proposed way out of it is to restrict the fields to a region with no singularity. These restriction are not natural from the point of quantum field theory and up to now nobody has succeeded in doing so consistently.

5. The CAT Model

We start with the following definitions:
$\omega \equiv \frac{Exp(-\Phi)}{k^{\frac{1}{2}}}$
$2\Omega(\omega) \equiv \omega(\omega^2-1)^{\frac{1}{2}} - Log(\omega + (\omega^2-1)^{\frac{1}{2}})$
for $\omega \geq 1$ (as in ref. 4) and
$2\Omega'(\omega) \equiv \omega(1-\omega^2)^{\frac{1}{2}} - ArcCos(\omega)$
for $\omega \leq 1$
$2\chi \equiv \rho + (\omega)^2$.
It should be noted that the value $\omega = 1$ corresponds to the critical value of the dilaton field.

The new idea is now the following: we shall formulate an effective theory with the explicit aim of describing physics around $\omega = 1$ by mean of a *non-local field redefinition*:

$$2\eta(\sigma) \equiv \Omega(\sigma+\epsilon) + \Omega'(\sigma-\epsilon)$$
$$2\xi(\sigma) \equiv \Omega(\sigma+\epsilon) - \Omega'(\sigma-\epsilon)$$

where $\sigma+\epsilon$, $\sigma-\epsilon$ are two nearby points and $\omega(\sigma+\epsilon) \sim 1^+$, $\omega(\sigma-\epsilon) \sim 1^-$.
The signature of the kinetic term thus obtained does not oscillate.

Now we have to construct interaction vertices in these new field variables with conformal weight $(1,1)$ and match them against the cosmological constant term in the classical limit.

The resulting action is

$$S_{cat} = \frac{1}{2\pi}\int d^2x[-k(\partial\chi)^2 + (\partial\eta)(\partial\xi) + \gamma_+ e^{A_+\chi} + \gamma_- e^{-A_-\chi + B\eta}] \tag{5.1}$$

where the parameters appearing in the action are adjusted by the request of conformal invariance[1].
S_{cat} describes an affine $\widehat{Sl(2)}$ Toda theory, previously studied by Babelon-Bonora in the framework of integrable models[3].
This is more evident if we make another field redefinition: $\varphi \equiv i\sqrt{2\kappa}\chi$. We shall refer to our model as the conformal affine Toda black hole (CATBH) model. The CATBH model allows a standard perturbative quantization with the aim of unveiling the quantum effects of the dilaton-graviton fields: as a matter of facts, CAT fields are suitable functions of the dilaton and of the Liouville fields.

We now want to consider the renormalization flow of the classical Babelon-Bonora action

$$S_{BB} = \frac{1}{2\pi}\int d^2x \left[\frac{1}{2}\partial_\mu\varphi\partial^\mu\varphi + \partial_\mu\eta\partial^\mu\xi - 2\left(e^{2\varphi} + e^{2\eta-2\varphi}\right)\right]. \tag{5.2}$$

At the quantum level one must implement wave and vertex function renormalizations so that in (5.2) one must introduce different bare coupling constants in front of the fields as well as in front of the vertex interaction terms. As a consequence one ends with the form (5.1). However, according to the general spirit of the renormalization procedure, all generally covariant dimension 2 counterterms are possible in (5.1) and hence also Feigin–Fuchs terms, i.e. the ones involving the $2D$-scalar curvature, are allowed. This ansatz is in agreement with the perturbative theory as one could show following Distler and Kawai. In our context the Feigin–Fuchs terms come naturally out if we look at the improved stress tensor . This leads us to consider the following quantum action:

$$S = \frac{1}{2\pi}\int d^2x\, \sqrt{g}\left[g^{\mu\nu}\left(\frac{1}{2}\partial_\mu\varphi\partial_\nu\varphi + \partial_\mu\eta\partial_\nu\xi\right) + \gamma_+ e^{i\lambda^{(+)}\varphi} + \gamma_- e^{i\delta\eta - i\lambda^{(-)}\varphi}\right. \\ \left. + iq_\varphi R\varphi + iq_\eta R\eta + iq_\xi R\xi\right]. \tag{5.3}$$

Let us now start with the renormalization procedure of (5.3) in a perturbative framework, in the hypothesis that $\lambda^{(-)^2} \sim \lambda^{(+)^2} \sim 4$. Notice that ξ plays the role of an auxiliary field; a variation with respect to ξ gives the on-shell equation of motion

$$\nabla_\mu\nabla^\mu\eta = iq_\xi R, \tag{5.4}$$

which in our perturbative scheme must be linearized around the flat space:

$$\partial_\mu\partial^\mu\eta = 0. \tag{5.5}$$

Following Ref.16, we define the renormalized quantities at an arbitrary mass scale μ by:

$$\begin{aligned} \varphi &= Z_\varphi^{\frac{1}{2}}\varphi_R, & \gamma_\pm &= \mu^2 Z_{\gamma_\pm}\gamma_{\pm R}, & q_\varphi^2 &= Z_\varphi^{-1}q_{\varphi R}^2, \\ \eta &= Z_\eta^{\frac{1}{2}}\eta_R, & \lambda^{(\pm)^2} &= Z_\varphi^{-1}\lambda^{(\pm)}{}_R^2, & q_\eta^2 &= Z_\eta^{-1}q_{\eta R}^2, \\ \xi &= Z_\eta^{-\frac{1}{2}}\xi_R, & \delta^2 &= Z_\eta^{-1}\delta_R^2, & q_\xi^2 &= Z_\eta q_{\xi R}^2. \end{aligned} \tag{5.6}$$

The following quantities are conserved through renormalization:

$$r = \tfrac{q_\varphi}{\lambda^{(-)}}, \quad l = \tfrac{\lambda^{(+)}}{\lambda^{(-)}}, \quad k = q_\xi\delta, \quad p = q_\xi q_\eta. \tag{5.7}$$

Following essentially the same procedures of Ref.16, with slightly modifications due to the presence of extra fields and by taking $\lambda^{(+)} \neq \lambda^{(-)}$, we find that the theory can be renormalized at one loop if we restrict ourself to the on-shell renormalization scheme, i.e. if we get rid of the terms in $\partial_\mu\partial^\mu\eta$ (produced by the renormalization) using (5.5). The true on-shell theory should rely on (5.4), but at this perturbative order curvature terms can be neglected: they are important only in the renormalization of $\gamma_\pm$. The curvature terms are taken into account[17] considering the modifications to the trace of the stress-energy tensor, calculated on-shell. We finally get the following β functions:

$$\begin{aligned}\beta_+ &:= \mu\frac{d\gamma_{+R}}{d\mu} = 2\gamma_{+R}\left(\frac{\lambda^{(+)}{}^2_R}{4} - 1 - q_{\varphi R}\lambda^{(+)}{}_R\right),\\ \beta_- &:= \mu\frac{d\gamma_{-R}}{d\mu} = 2\gamma_{-R}\left(\frac{\lambda^{(-)}{}^2_R}{4} - 1 + q_{\varphi R}\lambda^{(-)}{}_R - q_{\xi R}\delta_R\right),\end{aligned} \tag{5.8}$$

$$\begin{aligned}\beta_{\lambda^{(\pm)}} &:= \mu\frac{d\lambda^{(\pm)}{}^2_R}{d\mu} = \frac{1}{2}\gamma_{+R}\gamma_{-R}\frac{\lambda^{(+)}{}^2_R + \lambda^{(-)}{}^2_R}{2}\lambda^{(\pm)}{}^2_R,\\ \beta_\delta &:= \mu\frac{d\delta^2_R}{d\mu} = \frac{1}{2}\gamma_{+R}\gamma_{-R}\delta^4_R,\end{aligned} \tag{5.9}$$

$$\begin{aligned}\beta_{q_\varphi} &:= \mu\frac{dq_{\varphi}{}^2_R}{d\mu} = \frac{1}{2}\gamma_{+R}\gamma_{-R}\frac{\lambda^{(+)}{}^2_R + \lambda^{(-)}{}^2_R}{2}q^2_{\varphi R},\\ \beta_{q_\eta} &:= \mu\frac{dq_{\eta}{}^2_R}{d\mu} = \frac{1}{2}\gamma_{+R}\gamma_{-R}\delta^2_R q^2_{\eta R},\\ \beta_{q_\xi} &:= \mu\frac{dq_{\xi}{}^2_R}{d\mu} = -\frac{1}{2}\gamma_{+R}\gamma_{-R}\delta^2_R q^2_{\xi R}.\end{aligned} \tag{5.10}$$

Putting together the equations (5.9), we find that the following quantity is also conserved through renormalization.

$$d = \frac{1+l^2}{4}\frac{1}{\delta^2} - \frac{1}{\lambda^2}. \tag{5.11}$$

This result needs not to be valid beyond this perturbative order. Using the non-perturbative RG-invariants in (5.7) and dropping out the rather cumbersome R indices, we can rewrite the relevant β functions as:

$$\begin{aligned}\beta_+ &= 2\gamma_+\left(\frac{l^2-4lr}{4}\lambda^{(-)^2} - 1\right)\\ \beta_- &= 2\gamma_-\left(\frac{1+4r}{4}\lambda^{(-)^2} - 1 - k\right)\\ \beta_{\lambda^{(-)}} &= \frac{1+l^2}{4}\gamma_+\gamma_-\lambda^{(-)^4},\end{aligned} \tag{5.12}$$

Then we have to request that at a certain scale t_0, $t \equiv log(\mu)$, both vertex operators have a conformal weight $(1,1)$. As a consequence, $\gamma_\pm$ depend on the scale $t - t_0$

only through the cosh, which is an even function, and hence does not distinguish between IR and UV scales. The requirement of having vertex operators with the right conformal weight at the renormalization point t_0 forces the theory to be dual (under the exchange of the IR and UV scales). Explicitely we get:

$$\lambda^{(-)^2}(t) = \frac{8}{2+\kappa}\left\{1+k-\frac{2+(1-k)\kappa}{\kappa}\tanh\left[\frac{2+(1-k)\kappa}{\kappa}\frac{2+k}{1+k}(t-t_0)\right]\right\},$$
$$\gamma_-(t) = A\left\{\cosh\left[\frac{2+(1-k)\kappa}{\kappa}\frac{2+k}{1+k}(t-t_0)\right]\right\}^{-2\frac{1+k}{2+k}}. \tag{5.13}$$

This UV-IR duality implies that $\gamma_-(t)$ behaves in the same manner both at an UV and at an IR scale with the asymptotic behaviour:

$$\gamma_-(t) \overset{|t-t_0|\to\infty}{\sim} Ae^{-2s|t-t_0|}, \quad \text{with} \quad s = \tfrac{2+(1-k)\kappa}{\kappa}. \tag{5.14}$$

where $\kappa = (N-23)/12$ and s is always positive and close to 1.

The main result here is that the RG-analysis[1] shows that there exists an energy scale $t_{BC} \ll t_0$ at which our action "becomes" the de Alwis-Bilal-Callan action, in the sense that $\left|\frac{\gamma_-(t_{BC})}{\gamma_+(t_{BC})}\right| >> 1$. Notice that in the classical limit $e^\Phi \to 0$; our effective action in the IR-phase $t = O(t_{BC})$ becomes the one of ref. 4 since $\xi \to \eta \to \Omega$.

Our strategy at this point is to use the above running coupling constants γ_-, λ_- to define "effective" bh thermodynamic quantities at the scale t_{BC}. Indeed by a physical point of view the back-reaction should modify Hawking radiation emission and cause it to stop as soon as the bh has radiated away all its initial ADM mass, so that it should be reasonable to get a decreasing Hawking temperature rate at the end point of bh-evaporation. In our CATBH model the bh temperature is proportional to $\mu\gamma_-^{\frac{1}{2}}$. We can regard γ_- as a running coupling constant in terms of μ, which is roughly a measure of ADM mass. In particular, a reasonable ansatz for the bh solution formed by N infalling matter shock waves, allows us to find a simple link between the RG-scale t and the bh mass (that is the physical scale of our problem!). In fact a relation between the v. e. v. of the operator-valued scalar curvature $\sqrt{-g}R$ and the other CATBH running coupling constant λ_- can be obtained in the classical limit $e^\Phi \to 0$. If we consider the conformal gauge $\widehat{g_{\mu\nu}} = -\frac{1}{2}e^{2\lambda\rho}\eta_{\mu\nu}$, we get in the tree approximation:

$$< \sqrt{-\hat{g}}\hat{R} >_{tree} = 2\lambda\partial_\mu\partial^\mu < \rho >_{tree} . \tag{5.15}$$

In eq. (5.15) we may use the explicit solution describing the black hole formation by N-shock waves f_i, with $< \rho >_{tree}$ given by the CGHS classical solution:

$$e^{-2\lambda_-(t)<\rho(x^+)>_{tree}} = 1 - 2\lambda_-(t) < \rho(x^+) >_{tree} + O(\lambda_-^2)$$
$$= -\kappa a(x^+ - x_0^+)\theta(x^+ - x_0^+) - \gamma_-(t)x^+x^-, \tag{5.16}$$

where θ is the Heaviside function, $x^\pm \equiv x^0 \pm x^1$ and $a \equiv$ const. Therefore, at $x^+ = x_0^+$, where the f-waves are sitting, we have, using the light-cone coordinates $x^\pm$:$< \sqrt{-\hat{g}(x^+)}\hat{R}(x^+) >_{tree} \sim \gamma_-(t), x^+ \to x_0^+$.

Since here we have two scales x_0^+ and μ, it is reasonable to set (in $c = \hbar = 1$) $x_0^+ \equiv \frac{1}{\mu}$, since the natural scale which describes the black hole formation is x_0^+. Therefore we arrive to:

$$< [\sqrt{-\hat{g}}\hat{R}](x_0^+) > \sim \gamma_-[-log(x_0^+)]. \tag{5.17}$$

Furthermore in the CGHS solution, the mass m_{bh} of the bh created grows linearly with x_0^+, i.e. $m_{bh} \propto M_{Pl}^2 x_0^+$. Thus putting all together we get that the end point of the bh-evaporation is charaterized by:

$$< \sqrt{-\hat{g}}\hat{R} >_{tree} \to 0, m_{bh} \to 0, \tag{5.18}$$

$$T_{bh} \sim T_0(\frac{m_{bh}}{m_0})^{s-1} \to 0, m_{bh} \to 0. \tag{5.19}$$

Notice that for "astrophysical" bh, i.e. with large mass, one finds that

$$T_{bh} \sim T_0(\frac{m_0}{m_{bh}})^{s+1} \to 0, m_{bh} \to \infty. \tag{5.20}$$

The vanishing of the bh temperature both for small and large bh mass is a consequence of the duality between UV and IR scales observed above.
Then we see that the end-point of the bh-evaporation is charaterized by a regular geometry and an almost zero Hawking temperature.

There are various lines of developement of this model[2]; in particular we signal: i)the calculation of the geometry and dilaton associated to the solutions of our conformal affine Toda model; ii) the derivation of an (unitary?) exact S-matrix which is allowed (in principle) by the quantum integrability of (5.3).

References

1. F.Belgiorno, A.S. Cattaneo, M. Martellini and F. Fucito, *A Conformal Affine Toda Model of 2D Black Holes: a Quantum Study of the Evaporation End-Point* to appear in Phys.Lett.B.
2. F. Belgiorno, A.S. Cattaneo, M. Martellini and F. Fucito, *A Conformal Affine Toda Model of Two Dimensional Black Holes: the End Point State and the S-Matrix*, work in progress.
3. O. Babelon and L. Bonora , *Phys. Lett.* **B244**(90)220.
4. A. Bilal and C. Callan, *Liouville Models of Black Hole Evaporation*, Princeton preprint PUPT-1320, hept@xxx/9205089, May 1992.
5. B. Birnir, S.B. Giddings, J.A. Harvey and A. Strominger, *Quantum Black Holes*, Santa Barbara preprint UCSB-TH-92-08, hepth@xxx/9203042, March 1992.
6. C.G. Callan, S.B. Giddings, J.A. Harvey, and A. Strominger , *Phys. Rev.* **D45**(92)1005.
7. S.P. de Alwis, *Quantization of a Theory of 2d Dilaton Gravity* , Colorado preprint COLO-HEP-280, hepth@xxx/9205069, May 1992.
8. S.P. de Alwis, *Black Hole Physics from Liouville Theory*, Colorado preprint COLO-HEP-284, hepth@xxx/9206020, June 1992.
9. S.P. de Alwis, *Quantum Black Holes in Two Dimensions*, Colorado preprint COLO-HEP-288, hepth@xxx/9207095, Jul 1992.
10. S.B. Giddings and A. Strominger, *Quantum Theory of Dilaton Gravity*, UC Santa Barbara preprint, UCSB-TH-92-28, hepth@xxx/9207034, July 1992.
11. S.W. Hawking, *Comm.Math.Phys.* **43**(75)199.
12. S.W. Hawking, *Phys. Rev.* **D14**(76)2460.
13. S.W. Hawking, *Evaporation of Two Dimensional Black Holes*, CALTECH preprint CALT-68-1774, March 1992.
14. J.G. Russo, L. Susskind, and L. Thorlacius, *Black Hole Evaporation in* $1+1$ *Dimensions*, Stanford preprint SU-ITP-92-4, January 1992;
   L. Susskind and L. Thorlacius, *Hawking Radiation and Back-Reaction*, Stanford preprint SU-ITP-92-12, hepth@xxx/9203054, March 1992.
15. G. t'Hooft , *Nucl. Phys.* **B335**(90)138.
16. M.T.Grisaru, A.Lerda, S.Penati and D.Zanon *Nucl. Phys.* **B342**(90)564.
17. A.B.Zamolodchikov *JEPT Lett.* **43**(86)730.

STRINGY GRAVITATIONAL INSTANTONS, THE H-MAP AND N=4 MODULI DEFORMATIONS *

Marco Billó, Pietro Frè

SISSA - International School for Advanced Studies
Via Beirut 2, I-34100 Trieste, Italy
and I.N.F.N. sezione di Trieste

Luciano Girardello

Dipartimento di Fisica, Universitá di Milano
Via Celoria 16, I-20133 Milano, Italy
and I.N.F.N. sezione di Milano

Alberto Zaffaroni

SISSA - International School for Advanced Studies
Via Beirut 2, I-34100 Trieste, Italy
and I.N.F.N. sezione di Trieste

ABSTRACT

We report on recent results of ours on the problem of string propagation in a general instanton background for the case of the complete heterotic superstring. We have introduced the concept of generalized HyperKähler manifolds, and we have related it to (4,4) superconformal theories. Our proposal of a generalized h-map construction predicts a universal $SU(6)$ symmetry for the modes of the string excitations moving in an instanton background. We have also discussed the role of abstract N=4 moduli and, applying the method to the particular limit case of the solvable $SU(2) \times \mathrm{R}$ instanton found by Callan et al. we have shown that it admits deformations. As a consequence the $SU(2) \times \mathbf{R}$ instanton corresponds to a point in a 16-dimensional moduli space. The geometrical characterization of the other spaces in the same moduli-space remains an outstanding problem.

Gravitational instantons (for a review see [1,2]) have an intrinsic interest and furthermore may provide a mechanism for the non perturbative breaking of local supersymmetry [3]. Their study, in the context of string theory, is a part of a more general study, namely the identification of the Conformal Field Theories corresponding to the geometries under consideration and the study of their properties.

Significant advances have been recently made in identifying conformal field theories related with black-hole space-times in unphysical dimensions ($D=2$) [4] and some insights have also been obtained on the physical case [5]. A limit case of instanton conformal field theory has been discussed in a seminal paper by Callan, Harvey and Strominger [6].

So far this issue has not been addressed in a full-fledged heterotic superstring framework. Indeed, like the case of the Witten black-hole [4] the investigation has been limited to a discussion of the σ-model on the selected space-time, ignoring the internal degrees of freedom of the superstring and the question of modular invariance.

In a recent paper[7] we have considered the issue of gravitational instantons in the context of the complete heterotic superstring. Here we report on some advances we made in this field.

A basic point of our work is the proposal of a *generalized h-map*, according to which the propagation on an instanton background of a heterotic superstring compactified on a Calabi-Yau space is given by the tensor product of three conformal theories: a $c = 6$ (4,4)-theory, a $c = 9$ (2,2)-theory and the $c = 11$ right-moving current algebra of $SO(6) \times E_8'$. A general prediction of our framework is that all particle modes in any istantonic background of this type are classified by an $SU(6)$ symmetry group. This prediction is absolutely analogous to the prediction that, in Calabi-Yau compactifications, massless particles fall into E_6 representations.

In [7] we have studied in depth the relation between (4,4) world-sheet supersymmetry and the self-duality (respectively antiselfduality) of the curvatures $R(\omega_R \pm T)$, where T is the torsion and ω_R is the Riemannian connection. Spaces with this property correspond, as we show, to a *generalization of HyperKähler manifolds* and have the proper geometry to describe axionic-dilatonic instantons. These configurations were originally introduced by D'Auria and Regge [8] long time ago and were more recently rediscovered in string theory by Callan, Harvey and Strominger [6] and by Rey [9]. Their distinguished feature is asymptotic flatness which is an essential feature in order to utilize the instanton in any supersymmetry breaking mechanism à la Konishi et al. [10]. Indeed, for the instanton to contribute to a scattering amplitude, the asymptotic states must be the same in flat space and in the instanton background. Such asymptotic flatness requires an unsoldering of the Lorentz-bundle from the tangent bundle which is indeed what the axionic and dilatonic fields combined are able to realize.

Another basic point of our investigation has been the moduli-deformations of the (4,4)-theory. We have discussed the general characterization of the moduli within an arbitrary $c = 6$ (4,4)-theory and their use to construct the emission vertices of the particle zero-modes appearing in the spectrum of the heterotic superstring. Both HyperKähler and generalized HyperKähler manifolds lead to (4,4) world-sheet supersymmetry, so that the associated conformal field theory has the same structure and the same properties for both kind of manifolds. This is the very reason why we name generalized HyperKähler the manifolds under consideration in [7].

Considering the limit case where the manifold of the axion-dilaton instanton reduces to $SU(2)\times \mathrm{R}$ [6], case where asymptotic flatness is lost, but conformal solvability is gained, in [7] we have discussed the (4,4)-moduli of this specific model showing that they are four as for flat space. We have explicitely exhibited the infinitesimal deformations of the metric and of the torsion, finding that the $SU(2)\times \mathrm{R}$ instanton is a point in a 16-dimensional space. The global characterization of this moduli space and the geometric interpretation of the deformed space is therefore still an outstanding problem.

Let us shortly review the main points of the work we have just outlined.

We want to discuss a general framework for the investigation of gravitational instantons in the context of heterotic sting theory; the fundamental point is of course the characterization of the CFT representing the corrisponding complete quantum theory. We consider heterotic string theory with gauge group $E_8' \times E_8$ and we investigate the abstract structure of the consistent vacua corresponding to a target space o the form $\mathcal{M}_4^{instanton} \times \mathcal{M}_6^{CY}$ where $\mathcal{M}_4^{instanton}$ is the 4D manifold of a gravitational instanton and $\mathcal{M}_6^{CY}$ is a Calabi-Yau manifold. Particularly interesting are, in view of their possible relevance for the spontaneous breakdown of local susy[10], are the axionic-dilatonic instantons(instantons with torsion). Indeed, in order to contribute to the breakdown mechanism, the space must be asymptotically flat, and this requires an unsoldering of the principal Lorentz bundle from the tangent bundle[11]. This can be achieved by the combined action of axionic and dilatonic field.

As an example, we will treat the solution considered in [6], in the particular case in which it corresponds to an exactly solvable (4,4) supersymmetric conformal field theory (CFT) (loosing unfortunately asymptotic flatness). In particular, applying our general set-up, we are able to show how to deform it without spoiling its (4,4) supersymmetry, although the geometrical interpretation of the deformed theory is not obvious.

The intrinsic feature of gravitational instantons with torsion is the selfduality (resp. antiselfduality) of the curvatures for the two non-Riemannian connections $\omega_R - T, \omega_R + T$ (ω_R is the Riemannian connection, T the torsion); it is equivalent to the conditions for the σ-model based on this manifold to admit $(4,4)$ global SUSY of world-sheet. Let's see how. The (1,1)-supersymmetric action for the σ-model on a generic manifold is written, in 2nd order formalism, as follows [12]:

$$S = \frac{-1}{4\pi}\int_{\partial\mathcal{M}} dz\, d\bar{z} \left\{ V^a_z V^a_{\bar{z}} + 2i\lambda^a \overset{-}{\nabla}_z \lambda^a - 2i\mu^a \overset{+}{\nabla}_{\bar{z}} \mu^a - 4R^{ab}{}_{cd}\lambda^a\lambda^b\mu^c\mu^d \right\} + \\ + \frac{1}{4\pi}\int_{\mathcal{M}} H \tag{1}$$

where V^a are the target vielbein and where the left fermions λ^a are coupled via the left connection, $\omega_R - T$, and the right fermions μ^a via the right one, $\omega_R + T$; supersymmetry enforces the relation $T_{abc} = -3H_{abc}$ *.

* In [7] we reported the "local" version of the action, in which the two-dimensional gravitinos are switched on. In that context it is easy to retrieve the conditions for an extended (N,N) SUSY. From the local expression

As well known [13] the condition to admit (4,4) susy is the existence of two mutually commuting sets of 3 complex structures, each realizing a representation of the quaternionic algebra

$$\mathcal{J}^x \mathcal{J}^y = -\delta^{xy} + \epsilon^{xyz} \mathcal{J}^z \tag{2}$$

covariantly constant one with respect to the $\omega_R - T$ connection and the other with respect to the $\omega_R + T$ connection. For $T = 0$ these requirements force the space to be a HyperKähler manifold.
Any antisymmetric matrix can be expanded along the basis given by the two sets of 3 matrices

$$\begin{aligned} \hat{\mathcal{J}}^x_{ab} &= -(\delta_{a0}\delta_{bx} - \delta_{b0}\delta_{ax} + \epsilon_{xab}) \\ \tilde{\mathcal{J}}^x_{ab} &= (\delta_{a0}\delta_{bx} - \delta_{b0}\delta_{ax} - \epsilon_{xab}) \end{aligned} \tag{3}$$

respectively selfdual and antiselfdual, mutually commuting, each realizing the algebra (2). We can therefore expand on this basis the complex structures of the manifold. The requirement of mutual commutation implies then that one of them is selfdual and the other one antiselfdual.
Analogously the curvatures relative to $\omega_R \pm T$, regarded as antisymmetric matrices of two-forms, admit the expansion $R = A_x \hat{\mathcal{J}}^x + B_x \tilde{\mathcal{J}}^x$. The integrability condition $[R, \mathcal{J}^x] = 0$ for the covariant constancy of the relevant complex structures forces one of the curvatures to be selfdual and the other one antiselfdual. This coincides, as said before, with the characterizing feature of gravitational instantons with torsion.

Let us turn to the abstract construction of heterotic string vacua, compactified on a C.Y. manifold, where the usual $c = (6,4)$ CFT that represents 4-dimensional flat space is replaced by some new $c = (6,6)$ theory describing the string propagation on an instantonic 4-dimensional geometry. We consider an extension of the well-known h-map mechanism . We start from the CFT describing the heterotic σ-model with a 10-dimensional target space

$$(15, 26) = (6, 4) \otimes (9, 6) \otimes (0, 16)$$

where we have indicated the left and right central charges, and we have split the 10 dimensions into 6+4. The 32 r.m. fermions corresponding to the (0,16) theory generate the current algebra of $E_8' \times E_8$. The process of compactification and the insertion of an instantonic background into the 4-dim sector can be described as follows: let the four and six-dimensional theory "eat" 4 (resp. 6) heterotic fermions, and impose that the resulting c=(6,6), (c=(9,9)) left-right symmetric theories have (4,4) (resp.(2,2)) supersymmetries:

$$(15, 26) = (6, 6)_{4,4} \oplus (9, 9)_{2,2} \oplus (0, 11) \tag{4}$$

From the geometric point of view, indeed, we consider a target space of the form

$$\mathcal{M}_{10} = \mathcal{M}_6^{CY} \times \mathcal{M}_4^{instanton}$$

---

one can easily extract the classical expressions for the supercurrents. It is also worth to use first order formalism in which dilatonic contributions are explicitely exhibited. These conceptually straightforward but useful points are not reported in this communication.

where $\mathcal{M}_6^{CY}$ is a C.Y. manifold and thus its holonomy group is precisely $SU(3)$. The holonomy of a gravitational instanton (whose curvature is selfdual or antiselfdual) is contained in $SU(2)$. The ten-dimensional axion Bianchi Identity $dH = 0$ requires (at 1st order)

$$0 = \mathrm{Tr} F \wedge F - \mathrm{Tr} R_{(6)} \wedge R_{(6)} - \mathrm{Tr} R_{(4)} \wedge R_{(4)}$$

and this is obtained by embedding the holonomy groups of the two submanifolds into the gauge group. This latter is thus broken to $SU(6)$, as it follows from the maximal subgroups embedding

$$SU(6) \times SU(3) \times SU(2) \longrightarrow E_8 \tag{5}$$

The $SU(6)$ symmetry is abstractly reproduced by (4), since the $U(1)$ current of the $N{=}2$ algebra associated with $\mathcal{M}_6$, the $SU(2)$ currents of the $N{=}4$ associated with $\mathcal{M}_4$ and the $SO(6)$ currents of the heterotic fermions combine, together with suitable spin fields, to yeld the $SU(6)$ current algebra, according to the maximal embedding

$$SO(6) \times U(1) \times SU(2) \longrightarrow SU(6) \tag{6}$$

On this background the emission vertices for the particle modes are organized into $SU(6)$ representations as required by Kaluza-Klein analysis.

Modular invariance is easily enforced in a type II partition function. The h-map [20] converts modular invariant type II string vacua into consistent modular invariant heterotic vacua. It is based on the isomorphism between the $SO(2n)$ characters and those of $SO(2n{+}24)$ or $E_8' \times SO(2n{+}8)$. It is usually employed to replace in the flat type II theory the two transverse fermions present in light-cone quantization with 26 heterotic fermions generating $E_8' \times SO(10)$, ending with a consistent heterotic vacuum. In our case we are forced to use covariant quantization. However, it is well known [14] that the superghosts transform in such a way as to cancel the effects of two fermions; their subpartition functions can be organized so to have modular transformation obtained formally from those of the $SO(2n)$ characters by setting $n = -1$. We can therefore use the h-map to replace the superghosts with 22 heterotic fermions of the gauge group $E_8' \times SO(-2+8) = E_8' \times SO(6)$, reproducing a modular invariant vacuum which is just of the kind considered above.

The next important step is the construction of the vertex operators, that carry all the relevant physical information. In analogy with C.Y. compactification, we can put in corrispondence certain operators of the $(6,6)_{4,4}$ theory with "abstract cohomology classes" of $\mathcal{M}_4$; this is obtained comparing enumeration and group-theoretical indexing of the emission vertices for massless particle modes with the prediction of Kaluza-Klein analysis. The standard $N{=}4$ algebra contains the stress-energy tensor, the four supercurrents and three currents closing an $SU(2)_1$. In a (4,4) theory there is a realization of these operators both in the left and in the right sector. The fields of the theory are organized into representation of $SU(2)_L \times SU(2)_R$. We denote by $\Phi\begin{bmatrix} h,\tilde h \\ J,\tilde J\end{bmatrix}^{m,\tilde m}$ a primary field with left and right dimensione $h, \tilde h$ and isospins $J, \tilde J$, and with third components $m, \tilde m$. In each sector the $SU(2)_1$ can be bosonized in terms of a single free boson τ; the spectral flow of $N{=}2$ theories is extended to a "multiplet of spectral flows":

$$\Phi\begin{bmatrix} h \\ J \end{bmatrix}^m = e^{im\sqrt{2}\tau}\Phi\begin{bmatrix} h - m^2 \\ 0 \end{bmatrix}$$

As explained in [17], the N=4 analogues of the (c,c) and (c,a) fields of weights $(\frac{1}{2},\frac{1}{2})$ which play the role of abstract (1,1) and (2,1) forms of the $(9,9)_{2,2}$ theory are those primary fields of the $(6,6)_{4,4}$ which are of the form

$$\Psi_A \begin{bmatrix} 1/2, 1/2 \\ 1/2, 1/2 \end{bmatrix} \qquad A = 1, \ldots h^{1,1} \tag{7}$$

and that correspond to the lowest component of a short representation of the N=4 algebra. These fields are to be identified with the abstract (1,1) forms of the theory. Focusing on the left sector, a short representation is made of the following set of fields:

$$\Psi \begin{bmatrix} 1/2 \\ 1/2 \end{bmatrix}^a (z) \,,\; \Phi \begin{bmatrix} 1 \\ 0 \end{bmatrix} (z) \,,\; \Pi \begin{bmatrix} 1 \\ 0 \end{bmatrix} (z)$$

satisfying the OPEs

$$\mathcal{G}^a(z)\Psi^b(w) = \frac{\epsilon^{ab}\Phi(w)}{z-w} + reg. \qquad \overline{\mathcal{G}}^a(z)\Psi^b(w) = \frac{\delta^{ab}\Pi(w)}{z-w} + reg.$$

$$\overline{\mathcal{G}}^a(z)\Phi(w) = 2\epsilon^{ab}\partial\left(\frac{\Psi^b(w)}{z-w}\right) + reg. \qquad \mathcal{G}^a(z)\Pi(w) = -2\delta^{ab}\partial\left(\frac{\Psi^b(w)}{z-w}\right) + reg.$$

$$\mathcal{G}^a(z)\Phi(w) = \overline{\mathcal{G}}^a(z)\Pi(w) = 0$$

where $\mathcal{G}^a(z), \overline{\mathcal{G}}^a(z)$, $a = 1,2$ denote the supercurrents organized in two $SU(2)$ doublets. The fields Φ, Π, being last components of an N=4 rep., when added (combining left and right sectors) to the σ-model lagrangian preserve its N=4 invariance. We call these operators the N=4 moduli. It may happen that the particular $(6,6)_{4,4}$ theory allows left-right asymmetric fields of the form

$$\Psi_{\mathcal{A}} \begin{bmatrix} 0, 1/2 \\ 0, 1/2 \end{bmatrix} (z,\bar{z}) \qquad \mathcal{A} = 1, \ldots h^{0,1} \tag{9}$$

which in the right sector are lowest components of a short rep. They are then identified with the abstract (0,1) forms. The knowledge of an explicit $(6,6)_{4,4}$ CFT corresponding to a certain target manifold provides us with an "abstract Hodge diamond", that for zero torsion coincides with the usual Hodge diamond (at the same time, for zero torsion, generalized HyperKähler manifolds are just ordinary HyperKähler spaces) *.
When $\mathcal{M}_4$ is flat space, emission vertices have the form

$$V_\epsilon^{\cdot\cdot,\bullet}\,(k, z, \bar{z}) \;=\; \Phi_\epsilon(z,\bar{z})\, e^{[i\,k\cdot X(z,\bar{z})]}\, \Psi^{\cdot\cdot}(z,\bar{z})\, \Lambda^{\bullet}(z,\bar{z})$$

$\Phi_\epsilon(z,\bar{z})\, e^{[i\,k\cdot X(z,\bar{z})]}$ is a conformal field of the (6,4) theory, representing the CFT analogue of the pure-state wave function $\psi_{k,\epsilon}(X)$ satisfying $\Delta\psi_{k,\epsilon}(X) \;=\; m^2\, \psi_{k,\epsilon}(X)$ where Δ is

* See for example[15] where this fact is shown in the case of the K3 surface, whose corresponding CFT is given, in some points of its moduli space, by a tensor product of N=2 minimal models.

the relevant wave operator. It determines spin, polarization ϵ and momentum k which are good quantum numbers of the particle in flat space. $\Psi^{\cdot\cdot}(z,\bar{z})$ is a field of the $(9,9)_{2,2}$ and $\Lambda^{\bullet}(z,\bar{z})$ of the (0,13) theory: they determine the internal quantum numbers $(\cdot\cdot,\bullet)$.
If $\mathcal{M}_4$ is non-trivial, the eigenfunctions of the operators Δ are no longer plane waves and the spectrum is labeled by a new set of quantum numbers, correspondingly to the replacement of $\Phi_\epsilon(z,\bar{z})\, e^{[i\,k\cdot X(z,\bar{z})]}$ with suitable operators of the $(6,6)_{4,4}$ theory. A certain number of these operators correspond to zero modes of Δ and can be used to calculate the scattering of massless particles in the non-trivial background under consideration. The identificaton of such operators is made easier by recasting the flat space form of the vertices into an abstract $(6,6)_{4,4}$ language.
In this communication we focus only on the emission vertices for the fields of the gravitational multiplet. The construction of those for the other light fields of the effective theory * is quite straightforward and is reported in [7]; it exhibits the $SU(6)$ symmetry of the charged modes. CFT construction and Kaluza-Klein analysis perfectly agree.
In the $(6,6)_{4,4}$ theory describing flat space, the two sets of complex structures coincide (T=0)and can be chosen to be the $\hat{\mathcal{J}}^x$; the $SU(2)_1$ currents are made out of fermions only: $A^i = -\frac{i}{2}\psi^\mu \hat{\mathcal{J}}^x_{\mu\nu}\psi^\nu$. In each sector there are two short representations; the lowest components ($SU(2)$-doublets) are combinations of the ψ^μ; the last components Φ,Π are combinations involving the ∂X^μ. Multiplying the lowest components of the two sectors we find four abstract (1,1)-forms. It is also possible to construct two abstract (0,1)-forms $\mathbf{1}\left[\begin{smallmatrix}0\\0\end{smallmatrix}\right](z)\Psi\left[\begin{smallmatrix}\frac{1}{2}\\\frac{1}{2}\end{smallmatrix}\right](\bar{z})$. These numbers agree with the Hodge numbers for the compactified flat space. Of the four spin fields of the $SO(4)$ current algebra generated by the ψ^μ, the two with positive chirality are $SU(2)$ singlets and are the Ramond partners of the two fermion doublets; the two with negative chirality constitute an $SU(2)$ doublet related by spectral flow to the identity. For the usual flat space graviton and gravitino vertices we refer to [17,sec VI.10]. From the above observations, the generic $(6,6)_{4,4}$ expressions are easily recognized to be (the last field in the formulæbelow is the identity of the $(9,9)_{2,2}$ theory)

Graviton

$$e^{i\phi_{sg}(z)}\Phi_A\begin{bmatrix}1/2 & 1\\ 1/2 & 0\end{bmatrix}^a \mathbf{1}\begin{pmatrix}0 & 0\\ 0 & 0\end{pmatrix}$$

$$e^{i\phi_{sg}(z)}\Pi_A\begin{bmatrix}1/2 & 1\\ 1/2 & 0\end{bmatrix}^a \mathbf{1}\begin{pmatrix}0 & 0\\ 0 & 0\end{pmatrix}$$

$4h^{1,1}$ zero modes

Gravitino

$$e^{\frac{1}{2}\phi_{sg}(z)}\Phi_A\begin{bmatrix}1/4 & 1\\ 0 & 0\end{bmatrix}\mathbf{1}\begin{pmatrix}3/8 & 0\\ -3/2 & 0\end{pmatrix}$$

$$e^{\frac{1}{2}\phi_{sg}(z)}\Pi_A\begin{bmatrix}1/4 & 1\\ 0 & 0\end{bmatrix}\mathbf{1}\begin{pmatrix}3/8 & 0\\ -3/2 & 0\end{pmatrix}$$

---

* Recall that the matter content of the effective N=1, d=4 SUGRA is determined by the topology of the underlying CY manifold, and includes E_6-charged besides E_6-neutral fields

$$e^{\frac{1}{2}\phi_{sg}(z)}\mathbf{1}\begin{bmatrix}1/4\\1/2\end{bmatrix}^a \Phi_A \begin{bmatrix}1\\0\end{bmatrix}\mathbf{1}\begin{pmatrix}3/8 & 0\\3/2 & 0\end{pmatrix}$$

$$e^{\frac{1}{2}\phi_{sg}(z)}\mathbf{1}\begin{bmatrix}1/4\\1/2\end{bmatrix}^a \Pi_A \begin{bmatrix}1\\0\end{bmatrix}\mathbf{1}\begin{pmatrix}3/8 & 0\\3/2 & 0\end{pmatrix}$$

$2h^{1,1}$ zero modes of (+) chirality and $4h^{0,1}$ zero modes of (-) chirality

The counting agrees with the zero-modes counting à la Kaluza-Klein for the metric + torsion; $3h^{1,1} - 2 + (h^{1,1} + 2) = 4h^{1,1}$ and for the gravitino fields.

Now we turn to the instantonic solution of [6]. Our point of view is to consider solutions of the effective theory arising from the compactification. Without entering in details, we just recall that this theory is a particular kind of matter coupled $N{=}1$, $d{=}4$ supergravity [21], containing a linear multiplet $(\phi, \chi, B_{\mu\nu})$ which can be duality-rotated into a WZ multiplet [22] whose scalar and pseudoscalar fields are named dilaton and axion. This can be described in the New Minimal formulation [16] and it turns out [7] that the following configuration

$$\begin{aligned} ds^2 &= e^{-2\Phi}(dx)^2 & & & V^a &= e^{-\Phi}e^a \\ e^{-2\Phi} &= A + \frac{2k}{r^2} & &\leftrightarrow & \Phi &= \log\frac{r}{\sqrt{2k}} \end{aligned} \tag{10}$$

$$H_{abc} = \frac{1}{3}\epsilon_{abcd}\partial_d\Phi$$

is a solution of the New Minimal equations of motion. This configuration is the one proposed in [6]; it has the nice feature of being asymptotically flat. It has been argued [6] that there are no α' corrections to this background, which describes thus an exact (all-orders) CFT when inserted in a σ-model .

As noted in [6], in the particular case $A{=}0$ (in which asymptotic flatness is lost) the corresponding CFT is solvable and quite simple in fact. We write the flat metric as $dx^2 = dr^2 + (r^2/2)\Omega^i \otimes \Omega^i$, where the Ω^i are the Maurer-Cartan forms of $SU(2)$, such that $d\Omega^i = -(\epsilon^{ijk}/\sqrt{2})\Omega^j\,\Omega^k$; redefining the radial coordinate as $t = \sqrt{2k}\log(r/\sqrt{2k})$ the metric of eq.(9) becomes finally

$$ds^2 = dt^2 + k\Omega^i \otimes \Omega^i \tag{11}$$

while the dilaton is linear in the coordinate t:

$$\Phi = \frac{t}{\sqrt{2k}} \tag{12}$$

The vielbeins can therefore be chosen as

$$V^0 = dt \quad ; \quad V^i = -\sqrt{k}\,\Omega^i$$

We use the formalism of σ-model on group-manifolds [12,17]. We reduce all geometrical quantities to the left and right invariant one-forms of $SU(2) \times \mathbf{R}$

$$\Omega = g^{-1}dg = \Omega^a T^a \quad ; \quad \bar{\Omega} = dg\,g^{-1}\,\bar{\Omega}^a T^a$$

$$\tilde{\Omega}^b = \Gamma^{ba}\Omega^a$$

Antihermitean generators T^a are normalized to $\mathrm{Tr}T^a t^b = \delta^{ab}$, so that the structure constants f^{abc} vanish on zero index, and otherwise their value is $\sqrt{2}\epsilon^{ijk}$. The adjoint matrix $\Gamma^{ab} = -\mathrm{Tr}(T^a g T^b g^{-1})$ is such that $\Gamma^{00} = 1, \Gamma^{0i} = 0$, Γ^{ij} is the adjoint matrix of $SU(2)$ $(i,j{=}1,2,3)$. The Riemannian connection and the torsion are given by:

$$\omega_R^{abc} = T^{abc} = -\frac{1}{2}\sqrt{\frac{k}{2}}\epsilon^{abcd}abc \tag{13}$$

where we consider the tensor ϵ to vanish on zero indices.
We observe that the left connection $\omega_R - T$ is identically zero, and the curvatures of both the connections $\omega_R \pm T$ vanish. The σ-model equations of motion reduce to

$$\bar{\partial}\Omega^a_z = 0 \quad ; \quad \bar{\partial}\lambda^a = 0$$

$$\partial\tilde{\Omega}^a_{\bar{z}} = 0 \quad ; \quad \partial\tilde{\mu}^a = 0$$

Note that the right-moving fields of the theory are not directly those appearing in the lagrangian derived from (1), but are obtained from them through the action of the adjoint matrix. Finally, the σ-model looks like

$$S = -\frac{1}{4\pi}\int dzd\bar{z}((dt)^2 + k\Omega^i\Omega^i + 2i\lambda^a\partial_{\bar{z}}\lambda^a + 2i\tilde{\mu}^a\partial_z\tilde{\mu}^a) - \frac{1}{3}\frac{k}{4\pi}\int\frac{\epsilon^{ijk}}{\sqrt{2}}\Omega^i\Omega^j\Omega^k \tag{14}$$

We stress that the presence of the dilaton linear in t turns this latter into a Feigin-Fuchs boson with immaginary background charge $-i\sqrt{\frac{2}{k}}$ at classical level. In 1^{st}order formalism it is easy to work out dilaton contributions to the stress-energy tensor and classical supercurrents [7].
Conditions for (4,4) susy are matched: the left complex structures $\hat{\mathcal{J}}^x$ are of course covariantly constant w.r.t $\omega_R - T = 0$; it is easy to show that $\Gamma^T\tilde{\mathcal{J}}^x\Gamma$ are covariantly constant with respect to $\omega_r + T$ *.
The quantization of the susy WZW model on $SU(2)\times \mathrm{R}$ is now straightforward. Focusing on the left sector, using the notation

$$dt = i\sqrt{2}J^0 \quad ; \quad \Omega^i = i\frac{\sqrt{2}}{k}J^i \quad ; \quad \lambda^a = e^{i\pi/4}\psi^a$$

we find as the result of a standard procedure

$$J^i(z)J^j(w) = \frac{k}{2}\frac{\delta^{ij}}{(z-w)^2} + \frac{i\epsilon^{ijk}J^k}{z-w} \tag{15a}$$

* The analysis can be performed in purely algebraic terms [18] putting in correspondence complex structures and Cartan decompositions of the Lie Algebra of the group in question.

$$J^0(z)J^0(w) = \frac{1}{2(z-w)^2} \tag{15b}$$

$$\psi^a(z)\psi^b(w) = -\frac{1}{2}\frac{\delta^{ab}}{z-w} \tag{15c}$$

The N=4 superconformal theory involves the stress-energy tensor

$$T(z) = J^0J^0 + \frac{1}{k+2}J^iJ^i + \frac{i}{\sqrt{k+2}}\partial J^0 + \psi^a\partial\psi^a \tag{16a}$$

the four supercurrents

$$\begin{aligned} G^0(z) &= 2\left\{-ij^a\psi^a + \frac{1}{3}\frac{\epsilon_{ijk}}{\sqrt{k+2}}\psi^i\psi^j\psi^k + \frac{\partial\psi^0}{\sqrt{k+2}}\right\} \\ G^x(z) &= 2\left\{-ij^a\hat{\mathcal{J}}^x_{ab}\psi^b + \frac{\epsilon_{ijk}}{\sqrt{k+2}}(\hat{\mathcal{J}}^x\psi)^i\psi^j\psi^k + \frac{\partial(\hat{\mathcal{J}}^x\psi)^0}{\sqrt{k+2}}\right\} \end{aligned} \tag{16b}$$

(where we have used the notation $j^a = (J^0, \frac{J^i}{\sqrt{k+2}})$ and the $SU(2)$ currents

$$A^i(z) = -\frac{i}{2}\psi^a\hat{\mathcal{J}}^i_{ab}\psi^b = i(\psi^0\psi^i + \tfrac{1}{2}\epsilon^{ijk}\psi^j\psi^k) \tag{16c}$$

The N=4 algebra of the right sector is obtained by replacing everywhere non-tilded with tilded quantities and $\hat{\mathcal{J}}^x$ with $\bar{\mathcal{J}}^x$.
As apparent from (16a) at quantum level the F.F. central charge is shifted to $c_{FF} = 1 + (6/k+2)$,the only value for which the total central charge sums up to 6, the correct value for a four-dimensional supersymmetric solution:

$$c = c_{FF} + c_{WZW} + c_{free\ fermions} = 1 + \frac{6}{k+2} + \frac{3k}{k+2} + 2 = 6$$

This shift $k \to k+2$ is a known effect due to the fact that the background charge terms couple to the SWZW at a certain level κ and the super Kac-Moody at level κ correspond to an ordinary K.M. of level $\kappa - C_{\mathrm{V}} = \kappa - 2$ plus free fermions, that is our case for $\kappa - 2 = k$. Without the dilatonic contributions, N=4 would still be present, but the supercurrents would close the N=4 extended algebra, based on the group $SU(2)\times SU(2)\times U(1)$ [19]; the well-known way to reduce it to the standard one * is indeed to add a background charge term; this process is built-in in the solution we are considering.
Having organized the supercurrents in $SU(2)$ doublets as

$$\mathcal{G} = \frac{1}{\sqrt{2}}\begin{pmatrix} G^0 - iG^3 \\ -(G^2 + iG^1) \end{pmatrix} \quad ; \quad \overline{\mathcal{G}} = (\mathcal{G})^*$$

* We refer for the notations and normalizations of the standard N=4 to [17]

we search for short representations of the N=4 (i.e. solutions of the OPEs (8)). We find two representations with lowest components

$$\Psi_1(z) = e^{-\sqrt{\frac{2}{k+2}}t}\begin{pmatrix}\psi^0 + i\psi^3\\ \psi^2 + i\psi^1\end{pmatrix} \quad ; \quad \Psi_2(z) = e^{-\sqrt{\frac{2}{k+2}}t}\begin{pmatrix}\psi^2 - i\psi^1\\ -(\psi^0 - i\psi^3)\end{pmatrix} \tag{18}$$

These fields are extremely close to the analogous ones for flat space, except for the presence of the exponential term, which is an operator of conformal weight zero in the F.F. theory. This factor is the counterpart of the plane-wave factor $e^{ik\cdot X(z,\bar{z})}$ appearing in the flat space vertices; also in that case the exponential has conformal weight zero as $k^2 = 0$. We can say that $k_0 = \sqrt{\frac{2}{k+2}}$ is the energy component of the four-momentum. It is fixed to a constant value in terms of the "spatial" components **k**, but this latter is quantized to fixed values for the $SU(2) \times \mathbf{R}$ background, due to the compactness of the constant-time slices, which are S^3. Combining left and right sectors we find four abstract (1,1)-forms and also two (0,1)-forms, analogously to what happens in flat space (eqs.(7-8)). As an abstract $(6,6)_{4,4}$ theory this background has the same Hodge diamond as flat space, although $(T \neq 0)$ it does not coincide with the usual diamond.

The last point we want to discuss is the following. The existence of non-trivial N=4 moduli implies that a generalized HyperKähler manifold corresponding to a $(6,6)_{4,4}$ with $h^{1,1}$ abstract (1,1)-forms $\Psi_A\left[{}^{1/2,1/2}_{1/2,1/2}\right]$ is actually an element in a continuous family of such manifolds, with $4h^{1,1}$ parameters; indeed we can add to the σ-model Lagrangian all the operators of the form $(last\ comp.)_A\left[{}^{1,1}_{0,0}\right](z,\bar{z})$, where in each sector we can reach by susy the Φ or the Π field, making a total of four for each Ψ_A.
The geometrical interpretation of the moduli and the knowledge of the moduli space is very important because in the functional integration we are supposed to integrate over all geometries. Given a scattering process there is an expression for the emission vertices of the i-th particle in each conformally flat background corresponding to a specific $(6,6)_{4,4}$ theory. At every order in string perturbation theory the true scattering amplitude is obtained by calculating the correlators for a fixed background, by integrating over the Riemann surface and its moduli space and then by summing over the backgrounds. This last sum has a discrete (sum over the various topologies) and a continuous part (integration over the instanton moduli space).
For the $SU(2) \times \mathbf{R}$ background we conclude that it admits 16-parameters deformations preserving N=4 susy. Indeed we have four left moduli (the last components of the reps. (18)) whose bosonic part can be written (reshifting the background charge term to its classical value) as

$$\Phi_1(z) = (V_z^2 + iV_z^1)e^{-\sqrt{\frac{2}{k}}t} \qquad \Phi_2(z) = (V_z^0 + iV_z^3)e^{-\sqrt{\frac{2}{k}}t}$$
$$\Pi_1(z) = -(V_z^0 - iV_z^3)e^{-\sqrt{\frac{2}{k}}t} \qquad \Pi_2(z) = (V_z^2 - iV_z^1)e^{-\sqrt{\frac{2}{k}}t}$$

and four right ones, expressed in terms of the right vielbeins $\tilde{V}^a$. The result is that we can deform the σ-model by adding to the Lagrangian the expression

$$e^{-\sqrt{\frac{2}{k}}t(z,\bar{z})}V_z^a M_{ab}\Gamma_{bc}V_{\bar{z}}^c \tag{19}$$

The constant matrix M belongs to $GL(4,\mathbf{R})$ and contains the 16 parameters. Recasting the resulting σ-model in the generic form (1) one gets explicitely (at least for infinitesimal deformations) the expressions of the deformed metric and torsion (and of the two connections). It can be shown that, as expected, the deformed manifold is still generalized HyperKähler : the curvatures of $\omega_R \pm T$ are selfdual (resp. antiselfdual) and one can find the two sets of complex structures satisfying all the requirements for (4,4) supersymmetry. However, the deformed manifold has non-vanishing curvatures $R(\omega_R - T)$ and $R(\omega_R + T)$ and i is explicitely different from the original manifold. Indeed it is no longer a group-manifold. The intrinsic characterization of this deformed manifold is a problem for future investigation.

REFERENCES

[1] T.Eguchi, P.B.Gilkey, and A.J.Hanson, Phys. Rep. **66** (1980) 213.

[2] T.Eguchi, A.J.Hanson Phys. Lett. **74B**(1978) 249; Ann. Phys. **120** (1979) 82

[3] K.Konishi, N.Magnoli and H.Panagopoulis, Nucl. Phys. **B309** (1988) 201, Nucl. Phys **B323** (1989) 441.

[4] E.Witten, Phys. Rev **D44** (1991) 314; R.Dijkgraf, E.Verlinde and H.Verlinde *"String propagation in a Black Hole geometry"* PUPT-1252; S.Elitzur, A.Forge, E.Rabinovici *"Some Global aspects of String Compactifications"* RI-143/90; G.Mandal, A.Sengupta and S.Wadia *"Classical solutions of 2D string theory"* IASSNS-HEP-91/10; M.Rocek, K.Schoutens, A.Sevrin *"Off-Shell WZW Models in Extended Superspace"* IASSNS-HEP-91/14; A.Giveon Mod. Phys. Lett. **A6** No 31 (1991) 2843.

[5] M.Rocek, C.Ahn, K.Schoutens, A.Sevrin *"Superspace WZW Models and Black Holes"* IASSNS-HEP-91/69; P.Ginsparg, F.Quevedo, *"Strings on Curved Space-Time: Black Holes, Torsion, and Duality"* LA-UR-92-640, hepth@xxx/9202092;
C.Kounnas, D.Lust *"Cosmological String Backgrounds from Gauged WZW Models"* CERN-TH.6494/92

[6] C.G.Callan, J.A.Harvey, A.Strominger Nucl. Phys. **B359** (1991) 611 ; C.G.Callan, Lectures at Sixth J.A.Swieca summer school, PUPT-1278.

[7] M.Billò, P.Fré, L.Girardello, A.Zaffaroni SISSA/159/92/EP, submitted to IJMP

[8] R.d'Auria, T.Regge Nucl. Phys. **B195** 308

[9] S.J.Rey, Phys. Rev. **D43** (1991) 526; *"On String Theory and Axionic Strings and Instantons"* Slac-PUB-5659 (Sept. 1991).

[10] K.Konishi, N.Magnoli and H.Panagopoulis, Nucl. Phys. **B309** (1988) 201, Nucl. Phys **B323** (1989) 441.

[11] R.d'Auria, T.Regge Nucl. Phys. **B195** 308

[12] L.Castellani,R.d'Auria and D.Franco, Int.Jour.Mod.Phys. **A6** 4009

[13] L.Alvarez-Gaumé and D.Z.Freedman, Comm. Math. Phys. **80** (1981) 443; S.J.Gates, C.M.Hull, M.Rocek Nucl. Phys. **B248** (1984) 157; P.S.Howe and G.Papadopoulos, Nucl. Phys. **B289** (1987) 264, Class. Quant. Grav. **5** (1988) 1647

[14] E.Verlinde, H.Verlinde Nucl. Phys. **B288** (1987) 357.

[15] T.Eguchi, H.Ooguri, A.Taormina and S.K.Yang, Nucl. Phys. **B315** (1989) 193.

[16] V.P. Akulov, D.V.Volkov and V.A.Suroka Teor. Math. Phys. **31** (1977) 285; M.F.Sohnius, P.C.West Phys. Lett. **105B** (1981) 353; see also R.D'Auria, P.Fré, P.van Nieuwenhuizen , P.K.Townsend Ann. Phys **155** (1984) 423

[17] L.Castellani, R.d'Auria and P.Fré, *"Supergravity and Superstrings: A geometric Perspective"* (World Scientific ed.), 1990

[18] Ph.Spindel, A.Sevrin, W.Troost, A.Van Proeyen Phys.Lett. **B206** (1988) 71, Nucl.Phys. **B308** (1988) 662

[19] A.Sevrin,W.Troost and A.Van Proeyen, Phys.Lett. **B208** (1988) 447

[20] D.Gepner, Trieste Lectures 1989; Nucl. Phys. **B296** (1988) 757; Phys. Lett. **B199** (1987) 380.

[21] S.Cecotti, S.Ferrara and L.Girardello, Int. Jour. Mod. Phys. **A10** (1989) 2475; A.Strominger, Phys. Rev. Lett. **55** (1985) 2547; A.Strominger, E.Witten Comm. Mathy. Phys. **101** (1985) 341; P.Candelas Nucl. Phys. **B298** (1988) 458; L.Dixon, V.S.Kaplunovski, J.Louis Nucl. Phys. **B329** (1990) 27.

[22] R.D'Auria, P.Fré, G.De Matteis, I.Pesando, Int. Jour. Mod. Phys. **A4** No14 (1989) 3577

CONFINEMENT AS A DUAL MEISSNER EFFECT, MONOPOLES, MATRIX MODELS AND THE UNITARY GAUGE OF QUANTUM CHROMODYNAMICS (II)

M. BOCHICCHIO
Theory Division, Cern
1211 Geneva 23, Switzerland

ABSTRACT

Recently I have shown that the effective action at large distances of a pure $SU(N)$ theory in $d = 4$, in the gauge $F_{0i}^{ch} = 0$, is the action of a set of charged gluons and photons coupled to random magnetic fields or equivalently to closed monopoles lines lying in the (23) planes. A remarkable point is that this representation becomes exact in the large-N limit. I also discuss the reasons for the peculiarity of the gauge choice.

1. Universality

It is commonly believed that the only non-perturbative formulation of non-Abelian gauge theories is the lattice one. In addition the lattice regularization preserves exact gauge invariance [1]. However, since the old days of quantum field theory, it has been known that if the regulator violates a non-anomalous symmetry it is always possible to reach a symmetric theory at quantum level at the price of fine-tuning non-symmetric counter-terms in such a way as to recover, in the limit of a large cut-off, the formal Ward identities of the symmetric theory [2]. Therefore at least in principle, non-perturbative QCD could be defined by the following functional integral:

$$Z = \int \exp[-S + \text{n.g.i. regulator} + \text{n.g.i. counter-terms}]\, DA \quad . \tag{1}$$

The question then arises whether the continuous non-compact version of the theory given by Eq. (1) and the compact one on the lattice in $d = 4$ lie in the same universality class, that is, whether both confine and give the same confinement physics. While the lattice theory certainly confines at strong coupling, this does not help us to understand whether or why the two versions of the theory should confine at weak coupling, although we already know from numerical simulations that the lattice theory confines also at weak coupling [3]. The most popular conjecture about confinement was made by 't Hooft and Mandelstam [4] who argued that confinement could be explained by a dual mechanism of superconductivity: a magnetic condensate would squeeze electric fluxes into thin tubes in QCD in the same way as the electric condensate in a superconductor squeezes an external magnetic field (Meissner effect). However this idea alone is difficult to implement technically, because there is a topological obstruction to the existence of magnetic (string-like in $d = 4$) defects in a pure $SU(N)$ theory: $\pi_1(SU(N)) = \pi_2(SU(N)) = 0$. A possible way out was suggested by 't Hooft [5], who argued that monopoles might occur in the gauge:

$$X^{ch} = 0 \quad , \tag{2}$$

where X is anything which transforms in the adjoint representation and the superscript ^{ch} means projection onto the non-diagonal generators. This gauge leaves a residual $U(1)^{N-1}$ symmmetry and allows the existence of monopoles because $\pi_1(U(1)) = Z$. The gauge $X^{ch} = 0$ becomes singular where two eigenvalues of X coincide and 't Hooft argues that precisely at those points monopoles occur. Yet no monopoles have been found since the appearance of [5]. Recently I have shown [6] that a pure $SU(N)$ theory in the gauge $F_{01}^{ch} = 0$ possesses magnetic defects, which are monopoles lines in $d = 4$ in the (23) planes and monopoles points in $d = 3$. In this talk I would like to make some considerations about how and why this picture arises and to elaborate on some of its consequences.

2. Monopoles and scaling

The statement that some configuration is topologically allowed does not imply necessarily that it has an important role in the dynamics of a theory. For example continuous QED has a $U(1)$ gauge symmetry but no monopoles. According to 't Hooft, there are many gauges for $SU(N)$ that leave a residual $U(1)^{N-1}$ symmetry. Some examples are: the renormalizable gauge,

$$\begin{aligned} D_\mu(A^0)A_\mu^{ch} &= 0 \\ D_\mu(A^0) &= \partial_\mu - i[A_\mu^0, \ ] \quad , \end{aligned} \tag{3}$$

and the unitary gauge (I call it unitary because, as we will see, ghosts do not propagate in this gauge),

$$F_{01}^{ch} = 0 \quad . \tag{4}$$

Although from a topological point of view, in order to enhance our chances of finding monopoles, they have equal opportunities, from a dynamical point of view they are very different, as I will show by means of the following scaling argument. The (partially) gauge-fixed partition function corresponding to the renormalizable gauge is:

$$\begin{aligned} Z_{gf} = & \int \exp\Big[-\int \frac{N}{\lambda} Tr F_{\mu\nu}(x)^2 + \omega^{ch}(x) D_\mu(A^0)[D_\mu(A)\eta^{ch}(x)]^{ch}\, dx \Big] \times \\ & \times \delta(D_\mu(A^0)A_\mu^{ch})\; DA\; D\omega\; D\eta \quad , \end{aligned} \tag{5}$$

while the one corresponding to the unitary gauge is:

$$\begin{aligned} Z_{gf} &= \int \exp\Big[-\int \frac{N}{\lambda} Tr F_T(x)^2 \, dx\Big] \times \\ &\times \exp\Big[-\int \frac{N}{\lambda}\sum_i^N F_L^i(x)^2 \, dx\Big] \prod_{x,L}\prod_{i>j}^N [F_L^i(x) - F_L^j(x)]^2 \times \\ &\times \delta(F_L^{ch}(x)) \; DA \quad . \end{aligned} \tag{6}$$

By rescaling the fields A by a factor of $\sqrt{\lambda}$ and taking the limit $\lambda \to 0$, I see that, at lowest order, the renormalizable gauge describes massless gluons and the ghosts decouple. There are no monopoles. The unitary $F_{01}^{ch} = 0$ gauge, instead, describes even in the limit $\lambda \to 0$ a non-Gaussian interacting theory, given by the Gaussian integral of a polynomial that, in the thermodynamic limit, is of infinite degree. The result of the Gaussian functional integration is therefore represented, even at lowest order, by a sum of Feynman diagrams with infinite loops. In [6] I made the observation that the physical content of such a theory actually involves the interaction of $N-1$ kinds of random magnetic fields (or equivalently non-propagating magnetic monopoles) with the remaining elementary excitations: the photons associated with the Cartan subalgebra of $SU(N)$ and the charged gluons. Another remarkable point is that this picture becomes exact in the large-N limit. Incidentally this appears to be the first example of an idea which has been popular for some time, among string theorists, that some non-perturbative physics could be attached to diagrams with infinite loops [7]. The most common device in quantum field theory is the saddle-point method, which consists in approximating the result of functional integration by the value of the action at the minima times a factor that keeps track of the quadratic fluctuations:

$$\int \exp[-S(\varphi)] \, D\varphi = \sum_m \exp[-S(m)] \, Det(S''(m))^{-\frac{1}{2}} \quad . \tag{7}$$

However Eq. (7) is really equivalent to the following formula, in which no functional integration is actually performed:

$$\exp[-S(\varphi)] = \sum_m \exp[-S(m) - \frac{1}{2}S''(m)(\varphi - m)^2] \quad . \tag{8}$$

Such an estimate, which in statistical mechanics is called the Villain approximation, will be used in the following section, which is mainly a plagiarism of [6]. It is also clear that, in our case, the parameter that actually controls the fluctuations around the saddle points is N.

3. Random field and matrix models

A pure $SU(N)$ theory has a double expansion in Feynman graphs in powers of $\lambda = g^2N$ and N^{-1} [8]. A solution of the large-N limit of the theory is a very difficult task [9] and corresponds to summing all orders in λ at large N. It is commonly believed that such a limit is represented even in $d = 4$ by a string theory [10]. This is partially justified by the fact that the leading Feynman graphs are planar in the index (colour) space. This stringy behaviour would imply the area law for the Wilson loop and therefore quark confinement. However, this program has been completed only in $d = 2$ [11], [12]. I take here a different, less ambitious point of view: instead of trying to solve the theory for large N to all orders in λ, I use the large-N limit, in the gauge $F_{01}^{ch} = 0$, to show the existence of a new operator relevant in the infrared: the random field. After gauge-fixing, the gauge-fixed partition function for $U(N)$ becomes, in the continuum :

$$
\begin{aligned}
Z_{gf}(\mathcal{L}_{01}) &= Z_{gf}^{-1} \int \prod_{x,T} \exp\Big[-\frac{N}{\lambda} Tr F_T(x)^2 \Big] \times \\
&\times \prod_{x,L} \exp\Big[-\frac{N}{\lambda} \sum_i^N F_L^i(x)^2 + \sum_{i>j}^N \ln [F_L^i(x) - F_L^j(x)]^2 \Big] \times \\
&\times \delta(F_L^{ch}(x)) \; Tr \prod_{l \in \mathcal{L}_{01}} \exp(i e_l A_l) \; DA
\end{aligned} \tag{9}
$$

and on the lattice:

$$
\begin{aligned}
Z_{gf}(\mathcal{L}_{01}) &= Z_{gf}^{-1} \int \prod_{P_T} \exp\Big[\frac{2N}{\lambda} Tr \cos F(U_{P_T}) \Big] \times \\
&\times \prod_{P_L} \exp\Big[\frac{2N}{\lambda} \sum_i^N \cos F^i(U_{P_L}) + \sum_{i>j}^N \ln \sin^2 [\frac{F^i(U_{P_L}) - F^j(U_{P_L})}{2}] \Big] \times \\
&\times \delta(U_{P_L}^{nd}) \; Tr \prod_{l \in \mathcal{L}_{01}} U_l \; DU \quad .
\end{aligned} \tag{10}
$$

For $SU(N)$ there is the additional constraint $TrA = TrF = 0$. The superscript nd means projection onto the non diagonal matrices. With the normalization of Eqs. (9)-(10), the electric charge of the fundamental representation is 1. Let me notice that the longitudinal part of the gauge-fixed action as a function of F_L^0, the diagonal field strength, is the same as the one that occurs in the study of 0-dimensional matrix models [11] or of the reduced $d = 2$ $U(N)$ model [12]. This immediately suggests that in the large-N limit the longitudinal field strength will be concentrated more and more around the minima of the associated 0-dimensional model. However for $d \geq 2$ there is the important difference that the minima occur for every point or plaquette in the longitudinal directions. Therefore although fluctuations get more and more suppressed in the large-N limit this is compensated by the fact that there is more and more entropy due to the sum over the minima. Since these sums at different space-time points are not correlated, the net result is a random field. In $d = 2$ this random field is purely longitudinal, and the Wilson loop follows the area law because after integrating over the random field the usual $d = 2$ propagator is reproduced (see sect. 4). In $d = 3$ and $d = 4$, instead, the random field has a decomposition into

a longitudinal and a transverse part. The transverse part corresponds in $d = 3$ to monopoles points and in $d = 4$ to closed monopoles lines orthogonal to the longitudinal planes. Notice that the occurrence of monopoles is based on the dynamical contribution of the Faddeev-Popov determinant, as opposed as to the compactness of the lattice formulation as in ref. [13]. In $d = 2$ the a-priori distribution of the random field is given by the joint eigenvalues distribution of the associated 0-dimensional matrix model. In $d = 3$ and $d = 4$, instead, the a-priori random field distribution is renormalized by the addition of the effective action of the charged gluons (see sect. 4). This effective action is, at one-loop, the log of a functional determinant that is obtained from Eq. (22).
Since the longitudinal part of the exponential of the action is really a sum of delta functions at $N = \infty$ (see below):

$$\prod_{x,L} \exp\Big[-\frac{N}{\lambda}\sum_i^N F_L^i(x)^2 + \sum_{i>j}^N \ln[F_L^i(x) - F_L^j(x)]^2\Big] \rightarrow_{N\rightarrow\infty}$$
$$\rightarrow \prod_{x,L}\sum_{\sigma_x}\sum_{h^{\sigma_x(1)}<\ldots<h^{\sigma_x(N)}} \delta(F_L^i(x) - h^{\sigma_x(i)})\ldots\delta(F_L^j(x) - h^{\sigma_x(j)}) \qquad (11)$$

it is convenient to approximate it by a sum of Gaussians strongly peaked around the minima. This is the Villain approximation:

$$Z_{gf}^{Villain}(\mathcal{L}_{01}) = Z_{gf}^{Villain-1}\int\prod_{x,T}\exp\Big[-\frac{N}{\lambda}TrF_T(x)^2\Big]\times$$
$$\times\prod_{x,L}\sum_{\sigma_x}\sum_{h^{\sigma_x(1)}<\ldots<h^{\sigma_x(N)}}\exp\Big[-\frac{1}{2}\sum_{i,j}^N G_{ij}(\sigma_x)(F_L^i(x) - h^{\sigma_x(i)})(F_L^j(x) - h^{\sigma_x(j)})\Big]\times$$
$$\times\delta(F_L^{ch}(x))\ Tr\prod_{l\in\mathcal{L}_{01}}\exp(ie_lA_l)\ DA \quad ; \qquad (12)$$

σ_x is a permutation of $1\ldots N$, and $G_{ij}(\sigma_x)$ is the matrix of second derivatives at the minima $h^{\sigma_x(1)}\ldots h^{\sigma_x(N)}$ or better a matrix to be determined variationally; at every order in λ and for N large the result does not depend on the exact form of the matrix G_{ij} as long as it is of order N^2 (see below). The distribution of the h^i is given, in the large-N limit, by the function ρ :

$$\rho(x) = \lim_{N\rightarrow\infty} N^{-1}\sum_i^N \delta(x - h^i) \quad ; \qquad (13)$$

ρ is in the continuum [11]:

$$\rho(x) = \pi^{-1}\sqrt{\frac{2}{\lambda}}\sqrt{1-\frac{x^2}{2\lambda}}$$
$$-\sqrt{2\lambda} < x < \sqrt{2\lambda} \qquad (14)$$

and on the lattice [12]:

$$\rho(x) = \pi^{-1}\sqrt{\frac{2}{\lambda}}\cos(\frac{x}{2})\sqrt{1-\frac{2\sin^2(\frac{x}{2})}{\lambda}}$$
$$-2\arcsin\sqrt{\frac{\lambda}{2}} < x < 2\arcsin\sqrt{\frac{\lambda}{2}} \quad . \tag{15}$$

According to the fundamental laws of electrodynamics, $N-1$ monopoles magnetic currents occur (for every finite N) because the divergences of the dual of the $N-1$ neutral field strengths are non-vanishing at the minima. This corresponds to monopoles lines in $d=4$ (vector current) and to monopoles points in $d=3$ (scalar current). The magnetic current is given by:

$$\begin{aligned} j^i_\nu &= \partial_\mu \tilde{h}^{\sigma_x(i)}_{\mu\nu} \\ \tilde{h}_{\mu\nu} &= \frac{1}{2}\epsilon_{\mu\nu\rho\sigma}h_{\rho\sigma} \\ d &= 4 \\ j^i &= \partial_\mu \tilde{h}^{\sigma_x(i)}_\mu \\ \tilde{h}_\mu &= \epsilon_{\mu\nu\rho}h_{\nu\rho} \\ d &= 3 \\ h^i_{\mu\nu} &= h_L \quad (\mu\nu)=(01) \\ h^i_{\mu\nu} &= 0 \quad (\mu\nu)\neq(01) \quad . \end{aligned} \tag{16}$$

Since from Eqs. (14)-(15) it follows that $h^i = o(\frac{\sqrt{\lambda}}{N})$ and $G_{ij} = O(\frac{N^2}{\lambda})$, the Villain approximation improves in the large-N limit because fluctuations are suppressed. More technically I can justify the Villain approximation in the following way. The manifold of the field strength configurations can be triangulated in simplexes of the following form:

$$F^{\sigma_x(1)} < \ldots < F^{\sigma_x(N)} \quad . \tag{17}$$

Since the log term in the action diverges at coinciding points, the singularities of the action occur on the boundaries of the simplexes. The minima are in the interior regions, one minima for each region since $h^{\sigma_x(1)} < \ldots < h^{\sigma_x(N)}$, because of the infinite repulsion of the eigenvalues at coinciding points. In each region the potential can be bounded from below and from above by a quadratic form of order N^2, everywhere but for a region small at pleasure, which anyway gives a vanishing small contribution to the functional integral, since the potential is very large in it (indeed the exact potential has divergences at finite distance, while a quadratic form never diverges at finite distance, but can be chosen very large). The result, as we will see, is independent, to all orders in λ, from the actual form of the quadratic approximation, as long as the quadratic form is of order N^2 for N large. Therefore the Villain approximation is exact in the large-N limit and to all orders in λ. The Villain approximation is BRS-invariant in this gauge:

$$\delta^{BRS}\int \exp\left[-\frac{N}{\lambda}\sum_{T,a}\int F_T^{a^2}-S^{Villain}\right]\delta(F_L^{ch})\,DA=$$
$$=\delta^{BRS}\int \exp\left[-\frac{N}{\lambda}\sum_{T,a}\int F_T^{a^2}-\frac{N}{\lambda}\sum_i\int F_L^{i^2}\right]\delta(F_L^{ch})\,\Delta_{FP}\times$$
$$\times\frac{\exp[-S^{Villain}]}{\exp\left[-\frac{N}{\lambda}\sum_i\int F_L^{i^2}\right]\Delta_{FP}}\,DA=$$
$$=\int \exp[-S_{gf}]\delta^{BRS}\left[\frac{\exp[-S^{Villain}]}{\exp[-\frac{N}{\lambda}\sum_i\int F_L^{i^2}]\Delta_{FP}}\right]DA=0 \quad , \qquad (18)$$

since $\delta^{BRS}F_L^0=[F_L^{ch},\eta^{ch}]^0=0$, because of the gauge-fixing condition $F_L^{ch}=0$. This is not surprising since F_L^0 is actually gauge invariant in this gauge. The integral over the $N-1$ $U(1)$ gauge fields is performed by introducing the gauge-fixing term:

$$\int\prod_i Dc^i(x)\delta\left[c^i(x)-G_{ij}(\sigma_x)\partial_L A_L^j(x)-\frac{2N}{\lambda}\partial_T A_T^i(x)\right]\times$$
$$\times\exp\left[-\frac{1}{2}c^i(x)(G(\sigma_x)^{-1})^{ij}c^j(x)\right] \quad . \qquad (19)$$

Since in general each term in the sum of Eq. (12) is not translational invariant, but the entire sum is, it is convenient to substitute $G_{ij}(\sigma_x)$ with:

$$\tilde{G}_{ij} \;=\; \frac{1}{N!}\sum_\sigma G_{ij}(\sigma) \quad . \qquad (20)$$

The result is $\tilde{G}_{ij}$-independent in the limit of large $\tilde{G}_{ij}$, and the corresponding error is still of order N^{-1} because I can bound, in a way quite analogous to the preceding one, each quadratic term of the exponential in Eq. (12) by two different $\tilde{G}_{ij}$ of order N^2; the result is still $\tilde{G}_{ij}$-independent to order N^{-1} (this is explicitly checked below to leading λ order). It is convenient to use the representation:

$$\exp\left[-\frac{N}{\lambda}F_T^{ch2}\right]=\int D\Omega^{ch}\exp\left[i\sqrt{\frac{2N}{\lambda}}F_T^{ch}\Omega_T^{ch}-\frac{1}{2}\Omega_T^{ch2}\right] \qquad (21)$$

and to integrate over Ω^{ch} after the $U(1)^{N-1}$ integration. The result, to the lowest λ order, is:

$$Z_{gf}^{Villain}(\mathcal{L}_{01})=Z_{gf}^{Villain-1}\times$$
$$\times\int\prod_x\exp\left[-\frac{N}{\lambda}\sum_{(\mu\nu)=T,ch}(D_{[\mu}^{chch'}(H)A_{\nu]}^{ch'}(x))^2\right]$$

$$\times \exp\Big[-\frac{N}{\lambda}[A_T^i(x)(-\Delta\delta^{ij}+(\delta^{ij}-\frac{2N}{\lambda}(\tilde{G}^{-1})^{ij})\partial_T\partial_{T'})A_{T'}^j]\Big]$$
$$\times\delta(D_{[0}^{chch'}(H)A_{1]}^{ch'}(x))\ \delta(\sum_i^N H_{01}^i(x))\ \rho(H_{01}^1(x)\dots H_{01}^N(x))$$
$$\times\sum_k \exp[ie_\mu^k(\tilde{G}^{-1})^{kj}\partial_\nu\Delta^{-1}(\tilde{G})^{ji}H_{\mu\nu}^i]$$
$$\times[1+o(\sqrt{\lambda})+o(N^{-1})]DA_\mu^{ch}DA_T^iDH_{01}^i \tag{22}$$

$$\begin{aligned} D_\mu^{chch'}(H) &= \frac{\partial}{\partial x^\mu}\delta^{chch'} - i\sum_k\int dz f^{chkch'}(\tilde{G}^{-1})^{kj}\frac{\partial}{\partial x^\nu}\Delta^{-1}(\tilde{G})^{ji}(x-z)H_{\mu\nu}^i(z)\\ H_{\mu\nu}^i &= 0\ \ (\mu\nu)\neq(01)\\ \Delta^{-1}(\tilde{G})_{ji} &= (-\tilde{G}\Delta_L-\frac{2N}{\lambda}\Delta_T)_{ji}^{-1}\quad , \end{aligned} \tag{23}$$

where Δ_L is the Laplacian in the (01) plane and Δ_T the Laplacian in the (23) plane. I have performed the sum over permutations according to:

$$\begin{aligned}\rho(H^1(x)\dots H^N(x)) &= \lim_{N\to\infty}\frac{1}{N!}\sum_{\sigma_x}\sum_{h^{\sigma_x(1)}<\dots<h^{\sigma_x(N)}}\delta(H^1-h^{\sigma_x(1)})\dots\times\\ &\quad\times\delta(H^N-h^{\sigma_x(N)})\end{aligned} \tag{24}$$

which is the N-eigenvalues joint distribution of the associated 0-dimensional (matrix) model [11], [12], [14]. Equation (22) manifestly describes the coupling of the external current, e_μ, and of the charged gluons to a random magnetic field whose distribution is given, independently in each space-time point, by ρ. Notice that the dependence upon $\tilde{G}$ disappears in the $N\to\infty$ limit. An easy way for understanding Eq. (22) consists in noticing that, in the large N limit and at the lowest order in λ, according to Eq. (11), we can simply substitute F_L^i with h^i and A_L^i with the vector potential corresponding to h^i everywhere in Eq. (9) or Eq. (10). Then we must sum over all possible minima, h^i, according to Eq. (24). To summarize, Eq. (22) shows that in the large-N limit, in the gauge $F_{01}^{ch}=0$, although the fluctuations of A_L^i get suppressed, the coupling of the $U(1)$ currents to the non-trivial minima of the longitudinal field strength survives. The minima then possess enough entropy to fluctuate generating themselves a random field. It is easy to re-interpret this random field in terms of a set of $N-1$ non-propagating magnetic monopoles decomposing H_{01}^i into a longitudinal and a transverse part and replacing the DH_{01} integrals with a sum over an equally N-spaced partition of the support of $\rho(H^1)$ (which is the same to leading N^{-1} accuracy):

$$\begin{aligned} H_{\mu\nu}^i &= (H_{\mu\nu}^i)^L+(H_{\mu\nu}^i)^T\\ (H_{\mu\nu}^i)^L &= \partial_\mu n_\nu^i-\partial_\nu n_\mu^i\\ (H_{\mu\nu}^i)^T &= \epsilon_{\mu\nu\rho\sigma}(\partial_\rho\varphi_\sigma^i-\partial_\sigma\varphi_\rho^i)\\ q_\mu &= \partial_\nu\tilde{H}_{\mu\nu}\\ q_\mu &= (\Delta\delta_{\mu\nu}-\partial_\mu\partial_\nu)\varphi_\nu\quad . \end{aligned} \tag{25}$$

According to Eq. (22) and Eq. (25), the monopoles magnetic charge is the support of ρ divided by N times an integer. For large N the monopoles magnetic charge is $2\sqrt{2\lambda}N^{-1}\times$integer in the continuum and $4\arcsin(\sqrt{\frac{\lambda}{2}})N^{-1}\times$integer for $\lambda \leq 2$ and $2\pi N^{-1}\times$integer for $\lambda \geq 2$ on the lattice.

4. Random magnetic field and string tension

For a Wilson loop in the (01) plane:

$$\begin{aligned}&\langle Tr\ P\ \exp[i\int A_\mu\ dx^\mu]\rangle \simeq \\ &\simeq \langle Tr\ \exp[i\int F_{\mu\nu}\ dS^{\mu\nu}]\rangle = \\ &= \langle Tr\ \exp[i\int F^0_{\mu\nu}\ dS^{\mu\nu}]\rangle\end{aligned} \tag{26}$$

because of the gauge-fixing condition. I get from Eq. (22), at one-loop:

$$\begin{aligned}&Z \times \langle Tr\ \exp[i\int F_{\mu\nu}\ dS^{\mu\nu}]\rangle = \\ &= \int \prod_x \rho(H^1_{01}(x)\ldots H^N_{01}(x))\ \exp\left[-\frac{1}{2}\ln Det'(-D^2_\mu(H) + D_\mu(H)D_\nu(H))\right] \\ &\times \sum_k \exp[i\int H^k_{01}\ dS^{01}]\ DH^1_{01}\ldots DH^N_{01}\end{aligned} \tag{27}$$

$$\begin{aligned}&D_\mu(H) = \partial_\mu - i[\partial_\nu \Delta_L^{-1} H^i_{\mu\nu} T^i,\ \ ] \\ &Tr(T^i T^k) = \delta^{ik}\quad .\end{aligned} \tag{28}$$

The superscript $'$ means that the determinant must be restricted to the subspace $D_{[0}A^{ch}_{1]} = 0$. In $d = 2$, the functional determinant is absent, and the known result for the string tension is obtained, since, in each term of the sum in Eq. (27), $N-1$ fields H_{01} can be integrated away, giving the one-eigenvalue distribution [12]. In $d = 3$ and $d = 4$, instead, the functional determinant has to be evaluated.

5. Confinement versus symmetry breaking

Every renormalizable field theory depending only on a single dimensionless coupling constant and no mass scale at the classical level displays the mechanism of dimensional transmutation if a mass scale is dynamically generated. The existence of a dimensionful magnetic condensate in QCD:

$$\langle H^i(x)^2\rangle \neq 0 \tag{29}$$

implies that every dimensionful physical quantity must depend on the condensate in a way determined by dimensional analysis. This phenomenon is quite analogous to the case of a gauge theory coupled to a massless scalar, which is spontaneously broken by radiative corrections a' la Coleman-Weinberg [15]: in this case the mass ratio between the vector and the scalar particle (at least to lowest order) does not depend on the scalar coupling constant. Another feature that this model has in common with QCD is that, without a (electric) condensation, in the unitary gauge:

$$Im\ \varphi = 0 \tag{30}$$

the gluon propagator is not defined, since the quadratic part of the gauge fixed action still has zero modes corresponding to the linear part of a gauge transformation, even though the gauge freedom has already been fixed. In the pure gauge theory, in the unitary gauge $F_{01}^{ch} = 0$, both phenomena occur, dimensional transmutation and coupling of gluons to the magnetic condensate. Indeed the monopoles current couples universally to the charged currents via the $U(1)^{N-1}$ photons. In particular the propagator of the charged gluons becomes the propagator of a charged vector particle in a random magnetic field. I expect that monopoles condensation will generate the mass scale necessary to make the gluon propagator well defined. According to Eq. (22), the gluon propagator is the inverse of the following quadratic form:

$$\sum_{(\mu\nu)=T,ch} \langle (D^{chch'}_{[\mu}(H) A^{ch'}_{\nu]}(x))^2 \rangle_H \quad . \tag{31}$$

In addition, since there is no random flux flowing through a Wilson loop in the (23) plane, I also conjecture that the coupling of monopoles with the charged gluons will restore rotational invariance via dimensional reduction due to the random field [16]. This may be expected since the random field will tend to confine the charged gluons in the (23) planes.

6. References

1. K. Wilson, *Phys.Rev.* **D10** (1974) 2245
2. See for example, A. Borrelli, L. Maiani, G. C. Rossi, R. Sisto, M. Testa, *Nucl.Phys.* **B133** (1990) 335.
3. M. Creutz, *Phys.Rev.* **D21** (1980) 2308
4. G. 't Hooft, High Energy Physics EPSI Conference, Palermo (1975) edited by A. Zichichi.
   S. Mandelstam, *Phys.Rep.* **23C** (1976) 245.
5. G. 't Hooft, *Nucl.Phys.* **B190** (1981) 454.
6. M. Bochicchio, *Confinement as a dual Meissner effect, monopoles, matrix models and the unitary gauge of quantum chromodynamics (I).* Talk given at the conference Lattice 92, Amsterdam, 11-19 September 1992, to appear in *Nucl.Phys.* **B** *Suppl.*

7. D.Friedan, S.Shenker, *Phys.Lett.* **B175** (1986) 287.
8. G. 't Hooft, *Nucl.Phys.* **B72** (1976) 461.
9. V. Kazakov, A. Migdal, *Induced QCD at large N*, Princeton preprint PUPT-1322, May 1992.
10. See for example, A. M. Polyakov, *Gauge fields and strings*, (Harwood, Chur, 1987).
11. E. Brezin, C. Itzykson, G. Parisi, J. B. Zuber, *Comm.Math.Phys.* **59** (1978) 35.
12. D. Gross, E. Witten, *Phys.Rev.* **D21** (1980) 446.
13. M. Bochicchio, *Phys.Lett.* **B283** (1992) 97.
14. F. J. Dyson, *J.Math.Phys.* **3** (1962) 140, 157, 166.
15. S. Coleman, S. Weinberg, *Phys.Rev.* **D** (1973) 1888.
16. Y. Imry, S. K. Ma, *Phys.Rev.Lett.* **35** (1975) 139.
    G. Parisi, N. Sourlas, *Phys.Rev.Lett.* **43** (1979) 744.

RANDOM SURFACES AND QUANTUM GRAVITY: LOOKING FOR THE EMERGENCE OF CONTINUUM THEORIES FROM TRIANGULATED MANIFOLDS

MARK BOWICK, BERND BRÜGMANN, PAUL CODDINGTON, LEPING HAN, GEOFFREY HARRIS and ENZO MARINARI[1(*)]

Physics Department, Syracuse University
Syracuse, NY 13244 (USA)

(*) *and Dipartimento di Fisica, Università di Roma* Tor Vergata
V. Ricerca Scientifica, 00133 Roma (Italy)

(bowick brugman paulc han gharris marinari @ sccs.npac.syr.edu)

ABSTRACT

We investigate the possible existence of continuum theories originating from lattice theories defined on triangulated manifolds. We discuss the $2d$ case, considering theories with extrinsic curvature coupled to embedding bosonic matter fields, and the $4d$ case, where we find that the effect of a non-trivial measure is likely to be relevant.

1. Introduction

In this note we will discuss the output of two related fields of investigation [1, 2]. Both of them are based on the hope that a critical point may allow a continuum theory to emerge out of a discrete lattice theory defined on a (diffeormosphism invariant) triangulated lattice. In the $2d$ case one hopes that the addition to the *free* surface Hamiltonian of an extrinsic curvature dependent term could favour the emergence of a non-trivial continuum theory. In $d = 4$ the hope is that a non-trivial critical point could emerge already from the discretized form of the simple Einstein action. In this case maybe self-consistency of the theory (for example the simple demand of the emergence of a $4d$ manifold at large distances) could help in solving the key point of the determination of the correct measure factor. In both cases the tools used are from Statistical Mechanics combined with a very detailed analysis of the features of

[1]Talk presented by E. Marinari

the cross-over region.

2. 2*d* Random Surfaces with Extrinsic Curvature

String theory, in a number of guises, has been conjectured to describe the underlying fundamental physics of a wide variety of physical phenomena and models. These include the strong interaction at long distances, the three-dimensional Ising model and unified models incorporating gravity. In its simplest form, the bosonic string, it is a theory of free fluctuating surfaces. The functional integral for the Euclideanized bosonic string is just the partition function for an ensemble of random fluctuating fluid surfaces. Such surfaces are also ubiquitous in nature, being found for example in macro-emulsions and the lipid bilayers that form an important part of cell membranes [3]. These systems are fluid because their component 'molecules' are loosely bound. Their constituents are arranged so that the net surface tension (nearly) vanishes; thus these membranes are subject to large thermal fluctuations. In one important respect, however, these chemical/biological membranes differ fundamentally from the surfaces we discuss and simulate; they are self-avoiding. The worldsheets of the bosonic string, in contrast, generically self-intersect.

The bosonic string for surfaces embedded in 26 dimensional space has been studied extensively. Much progress in numerically simulating strings has been made through the use of Dynamically Triangulated Random Surfaces (DTRS)[6, 7]. For theories of surfaces embedded in $D \leq 1$, analysis of the continuum Liouville theory [4, 5] and of matrix models has led to consistent predictions for critical exponents and correlation functions. In the 'double scaling limit', in fact, exact nonperturbative solutions [8] have been found from these matrix models; the functional integral over surfaces has been essentially summed over all genera.

These analytical techniques have failed, for the most part, in probing the theory of random surfaces in the more interesting embedding regime $D > 1$. Indeed the formulas for critical exponents computed in Liouville theory give complex results when continued to the range $1 < D < 25$.[2] The matrix models describing $D > 1$ strings have so far been too complicated to solve.

Monte Carlo simulations for $D > 1$ [10, 11, 12, 13] indicate that these theories do not appear to describe the fluctuations of two dimensional *smooth* surfaces in the continuum limit. Extremely spiky, branched-polymer-like configurations with high (perhaps infinite) Hausdorff dimension dominate the functional integral[3].

Evidence for this pathology was obtained, for instance, when it was shown that the bare string tension, essentially the amount of work per unit area needed to perturb a

[2]More precisely, one encounters these instabilities in Liouville theory when the quantity $c-24\Delta > 1$, where c denotes the central charge of the matter theory which describes the embedding of the surfaces and Δ is the conformal weight of the lowest weight state in this theory [9]. Since here we are considering flat space, $c = D$ and $\Delta = 0$.

[3]In the same way singular configurations dominate the Gaussian theory, which is essentially a theory of free random walks.

boundary loop on these configurations, cannot vanish at the critical point [14]. (This implies that the renormalized string tension diverges in the continuum limit.) It has been speculated in ref. [15] that the proliferation of vortex configurations of the internal geometry (the 'Liouville mode') induces the degeneration of these surfaces in the embedding space. A related explanation that is often proferred is that a negative mass-squared particle[4], which comes on shell in the string theory for $D > 1$, creates instabilities which are made manifest by these singular configurations.

The tachyon, and apparently these related instabilities, can be eliminated in particular cases by introducing fermionic coordinates and supersymmetry on the worldsheet, and implementing an appropriate projection of states. Presumably, the fermions effectively smooth out the surfaces. This would be consistent with what has been observed for one-dimensional geometries; the random walk of a spin one-half particle has Hausdorff dimension one and thus appears to be smooth [16]. Many authors have proposed an alternative modification of the string action [17, 18, 19, 20] via the addition of a term that directly suppresses extrinsic curvature[5].

To write down our action we introduce an explicit parametrization of a generic surface $\mathcal{M}$ in R^3 with coordinates (σ_1, σ_2) and the embedding $X^\mu(\sigma_i)$. μ runs from 1 to 3 (since we only study the case of a $3d$ embedding space). The induced metric (the pullback of the Euclidean R^3 metric via the embedding) is given by

$$h_{ij} = \partial_{\sigma_i} X^\mu \partial_{\sigma_j} X_\mu \ . \tag{1}$$

We will use Greek letters for the embedding space indices; they can be raised and lowered at will since our background space is flat. Associated with each point in $\mathcal{M}$ are tangent vectors $(t^\mu_i \in T\mathcal{M})$ and a normal vector $n^\mu \in T\mathcal{M}^\perp$. The extrinsic curvature matrix K_{ij} (the second fundamental form) can be defined by

$$\partial_i n^\mu = -K_{ij} t^{\mu j} \ . \tag{2}$$

The eigenvalues of this matrix are the inverses of the radii of curvature of $\mathcal{M}$. One usually describes the geometry of these surfaces in terms of the mean curvature [22, 23]

$$H = \frac{1}{2} h^{ij} K_{ij} \ , \tag{3}$$

and the Gaussian curvature

$$K = \epsilon^{ik} \epsilon^{jl} K_{ij} K_{kl} \ . \tag{4}$$

One can show that the Gaussian curvature can be computed solely from the metric h_{ij}, while the mean curvature depends explicitly on the embedding X^μ.

We shall be concerned primarily with the Polyakov form of the string action [24], in which an additional *intrinsic metric* g_{ij} is introduced to describe the surface geometry.

[4] In some contexts, this is referred to as the 'tachyon'.

[5] In fact, integrating the fermions out of the Green-Schwarz superstring yields an action similar to the one we consider, but with the addition of a complex Wess-Zumino type term [21].

We discretize our model by triangulating surfaces. In this construction, each triangle is equilateral with area 1 in the intrinsic metric; the coordination number at each vertex determines the intrinsic curvature of the surface. The coordinates i label the vertices of the triangulation. Then the discrete analogue of the intrinsic metric is the adjacency matrix C_{ij} whose elements equal 1 if i and j label neighbouring nodes of the triangulation, and vanish otherwise. Two-dimensional diffeomorphism invariance reduces to the permutation symmetry of the adjacency matrix at this discrete level. One of the keys, in fact, to the power of this construction is the preservation of this symmetry. Each vertex of the triangulation is embedded in R^3 via the mapping X_i^μ. Given the embedding X, we can also associate a unit normal vector $(n^\mu)_{\dot{k}}$ with each triangle on the surface (dotted Roman indices label the triangles). Note that all of the surface curvature of our triangulations is concentrated along the links and vertices. The surface is still flat in the direction tangent (but not transverse) to each link, so that the mean curvature has support on the links, while the Gaussian curvature is non-zero only at the vertices. The intrinsic curvature R_i at vertex i is given by the deficit angle determined solely by the triangulation

$$R_i = \pi \frac{(6 - q_i)}{q_i}, \tag{5}$$

where q_i denotes the connectivity of the lattice at vertex i. The Gaussian curvature K on the other hand is expressed in terms of the deficit angle in the embedding space.

We shall study the theory defined by the action

$$S = S_{Gauss} + \lambda S_E = \sum_{i,j,\mu} C_{ij}(X_i^\mu - X_j^\mu)^2 + \lambda \sum_{\dot{k},\dot{l},\mu} C^{\dot{k}\dot{l}}(1 - n_{\dot{k}}^\mu \cdot n_{\dot{l}}^\mu) . \tag{6}$$

Thus, for $\lambda > 0$, we have introduced a ferromagnetic interaction in the surface normals. The model defined by this action has been studied in [25, 26, 27, 28, 29, 30] and references therein.

From (2) and the definition of the induced metric, it follows that this is a discretization of the continuum action

$$S = \int \sqrt{|\det g|}(g^{ij}\partial_i X^\mu \partial_j X^\mu + \frac{\lambda}{2} g^{ij} h^{kl} K_{ik} K_{jl}) . \tag{7}$$

Note that the second term in this action is manifestly positive, Weyl and reparametrization invariant, and that λ is a dimensionless coupling. So, naively, it is not clear whether it is relevant or not. If it were relevant, one would then anticipate that (since it obeys all of the appropriate symmetries) it should be effectively generated in any string action, and that it should engender ordering of the normals. It should then lead to another RG fixed point at a finite value of λ, which would characterize a phase transition between the crumpled phase (observed when $\lambda = 0$) and a 'smooth(er)' phase.[6]

[6]The extrinsic curvature term is also higher-derivative, indicating that the field theory described

A renormalization group analysis [18, 32, 33, 19] indicates, however, that there should be no phase transition at finite coupling when such extrinsic curvature dependent operators are added to the action. The computations of refs. [18, 32, 33] use the action

$$S = \int d^2\sigma \, (\mu_0\sqrt{\det h} + \frac{1}{\alpha}\sqrt{\det h}(h^{ij}K_{ij})^2) , \tag{8}$$

in the regime in which the string tension μ_0 is small (unlike the usual particle physics limit of string theory, which is characterized by large μ_0). After integrating out fluctuations of the embedding X^μ between momentum scales Λ and $\tilde{\Lambda}$, it is found that the renormalization of the extrinsic curvature coupling is given to one-loop order by

$$\beta(\alpha) \equiv \Lambda\frac{d\alpha}{d\Lambda} = -\frac{3}{4\pi}\alpha^2 , \tag{9}$$

so that α is driven to infinity in the infra-red. This theory thus exhibits asymptotic freedom. Surfaces are smooth (the normals are correlated) below a persistence length [34]

$$\xi_p \sim \exp(\frac{4\pi}{3\alpha_{bare}}) , \tag{10}$$

and are disordered above this scale. Some intuition into this result can be gained by observing that this theory is similar to the $O(3)$ sigma model, which is asymptotically free [16]. The normals to M are the analogues of $O(3)$ vectors, though in this case they are constrained to be normal to a surface governed by the action (7).

Without the extrinsic curvature term, (8) is the Nambu-Goto action, while (7), which we use in our simulations, is based on the action quantized by Polyakov. Classically (when the equations of motion for the Polyakov action are solved and substituted back into the action), the two actions are equivalent. It has also been demonstrated [35] that the two quantizations are equivalent in the critical dimension $D = 26$. In lower dimensions (note that the Nambu-Goto action clearly does not make sense for $D < 2$), it is not so clear that quantizations 'based' on the two actions are indeed the same. The work of Polchinski and Strominger [36] suggests that there are alternative quantizations. Distler ([37]) has also questioned the equivalence of these quantizations in $D = 3$. Indeed, even if the two quantizations are equivalent, it does not automatically follow that the two theories are still the same once an extrinsic curvature dependent term has been added.

In fact, Polyakov in [19] uses a hybrid form of the action (8) and still obtains the same result for the beta function. He introduces an intrinsic metric g_{ij}, chooses the

---

by this action is non-unitary. This fact alone does not imply that the associated string-scattering amplitudes do not satisfy unitarity. Polchinski and Yang [31] do, however, contend that in this case the string theory will not be unitary. Even if this were so, this model could still be an appropriate description of the statistical mechanics of fluctuating surfaces, although not one corresponding to a physical fundamental string theory.

conformal gauge $g_{ij} = \rho\delta_{ij}$ and considers

$$S = \frac{1}{2\alpha}\int d^2\sigma(\mu_o\rho + \rho^{-1}(\partial^2 X^\mu)(\partial^2 X^\mu) + \lambda^{ij}(\partial_i X^\mu \partial_j X^\mu - \rho\delta_{ij})) . \tag{11}$$

Classically, the Lagrange multiplier λ^{ij} constrains the intrinsic metric to equal the induced metric (this equality is not enforced by the classical equations of motion for the original Polyakov action). This constraint should be relaxed quantum mechanically if, as Polyakov [16] argues, the condensate of this Lagrange multiplier assumes a value of the order of the momentum cutoff. If this dynamical assumption is correct, then one can essentially derive the equivalence of this Nambu-Goto-like and the original Polyakov quantizations. In the large D (embedding dimension) limit, saddle point calculations [38] show that λ indeed does acquire a large expectation value, and that for small values of the string tension μ_o, the coupling α is asymptotically free, as the RG calculations suggest.

There are, however, a couple of caveats and suggestions in the analytic literature that do allow for the existence of a crumpling transition for fluid surfaces. Polyakov remarks that if, in the infrared region, fluctuations of the internal geometry (ρ) are suppressed relative to fluctuations of the extrinsic metric, then the beta function is proportional to α and hence the continuum limit of the theory exhibits non-trivial scaling behaviour; this presumably cannot be the case in the large D limit. Another RG calculation, performed by Yang [39] using the Polchinski-Strominger action [36] with an extrinsic curvature dependent term, indicates that the two-loop correction (which is proportional to α^3) might be large enough to yield a zero of the beta function, and thus a non-trivial IR fixed point. The Polchinski-Strominger action is based on the assumption that the Liouville mode ρ effectively decouples (its mass is much greater than the momentum scale set by the string tension); it is not clear why this assumption should hold for the model that we simulate. Finally, note that these computations are perturbative (in $1/D$ or α). It is possible that non-perturbative effects could drive a crumpling transition.

Monte Carlo simulations of the action (7) on dynamically triangulated random surfaces (DTRS) were first performed by Catterall [25], and shortly thereafter by Baillie, Johnston, and Williams [26, 27] and Catterall, Kogut and Renken [30]. They simulated triangulations with the topology of the sphere, and measured the specific heat

$$C(\lambda) \equiv \frac{\lambda^2}{N}(< S_E^2 > - < S_E >^2) , \tag{12}$$

on surfaces with up to $N = 144$ nodes (and $N = 288$ nodes in the last reference). They found a peak in the specific heat; the peak size appeared to grow with N. A similar model that can be vectorized rather straightforwardly was also considered; the set of planar ϕ^3 graphs was simulated [30, 40]. Each vertex of these ϕ^3 graph was embedded in R^3 and the action (7) was used; graphs of up to 1000 nodes were simulated (these would be dual to 500 node triangulations). It was found that the specific heat peak grew with N, albeit slowly, as

$$C_{max} = AN^{\omega} + B \ , \tag{13}$$

with $\omega = 0.185(50)$. Further work in [28, 29], using dynamical triangulations with the topology of the torus and lattices with up to $N = 576$ nodes, indicated that the rate of increase of the peak height severely diminishes with increasing N. The data strongly suggests that in fact the specific heat peak height does **not** diverge as $N \to \infty$. These authors also measured the bare string tension and mass gap, by embedding the torus in a background toroidal space spanned by a loop, and measuring the dependence of the free energy on the loop size. They found that these measurements (when taken for λ values near the peak position) are consistent with the appropriate scaling relations (with vanishing bare string tension and mass gap) that should characterize a phase transition to smooth surfaces. This measurement, which may turn out to be quite important (since it constitutes the best evidence we have, up to now, that there could really be a phase transition for $\lambda = \lambda_c$), is still quite an indirect way of measuring correlation functions. As we will discuss, these scaling relations could contradict other observed phenomena (such as the absence of diverging correlation times and increasing finite size effects at the pseudo-critical point).

Thus it appears that numerical evidence could allow for the existence of a crumpling transition (most probably of higher order), while analytical calculations generally indicate that no such transition should occur.

In [41] the peak was measured in a DTRS simulation that incorporated self-avoidance and the extrinsic curvature term S_E, with a solid-wall potential substituted for the Gaussian term in the action. The results for the specific heat turned out to be very similar to those found in the simulations we have just discussed, for example, in [29]. The specific heat peak is, in this context, considered to be a lattice artifact, because the peak height levels off with large N (of order 500). These simulations included a crude block-spin measurement that suggests that the renormalization group flow of λ is consistent with the analytical result of asymptotic freedom.

Simulations using other discretizations for the extrinsic curvature dependent term have yielded somewhat different results [25, 26]. The specific heat peak, measured in simulations employing what is referred to as the 'area discretization', is rather feeble, and levels off for small values of N (by $N = 72$) (the authors interpret this as being indicative of perhaps a 'third' order transition). Actions based on these various discretizations have been simulated for fixed, triangular meshes. These systems model tethered or crystalline membranes, in which the constituent molecules are tightly bound together. In the tethered case, the specific heat peak obtained from simulations of the edge action (7) grows vigorously as a function of N for very large (128×128) lattices [42]. This is strong evidence for the existence of a second order transition which, in this case, is in accord with the analytic results – these calculations are reviewed by Nelson [43] and David [3, 44] and involve mean field and large D computations which suggest that the β function is linear at leading order, with a zero for finite α, i.e. a UV fixed point. When the alternate area discretization is used in the tethered case, the specific heat peak again stops growing. Recent work has demonstrated that this other discretization is pathological in the tethered case; the class of 'corrugated' surfaces, which are singular in one direction and smooth in the other, then dominates the path integral [42].

Thus, given the muddle of somewhat contradictory evidence, it is unclear whether or not a crumpling transition exists for fluid surfaces. We have pursued this question by taking high statistics measurements of the specific heat peak, and by measuring other observables describing the geometry of these surfaces, since observables with different quantum numbers can give quite different information. For example, in the Ising model the magnetization behaves quite unlike the internal energy (which is invariant under the standard Z_2 transformation).

To analyze and interpret this data, we have applied insights gained from work on better understood systems, primarily spin models and lattice gauge theories. Issues of the equivalence of the Nambu-Goto and Polyakov quantizations have also motivated us to compare the intrinsic and induced geometry of the surfaces that we simulate.

3. Numerical Simulations in $d = 2$

To minimize finite size effects, we have considered triangulations with the topology of the torus. The action (7) was used, with the BRST invariant measure utilized also by Baillie, Johnston, and Williams [26], so that

$$Z = \sum_{G \in T(1)} \int \prod_{\mu,i} dX_i^\mu \prod_i q_i^{\frac{d}{2}} \exp(-S_{Gauss} - \lambda S_E) , \tag{14}$$

where $d = 3$, q_i is the connectivity of the ith vertex, and $T(1)$ refers to the set of triangulations of genus 1. The authors of [28, 29] do not include this connectivity dependent term in their measure. The long-distance physics of the simulations is presumably insensitive to the presence of this term. Because we have chosen a different measure, though, our quantitative results cannot be precisely compared with theirs.

We measured a variety of quantities that characterize the extrinsic and intrinsic geometry of these surfaces. These observables include:

1. The edge curvature S_E and the associated specific heat $C(\lambda)$, which is a sensitive indicator of the presence of a phase transition.

2. The squared radius of gyration R_G;

$$R_G \equiv \frac{1}{N} \sum_{i,\mu} (X_i^\mu - X_{\text{com}}^\mu)^2 , \tag{15}$$

where the com subscript refers to the center of mass of the surface. By measuring the N dependence of the gyration radius, we can extract a value for the extrinsic

Hausdorff dimension, which is given by

$$R_G \sim N^{\nu} \sim N^{\frac{2}{d_{extr}}} \ . \tag{16}$$

3. The magnitude of the extrinsic Gaussian curvature. We measure a discretization of $\int \mid K \mid \sqrt{\mid h \mid}$, with

$$\mid \mathcal{K} \mid = \frac{1}{N} \sum_i \mid 2\pi - \sum_j \phi_i^j \mid \ . \tag{17}$$

   Here ϕ_i^j denotes the angle subtended by the jth triangle at the ith vertex. This quantity, therefore, measures the magnitude of the deficit angle in the embedding space averaged over all vertices. We also record the mean square fluctuation of $\mid \mathcal{K} \mid$, denoted by $F[\mid \mathcal{K} \mid]$.

4. The corresponding intrinsic quantity, $\mid \mathcal{R} \mid$, given by

$$\mid \mathcal{R} \mid = \frac{\pi}{3N} \sum_i \mid 6 - q_i \mid \ , \tag{18}$$

   and its fluctuations. When the intrinsic and extrinsic metrics are equal, the intrinsic and extrinsic deficit angles are identical, and $K = R/2$.

5. To study the correlation between intrinsic and extrinsic geometry, we also measure the quantity which we refer to as $\mathcal{K} * \mathcal{R}$:

$$\mathcal{K} * \mathcal{R} \equiv \frac{\int KR}{\sqrt{\int K^2 \int R^2}} = \frac{\sum_i (2\pi - \sum_j \phi_i^j)(6 - q_i)}{\sqrt{\sum_i (2\pi - \sum_j \phi_i^j)^2 \sum_i (6 - q_i)^2}} \ . \tag{19}$$

   This quantity is 1 when the metrics are equal, 0 if they are un-correlated, and negative when these curvatures are anti-correlated.

6. We measure, finally, the average maximum coordination number of the surface vertices, $\max_i q_i$.

In our simulations we have used the standard Metropolis algorithm to update the embedding fields X_i^{μ}. To sweep through the space of triangulations we performed flips (see reference [12]) on randomly chosen links. Flips were automatically rejected if they yielded a degenerate triangulation; i.e. one in which a particular vertex has fewer than three neighbors, or in which a vertex is labeled as its own neighbor, or where more than one link connects two vertices. (It has been proven in ref. [12, 13] that the entire space of graphs of a given topology can be spanned by only performing

N=144	.8	1.25	1.35	1.40	1.45	1.50	2.0	λ
	3	3	3	3	3	3	3	$\times 10^6$ sweeps
N=288	.8		1.375	1.40	1.425	1.475	2.0	λ
	14.4		21.0	15.0	16.2	13.5	14.4	$\times 10^6$ sweeps
N=576	.8	1.325	1.375	1.40	1.425	1.475	2.0	λ
	12.0	27.0	27.0	27.0	27.0	27.0	9.6	$\times 10^6$ sweeps

Table 1: A record of the number of sweeps performed at each different λ value for 3 different lattice sizes.

these flips.) After a set of $3N$ flips was performed, $3N$ randomly selected embedding coordinates were updated via random shifts from a flat distribution,

$$X^\mu \rightarrow X^\mu + \delta X^\mu \ . \tag{20}$$

The mean magnitude of these shifts

$$< \delta X^\mu \delta X^\mu > \tag{21}$$

was chosen so that the acceptance rate for updates of the X^μ was roughly 50 percent. Most of the Monte Carlo simulations were performed on HP-9000 (720 and 750 series) workstations; we also collected some data by simulating lattices on each of the 32 nodes of a CM-5. Our code was in Fortran, with a Fibonacci random number generator.

In Table 1 we summarize our runs. Note that we have performed quite long runs on the larger lattice sizes. We will discuss later why we believe runs of this length are just sufficient to yield accurate values of the observables for the largest lattice size ($N = 576$).

In all of our figures the different dots will be printed with their associated statistical error (sometimes too small to be visible). The statistical error is computed by means of a standard binning procedure. We will explicitly discuss the cases in which our estimator for the statistical error is not asymptotic.

The lines in these figures are from a histogram reconstruction (see for example [45, 46]). We patch different histograms [47, 48, 49] by weighting them with the associated statistical indeterminacy (which we estimate by a jack-knife binned procedure); this procedure seems to be very effective and reliable. All of the reconstruction curve sets (3: dotted, dashed and continuous for 3 surface sizes on each figure) consist of 3 curves (which sometimes appear as a single one). The middle curve is the histogram reconstruction, and the upper and the lower ones bound the data within the errors obtained by the procedure we have just described.

For $N = 144$ we have patched the four histograms originating from $\lambda = 1.35$, 1.40, 1.45, 1.50. For $N = 288$ we have used $\lambda = 1.375$, 1.40, 1.425 and 1.475. For $N = 576$, we chose $\lambda = 1.375$, 1.40 and 1.425.

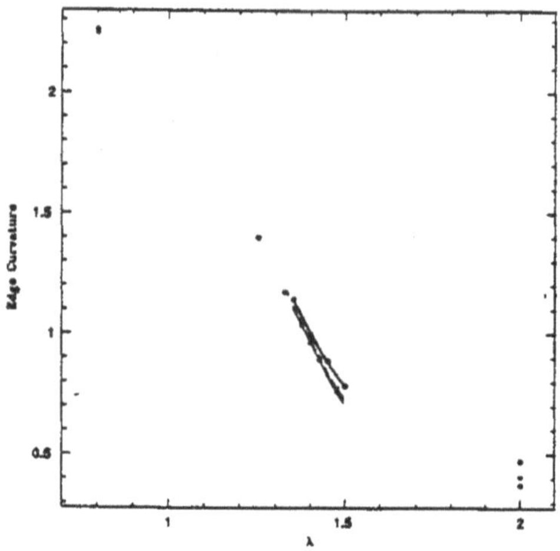

Figure 1: The edge curvature S_E as a function of λ. As in all other pictures, filled circles and a dotted line correspond to $N = 144$, crosses and a dashed line indicate $N = 288$, and empty squares and a solid line represent $N = 576$.

We have only drawn the reconstructed, patched curves (with their reliable errors) in the regions where we trust them. For example, close to the pseudo-critical region we can trust a peak pattern only when we can reconstruct the peak by starting from both sides of the transition (without multi-histogram patching). So we have always used single histogram reconstructions to check these criteria, before constructing the final, multi-histogram data.

We have measured a large number of local observables. We will see that a mixed picture emerges from these measurements. For example the observables related to the dynamical triangulations exhibit a characteristic pattern, to be discussed in detail below.

We start by showing, in Fig. 1, the edge curvature S_E as a function of λ. The crossover region is around $\lambda \simeq 1.4$. For small values of λ, the surface is crumpled (see the latter part of this section). In this region, finite size effects are already negligible for our lattice sizes, and our 3 data points are on top of each other. We can see weak finite size effects by comparing the continuous lines in the transient region. The $N = 144$ dotted line is far from the ones of the two larger lattices, which lie, on the contrary, on top of each other. Finite size effects are larger in the large λ phase. One would expect, close to a phase transition with a diverging correlation length, an increase of the finite size effects which we do not observe here. The lattice should feel the presence of the zero mass excitation, and the finite size corrections should be larger than everywhere else (in the case of periodic boundary conditions they would obey a power-law, rather than decaying exponentially with size). This is surely not

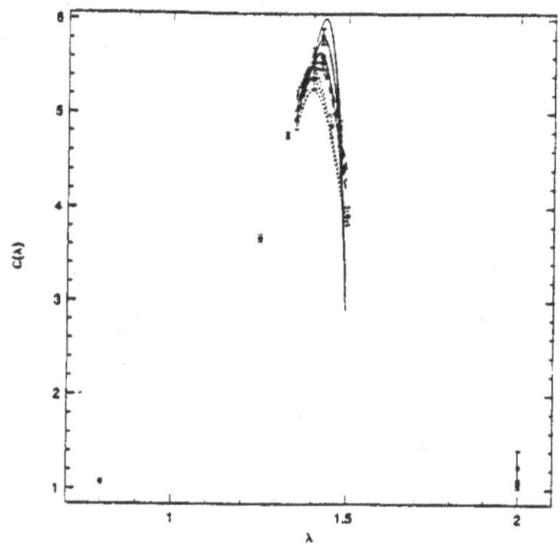

Figure 2: The edge curvature specific heat, $C(\lambda)$. The edge curvature S_E as a function of λ. As in all other pictures, filled circles and a dotted line correspond to $N = 144$, crosses and a dashed line indicate $N = 288$, and empty squares and a solid line represent $N = 576$.

N	$C(\lambda)^{\max}$	λ_c
144	5.37(14)	1.395(30)
288	5.55(7)	1.410(25)
576	5.81(17)	1.425(30)

Table 2: The maximum of the specific heat and its position, with their errors, for 3 different lattice sizes.

firm evidence against the presence of a phase transition, but it does show that the putative critical behaviour is atypical.

The errors in the 'flat phase' ($\lambda = 2.0$) are not under control. Our estimators do not plateau under repeated iterations of the binning procedure. In this regime, correlation times are large, as we will discuss in next section. This *caveat* holds for this figure and for all the quantities we have measured.

In Fig. 2 we show the related specific heat $C(\lambda)$, in the same λ region. In Fig. 3 we enlarge the pseudo-critical λ region, in order to show the reconstructed peak of the specific heat. As already noted our reconstruction procedure is quite reliable here.

In Table 2 we give the maximum of the edge curvature specific heat and its location for the 3 different lattice sizes. We can extract from these data a specific heat exponent $\omega = .06 \pm .05$, with ω defined as in equation (13), and the constant

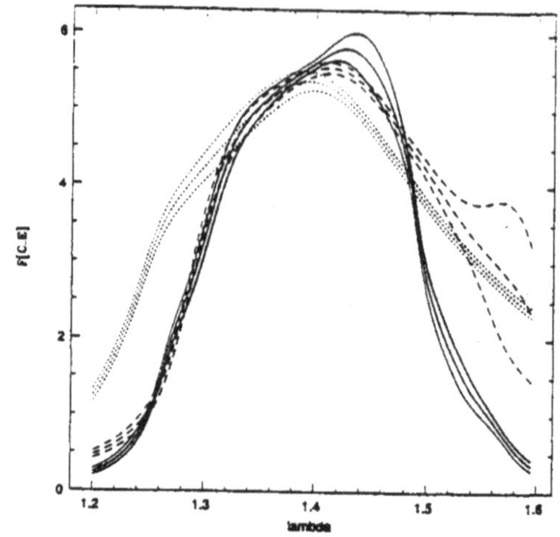

Figure 3: As in Fig. 1, but with the multi-histogram reconstruction in the transient region. The edge curvature S_E as a function of λ. As in all other pictures, filled circles and a dotted line correspond to $N = 144$, crosses and a dashed line indicate $N = 288$, and empty squares and a solid line represent $N = 576$.

B set to zero. If we estimate an effective exponent from the two smaller lattice sizes we get $.05 \pm .06$, and from the two larger ones we get $.07 \pm .06$; this demonstrates that we do not see, within our statistical precision, any sign of a non-pure-power, non-asymptotic behavior. Note that if the constraint that B vanishes is relaxed, our data is not accurate enough to yield a meaningful fit to equation (13). A very small (asymptotically finite) correlation length is sufficient to produce such a small effect on our quite small lattice sizes. These results appear to be consistent with those of the Copenhagen group [29], and they are not so far from the ones of the Urbana group [30, 40].

The critical value of λ shifts (very slowly, i.e. in a way compatible with no increase given our statistical precision) to higher values for increasing N. Also the shape of the specific heat (the width, for example), is basically unchanged as we go to larger lattices. From Figs. 2 and 3 we do not infer any evidence of criticality.

In Fig. 4 we show the radius of gyration of the surface, R_G, as defined in (15). Here obviously the volume scaling is non-trivial: larger surfaces have larger radius. The histogram reconstruction ceases to work already for quite low values of λ for the larger lattice. This effect could be related to the interesting finite size scaling behavior of this quantity, which we illustrate in better detail in Fig. 5 and 8.

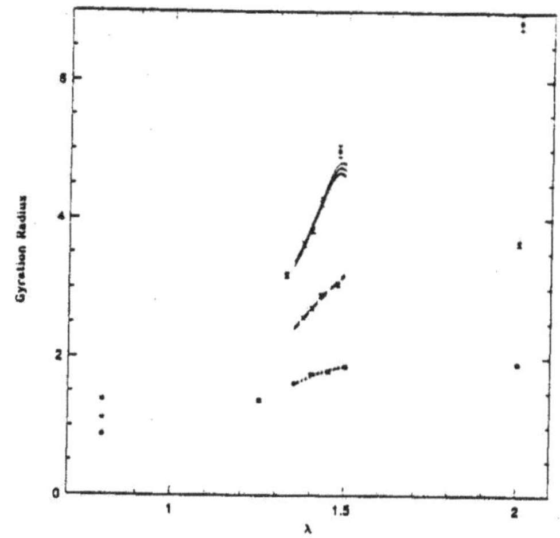

Figure 4: The gyration radius R_G defined in (15), plotted as in Fig. 1. The edge curvature S_E as a function of λ. As in all other pictures, filled circles and a dotted line correspond to $N = 144$, crosses and a dashed line indicate $N = 288$, and empty squares and a solid line represent $N = 576$.

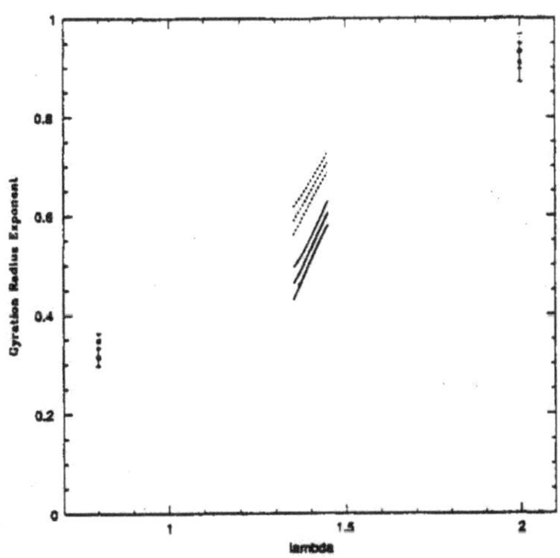

Figure 5: The effective inverse Hausdorff dimension ν as a function of λ, as defined in (16). The filled dots and the dashed curve are from a fit to the $N = 288$ and $N = 144$ data, while the empty dots and solid curve represent the fit to $N = 576$ and $N = 288$.

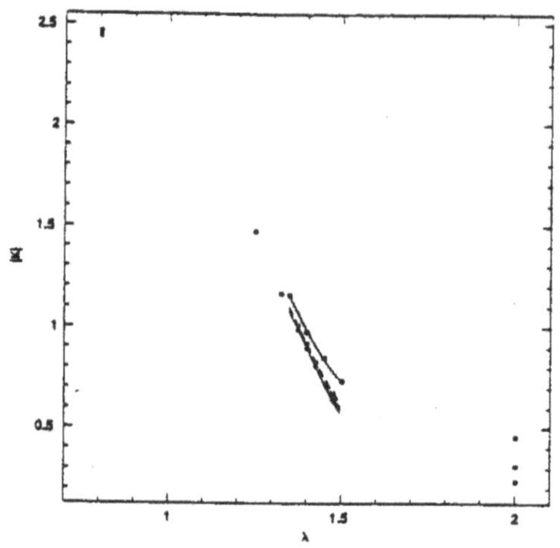

Figure 6: The extrinsic Gaussian curvature $| \mathcal{K} |$ defined in (17), plotted as in Fig. 1.

Here we plot

$$\nu(N) \equiv \frac{\log \frac{R(N)}{R(\frac{N}{2})}}{\log(2)} \ . \tag{22}$$

This is an effective inverse Hausdorff dimension, which is a function of λ. In the large λ limit $\nu \rightarrow 1$ and $d_{extr} \rightarrow 2$, as expected for flat surfaces. In the low λ limit d_{extr} becomes very large. In the pseudo-critical region ν is a linear function of λ. Curiously enough, the latter curve yields a Hausdorff dimension of 4, a value characteristic of branched polymers, near the location of the specific heat peak. Due to finite-size effects, this value is not particularly reliable, though (and a value changing rapidly from 2 to a large number must to pass through 4 somewhere in the crossover region). In ref. [29] a value compatible with ours is quoted for the critical theory ($D_H(\lambda_c) > 3.4$), Which is compatible with our findings. We stress however (also here, in complete agreement with remarks of [29]) the fact that the Hausdorff dimension in the pseudo-critical region depends heavily and in a quite unusual manner on N.

In both high and low λ regions finite size effects are quite small (compatible with zero in one standard deviation). On the contrary, in the pseudo-critical region, finite size effects are strong. This effect cannot be explained by the shift in λ which one gets from the shift of the peak of the specific heat, which is far too small. This behaviour is very different from the one which we have discussed for S_E, and it seems to indicate the possibility of some sort of critical behavior close to $\lambda = 1.4$.

In Fig. 6 we plot the expectation values of the magnitude of the extrinsic Gaussian curvature $| \mathcal{K} |$. If the induced metric is equal to the intrinsic metric, then $| \mathcal{K} |= \frac{|\mathcal{R}|}{\epsilon}$. This plot is not substantially different from that of S_E. We note that finite size effects

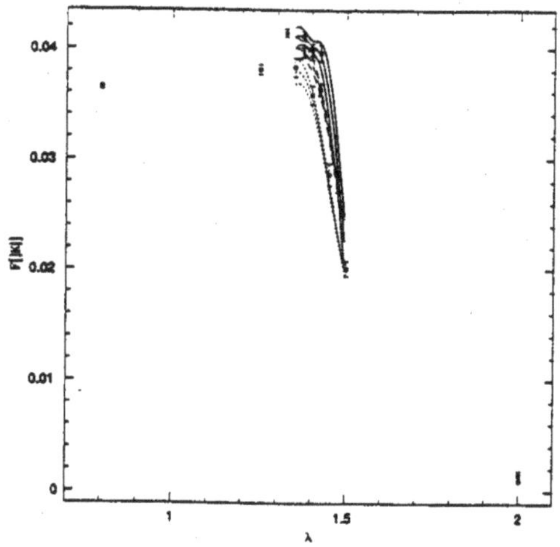

Figure 7: The fluctuations of $|\mathcal{K}|$.

are somewhat larger in this case than for the edge action, but they follow the same pattern (exhibiting a big increase in the flat phase).

The plot of the fluctuations of the extrinsic Gaussian curvature, $F[\mathcal{K}]$ which we present in Fig. 7, shows something very new. A very sharp crossover, with perhaps a peak developing for large N, dominates the pseudo-critical behavior. Fluctuations do not seem to depend on λ in the crumpled phase, while they drop dramatically, in a very small λ interval, in the flat region. Here, again, finite size effects are sizeable in the pseudo-critical region. The position of the crossover does not depend sensitively on N, while the detailed shape at λ_c seems to change slightly with N. It is difficult to give a precise interpretation of a plot like this, but, as we said, the change of regime is very clear here.

In Fig. 8 we give the intrinsic curvature $\mathcal{R}$, and in Fig. 9 its fluctuations. Both plots are very similar to the related, extrinsic curvature $\mathcal{K}$ plots. $|\mathcal{R}|$ drops off rapidly, just as $|\mathcal{K}|$ also does. Through the peak region, though, $|\mathcal{K}|$ decreases by about a factor of 5 while $|\mathcal{R}|$ diminishes to only about .6 of its value on the left-hand side of the peak. Since the action explicitly suppresses mean curvature, and the mean and extrinsic Gaussian curvature are closely related (for instance, $H^2 > \frac{K}{2}$), we would expect that for large λ extrinsic fluctuations would be suppressed much more than fluctuations of intrinsic geometry.

In Fig. 10 we plot the intrinsic extrinsic curvature correlation. The plot of $\mathcal{K}^*\mathcal{R}$ indicates that intrinsic and extrinsic geometry are strongly correlated for small λ, but as one passes through the peak region, they become decorrelated. This is not particularly surprising, given that the action directly suppresses only extrinsic fluctuations.

In Fig. 11 we plot the expectation value of the maximum correlation number,

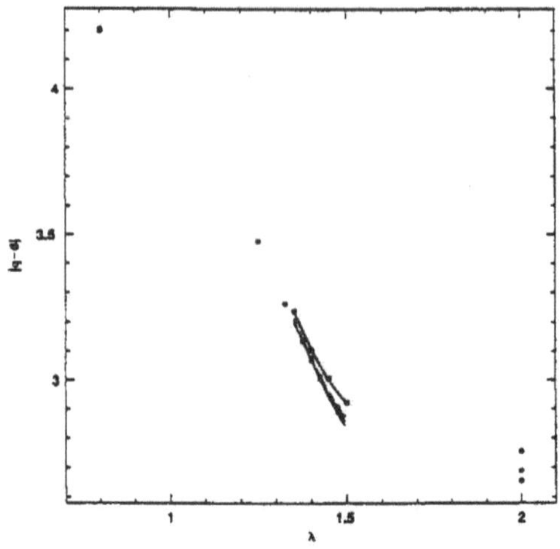

Figure 8: The intrinsic curvature $|\mathcal{R}|$ defined in (18), plotted as in Fig. 1.

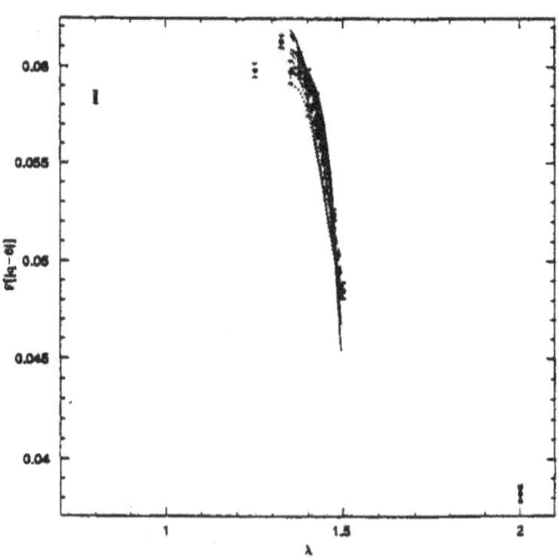

Figure 9: The fluctuations of $|\mathcal{R}|$.

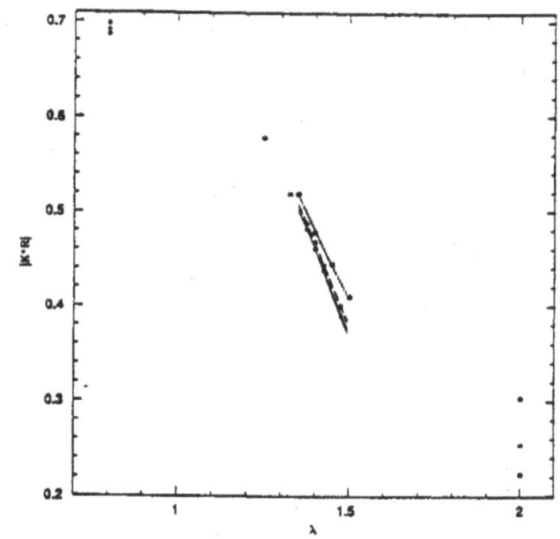

Figure 10: The intrinsic extrinsic curvature correlation, as defined in (19), plotted as in Fig. 1.

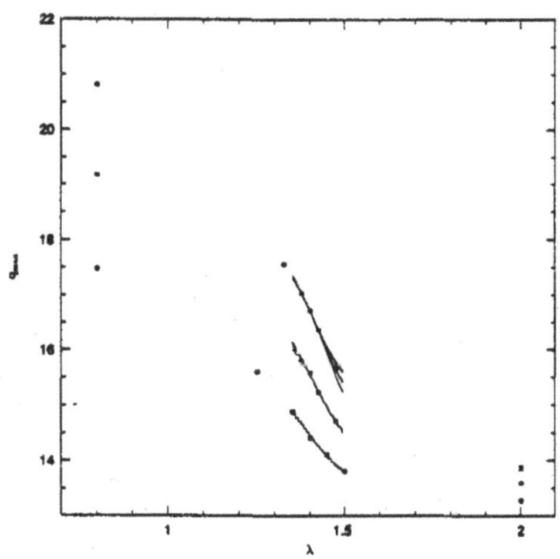

Figure 11: The average maximum connectivity number of the surface vertices, $\max_i q_i$, plotted as in Fig. 1.

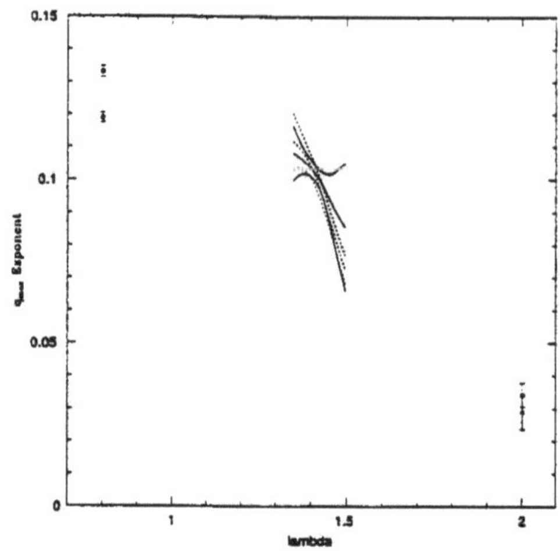

Figure 12: The scaling exponent of $\max_i q_i$, plotted as in Fig. 1.

which has non-trivial scaling behavior. In Fig. 12 we give its scaling exponent, defined analogously to the exponent we have exhibited for the gyration radius. In the pseudo-critical region q_{max} scales (for our 3 lattice sizes) as a power, with an exponent close to 0.1; we do not know if this scaling is meaningful.

We have also computed correlation times for different observables. As we already noticed, correlation times become very large in the large λ region. In agreement with ref. [29] we do not see any increase of the correlation times close to the pseudo-critical point.

We will not present precise estimates of correlation times (exponential or integrated): they are too large to get precise estimates. We will limit ourselves to a discussion of a few figures, which give quite a clear idea of what is happening. The comparison with ref. [29] cannot be very direct, since our action is different, and because their dynamics may be more effective than ours. Still, the comparison is quite puzzling, since we estimate and exhibit correlation times which are much (orders of magnitude) larger than the ones of [29]. Applying usual methods to estimate τ_{int} can lead to an underestimate of short correlation times if more than one time scale is present (that does surely happen to our data if we integrate our data on a window of a reasonable size).

In Fig. 13 we plot S_E for $N = 144$, $\lambda = 1.4$, and in Fig. 14 the gyration radius for these values (with a different time scale). Clearly, the correlation time is at least of order 40,000 sweeps in the first case and 100,000 sweeps in the second one. In Figs. 15 and 16, we plot the same quantities for $\lambda = 1.5$. Here correlation times are larger, of order 50,000 steps for S_E and larger than 150,000 steps for R_G. In Figs. 17 and 18 we draw the same plot on the largest lattice we study ($N = 576$) for $\lambda = 1.4$. Here

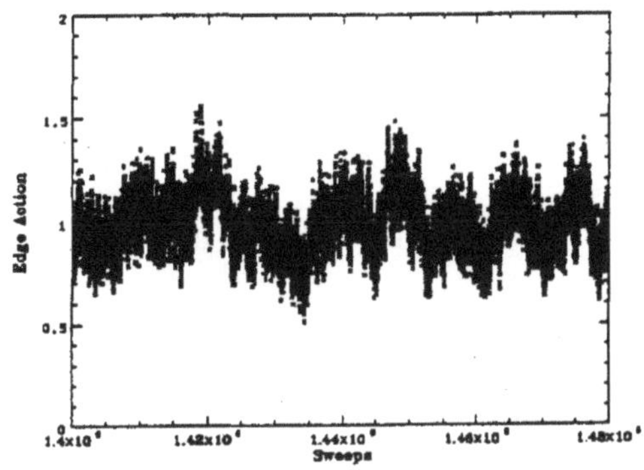

Figure 13: S_E as a function of Monte Carlo time (80,000 steps) for $N = 144$, $\lambda = 1.4$.

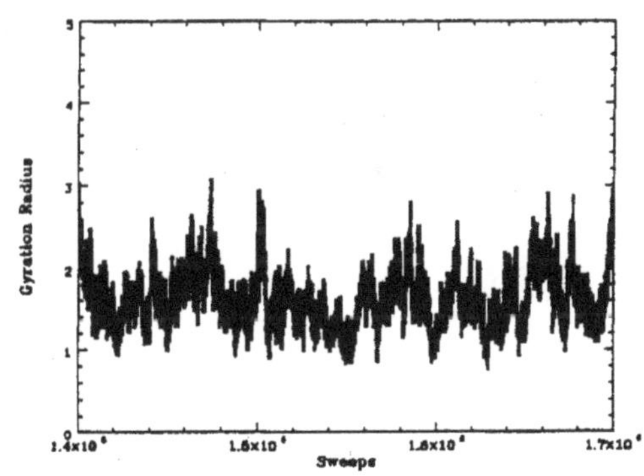

Figure 14: R as a function of Monte Carlo time (300,000 steps) for $N = 144$, $\lambda = 1.4$.

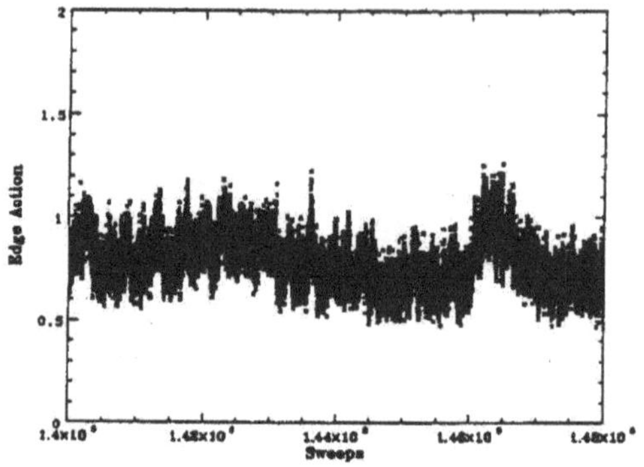

Figure 15: S_E as a function of Monte Carlo time $(80,000$ steps) for $N = 144$, $\lambda = 1.5$.

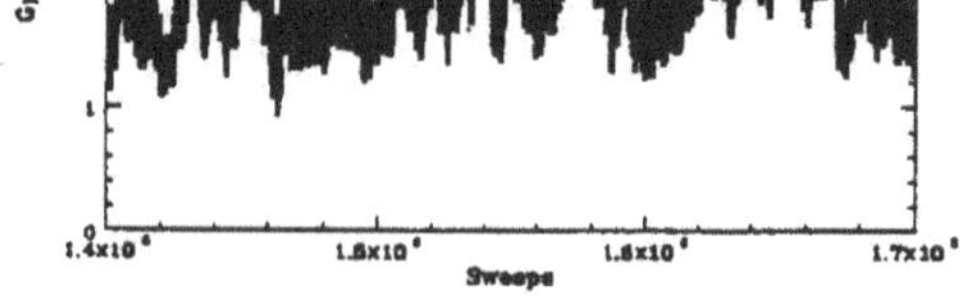

Figure 16: Fig. 16 R as a function of Monte Carlo time $(300,000$ steps) for $N = 144$, $\lambda = 1.5$.

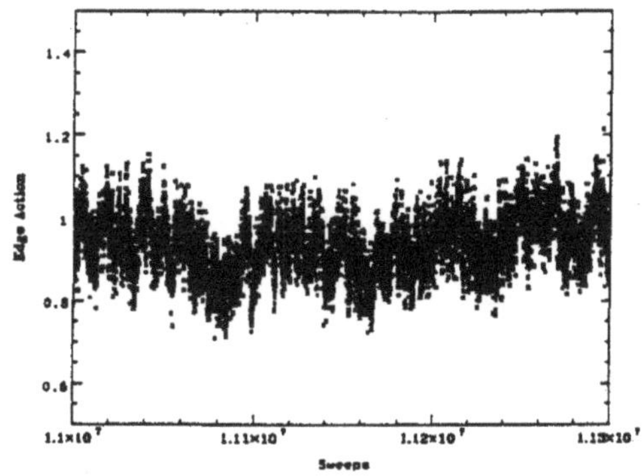

Figure 17: S_E as a function of Monte Carlo time $(300,000$ steps) for $N = 576$, $\lambda = 1.4$.

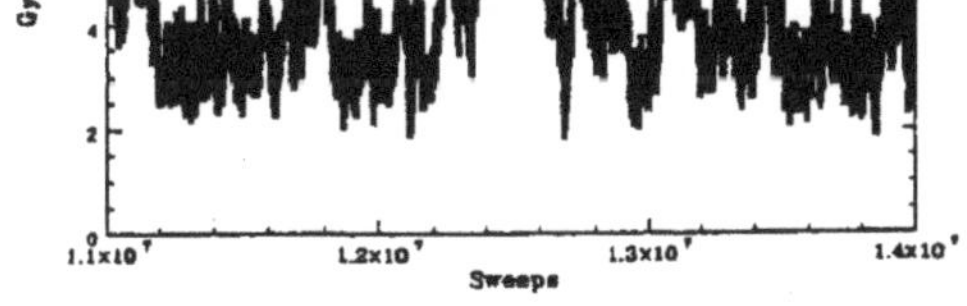

Figure 18: R as a function of Monte Carlo time $(3,000,000$ steps) for $N = 576$, $\lambda = 1.4$.

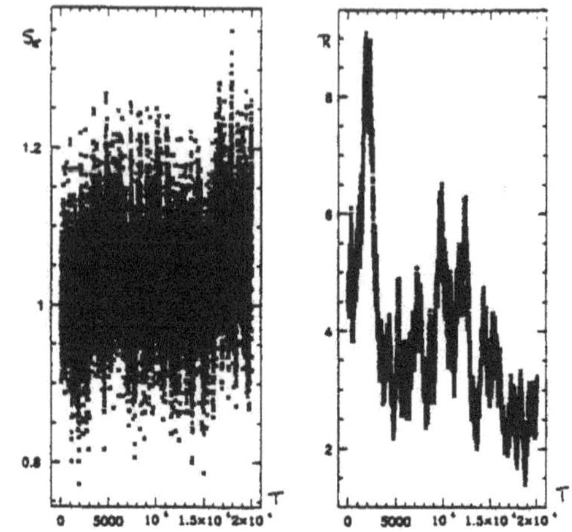

Figure 19: S_E and R from the same Monte Carlo run, $N = 576$, $\lambda = 1.325$, 20,000 steps.

we can see dramatic correlations, with times of at least 100,000 steps for S_E and of at least 1,000,000 steps for R_G.

In Fig. 19 we plot, for the same time history and on the same scale, both S_E and R_G. This figure shows a clear anticorrelation: larger surfaces are flatter, and have smaller curvature (this is apparent in the region close to the 2000th step).

4. A Discussion of the 2*d* Results

We have shown that the transition from crumpled to flat surfaces with increasing λ is quite complex. Our interpretation of what is going on is, at present, mainly based on these data and on the results of ref. [29].

On lattices with up to 576 vertices we can clearly see a sharp crossover, but the absence of a diverging specific heat, of diverging correlation times and of strong finite size effects suggests that we are not observing a usual second order phase transition. On the other hand, quantities like the intrinsic curvature, or the mass measurements of [29] show that something non-trivial is happening.

Let us review the crux of our observations again. This model of crumpled surfaces appears to exhibit sharp crossover behavior in the region around $\lambda = 1.4$. The sharp growth in the gyration radius and the suppression of curvature fluctuations indicate that the normals acquire long-range correlations, up to the size of the systems we examine. Presumably the zero string tension measurement of [29] also shows that the disordered regime differs from the regime in which the surfaces are ordered (up to scale of the lattices that are simulated) by only a small shift in λ. This evidence might indicate the presence of a phase transition at this point. If so, it is very likely to be of order higher than 2 (or, rather implausibly, it could be second order with an extremely low negative specific heat exponent; our lattices are much too small for us to confidently extrapolate the value of the specific heat exponent as $N \to \infty$).

If the transition were higher order, the peak should exhibit a cusp, but we would need far more accurate data to detect this. The existence of this phase transition would then suggest the existence of a new continuum string theory, though many other issues would have to be resolved (e.g. unitarity) to determine if such a theory is physically desirable.

There are other possible interpretations of our data. We need to consider the influence of finite-size effects, since the surfaces which we simulate are quite small, even smaller than one might naively assume because they are not intrinsically smooth. For instance, random surfaces characteristic of $D = 0$ gravity have a Hausdorff dimension of roughly $d_{intr} = 2.8$ [50, 51]; it has been predicted that surfaces embedded in 1 dimension have Hausdorff dimension $2 + \sqrt{2}$ [50]. Thus, for instance, if the surfaces in our simulations had an intrinsic dimension of 3, they would have a linear size of fewer than 9 lattice spacings[7].

Perhaps the simplest alternative explanation for the presence of this peak is suggested by the arguments of Kroll and Gompper [41]. They argue that the peak occurs when the persistence length of the system approaches the size of the lattice ($\xi_p \sim N^{\frac{1}{d}}$)[8]. For larger λ, fluctuations on a larger scale become more important, but when this scale is greater than the lattice size, these fluctuations are suppressed. Thus one might surmise that the specific heat will drop for large λ. (It clearly goes to zero for small λ because of the presence of the prefactor λ^2; the lattice implements a ultraviolet cutoff that freezes out very short-range fluctuations.)

The one-loop renormalization group calculation (10) predicts that the persistence length grows as $\xi_p \sim \exp(C\lambda)$; C is inversely proportional to the leading coefficient of the beta function. We would expect that the peak position should shift to the right with increasing N in this scenario as

$$\Delta = \frac{\delta \ln N}{d_{intr} C}. \tag{23}$$

Quite a large value of C is needed to explain the rapid crossover; roughly values of $C \sim 10, d_{intr} \sim 3$ are more or less consistent with the magnitude of the peak shift and crossover width. The RG calculations using different forms of the action yield $C = \frac{4\pi}{3}$ (see equation 10), but this may not apply to the action we simulate.

This reasoning also indicates that the peak should widen as the lattice size increases; we do not observe this at all. It seems plausible though that these arguments, based only on the leading term of the high lambda expansion, are too naive.

[7] Of course, our lattices are too small, by one or two orders of magnitude, to really exhibit a convincing fractal structure.

[8] For couplings above this point, our simulations would simply be measuring finite size effects.

An alternative scenario, which builds on the ideas in the above paragraph, is suggested by the tantalizing similarities between the results of our fluid surface simulations and what has been observed for the $d = 4$ $SU(2)$ Lattice Gauge Theory [45] and for the $d = 2$ $O(3)$ model. Let us discuss the case of the $O(3)$ model.

The $O(3)$ model, which is asymptotically free, exhibits a specific heat peak near $\beta = 1.4$. This peak was first measured via Monte Carlo simulations by Colot [53]. It can also be obtained by differentiating the energy data measured by Shenker and Tobochnik [54, 55]. The origin of this peak is understood [55, 56]; it is due to the fluctuations of the sigma particle, a low-mass bound state of the massless $O(3)$ pions. The sigma induces short-range order, and contributes to the specific heat as a degree of freedom only at high temperatures (when the correlation length in the system becomes smaller than its inverse mass). The peak thus occurs at the beginning of the crossover regime, when the correlation length is several lattice spacings.

According to the low temperature expansion, the correlation length grows as $\xi \sim \exp(2\pi\beta)/\beta$. Thus one would expect a fairly rapid crossover in the $O(3)$ model; the correlation length should increase by roughly a factor of 9 when β is shifted by about .35[9]. Such a crossover is indeed observed, though it is somewhat less dramatic than the behavior observed for fluid surfaces; the specific heat peak we measure is roughly half the width of the peak for the $O(3)$ model.

Recent simulations of the $O(3)$ model [52] indicate that the specific heat peak grows significantly when the lattice size L is increased from 5 to 15, and that virtually no growth in peak height is evident as L is increased further up to 100. Also, the peak position shifts to the right as L grows, and then appears to stabilize for large L. This is more or less what we observe in our simulations of fluid surfaces, on lattices of small size. We point out these similarities largely to emphasize that there does exist an asymptotically free theory (with low mass excitations) which exhibits crossover behavior qualitatively similar to that observed in our simulations.

The analogy is perhaps deeper, though, since the fluid surface action (with extrinsic curvature) in certain guises looks like a sigma model action. So, perhaps it would not be so surprising from this point of view to find a sigma particle in these theories perhaps associated with $(\hat{n}^2 - 1)$, in which $\hat{n}$ denotes the unit normal to our surfaces.

Another additional possibility is that fluctuations of the intrinsic geometry (the Liouville mode) are responsible for short-range order and contribute to the specific heat peak.

There remains much to be done to clarify whether or not a crumpling transition occurs for a finite value of the extrinsic curvature coupling λ. It would be interesting (and probably a fair amount of work) to apply Wilson renormalization group techniques to the actual action (7) which we simulate, to determine the leading coefficient

[9] In fact, the presence of the sigma significantly modifies this low-temperature expansion result [56] in this intermediate regime, but does not qualitatively destroy the rapidity of the crossover. Indeed, despite heroic efforts, it has been impossible to extend computationally beyond this regime and precisely verify the asymptotic low-temperature relation for the correlation length [57, 58].

of the beta function. Additionally, perhaps a calculation of $1/D$ corrections to the large D computations already performed could unearth evidence of a sigma-type excitation in these theories (the effects of the sigma appear as $1/N$ corrections in the $O(N)$ model).

We also are histogramming our data to examine the behavior of complex zeroes (in complex λ space) of the partition function of our simulations [59]. It has been shown (in the case of $SU(2)$ lattice gauge theory) that such zeroes, when they are near but do not approach the real axis in the infinite volume limit, occur in theories which exhibit specific heat peaks with no associated phase transition [45]. High temperature expansions also indicate that the $O(3)$ model susceptibility has a complex singularity near the real axis [60]– presumably this corresponds to a zero of the partition function and is a manifestation of the sigma.)

Of course, simulations on large lattices, with better statistics, should also help us evaluate whether a crumpling transition exists. We are testing algorithms, such as simulated tempering [61], in order to evade the long auto-correlation times that have characterized our simulations so far.

Even if no such transition exists for finite λ, one could still attempt to study a continuum theory in the strong coupling limit, as is done for QCD, for instance. To do so, we would like to examine global quantities, such as masses extracted from normal-normal correlation functions, rather than just the local quantities (energy, e.g.) that we have measured. Measuring these correlations requires a definition of distance on these triangulated lattices; the most successful definition of the metric is based on the propagation of massive particles (via inversion of the Laplacian) on these lattices [62].

We have observed dramatic crossover behavior for particular observables in our Monte Carlo simulations, but on the other hand, the correlation times and certain finite-size effects do not behave as one would expect in the presence of a phase transition. The behavior of other lattice models also indicates that it is possible that we are observing the effects of finite-mass excitations on small lattices, rather than a phase transition. We hope that future work will clarify this murky state of affairs, to determine if there indeed exists a crumpling transition for fluid surfaces.

5. 4*d* Triangulated Manifolds and Quantum Gravity

Dynamically triangulated random surfaces (DTRS) [6] play an important role in the efforts to develop a coheerent description of quantum gravity. The (euclidean) space-time is approximated by a d-dimensional simplicial triangulation, where the link length is constant, equal to 1, but the connectivity matrix is a dynamical variable.

The most important advances have been obtained in two-dimensional quantum gravity, where DTRS are simplicial triangulations of a 2d manifolds. The analytic success of matrix models, which can be for example exactly solved in the case of pure 2d gravity [8], has strongly encouraged this approach. The results obtained in the triangulated approach and in the continuum lead to consistent predictions for correlation functions and critical exponents.

Dynamical triangulations are also potentially relevant in four dimensions. One can hope that a sensible, non-perturbative definition of the quantum gravity theory can be obtained in some *scaling* limit of the theory of $4d$ hyper-tetrahedra. This approach has much in common with Regge calculus, where the connectivity is fixed but the functional integration runs over the link lengths. The underlying principle is clearly very similar, and one could say that DTRS have the status of an improved Regge calculus. We face the usual problem inherent in discretizing a theory, i.e. the discretization scheme can break some of the continuous symmetries, which will have to be recovered in the continuum limit (if there is one). Indeed, Wilson lattice gauge theories have taught us an important lesson. The fact that gauge invariance is exactly conserved in the lattice theory, for all values of the lattice spacing a, is in that case crucial: it would have been very difficult to establish firm numerical results if one would have had to care about the presence of non gauge-invariant correction, which would disappear only in the $a \to 0$ limit. In the case of quantum gravity, diffeomorphism invariance plays such a crucial role, and DTRS are diffeomorphism invariant by construction, at least on the space of piecewise flat manifolds. Hence part of the difficulties Regge calculus has in forgetting about the lattice structure are eliminated a priori in the DTRS lattice approach.

There are two more important points to stress. The first one is that in the DTRS approach in $3d$ and $4d$, as opposed to $2d$, we can try to make sense out of the pure Einstein action, without, for example, curvature squared terms. Even though the partition function formally diverges, at fixed volume the local curvature is bounded both from below and from above. Therefore we can study the theory at fixed (or better quasi-fixed, see later) volume, and look for the existence of a stable fixed point in the large volume limit. A second order phase transition with diverging correlation lengths, in the statistical mechanics language, would allow us to define a continuum limit which is universal and is not influenced by the details of the underlying discrete lattice structure. Precisely this scenario constitutes one of the best hopes we have to find a consistent quantum theory of gravity. If euclidean quantum gravity based on the Einstein action does have non-perturbative meaning, then we can exhibit it in this way.

The second problem is to determine the measure one should use to define the quantum theory. This problem, far from being solved in the continuum, is completely open in the lattice approach (for a good review see for example [63]). In quantum Regge calculus the influence of the measure has been examined in [64] but there is no direct relation to DTRS. The problem of the measure is the main point we address here, and we want to suggest that the DTRS approach may be powerful enough to solve it.

Recently two groups have pioneered Monte Carlo simulations of DTRS [65, 66, 67]. Numerical simulations (even if on quite a small scale) turned out to be feasible, and lead to very non-trivial results. One clearly observes a phase transition structure. Although quantitative statements are not easy to make, given the limited statistics and the small lattice size, it is clear that the situation is different from the $3d$ case, where the phase transition is manifestly of first order, and there is no continuum theory. In $4d$ [65] there is an open possibility that the transition is second order (although that cannot be claimed without a much more detailed finite size study). One does not observe hysteresis cycles, and the crossover is less sharp then in $3d$. More involved statements about critical exponents have to be taken at this point, we believe, *cum grano salis*, but there is evidence for the possible presence of a critical point with a non-trivial continuum theory in the phase diagram for the euclidean Einstein action. This observation certainly warrants further careful investigations.

These first simulations have been run with uniform measure, where all triangulations have the same weight in the sum which defines the path integral of the quantum theory. There are no particularly good reasons for this choice to be the correct one. In fact, on general grounds one expects the measure to be non-uniform [63, 70]. In the following we investigate the changes introduced by defining simplicial quantum gravity with a non-trivial measure.

One of us has described in [68] the structure of the Monte Carlo simulations, and how the efficiency of the numerical procedure can be optimized. The programming of dynamical triangulations is difficult since a dynamical data structure is required, but it is very relevant in many practical applications. For the same reason the implementation of DTRS on parallel computers is a hard problem. Since a set of ergodic moves preserving the volume for canonical simulations is not known, one has to consider a Markov chain which sweeps out the space of different volume simplicial manifolds. We have used the quasi-canonical method introduced in [69], which allows us to control the systematic distortion arising from the *ad hoc* potential that keeps the system close to a given number of 4-simplices [68]. All the details about the system we study and about the numerical procedures are given in [68].

In the continuum the euclidean Einstein-Hilbert action for a metric $g_{\mu\nu}$ has the form

$$S_E[g] = \int d^4x\sqrt{g}(\lambda - \frac{R(g)}{G}) , \tag{24}$$

where $R(g)$ is the Ricci scalar and λ and G are the cosmological and gravitational constants, respectively. We consider a fixed S^4 topology. On a triangulation T we discretize according to

$$V = \int d^4x\sqrt{g} \rightarrow N_4[T] , \tag{25}$$

$$R = \int d^4x\sqrt{g}R(g) \rightarrow \frac{2\pi}{\alpha}N_2[T] - 10N_4[T] , \tag{26}$$

where $\alpha \simeq 1.318$, and $N_i[T]$ is the number of i-simplices of the triangulation T. The discrete action is then

$$S_E[T] = k_4N_4[T] - k_2N_2[T] , \tag{27}$$

where $k_4 = \lambda + 10/G$ and $k_2 = 2\pi/\alpha G$.

In the discrete quantum theory there exists a critical line $k_4 = k_4^c(k_2)$ such that if k_4 is different from k_4^c for a given value of k_2 then the random walk tends to either zero or infinite volume. All measurements are made for $k_4 = k_4^c(k_2)$.

We have selected not one but a family of measures in order to investigate the influence of the measure in a rather general setting. Our choice is guided by diffeomorphism invariance of the measure [70] but ignores more sophisticated arguments like BRST invariance. We have studied, as a function of n, the uniform measure times the weight $g^{n/2}$,

$$\prod_x g^{n/2} \prod_{\mu<\nu} dg_{\mu\nu}. \tag{28}$$

In the triangulated theory $S_E[T]$ is replaced by $S[T] = S_E[T] + S_M[T]$, where

$$S_M = -n \sum_a \log \frac{o(a)}{5} . \tag{29}$$

The sum runs over all 0-simplices (sites) of the manifold, and $o(a)$ is the number of 4-simplices which include the site a. We considered n in the interval from -5 to 5. The case $n = 0$ repeats simulations with the trivial, uniform measure, which can be compared with previous results.

In 2d DTRS it has been already observed that on finite lattices, as expected, the phase diagram depends on the measure coupling (e.g. [71]). However, in this case the measure term is naively an irrelevant operator, and once an extrinsic curvature term has been added it has been shown that the phase diagram [1, 29] is insensitive to this term in the large lattice limit (also at the crossover point). In the 4d case the phase diagram is also expected to depend on n; if one insists on sending n to $\pm\infty$, then the measure term would dominate the action. It is an interesting question to ask whether for some reasonable values of n one obtains different universality classes. The choice of uniform measure in [65, 66, 67] would be valid if this is not the case.

Let us summarize our results. We confirm the fact that the phase transition can be of second order, and that it is plausible that we will be able to define a sensible theory. We find that the measure factor plays an important role, and that the critical behavior does depend on n. Varying n does not only change non-universal quantities, like for example the value of the critical coupling, but changes the actual (pseudo-)critical behavior.

In figure 20 we plot the average curvature R/V for $V \equiv N_4 = 4000$ as a function of the coupling k_2 for different values of n, $n = -5$ for the lowest curve, then $n = -1$, 0, 1 and $n = 5$ for the upper curve. In figure 21 we plot the average distance (in the internal space) of two 4-simplices. We count the minimum number of steps from 4-simplex to 4-simplex across 3-simplex faces that connect a pair of 4-simplices and average over all 4-simplices and random manifolds.

Both figures show that the measure operator has a pronounced effect. Increasing the coupling of the measure term leads to a continuous, monotonous deformation of

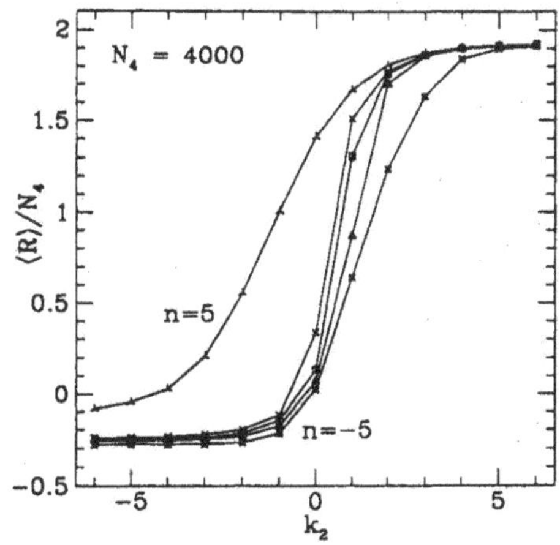

Figure 20: Expectation value of R/V for $n = -5, -1, 0, 1, 5$. The curves are ordered from the smallest n at the bottom upwards.

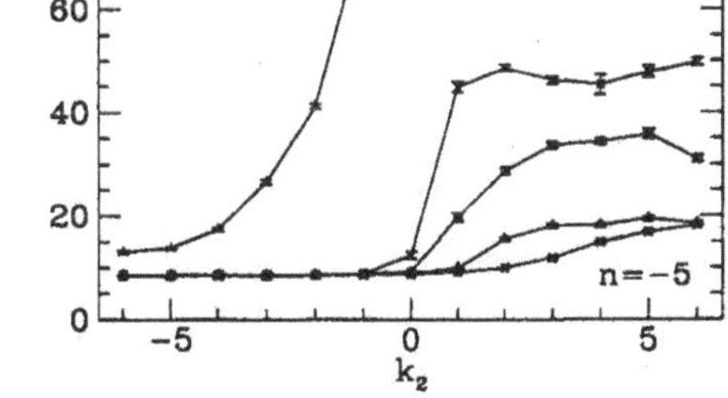

Figure 21: Expectation value of d for $n = -5, -1, 0, 1, 5$, $N_4 = 4000$; curves ordered as in figure 20.

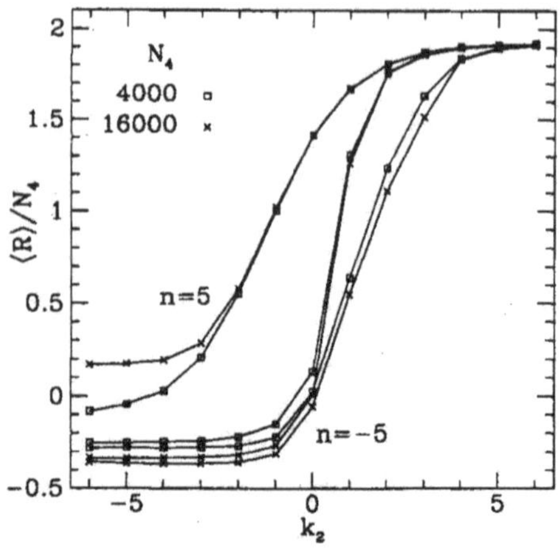

Figure 22: Expectation value of R/V for $n = -5, 0, 5$, $N_4 = 4000, 16000$; curves ordered as in figure 20.

the curves. Notice that the curves are not just shifted. In the case of R/V, the singularity seems stronger for $n \simeq 0$, where the jump in R/V is quite sharp. The distance d has a sharper jump for $n = 1$, where it seems to jump from one constant value to another constant. Smaller values of n show a slower increase in d.

For large absolute values of n, especially for $n = -5$, the plots show a weaker singularity. The profile of R/V hints less at a sharp jump than the former cases, and the distance increases very smoothly from a critical value of k_2, $k_2^c(n)$ on. When n increases to the value of 5 the system seems to loose criticality on an absolute scale. Its behavior through the crossover is quite smooth.

A critical value of k_2^c can be defined, for example, as the point where the distance value starts to change. But for the $n = 5$ case the transition point is not very clear. Let us note that such a value of k_2^c changes its sign as a function of n.

Figure 22 and 23 show results for $N_4 = 16000$ to indicate what kind of finite size effects are present. As evident from figure 22, larger volumes amplify the effect of large absolute n for $k_2 < k_2^c$. Figure 23 for the average distance d displays the same qualitative behaviour as figure 21 for $k_2 < k_2^c$, but for large k_2 the average distance does not remain constant, which might already have been guessed in figure 21. The explanation is that the measure term S_M for positive n introduces a bias towards smaller $o(a)$ which increases d, but for large enough k_2 and k_4 the contribution of S_E dominates and d approaches its $n = 0$ value. This may be the case for $k_2 > k_2^c$, while for $k_2 > k_2^c$ the critical value $k_4^c(k_2)$ is such that $S_E \approx 0$, and at least for the range of negative k_2 considered here S_M is relevant and d remains large for $n = 5$. The same

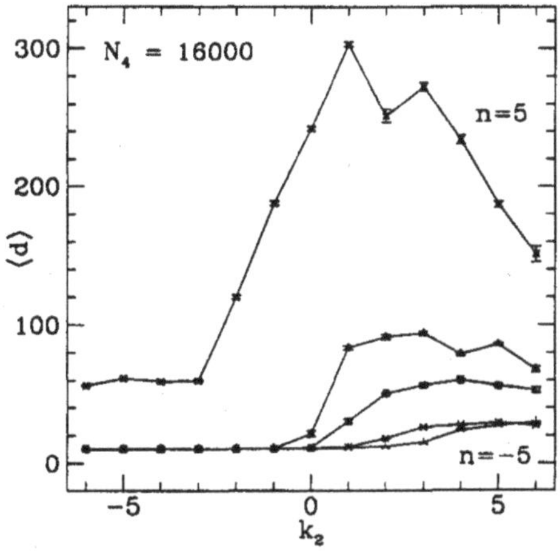

Figure 23: Expectation value of d for $n = -5, -1, 0, 1, 5$, $N_4 = 16000$; curves ordered as in figure 20.

holds for R/V when $n = 5$.

Our conclusion has been that the measure term has a strong effect, which seems difficult to reabsorb in a simple renormalization of the critical coupling. Always keeping in mind that a precise finite size study is required before making quantitative statements [72], we believe there are two basic possibilities. The first possibility is that there is only one universality class, and that all the theories we have studied do asymptotically show the same critical behavior. In this case the rate of approach to the continuum limit is strongly influenced by n. We will select the theory with faster convergence to the continuum.

The second possibility (which is the most interesting one) is that the measure factor changes the universality class. Our results, albeit preliminary, seem to hint in this direction. In this case we could have a critical value of n_c, and transitions belonging to different universality classes. This is a very appealing scenario, and here the lattice discrete theory could make its own original contribution. It could be possible to pick out the correct measure, on the lattice, by requiring a particular expectation value and scaling behavior of some physical observable. Such a prescription would be a powerful tool, turning the discrete version of the theory from a source of indetermination into a completely determined scheme.

Acknowledgements

We are much indebted to Geoffrey Fox and NPAC, for making this work possible.

References

1. M. Bowick, P. Coddington, L. Han, G. Harris and E. Marinari, *The Phase Diagram of Fluid Random Surfaces with Extrinsic Curvature,* preprint hep-lat/9209020 (September 1992).
2. B. Brügmann and E. Marinari, *4d Simplicial Quantum Gravity with a Non-Trivial Measure,* preprint hep-lat/9210002 (October 1992).
3. F. David, *Introduction to the Statistical Mechanics of Random Surfaces and Membranes*, Saclay preprint SPhT/91-113.
4. V. G. Knizhnik, A. M. Polyakov and A. B. Zamolodchikov, *Mod. Phys. Lett.* **A3** (1988) 819.
5. F. David, *Mod. Phys. Lett.* **A3** (1988) 1651; J. Distler and H. Kawai, Nucl. Phys. **B321** (1989) 509.
6. D. Weingarten, Phys. Lett. **B90** (1980) 285, Nucl. Phys. **B210** [FS6] (1982) 229; V. A. Kazakov, *Phys. Lett.* **150B** (1985) 282; F. David, *Nucl. Phys.* **B257** (1985) 45; J. Ambjørn, B. Durhuus and J. Frölich, Nucl. Phys. **B257** (1985) 433.
7. V. A. Kazakov and A. A. Migdal, *Nucl. Phys.* **B311** (1988/89) 171.
8. E. Brézin and V.A. Kazakov, *Phys. Lett.* **236B** (1990) 144; M.R. Douglas and S.H. Shenker, *Nucl. Phys.* **B335** (1990) 635; D. J. Gross and A. A. Migdal, *Phys. Rev. Lett.* **64** (1990) 127;
9. D. Kutasov and N. Seiberg, *Nucl. Phys.* **B358** (1991) 600.
10. A. Billoire and F. David, *Phys. Lett.* **168B** (1986) 87; *Nucl. Phys.* **B275** (1986) 617.
11. J. Ambjørn, B. Durhuus and J. Fröhlich, *Nucl. Phys.* **B275** (1986) 161; J. Ambjørn, B. Durhuus, J. Fröhlich and P. Orland, *Nucl. Phys.* **B270** (1986) 457.
12. D. Boulatov, V. Kazakov, I. Kostov and A.A. Migdal, *Phys. Lett.* **157B** (1985) 295.
13. D. Boulatov, V. Kazakov, I. Kostov and A.A. Migdal, *Phys. Lett.* **174B** (1986) 87; *Nucl. Phys.* **B275** (1986) 641.
14. J. Ambjørn and B. Durhuus, *Phys. Lett.* **188B** (1987) 253.
15. M.E. Cates, *Europhys. Lett.* **7** (1988) 719.

16. A. M. Polyakov, *Gauge Fields and Strings* (Harwood Academic Publishers, Chur, Switzerland 1987).
17. P. B. Canham, *J. Theor. Biol.* **26** (1970) 61.
18. W. Helfrich, *J. Naturforsch.* **28C** (1973) 693.
19. A. M. Polyakov, *Nucl. Phys.* **B268** (1986) 406.
20. H. Kleinert, *Phys. Lett.* **174B** (1986) 335.
21. P. Wiegmann, *Nucl. Phys.* **B323** (1989) 330.
22. Y. Choquet-Bruhat, C. DeWitt-Morette and M. Dillard-Bleick, *Analysis, Manifolds, and Physics* (Elsevier, Amsterdam 1987).
23. M. P. Do Carmo, *Differential Geometry of Curves and Surfaces* (Prentice-Hall, Englewood, N.J., USA 1976).
24. A. M. Polyakov, *Phys. Lett.* **103B** (1981) 207.
25. S. Catterall, *Phys. Lett.* **220B** (1989) 207.
26. C. Baillie, D. Johnston and R. Williams, *Nucl. Phys.* **B335** (1990) 469.
27. C. Baillie, S. Catterall, D. Johnston and R. Williams, *Nucl. Phys.* **B348** (1991) 543.
28. J. Ambjørn, J. Jurkiewicz, S. Varsted, A. Irbäck and B. Petersson, *Phys. Lett.* **275B** (1992) 295.
29. J. Ambjørn, A. Irbäck, J. Jurkiewicz and B. Petersson, *The Theory of Dynamical Random Surfaces with Extrinsic Curvature*, Niels Bohr Institute Preprint NBI-HE-92-40.
30. S. Catterall, J. Kogut and R. Renken, *Nucl. Phys.* **B** (Proc. Suppl.) **99A** (1991) 1.
31. J. Polchinski and Z. Yang, U. Texas and Rochester preprint UTTG-08-92/UR-1254/ER-40685-706.
32. H. Kleinert, *Phys. Lett.* **114A** (1986) 263.
33. D. Forster, *Phys. Lett.* **114A** (1986) 115.
34. L. Peliti and S. Leibler, *Phys. Rev. Lett.* **54** (1985) 1690.
35. T. Morris, *Nucl. Phys.* **B341** (1990) 443; J. Govaerts, *Int. J. Mod. Phys.* **A4** (1989) 173.
36. J. Polchinski and A. Strominger, *Phys. Rev. Lett.* **67** (1991) 1681.
37. J. Distler, *A note on the 3D Ising Model as a String Theory*, Princeton University preprint PUPT-1324 (hep-th-9205100).
38. F. David, *Europhys. Lett.* **2** (1986) 577; F. David and E. Guitter, *Europhys. Lett.* **3** (1987) 1169; *Nucl. Phys.* **B295** (1988) 332; F. Alonso and D. Espriu, *Nucl. Phys.* **B283** (1987) 393.
39. Z. Yang, Univ. of Rochester preprint, UR-1136/ER-13065-678.

40. S. Catterall, D. Eisenstein, J. Kogut and R. Renken, *Nucl. Phys.* **B366** (1991) 647.

41. D. Kroll and G. Gompper, *Science* **255** (1992) 968; *Europhys. Lett.* **19** (1992) 581; *The Scaling Behavior of Randomly Triangulated Self-Avoiding Surfaces, Phys. Rev.* **A**; to appear.

42. R. Harnish and J. Wheater, *Nucl. Phys.* **B350** (1991) 861.

43. D. Nelson, in *Statistical Mechanics of Membranes and Surfaces*, Jerusalem Winter School, Vol. **5**, edited by D. Nelson, T. Piran and S. Weinberg (World Scientific, Singapore 1989) 167.

44. F. David, in *Statistical Mechanics of Membranes and Surfaces*, Jerusalem Winter School, Vol. **5**, edited by D. Nelson, T. Piran and S. Weinberg (World Scientific, Singapore 1989) 158.

45. M. Falcioni, E. Marinari, M. L. Paciello, G. Parisi and B. Taglienti, *Phys. Lett.* **102B** (1981) 270; Nucl. Phys. **B190** (1981) 782; *Phys. Lett.* **108B** (1982) 331; E. Marinari, *Nucl. Phys.* **B235** (1984) 123.

46. A. M. Ferrenberg and R. H. Swendsen, *Phys. Rev. Lett.* **61** (1988) 2635 and *Erratum*, ibid. **63** (1989) 1658.

47. A. M. Ferrenberg and R. H. Swendsen, *Phys. Rev. Lett.* **63** (1988) 1196.

48. N. A. Alves, B. A. Berg and S. Sanielevici, *Spectral Density Study of the SU(3) Deconfining Phase Transition*, to be published in *Nucl. Phys.* **B**.

49. R. H. Swendsen, summary talk given at the 1992 Berlin Statistical Physics Conference STATPHYS 18.

50. H. Kawai and M. Ninomiya, *Nucl. Phys.* **B336** (1990) 115.

51. A. Migdal, talk given at Syracuse University, April 1992.

52. K. Anagnostopoulos, private communication.

53. J. L. Colot, *J. Phys.* **A16** (1983) 4423.

54. S. Shenker and J. Tobochnik, *Phys. Rev.* **B22** (1980) 4462.

55. R. Brout, W. Deans and A. Silovy, *Phys. Rev.* **B27** (1983) 5813.

56. R. Brout and W. Deans, *Nucl. Phys.* **B215** (1983) 407; J. Orloff and R. Brout, *Nucl. Phys.* **B270** (1986) 273.

57. J. Apostolakis, C. Baillie and G. Fox, *Phys. Rev.* **D43** (1991) 2687.

58. P. Hasenfratz and F. Niedermayer, *Phys. Lett.* **B245** (1989) 522.

59. M. Bowick, P. Coddington, L. Han, G. Harris and E. Marinari, work in progress.

60. P. Butera, M. Comi and G. Marchesini, *Nucl. Phys.* **B300** (1989) 1.

61. E. Marinari and G. Parisi, *Europhys. Lett.* **19** (1992) 451.

62. M. Agishtein, R. Benav, A. Migdal and S. Solomon, *Mod. Phys. Lett.* **A6** (1991) 1115.

63. P. Menotti, Nucl. Phys. **B** (Proc. Suppl.) **17** (1990) 29.

64. See for example H. Hamber, talk given at Lattice 92 (Amsterdam), and references therein; W. Beirl, E. Gerstenmayer and H. Markum, Phys. Rev. Lett. **69** (1992) 713

65. J. Ambjørn and J. Jurkiewicz, Phys. Lett. **B 278** (1992) 42.

66. M. E. Agishtein and A. A. Migdal, Mod. Phys. Lett. **A7** (1992) 1039.

67. M. E. Agishtein and A. A. Migdal, Nucl. Phys. B385 (1992) 395

68. B. Brügmann, *Non-Uniform Measure in Four-Dimensional Simplicial Quantum Gravity,* Syracuse preprint SU-GP-92/9-1, September 1992, submitted to Phys. Rev. **D**.

69. B. Baumann, Nucl. Phys. **B285** (1987) 391.

70. N. Konopleva and V. Popov, *Gauge Fields* (Harwood, New York 1979); H. Leutwyler, Phys. Rev. 134 (1964) B1155; E. Fradkin and G. Vilkovisky, Phys. Rev. D8 (1973) 4241

71. F. David, J. Jurkiewicz, A. Krzywicki and P. Petersson, Nucl. Phys. B290 (1987) 218

72. B. Brügmann and E. Marinari, work in progress.

ENTROPY GENERATION IN QUANTUM GRAVITY AND BLACK HOLE REMNANTS*

A. CASHER and F. ENGLERT†

Service de Physique Théorique
Université Libre de Bruxelles, Campus Plaine, C.P.225
Boulevard du Triomphe, B-1050 Bruxelles, Belgium
and
School of Physics and Astronomy
Raymond and Beverly Sackler Faculty of Exact Sciences
Tel-Aviv University, Ramat-Aviv, 69978 Tel-Aviv, Israel

Abstract

The area entropy $A/4$ and the related Hawking temperature in the presence of event horizons are rederived, for de Sitter and black hole topologies, as a consequence of a tunneling of the wave functional associated to the classical coupled matter and gravitational fields. The extension of the wave functional outside the barrier provides a reservoir of quantum states which allows for an additive constant to $A/4$. While, in a semi-classical analysis, this gives no new information in the de Sitter case, it yields an infinite constant in the black hole case. Evaporating black holes would then leave residual "planckons" - Planckian remnants with infinite degeneracy. Generic planckons can neither decay into, nor be directly formed from, ordinary matter in a finite time. Such opening at the Planck scale of an infinite Hilbert space is expected to provide the ultraviolet cutoff required to render the theory finite in the sector of large scale physics.

⋆ **presented by F.Englert**

1. Introduction

Tunneling in quantum gravity can generate entropy[1],[2]. To understand how such an apparent violation of unitarity may arise, let us first consider a classical spacetime background geometry with compact Cauchy hypersurfaces. If quantum fluctuations of the background are taken into account, quantum gravity leaves no "external" time parameter to describe the evolution of matter configurations in this background. Indeed, the solutions of the Wheeler-De Witt equation[3]

$$\mathcal{H}|\Psi\rangle = 0 \tag{1}$$

where $\mathcal{H}$ is the Hamiltonian density of the interacting gravity-matter system can contain no reference to such time when there are no contribution to the energy from surface terms at spatial infinity[4]. This is due to the vanishing of the time displacement generator and even though the theory can be unambiguously formulated only at the semi-classical level, such a consequence of reparametrization invariance should have a more general range of validity.

To parametrize evolution, one then needs a "clock" which would define time through correlations[5], namely correlations of matter configurations with ordered sequences of spatial geometries. If quantum fluctuations of the metric field can be neglected, the field components g_{ij} at every point of space can always be parametrized by a classical time parameter, in accordance with the classical equations of motion. This classical time, which is in fact a function of the g_{ij}, can be used to describe the evolution of matter and constitutes thus such a dynamical "clock" correlating matter to the gravitational field[6]. This description is available in the Hamilton-Jacoby limit of (1) where the classical background evolving in time is represented by a coherent superposition of "forward" waves formed from eigenstates of (1). When quantum metric field fluctuations are taken into account, "backward" waves, which can be interpreted as flowing backwards in time, are unavoidably generated from (1) and the operational significance of the metric clock gets lost outside the domain of validity of the Hamilton-Jacoby limit. Nevertheless, in domains of metric field configurations where both forward and backward waves are present but where quantum fluctuations are sufficiently small, interferences with such "time reversed" semi-classical solutions will in general be negligible‡. Projecting then out the backward waves restores the operational significance of the metric clock but the evolution marked by the correlation time is no more unitary:

‡ For a recent discussion of related problems see reference [7].

information has been lost in projecting these backward waves stemming from regions where quantum fluctuations of the clock are significant. This is only an apparent violation of unitarity which would be disposed of if the full content of the theory would be kept, perhaps eventually by reinterpreting backward waves in terms of the creation of "universe" quanta through a further quantization of the wavefunction (1).

This apparent violation of unitarity is particularly marked if the gravitational clock experiences the strong quantum fluctuations arising from a tunneling process. This can be illustrated from the simple analogy, represented in Fig.1, offered by a nonrelativistic closed system of total fixed energy where a particle in one space dimension x moving in a potential $U(x)$ plays the role of a clock for surrounding matter and tunnels through a large potential barrier. Outside the barrier, the clock is well approximated by semi-classical waves, but if on the left of the turning point a one would take only forward waves, one would inevitably have on the right of the other turning point b both forward and backward waves with large amplitudes compared with the original ones. The ratio between the squares of the forward amplitudes on the right and on the left of the barrier for a component of the clock wave with given clock energy E_c is the inverse transmission coefficient $N_0(E_c)$ through the barrier and provides a measure of the apparent violation of unitarity. In the Hamilton-Jacoby limit of quantum gravity, the characterisation of tunneling amplitudes by inverse transmission coefficients N_0 will appear as the natural one to compute the entropy transferable reversibly between the metric clock and matter. More precisely, we shall see that, in this limit, spherically symmetric spacetimes bounded by event horizons are in general connected by tunneling to another manifold and that the entropy gained by tunneling from the latter to the former is, for large barriers, $\log N_0$. Explicit evaluation of this tunneling entropy yields $\ln N_0 = A/4$ where A is the area of the event horizon. In this way the horizon thermodynamics of Gibbons and Hawking[8] is recovered. But the present approach has potentially additional information.

The tunneling entropy $\ln N_0$ is in last analysis an effect of quantum fluctuations in quantum gravity. Therefore, despite the fact that no violation of unitarity would appear in a complete description including backward waves, this entropy should be expressible in terms of density of states of matter and gravity. Tunneling offers an interesting perspective in this direction because it enlarges the semi-classical wave function of spacetime to include in its description the other side of the barrier. This can yield a reservoir of quantum states which may provide, in addition to the $\exp(A/4)$ states building the entropy, residual states which would be expressed as an "integration constant" in the total entropy $\mathcal{S}$ of spacetime. Thus we shall write

$$\mathcal{S} = A/4 + C \tag{2}$$

and try to get some information about the constant C by analysing both sides of the barrier.

The knowledge of C is crucial, in particular for the understanding of the black hole behaviour at the final stage of evaporation.

A infinitely large value of C would indeed indicate that the evaporation can only radiate a finite number of "surface" states of order $\exp(A/4)$ out of an infinite set of available internal states. This mismatch would entirely modify the black hole evaporation process at its last stage and bring the decay to a halt. Indeed when the black hole evaporates to the Planck scale, it becomes, if C is infinite, a "planckon"[9], that is a remnant with infinite degeneracy. Causality and unitarity prevent the decay and the production in a finite time of planckons directly out of ordinary matter for nearly all such states[9]. Namely, if a planckon state $|A_i\rangle$ of finite size and mass m decays (or is produced) within a finite time τ in an approximately flat space-time background, the total number of possible final (or initial) states is limited through causality by the number $\mathcal{N}(\tau)$ of orthogonal states with total mass m in a volume τ^3. Assuming the number of quantum fields which describe physics at scales large compared to the Planck scale to be finite, $\mathcal{N}(\tau)$ is a finite number. Unitarity then implies that if the dimension $\nu(m)$ of the Hilbert space spanned by the degenerate states $|A_i\rangle$ becomes greater than $\mathcal{N}(\tau)$, a subspace of planckon states whose dimension is $\nu(m) - \mathcal{N}(\tau)$ will be, for times smaller than τ, orthogonal to the Hilbert space of states formed by these quantum fields. Thus, when $\nu(m) \to \infty$, generic planckons cannot decay (nor be formed) in a finite time. Of course the above argument does not preclude the very formation of the *finite* number of distinct planckons which can be generated in a finite time as remnants of macroscopic decaying black holes. This time is however unrelated to the time for their decay (creation) directly at the Planck size into (from) ordinary matter quanta; the latter time is generically infinite.

On the other hand, a zero or finite value of C would lead to the disappearance of the hole at the end point of evaporation and hence probably imply a genuine violation of unitarity within our universe*.

★ For a comprehensive review on recent attempts to solve the black hole unitarity puzzle, see reference [10]. See also reference [11].

The main content of our work is that, in an asymptotically flat background, the constant C for black holes, as deduced from a WKB analysis of tunneling, is in fact infinite. Hence the present approach indicates that the solution of the unitarity problem posed by the black hole decay is provided by planckon remnants. This conclusion is however contingent upon the limitation of the semi-classical approach to quantum gravity used here and remains therefore a tentative one.

We shall first review the computation of tunneling amplitudes in quantum gravity through static barriers[2] and compute the tunneling entropy for the case of de Sitter spacetime topologies. However the estimation of C appears in this case intractable within the semi-classical approximation. We then shall analyse in similar terms the black hole geometries[12]. The above-mentioned results will be derived and discussed.

2. Tunneling amplitudes in quantum gravity

Our basic action in four dimensional Minkowski space-time will be

$$S = S_{grav} + S_{matter} \tag{3}$$

where S_{grav} has the conventional form ($G = 1$) :

$$S_{grav} = -\frac{1}{16\pi}\int \sqrt{-g}R\, d^4x \tag{4}$$

and S_{matter} contains sufficiently many free parameters to allow for the stress tensors considered below. A possible cosmological constant term can be included in the matter action.

Consider (Fig.2) in general two spacelike hypersurfaces Σ_1 and Σ_2 which are turning points in superspace (or turning hypersurfaces) along which solutions of the Minkowskian classical equations of motion for gravity and matter meet a classical solution of their Euclidean extension. Σ_1 and Σ_2 are thus the boundaries of a region $\mathcal{E}$ of Euclidean space-time defined by the Euclidean solution. If $\mathcal{E}$ can be continuously shrunk to zero one can span $\mathcal{E}$ by a continuous set of hypersurfaces τ_e = constant such that $\tau_e \equiv \tau_{e,1}$ on Σ_1 and $\tau_e \equiv \tau_{e,2}$ on Σ_2. These τ_e = constant surfaces define a Euclidean coordinate system which we shall call synchronous; the Euclidean metric in $\mathcal{E}$ can be written in the form

$$ds^2 = N^2(\tau_e, x_k)\, d\tau_e^2 + g_{ij}(\tau, x_k)\, dx^i\, dx^j \tag{5}$$

where $N(\tau_e, x_k)$ is a lapse function. The Euclidean action S_e over $\mathcal{E}$, from Σ_1 to Σ_2, is obtained by analytic continuation from the Minkowskian action (3) and can be written as

$$S_e(\Sigma_2, \Sigma_1) = \int_{\mathcal{E}} \Pi^{ij} g'_{ij}\, d^4x + \int_{\mathcal{E}} \Pi^a \Phi'_a\, d^4x - \int_{\mathcal{E}} (g_{ij}\Pi^{ij})'\, d^4x \\ - \frac{1}{8\pi} \int_{\mathcal{E}} \partial_k[(\partial_j N) g^{kj} \sqrt{g^{(3)}}]\, d^4x. \tag{6}$$

Here Π^{ij} and Π^a are the Euclidean momenta conjugate to the gravitational fields g_{ij} and to the matter fields ϕ_a; $g^{(3)}$ is the three dimensional determinant and the $'$ symbol indicates a derivative with respect to τ_e. In the gauges (5), Π^{ij} is expressed as

$$\Pi^{ij} = \frac{\sqrt{g^{(3)}}}{32\pi N}[g^{im}g^{jn} - g^{ij}g^{mn}]g'_{mn}, \tag{7}$$

On the turning hypersurfaces Σ_1 and Σ_2, all field momenta (Π^{ij}, Π^a) are zero in the synchronous system and the third term in (6) vanishes. The last term in (6) also vanishes if the hypersurfaces Σ_1 and Σ_2 are compact but may receive contributions from infinity otherwise. In this case, we shall assume that turning hypersurfaces merge at infinity sufficiently fast so that the Euclidean action $S_e(\Sigma_2, \Sigma_1)$ does not get contributions from the last term in (6). The classical Minkowskian solution in the space-time $\mathcal{M}_1$ containing Σ_1 can be represented quantum mechanically by a "forward wave" solution $\Psi(g_{ij}, \phi_a)$ of the Wheeler-De Witt equation (1) in the Hamilton-Jacoby limit. At Σ_1, this wave function enters, in the WKB limit, the Euclidean region $\mathcal{E}$ and leaves it at Σ_2 to penetrate a new Minkowskian space-time $\mathcal{M}_2$. The tunneling of $\Psi(g_{ij}, \phi_a)$ through $\mathcal{E}$ engenders in addition to the "forward wave" solution a time reversed "backward wave". The inverse transmission coefficient N_0 through the barrier measures the ratio of the norms of the forward waves at Σ_2 and Σ_1. For large N_0 one may write in the synchronous system

$$N_0 = \exp\left[-2\left(\int_{\mathcal{E}} \Pi^{ij} g'_{ij}\, d^4x + \int_{\mathcal{E}} \Pi^a \phi'_a\, d^4x\right)\right]. \tag{8}$$

As all surface terms in (6) vanish in this system, (8) can be rewritten in the coordinate invariant form

$$N_0 = \exp\left[2S_e(\Sigma_1, \Sigma_2)\right]. \tag{9}$$

Let us examine the case where the Euclidean manifold $\mathcal{E}$ is static in the sense that it admits a Killing symmetry. We can take advantage of the covariance of the action S_e and express it in terms of a new "static" coordinate system, possibly singular, with momenta everywhere vanishing in $\mathcal{E}$. In this way, momenta in (8) get squeezed into the last surface term of (6) and one gets

$$N_0 = \Delta t_e \frac{1}{4\pi} \int_{\mathcal{E}} \partial_k[(\partial_j N) g^{kj} \sqrt{g^{(3)}}]\, d^3x \tag{10}$$

where Δt_e is the Euclidean time needed to span $\mathcal{E}$ in the static system. The tunneling amplitude will then be computable from this surface term only, even when the static parametrisation is singular.

3. The tunneling entropy in de Sitter spacetime topologies

Let us first illustrate the equivalence implied by (9) of the expressions (8) and (10) for the de Sitter spacetime which is the classical solution of pure gravity in the presence of a cosmological constant Λ. In the present formalism, Λ should be viewed as the Lagrangian density of the matter action in (2); it plays the role of a matter distributed with rest energy density $\sigma = \Lambda$ and obeying the equation of state $\sigma = -p$ where p is a (negative) pressure. The full Minkowskian solution is the 4-hyperboloid (Fig.3) which can be parametrized by the minisuperspace metric

$$ds^2 = d\tau^2 - a^2 d\sigma^2; \quad a = r_h \cosh \frac{\tau}{r_h} \tag{11}$$

where $r_h = (3/8\pi\Lambda)$. The hypersurface $\tau = 0$ is a turning hypersurface connecting the hyperboloid to the Euclidean solution consisting of the 4-sphere which can be described in a synchronous system by replacing in (11) τ by $-i\tau_e$. This yields the Euclidean scale factor

$$a_e = r_h \cos \frac{\tau_e}{r_h}. \tag{12}$$

The half-sphere delimited by $-\pi r_h/2 \le \tau_e \le 0$ has another turning point at, say, the south pole $\tau_e = -\pi r_h/2$ where $a_e = 0^\star$: in the above synchronous system the space integral of the momenta in a $\tau_e =$ constant hypersurface vanishes in the vicinity of this point and so does the third term in (6). The half-sphere considered constitute the domain $\mathcal{E}$ through which a "wormhole" at, say, $a_e = 0$ is connected by tunneling to the de Sitter spacetime. The inverse

⋆ In fact, any point on the half-sphere can be taken as a turning hypersurface.

transmission coefficient N_0 can be straightforwardly computed from (8) using (7) and one gets

$$N_0 = \exp\left[\frac{3\pi}{2}\int\limits_{-\pi r_h/2}^{+\pi r_h/2} a_e \left(\frac{da_e}{d\tau_e}\right)^2 d\tau_e\right] = \exp(\pi r_h^2) = \exp(A/4) \tag{13}$$

where A is the area of the event horizon.

The significance of this result is best appreciated when the 4-sphere is described in static coordinates:

$$ds^2 = (1 - r^2/r_h^2)\, dt_e^2 + (1 - r^2/r_h^2)^{-1}\, dr^2 + r^2\, d\Omega^2. \tag{14}$$

In this static frame, all momenta vanish everywhere on the sphere and the tunneling is expressible by the surface term (10) only where the radial integration is carried from $r = 0$ to $r = r_h$. The Euclidean time is periodic with period $\mathcal{T}^{-1} = 2\pi r_h$. Using (10) with $\Delta t_e = (1/2)\mathcal{T}^{-1} = \pi r_h$, one recovers the result (13).

It is now easy to verify that the equality between the inverse transmission coefficient and $\exp(A/4)$ is maintained when the de Sitter spacetime is perturbed by spherically symmetric static matter distributions[2]. This establishes the validity of (13) for these generalized de Sitter spacetimes.

Let us tentatively take boundary conditions in field space by assigning pure forward waves at the wormhole turning point. The probability of finding an expanding generalized de Sitter spacetime for a corresponding wormhole state is then N_0, since in the classical limit interferences between spaces evolving forward or backward in time must be negligible. Assuming that all wormhole states are equally probable, we get from (13) that the relative probability of finding two matter configurations in the generalized de Sitter spacetimes is

$$\frac{N_0^{(1)}}{N_0^{(2)}} = \exp\left[\frac{A^{(1)}}{4} - \frac{A^{(2)}}{4}\right]. \tag{15}$$

Integrating the constraint equation $\mathcal{H} = 0$ over a static domain of the Minkowskian spacetime one gets

$$\frac{1}{16\pi}\int \sqrt{-g}R\, d^3x + H_{matter} + \frac{\mathcal{T}A}{4} = 0 \tag{16}$$

where H_{matter} is the total matter energy. The variation of (16) yields

$$-\frac{\delta A}{4} = \mathcal{T}^{-1}\delta_\lambda H_{matter} \tag{17}$$

where λ labels the *explicit* dependence of H_{matter} on all other (non gravitational) "external" parameters. Equation (17) is the differential Killing identity of reference [13].

It now follows from (15) and (17) that matter configurations with neighbouring energies in a static patch of a generalized de Sitter spacetime would be Boltzmann distributed at the global temperature $\mathcal{T}$ provided our ignorance about wormhole states allows to take them to be equally probable. Thus the temperature of the static patch is $\mathcal{T}$ and therefore (17) also implies that $A/4$ is (up to an integration constant $C_{deSitter}$) the entropy of spacetime and that the latter is in thermal equilibrium with the surrounding matter. As the entropy must be an intrinsic property of spacetime, not only is equilibrium a consequence of the chosen boundary conditions in field space but the converse is also true: the temperature obtained directly from (17) with the spacetime entropy identified as $A/4 + C_{deSitter}$ must agree at equilibrium with the thermal distribution generated from the field boundary conditions. This justifies a posteriori the above choice of boundary conditions*.

The tunneling approach to the horizon entropy and temperature[1],[2] used here differs from the analysis based on the Euclidean periodicity of Green's functions[8] in two respects. On the one hand, the present approach yields the thermal spectrum, and then the entropy, from the *backreaction* of the thermal matter on the gravitational field, in contradistinction to the Green's function approach. On the other hand however, the thermal matter considered here is taken in the classical limit while the Green's function method describes genuine quantum radiation. Both methods fall short of a fully consistent quantum treatment of the backreaction. But as stated in the introduction the present approach may uncover from the

hidden side of the barrier a density of state building the full entropy. Unfortunately, for the de Sitter spacetimes considered above, the hidden side is a wormhole whose description cannot be achieved in our semi-classical approach. Hence, for de Sitter spacetimes, we do not gain at this stage any information on the integration constant $C_{deSitter}$ which measures the

density of states left when the full spacetime reduces to the (planckian) wormhole. As we shall now see the situation appears quite different in the case of black hole geometries.

4. The tunneling entropy of black holes

* up to changes which would not alter the probability ratios in the large N_0 limit.

A Schwarzschild static patch of an eternal black hole of mass m_0 is described by the metric

$$ds^2 = (1 - \frac{2m_0}{r})\, dt^2 - (1 - \frac{2m_0}{r})^{-1}\, dr^2 - r^2\, d\Omega^2. \tag{18}$$

Surrounding the black hole by static matter generalizes (18) to

$$ds^2 = g_{00}(r)\, dt^2 - g_{11}(r)\, dr^2 - r^2\, d\Omega^2 \tag{19}$$

where in absence of outer horizon one has

$$r \to \infty : g_{00}(r) = g_{11}^{-1}(r) \to 1 - \frac{2M}{r}. \tag{20}$$

Here $M = M(\infty)$ is the total mass and

$$\begin{aligned} M(r) &= m_0 + \int_{2m_0}^{r} 4\pi z^2 \sigma(z)\, dz \\ g_{11}^{-1}(r) &= 1 - \frac{2M(r)}{r} \\ g_{00}(r) &= \left(1 - \frac{2M(r)}{r}\right) \exp\left[- \int_{r}^{\infty} (\sigma + p_1) 8\pi z g_{11}(z)\, dz \right] \end{aligned} \tag{21}$$

The metric (19) can be extended to the four quadrants of a Kruskal space and we choose identical matter distributions in the two Schwartzschild patches to keep a twofold symmetry around the Kruskal time axis. The Kruskal diagram is depicted in Fig.4 where we have also indicated its Euclidean extension $T_e = iT$ resulting from the analytic continuation of the static metric (19) to the periodic time $t_e = it$. The Euclidean periodicity is

$$\mathcal{T} = \frac{1}{4\pi} [g_{00}(2m_0)\, g_{11}(2m_0)]^{-1/2} \frac{dg_{00}(r)}{dr}\Big|_{r=2m_0} \tag{22}$$

or from (21)

$$\mathcal{T} = \frac{1}{8\pi m_0} \exp\left[- \int_{2m_0}^{\infty} (\sigma + p_1) 4\pi z g_{11}\, dz \right]. \tag{23}$$

The Euclidean extension of the black hole surrounded by static matter is represented in Fig.5.

In contradistinction to the de Sitter case, there is clearly no WKB tunneling from a wormhole to a black hole because of the mismatch in topologies. One is therefore lead to investigate possible tunnelings between two black holes geometries $(B.H.)_1$ and $(B.H.)_2$ constituted respectively by black holes of mass m_0 and m ($m > m_0$) surrounded by matter. The Euclidean sections of $(B.H.)_1$ and $(B.H.)_2$, depicted in Fig.6, are engendered by a rotation of half a Euclidean period of the hypersurfaces a_1 and a_2 labeled by $T = 0$ in their Kruskal diagrams. These are turning hypersurfaces along which Minkowskian and Euclidean black holes meet. We now search for two black holes such that a_1 and a_2 are also the boundaries of an Euclidean solution $\mathcal{E}$ of the Euclidean equations of motion through which tunneling can take place from one Minkowskian black hole to the other. A necessary condition for this to happen is that the total mass M of the two black hole-matter systems and their Euclidean period $\mathcal{T}^{-1}$ be the same, so that the turning hypersurfaces a_1 and a_2 of the two geometries merge at spatial infinity.

Let us choose identical matter distributions outside a radius $r_c = 2m + \eta$. η is positive and such that the mass of the matter between the horizon and r_c is 0 for $(B.H.)_2$ and thus $m - m_0$ for $(B.H.)_1$. Keeping m and the matter distribution outside r_c fixed, we now decrease m_0 towards 0. From (23), in order to keep the Euclidean period $\mathcal{T}^{-1}$ constant for $(B.H.)_1$, we have also to decrease η towards 0. As $\eta \to 0$ the mass $m - m_0$ surrounding the infinitesimal mass m_0 in $(B.H.)_1$ approaches its own Schwartzschild radius. It then follows from (21) that $g_{00}(r)$ tends to zero for the whole interval $2m_0 < r < 2m$. In other words any frequency stemming from the neighbourhood of the small mass hole is infinitely redshifted by the matter in that interval, as encoded in the damping exponentials in equations (21) and (23).

It is clear, from the static coordinate description (19) extended to Euclidean times $t_e = it$, that the Euclidean sections of $(B.H.)_1$ and $(B.H.)_2$ coincide for $r > 2m + \eta$ but, while for $(B.H.)_2$ the Euclidean section terminates at $r = 2m$, $(B.H.)_1$ presents an extra "needle" in the region $2m_0 < r < 2m$ whose 4-volume is vanishingly small when $\eta \to 0$. As we now show, this is where tunneling between $(B.H.)_1$ and $(B.H.)_2$ occurs.

To this effect, following the notations of section 2, we identify at finite η the first turning hypersurface Σ_1 through which tunneling takes place with a_1 and consider instead of a second turning hypersurface Σ_2 a hypersurface a'_2 which lies in the Euclidean section of both $(B.H.)_1$ and $(B.H.)_2$; thus r is greater than $2m + \eta$ everywhere on a'_2.

When $\eta \to 0$, we can choose a'_2 arbitrarily close to a_2. One can then prove[12] that all gravitational momenta vanish in this limit on a_2' in a synchronous system. We may then identify a_2' with Σ_2. The region $\mathcal{E}$ is thus contained in the needle $2m_0 < r < 2m+\eta$. Because of the Kruskal twofold symmetry a_1 is mapped onto itself by a Euclidean time rotation of half a period and thus $\mathcal{E}$ spans only half the needle 4-volume. ¿From (9), we learn that the inverse transmission coefficient N_0 is simply the exponential of the total Euclidean action of the needle. Although the limiting 4-volume of the needle vanishes, the action will turn out to be finite. It is in fact computable as the difference between the Euclidean actions of the two black holes cut off at the arbitrary radius r_c greater than $2m$ because the two geometries and the two actions coincide for all $r > r_c$.

We thus write

$$N_0 = \exp[S_e^{(B.H.)_2} - S_e^{(B.H.)_1}]. \tag{24}$$

To evaluate these actions we take advantage of the covariance to express them in terms of the static coordinate system (19) with $t = -it_e$. Using equations (10) and (22) and the fact that the integrand in (10) is the same at r_c for $S_e^{(B.H.)_1}$ and for $S_e^{(B.H.)_2}$, we get

$$N_0 = \exp\left[4\pi m^2 - 4\pi m_0^2(\eta \to 0)\right] \tag{25}$$

or, as m_0 vanishes in the limit,

$$N_0 = \exp A/4 \tag{26}$$

where $A = 16\pi m^2$ is the area of the event horizon of the black hole.

We have thus learned that black holes are related by quantum tunneling to another classical solution for gravity and matter, namely to a "germ black hole" of infinitesimal mass determining the spacetime topology surrounded by a static distribution of matter characterized by a vanishing $g_{00}(r)$. This domain of space-time is characterized by a limiting light-like Killing vector. When the space-time geometry presents a 4-domain endowed with such a Killing vector, we shall call the domain an "achronon". All spherically symmetric achronon configurations will exhibit an infinite time dilation in the Schwarzschild time t, or equivalently massless modes emitted by the achronon are infinitely redschifted. Classically, the achronon has the "frozen" appearance of a collapse at infinite Schwarzschild time. The difference is that it is also frozen in space-time.

To see that achronons can indeed be constructed, at least in a phenomenological fluid model, we shall build a shell model with the required properties.

Let us consider a static spherically symmetric distribution of matter surrounded by an extended shell comprised between two radii r_a and r_b. We define

$$\hat{\sigma} \equiv \int_{r_a}^{r_b} \sigma g_{11}^{1/2} dr, \quad \hat{p}_\theta \equiv \int_{r_a}^{r_b} p_\theta g_{11}^{1/2} dr, \quad \hat{p}_1 \equiv \int_{r_a}^{r_b} p_1 g_{11}^{1/2} dr \tag{27}$$

where $p_\theta = -T_\theta^\theta$ and $p_\phi = -T_\phi^\phi$.

Assuming $p_1 = 0$, one may perform the thin shell limit $r_b \to r_a = R$ in these integrals using (21) and the Bianchi identity

$$p_\theta = p_\phi = \frac{1}{4}(\sigma + p_1)\frac{8\pi r^2 p_1 + 2M(r)/r}{1 - 2M(r)/r} + \frac{1}{2} r p_1' + p_1. \tag{28}$$

One then gets

$$4\pi R\hat{\sigma} = (1 - 2m^-/R)^{\frac{1}{2}} - (1 - 2m/R)^{\frac{1}{2}} \tag{29}$$

$$8\pi R\hat{p}_\theta = \frac{1 - m/R}{(1 - 2m/R)^{\frac{1}{2}}} - \frac{1 - m^-/R}{(1 - 2m^-/R)^{\frac{1}{2}}}; \quad \hat{p}_1 = 0 \tag{30}$$

where m and m^- are the values of $M(r)$ respectively at r_b and r_a and $m_s = m - m^-$ is thus the mass of the shell. Equations (29) and (30) are the standard result[14]. As the radius R approaches $2m$, these solutions become physically meaningless when $\hat{p}_\theta$ becomes greater than $\hat{\sigma}$: this violates the "dominant energy condition"[15], implying the existence of observers for which the momentum flow of the classical matter becomes spacelike; in fact, the shell is mechanically unstable even before this condition is violated[16].

The divergence of $\hat{p}_\theta$ when $R \to 2m$ appears in (30) because of the vanishing denominator in (28). Equation (30) depends however crucially on the radial pressure being zero inside the shell. Relaxing this condition it is possible to avoid all singularities of the stress tensor as $R \to 2m$ by requiring p_1 inside the shell to satisfy, before performing the thin shell limit,

$$4\pi r^2 p_1 + \frac{M(r)}{r} = 0. \tag{31}$$

This solution is unsatisfactory if the (extended) shell sits in an arbitrary background because of the finite discontinuity of the radial pressure across the shell boundaries which would lead to singularities in p_θ. We may ensure continuity of the radial pressure by immersing the shell in suitable left and right backgrounds. To avoid reintroducing stress divergences when

r^b approaches $2M(r^b)$ these should satisfy $(\sigma + p_1) = 0$ at the shell boundaries. One can now perform the thin shell limit. The finite discontinuity of $p_1(r)$ at $r = R$ leads now to

$$\hat{p}_\theta = -\frac{\hat{\sigma}}{2}, \quad \hat{p}_1 = 0 \tag{32}$$

instead of (30), while $\hat{\sigma}$ is still given by (29). The dominant energy condition is now satisfied everywhere and provided the background is smooth enough in the neighbourhood of the shell, no stress divergences will appear when it approaches the Schwarzschild radius. Performing the explicit integration over the shell in the exponential term in (21), one gets for $g_{00}(r)$in the region $0 \leq r < 2m$,

$$g_{00}(r) = (1 - \frac{2M(r)}{r}) \left[\frac{R - 2m}{R - 2m^-}\right] \exp\left[-\mathcal{R}\int_r^\infty (\sigma + p_1) 8\pi z g_{11}\, dz\right]. \tag{33}$$

Here the radius R of the shell is taken at $R = 2m + \eta$ where η is a positive infinitesimal and the symbol $\mathcal{R}$ means that the integral is carried over the regular matter contribution only. Clearly, $g_{00}(r) = O(\eta)$ for $0 \leq r < 2m$, t arbitrary and the above matter distribution constitutes indeed an achronon. These phenomenological shell models are designed only to illustrate achronon properties but are hardly physically relevant as such and we shall tentatively assume that realistic field theoretic achronons do exist. This is of course a crucial issue which deserves further study.

We now relate in general the tunneling as encoded by (26) to the black hole entropy. This can be done following the analysis of the de Sitter case. Assume all states formed by achronons of mass m surrounded by matter configurations of mass $M - m$, M fixed, to be equally probable. This amounts here to assume the validity of the microcanonical ensemble as achronons can be viewed just as lumps of ordinary matter taken out from the surroundings. The relative probability of finding two black hole geometries for a given total mass M is then given by (15) with $A^{(1)}$ and $A^{(2)}$ identified here with the black hole areas. The differential Killing identity (17) follows as before from the integrated constraint equation, the only difference being in general an additional term δM on the right hand side arising from a surface term at spatial infinity. As M is kept fixed, this term plays no role and (17) remains valid as such. Therefore, in analogy with the de Sitter case, matter configurations with neighbouring energies in a static Schwartszchild patch of an eternal black hole surrounded by matter have a Boltzmann distribution at a global temperature $\mathcal{T}$. The latter now coincides with the local temperature at spatial infinity. $\delta A/4$ is the differential entropy of the hole and $A/4$ is the amount of entropy transferable to matter reversibly. The total black hole entropy is

$$S = A/4 + C_{B.H.} \tag{34}$$

where $\exp C_{B.H.}$ measures the number of quantum states of a residual planckian black hole. The boundary condition in field space at equilibrium are such that the wave functional of an eternal black hole has a small amplitude describing an

achronon configuration whose relative weight with respect to the classical black hole configuration is of order $\exp -(A/8)$.

5. From achronon to planckons

We now discuss the nature and the significance of $C_{B.H.}$. The entropy $A/4$ which can be exchanged reversibly from a black hole to ordinary matter was rederived in the preceding section from the existence of a "potential barrier" between a black hole of mass m and an achronon of the same mass. This was done in the context of eternal black holes admitting a Kruskal twofold symmetry, so that there are in fact two achronons imbedded in two causally disconnected static spaces. Within each space black hole-achronon states are in thermal equilibrium with their surroundings. We are therefore led to picture in such a space a quantum black hole state , in the semi-classical limit, as a quantum superposition of two coherent (normalized) states, $|B.H.\rangle$ and $|A\rangle$ representing respectively a classical black hole and a classical achronon. The relative weight of the two states is approximately, up to a phase, $\exp(-A/8)$. It follows from detailed balance at equilibrium between radiated matter and the black hole that the same superposition should hold for a the black hole who would only emit (and not receive) thermal radiation at the equilibrium temperature. As a black hole formed from collapse would asymptotically emit such a thermal flux, we infer that the collapsing black hole approaches a state $|C\rangle$ containing an achronon component with the same weight as in thermal equilibrium. We thus write

$$|C\rangle \simeq |B.H.\rangle + \exp(-A/8)|A\rangle. \tag{35}$$

To a classical single black hole configuration one may associate many distinct classical achronon configurations. In the shell model, for instance, there are infinitely many distinct classical matter configurations of the same total mass m. The argument is however much more general and infinite quantum degeneracy of the achronon is a direct consequence of the infinite time dilation. Indeed, the Hamiltonian H_{matter} is of the form

$$H_{matter} = \int \sqrt{g_{00}} K(\phi_a, g_{ij}, \Pi_a,)\, d^3x \tag{36}$$

and all its eigenvalues are squashed towards zero by the Schwarzschild time dilation factor $\sqrt{g_{00}}$, thus generating an infinite number of orthogonal zero energy modes on top of the original achronon.

The infinity of zero energy modes around any background implies an infinite degeneracy of achronons of given mass and thus an infinity of distinct quantum black hole states of the same mass differing by the achronon component of their wave function. This infinite degeneracy of the quantum black hole provides the reservoir from which are taken the finite number of "surface" quantum states $\exp A/4$ counted by the area entropy $A/4$ transferable reversibly to outside matter.

Except for providing a rational for the large but finite testable entropy of the black hole, achronons do not modify the behaviour of large macroscopic black holes. However when their mass is reduced by evaporation and approaches the Planck mass the barrier disappears and quantum superposition completely mixes the two components. Of course, this means that both the description in terms of semiclassical configurations and of tunneling disappears. What remains however as a consequence of unitarity, is the infinity of distinct orthogonal quantum states available which have no counterpart in the finite number of decayed states. The quantum black hole has become a planckon[12], that is a Planckian mass object with infinite degeneracy. In terms of equation (34), this means that in a asymptotically flat background, the integration constant of the black hole entropy $C_{B.H.}$ is infinite. As discussed in the introduction, this means that in such a background a generic planckon cannot decay nor be formed in a finite time.

This conclusion however is contingent upon the validity at the qualitative level of our semi-classical approach and upon the possibility of building achronons from genuine field theory. The main question is whether or not the *quantum* backreaction of the matter on the metric removes the infinite degeneracy. Answering it requires further analysis.

Finally, it is of interest to note that the planckon solution to the unitarity problem posed by the evaporating black hole would have, at a fundamental level, far reaching implications on the spectrum of quantum gravity. The opening at the Planck size of an infinite number of states, an unavoidable consequence of the existence of planckons, may appear as a horrendous complication which could make quantum gravity definitely unmanageable but hopefully the converse may be true. Indeed planckons should make quantum gravity ultraviolet finite. The Hilbert space of physical states available to macroscopic observer must be orthogonal to the infinite set of states describing Planckian bound states. Their wave function at Planckian scales where planckon configurations are concentrated are therefore expected to be vanishingly small. In this way, planckons would provide the required short distance cut-off for a consistent field theoretic description of quantum gravity within our universe while leaving the largest part of its information content hidden at the Planck scale.

An operational formulation of quantum gravity applicable within our universe and based on conventional four dimensional gravity might thus well be within reach. Nevertheless, the sudden widening of the spectrum of physical states at the Planck scale and the relative scarcity of states which describe large distance physics suggest that a fully consistent theory cannot be formulated in terms of only long range quantum fields (including the metric), and a larger scheme may be required to cope with the infinite amount of information relegated to the Planck scale.

Acknowledgements

We are very grateful to R. Balbinot, R. Brout, J. Katz, J. Orloff, R. Parentani and Ph. Spindel for most enjoyable, stimulating and clarifying discussions.

References

[1] F. Englert, "From Quantum Correlations to Time and Entropy" in "The Gardener of Eden" Physicalia Magazine (special issue in honour of R. Brout's birthday), (1990) Belgium, Ed. by P. Nicoletopoulos and J. Orloff.

[2] A. Casher and F. Englert, Class. Quantum Grav. **9** (1992) 2231.

[3] B. De Witt, Phys. Rev. **160** (1967) 113.

[4] T. Regge and C. Teitelboim, Annals of Physics **88** (1974) 286.

[5] D. Page and K.W. Wooters, Phys. Rev. **D27** (1983) 2885.

Y. Aharonov and T. Kaufherr, Phys. Rev. **D30** (1984) 368.

J.B. Hartle, Phys. Rev. **D37** (1988) 2818; Phys. Rev. **D38** (1988) 2985

[6] T. Banks, Nucl. Phys. **249** (1985) 332,
R. Brout, Foundations of Physics **17** (1987) 603,
R. Brout, G. Horwitz and D. Weil, Phys. Lett. **B192** (1987) 318.

[7] W. Unruh and W.H. Zurek, Phys. Rev. **D40** (1989) 1064,
J.J. Halliwell, Phys. Rev. **D39** (1989) 2912.

[8] G. Gibbons and S. Hawking, Phys. Rev. **D15** (1977) 2738; 2752.

[9] Y. Aharonov, A. Casher and S. Nussinov, Phys. Lett. **B191** (1987) 51.

[10] J.A. Harvey and A. Strominger, "Quantum Aspects of Black Holes", Preprint EFI-92-41; hep-th/9209055.

[11] T. Banks, M.O'Loughlin and A .Strominger , "Black Hole Remnants and the Information Puzzle" Preprint RU-92-40; hep-th/9211030.

[12] A. Casher and F. Englert, "Black Hole Tunneling Entropy and the Spectrum of Gravity" Preprint ULB-TH 8/92; TAUP 2017-92; gr-qc/9212010.

[13] J.M. Bardeen, B. Carter and S.W. Hawking, Comm. Math. Phys. **31** (1973) 161.

[14] J. Frauendiener, C. Hoenselaers and W. Conrad, Class. Quantum Grav. **7** (1990) 585.

[15] S.W. Hawking and G.F.R. Ellis, "The Large Scale of Space-Time", (1973), (Cambridge University Press, Cambridge, England).

[16] P.R. Brady, J Louko and E. Poisson, Phys. Rev. **D44** (1991) 1891.

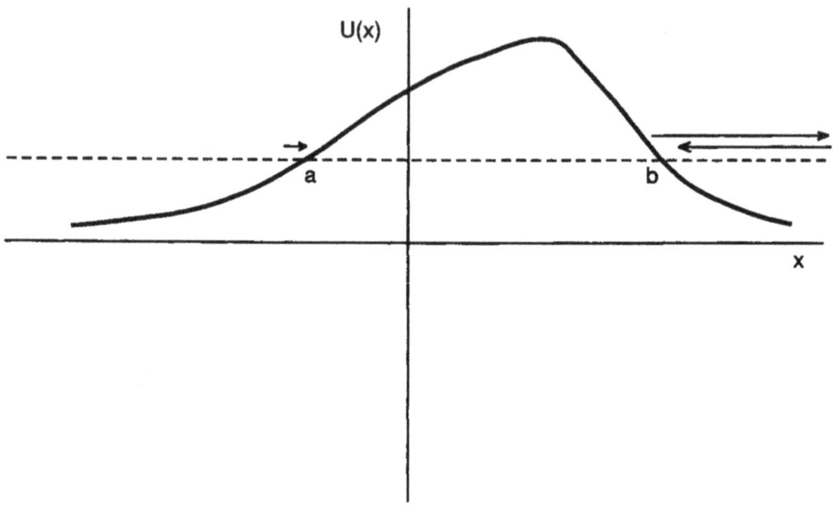

Figure 1. Tunneling of a nonrelativistic "clock".

The energy of the clock E_c is represented by the dashed line. On the left of the turning point a the clock is well represented by a forward wave depicted here by a single arrow. On the right of the turning point b the amplification of the forward wave and the large concomitant backward wave are indicated.

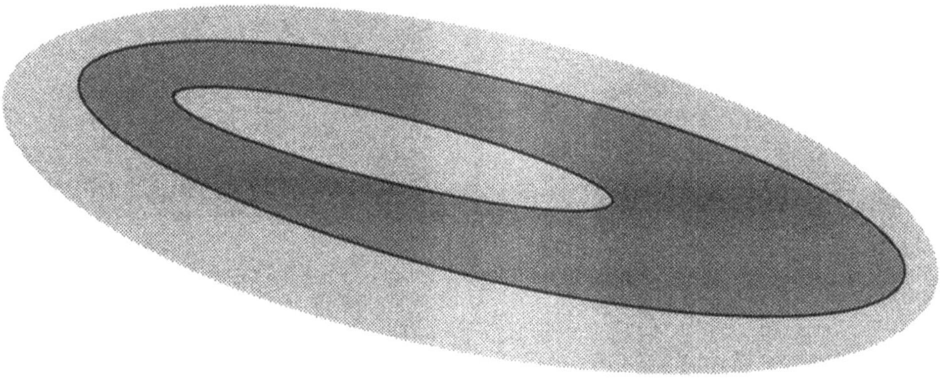

Figure 2. Tunneling in quantum gravity.

The two solid curves represent turning hypersurfaces Σ_1 and Σ_2 separating the dark gray Euclidean region $\mathcal{E}$ from two Minkowskian spacetimes depicted in light gray.

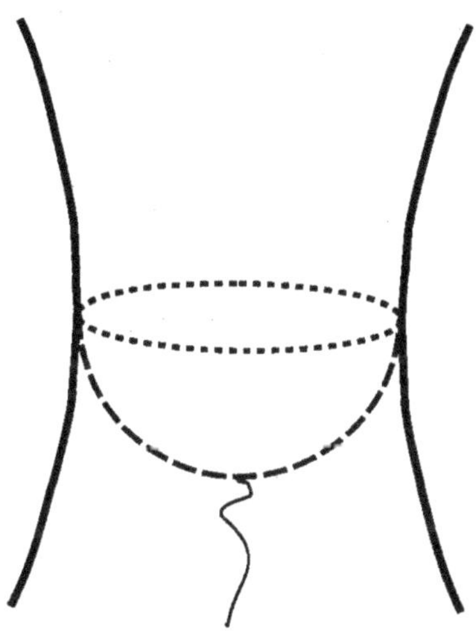

Figure 3. Tunneling in de Sitter topology

The heavy solid line delineates a 4-hyperboloid and the thin one a wormhole. The Euclidean domain $\mathcal{E}$ constituted by a half 4-sphere is delineated by a dashed line. The dotted circle is the turning hypersurface $\tau = 0$.

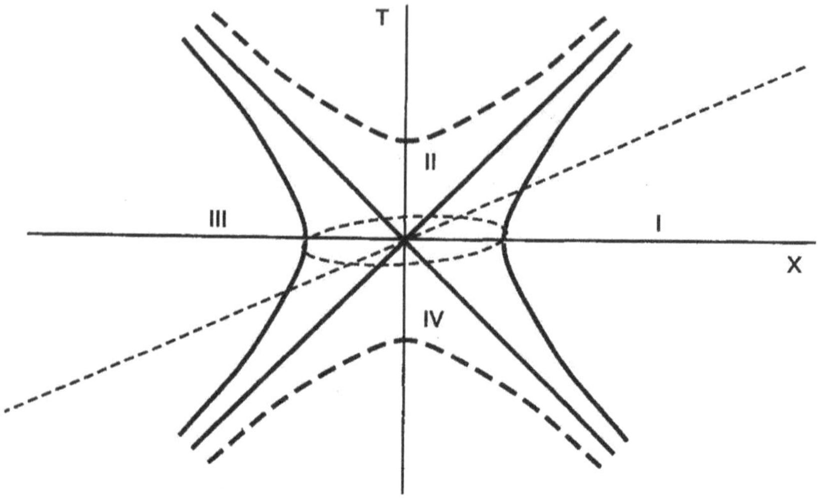

Figure 4. The Kruskal representation of a black hole, eventually surrounded by static matter.

The dashed straight line is the Euclidean axis T_e and the dashed circle is the analytic continuation in Euclidean time of the solid hyperbolae representing trajectories r =constant in the static patches I and III. These are separated from the dynamical regions II and IV by the horizons $r = 2m_0$ where lay the past and future singularities $r = 0$ depicted by the dashed hyperbolae. The Schwarzschild time t run on opposite directions on the two hyperbolae r =constant and the Euclidean time t_e spans the period $\mathcal{T}^{-\infty}$ on the analytically continued circle.

Figure 5. Euclidean black hole surrounded by static matter.

Each point is a 2-sphere and the circles span the Euclidean time t_e. The heavy solid line is the turning hypersurface described in Kruskal time by $T = 0$.

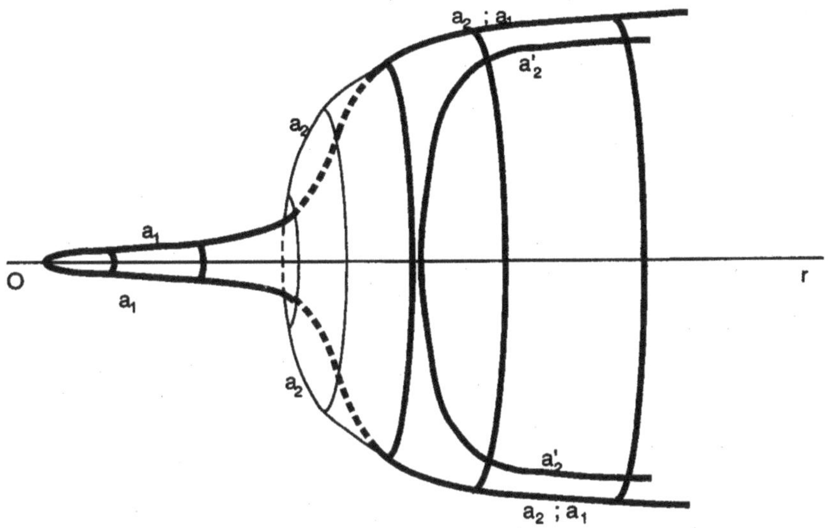

Figure 6. Black hole tunneling.

The figure represents the Euclidean sections of the two black hole geometries $(B.H.)_1$ and $(B.H.)_2$. The $(B.H.)_1$ geometry is depicted by thick lines and the $(B.H.)_2$ geometry by thin lines in the region where it differs from the first.

The curve a_1 represents a turning hypersurface of $(B.H.)_1$, to be identified with Σ_1. The curve a_2 represents a turning hypersurface of $(B.H.)_2$. The curve a_2' represents a hypersurface which lays in the intersection of the Euclidean sections of $(B.H.)_1$ and $(B.H.)_2$ and tends to Σ_2 in the limit $m_0 \to 0$.

Comments on Fractional Superstrings

Gerald B. Cleaver
California Institute of Technology
Pasadena, CA 91125, USA

Abstract

Using the factorization approach of Gepner and Qiu, I derive the closed fractional superstring partition functions for $K = 4$, 8, and 16. The relationship between the massless graviton/gravitino sector and the purely massive sectors of these theories is explored. Properties of the massive sectors are investigated. A twist current in these models is found responsible for spacetime supersymmetry. Specifically, this current is shown to transform bosonic (fermionic) projection states into fermionic (bosonic) non-projection states and vice-versa.

Section 1: Brief Review of Fractional Superstrings

In the prior talk, Keith Dienes presented a general review of fractional superstrings with critical dimensions of six, four, and three. These fractional superstrings are based on $SU(2)_K/U(1)$ coset (a.k.a. parafermion) conformal field theories. The relationship between critical dimension, D, and the level, K, of the $SU(2)_K/U(1)$ theory is,

$$D = 2 + 16/K \ . \qquad (1.1) \ .$$

Thus integer* $D = 6, 4, 3$ corresponds to $K = 4, 8, 16$, respectively. In the present talk, I will focus specifically on the associated partition functions, how we can derive these from "first principles", and what is learned from this. These closed string partition functions, $Z(\text{level}-K)$, have the general form

$$Z(\text{level}-K) = |A_K|^2 + |B_K|^2 + |C_K|^2 \ . \qquad (1.2)$$

The A_K term in Eq. (1.2) contains the massless graviton and gravitino. These $D < 10$ fractional superstrings have a new feature not present in the standard $D = 10$ superstrings (which correspond to K=2). This feature is the existence of the massive B_K– and C_K–sectors. These additional sectors were originally derived by the authors of refs. 2–4 by applying S transformations to the A_K–sector and then demanding modular invariance of the theory. I will remark (1) on new aspects of the relationship between the B_K– and C_K–sectors and the A_K–sector and (2) on the presence of spacetime supersymmetry (SUSY) in all sectors. Following this, I will demonstrate that spacetime SUSY results from the action of a twist current used in the derivation of these partition functions. Only by this twisting can cancellation between bosonic and fermionic terms occur at each mass level in the A_K– and B_K–sectors. The same twisting results a in "self-cancellation" of terms in the C_K–sector (which exists only in the four- and three-dimensional models). This self-cancellation may suggest an anyonic interpretation of the C_K–sector states.†

Before I rederive the fractional superstring partition functions for each critical dimension, let us review the partition function $Z(\phi^j_m)$ for the Verma module, $[\phi^j_m(\tau)]$,‡ containing a single (holomorphic) parafermion primary field $\phi^j_m(z)$ and its

---

*In ref. 1, I show that fractional superstrings with non-integer dimensions cannot have spacetime supersymmetry.

†Uncompactified spacetime anyons can presumably exist only in three or less dimensions. This would seem to contradict a claim that the $D = 4$ model may contain spacetime anyons. I argue shortly that one dimension of the $D = 4$ fractional string is probably compactified. Examination of the B_K–sector in the $D = 4$ model further suggests this. Anyonic interpretation of the C_K–sector fields was first proposed in ref. 2.

‡From here on, I do not distinguish between the primary field ϕ^j_m and its complete Verma module $[\phi^j_m]$. Instead, ϕ^j_m represents both where meaningful.

parafermion descendents. The form of the character is

$$Z(\phi^j_m(z)) = \eta(\tau) c^{2j}_{2m}(\tau)\,, \tag{1.3}$$

where η is the Dedekind eta-function,

$$\eta = q^{1/24} \prod_{n=1}^{\infty} (1 - q^n) \tag{1.4}$$

with $q = e^{2\pi i \tau}$, and c^{2j}_{2m} is a string function[5] defined by

$$c^{2j}_{2m} = \frac{1}{\eta^{-3}(\tau)} \sum_{x,y} \mathrm{sign}(x) q^{x^2(K+2) - y^2 K} \tag{1.5}$$

where in (1.5) the conditions
1. $-|x| < y \leq |x|$,
2. either $x = \frac{l+1}{2(K+2)} \pmod 1$ or $(\frac{1}{2} - x) = \frac{l+1}{2(K+2)} \pmod 1$; and
3. either $y = \frac{n}{2K} \pmod 1$ or $(\frac{1}{2} + y) = \frac{n}{2K} \pmod 1$

must be met simultaneously.[2] These string functions obey the same equivalences as their associated primary fields ϕ^j_m:

$$c^{2j}_{2m} = c^{2j}_{2m+2K} = c^{K-2j}_{2m-K}\,. \tag{1.6a}$$

Additionally, since $\phi^j_m = (\phi^j_{-m})^\dagger$

$$c^{2j}_{2m} = c^{2j}_{-2m}\,. \tag{1.6b}$$

For each spacetime dimension in these theories, a partition function (character) term of the form (1.3) is tensored with the partition function $Z\,(X)$ for an uncompactified chiral boson $X(z)$. Since

$$Z\,(X) \propto \frac{1}{\eta(\tau)}\,, \tag{1.7}$$

the $\eta(\tau)$ factors cancel out in $Z(\phi^j_m) \times Z(X)$. Similar cancellation of $\bar{\eta}(\bar{\tau})$ occurs in the antiholomorphic sector. In the following partition function equations, I suppress the trivial factor of $(\mathrm{Im}\,\tau)^{-8/K}$ contributed by the $D-2$ holomorphic and anti-holomorphic world sheet boson partition functions together.

Section 2: Fractional Superstring Partition Functions

By using the string function equivalences (1.6a-b), the partition functions for the level-K fractional superstrings in refs. 2–4,6, with critical spacetime dimensions $D = 2 + \frac{16}{K} = 10, 6,\ 4,$ and 3, can be written (in light-cone gauge) as:

$D=10 \quad (K=2)$: $\qquad Z=|A_2|^2$, where

$$\begin{aligned} A_2 &= \frac{1}{2}\left\{(c_0^0+c_0^2)^8-(c_0^0-c_0^2)^8\right\}_{\text{boson}}-8(c_1^1)^8_{\text{fermion}} \\ &= 8\left\{(c_0^0)^7c_0^2+7(c_0^0)^5(c_0^2)^3+7(c_0^0)^3(c_0^2)^5+c_0^0(c_0^2)^7\right\}_{\text{boson}}-8(c_1^1)^8_{\text{fermion}} \end{aligned} \tag{2.1}$$

$D=6 \quad (K=4)$: $\qquad Z=|A_4|^2+3|B_4|^2$, where

$$\begin{aligned} A_4 = {} & 4\left\{(c_0^0+c_0^4)^3(c_0^2)-(c_0^2)^4\right\} \\ & +4\left\{(c_2^0+c_2^4)^3(c_2^2)-(c_2^2)^4\right\} \end{aligned} \tag{2.2a}$$

$$\begin{aligned} B_4 = {} & 4\left\{(c_0^0+c_0^4)(c_2^0+c_2^4)^2(c_0^2)-(c_0^2)^2(c_2^2)^2\right\} \\ & +4\left\{(c_2^0+c_2^4)(c_0^0+c_0^4)^2(c_2^2)-(c_2^2)^2(c_0^2)^2\right\} \end{aligned} \tag{2.2b}$$

$D=4 \quad (K=8)$: $\qquad Z=|A_8|^2+|B_8|^2+2|C_8|^2$, where

$$\begin{aligned} A_8 = {} & 2\left\{(c_0^0+c_0^8)(c_0^2+c_0^6)-(c_0^4)^2\right\} \\ & +2\left\{(c_4^0+c_4^8)(c_4^2+c_4^6)-(c_4^4)^2\right\} \end{aligned} \tag{2.3a}$$

$$\begin{aligned} B_8 = {} & 2\left\{(c_0^0+c_0^8)(c_4^2+c_4^6)-(c_0^4c_4^4)\right\} \\ & +2\left\{(c_4^0+c_4^8)(c_0^2+c_0^6)-(c_4^4c_0^4)\right\} \end{aligned} \tag{2.3b}$$

$$\begin{aligned} C_8 = {} & 2\left\{(c_2^0+c_2^8)(c_2^2+c_2^6)-(c_2^4)^2\right\} \\ & +2\left\{(c_2^0+c_2^8)(c_2^2+c_2^6)-(c_2^4)^2\right\} \end{aligned} \tag{2.3c}$$

$D=3 \quad (K=16)$: $\qquad Z=|A_{16}|^2+|C_{16}|^2$, where

$$\begin{aligned} A_{16} = {} & \left\{(c_0^2+c_0^{14})-c_0^8\right\} \\ & +\left\{(c_8^2+c_8^{14})-c_8^8\right\} \end{aligned} \tag{2.4a}$$

$$\begin{aligned} C_{16} = {} & \left\{(c_4^2+c_4^{14})-c_4^8\right\} \\ & +\left\{(c_4^2+c_4^{14})-c_4^8\right\}\ . \end{aligned} \tag{2.4b}$$

The $D=10$ partition function, written in terms of string functions, was originally obtained by the authors of refs. 2,4 using the equivalences between the $K=2$ string functions and the non-zero Jacobi ϑ-functions.[§] The partition functions for $K>2$

[§]The $K=2$ parafermion $c=\frac{1}{2}$ CFT corresponds to an Ising (free fermion) model.

were derived by demanding a massless spin-two particle, modular invariance, and no tachyons. In each model, the massless spin-two particle and its supersymmetric partner arise from the A_K-sector. The B_K- and C_K-sectors were obtained by acting on the A_K-sector with the $PSL(2,\mathbf{Z})$ modular group generators, $S:\tau \to -1/\tau$, and $T:\tau \to \tau+1$. At each level, K, the contribution of each sector is separately zero. This is consistent with spacetime SUSY and suggests cancellation between bosonic and fermionic terms at each mass level. This leads to the following identities[2]:

$$A_2 = A_4 = B_4 = A_8 = B_8 = C_8 = A_{16} = C_{16} = 0\,. \tag{2.5}$$

The factorization method of Gepner and Qiu[7] for string function partition functions allows us to rederive the fractional superstring partition functions in this new form systematically. Using this approach we can express a general parafermion partition function (with the level, K, of the string functions henceforth suppressed),

$$Z = |\eta|^2 \sum N_{l,n,\bar{l},\bar{n}} c^l_n \bar{c}^{\bar{l}}_{\bar{n}}\,, \tag{2.6a}$$

as

$$Z = |\eta|^2 \sum \frac{1}{2} L_{l,\bar{l}} M_{n,\bar{n}} c^l_n \bar{c}^{\bar{l}}_{\bar{n}}\,, \tag{2.6b}$$

(with $c^{l=2j}_{n=2m} = 0$ unless $l-n \in 2\mathbf{Z}$ since $\phi^j_m = 0$ for $j-m \notin \mathbf{Z}$). As a result of the factorization,

$$N_{l,n,\bar{l},\bar{n}} = \frac{1}{2} L_{l,\bar{l}} M_{m,\bar{m}}\,, \tag{2.7}$$

we can construct all modular invariant partition functions (MIPF's) for parafermions from a tensor product of modular invariant solutions for the $(l,\bar{l})$ and $(n,\bar{n})$ indices separately. This results from the definition of a level-K string functions c^l_n in terms of the $SU(2)_K$ characters χ_l and the Jacobi theta-function $\vartheta_{n,K}$:[◁]

$$\chi_l(\tau) = \sum_{n=-K+1}^{K} c^l_n(\tau)\vartheta_{n,K}(\tau)\,, \tag{2.8}$$

where the theta-function is defined by

$$\vartheta_{n,K}(\tau) = \sum_{p\in \mathbf{Z}+\frac{n}{2K}} e^{2\pi i K p^2 \tau}\,, \tag{2.9}$$

[◁]The associated relationship between the level-K $SU(2)$ primary fields Φ^j and the parafermion ϕ^j_m is

$$\Phi^j = \sum_{m=-j}^{j} \phi^j_m \,:\exp\left\{i\frac{m}{\sqrt{K}}\varphi\right\}:$$

where φ is the $U(1)$ boson field of the $SU(2)$ theory.

and χ_l is the character for the spin-$\frac{l}{2}$ representation of $SU(2)_K$,

$$\chi_l(\tau) = \frac{\vartheta_{l+1,K+2}(\tau) - \vartheta_{-l-1,K+2}(\tau)}{\vartheta_{1,2}(\tau) - \vartheta_{-1,2}(\tau)} \ . \tag{2.10}$$

This factorization is clearly seen in the transformation properties of c^l_n under the modular group generators S and T,

$$S : c^l_n \rightarrow \frac{1}{\sqrt{-i\tau K(K+2)}} \sum_{l'=0}^{K} \sum_{\substack{n'=-K+1 \\ l'-n' \in 2Z}}^{K} \exp\left\{\frac{i\pi n n'}{K}\right\} \sin\left\{\frac{\pi(l+1)(l'+1)}{K+2}\right\} c^{l'}_{n'} \tag{2.11a}$$

$$T : c^l_n \rightarrow \exp\left\{2\pi i \left(\frac{l(l+2)}{4(K+2)} - \frac{n^2}{4K} - \frac{K}{8(K+2)}\right)\right\} c^l_n \ . \tag{2.11b}$$

Thus, (2.6b) is modular invariant if and only if the $SU(2)$ affine partition function

$$W = \sum_{l,\bar{l}=0}^{K} L_{l,\bar{l}} \chi_l(\tau) \bar{\chi}_{\bar{l}}(\bar{\tau}) \tag{2.12}$$

and the $U(1)$ partition function

$$V = \frac{1}{|\eta(\tau)|^2} \sum_{n,\bar{n}=-K+1}^{K} M_{n,\bar{n}} \vartheta_{n,K} \bar{\vartheta}_{\bar{n},K} \tag{2.13}$$

are simultaneously modular invariant. That is, $N_{l,n,\bar{l},\bar{n}} = \frac{1}{2} L_{l,\bar{l}} M_{n,\bar{n}}$ corresponds to a MIPF (2.6a) if and only if $L_{l,\bar{l}}$ and $M_{n,\bar{n}}$ correspond to MIPF's of the forms (2.12) and (2.13), respectively.

This factorization is also possible for parafermion tensor product theories, with matrices $\boldsymbol{L}$ and $\boldsymbol{M}$ generalized to tensors. Any tensor $\boldsymbol{M}$ corresponding to a MIPF for p factors of $U(1)$ CFT's can be written as a tensor product of p independent matrix $\boldsymbol{M}$ solutions to (2.13) twisted by simple currents $\mathcal{J}$.[8] This approach greatly simplifies the derivation of the fractional superstring partition functions, while simultaneously suggesting much about the meaning of the different sectors, the origin of spacetime supersymmetry, and related "projection" terms.

2.1: *Affine Factor and "W" Partition Function*

In the A_K–sectors defined by Eqs. (2.2a, 2.3a, 2.4a) the terms inside the first (upper) set of brackets, which carry "$n \equiv 2m = 0$" subscripts, correspond to spacetime bosons and the terms inside the second (lower) set, carrying "$n = K/2$" correspond to spacetime fermions. Expressing the A_K–sectors in this form makes a one–to–one correspondence between bosonic and fermionic states in the A_K–sector manifest. If I remove the subscripts on the string functions in the bosonic and fermionic subsectors (which is analogous to replacing c_n^l with χ_l) we find the subsectors become equivalent. In fact, under this operation of removing the "n" subscripts (which I will denote by $\stackrel{\text{affine}}{\Longrightarrow}$), all sectors become the same up to an integer coefficient:

$D = 6 \qquad (K = 4)$:

$$A_4, B_4 \qquad \stackrel{\text{affine}}{\Longrightarrow} \qquad A_4^{\text{aff}} \equiv (\chi_0 + \chi_K)^3 \chi_{K/2} - (\chi_{K/2})^4 \tag{2.14a}$$

$D = 4 \qquad (K = 8)$:

$$A_8, B_8, C_8 \qquad \stackrel{\text{affine}}{\Longrightarrow} \qquad A_8^{\text{aff}} \equiv (\chi_0 + \chi_K)(\chi_2 + \chi_{K-2}) - (\chi_{K/2})^2 \tag{2.14b}$$

$D = 3 \qquad (K = 16)$:

$$A_{16}, C_{16} \qquad \stackrel{\text{affine}}{\Longrightarrow} \qquad A_{16}^{\text{aff}} \equiv (\chi_2 + \chi_{K-2}) - \chi_{K/2} \,. \tag{2.14c}$$

Eqs. (2.13a-c) all have the general form

$$A_K^{\text{aff}} \equiv (\chi_0 + \chi_K)^{D-3} (\chi_2 + \chi_{K-2}) - (\chi_{K/2})^{D-2} \,. \tag{2.15}$$

Thus,

$$W(\text{level}-K) = |A_K^{\text{aff}}|^2 \,. \tag{2.16}$$

(Note that the modular invariance of W requires that A_K^{aff} transforms back into itself under S.)

The class of partition functions (2.16) is indeed modular invariant and possesses special properties. This is easiest to show for $K = 16$. The $SU(2)_{16}$ MIPF's for $D = 3$ are trivial to classify, since at this level the A–D–E classification forms a complete basis set of modular invariants, even for MIPF's containing terms with negative coefficients. The only free parameters in $K = 16$ affine partition functions $Z\left(SU(2)_{16}\right)$ are integers a, b, and c, where

$$Z\left(SU(2)_{K=16}\right) = a \times Z(\mathrm{A}_{17}) + b \times Z(\mathrm{D}_{10}) + c \times Z(\mathrm{E}_7) \,. \tag{2.17}$$

Demanding that neither a left- nor a right-moving tachyonic state be in the Hilbert space of states in the $K = 16$ fractional superstring when the intercept v, defined by

$$L_0|\text{physical}\rangle = v|\text{physical}\rangle \,, \tag{2.18}$$

is positive, removes these degrees of freedom and requires $a = -(b+c) = 0$, independent of the possible $U(1)$ partition functions. These specific values for a, b, and c give us (2.16) for this level:

$$W(K=16) = Z(\mathrm{D}_{10}) - Z(\mathrm{E}_7) = |A^{\mathrm{aff}}_{16}|^2 \,. \tag{2.19}$$

Though not quite as straightforward a process, we can also derive the affine partition functions $W(K)$ for the remaining levels. The affine factors in the $K=4$ and 8 partition functions involve twisting by a non-simple current.* These cases correspond to theories that are the difference between a $\bigotimes_{\substack{D-2 \\ \text{factors}}} \mathrm{D}_{\frac{K}{2}+2}$ tensor product model and a $\bigotimes_{\substack{D-2 \\ \text{factors}}} \mathrm{D}_{\frac{K}{2}+2}$ tensor product model twisted by the affine current°

$$J^{K,\,\text{affine}}_{\text{non-simple}} = (\Phi^{\frac{K}{4}})^{D-2}\bar{\Phi}^1(\bar{\Phi}^0)^{D-3} \,. \tag{2.20}$$

The equivalent parafermion twist current is obvious,

$$J^{K,\,\text{parafermion}}_{\text{non-simple}} = (\phi_0^{\frac{K}{4}})^{D-2}\bar{\phi}_0^1(\bar{\phi}_0^0)^{D-3} \,. \tag{2.21}$$

(This derivation applies to the $K=16$ case also .)

---

*Models involving non-simple current twisting is discussed briefly in the appendix. See also refs. 9 and 10 for a discussion of simple and non-simple currents and their "twisting" effects. Multiloop modular invariance of $SU(2)_K$ tensor product models with non-simple current twisting is an issue currently under my investigation.

°I have left off the spacetime indices on most of the following currents and fields. I am working in light-cone gauge so only indices for transverse modes are implied. The $D-2$ transverse dimensions are assigned indices in the range 1 to $D-2$ (and are generically represented by lowercase Greek superscripts.) When spacetime indices are suppressed, the fields, and their corresponding partition functions, acting along directions 1 to $D-2$ are ordered in equations from left to right, respectively, for both the holomorphic and antiholomorphic sectors separately. Often I will be still more implicit in my notation and will express r identical factors of ϕ^j_m along consecutive directions (when these directions are either all compactified or uncompactified) as $(\phi^j_m)^r$. Thus, Eq. (2.21) for $K=8$ means

$$J^{K=8,\,\text{parafermion}}_{\text{non-simple}} \equiv (\phi_0^{K/4})^{\mu=1}(\phi_0^{K/4})^{\nu=2}(\bar{\phi}_0^1)^{\bar{\mu}=1}(\bar{\phi}_0^0)^{\bar{\nu}=2} \,.$$

2.2: *Theta-Function Factor and the "V" Partition Function*

I now consider the theta-function (*i.e.*, $U(1)$) factors, $\boldsymbol{M}$, carrying the $(n,\bar{n})$ indices in the fractional superstring partition functions. Since all A_K–, B_K–, C_K–sectors in the level-K fractional superstring partition function (and even the boson and fermion subsectors separately in A_K) contain the same affine factor, it is clearly the choice of the theta-function factor which determines the spacetime supersymmetry of the fractional superstring theories. That is, spacetime spins of particles in the Hilbert space of states depend upon the $\boldsymbol{M}$'s that are allowed in tensored versions of Eq. (2.13). In the case of matrix $\boldsymbol{M}$, rather than a more complicated tensor, invariance of (2.13) under S requires that the components $M_{n\bar{n}}$ be related by

$$M_{n',\bar{n}'} = \frac{1}{2K}\sum_{n,\bar{n}} M_{n,\bar{n}} e^{i\pi nn'/K} e^{i\pi\bar{n}\bar{n}'/K} \, , \tag{2.22a}$$

and T invariance demands that

$$\frac{n^2-\bar{n}^2}{4K} \in \mathbf{Z} \, , \quad \text{if } M_{n,\bar{n}} \neq 0 \, . \tag{2.22b}$$

At every level, K, there is a unique modular invariant function corresponding to each factorization[7], $\alpha \times \beta = K$, where α, $\beta \in \mathbf{Z}$. Denoting the matrix elements of $\boldsymbol{M}^{\alpha,\beta}$ by $M^{\alpha,\beta}_{n,\bar{n}}$, they are given by[▹]

$$M^{\alpha,\beta}_{n,\bar{n}} = \frac{1}{2}\sum_{\substack{x\in\mathbf{Z}_{2\beta}\\ y\in\mathbf{Z}_{2\alpha}}} \delta_{n,\alpha x+\beta y}\,\delta_{\bar{n},\alpha x-\beta y} \, . \tag{2.23}$$

Thus, for $K=4$ the two distinct choices for the matrix $\boldsymbol{M}^{\alpha,\beta}$ are $\boldsymbol{M}^{1,4}$ and $\boldsymbol{M}^{2,2}$; for $K=8$, we have $\boldsymbol{M}^{1,8}$ and $\mathbf{M}^{2,4}$; and for $K=16$, the three alternatives are $\boldsymbol{M}^{1,16}$, $\boldsymbol{M}^{2,8}$, and $\boldsymbol{M}^{4,4}$. $\boldsymbol{M}^{1,K}$ represents the level-K diagonal, *i.e.*, $n=\bar{n}$, partition function. $\boldsymbol{M}^{\alpha,\beta=\frac{K}{\alpha}}$ corresponds to the diagonal partition function twisted by a $\mathbf{Z}_\alpha$ symmetry. (Twisting by $\mathbf{Z}_\alpha$ and $\mathbf{Z}_{K/\alpha}$ produce isomorphic models.) Simple tensor products of these $\boldsymbol{M}^{\alpha,\beta}$ matrices are insufficient for producing fractional superstrings

---

▹By Eq. (2.23), $M^{\alpha,\beta}_{n,\bar{n}} = M^{\beta,\alpha}_{n,-\bar{n}}$. Hence, $\boldsymbol{M}^{\alpha,\beta}$ and $\boldsymbol{M}^{\beta,\alpha}$ result in equivalent fractional superstring partition functions. To avoid this redundancy, I choose $\alpha \leq \beta$.

Throughout this subsection I will view the n as representing, simultaneously, the holomorphic $\vartheta_{n,K}$ characters for $U(1)$ theories and, in some sense, the holomorphic string functions, c^0_n, for parafermions. ($\bar{n}$ represents the antiholomorphic parallels.) However, I do not intend to imply that the string functions can actually be factored into $c^l_0 \times c^0_n = c^l_n$. Rather, I mean to use this in Eqs. (2.25b, 2.27b, 2.29b) only as an artificial construct for developing a deeper understanding of the function of the parafermion primary fields (Verma modules) ϕ^0_m in these models. In the case of the primary fields, ϕ^j_m, factorization is, indeed, valid: $\phi^j_0 \otimes \phi^0_m = \phi^j_m$ (for integer j, m).

with spacetime SUSY (and, thus, no tachyons). I have found that twisting by a special simple $U(1)$ current is required to achieve this. Of the potential choices for the $U(1)$ MIPF's, $V(\text{level}-K)$, the following are the only ones that produce numerically zero fractional superstring partition functions:

$D=6 \qquad (K=4)$:

The $\boldsymbol{M}=\boldsymbol{M}^{2,2}\otimes\boldsymbol{M}^{2,2}\otimes\boldsymbol{M}^{2,2}\otimes\boldsymbol{M}^{2,2}$ model twisted by the simple $U(1)$ current**

$$\mathcal{J}_4 \equiv \phi^0_{K/4}\phi^0_{K/4}\phi^0_{K/4}\phi^0_{K/4}\bar{\phi}^0_0\bar{\phi}^0_0\bar{\phi}^0_0\bar{\phi}^0_0 \tag{2.24}$$

results in the following $U(1)$ partition functions:

$$\begin{aligned} V(K=4) = \; & [(\vartheta_{0,4}+\vartheta_{4,4})^4(\bar{\vartheta}_{0,4}+\bar{\vartheta}_{4,4})^4+(\vartheta_{2,4}+\vartheta_{-2,4})^4(\bar{\vartheta}_{2,4}+\bar{\vartheta}_{-2,4})^4 \\ & +(\vartheta_{0,4}+\vartheta_{4,4})^2(\vartheta_{2,4}+\vartheta_{-2,4})^2(\bar{\vartheta}_{0,4}+\bar{\vartheta}_{4,4})^2(\bar{\vartheta}_{2,4}+\bar{\vartheta}_{-2,4})^2 \\ & +(\vartheta_{2,4}+\vartheta_{-2,4})^2(\vartheta_{0,4}+\vartheta_{4,4})^2(\bar{\vartheta}_{2,4}+\bar{\vartheta}_{-2,4})^2(\bar{\vartheta}_{0,4}+\bar{\vartheta}_{4,4})^2]_{\text{untwisted}} \end{aligned} \tag{2.25a}$$

$$\begin{aligned} & +[(\vartheta_{2,4}+\vartheta_{-2,4})^4(\bar{\vartheta}_{0,4}+\bar{\vartheta}_{4,4})^4+(\vartheta_{4,4}+\vartheta_{0,4})^4(\bar{\vartheta}_{2,4}+\bar{\vartheta}_{-2,4})^4 \\ & +\;(\vartheta_{2,4}+\vartheta_{-2,4})^2(\vartheta_{4,4}+\vartheta_{0,4})^2(\bar{\vartheta}_{0,4}+\bar{\vartheta}_{4,4})^2(\bar{\vartheta}_{2,4}+\bar{\vartheta}_{-2,4})^2 \\ & +\;(\vartheta_{4,4}+\vartheta_{0,4})^2(\vartheta_{2,4}+\vartheta_{-2,4})^2(\bar{\vartheta}_{2,4}+\bar{\vartheta}_{-2,4})^2(\bar{\vartheta}_{0,4}+\bar{\vartheta}_{4,4})^2]_{\text{twisted}}\;. \end{aligned}$$

Writing this in parafermion form, and then using string function identities, followed by regrouping according to A_4 and B_4 components, results in

$$Z(\text{theta factor}, K=4) = |(c^0_0)^4+(c^0_4)^4|^2_{(A_4)} + |(c^0_0)^2(c^0_4)^2+(c^0_4)^2(c^0_0)^2|^2_{(B_4)}\;. \tag{2.25b}$$

**Recall that the parafermion primary fields ϕ^0_m have simple fusion rules,

$$\phi^0_m\otimes\phi^0_{m'} = \phi^0_{m+m'} \quad (\text{mod } K)$$

and form a $\boldsymbol{Z}_K$ closed subalgebra. This fusion rule, likewise, holds for the $U(1)$ fields : $\exp\{i\frac{m}{K}\varphi\}$: . This isomorphism makes it clear that any simple $U(1)$ current, $\mathcal{J}_K$, in this subsection that contains only integer m can be expressed equivalently either in terms of these parafermion fields or in terms of $U(1)$ fields. (We specify integer m since $\phi^0_m=0$ for half-integer m.) In view of the following discussion, I define all of the simple twist currents, $\mathcal{J}_K$, as composed of the former. Please note, to distinguish between simple $U(1)$ currents and affine currents, the former appear in calligraphic style.

$D = 4 \qquad (K = 8)$:

The $\boldsymbol{M} = \boldsymbol{M}^{2,4} \otimes \boldsymbol{M}^{2,4}$ model twisted by the simple $U(1)$ current

$$\mathcal{J}_8 \equiv \phi^0_{K/4}\phi^0_{K/4}\bar{\phi}^0_0\bar{\phi}^0_0 \tag{2.26}$$

results in

$$\begin{aligned} V(K=8) = \; & [(\vartheta_{0,8} + \vartheta_{8,8})(\bar{\vartheta}_{0,8} + \bar{\vartheta}_{8,8}) + (\vartheta_{4,8} + \vartheta_{-4,8})(\bar{\vartheta}_{4,8} + \bar{\vartheta}_{-4,8})]^2_{\text{untwisted}} \\ & + [(\vartheta_{2,8} + \vartheta_{-6,8})(\bar{\vartheta}_{2,8} + \bar{\vartheta}_{-6,8}) + (\vartheta_{-2,8} + \vartheta_{6,8})(\bar{\vartheta}_{-2,8} + \bar{\vartheta}_{6,8})]^2_{\text{untwisted}} \\ & + [(\vartheta_{4,8} + \vartheta_{-4,8})(\bar{\vartheta}_{0,8} + \bar{\vartheta}_{8,8}) + (\vartheta_{0,8} + \vartheta_{8,8})(\bar{\vartheta}_{4,8} + \bar{\vartheta}_{-4,8})]^2_{\text{twisted}} \\ & + [(\vartheta_{6,8} + \vartheta_{-2,8})(\bar{\vartheta}_{2,8} + \bar{\vartheta}_{-6,8}) + (\vartheta_{2,8} + \vartheta_{-6,8})(\bar{\vartheta}_{-2,8} + \bar{\vartheta}_{6,8})]^2_{\text{twisted}} \, . \end{aligned} \tag{2.27a}$$

Hence,

$$Z(\text{theta factor}, K=8) = |(c^0_0)^2 + (c^0_4)^2|^2_{(A_8)} + |(c^0_0)(c^0_4) + (c^0_4)(c^0_0)|^2_{(B_8)} + 4|(c^0_2)^2|^2_{(C_8)} \, . \tag{2.27b}$$

$D = 3 \qquad (K = 16)$:

The $\boldsymbol{M} = \boldsymbol{M}^{4,4}$ model twisted by the simple $U(1)$ current

$$\mathcal{J}_{16} \equiv \phi^0_{K/4}\bar{\phi}^0_0 \tag{2.28}$$

produces,

$$\begin{aligned} V(K=16) = \; & |(\vartheta_{0,16} + \vartheta_{16,16}) + (\vartheta_{8,16} + \vartheta_{-8,16})|^2_{\text{untwisted}} \\ & + |(\vartheta_{4,16} + \vartheta_{-4,16}) + (\vartheta_{12,16} + \vartheta_{-12,16})|^2_{\text{untwisted}} \, . \end{aligned} \tag{2.29a}$$

Thus,

$$Z(\text{theta factor}, K=16) = |c^0_0 + c^0_8|^2_{(A_{16})} + 4|c^0_4|^2_{(C_{16})} \, . \tag{2.29b}$$

(In this case the twisting is trivial since J_{16} is in the initial untwisted model.)

All of the above simple $U(1)$ twist currents are of the general form

$$\mathcal{J}_K = (\phi^0_{K/4})^{D-2}(\bar{\phi}^0_0)^{D-2} \quad \text{for } K > 2 \, . \tag{2.30}$$

I believe this specific class of twist currents is the key to spacetime supersymmetry in the parafermion models.[††] Without its twisting effect, numerically zero fractional superstring MIPF's in three, four, and six dimensions cannot be formed and, thus, spacetime SUSY would be impossible. This twisting also reveals much about the necessity of non-A_K-sectors. Terms from the twisted and untwisted sectors of these models become equally mixed in the $|A_K|^2$, $|B_K|^2$, and $|C_K|^2$ contributions to the level-K partition function. Further, this twisting keeps the string functions with $n \not\equiv 0, K/2 \pmod{K}$ from mixing with those possessing $n \equiv 0, K/2 \pmod{K}$. This is especially significant since I believe the former string functions in the C_K-sector likely correspond to spacetime fields of fractional spin-statistics (*i.e.*, anyons) and the latter in both A_K and B_K to spacetime bosons and fermions. If mixing were allowed, normal spacetime SUSY would be broken, to be replaced by a fractional supersymmetry, most-likely ruining Lorentz invariance for $D > 3$.

Since in the antiholomorphic sector $\mathcal{J}_K$ acts as the identity, I will focus on its effect in the holomorphic sector. In the A_K-sector the operator $(\phi^0_{K/4})^{D-2}$ transforms the bosonic (fermionic) nonprojection[‡‡] fields into the fermionic (bosonic) projection fields and vice-versa. For example, consider the effect of this twist current on the fields represented in

$$A_4 \equiv A_4^{\text{boson}} - A_4^{\text{fermion}} \, . \tag{2.31a}$$

where

$$A_4^{\text{boson}} = 4\left\{(c^0_0 + c^4_0)^3(c^2_0) - (c^2_0)^4\right\} \tag{2.31b}$$

$$A_4^{\text{fermion}} = 4\left\{(c^2_2)^4 - (c^0_2 + c^4_2)^3(c^2_2)\right\} \, . \tag{2.31c}$$

Twisting by $(\phi^0_{K/4})^{D-2}$ transforms the related fields according to

$$(\phi^0_0 + \phi^2_0)^3(\phi^1_0) \quad \overset{(\phi^0_{K/4})^{D-2}}{\Longleftrightarrow} \quad (\phi^0_1 + \phi^2_1)^3(\phi^1_1) \tag{2.32a}$$

$$(\phi^1_0)^4 \quad \overset{(\phi^0_{K/4})^{D-2}}{\Longleftrightarrow} \quad (\phi^1_1)^4 \, . \tag{2.32b}$$

Although the full meaning of the projection fields is not yet understood, the authors of refs. 3 and 6 argue that the corresponding string functions should be

---

[††] $\bar{\mathcal{J}}_K = (\phi^0_0)^{D-2}(\bar{\phi}^0_{K/4})^{D-2}$ is automatically generated as a twisted state.

[‡‡] We use the same language as the authors of refs. 2–4, 6. Nonprojection refers to the bosonic and fermionic fields in the A_K^{boson} and A_K^{fermion} subsectors, respectively, corresponding to string functions with positive coefficients, whereas projection fields refer to those corresponding to string functions with negative signs. With this definition comes an overall minus sign coefficient on A_K^{fermion}, as shown in Eq. (2.31a). For example, in (2.31b), the bosonic non-projection fields are $(\phi^0_0 + \phi^2_0)^3(\phi^1_0)$ and the bosonic projection is $(\phi^1_0)^4$. Similarly, in (2.31c) the fermionic non-projection field is $(\phi^1_1)^4$ and the projections are $(\phi^0_1 + \phi^2_1)^3(\phi^1_1)$.

interpreted as "internal" projections, *i.e.*, cancellations of degrees of freedom in the fractional superstring models. Relatedly, the authors show that when the A_K–sector is written as $A_K^{\rm boson} - A_K^{\rm fermion}$, as above, the q-expansions of both $A_K^{\rm boson}$ and $A_K^{\rm fermion}$ are all positive. Including the fermionic projection terms results in the identity

$$\eta^{D-2} A_K^{\rm fermion} = (D-2)\left(\frac{(\vartheta_2)^4}{16\eta^4}\right)^{(D-2)/8} . \tag{2.33a}$$

Eq. (2.33a) is the standard theta-function expression for $D-2$ world sheet Ramond Majorana-Weyl fermions. Further,

$$\eta^{D-2} A_K^{\rm boson} = (D-2)\left(\frac{(\vartheta_3)^4 - (\vartheta_4)^4}{16\eta^4}\right)^{(D-2)/8} . \tag{2.33b}$$

Now consider the B_K–sectors. For $K=4$ and 8, the operator $(\phi^0_{K/4})^{D-2}$ transforms the primary fields corresponding to the partition functions terms in the first set of brackets on the RHS of Eqs. (2.2b, 2.3b) into the fields represented by the partition functions terms in the second set. For example, in the $K=4$ ($D=6$) case

$$(\phi^0_0 + \phi^2_0)(\phi^1_0)(\phi^0_1 + \phi^2_1)^2 \overset{(\phi^0_{K/4})^{D-2}}{\Longleftrightarrow} (\phi^0_1 + \phi^2_1)(\phi^1_1)(\phi^2_0 + \phi^0_0)^2 \tag{2.34a}$$

$$(\phi^1_0)^2(\phi^1_1)^2 \overset{(\phi^0_{K/4})^{D-2}}{\Longleftrightarrow} (\phi^1_1)^2(\phi^1_0)^2 . \tag{2.34b}$$

Making an analogy with what occurs in the A_K–sector, I suggest that $(\phi^0_{K/4})^{D-2}$ transforms bosonic (fermionic) nonprojection fields into fermionic (bosonic) projection fields and vice-versa in the B_K–sector also. Thus, use of the twist current $\mathcal{J}_K$ allows for bosonic and fermionic interpretation of these fields§§:

$$B_4 \equiv B_4^{\rm boson} - B_4^{\rm fermion} , \tag{2.35a}$$

where

$$B_4^{\rm boson} = 4\left\{(c^0_0 + c^4_0)(c^2_0)(c^0_2 + c^4_2)^2 - (c^2_0)^2(c^2_2)^2\right\} \tag{2.35b}$$

$$B_4^{\rm fermion} = 4\left\{(c^2_2)^2(c^2_0)^2 - (c^0_2 + c^4_2)(c^2_2)(c^0_0 + c^4_0)^2\right\} . \tag{2.35c}$$

What appears as the projection term, $(c^2_0)^2(c^2_2)^2$, for the proposed bosonic part acts as the nonprojection term for the fermionic half, when the subscripts are reversed. One interpretation is that this implies a compactification of two transverse dimensions.◁◁

§§Similar conclusions have been reached by K. Dienes and P. Argyres for different reasons. They have, in fact, found theta-function expressions for the $B_K^{\rm boson}$ and $B_K^{\rm fermion}$ subsectors.[11]

◁◁This was also suggested in ref. 3 working from a different approach.

The spin-statistics of the physical states of the $D = 6$ model as observed in four-dimensional uncompactified spacetime would be determined by the (matching) n subscripts of the first two string functions** (corresponding to the two uncompactified transverse dimensions) in each term of four string functions, $c_n^{l_1} c_n^{l_2} c_{n'}^{l_3} c_{n'}^{l_4}$. The B_8 terms can be interpreted similarly when one dimension is compactified.

However, the C_K-sectors are harder to interpret. Under $(\phi^0_{K/4})^{D-2}$ twisting, string functions with $K/4$ subscripts are invariant, transforming back into themselves. Thus, following the pattern of A_K and B_K we would, for example, express C_{16} as

$$C_{16} = C^a_{16} - C^b_{16} \tag{2.36a}$$

where,

$$C^a_{16} = (c^2_4 + c^{14}_4) - c^8_4 \tag{2.36b}$$

$$C^b_{16} = c^8_4 - (c^2_4 + c^{14}_4) \ . \tag{2.36c}$$

The transformations of the corresponding primary fields are not quite as trivial, though. $(\phi^1_2 + \phi^7_2)$ is transformed into its conjugate field $(\phi^7_{-2} + \phi^1_{-2})$ and likewise ϕ^4_2 into ϕ^4_{-2}, suggesting that C^a_{16} and C^b_{16} are the partition functions for conjugate fields. Remember, however, that $C_{16} = 0$. Even though we may interpret this sector as containing two conjugate spacetime fields, this (trivially) means that the partition function for each is identically zero. I refer to this effect in the C_K-sector as "self-cancellation." One interpretation is that there are no states in the C_K-sector of the Hilbert space that survive all of the internal projections. If this is correct, a question may arise as to the consistency of the $K = 8$ and 16 theories. Alternatively, perhaps anyon statistics allow two (interacting?) fields of either identical fractional spacetime spins $s_1 = s_2 = \frac{2m}{K}$, or spacetime spins related by $s_1 = \frac{2m}{K} = 1 - s_2$, where in both cases $0 < m < \frac{K}{2}$ (mod 1), to cancel each other's contribution to the partition function.

Conclusions

A viable and consistent generalization of the superstring would be significant. I have shown that the partition functions for fractional superstrings have simple origins when derived systematically through the factorization approach of Gepner and Qiu. Further, using this affine/theta-function factorization of the parafermion partition functions, I have related the A_K-sector, containing the graviton and gravitino, with the massive sectors B_K and C_K. A bosonic/fermionic interpretation of the B_K-subsectors was given. Apparent "self-cancellation" of the C_K-sector was shown. I am currently investigating the meaning of this last issue.

**Using the subscripts n' of last two string functions to define spin-statistics in four-dimensional uncompactified spacetime corresponds to interchanging the definitions of B_4^{boson} and B_4^{fermion}.

Appendix: Simple and Non-Simple Currents

A simple current, J_s, is a primary field of a conformal field theory (CFT) which, when fused with any other primary field (including itself) $\Phi_{\mathbf{l}}$, where $\mathbf{l}$ labels a primary field of the CFT, produces only a single primary field as a product state:

$$J_s \otimes \Phi_{\mathbf{l}} = \Phi_{\mathbf{l}'} \,. \tag{A.1}$$

A non-simple current J_{ns}, when fused with at least one other primary field (possibly itself), produces more than one term:

$$J_{ns} \otimes \Phi_{\mathbf{l}} = \sum_{\mathbf{l}'} \Phi_{\mathbf{l}'} \,. \tag{A.2}$$

Schellekens *et. al,* [12] have shown that twisting a known modular invariant model by a simple current results in a new modular invariant. Let N be the order of the simple current. That is, N is the smallest positive integer such that

$$(J_s)^N \otimes \Phi_{\mathbf{l}} = \Phi_{\mathbf{l}} \tag{A.3}$$

for all primary fields $\Phi_{\mathbf{l}}$ in the CFT. The "untwisted sector" of the new model for this CFT is formed by those primary fields in the original, which when fused with any power of the current $(J_s)^i$, $i = 1$ to $N-1$, only produce T invariant states. The "twisted sectors" of the new model are formed by the sets of (T invariant) fusion products formed from each $(J_s)^i$ acting on the untwisted sector of the new model.

Deriving new models using non-simple currents J_{ns} is more difficult. The general requirement that a primary field of the original model remain in the untwisted sector of the twisted model appears to be that there is at least one T invariant in the fusion product of that field and J_{ns}. The simplest type of model involving non-simple currents is a class formed using a closed subalgebra of non-simple currents, $\{J_{ns,a}\}$, where a labels the currents. For an investigation into models formed from non-simple twistings see ref. 9.

References:

1. G. Cleaver and P. Rosenthal, CALT-68-1756.
2. K. Dienes and S.-H. Tye, *Nucl. Phys.* **B376** (1992) 297.
3. P. Argyres and S.-H. Tye, *Phys. Rev. Lett.* **67** (1991) 3339.
4. P. Argyres, K. Dienes and S.-H. Tye, CLNS 91/1113; McGill-91-37.
5. V. Kač, *Adv. Math.* **35** (1980) 264;
   V. Kač and D. Peterson, *Bull. AMS* **3** (1980) 1057;
   __________, *Adv. Math.* **53** (1984) 125.

6. P. Argyres, E. Lyman, and S.-H. Tye, CLNS 91/1121.
7. D. Gepner and Z. Qiu, *Nucl. Phys.* **B285** (1987) 423.
8. G. Cleaver, Unpublished research.
9. G. Cleaver and D. Lewellen, *Phys. Lett.* **B200** (1993) 354.
10. G. Cleaver, Ph.D. Thesis.
11. K. Dienes, Private communications.
12. A.N. Schellekens and S. Yankielowicz, *Nucl. Phys.* **B327** (1989) 3;
    A.N. Schellekens, *Phys. Lett.* **244** (1990) 255;
    B. Gato-Rivera and A.N. Schellekens, *Nucl. Phys.* **B353** (1991) 519.
    ___________, *CERN-TH.6056/91.*

PICARD–FUCHS MONODROMY GROUPS AND QUANTUM SYMMETRIES OF STRING EFFECTIVE ACTIONS

R. D'AURIA
Department of Physics, Politecnico di Torino and INFN Sezione di Torino, Torino, Italy

S. FERRARA
CERN, 1211 Geneve 23, Switzerland

ABSTRACT

Local and global properties of the moduli space of Calabi–Yau type compactifications determine the low energy parameters of the string effective action. We show that the moduli space geometry is entirely encoded in the Picard–Fuchs equations for the periods of the Calabi–Yau $H^{(3)}$–cohomology.

1. Introduction

It is well known that among the possible vacua of heterotic string theory the (2,2), c=9 world–sheet superconformal field theories (SCFT) are of prominent importance since they give rise to N=1 supersymmetry in four dimensions, which is currently believed to be a necessary ingredient of the low–energy effective Lagrangian [1].

In general it is quite difficult to compute the correlators of a Superconformal Field Theory (SCFT) and usually more insight into their structure can be gained by using the Landau–Ginzburg approach to N=2 SCFT [2] or else in the context of the N=2 topological field theory [3]. The description of the (2,2)–string vacua in terms of the Landau–Ginzburg superpotential (possibly in its twisted topological version) is strictly related to the geometrical description in terms of the compactification of the heterotic string on a Calaby–Yau 3–fold (C.Y.) and in fact the determination of the physical low–energy parameters in both frameworks can be performed by using the same techniques of algebraic geometry [4,5]. One is actually faced with the problem of computing such low–energy quantities in terms of the properties of the moduli space of a given SCFT. Moduli space is the space of all the marginal deformations of the underlying N=2 SCFT or, in the geometrical picture of Calabi–Yau compactifications, is the space of the parameters φ^{α}, ψ^{a} describing respectively the deformations of the Kähler class and/or of the complex structure of the C.Y.

manifold [6]. In the low energy theory the moduli parameters $\varphi^\alpha(x)$, $\psi^a(x)$ appear as neutral massless scalar fields with vanishing potential.

The low energy objects that we are interested in as functions of the moduli are: i) the moduli and family metrics; ii) the Yukawa couplings. Another important moduli dependent effect is: iii) the running of the gauge couplings due to one–loop stringy effects.

These quantities appear in the effective N=1 Lagrangian in the following form:

i)

$$g_{i\bar{j}}\partial_\mu M^i \partial_\nu \bar{M}^{\bar{j}} G^{\mu\nu} \tag{1.1}$$

$$g^{(27)}_{\alpha\bar{\beta}}\partial_\mu \varphi^\alpha \partial_\nu \bar{\varphi}^{\bar{\beta}} G^{\mu\nu} \tag{1.2}$$

$$g^{(\overline{27})}_{a\bar{b}}\partial_\mu \psi^a \partial_\nu \bar{\psi}^{\bar{b}} G^{\mu\nu} \tag{1.3}$$

ii)

$$W_{abc}(\psi)\phi^a_{27}\phi^b_{27}\phi^c_{27} \tag{1.4}$$

$$W_{\alpha\beta\gamma}(\varphi)\phi^\alpha_{\overline{27}}\phi^\beta_{\overline{27}}\phi^\gamma_{\overline{27}} \tag{1.5}$$

iii)

$$\frac{1}{g^2(S,\varphi,\psi)} T_r F_{\mu\nu}F^{\mu\nu} + \theta(S,\varphi,\psi) T_r F_{\mu\nu}\tilde{F}^{\mu\nu} \tag{1.6}$$

where M^i denote any of the two sets of moduli φ^α, ψ^a, $S = D + ia$ is the dilaton–axion scalar field, $g_{i\bar{j}}, g^{(27)}_{\alpha\bar{\beta}}, g^{(\overline{27})}_{a\bar{b}}$ are the metrics for moduli, families and antifamilies respectively, W_{abc}, $W_{\alpha\beta\gamma}$ are the Yukawa couplings, g^2 and θ are the gauge coupling and the θ–angle for the gauge field strength $F_{\mu\nu}$ ($\tilde{F}_{\mu\nu}$ is the dual of $F_{\mu\nu}$).

We note that the moduli space $\mathcal{M}$ has the structure of a product space

$$\mathcal{M} = \mathcal{M}_1 \times \mathcal{M}_2 \tag{1.7}$$

where $\mathcal{M}_1, \mathcal{M}_2$ are parametrized by the φ^α and ψ^a, respectively. This can be proven either in SCFT or by supersymmetry arguments [7]. The two spaces are actually related by the mirror hypothesis [8] which establishes that exchanging the role of the two parameters φ^α, ψ^a associated to the Kähler class and complex structure deformations of the C.Y. 3–fold, does not change the underlying SCFT and therefore also the physical low–energy parameters. In the following we shall restrict our attention only to the manifold $\mathcal{M}$ of the complex structure moduli. The results thus obtained are also valid for the moduli space of the Kähler class deformations of the associated mirror C.Y. manifold.

The structure of the quantities appearing in eq.s (1.1–6) is completely determined by the local and global properties of the moduli space. These properties in turn are encoded in the Picard–Fuchs equations for the periods $X^A(\psi), F_A(\psi)$ of the Calabi–Yau family parametrized by the $\psi^{a'}$s. Once we know the periods we may reconstruct the Kähler potential of the moduli space and the Yukawa couplings as follows [6]:

$$K(\psi,\bar{\psi}) = -\log\ i(X^A\bar{F}_A - \bar{X}^A F_A) \tag{1.8}$$

$$W_{abc} = \frac{\partial X^A}{\partial\psi^a}\frac{\partial X^B}{\partial\psi^b}\frac{\partial X^C}{\partial\psi^c}\frac{\partial^3 F}{\partial X^A\partial X^B\partial X^C} \tag{1.9}$$

where $F = F(X^A)$ is a degree two homogeneous function such that

$$X^A F_A = 2F \quad \rightarrow \quad F_A \equiv \frac{\partial F}{\partial X^A} \tag{1.10}$$

Notice that the characterization of the local geometry of the moduli space given by eq.s (1.8–1.10) coincides with the definition of Special Kähler [9,10,11] geometry and indeed the periods (X^A, F_A) of the Calabi–Yau 3–fold can be indentified with a set of globally defined holomorphic sections in terms of which the special Kähler geometry can be defined. In this lecture however we will be mainly concerned with the global properties of the moduli space $\mathcal{M}$ which are consequence of the fact that $\mathcal{M}$ possesses a group Γ of discrete isometries which is generally referred to as the target space duality group, or modular group [12]. The duality group describes quantum symmetries of the string effective actions and is the discrete version of the non–compact symmetries of old supergravity Lagrangians (no–scale supergravities). The most celebrated example is the effective Lagrangian obtained by compactification of the heterotic string on a 6–torus, in the limit of large radius of the compactified space. The Kähler class modulus $t = 2(R^2 + i\sqrt{|g|})$ parametrizes the homogeneous space [13]:

$$\frac{SU(1,1)}{U(1)} \tag{1.11}$$

The Kähler potential of the scalar fields t, ϕ_{27} is

$$K = -\log\ [i(t-\bar{t}) - \phi_{27}\bar{\phi}_{\overline{27}}] \tag{1.12}$$

and the superpotential $\mathcal{W}$ is given by $\mathcal{W} = C\phi_{27}^3$ corresponding to a constant Yukawa coupling, $W = C$. The theory is obviously invariant under

$$t' = \frac{at+b}{ct+d} \quad , \quad \phi'_{27} = -\frac{1}{ct+d}\phi_{27} \quad , \quad ad - bc = 1 \tag{1.13}$$

and the duality group is given in this case by $SL(2,\mathbf{Z})$ with generators

$$t \rightarrow t+1 \qquad\qquad t \rightarrow \frac{1}{t} \tag{1.14}$$

For Calabi-Yau manifolds W is no longer a constant, but in general it depends on the moduli : $W_{\alpha\beta\gamma} = W_{\alpha\beta\gamma}(M)$ because of the instanton corrections to the σ-model perturbative result. In the one modulus case, however, it turns out that $t \rightarrow t+1$ is still an exact symmetry if t is the special coordinate of the Kähler Special Geometry, or, equivalently, the flat coordinate of the associated topological field theory. Actually one is led to consider the possibility that the translational symmetry is exact for any number of moduli since invariance under $t^a \rightarrow t^a + 1$, $a = 1, ..., n$ has its stringy origin in the presence of the antisymmetric axion field B_{ij} in the σ-model action, and means that the t^a-variables are periodic. For example the Yukawa coupling in the one modulus case can be written as follows

$$W(t) = \sum_{n=0}^{\infty} d_n \; e^{2\pi i n t} \tag{1.15}$$

and for many variables we have analogous expansions. In the large radius limit, $t \rightarrow i\infty$, eq. (1.15) gives

$$W(t) = d_0 = const. \tag{1.16}$$

which implies that $\mathcal{F}(t) \equiv |X^0|^{-2}F(X^A)$ is cubic in t. Actually the most general prepotential such that $W(t)$ and the Kähler potential (1.8) are invariant under $t \rightarrow t+1$ is (Cecotti et al.[7])

$$F(X^A) = (X^0)^2[t^3 + \lambda(t) + P_2(t)] \tag{1.17}$$

where $P_2(t)$ is a polynomial of degree two with purely imaginary coefficients and $\lambda(t)$ a periodic function of t. If instead invariance under arbitrary shifts is required, $t \rightarrow t + c$, $c \in \mathbf{R}$, then λ is a constant.

Under the inversion $t \rightarrow \frac{1}{t}$ the Kähler potential transforms as follows

$$K \rightarrow K + f(t) + f(\bar{t}) \tag{1.18}$$

so that W, which has a non trivial Kähler weight undergoes a non trivial transformation.

In the next section we will show that the generalization of the inversion for a generic duality group is given by

$$t^a \rightarrow f^a(t^a, \partial_a\mathcal{F}, \mathcal{F}; A, B, C, D) \qquad a = 1, ..., n \tag{1.19}$$

where n is the number of moduli, $\mathcal{F}(t^a) = (X^0)^{-2}F(X^A)$ and A, B, C, D are $(n+1) \times (n+1)$ matrices parametrizing a generic $Sp(2n+2; \mathbf{Z})$ transformation M:

$$M = \begin{pmatrix} A & B \\ C & D \end{pmatrix} \quad \in \quad Sp(2n+2; \mathbf{Z}) \tag{1.20}$$

For example in the case of the quintic studied by Candelas et al. the duality group acting on the single modulus t is generated by two transformations:

$$t \to t+1 \tag{1.21}$$

$$t \to \frac{t}{t\mathcal{F}'_t - 2\mathcal{F} + 1} \tag{1.22}$$

The knowledge of the duality group is very important since the physical quantities appearing in the effective Lagrangian must transform in a definite way under the group: for example the gauge coupling $g_a^{-2}(t,\bar{t})$ is a real function which must be modular invariant in $t, \bar{t}$.

Moreover some of the duality transformations may induce target–space anomalies in the effective Lagrangian [14]. These anomalies arise when the Kähler potential is not invariant under the duality transformations, but undergoes the transformation (1.18). In this case all the Kähler gauge dependent quantities get transformed and anomalies can be generated. From the above discussion we see that in general possibly anomalous transformations belong to the coset $\frac{\Gamma}{T}$ where Γ is the duality group and T generates the discrete translations under which the Kähler potential is invariant. The modular group described here acts on the special coordinates t^a which are the coordinates which flatten the holomorphic connection of the associated topological field theory. The modular coordinate γ on which the two transformations described earlier act as $SL(2,R)$ transformations (for any F–function) is described in the next section and the relation of this coordinate with the special coordinate as well as with the Landau–Ginzburg coordinate of the defining polynomial in CP$(d+1)$ will also be given. In this respect the target space duality group Γ is closely related to the monodromy group Γ_M of the Picard–Fuchs equations which determine the period matrix for Calabi–Yau manifolds, the precise relation being discussed in the following.

2. Monodromy of the Picard-Fuchs equations and the quantum modular group

The periods (X^A, F_A) of a given family of Calabi–Yau d–folds are known to satisfy a coupled set of linear partial differential equations called Picard–Fuchs equations (PFE). A general property of these equations is that they are of Fuchsian type, that is they only have regular singular points. They can be derived from the defining polynomial equation $W(y^i, \psi^\alpha) = 0$ of the Calabi–Yau manifold by using simple algorithms which have been described in ref. [15]. If one solves the PFE for the periods and uses eq.s (1.8) and (1.9), then one reconstructs the Kähler potential of the moduli space and the Yukawa couplings associated to that particular class of C.Y. compactifications. Furthermore the PFE also encode essential informations one the global structure of the moduli space through their monodromy group.

Let us denote by Γ the target space duality group (quantum modular group) and by Γ_W the group of invariance of the superpotential $W(y^i, \psi^a)$. Γ_W consists of these diffeomorphisms of the moduli ψ^α which leave $W = 0$ invariant except for a (quasi)–homogeneous change of the $CP(d+1)$ coordinates:

$$W(y^i, \psi^\alpha) = 0 \quad \underset{\Gamma_W}{\longrightarrow} \quad W(\tilde{y}^i(y); \tilde{\psi}^\alpha(\psi)) = 0 \tag{2.1}$$

where $\tilde{y}^i = U^i_j y^j$ and i, j run over all chiral fields with same $U(1)$ charge.

Finally let us denote by Γ_M the monodromy group of the PFE's. To define it in the simplest way we restrict our attention to the case of one single modulus, in which case the PFE's are ordinary differential equations.

Then, if we denote by $(f_1(z), ..., f_n(z))$ a basis of solutions of the differential equation at a point z, by analytically continuing $(f_1, ..., f_n)$ along a closed loop around a singularity z_1 of the equation we arrive at a new solution at z_1 which must therefore be expressible as a linear combination of the basis $(f_1, ..., f_n)$:

$$(f_1, ..., f_n) \rightarrow (\hat{f}_1, ..., \hat{f}_n) = (f_1, ..., f_n)\, A_{z_1} \tag{2.2}$$

where the $n \times n$ non singular matrix A_{z_1} defines the monodromy around z_1. If the equation has r singular points we obtain r monodromy matrices $A_{z_1}, ..., A_{z_r}$, and if we compose closed loops around z_i and z_j in the usual way it is clearly seen that to the loop $\gamma_i \circ \gamma_j \equiv \gamma_{ij}$ encircling z_i and z_j corresponds the monodromy matrix $A_{z_j} \cdot A_{z_i}$, and that more generally $A_{z_1}, ..., A_{z_r}$ generate a group, the monodromy group of the differential equation (here the inverse $A^{-1}_{z_i}$ is the matrix obtained by running around z_i in the opposite direction, and $\mathbf{1}$ corresponds to a circuit contractible to a point).

It turns out that in the known cases the monodromy group Γ_M is a normal (generally infinite) subgroup of Γ and that

$$\Gamma/\Gamma_M \simeq \Gamma_W \tag{2.3}$$

Following the proposal of Lerche et al.[15] we assume that (2.3) is true in general, possibly also for the case of many moduli. Equation (2.3) suggests that in order to reconstruct Γ we can compute the monodromy group of the PFE's and the invariance group of $W = 0$, so that $\Gamma \simeq \Gamma_W \not> \Gamma_M$. In this section we give two explicit examples of such construction: the 1–dimensional C.Y. manifold described by a cubic polynomials in CP(2):

$$W = \frac{1}{3}(y_1^3 + y_2^3 + y_3^3) - \psi y_1 y_2 y_3 = 0 \tag{2.4}$$

and the 3–dimensional C.Y. manifold describe by a quintic polynomial in CP(4):

$$W = \frac{1}{5}(y_1^5 + y_2^5 + y_3^5 + y_4^5 + y_5^5) - \psi y_1 y_2 y_3 y_4 y_5 = 0 \tag{2.5}$$

where $y^i, i = 1, ..., d+2$ are homogeneous coordinates in CP(d+1) d is the complex dimension equal to 1 or 3 in our case and ψ is a single modulus parametrizing the complex structure deformations of the hypersurface. Note that while in the case of the torus the space of complex structure deformations is one–dimensional, in the case of the quintic $W_0 = \frac{1}{5}(y_1^5 + y_2^5 + y_3^5 + y_4^5 + y_5^5)$ there are 101 indipendent complex structure deformations so that the moduli space is 101–dimensional. The particular 1–dimensional subspace described by the simple deformation (2.5) is such that $1, y_1 \dots y_5, (y_1 \dots y_5)^2, (y_1 \dots y_5)^3$ close a subring of the chiral ring of all the marginal operators associated to the deformations of the quintic.

Let us begin to study the case of the torus. From (2.4), using known algorithms, (see ref. [15]),one obtains the following PFE's:

$$\frac{d}{d\psi}\begin{pmatrix}\omega_0\\ \omega_1\end{pmatrix} = \begin{pmatrix}0 & 1\\ \frac{\psi}{1-\psi^3} & \frac{3\psi^2}{1-\psi^3}\end{pmatrix}\begin{pmatrix}\omega_0\\ \omega_1\end{pmatrix} \tag{2.6}$$

This can be traded for a single 2nd-order differential equation for ω_0

$$\left(\frac{d^2}{d\psi^2} - \frac{3\psi^2}{1-\psi^3}\frac{d}{d\psi} - \frac{\psi}{1-\psi^3}\right)\omega_0 = 0 \tag{2.7}$$

which exhibits four regular singular points at $\psi^3 = 1, \psi = \infty$.

The monodromy group of this equation can be studied as follows. First of all we note that it is sufficient to compute the monodromy matrix T_0 around $\psi = 1$. In fact the effect of a closed loop around $\psi = \alpha$ and $\psi = \alpha^2$ $(\alpha = e^{2\pi i/3})$ can be computed from the monodromy matrix T_0 around $\psi = 1$ by conjugation with $\mathcal{A}$, where $\mathcal{A}$ represents the operation $\psi \to \alpha\psi$:

$$\begin{aligned} T_1 &= \mathcal{A}^1 T_0 \mathcal{A}^{-1}\\ T_2 &= \mathcal{A}^2 T_0 \mathcal{A}^{-2} \end{aligned} \tag{2.8}$$

Furthermore a closed loop which encloses all the singular points, including ∞, is contractible and therefore

$$T_\infty T_2 T_1 T_0 = 1 \quad \to \quad T_\infty = (T_2 T_1 T_0)^{-1} \tag{2.9}$$

To compute T_0 it is convenient to perform the substitution $z = \psi^3$ in the differential equation (2.7). We obtain

$$\left\{9z(1-z)\frac{d^2}{dz^2} + (6 - 15z)\frac{d}{dz} - 1\right\}\omega = 0 \tag{2.10}$$

This is a hypergeometric equation of parameters $a = b = 1/3, c = 2/3$ and therefore a set of independent solutions around $z \equiv \psi^3 = 0$ is given by

$$\begin{cases} U_1 = \frac{\Gamma^2(1/3)}{\Gamma(2/3)} F(1/3, 1/3, 2/3; \psi^3)\\ U_2 = \frac{\Gamma^2(2/3)}{\Gamma(4/3)} \psi F(2/3, 2/3, 4/3; \psi^3) \end{cases} \tag{2.11}$$

where $F(a, b, c; z)$ is the hypergeometric functions.

These 2 solutions can be continued around $\psi^3 = 1$ by known formulae [16]: one finds

$$U_1 = -\log(1-z)F(1/3,1/3,1;1-\psi^3) + B_1(1-\psi^3)$$
$$U_2 = -\log(1-z)F(1/3,1/3,1;1-\psi^3) + B_2(1-\psi^3) \tag{2.12}$$

where B_1 and B_2 are regular series around $\psi^3 = 1$. (The appearance of the logarithmic factor in (2.12) is traceable to the equality of the roots of the indicial equation around $z \equiv \psi^3 = 1$). A closed loop around $\psi = 1$ gives

$$\begin{pmatrix} U_1 \\ U_2 \end{pmatrix} \to \begin{pmatrix} U_1' \\ U_2' \end{pmatrix} = \begin{pmatrix} U_1 \\ U_2 \end{pmatrix} - 2\pi i\, F(\frac{2}{3},\frac{2}{3},1;1-\psi^3)\begin{pmatrix} 1 \\ 1 \end{pmatrix} \tag{2.13}$$

The Kummer relations [16] among hypergeometric functions allow us to reexpress $F(\frac{2}{3},\frac{2}{3},1;1-\psi^3)$ in terms of the original basis (U_1, U_2) around $\psi = 0$,

$$F\Big(\frac{1}{3},\frac{1}{3},1;1-z\Big) = \frac{\Gamma(\frac{1}{3})}{\Gamma^2(\frac{2}{3})}F\Big(\frac{1}{3},\frac{1}{3},\frac{2}{3};z\Big) + \frac{\Gamma(-\frac{1}{3})}{\Gamma^2(\frac{1}{3})}F\Big(\frac{2}{3},\frac{2}{3},\frac{4}{3};z\Big) \tag{2.14}$$

Therefore, using the relation $\Gamma(z)\Gamma(1-z) = \frac{\pi}{\sin \pi z}$ one obtains

$$\begin{pmatrix} U_1' \\ U_2' \end{pmatrix} = \begin{pmatrix} 1 + i\, tg\frac{2\pi}{3} & i\, tg\frac{2\pi}{3} \\ i\, tg\frac{2\pi}{3} & i - i\, tg\frac{2\pi}{3} \end{pmatrix} \begin{pmatrix} U_1 \\ U_2 \end{pmatrix} \tag{2.15}$$

that is the monodromy matrix around $\psi = 1$ is

$$T_0 = \begin{pmatrix} 1 - i\sqrt{3} & i\sqrt{3} \\ -i\sqrt{3} & 1 + i\sqrt{3} \end{pmatrix} \tag{2.16}$$

To find T_1, T_2 we need to represent $\mathcal{A} : \psi \to \alpha\psi$ on U_1, U_2. From (2.6), (2.7) and (2.11) we see that under $\psi \to \alpha\psi$ the differential operator is invariant while

$$\begin{pmatrix} U_1 \\ U_2 \end{pmatrix} \to \begin{pmatrix} 1 & 0 \\ 0 & \alpha \end{pmatrix} \begin{pmatrix} U_1 \\ U_2 \end{pmatrix} \tag{2.17}$$

Since we are interested in the projective representation of the monodromy group we may rescale our basis in such a way that $\det \mathcal{A} = 1$ (note that T_0 already satisfies $\det T_0 = 1$). Hence we have

$$\mathcal{A} = \begin{pmatrix} \alpha^{-1/2} & 0 \\ 0 & \alpha^{1/2} \end{pmatrix} \tag{2.18}$$

and from (2.8)

$$T_1 = \begin{pmatrix} 1 - i\sqrt{3} & \alpha^{-1}i\sqrt{3} \\ -\alpha i\sqrt{3} & 1 + i\sqrt{3} \end{pmatrix} \qquad T_2 = \begin{pmatrix} 1 - i\sqrt{3} & \alpha^{-2}i\sqrt{3} \\ -\alpha^2 i\sqrt{3} & 1 + i\sqrt{3} \end{pmatrix} \tag{2.19}$$

Let us now recall that the modular group is given by the group of transformations on the variable ψ which leaves the theory invariant. The monodromy group Γ_M of the PFE's must therefore be a subgroup of the modular group. In our case the modular group of the torus is known a priori to be $\Gamma = SL(2;\mathbf{Z})$ and therefore it should be possible to perform a change of basis on the periods U_i such that the entries of the generators T_0, T_1, T_2 are integer numbers. Actually it is known since the last century that Γ_M is isomorphic to $\Gamma(3)$, where $\Gamma(3)$ is the group of matrices equivalent to the identity modulo 3. The basis $(\mathcal{F}_1, \mathcal{F}_2)$ where $\Gamma_M \simeq \Gamma(3)$ is obtained by the following linear transformation [17]

$$\begin{pmatrix} \mathcal{F}_1 \\ \mathcal{F}_2 \end{pmatrix} = \frac{1}{3(1+\alpha^{-1/2})} \begin{pmatrix} 3\alpha^{1/2} & -3 \\ 1+\alpha^{1/2} & \alpha^2 - 1 \end{pmatrix} \begin{pmatrix} U_1 \\ U_2 \end{pmatrix} \tag{2.20}$$

The transformed Γ_M generators $\hat{T}_i$ take the following form:

$$\hat{T}_0 = \begin{pmatrix} 1 & 3 \\ 0 & 1 \end{pmatrix} \;;\hat{T}_1 = \begin{pmatrix} -5 & 12 \\ -3 & 7 \end{pmatrix} \;;\hat{T}_2 = \begin{pmatrix} 1 & 0 \\ -3 & 1 \end{pmatrix}$$

$$\hat{T}_\infty \equiv (\hat{T}_2\hat{T}_1\hat{T}_0)^{-1} = \begin{pmatrix} 1 & 0 \\ 9 & 1 \end{pmatrix} \tag{2.21}$$

The transformation $\mathcal{A} : \psi \to \alpha\psi$ is obviously an invariance of $W = 0$ (and of the differential operator (2.17)) since it can be undone by the coordinate transformation $y^i \to \alpha^{-1/3} y^i$ Less evident is the invariance under the transformation $\mathcal{B}$:

$$\mathcal{B} : \psi' = -\frac{\psi + 2}{1 - \psi} \tag{2.22}$$

which can be undone by the change of coordinates

$$\begin{pmatrix} y_1' \\ y_2' \\ y_3' \end{pmatrix} = \frac{i}{\sqrt{3}} \begin{pmatrix} 1 & 1 & 1 \\ 1 & \alpha & \alpha^2 \\ 1 & \alpha^2 & \alpha \end{pmatrix} \begin{pmatrix} y_1 \\ y_2 \\ y_3 \end{pmatrix} \tag{2.23}$$

In the basis $(\mathcal{F}_1, \mathcal{F}_2)$ A and B take the form

$$\hat{A} = \begin{pmatrix} 1 & -3 \\ 1 & -2 \end{pmatrix} \;;\; \hat{B} = \begin{pmatrix} 1 & -1 \\ 0 & 1 \end{pmatrix} \tag{2.24}$$

We note that the Γ_W generators A, B satisfy the relation $A^3 = B^2 = 1$, $(AB)^3 = 1$ which are the defining relations of the tetrahedal group Δ. Indeed $\Gamma(3)$ is a normal subgroup of $\Gamma \equiv SL(2,\mathbf{Z})$, the modular group of the torus ,and $\Gamma/\Gamma(3) \equiv \Gamma_W$.

We have thus verified that the relation $\Gamma/\Gamma_M \simeq \Gamma_W$ actually holds in the case of the torus.

Let us now consider the case of the quintic (2.5) describing a C.Y. 3–fold in CP(4). The PFE for the periods is given by the following fourth–order equation [4,5,15]:

$$\frac{d^4V}{d\psi^4} - \frac{10\psi^4}{1-\psi^5}\frac{d^3V}{d\psi^3} - \frac{25\psi^3}{1-\psi^5}\frac{d^2V}{d\psi^2} - \frac{15\psi^2}{1-\psi^5}\frac{dV}{d\psi} - \frac{\psi}{1-\psi^5}V = 0 \tag{2.25}$$

The four independent solutions of this equation represent the four periods of the uniquely defined (3,0)–form Ω which always exists on a C.Y. space; in our case it can be explicit computed from the defining polynomial (2.5), but we do not need its explicit form in the following. The periods are defined by integrating $\Omega(y,\psi)$ on a basis (γ_A, γ^B) of 3–cycles of W satisfying

$$\gamma^A \cap \gamma_B = -\gamma_B \cap \gamma^A = \delta^A_B \qquad ; \qquad \gamma^A \cap \gamma^B = \gamma_A \cap \gamma_B = 0 \tag{2.26}$$

The basis (2.26) is defined only up $Sp(4;\mathbf{Z})$ transformations, which leave the intersection properties (2.26) invariant. We define

$$V = (X^0, X^1, F_0, F_1) \equiv (X^A, F_A) \tag{2.27}$$

where

$$\int_{\gamma_A} \Omega = X^A(\psi)\,; \; \int_{\gamma^A} \Omega = F_A(\psi) \tag{2.28}$$

Actually on a C.Y. 3–fold the four periods V are not functionally independent, but satisfy the relation (1.10) where F is a homogeneous function of degree two in X^0, X^1. That has its counterpart in a functional relation satisfied by the coefficients of the differential equation (2.25), namely[18,19]

$$W_3 = a_1 - \frac{da_2}{d\psi} - \frac{1}{2}\frac{d^2 a_3}{d\psi^2} = 0 \tag{2.29}$$

where a_i is the coefficients of $\frac{d^i}{dz^i}$ in the differential equation (2.25).

The reason why we have called W_3 the l.h.s. of eq. (2.29) is that it coincides with the W_3 generator of the W_n–algebra associated to a generic n–th order linear differential operator (2.25). Actually $W_3 = 0$ is an invariant statement since W_3 transforms as a covariant tensor of order 3 under ψ–reparametrizations

$$W_3'(\psi') = J^{-3} W_3(\psi) \tag{2.30}$$

where J is the Jacobian of the transformation. The fact the $W_3 = 0$ for the Picard–Fuchs equation of the quintic was first noted by Lerche et al. [15]. Actually this constraint has been proven to hold quite generally as a consequence of the fact that the moduli space geometry is a special Kähler geometry [18,19]. Indeed one finds that in the one modulus case $W_3 = 0$ is equivalent to the statement that the associated 4×4 linear system

$$\left(\frac{d}{d\psi} + A\right) \hat{V} = 0 \tag{2.31}$$

where $\hat{V} \equiv (V, V', V'', V''')^t$, has a Drinfeld–Sokolov connection A which is gauge equivalent to a $Sp(4)$–connection [20].

The gauge group acting on (2.31) is the group generated by the strictly lower triangular matrices which leave invariant the top component of $\hat{V}$ and therefore also the differential equation (2.26).* Note that in the gauge where A is valued in the Lie Algebra of $Sp(4)$ the matrix of solutions of (2.31) is an element of $Sp(4)$

$$\hat{V} = \begin{pmatrix} V_1 & V_2 & V_3 & V_4 \\ \times & \times & \times & \times \\ \times & \times & \times & \times \\ \times & \times & \times & \times \end{pmatrix} \in Sp(4) \tag{2.32}$$

In (2.32) the top row is the set of gauge invariant solutions of (2.25). Of course the gauge (2.32) is not completely fixed since arbitrary $Sp(4)$ gauge transformations on (2.31) keep the matrix $\hat{V}$ in the symplectic gauge. Another consequence of the relation of the PFE's with Special Geometry is that the coefficient a_3 of $\frac{d^3}{dz^3}$ has always the following form [18]:

$$a_3 = -W^{-1}\frac{dW^{-1}}{dz} \tag{2.33}$$

where $W(\psi)$ is the Yukawa coupling of the effective $N = 1$ theory obtained by compactification on the given C.Y. 3–fold. As an example in the quintic case by comparing (2.33) with (2.25) we find

$$W(\psi) = \frac{1}{1-\psi^5} \tag{2.34}$$

The important thing to observe is that many of the properties that we discussed in the one modulus case can be easily extended to the case of n moduli, where the PFE's are partial differential equations. For more details see Ceresole et al.[19].

Let us now discuss the modular group Γ of the moduli space of the quintic polynomial (2.5).

Let us first consider the duality group Γ_W of the defining polynomial equation $W = 0$ given by (2.5). It is obvious that $\mathcal{A} : \psi \to \alpha\psi$, where $\alpha \equiv e^{2\pi i/5}$, is a symmetry of $W = 0$ since it can be undone by a rescaling of the CP (4) homogeneous coordinates: $(y_1, y_2, y_3, y_4, y_5) \to (\alpha^{-1}y_1, y_2, y_3, y_4, y_5)$. Obviously $\mathcal{A}^5 = 1$ and this excludes a priori the possibility that we can represent the modular group as a subgroup of $SL(2;\mathbf{Z})$ acting projectively on a function of ψ, since $SL(2,\mathbf{Z})$ does not possess elements of order 5. Since there are apparently no other ψ–transformations which can be undone by linear transformations of the y_i's, Γ_W is simply the cyclic group $\mathbf{Z}_5$. According to our previous discussion to reconstruct the full modular group we must now compute the monodromy group Γ_M of eq. (2.25).

Γ_M will be represented by 4×4 matrices on the four periods (V_1, V_2, V_3, V_4) solutions of (2.25). The same is true for Γ_W since $\mathcal{A} : \psi \to \alpha\psi$ leaves invariant the differential operator of eq. (2.25) and therefore induces just a linear combinations

* If one allows for rescaling of the solutions of (2.25) then the gauge group can be extended to the Borel subgroup of $GL(4)$.

of the periods. Furthermore by using the gauge (2.32) we may represent Γ_M and Γ_W by $Sp(4;\mathbf{Z})$,–matrices.

Let us first compute Γ_M. We sketch briefly the procedure, for further details see ref. [4].

The differential equation (2.26) is a Fuchsian equation with regular singular points at $\psi = \alpha^k$, $(k = 0, 1, ..., 4)$, $\alpha = e^{2\pi i/5}$, and $\psi = \infty$. As in the case of the torus it is sufficient to study the monodromy matrix T_0 around $\psi = 1$, since around $\psi = \alpha^k$ the corresponding monodromy matrices T_k, $k = 1, 2, 3, 4$ are given by

$$T_0 \to T_k = A^k T_0 A^{-k} \tag{2.35}$$

where A represents $\psi \to \alpha\psi$. The monodromy around $\psi = \infty$ depends on the other generators around $\psi = \infty$ through the relations $T_\infty T_4 T_3 T_2 T_1 T = 1$.

As in the torus case it is convenient to transform eq. (2.25) into a generalized hypergeometric equation through the substitution $z = \psi^{-5}$. We obtain

$$\begin{aligned}\Big\{\frac{d^4}{dz^4} - \frac{2(4z-3)}{z(1-z)}\frac{d^3}{dz^3} - \frac{72z-35}{5z^2(1-z)}\frac{d^2}{dz^2} - \frac{24z-5}{5z^3(1-z)}\frac{d}{dz} - \\ - \frac{24}{625z^3(1-z)}\Big\}V(z) = 0\end{aligned} \tag{2.36}$$

which has singular Fuchsian points at $z = 0, 1, \infty$ with associated Riemann P–symbol

$$P\left\{\begin{matrix} 0 & \infty & 1 & \\ 0 & 1/5 & 0 & \\ 0 & 2/5 & 1 & ;\psi^{-5} \\ 0 & 3/5 & 2 & \\ 0 & 4/5 & 1 & \end{matrix}\right\} \tag{2.37}$$

We notice that in the variable $z = \psi^{-5}$ we have introduced a new singular point around $\psi = 0$ so that the monodromy around $z = \infty$ corresponds exactly to the representation of $\psi \to \alpha\psi$ on the periods. In other words the duality generator A becomes part of the monodromy generators of the new equation (2.36).* A solution of (2.36) around $z = 0$ $(\psi = \infty)$ is given by

$$\omega_0(\psi) = {}_4F_3\left(\frac{1}{5}, \frac{2}{5}, \frac{3}{5}, \frac{4}{5}; 1, 1, 1; \psi^{-5}\right) \tag{2.38}$$

In order to represent A in a simple way we may construct a basis of solutions around $\psi = 0$ as follows. We first continue $\omega_0(\psi)$ around $\psi = 0$ by using a Barnes–type integral representation and we obtain:

$$\omega_0(\psi) = -\frac{1}{5 \cdot 16\pi^4} \sum_{n=0}^{\infty} \frac{\Gamma^5(\frac{n}{5})}{\Gamma(n)} (\alpha^n - 1)^4 (5\psi)^n \qquad (|\psi < 1) \tag{2.39}$$

* The same observation can be done in the case of the torus previously discussed.

Then we recall that $\psi \to \alpha\psi$ leaves the differential operator (2.25) invariant so that

$$\omega_j(\psi) \doteq \omega_0(\alpha^j \psi) \qquad j = 0, 1, 2, 3, 4 \tag{2.40}$$

are also solutions of (2.38). The five functions ω_j are subject to the linear relations $\sum_{n=0}^{4} \omega_j = 0$, as it follows from their explicit expression by the power series (2.39) and the analogous ones derived from (2.40).

If we take $\omega_0, \omega_1, \omega_2, \omega_4$ as a basis of solutions around $\psi = 0$ it follows immediately that $\psi \to \alpha\psi$ is represented on $(\omega_2, \omega_1, \omega_0, \omega_4)^t$ as follows

$$A = \begin{pmatrix} -1 & -1 & -1 & -1 \\ 1 & 0 & 0 & 0 \\ 0 & 1 & 0 & 0 \\ 0 & 0 & 1 & 0 \end{pmatrix} \tag{2.41}$$

Next one examines the monodromy T_0 around $\psi = 1$. For this purpose one observes that since $z = 1$ has the double root $\rho = 1$ for the indicial equation, the continuation of the series $\omega_j(\psi), |\psi| < 1$, to the neighbourhood $|\psi - 1| < 1$ will contain logarithms. Indeed one can write

$$\omega_j(\psi) = \frac{1}{2\pi i} c_j \tilde{\omega}(\psi) \log(\psi - 1) + reg. \tag{2.42}$$

where $\tilde{\omega}$ is a linear combination of regular solutions around $\psi = 1$. It turns out that

$$\tilde{\omega}(\psi) = -\frac{1}{c_1}(\omega_1(\psi) - \omega_0(\psi)) = \frac{1}{c_1}(\psi - 1) + O(\psi - 1)^2) \tag{2.43}$$

It follows

$$\begin{aligned} \frac{d\omega_j}{d\psi} &= \frac{1}{2\pi i} c_j \left(\frac{d\tilde{\omega}}{d\psi} \log(\psi - 1) + \tilde{\omega}(\psi) \frac{1}{\psi - 1} + \ldots\ldots \right) \\ &=_{\psi \to 1} \frac{1}{2\pi i} \frac{c_j}{c_1} \log(\psi - 1) + \cdots \end{aligned} \tag{2.44}$$

We see that in order to compute the monodromy coefficients c_j one has to compute the asymptotic behaviour of $\frac{d\omega_j}{d\psi}$ and look at the coefficient of $\log(\psi - 1)$. Using the series expansion for ω_j derived from (2.39) and (2.40) one finds (see [4] for details)

$$c_j = (1, 1, -4, 6, -4) \tag{2.45}$$

From eqs. (2.42–45), one easily finds that the monodromy matrix around $\psi = 1$ acting on the basis $(\omega_2, \omega_1, \omega_0, \omega_4)^t$ is given by

$$T_0 = \begin{pmatrix} 1 & 4 & -4 & 0 \\ 0 & 0 & 1 & 0 \\ 0 & -1 & 2 & 0 \\ 0 & 4 & -4 & 1 \end{pmatrix} \tag{2.46}$$

The matrices A and T_0 given by eqs. (2.41) and (2.46) are integer valued, but not symplectic, since the ω_j–basis is not a symplectic basis. According to our previous discussion there must exist a matrix m such that

$$\hat{T}_0 = mT_0m^{-1} \quad ; \quad \hat{A} = mAm^{-1} \tag{2.47}$$

are not only integer–valued but also symplectic.

A solution for m has been found in ref. [4] which is unique up to $Sp(4;\mathbf{Z})$ transformations.

We choose the following solution:

$$\hat{A} = \begin{pmatrix} -1 & 3 & 5 & 3 \\ 1 & -4 & 8 & -5 \\ 0 & -1 & 1 & -1 \\ 0 & 1 & 0 & -1 \end{pmatrix} \quad ; \quad \hat{T}_0 = \begin{pmatrix} 1 & 0 & 0 & 0 \\ 0 & 1 & 0 & 0 \\ 0 & 0 & 1 & 0 \\ 0 & -1 & 0 & 1 \end{pmatrix} \tag{2.48}$$

The other monodromy generators $\hat{T}_k$, $\hat{T}_\infty$ around $\psi = \alpha^k$ and $\psi = \infty$ are finally computed from eqs. (2.35) and $T_\infty = (T_4T_3T_2T_1T_0)^{-1}$.

The conclusion is the following: the duality group Γ of the moduli space of the C.Y. 3–fold (2.5) can be given a 4×4 representation on the integer valued and symplectic basis of the periods. Γ is a subgroup of $Sp(4;\mathbf{Z})$ generated by the matrices $\hat{A}$, $\hat{T}_k(k = 0,1,2,3,4)$, where $\hat{A}$ is a representation of the $\mathbf{Z}_5$ subgroup of $Sp(4;\mathbf{Z})$ which leaves $W = 0$ invariant and the $\hat{T}_k$'s generate the monodromy group of the PFE (2.25).

We have already observed that the group Γ must be a subgroup of $SL(2,\mathbf{Z})$ since $SL(2,\mathbf{Z})$ does not contain elements of order 5. We may however represent Γ as a subgroup of $SL(2\mathbf{R})$ since $\mathbf{Z}_5 \in SL(2,\mathbf{R})$. To find the representation we need to a variable $\gamma(\psi)$ such that $\mathcal{A} : \psi \to \alpha\psi$ and the transport around $\psi = 1$ are represented as $PSL(2,\mathbf{R})$ transformations on $\gamma(\psi)$. The determination of $\gamma(\psi)$ and the associated 2×2 representation of $\Gamma \in SL(2,\mathbf{R})$ has been given in ref. [4] by requiring that $\gamma(\psi)$ be a modular parameter on which A and T_0 act as transformations of order five and infinity, respectively, on the upper γ–plane. Standard formulae of the theory of the automorphic functions then determine $\gamma(\psi)$. We give here a different, but closely related derivation, which is based only on the structure of the PFE (2.25).

Let us observe that if we denote by $V(\psi(\gamma))$ the four–dimensional row vector of the periods as a function of γ, then we must have:

$$V\Big(\psi\Big(\frac{a_i\gamma + b_i}{c_i\gamma + d_i}\Big)\Big) = V(\psi(\gamma))\Gamma_i \tag{2.49}$$

where Γ_i is any of the matrices $\hat{A}, \hat{T}_k$ and $S_i \equiv \frac{a_i\gamma+b_i}{c_i\gamma+d_i}$ the corresponding 2–dimensional action or γ. If V is required to be a uniform function of γ, then $\psi = \psi(\gamma)$ must be uniform and such that the entire ψ^5–plane is mapped into a fundamental region of the γ–plane for the group $\Gamma \equiv \{S_i\}$. That amounts to say that γ is a modular variable and ψ is an automorphic function of γ with respect to

Γ. There is a general procedure to construct the uniformizing variable γ directly from the PFE for V. It consists in associating to the main differential equation, eq. (2.36) in our case, a second order differential equation with the same singular points ($z = 0, 1, \infty$ in our case) and with exponents determined as follows.

If all the integrals of the main equation are regular around the given singularity (no two roots of the indicial equation differ by integers) and if all the roots are commensurable quantities multiple of $1/k$ (k integer), then the difference of the roots of the indicial equations of the associated 2nd–order equation is taken equal to $1/k$. In all the other cases the difference of roots is taken equal to zero. The uniformizing variable γ is then given by the ratio of two solutions of the associated 2nd–order equation. Let us see how this works in our case. To adhere to the same notations as in [4] we perform the substitution $z \to \frac{1}{z}$ in the equation (2.36). The P–Riemann symbol (2.37) becomes transformed into

$$P\begin{pmatrix} 0 & \infty & 1 & \\ 1/5 & 0 & 0 & \\ 2/5 & 0 & 1 & ;\psi^5 \\ 3/5 & 0 & 2 & \\ 4/5 & 0 & 1 & \end{pmatrix} \tag{2.50}$$

From (2.50) we see that at $z^{-1} \equiv \psi^5 = 0$ all the roots are multiple of $\frac{1}{k} \equiv \frac{1}{5}$ and do not differ by integers. At $z^{-1} = \infty$ and $z^{-1} = 1$ instead we have at least two coincident roots. Therefore calling λ, μ, ν the differences of the roots of the indicial equation for the associated 2nd–order equation we have

$$\lambda = 1/5 \qquad ; \qquad \mu = \nu = 0 \tag{2.51}$$

Given the exponent we can immediately write down the associated 2nd–order equation, which, having regular singular points at $z = 0, 1, \infty$ is a hypergeometric equation of parameters

$$a = \frac{1}{2}(1 - \lambda - \mu + \nu) = \frac{2}{5};\ b = \frac{1}{2}(1 - \lambda - \mu - \nu) = \frac{2}{5};\ c = 1 - \lambda = \frac{2}{5} \tag{2.52}$$

that is

$$z(1-z)\mathcal{F}'' + \left(\frac{4}{5} - \frac{7}{5}z\right)\mathcal{F}' - \frac{4}{25}\mathcal{F} = 0 \tag{2.53}$$

The uniformizing variable is then given by

$$\gamma = \frac{\mathcal{F}_1}{\mathcal{F}_2} \tag{2.54}$$

where

$$\mathcal{F}_1 \equiv \frac{\Gamma^2(\frac{2}{5})}{\Gamma(\frac{4}{5})} F(2/5, 2/5, 4/5; \psi^5) \tag{2.55}$$

$$\mathcal{F}_2 \equiv \frac{\Gamma^2(\frac{3}{5})}{\Gamma(\frac{6}{5})} F(3/5, 3/5, 6/5; \psi^5) \tag{2.56}$$

are two linearly independent solutions of (2.53). From the theory of the automorphic functions we know that γ maps the ψ^5–plane onto a couple of adjacent triangles inside the circle $|\gamma|^2 = 1$ with internal angles $(0, 0, \pi/5)$; they constitute a fundamental region for the projective action of the modular group Γ, and the inverse function $\psi = \psi(\gamma)$ is automorphic with respect to Γ.

It is now easy to derive the explicit representation of A and T_0 as a subgroup of $SL(2, \mathbf{R})$ on $\mathcal{F}_1, \mathcal{F}_2$: one has simply to perform the study of the monodromy group of the differential equation (2.53) in exactly the same way as we did it in the case of the torus, eq. (2.10). One obtains in this case, with completely analogous calculations:

$$A = \begin{pmatrix} e^{-i\pi/5} & 0 \\ 0 & e^{i\pi/5} \end{pmatrix} \quad ; \quad T_0 = \begin{pmatrix} 1 - itg\frac{2\pi}{5} & itg\frac{2\pi}{5} \\ -itg\frac{2\pi}{5} & 1 + itg\frac{2\pi}{5} \end{pmatrix} \tag{2.57}$$

and

$$T_k = A^k T_0 A^{-k} \quad (k = 1, 2, 3, 4); \quad T_\infty = (T_4 T_3 T_2 T_1 T_0)^{-1} \equiv (AT)^{-5} \tag{2.58}$$

Note that the matrices quoted in the Candèlas et al. [4] are related to those given in (2.57) by the change of basis

$$\begin{pmatrix} Z_1 \\ Z_2 \end{pmatrix} = \begin{pmatrix} i & -i\alpha^2 \\ 1 & -\alpha^2 \end{pmatrix} \begin{pmatrix} \mathcal{F}_1 \\ \mathcal{F}_2 \end{pmatrix} \tag{2.59}$$

which map the interior of the circle $|\gamma|^2 = 1$ into the upper half–plane $Im\gamma > 0$.

Till now we have considered the representation of the modular group Γ on the periods as 4×4 $Sp(4; \mathbf{Z})$–valued matrices, or as $SL(2; \mathbf{R})$–matrices acting projectively on the unformizing variable γ. There is however another important variable, the special variable t, in terms of which we may give an interesting representation of Γ which we now discuss.

To introduce it let us recall that in general, for any number n of moduli, a symplectic transformation on the period vector $V(X^A, F_A) \equiv (X^0, X^a, F_0, F_a)$, $a = 1, ..., n$, induces a reparametrization on the special coordinates $t^a = \frac{X^a}{X^0}$. Indeed from

$$\begin{aligned} \tilde{V} &= VM \\ M &= \begin{pmatrix} A & C \\ B & D \end{pmatrix} \quad \in Sp(2n+2, \mathbf{R}) \end{aligned} \tag{2.60}$$

where A, B, C, D are $(n+1) \times (n+1)$ matrices obeying

$$\begin{aligned} A^t B &= B^t A \\ C^t D &= D^t C \\ A^t D + B^t C &= \mathbf{1} \end{aligned} \tag{2.61}$$

we readily obtain [19]

$$\tilde{t}^a = \frac{\tilde{X}^a}{\tilde{X}^0} = \frac{A^a_B X^B + B^a F_B}{A^0_B X^B + B^{0B} F_B} \tag{2.62}$$

Recalling that $F_A = \frac{\partial F}{\partial X^A}$ where $F(X)$ is a homogeneous function of degree two we have

$$\begin{aligned} F(X^A) &= (X^0)^2 \mathcal{F}(t^a) \\ F_0(X^A) &\equiv \frac{\partial F}{\partial X^a} = X^0[2\mathcal{F}(t^a) - t^a \partial_a \mathcal{F}(t^a)] \\ F_a(X^A) &\equiv \frac{\partial F}{\partial X^0} = X^0 \partial_a \mathcal{F}(t^a) \end{aligned} \tag{2.63}$$

and substituting in (2.62) we find

$$\tilde{t}^a = \frac{A^a_b t^b + A^a_0 + B^{ab}\mathcal{F}_b + B^{a0}(2\mathcal{F} - t^b\mathcal{F}_b)}{A^0_b t^b + A^0_0 + B^{ab}\mathcal{F}_b + B^{00}(2\mathcal{F} - t^b\mathcal{F}_b)} \tag{2.64}$$

where $\mathcal{F}_a \equiv \partial_a \mathcal{F}$. If we now restrict the generic $Sp(2n+2, \mathbf{R})$ matrix M to belong to $\Gamma \subset Sp(2n+2, \mathbf{Z})$ we obtain the representation of Γ on the variables t^a. In particular the subgroup of $Sp(2n+2; \mathbf{R})$ consisting of matrices of the form

$$\begin{pmatrix} A & C \\ 0 & D \end{pmatrix} \qquad \begin{aligned} D &= (A^t)^{-1} \\ C &= AC^t(A^t)^{-1} \end{aligned} \tag{2.65}$$

act on the $t^{a'}$s as a group of linear fractional transformations. It is compelling to assume that the subgroup $T \subset PSL(2, \mathbf{Z})$ of discrete integer translations

$$t^a \to t^a + n^a \quad ; \quad n^a \in \mathbf{Z}^n \tag{2.66}$$

is always contained in Γ. Indeed the symmetry has its stringy origin in the Wess–Zumino term for the B_{ij} axion 2–form in the σ–model.

Indeed the σ–model term

$$\int_{W.S.} d^2\sigma \partial_\alpha Y^i \partial_\beta \bar{Y}^{\bar{j}} B_{i\bar{j}} \epsilon^{\alpha\beta} = \int_T B_{ij} dY^i \wedge d\bar{Y}^{\bar{j}} \tag{2.67}$$

where T is the image of the world-sheet (W.S.) in the C.Y. 3–fold, is topological in nature and the B_{ij} are analogous to the θ–parameters of Q.C.D. If the complexified Kähler (1,1)–form is parametrized as

$$g_{i\bar{j}} + iB_{i\bar{j}} = -i \sum_{a=1}^{n} t^a L^a_{ij} \tag{2.68}$$

where the $L^a_{i\bar{j}}$ are a basis of the (1,1)–cohomology, a shift $t^a \to t^a + n^a, n^a \in \mathbf{Z}$ induces a topologically non trivial mapping from the world sheet to the C.Y. and corresponds to an instanton correction to the σ–model perturbative result. Such integral shift is an invariance of the quantum action.

In the one modulus case the existence of the translation symmetry $t \to t+1$ has been verified in the case of the quintic by Candelas et al.[4] and proven for a large class of 3–fold by Morrison [5]. Let us verify it for the quintic. From the explicit form of the two generator A, T_0 given by eq. (2.48) we find that on the (X^0, X^1, F_0, F_1)–basis

$$(AT_0)^{-1} = \begin{pmatrix} 1 & 1 & -5 & -8 \\ 0 & 1 & -3 & -5 \\ 0 & 0 & 1 & 0 \\ 0 & 0 & -1 & 1 \end{pmatrix} \tag{2.69}$$

so that

$$\tilde{t} \equiv \frac{\tilde{X}^1}{\tilde{X}^0} = \frac{X^1}{X^0} + 1 \equiv t+1 \tag{2.70}$$

The t–transformations realized by A and T_0 are instead non linear:

$$A : \tilde{t} = \frac{X^1 - (X^0 + F^0)}{X^0 + F^0} = \frac{t - 1 - (2\mathcal{F} - t\mathcal{F}')}{2\mathcal{F} - t\mathcal{F}'} \tag{2.71}$$

$$T_0 : \tilde{t} = \frac{X^1}{X^0 - F_0} = \frac{t}{1 - 2\mathcal{F} + t\mathcal{F}'} \tag{2.72}$$

We note that while $(AT_0)^{-1}$ corresponds to a circuit around $z = 0$, the monodromy around $\psi = \infty$ is represented by $(AT_0)^{-5}$ so that

$$(AT_0)^{-5} : \qquad \tilde{t} \to t+5 \tag{2.73}$$

The transformations (2.73) and (2.71) or (2.72) generate the whole modular group on the t–modulus.

Coming back to the case of n moduli t^a, we may also write down the transformation law of the prepotential $F(X)$ in the general case. Indeed recalling the homogeneity relation $2F(X) = X^A F_A$ one finds for a generic $Sp(4)$–transformations (2.60) [21]:

$$\begin{aligned} 2\tilde{F}(\tilde{X}) &= (X^A, F_A) \begin{pmatrix} AC^t & AD^t \\ BC^t & BD^t \end{pmatrix} \begin{pmatrix} X^A \\ F_A \end{pmatrix} \\ &= 2F(X) + 2F_A (BC^t)^A_B X^B + X^A (AC^t)_{AB} X^B + F_A (BD^t)^{AB} F_B \end{aligned} \tag{2.74}$$

where we have used the conditions of symplecticity of the transposed matrix M^t in order to reconstruct $F(X)$ on the r.h.s. of (2.74).

If $M \subset \Gamma$, then $\tilde{F} = F$ since a modular transformation is a discrete isometry. In particular for a translation (2.74) gives

$$F(X^B A^A_B) = F(X^A) + X^A (AC^t)_{AB} X^B \tag{2.75}$$

since $B = 0$ and $A = \begin{pmatrix} 1 & n^a \\ 0 & \delta^a_b \end{pmatrix}$. In the t^a–variables eq. (2.75) becomes

$$\mathcal{F}(t^a + n^a) = \mathcal{F}(t^a) + (AC^t)_{ab} t^a t^b + 2(AC^t)_{0b} t^b + (AC^t)_{00} \tag{2.76}$$

Thus F (or $\mathcal{F}$) is periodic in the X^A (or t^a) up to quadratic additions. In particular the Yukawa coupling $W_{abc} = \frac{\partial^3 \mathcal{F}}{\partial t^a \partial t^b \partial t^c}$ is periodic, $W_{abc}(t^a + n^a) = W_{abc}(t^a)$ and can be expanded in a multiple Fourier series:

$$W_{abc}(t^a) = \sum_{n_a, n_b, a_c} d_{abc}(n_a, n_b, n_c) e^{2\pi i \vec{n} \cdot \vec{t}} \tag{2.77}$$

In the limit $t^a \to i\infty, \forall a$, we get a constant W_{abc} which is the perturbative σ–model result (note that $t^a \to i\infty$ means large radius limit of the C.Y.). Thus the vector $\vec{n}$ identifies the instanton corrections to the perturbative Yukawa-coupling value $W_{abc} = d_{abc}(0,0,0)$.

In the quintic case the numbers $d(n)$ have been identified as the number of rational curves of degree n existing on the quintic [4].

The quadratic addition (2.76) can also be evaluated explicitly for the quintic, using the translation matrix (2.69). One finds

$$\mathcal{F}(t+1) = \mathcal{F}(t) - (5t^2 + 16t + 3). \tag{2.78}$$

3. Conclusions

In this lecture we have described at some length the mathematical origin of target space duality symmetry, its relation to $N = 2$ Superconformal field theories, algebraic geometry and effective actions described by Supergravity Lagrangians.

The important discovery of mirror symmetry in the case of Calabi–Yau string compactifications allows one to compute instanton non–perturbative σ–model corrections to the effective action purely in terms of algebraic geometrical methods supplemented with the notion of special geometry. When applied to space–time fermions, target space duality, which is a discrete isometry of the moduli space, is seen as a compensating (field dependent) holonomy transformation. Since fermions live in a complex representation of the holonomy group $H \in U(N)$ (N is the dimension of the Kähler manifold), target space duality transformations induce σ–model types of anomalies and in particular mixed $U(1)$ gauge anomalies [14]. In supersymmetric gauge theories derived from strings the anomaly cancellation mechanism which takes place originated from a 4D Green–Schwarz mechanism for the 4D antisymmetric tensor $b_{\mu\nu}$ as well as by the existence of local Wess–Zumino terms which involve automorphic functions of the target space modular group. By supersymmetry they induce a moduli dependence on the gauge coupling constant, Δ_a, with important effects for the running toward the string unification scale:

$$\frac{1}{g_a^2(\mu)} = k_a \frac{1}{g_{string}^2} + \frac{b_a}{16\pi^2} \log \frac{M_{string}^2}{\mu^2} + \Delta_a \tag{3.1}$$

where k_a are the levels of the Kac–Moody algebra for the gauge group factors G_a, $\frac{1}{g_{string}^2}$ is the v.e.v. of the dilaton field, M_{string}^2 is the string mass scale, $M_{string}^2 = a\alpha'^{-1}$ (a numerical constant), and b_a is the 1–loop β–function for the gauge group factor G_a

$$b_a = -3T(G_a) + \sum_R T(R) \tag{3.2}$$

($T(G_a)$ is the Casimir of the adjoint and $T(R)$ are the quadratic Casimirs of the representations R of chiral multiplets).

For a C.Y. manifold with $G = E_6 \oplus E_8$ the M_{GUT}–scale occur then at the value

$$M_X^2 = M_{string}^2 \left(\frac{Y_8}{Y_6}\right)^{\frac{1}{b_8 - b_6}} \qquad (\Delta_a = \frac{1}{16\pi^2} \log Y_a) \tag{3.3}$$

where Y_a are automorphic functions of the target space duality group Γ. The non–harmonic part of Δ_a can be entirely computed from quantities in special geometry constructed out of the holomorphic period vector (X^A, F_A), namely the norm $X^A \bar{F}_A - \bar{X}^A F_A$ and $\det\left(\frac{\partial^2 F}{\partial X^A \partial X^B}\right)$.

Threshold effects at the string scale, due to moduli dependence, may play a crucial role for the discussion of supersymmetry breaking when non–perturbative stringy effects, such as gaugino condensation, are suitably incorporated.

Acknowledgements

We would like to thank P.Candelas,A. Ceresole,P.Fre',W.Lerche,J.Louis and P.Soriani for enlightening discussion.

References

1. P. Candelas, G. Horovitz, A. Strominger and A. Witten, *Nucl. Phys.* B268 (1985) 46. See also M. Green, J. Schwarz and E. Witten, *Superstring Theory* (Cambridge, Univ. Press, 1987).
2. E. Martinez, *Phys. Lett.* B217 (1989) 431;
   C. Vafa and N.P. Warner, *Phys. Lett.* B218 (1989) 51;
   W. Lerche, C. Vafa and N. Warner, *Nucl. Phys.* B324 (1989) 427;
   D. Gepner, *Phys. Lett.* B222 (1989) 207;
   P. Howe and P. West, *Phys. Lett.* B223 (1989) 377;

S. Cecotti, L. Girardello and A. Pasquinucci, *Nucl. Phys.* B328 (1989) 701, *Int. J. Mod. Phys.* A6 (1991) 2427;
C. Vafa, *Int. J. Mod. Phys.* A6 (1991) 2829;
K. Intrilligator and C. Vafa, *Nucl. Phys.* B339 (1990) 95;
C. Vafa, *Mod. Phys. Lett.* A4 1615 and *Mod. Phys. Lett.* A4 (1989) 1169;
S. Cecotti, *Int. J. Mod. Phys.* A6 (1991) 1749 and *Nucl. Phys.* B355 (1991) 755;
A. Giveon and D.J. Smit, *Mod. Phys. Lett.* A6 24 (1991) 2211.

3. E. Witten, *Commun. Math. Phys.* 117 (1988) 353; 118 (1988) 411 and *Nucl. Phys.* B340 (1990), 281;
T. Eguchi and S.K. Yang, *Mod. Phys. Lett.* A5 (1900) 1693;
C. Vafa, *Mod. Phys. Lett.* A6 (1991) 337;
K. Li, *Nucl.Phys.* B354 (1991) 711;
B. Blok and A. Varchenko, *Int.Jou. Mod. Phys.* A7 (1992) 1467;
R. Dijkgraaf, E. Verlinde and H. Verlinde, *Nucl. Phys.* B348 (1991) 435 and B352 (1991) 59;
A. Giveon and D.J. Smit, *Progr. Theor. Phys. Suppl.* 102 (1990) 351; *Mod. Phys. Lett.* A6 (1991) 2211; *Int.Jou.Mod. Phys.* A7 (1992) 973;
S. Cecotti and C. Vafa, *Nucl. Phys.* B367 (1991) 359.

4. P. Candelas, X.C. de la Ossa, P.S. Green and L. Parkes, *Phys. Lett.* 258B (1991) 118;
P. Candelas, X.C. de la Ossa, P.S. Green and L. Parkes, *Nucl. Phys.* B359 (1991) 21.

5. D. Morrison, *Essays on Mirror manifolds*,S.T.Yau Editor, Intenational Press (1992);
A. Font, *Periods and Duality Symmetries in Calabi–Yau Compactifications* (preprint UCVFC/DF-1-92);
A. Klemm and S. Theisen, *Considerations of one modulus Calabi–Yau compactifications: Picard–Fuchs equations, Kähler potentials and mirror maps* (Karlsruhe preprint KA-THEP-03-92).

6. P. Candelas and X.C. de la Ossa, *Nucl. Phys.* B355 (1991) 455.

7. S. Cecotti, S. Ferrara and L. Girardello, *Int. Mod. J. Phys.* A4 (1989) 2475;
S. Cecotti, S. Ferrara and L. Girardello, *Phys. Lett.* B213 (1988) 443;
L.J. Dixon, V.S. Kaplunovsky and J. Louis, *Nucl. Phys.* B329 (1990) 27.

8. P. AAspinwall and D. Morrison, *Topological field theory and rational curves*, preprint DUK-M-91-12;
E. Witten, *Essays on Mirror manifolds* , S.T.Yau Editor,International Press (1992)
B.R. Greene and M.R. Plesser, *Nucl. Phys.* B338 (1990) 15;
P. Candelas, M. Linker and R. Schimmrigk, *Nucl. Phys.* B341 (1990) 383;
P. Aspinwall, C.A. Lütken and G.G. Ross, *Phys. Lett.* B241 (1990) 373;
P. Aspinwall, C.A. Lütken, *Nucl. Phys.* B353 (1991) 427 and B355 (1991) 482; See also contributions in "Essay on Mirror Manifolds" edited by S.T. Yau, International Press (1992).

9. B. de Witt and A. Van Proeyen, *Nucl. Phys.* B245 (1984) 89
B. de Witt, P. Lauwers and A. Van Proeyen, *Nucl. Phys.* B255 (1985) 569;
E. Cremmer, C. Kounnas, A. Van Proeyen, J.P. Deredinger, S. Ferrara, B. de Witt and L. Girardello, *Nucl. Phys.* B250 (1985) 385.

10. A. Strominger, *Commun. Math. Phys.* 133 (1990) 163.

11. L. Castellani, R. D'Auria and S. Ferrara, *Phys. Lett.* B241 (1990) 57;
L. Castellani, R. D'Auria and S. Ferrara, *Class. Quant. Grav.* 1 (1990) 317;
R. D'Auria, S. Ferrara and P. Frè, *Nucl. Phys.* B359 (1991) 705.

12. K. Kikkawa and M. Yamasaki, *Phys. Lett.* 149B (1984) 357;
N. Sakai and L. Senda, *Progr. Theor. Phys.* 75 (1986) 692;
V.P. Nair, A. Shapere, A. Strominger and F. Wilczek, *Nucl. Phys.* B287 (1987) 402;
A. Giveon, E. Rabinovici and G. Veneziano, *Nucl. Phys.* B322 (1989) 167;
A. Shapere and F. Wilczek, *Nucl. Phys.* B320 (1989) 167;
M. Dine, P. Huet and N. Seiberg, *Nucl. Phys.* B322 (1989) 301;
J. Molera and B. Ovrut, *Phys. Rev.* D40 (1989) 1150; J. Lauer, J. Maas and H.P. Nilles, *Phys. Lett.* B226 (1989) 251 and *Nucl. Phys.* B351 (1991) 353;
W. Lerche, D. Lüst and N.P. Warner, *Phys. Lett.* B231 (1989) 418;
M. Duff, *Nucl. Phys.* B335 (1990) 610;
A. Giveon and M. Porrati, *Phys. Lett.* B246 (1990) 54 and *Nucl. Phys.* B355 (1991) 422;
Giveon, N. Malkin and E. Rabinovici, *Phys. Lett.* B238 (1990) 57;
J. Erler, D. Jungnickel and H.P. Nilles, *MPI-Ph/91-90*;
S. Ferrara, D. Lüst, A. Shapere and S. Theisen, *Phys. Lett.* B233 (1989) 147;
J. Schwarz, *Caltech preprints CALT-6S-1581* (1990) *CALT-68-1728* (1991) and *CALT-68-1740* (1991);
J. Erler, D. Jungnickel and H.P. Nilles, *MPI-Ph/91-81.*

13. E. Witten, *Phys. Lett.* B155 (1985) 151.

14. L.J. Dixon, V.S. Kaplunovsky and J. Louis, *Nucl. Phys.* B355 (1991) 649;
J.P. Deredinger, S. Ferrara, C. Kounnas and F. Zwirner, *Nucl. Phys.* B372 (1992) 145;
J. Louis, *PASCOS*, (1991), Proceedings, P. Nath Editor, World Scientific 1991;
G.L. Cardoso and B. Ovrut, *Nucl. Phys.* B369 (1992) 351.

15. A. Cadavid and S. Ferrara, *Phys. Lett.* B267 (1991) 193;
W. Lerche, D. Smit and N. Warner, *Nucl. Phys.* B372 (1992) 87;

16. A. Erdélyi, F. Oberhettinger, W. Magnus and F.G. Tricomi, *Higher Trascendental Functions* Mac Graw–Hill, New York, 1953.

17. P. Frè and P. Soriani, private communication, see also P. Soriani, SISSA Ph.D–thesis.

18. S. Ferrara and J. Louis, *Phys. Lett.* B278 (1992) 240.

19. A. Ceresole, R. D'Auria, S. Ferrara, W. Lerche and J. Louis, *Picard–Fuchs equations and special geometry*, *Int.Jou.Mod.Phys.* A8 (1993) 79;

20. P. Di Francesco, C. Itzynkson and J.B. Zuber, *Commun. Math. Phys.* 140 (1991) 543.

21. M. Villasante, *Phys. Rev.* D45 (1992) 1831.

FRACTIONAL SUPERSTRINGS WITH CRITICAL SPACETIME DIMENSIONS FOUR AND SIX: A STATUS REPORT

Keith R. Dienes*

Department of Physics, McGill University
3600 University St., Montréal, Québec H3A-2T8 Canada

ABSTRACT

We provide a short non-technical survey of the fundamental issues involved in the recently-proposed fractional superstring theories. After introducing the basic ideas which underlie these new string theories, we review their early successes and outline some related outstanding questions.

1. Fractional Superstrings: The Basic Ideas

Fractional superstrings are non-trivial generalizations of ordinary superstrings and heterotic strings, and have critical spacetime dimensions which are less than ten. In this paper, I aim to provide a self-contained introduction to these new string theories, giving an overview of the entire field as it currently stands rather than focusing exclusively on any particular aspect. The first half of this paper, therefore, will be entirely non-technical, and will summarize the basic idea behind the fractional superstrings as well as their early successes and some related outstanding issues. The second half will then provide more details concerning the construction of the fractional-superstring worldsheet theories, their partition functions, and their various possible spacetime interpretations.

1.1. *Motivation: Why new string theories?*

There have been, historically, a variety of different classes of string theories which have been constructed and analyzed. The ***bosonic string*** is characterized by a conformal symmetry on the worldsheet, and contains worldsheet fields which are the coordinate bosons X^μ. The critical dimension of this theory, however, is 26, and the theory suffers from spacetime tachyons and the absence of spacetime fermions. The ***superstring*** augments the conformal symmetry on the worldsheet by adding a worldsheet supersymmetry (resulting in a superconformal symmetry), and has worldsheet fields which are the bosons X^μ and their fermionic superpartners ψ^μ. While this removes the spacetime tachyons, gives rise to spacetime fermions, and causes the critical dimension to fall to 10, it is not clear how to incorporate gauge groups into this theory in a natural manner or obtain a reasonable spacetime phenomenology. The (closed) *heterotic string*, on the

*E-mail address: dien@hep.physics.mcgill.ca

other hand, contains a superconformal worldsheet symmetry for the right-movers and a conformal symmetry for the left-movers. Although the critical dimension of this theory remains 10, gauge groups appear in a natural manner and indeed realistic low-energy phenomenologies can emerge. Thus the heterotic string has proven to be a fruitful starting point in string-theoretic model-building. Even though the critical spacetime dimension of the heterotic string is 10, this can be overcome by either *compactifying* the six undesired dimensions (*e.g.*, on a variety of different orbifolds or Calabi-Yau manifolds) or *replacing* them with alternative but "equivalent" conformal field theories (*e.g.*, those involving a sufficient number of free bosons or fermions to achieve conformal anomaly cancellation).

The problem with this dimensional-reduction "solution," from a phenomenological point of view, is that there are *many* different ways of passing from the critical dimension $D_c = 10$ to $D = 4$, and each of these thus gives rise to a different model-construction procedure yielding an entirely new class of models. This great freedom in constructing models would not in itself be a problem, of course, were it not for the fact that each resulting model has a unique spacetime phenomenology, with its own set of particles, gauge groups, and interactions. Thus, the physical properties of the resulting theory depend crucially on the particular model chosen. Even this would not be a problem, however, were it not for the observation that the original choice is completely arbitrary: string theory, as presently formulated, leaves all of these choices equally likely, classically degenerate in the space of possible string field theory vacua. While this degeneracy might hopefully be lifted by string-field-theoretic effects, such a string field theory has yet to be formulated.

Our goal, then, is clear. Since the source of this problem was the great freedom in passing from the critical dimension to the actual spacetime dimension, we would like to construct new types of string theories which have *critical* spacetime dimension $D_c = 4$. We would then have no need for compactification, have fewer extra degrees of freedom to arbitrarily choose or adjust, and might possibly be able to construct a truly unique string theory directly in four dimensions. There are also, however, less phenomenological reasons for attempting to construct such theories. String theories are thought to embody a symmetry principle that is a generalization of the usual field-theory notion of gauge invariance to include higher-spin fields and non-linear interactions. Our understanding of these spacetime string symmetries remains sketchy, however, since their connection to structures and symmetries on the worldsheet is as yet somewhat mysterious. Another motivation, then, for searching for new such string theories is to possibly shed light on this connection between worldsheet and spacetime physics.

1.2. The Fractional Superstring

How might we proceed to construct theories with smaller critical dimensions? We already have some indications: we saw that if there is *no* worldsheet supersymmetry (as in the bosonic string), the critical dimension is $D_c = 26$, whereas if we add a single worldsheet supersymmetry (as in the super- or heterotic string), the critical dimension falls to $D_c = 10$. A natural guess, then, is to increase the symmetry on the worldsheet. However, one finds that putting an *additional* supersymmetry on the worldsheet yields a theory with $D_c = 2$,* whereas the addition of extra worldsheet supersymmetries turns out not to further change the critical dimension. Thus, it is clear that we need something intermediate – a *fractional* supersymmetry. It turns out that such fractional supersymmetries can indeed be formulated on the two-dimensional worldsheet, and the strings based on such worldsheet symmetries are called "fractional superstrings".

We can thus add a new type of string to the previous list: the fractional superstring. The worldsheet symmetry of this theory is therefore the usual conformal symmetry plus a fractional supersymmetry, together forming what is called a fractional superconformal symmetry. Correspondingly, the worldsheet fields in this theory consist not only of the bosons X^μ, but also their *fractional* superpartners, the *para*fermions ϵ^μ. These parafermionic fields have fractional spin on the worldsheet, but we stress that in two dimensions such fields are allowed; in particular, the presence of such fields on the worldsheet does not necessarily imply the appearance of fractional-spin particles in spacetime. Constructing such theories, then, we find that their critical spacetime dimensions are indeed less than ten, and the precise value depends on the fraction chosen for the supersymmetry. In particular, it turns out that we can parametrize this fraction by an integer $K \geq 2$ (the degenerate case $K = 1$ corresponds to the absence of any supersymmetry). The special case $K = 2$ corresponds to ordinary supersymmetry, which, we recall, transforms a worldsheet boson X of spin zero into a worldsheet fermion ψ of spin $\Delta(\psi) = 1/2$ via the action of a supercurrent of the form $J \sim \psi\partial X$ with spin $1 + \Delta(\psi) = 3/2$. The cases with $K > 2$ then correspond to our fractional supersymmetries: for each such K, these symmetries transform a worldsheet boson X of spin zero into a so-called "$\mathbf{Z}_K$ parafermion" ϵ of spin $\Delta(\epsilon) = 2/(K+2)$ via the action of a worldsheet fractional supercurrent of the form $J \sim \epsilon\partial X + ...$ with spin $1 + \Delta(\epsilon)$. Given this parametrization, then, one finds the corresponding critical spacetime dimensions:

$$D_c = \begin{cases} 26 & \text{for } K = 1 \\ 2 + 16/K & \text{for } K \geq 2. \end{cases} \tag{1}$$

We therefore have the following cases with integer values of K and D_c:

$$\begin{array}{rll} K = 1: & D_c = 26 & \\ K = 2: & D_c = 10, & \text{spin}(\psi) = 1/2 \\ K = 4: & D_c = \ \ 6, & \text{spin}(\epsilon) = 1/3 \\ K = 8: & D_c = \ \ 4, & \text{spin}(\epsilon) = 1/5 \\ K = 16: & D_c = \ \ 3, & \text{spin}(\epsilon) = 1/9\ . \end{array} \tag{2}$$

Given these fractional supersymmetries, it is possible to generate many new classes of fractional superstring theories with various critical spacetime dimensions. Since we are considering only closed strings, the left- and right-moving symmetries need not be the same, and thus we can label our (K_L, K_R) theories according to the fractions of their left- and right-moving supersymmetries. In this new language, we see that the ordinary bosonic string is simply the $(1,1)$ special case of the fractional superstring, the ordinary superstring (*e.g.*, the Type II string) is simply the $(2,2)$ special

*These are actually two *complex* dimensions, implying a 4D spacetime with $(2,2)$ signature.

case, and the ordinary heterotic string can be identified simply as the $(1,2)$ string. The more general (K_L, K_R) strings, however, will have critical spacetime dimensions less than ten. In particular, the (K,K) strings have the same fractional supersymmetry for the left- and right-moving sectors, and are the natural generalizations of the closed superstring. Similarly, the $(1,K)$ strings are the natural generalizations of the heterotic string.

1.3. Overview of Results and Open Questions

Since the original fractional-superstring proposal,[1] there has been much progress in attempting to demonstrate the self-consistency of these new string theories and in understanding much of the new physics they contain.[2–7] We shall defer a detailed examination of some of these issues to Sect. 2, and instead now proceed to give a quick overview of the results achieved as well as a brief discussion of the many issues which remain to be completely resolved.

There have been a number of early successes which indicate that the fractional superstring is likely to be a consistent theory both on the worldsheet and in spacetime. First, of course, this approach provides a natural generalization of the more traditional string theories, establishing (via the use of two-dimensional fractional supersymmetries and parafermions) a uniform framework for examining a variety of different string theories.[1,2] Secondly, as mentioned, these strings have critical spacetime dimensions which are less than ten, and in particular the phenomenologically interesting dimensions $D_c = 4, 6$ are allowed values. Third, the low-lying spacetime spectra of these theories have been examined in some detail,[4] and indeed for each relevant value of K one finds a tachyon-free spectrum containing a graviton, gauge bosons, and massless spacetime fermions; indeed, these latter spacetime spinorial states satisfy the Dirac equation by virtue of a parafermion zero-mode Clifford algebra, in complete analogy to what occurs in the ordinary superstring. Fourth, it appears that these new string theories have a *spacetime* supersymmetry:[3] this symmetry appears naturally in these theories, evidenced by equal numbers of spacetime bosonic and fermionic states at all mass levels, and is consistent with the appearance of massless spin-3/2 particles (gravitinos). Fifth, it has been possible to construct modular-invariant partition functions for these theories which are consistent with each of the above properties,[1,2] and these in turn have been used profitably to obtain much insight into some of the more difficult and unique aspects of these theories.[3,5,7] Sixth, promising first steps towards fractional-superstring *model-building* have been taken,[2] and as expected one finds that the set of $K > 2$ models is indeed orders of magnitude smaller than the analogous set of $K = 2$ models. These successes are all, therefore, signs that the fractional superstring theories will ultimately prove both consistent and fruitful. In fact, work in this area has even led to results which are independent of fractional superstrings, such as whole classes of new identities relating the characters of $\mathbf{Z}_K$ parafermions for different values of K.[3,5,6]

There remain, however, many different aspects of the fractional superstrings which still need to be investigated. While some of these will be necessary for proving the ultimate consistency of these theories, others will be important for uncovering the intrinsically new physics these theories contain. These different issues manifest themselves in the worldsheet physics, in the spacetime physics, and in the mapping between the two.

On the worldsheet, for example, it will be necessary to have a more complete understanding of the two-dimensional fractional superconformal symmetries and algebras which are the fundamental worldsheet building blocks of the fractional superstrings. Since these algebras involve fields which have fractional spin, the field-operator product expansions which comprise these algebras contain *cuts* rather than simple poles, and thus these algebras are necessarily non-local (as well as non-linear). Most of the difficulty in dealing with these algebras stems from this simple fact. For example, one important issue which has yet to be resolved is the fractional superghost system – one will need to understand the ghosts arising from fractional superconformal gauge-fixing in order to "properly" derive the critical dimension D_c from conformal anomaly cancellation and thereby demonstrate the consistency of this aspect of the fractional superstring. This non-linearity of the fractionally-supersymmetric boson plus parafermion (X, ϵ) theory may also be the key to many of the intrinsically *new* features of these strings. For example, it is tacitly assumed in the fractional superstring construction that one can start with a single such (X, ϵ) theory corresponding to one given spacetime dimension, and then simply tensor together D_c copies of this theory to produce the full D_c-dimensional worldsheet theory; certainly this would be the straightforward generalization of the bosonic and superstring constructions. However, each individual (X, ϵ) theory is non-linear and non-local, while a tensor product is necessarily a linear, local combination (*i.e.*, parafermions corresponding to different dimensions in the tensor product do not "feel" each other in a non-local way, while those corresponding to the same dimension do). Whether such a straightforward tensor-product construction can produce a fully self-consistent Lorentz-invariant theory remains to be seen.[4]

A similar situation exists for the spacetime physics of the fractional superstring. Here too there are a variety of issues which still must be examined in order to prove the physical consistency of the fractional superstring (*e.g.*, one must develop formalisms for calculating scattering amplitudes and demonstrating unitarity, as well as, eventually, for *bona-fide* model-building and examining the resulting spacetime phenomenologies[2,6]), but once again there are intrinsically new features which appear and require closer analysis. One of these issues is spacetime Lorentz invariance: while it appears that certain "sectors" of the theory are fully consistent with Lorentz invariance, other new Planck-scale sectors may be invariant only under *subgroups* of the Lorentz group.[2,3] This issue is undoubtedly the spacetime reflection of the worldsheet tensor-product issue discussed above, and whether the appearance of such new sectors indicates a Planck-scale violation of Lorentz invariance (which in itself is not ruled out by experiment) or a highly non-trivial self-induced spacetime compactification remains to be seen. These additional sectors also appear to contain Planck-scale particles which seem to violate the spin-statistics connection.[2,3] Properly interpreting these states and their roles in spacetime physics will be an important endeavor.

Finally, there are various outstanding issues concerning the *mapping* between the worldsheet theory and the spacetime physics. In order to map between these different spaces, one starts with a given worldsheet conformal-field-theory vacuum and examines the entire Fock space of excitations on that vacuum; those states in the Fock space satisfying certain physical-state conditions are then interpreted as particles in spacetime. For the fractional superstrings, it is still an open question as to what the complete set of physical-state conditions are: although a subset of these conditions are known,[4] there are many indications that further conditions having no analogues in the ordinary string theories remain to be found. It will also then be necessary to develop a no-ghost theorem for the fractional-superstring Fock space – *i.e.*, to demonstrate that the all of the states in the resulting space have non-negative norm. Additionally, there are various "projections" which occur in the fractional-superstring Fock space and which need to be more adequately understood. While some of these projections are analogous to the ordinary GSO projection which removes tachyonic states from the spacetime spectrum of the superstring, other so-called "internal projections" have no analogue in the superstring and turn out to be responsible for removing exponentially large numbers of states from the theory (thereby changing its Hagedorn temperature and effective central charge).[3–5]

Thus, while the fractional superstring has seen many successes, it also provides an arena in which many qualitatively new issues can and must be explored. Work in all of these areas is actively being pursued.

2. Fractional Superstrings: Some Details

In order to be more specific concerning some of these outstanding issues, we must first examine these fractional superstring theories in more detail. In this section, therefore, we shall provide a short tutorial in the fractional-superstring fundamentals, examining the construction of the fractional-superstring worldsheet theories, their corresponding partition functions, and the resulting spacetime physics.

2.1. Constructing the Worldsheet Theory

Before discussing the worldsheet theory of the fractional superstring, it proves instructive to re-examine that of the ordinary superstring (which is the $K = 2$ special case). The worldsheet fields of the superstring are of course the free spin-0 bosons X^μ and their superpartners, the free Majorana spin-1/2 fermions ψ^μ. Here $\mu = 0, 1, ..., D_c - 1$, and we therefore have D_c copies of the fundamental supersymmetric (X, ψ) theory, one copy corresponding to each spacetime dimension. In order to fully specify this theory, however, we must choose for our fields any of four different boundary conditions around the two non-contractible cycles (spacelike and timelike, respectively) of the torus. The bosons X^μ must always have periodic/periodic boundary conditions around these two cycles (in keeping with their interpretation as spacetime coordinates), but the fermions ψ^μ, which always appear in physical quantities through $\psi\psi$ bilinears, may have periodic or antiperiodic boundary conditions. Choosing (AP) or (AA) boundary conditions for the fermions puts us in the Neveu-Schwarz sector of the theory, and all excitations in this sector correspond to spacetime bosons. Choosing (PP) or (PA) boundary conditions, however, puts us in the Ramond sector of the theory, and all excitations in this sector correspond to spacetime fermions.

It turns out that this free Majorana fermion theory which appears on the superstring worldsheet can be rewritten in a variety of ways which are useful for later generalization. One way is to think of this theory as the Ising model with central charge $c = 1/2$. The fields in the Ising model are $\{\mathbf{1}, \psi, \sigma, \sigma^\dagger\}$, where $\mathbf{1}$ is the identity with highest weight $h = 0$, ψ is the fermion with $h = 1/2$, and σ and $\sigma^\dagger$ are the spin field and its conjugate with $h = 1/16$. These four fields are, of course, linear combinations of the previous Majorana fermion with its four possible boundary conditions: indeed, the $\mathbf{1}$ and ψ fields combine to yield the Neveu-Schwarz sector, whereas the spin fields σ and $\sigma^\dagger$ combine to yield the Ramond sector. A second useful way to think of the Majorana-fermion theory is as the Kač-Moody coset theory $SU(2)_2/U(1)$: here the $SU(2)$ is taken at Kač-Moody level 2 (as indicated by the subscript), and after modding out by $U(1)$ the resulting quotient theory is what can be called a "$\mathbf{Z}_2$ parafermion theory". We shall describe such quotient theories in more detail in the next subsection, but the basic idea is that the fields ϕ^j_m in such a theory fall into representations labeled by quantum numbers j and m inherited from the underlying $SU(2)$ structure. For this particular theory at level 2, there are only the four fundamental fields $\{\phi^0_0, \phi^1_0, \phi^{1/2}_{\pm 1/2}\}$, and these are again in direct one-to-one correspondence with the four fields of the Ising model, with ϕ^0_0 and ϕ^1_0 identified as the $\mathbf{1}$ and ψ fields respectively, and $\phi^{1/2}_{\pm 1/2}$ identified as the spin fields. Since the contribution to the worldsheet supercurrent from the fields corresponding to any one spacetime dimension in the superstring is of course $J = \psi\partial X$, we may rewrite this in terms of the quotient theory as $J = \phi^1_0 \partial X$.

Now that we have written the superstring ($K = 2$) worldsheet theory in this manner, it is straightforward to generalize to higher values of K. To construct the worldsheet theory of the fractional superstring for general $K \geq 2$, we begin with a single copy of the *fractionally*-supersymmetric (X, ϕ) theory, where ϕ now denotes the so-called "$\mathbf{Z}_K$ parafermion theory" – *i.e.*, the Kač-Moody coset theory $SU(2)_K/U(1)$, which will be discussed below. We then tensor together D_c copies of this theory, obtaining (X^μ, ϕ^μ) where the spacetime index $\mu = 0, 1, ..., D_c - 1$ is understood to be contracted with the Minkowski metric. Similarly, our worldsheet *fractional* supercurrent per dimension then takes the analogous form $J \sim \phi^1_0 \partial X + \ldots$, where ϕ^1_0 is now a field in the general $\mathbf{Z}_K$ parafermion theory.

2.2. $\mathbf{Z}_K$ Parafermion Theories: Basic Facts

These $\mathbf{Z}_K$ parafermion theories, although complicated, can be easily summarized by a few simple facts. For any K, the $\mathbf{Z}_K$ parafermion theories are conformal field theories with central charges $c = (2K - 2)/(K + 2)$, and their primary fields ϕ^j_m can be labeled and organized by their $SU(2)$ quantum numbers j and m, where $|m| \leq j$, $j - m \in \mathbf{Z}$, and where $j \geq 0$. One new feature, however, is the fact that j is also bounded from above, $j \leq K/2$, and thus these theories contain only a finite number of fundamental fields, this number growing with increasing K. The spins of these fields have an expected $j(j+1)$ dependence, as is usual for $SU(2)$, but once again there is a novelty, an m-dependence:

$$\Delta^j_m = \frac{j(j+1)}{K+2} - \frac{m^2}{K} . \tag{3}$$

The fusion rules of these fields also resemble those of an $SU(2)$ tensor product:

$$[\phi^{j_1}_{m_1}] \otimes [\phi^{j_2}_{m_2}] = \sum_{J=|j_1-j_2|}^{J_{\max}} [\phi^J_{m_1+m_2}] , \tag{4}$$

except that here $J_{\max}$ is not merely $j_1 + j_2$ but rather the minimum of this quantity and $K - j_1 - j_2$. Indeed, this general new feature, this "reflection" symmetry $j \leftrightarrow K/2 - j$, allows us to consistently identify the fields $\phi^j_m = \phi^j_{m+K} = \phi^{K/2-j}_{-(K/2-m)}$ for values outside the range $|m| \leq j$.

For any value of K, there are certain fields in the $\mathbf{Z}_K$ parafermion theory which will play a special role in the fractional superstring. The ϕ^0_0 field, as we have seen, is the identity, and will produce the Neveu-Schwarz vacuum state. The $\phi^{K/4}_{\pm K/4}$ fields, by contrast, are the analogues of the $K = 2$ spin fields, and they will correspondingly produce the Ramond vacuum state. A third special field, as we have seen, is the ϕ^1_0 field, often simply denoted ϵ. This is the energy operator in the $\mathbf{Z}_K$ theories, and is the field which appears in the supercurrent and through which Fock-space excitations take place.

These parafermion theories are of course more complicated than sketched here. In particular, as we have stated, the spins Δ^j_m are not simple half-integers, and thus the OPEs between these fields will involve branch *cuts* rather than simple poles. These theories are hence non-local on the two-dimensional worldsheet. This also implies that the parafermion mode algebras will not involve simple commutation or anticommutation relations, but rather *generalized* commutation relations. A discussion of these issues in the context of fractional superstrings, particularly for the $K = 4$ case, can be found in [4]. What makes the $K = 4$ case particularly tractable, it turns out, is the equivalence of the $\mathbf{Z}_4$ parafermion theory to that of a free boson compactified on a circle of radius $R = \sqrt{6}$. Furthermore, although these theories are non-local, the $K = 4$ theory is *abelianly braided* (*i.e.*, involves only one type of cut in the OPEs). The $K = 8$ and $K = 16$ cases, by contrast, do not share this property.

2.3. *The Critical Dimension*

Our discussion of the construction of the fractional-superstring worldsheet theory indicated that we start with a single fractionally-supersymmetric (X, ϕ) theory (where ϕ indicates the $\mathbf{Z}_K$ parafermion theory), and then tensor together D_c copies to form the full worldsheet theory. It is necessary, however, to determine the critical dimensions D_c for each value of K – *i.e.*, to derive the result quoted Eq. (1). The standard method for obtaining this would normally be to calculate the contribution to the conformal anomaly from the ghosts arising from fractional-superconformal gauge fixing, and then choose D_c so that $D_c c_0$ cancels this quantity [where $c_0 = 3K/(K + 2)$ is the total central charge of each (X, ϕ) theory]. The problem, however, is that the fractional superghost system is not yet understood, and the central charge of this system has not yet been calculated. We therefore choose an alternative path, demanding that our resulting theory indeed be a string theory including gravity and hence containing a massless graviton.

This argument, which is a straightforward generalization of an old argument due to Brink and Nielsen, proceeds in two steps. First we calculate the vacuum energy (or intercept $-v$) of the theory. Since we know that conformal invariance requires that $v = c_{\text{eff}}/24$ where c_{eff} is the effective central charge of *propagating* degrees of freedom in the theory, and since (as usual) only the $D_c - 2$ transverse dimensions worth of states yield physically propagating states, we see that $c_{\text{eff}} = (D_c - 2)c_0$. Thus in general we have the intercept $v = (D_c - 2)c_0/24$. Our second step is to identify the graviton as a state excited from this vacuum, and then adjust v (and thereby D_c) so that this excited state is massless. For example, in the $K = 1$ bosonic string, the graviton is $X^{\mu}_{-1}|p\rangle_R \otimes \tilde{X}^{\nu}_{-1}|p\rangle_L$ where $|p\rangle_{L,R}$ indicate these left- and right-moving vacuum states and where the lowest excitation mode of the X-field increases the energy by one unit (note that the spinless X^{μ}-field must have integer moding). Thus the resulting graviton state will be massless if $v = 1$, and since $c_0 = 1$ for this theory, we obtain $D_c = 26$. For the $K = 2$ superstring, the situation is similar. Here the graviton state is $b^{\mu}_{-1/2}|p\rangle_R \otimes \tilde{b}^{\nu}_{-1/2}|p\rangle_L$ where the excitation operators are now modes of ψ; since ψ has spin-1/2, it has half-integer moding and the lowest excitation adds a half-unit of energy. Thus the graviton will be massless if $v = 1/2$, and since $c_0 = 3/2$ for the (X, ψ) theory, we obtain $D_c = 10$.

The fractional-superstring case is analogous. Here the graviton state is given by $\epsilon^{\mu}_{-2/(K+2)}|p\rangle_R \otimes \tilde{\epsilon}^{\nu}_{-2/(K+2)}|p\rangle_L$, where once again the moding of the $\epsilon \equiv \phi^1_0$ field follows from its spin. This state will thus be massless if $v = 2/(K+2)$, and since $c_0 = 3K/(K+2)$, we obtain the result quoted in Eq. (1). Note that this set of critical dimensions also guarantees the existence of *massless* spacetime vectors (*i.e.*, gauge bosons) in the spectra of heterotic $(1, K)$ fractional superstrings, and also assures us that the chiral light-cone gauge Ramond vacuum states $\prod_{\mu=1}^{D_c-2}(\phi^{K/4}_{K/4})^{\mu}|p\rangle_{L,R}$ will be massless as well [since $(D_c - 2)\Delta^{K/4}_{K/4} - v = 0$]. This implies that these theories will in general contain a massless *gravitino* state, whereupon consistency requires that they exhibit full spacetime supersymmetry. We will see below that this is indeed the case.

2.4. Partition Functions

One method of efficiently surveying the *entire* spectrum of states (rather than merely those that are massless) is to examine the one-loop partition functions $\mathcal{Z}(\tau)$ that these theories can have; indeed, $\mathcal{Z}(\tau)$ receives contributions from states at all mass levels: $\mathcal{Z} \equiv \text{Tr}(-1)^F q^H \overline{q}^{\tilde{H}}$ where τ is the torus modular parameter, $q \equiv e^{2\pi i\tau}$, and the trace is over the complete Fock space of states with (left,right) energies $(H, \tilde{H})$ and fermion number F. In order to calculate such partition functions, we must multiply the characters corresponding to each of the worldsheet fields. The character of each worldsheet boson is given by $(\sqrt{\tau_2}\eta\overline{\eta})^{-1}$ where $\tau_2 \equiv \text{Im}\,\tau$ and where $\eta \equiv q^{1/24}\prod_{n=1}^{\infty}(1-q^n)$ is the Dedekind η-function; likewise, the character of each complex (left-moving) worldsheet fermion generally takes the form ϑ/η where ϑ represents one of the four classical Jacobi ϑ_i-functions, depending on the boundary conditions of the fermion. Similarly, the character of each (left-moving) *para*fermion ϕ^j_m is given by ηc^{2j}_{2m}, where the c's denote the level-K *string functions*; these functions are the higher-K generalizations of the Jacobi ϑ-functions, and reduce to them in the $K = 2$ special case. (Note that the super- and sub-scripts on the c's are defined as *twice* those on the corresponding ϕ^j_m.) Since the worldsheet field content of the (K, K) fractional superstrings consists (in light-cone gauge) of $D_c - 2$ copies of the (X, ϕ) theory, their partition functions must take the forms $\mathcal{Z}_K = {\tau_2}^k \sum(\overline{c})^{D_c-2}(c)^{D_c-2}$ where $k = 1 - D_c/2$. Here the antiholomorphic components come from the right-moving components of the worldsheet fields, and the η-functions have cancelled between the boson and parafermion characters.

Any linear combinations of the form $\sum(c)^{D_c-2}$ which can appear in these partition functions $\mathcal{Z}_K$ must satisfy several conditions. First, if we q-expand these combinations as $\sum a_n q^n$, we must have $a_n = 0$ for all $n < 0$ (absence of physical tachyons) as well as $a_0 \neq 0$ for some *subset* of terms within the linear combinations (corresponding to a massless sector). We must also demand these combinations close under the S modular transformation, and that our total partition functions $\mathcal{Z}_K$ be modular invariant. Remarkably, there exists only a single unique solution satisfying all of these constraints for each value $K = 2, 4, 8$, and 16. These solutions for ${\tau_2}^{-k}\mathcal{Z}_K$ are $|A_2|^2$ for $K = 2$; $|A_4|^2 + 3|B_4|^2$ for $K = 4$; $|A_8|^2 + |B_8|^2 + 2|C_8|^2$ for $K = 8$; and $|A_{16}|^2 + 4|C_{16}|^2$ for $K = 16$, where

$$
\begin{aligned}
A_2 &= [8(c_0^0)^7(c_0^2) + 56(c_0^0)^5(c_0^2)^3 + 56(c_0^0)^3(c_0^2)^5 + 8(c_0^0)(c_0^2)^7] \;-\; [8(c_1^1)^8] \quad \sim q^0 \\
A_4 &= [4(c_0^0 + c_0^4)^3(c_0^2) - 4(c_0^2)^4] \;-\; [4(c_2^2)^4 - 32(c_2^2)(c_2^4)^3] \quad \sim q^0 \\
B_4 &= [8(c_0^0 + c_0^4)^2(c_2^2)(c_2^4) - 4(c_0^2)^2(c_2^2)^2] - [4(c_0^2)^2(c_2^2)^2 - 16(c_0^0 + c_0^4)(c_0^2)(c_2^4)^2] \sim q^{1/2} \\
A_8 &= [2(c_0^0 + c_0^8)(c_0^2 + c_0^6) - 2(c_0^4)^2] \;-\; [2(c_4^4)^2 - 8(c_4^6 c_4^8)] \quad \sim q^0 \\
B_8 &= [4(c_0^0 + c_0^8)(c_4^6) - 2(c_0^4 c_4^4)] \;-\; [2(c_0^4 c_4^4) - 4(c_0^2 + c_0^6)(c_4^8)] \quad \sim q^{1/2} \\
C_8 &= 4(c_2^2 + c_2^6)(c_2^0 + c_2^8) - 4(c_2^4)^2 \quad \sim q^{3/4} \\
A_{16} &= [c_0^2 + c_0^{14} - c_0^8] \;-\; [c_8^8 - 2c_8^{14}] \quad \sim q^0 \\
C_{16} &= c_4^2 + c_4^{14} - c_4^8 \quad \sim q^{3/4} \, .
\end{aligned}
\tag{5}
$$

Certain features of these partition functions can be easily interpreted. The A-terms, whose expansions begin with q^0, contain the contributions from the massless states in these theories, and indeed we find that each of these A_K expressions contains a term $+(D_c-2)(c_0^0)^{D_c-3}(c_0^2)$ corresponding to the massless vector state $\epsilon^\mu|p\rangle$ (recall that $\epsilon \equiv \phi_0^1$). Similarly, each A_K expression also contains a term $-(D_c-2)(c_{K/2}^{K/2})^{D_c-2}$ corresponding to the massless fermion state (the Ramond vacuum); the minus sign indicates the spacetime statistics of this state, and once again the coefficient indicates the number of degrees of freedom appropriate for such spinors. In fact, it is straightforward to identify those terms within each A_K which correspond to spacetime bosonic or fermionic states of *any* mass. Since the bosonic Neveu-Schwarz vacuum state (respectively the fermionic Ramond vacuum state) has its m-quantum number equal to zero (respectively $K/4$) for each of the $D_c - 2$ transverse directions, and since all excitations on these vacua take place through the ϕ_0^1 field with vanishing m-quantum number, we see from the fusion rules (4) that *any* excited states must have the same m-quantum numbers as their underlying vacua. Indeed, since m-quantum numbers are additive modulo $K/2$ under fusion, this yields the usual boson/fermion fusions $B \otimes B = F$, $B \otimes F = F$, and $F \otimes F = B$. These observations thus enable us to split each expression A_K into its separate bosonic and fermionic components, with the first square-bracketed piece within each A_K in Eq. (5) arising from spacetime bosons, and the second piece (preceded by an overall minus sign) arising from spacetime fermions.

Another feature concerns spacetime supersymmetry. For $K = 2$, we can re-express A_2 in terms of the equivalent ϑ-functions to find that $A_2 = \frac{1}{2}\eta^{-12}J$ where

$J \equiv \vartheta_3{}^4 - \vartheta_2{}^4 - \vartheta_4{}^4$; thus $\mathcal{Z}_2$ is indeed recognized as the partition function of the ordinary superstring, and the Jacobi identity $J = 0$ (or $A_2 = 0$) indicates the vanishing of $\mathcal{Z}_2$ (*i.e.*, the exact level-by-level cancellation of bosonic and fermionic states). This is of course the partition-function reflection of the spacetime supersymmetry of this theory. Remarkably, however, this property extends to higher K as well, for it can be proven[3] that *each* of the combinations listed in Eq. (5) vanishes identically: $A_K = B_K = C_K = 0$. These resulting new identities, which are the higher-K generalizations of the $K = 2$ Jacobi identity $A_2 = 0$, can therefore be taken as evidence of spacetime supersymmetry in the *fractional* superstrings.

2.5. *Open Questions: New Vacua and Internal Projections*

There are, however, a number of issues concerning these partition functions which are yet to be resolved and which are undoubtedly at the heart of the intrinsically new physics of the fractional superstrings. These are, essentially, the questions of new vacuum states and of "internal projections."

While it was straightforward to identify each term within the A_K-sectors as arising from either a Neveu-Schwarz or Ramond vacuum state, it is clear that this is not the case for the B_K or C_K sectors. Of course, these sectors have q-expansions beginning only at q^h with $h > 0$, and thus they do not contribute to massless (*i.e.*, observable) physics except through loop effects suppressed by powers of the Planck mass. Nevertheless, although such sectors do not appear in the ordinary $K = 2$ superstring case, they do appear as fundamentally new features of the fractional superstrings, and their roles (and spacetime-physics properties) are not yet understood. The states in the B_K-sectors, for example, appear to originate from "mixed" Neveu-Schwarz/Ramond vacuum states in which *half* of the transverse dimensions contribute Neveu-Schwarz vacua and the other half Ramond: $(\phi_0^0 \phi_{K/4}^{K/4})^{(D_c-2)/2}|p\rangle$. The states in the C_K-sectors, by contrast, appear to originate from vacuum states $(\phi_{K/8}^{K/8})^{D_c-2}|p\rangle$ which are "pure" (*i.e.*, the same in each transverse dimension), but of fractional spin (because the fusion of *two* such vacuum states seems necessary to produce the Ramond vacuum). One possible interpretation of the roles of these new Planck-scale states is as follows.[2] Although the states in the B_K-sectors appear to violate full D_c-dimensional Lorentz invariance (because their underlying vacua are not invariant under the permutations of all of the D_c-2 transverse spacetime directions), they may well be invariant under D-dimensional ***subgroups*** of the Lorentz group, with $D < D_c$. The appearance of such states would then provide an intrinsically new mechanism for Lorentz symmetry breaking — *i.e.*, for spontaneous spacetime self-compactification. Indeed, in the $K = 4$ case, the form of B_4 suggests that in order for each corresponding state to be interpretable as arising from an underlying Neveu-Schwarz or Ramond vacuum, the resulting spacetime dimension D can be at most four (*i.e.*, with only *two* transverse dimensions); the two remaining string-function factors would then correspond to the internal CFT induced by the compactification. Similarly, in the $K = 8$ case, we find that D can be at most three, in which case the appearance of the fractional-spin C_8 sector would be fully consistent with the spin-statistics connection. [Note that the $K = 16$ theory, which is already has $D_c = 3$, requires (and receives) no additional compactification from any B-type sector, and the presence of a C_{16} sector is already non-problematic.] Furthermore, it turns out that such self-induced compactifications would naturally provide a mechanism for the appearance of phenomenologically necessary chiral spacetime fermions in the fractional-superstring spectra.[2] Of course, detailed analyses are necessary in order to verify these ideas.

A more concrete new issue, perhaps, is the question of the "internal projections" which arise in the $K > 2$ fractional superstrings and which are also reflected in the above partition functions. We have already seen that we could split each expression A_K into separate (though cancelling) bosonic and fermionic components $A_K = A_K^b - A_K^f$, and while the relative minus sign *between* A_K^b and A_K^f is of course attributed to spacetime statistics [*i.e.*, to the $(-1)^F$ factor in the definition of $\mathcal{Z}$], the minus signs appearing *within* each of these components for $K > 2$ clearly cannot be understood as arising due to spacetime statistics. Rather, they must be interpreted as fundamentally new "internal" projections which, like the GSO projection, act between states of the same spacetime statistics. One might of course worry that in the individual component q-expansions $A_K^{b,f} = \sum a_n q^n$, these extra minus signs might cause some coefficients a_n to become negative (thereby ensuring that A_K^b and A_K^f could not be the interpreted as the characters of bosonic or fermionic highest-weight sectors of an underlying post-projection conformal field theory). However, it has been proven[3] that *each* of these components individually satisfies the remarkable identity

$$A_K^f = (D_c - 2)\left(\prod_{n=1}^{\infty} \frac{1+q^n}{1-q^n}\right)^{D_c-2} = (D_c-2)\left[\frac{\vartheta_2(\tau)}{2\eta^3(\tau)}\right]^{(D_c-2)/2} . \tag{6}$$

This result thus not only guarantees that the coefficients a_n are all non-negative, but in fact indicates that the residual conformal field theories which underlie the A_K^f sectors of the fractional superstrings *after the internal projections* are (in light-cone gauge) nothing but those of $D_c - 2$ free bosons and Ramond fermions, with apparently no trace remaining of the original *para*fermions! This implies that the internal projection reduces the A-sector theory of the fractional superstrings to that of the ordinary Green-Schwarz string formulated in $D_c = 3, 4, 6$, or 10 spacetime dimensions. Now while these are of course precisely the dimensions in which the *classical* Green-Schwarz string can be formulated, it is well-known that this string is quantum-mechanically self-consistent only in 10 dimensions. Presumably, then, it is the role of the new fractional-superstring B-sectors — which appear only for $K > 2$ (or $D_c < 10$) — to restore this self-consistency and cancel the quantum anomalies. It turns out that there exist arguments[5,7] indicating that proper splittings for the B_K-sectors are those indicated by the square-brackets within the B_K-expressions in Eq. (5), and indeed one finds[5] a similar remarkable identity for the B-sectors:

$$B_K^f = (D_c - 2)\left[\frac{\vartheta_2(\lambda\tau)}{2\eta^3(\tau)}\right]^{(D_c-2)/2} \tag{7}$$

where $\lambda = [\Delta(\epsilon)]^{-1} = \frac{1}{2}(K+2)$. This implies that the internally-projected B-sectors of the fractional superstrings also resemble Green-Schwarz strings in $D_c < 10$, except with fermions formulated on *rescaled* momentum lattices; indeed, the underlying parafermionic spins now appear as the relevant scaling factors. The C-sectors, by contrast, appear to be entirely removed by the internal projections, and thus may contain no physical states of any spacetime statistics.[5,7]

The effects of these internal projections, as indicated by Eqs. (6) and (7), are quite profound. What had begun (in light-cone gauge) as D_c-2 copies of a fractionally-supersymmetric boson plus $\mathbf{Z}_K$ parafermion conformal field theory with total central charge $(D_c-2)c_0 = 48/(K+2)$ has now been internally projected down to D_c-2 copies of a boson plus (rescaled) fermion theory with total central charge $3(D_c-2)/2 = 24/K$. Thus exponentially large numbers of states at each mass level of the Fock space are being projected out of the spectrum by these internal projections, enough to reduce the effective central charges (and thereby alter such features of these theories as their Hagedorn temperatures, which depend only on how the numbers of states diverge at infinite mass.) Indeed, these internal projections indicate that our original parafermionic Fock space is too large, and that although the fractional superstrings can be formulated in terms of these parafermionic CFTs *with internal projections*, an alternative (and smaller) CFT might exist which would more efficiently describe the residual physical states surviving the internal projections. Restoring the contributions from longitudinal and timelike directions, we see that such residual covariant CFTs would have to have total central charges $c = 24/K + 6K/(K+2)$, and remarkably this is the spectrum of critical central charges for proposed string theories[8] based upon spin-$(K+4)/(K+2)$ algebras (*e.g.*, the so-called "spin-4/3 algebra" for $K=4$, which has no linear tensor-product formulation). An important goal for the fractional-superstring theories, then, is to construct suitable representations of these algebras with precisely this spectrum of critical central charges.

It is a pleasure to thank Philip Argyres and Henry Tye, with whom I collaborated on much of the work described herein, for many useful discussions. I would also like to acknowledge the hospitality of the Aspen Center for Physics, where portions of this work were completed. This work was supported in part by the Natural Sciences and Engineering Research Council of Canada and by les Fonds FCAR du Québec.

Fractional Superstring References

[1] P.C. Argyres and S.-H.H. Tye, *Phys. Rev. Lett.* **67** (1991) 3339.

[2] K.R. Dienes and S.-H.H. Tye, *Nucl. Phys.* **B376** (1992) 297.

[3] P.C. Argyres, K.R. Dienes, and S.-H.H. Tye, *New Jacobi-like Identities for* $\mathbf{Z}_K$ *Parafermion Characters*, CLNS 91/1113, McGill/91-37, hepth/9201078 (January 1992).

[4] P.C. Argyres, E. Lyman, and S.-H.H. Tye, *Low-lying States of the Six-Dimensional Fractional Superstring*, CLNS 91/1121, hepth/9205113 (May 1992).

[5] P.C. Argyres and K.R. Dienes, *Towards an Unraveling of the Massive Sectors and Internal Projections of the Fractional Superstring*, CLNS 92/1168, McGill/92-41 (to appear).

[6] P.H. Frampton and J.T. Liu, University of North Carolina preprint IFP-440-UNC, hepth/9210049 (September 1992).

[7] G. Cleaver and P. Rosenthal, Caltech preprint (to appear).

[8] P.C. Argyres, A. LeClair, and S.-H.H. Tye, *Phys. Lett.* **B253** (1991) 306.

INTEGRABLE HIERARCHIES AND $c = 1$ STRING THEORY

ROBBERT DIJKGRAAF
*Department of Mathematics, University of Amsterdam, Plantage Muidergracht 24
1018 TV Amsterdam, The Netherlands*

ABSTRACT

We review the relation between non-critical string theories and integrable hierarchies with particular emphasis on the $c = 1$ model. Here the interpretation of integrability in terms of space-time physics is explained.

1. Non-Critical Strings

Recent years have seen an enormous increase in our understanding of a particular class of string theories – the so-called non-critical bosonic strings. Most of this progress is due to the advent of matrix models [1] and their solutions in the double-scaling limit [2], see *e.g.* the reviews in [3]. These models can be completely solved within the framework of integrable hierarchies [4, 5]. A special role is played by the so-called $c = 1$ model. In many aspects this is the most physical and richest non-critical model. It also can be solved using matrix model techniques [6], but the determination of its correlation functions has been rather complicated [7, 8, 9]. In [10] the explicit construction of these scattering amplitudes has been clarified and the relation between integrable hierarchies and the $c = 1$ model has been explained, thereby completing in a sense the unification of all solvable non-critical string models. In these notes we will give a more leisurely account of all this.

1.1. Elementary introduction to string theory

Before we plunge into the details of the explicit solutions of these string theories, let us start with some very general remarks to place this work in the relevant context. Recall that from a perturbative point of view a string theory is a quantum theory of excitations (particles) in some space-time X that has an interpretation in terms of surfaces, *i.e.* in terms of maps

$$x : \Sigma \to X, \tag{1}$$

with Σ a surface of a particular genus. Indeed the 'central dogma' of string theory (see *fig.* 1.) equates (connected) S-matrix elements in space-time with correlation functions in a two-dimensional field theory

$$\langle out|S|in\rangle_c \sim \sum_{g=0}^{\infty} \lambda^{2g-2} \langle \prod V^{out} \prod V^{in} \rangle_g \,. \tag{2}$$

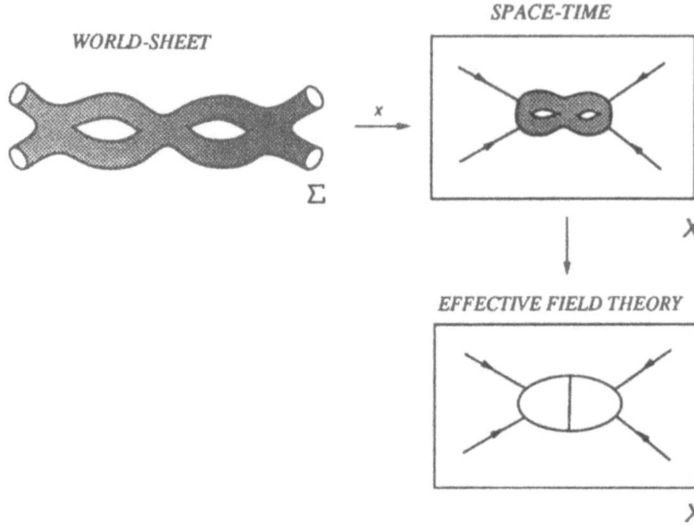

Fig. 1: The 'central dogma' of string theory expresses the space-time scattering amplitudes in terms of correlation functions in a quantum field theory defined on a world-sheet surface. In the effective field theory limit, these string perturbation contributions correspond to (sums of) point-particle Feynman diagrams.

Here the object on the LHS is a scattering amplitude in the space-time X whose expansion in terms of the string coupling constant λ is given by correlation functions of vertex operators V^{out}, V^{in} in a two-dimensional quantum field theory defined on a surface Σ of genus g — the world-sheet. Since the correlation functions on the RHS should not depend on intrinsic properties of the surface, like the metric and the position of the vertex operators, the world-sheet field theory should include in some form an integral over all metrics on the surface, and thus we are naturally lead to consider two-dimensional quantum gravity.

The analogue idea in quantum field theory is well-known: amplitudes in field theory can be written as correlation functions in a one-dimensional field theory, *i.e.* quantum mechanics. For example, the usual propagator of a massive scalar field $\phi(x)$ on $X = \mathbf{R}^d$ with Lagrangian

$$L = (\nabla\phi)^2 + m^2\phi^2 \tag{3}$$

is given by the amplitude of a particle to propagate from x to y, and can be written as one-dimensional path-integral over a field $x(t) \in \mathbf{R}^d$ and a metric $g(t)$ on the world-line

$$x \; — \; y \; = \; \langle x|\frac{1}{\partial^2+m^2}|y\rangle$$
$$= \int [dx][dg] e^{-\int dt \sqrt{g}(g^{-1}\dot{x}^2+m^2)} \, . \tag{4}$$

The path-integral is subject to the boundary conditions $x(0) = x$ and $x(1) = y$. The integral over the one-dimensional metric $g(t)$ modulo diffeomorphisms is not very interesting, and reduces to an integral over the total length T of our one-dimensional world. In this way we recover the familiar Schwinger parametrization of the propagator

$$\int [dx][dg] e^{-\int dt \sqrt{g}(g^{-1}\dot{x}^2+m^2)} = \int_0^\infty dT \, \langle x|e^{-T(\partial^2+m^2)}|y\rangle \, . \tag{5}$$

More complicated Feynman diagrams are treated by considering maps of graphs Γ into space-time, with overlap integrals for the vertices.

In string theory we consider maps $x : S^1 \to X$ instead of points $x \in X$. The analogue of the propagator is the two point function

$$x(\sigma) \; y(\sigma) = \int [dx][dg] e^{-\int d^2z \sqrt{g}(g^{\mu\nu}\partial_\mu x \partial_\nu x + \mu)} \, , \tag{6}$$

where $x(\sigma,\tau)$ and $g_{\mu\nu}(\sigma,\tau)$ are now a field and a metric on a two-dimensional surface, which for a propagator has the topology of a cylinder. We also introduced a two-dimensional cosmological constant μ that is in some sense the analogue of the mass of the string field.

This definition can now be very naturally extended to include all interactions, just by considering the two-dimensional field theory on surfaces Σ with more complicated topology

$$\Sigma \; = \int [dx][dg] e^{-\int_\Sigma d^2z \sqrt{g}(g^{\mu\nu}\partial_\mu x \partial_\nu x + \mu)} \tag{7}$$

So, in string theories the two-dimensional metric $g_{\mu\nu}$ is a fundamental variable and the path-integral should include an integral over all inequivalent Riemannian structures, *i.e.* metrics modulo diffeomorphisms. Although this space is infinite dimensional, it can be contracted to the moduli space $\mathcal{M}_g$ of inequivalent complex structures on the surface. One understands this fact by the result that any metric can be written as

$$g_{\mu\nu} = e^{2\phi} h_{\mu\nu}, \tag{8}$$

where $h_{\mu\nu}$ is a fixed representative of a conformal class, for example a metric of constant curvature $R_h = -1$, which uniquely determines a complex structure through

$$J_\mu{}^\nu = \sqrt{g}\epsilon_{\mu\lambda}g^{\lambda\nu}. \tag{9}$$

$\mathcal{M}_g$ is a rich mathematical object that plays a crucial role in our understanding of string theory. Our knowledge of this space has also benefited from the recent developments in string theory [11, 12], see *e.g.* the review [13].

The string theories with target space $\mathbf{R}^d$ (with a flat metric) can be most easily quantized in conformal gauge. One finds the action

$$S = \int \frac{d^2z}{2\pi}(\partial x^i \overline{\partial} x^i + \alpha_1 \partial\phi\overline{\partial}\phi + \alpha_2 R_h \phi + \alpha_3 \mu e^{2\phi} + b\overline{\partial}c + \overline{b}\partial\overline{c}) \tag{10}$$

with some appropriate constants α_i [14] chosen such that the total central charge of the x-field and the Liouville field equals 26 and thereby cancels the central charge $c = -26$ of the ghosts b, c. For d arbitrary the Liouville field does not decouple and represents an extra degree of freedom. It is a remarkable effect that the Liouville field forms essentially an extra embedding dimension if $d \neq 26$. Although this effect would be an embarrassment for, say, a four-dimensional string describing QCD flux tubes, it is a very welcome phenomenon in the case $d = 1$ that we will be interested in the subsequent. For $d = 26$ the Liouville field decouples, $\alpha_i = 0$, and we obtain a critical string.

1.2. $c = 1$ *string theory*

The $c = 1$ string is perhaps the most simple bosonic string theory, and and yet it has a very rich structure. We are naturally led to this model, since the class of bosonic strings that we wrote down in (10) with target space $\mathbf{R}^d$ is actually inconsistent if $d > 1$. This is a consequence of the famous tachyon instability. The zero-mode of the string

$$x(\tau) = \int d\sigma\, x(\sigma, \tau) \tag{11}$$

gives rise to a scalar particle whose mass can be calculated to be

$$m^2 = \frac{1-d}{24}. \tag{12}$$

This is indeed a tachyon for dimensions larger than one. The world-sheet interpretation of this instability is conjectured to be the formation of long, thin tree-like surfaces (branched polymers). However we see that this tachyon field is actually massless in dimension $d = 1$ and massive in $d = 0$. So here we stand a chance of constructing consistent bosonic string theories. We will concentrate on the case $d = 1$ where the space-time physics is more clear.

This model consists of one free field x that we couple to two-dimensional gravity. So our action reads

$$S = \int \frac{d^2z}{2\pi} \sqrt{g} g^{\mu\nu} \partial_\mu x \partial_\nu x. \tag{13}$$

Here the field x might be compactified, *i.e.*, we also consider the case where x is a periodic field

$$x \cong x + 2\pi\beta. \tag{14}$$

Since one scalar field gives a conformal field theory of central charge $c = 1$, we will refer to this model as the $c = 1$ string.

As explained above, the Liouville field ϕ does not decouple in this case. With the appropriate constants the action reads (not including the ghosts)

$$S = \int \frac{d^2z}{2\pi}\sqrt{h}\left[\tfrac{1}{2}h^{\mu\nu}(\partial_\mu x\partial_\nu x + \partial_\mu\phi\partial_\nu\phi) + R_h\phi + \mu e^{2\phi}\right] \tag{15}$$

This Lagrangian effectively describes a sigma model with a two-dimensional target space, with the Liouville field regarded as a spontaneously generated extra dimension. So we are dealing with a *two-dimensional critical string* (critical in the sense that the total matter central charge is 26). Of course, because of the background charge and the exponential potential, we do not have translation invariance in the ϕ-direction. The Hilbert space of the model is simply given by the tensor products of the state spaces of the free field x, the Liouville field ϕ and the ghosts b, c.

Physical modes in this model are limited. Since for a string in n dimensions we expect $n-2$ transverse oscillations, naively only the zero-mode of a two-dimensional string theory can be expected to lead to a physical excitation. This field is the massless scalar field known as the tachyon. The world-sheet vertex operators that describe the emission/absorption of tachyons of energy ω simply read (here and in the subsequent $\omega > 0$)

$$T_\omega^\pm = \frac{\Gamma(\omega)}{\Gamma(-\omega)}\mu^{1-\omega/2}e^{\pm i\omega x}e^{(2-\omega)\phi}. \tag{16}$$

With the explicit factors of μ inserted, we see that a genus g correlation function

$$\langle T_{\omega_1}^\pm \cdots T_{\omega_n}^\pm\rangle_g \tag{17}$$

scales simply as μ^{2-2g}. (This can be seen most easily by shifting the Liouville field by $\phi \to \phi - \frac{1}{2}\log\mu$.) Consequently we can identify the string coupling constant λ with the inverse of the cosmological constant

$$\lambda = \frac{1}{\mu}. \tag{18}$$

If we compactify x, the energies ω will be quantized

$$\omega_n = \frac{n}{\beta}, \qquad n \in \mathbf{Z}^+. \tag{19}$$

We will write T_n instead of T_{ω_n}. The point $\beta = 1$ is very special, since it is the invariant point under the duality transformation

$$\beta \to 1/\beta, \tag{20}$$

that interchanges a large and a small radius of compactification.

Clearly the 'holy grail' in $c = 1$ string theory would be the generating functional of all connected scattering amplitudes. For the compactified case it would read

$$F(t,\bar{t}) \sim \sum_{g=0}^{\infty} \lambda^{2g-2} \langle \exp \sum_n (t_n T_n^+ + \bar{t}_n T_n^-) \rangle_g \tag{21}$$

In terms of space-time field theory we would have incoming states $|k_1, k_2, \ldots\rangle$ with k_i tachyons of energy ω_i and corresponding outgoing states $\langle \bar{k}_1, \bar{k}_2, \ldots |$, and $\tau = e^F$ would be the generating function of all matrix elements

$$\tau(t,\bar{t}) = \sum_{\{k_i, \bar{k}_i\}} \prod_{i,j} \frac{t_i^{k_i}}{k_i!} \frac{\bar{t}_i^{\bar{k}_i}}{\bar{k}_i!} \langle \bar{k}_1, \bar{k}_2, \ldots | S | k_i | k_1, k_2, \ldots \rangle \ . \tag{22}$$

Actually there are more degrees of freedom than just the tachyons, but these so-called discrete states appear at very particular energy and momentum [15]. Consequently, they do not have an interpretation as propagating particles, but are more analogous to quantum mechanical degrees of freedom.

1.3. Minimal models

Before we go on and discuss the solution of the $c = 1$ string theory, we want to turn briefly to another interesting class of models, the so-called $c < 1$ models. These models are the unique CFTs with a finite number of irreducible representations of the Virasoro algebra. They are labeled by two integers (p, q) which should be coprime. It is rather difficult to write a Lagrangian for these models, but roughly they can be described as a $c = 1$ model with an extra background charge for the scalar field x

$$S = \int \partial x \bar{\partial} x + QRx. \tag{23}$$

So we loose all translation invariance. Furthermore, the occurrence of screening operators and an extra BRST charge make the analysis of the correct physical states much more complicated. Nevertheless, these string theories have been completely solved, due to the magic of matrix models. Instead of discussing this approach to two-dimensional quantum gravity here (see for example [3]), we just give a brief account of the nature of the solution as found in [4, 5].

First of all, the spectrum is extremely simple and universal for all (p, q) models and the massive deformation that can be obtained of them. For every positive integer we have a physical state, so we have vertex operators

$$V = \mathcal{O}_1, \mathcal{O}_2, \mathcal{O}_3, \ldots \tag{24}$$

According to our discussion above, the string theory correlation functions of these operators are on general grounds given by a (complicated and in general unknown) integral over moduli space, where the volume form is a correlator in a CFT consisting of the minimal model coupled to ghosts and Liouville theory

$$\langle \mathcal{O}_{n_1} \cdots \mathcal{O}_{n_s} \rangle_g = \int_{\mathcal{M}_{g,s}} \langle \ldots \rangle_{CFT} \tag{25}$$

The matrix model techniques give an alternative route to the calculation of these correlators, and this result can be summarized as follows. The generating functional of correlation functions for the (p, q) model

$$\tau_{p,q}(t) = \exp \sum_{g=0}^{\infty} \lambda^{2-2g} \langle \exp \sum_n t_n \mathcal{O}_n \rangle_g \tag{26}$$

is a tau-function of the KP hierarchy and all (p, q)-generating functions for fixed p lie on one orbit. More precisely, relative to a convenient choice of origin, the (p, q) model is obtained as

$$\tau_{p,q}(t_k) = \tau_p(t_k + \delta_{k,p+q}). \tag{27}$$

Furthermore, the KP hierarchy reduces to the pth KdV hierarchy, which implies that all correlation functions of the operators $\mathcal{O}_n$ with $n \equiv 0 \pmod p$ vanish.

At the point $(p, 1)$ we also have a very explicit representation of the τ-function due to Kontsevich [12] (for more details see *e.g.* [13, 16]). Let Z be an $N \times N$ Hermitian matrix, and choose particular values for the KP times t_n given by

$$t_n = \frac{\lambda}{n} Tr Z^{-n} . \tag{28}$$

With this parametrization, the τ-function is given by the following matrix integral, where Y is also an $N \times N$ Hermitian matrix

$$\tau_{p,1}(t) = c^{-1} \cdot \int dY \cdot e^{i\mathrm{Tr}(Z^p Y - \frac{Y^{p+1}}{p+1})/\lambda} . \tag{29}$$

Here the normalization constant is given by

$$c = (2\pi i/\lambda)^{\frac{N^2}{2}} \det(Z^{\frac{1-p}{2}}) \frac{\Delta(z)}{\Delta(z^p)} e^{i\frac{p}{p+1} Tr Z^{p+1}/\lambda} . \tag{30}$$

and $\Delta(z)$ is the Vandermonde determinant

$$\Delta(z) = \det z_i^{j-1} = \prod_{i>j} (z_i - z_j). \tag{31}$$

This rather formidable looking expression for the normalization constant is actually very natural. The integral (29) can be expanded around the critical point $Y = Z$ of the action

$$\mathrm{Tr}\,(Z^p Y - \frac{Y^{p+1}}{p+1}). \tag{32}$$

The normalization constant is now just the classical value of the action plus the gaussian (one-loop) contribution to the integral

$$c = e^{i\frac{p}{p+1} Tr Z^{p+1}/\lambda} \cdot c', \tag{33}$$

where c'_p is defined by the Gaussian integral

$$\begin{aligned} c' = \int dY \cdot e^{iS(Y,Z)/\lambda} &= (2\pi i/\lambda)^{\frac{N^2}{2}} \prod_{i,j} \sqrt{\frac{z_i - z_j}{z_i^p - z_j^p}} \\ &= (2\pi i/\lambda)^{\frac{N^2}{2}} \det Z^{\frac{1-p}{2}} \frac{\Delta(z)}{\Delta(z^p)}, \end{aligned} \tag{34}$$

and

$$S(Y,Z) = \sum_{k=0}^{p-1} \mathrm{Tr}\,[Y Z^k Y Z^{p-1-k}]. \tag{35}$$

This representation makes clear that $\tau_{p,1}(t)$ can indeed have an asymptotic expansion in the variables t_n.

2. Integrable Hierarchies and Quantum Field Theory

In order to understand better the integrable hierarchies that naturally appear in non-critical string theory and in particular the remarkable integral representation (29), we will now review some relations between quantum field theory and integrable systems. We follow the standard exposition of the KP hierarchy [17].

2.1. 2d Chiral quantum field theory

Consider a two-dimensional free chiral scalar field $\varphi(z)$, with the usual mode expansion

$$\partial\varphi(z) = \sum_n \alpha_n z^{-n-1} . \tag{1}$$

Here $z = e^{ix-\tau}$ is a coordinate on a cylinder, with x the periodic space variable and τ Euclidean time. We have standard commutation relations $[\alpha_n, \alpha_m] = n\delta_{n+m}$, and a vacuum $|0\rangle$ satisfying $\alpha_n|0\rangle = 0$, $n \geq 0$. The Hilbert space $\mathcal{H}$ can be considered as the completion of the Fock space spanned by states $\alpha_{-n_1} \ldots \alpha_{-n_s}|0\rangle$. The reader is encouraged to think about this two-dimensional scalar field as the tachyon field living in the space-time with coordinates (x, ϕ). Here we roughly have $\phi = \tau$.

As in the case of a harmonic oscillator one can consider coherent states,

$$|t\rangle = \exp\left(\sum_{n=1}^{\infty} it_n\alpha_{-n}\right)|0\rangle, \tag{2}$$

and their Hermitian conjugates

$$\langle t| = \langle 0|\exp\left(\sum_{n=1}^{\infty} -it_n\alpha_n\right) . \tag{3}$$

Now to any state $|W\rangle$ in the Hilbert space $\mathcal{H}$ we can associate a coherent state wave-function $\tau_W(t)$ by considering the inner product

$$\tau_W(t) = \langle t|W\rangle . \tag{4}$$

This function is a tau-function of the KP hierarchy if and only if the state $|W\rangle$ lies in the so-called Grassmannian.

To explain the concept of the Grassmannian we have to turn to the alternative description of the chiral boson in terms of chiral Weyl fermions $\psi(z)$, $\overline{\psi}(z)$ by means of the well-known bosonization formulas

$$i\partial\varphi = \overline{\psi}\psi, \quad \psi = e^{i\varphi}, \quad \overline{\psi} = e^{-i\varphi} . \tag{5}$$

These free fermions have mode expansions

$$\psi(z) = \sum_{a\in\mathbf{Z}+\frac{1}{2}} \psi_a z^{-a-\frac{1}{2}}, \tag{6}$$

$$\overline{\psi}(z) = \sum_{a\in\mathbf{Z}+\frac{1}{2}} \overline{\psi}_a z^{-a-\frac{1}{2}}, \tag{7}$$

and canonical anti-commutation relations

$$[\psi_a, \overline{\psi}_b]_+ = \delta_{a+b}, \qquad [\psi_a, \psi_b]_+ = [\overline{\psi}_a, \overline{\psi}_b]_+ = 0. \tag{8}$$

Loosely speaking, the Grassmannian can be defined as the collection of all *fermionic* Bogoliubov transforms of the vacuum $|0\rangle$. That is, the state $|W\rangle$ belongs to the Grassmannian if it is annihilated by particular linear combinations of the fermionic creation and annihilation operators.

$$(\psi_{n+\frac{1}{2}} - \sum_{m=1}^{\infty} A_{nm}\psi_{-m+\frac{1}{2}})|W\rangle = 0, \qquad n \geq 0\ , \tag{9}$$

or equivalently,

$$|W\rangle = S|0\rangle, \qquad S = \exp\left(\sum_{n,m} A_{nm}\overline{\psi}_{-n-\frac{1}{2}}\psi_{-m+\frac{1}{2}}\right). \tag{10}$$

Note that the operator S can be considered as an element of the infinite-dimensional linear group, $S \in GL(\infty, \mathbf{C})$.

By replacing the vacuum $|0\rangle$ by the state $|W\rangle$, we simply made another decomposition into positive and negative energy states, and filled these new negative energy states. The positive energy wave-functions are no longer given by

$$z^0, z^1, z^2, \ldots \tag{11}$$

but are now replaced by the functions

$$v_0(z), v_1(z), v_2(z), \ldots \tag{12}$$

with

$$v_n(z) = z^n - \sum_{m=1}^{\infty} A_{nm} z^{-m}\ . \tag{13}$$

If one prefers the language of semi-infinite differential forms, we have a formula

$$|W\rangle = v_0 \wedge v_1 \wedge \ldots \tag{14}$$

which should be contrasted with

$$|0\rangle = z^0 \wedge z^1 \wedge \ldots \tag{15}$$

We want to mention at this point one important generalization. In the above fashion one generates solutions to the KP hierarchy. This construction can be extended to give a tau-function for the two-dimensional Toda Lattice hierarchy by considering a second set of times $\bar{t}_k$, as discussed in detail in [18]. In terms of our conformal field theory, the *Toda tau-function* is simply obtained as

$$\tau(t,\bar{t}) = \langle t|S|\bar{t}\rangle \ , \tag{16}$$

with $|\bar{t}\rangle$ and $\langle t|$ the coherent states (2) and (3) and S a general $GL(\infty, \mathbf{C})$ element, *i.e.* an exponentiated fermion bilinear of type (10). We will return to the Toda hierarchy in §3.

Instead of taking the inner product of the state $|W\rangle$ with a coherent bosonic state, one can also consider *fermionic* N-point functions. In fact, one finds in this way a simple expression in terms of an $N \times N$ determinant of the wave-functions (13) [13, 16]

$$\langle N|\psi(z_1)\ldots\psi(z_N)|W\rangle = \det v_{j-1}(z_i) \ . \tag{17}$$

Using the bosonization formulas, one recognizes this correlation function as a special coherent state wave-function $\langle t|W\rangle$ where the parameters t_n are given by

$$t_n = \sum_{i=1}^{N} \frac{1}{n} z_i^{-n} \ . \tag{18}$$

With this choice of parameterization, and after taking into account a normal ordering contribution, the tau-function can be written as

$$\tau(t) = \frac{\det v_{j-1}(z_i)}{\Delta(z)} \ , \tag{19}$$

with $\Delta(z)$ the Vandermonde determinant.

2.2. Kontsevich integrals and the $c < 1$ models

We are now in a position to understand the integral representation that we gave for the minimal models partition functions $\tau_{p,q}(t)$. The state $|W\rangle$ corresponding to this orbit is most simply described at the $(p,1)$ point, where a description in terms of topological field theory can be given. In the Lax operator description of the KdV hierarchies [17], this point can be very efficiently characterized by the following initial value for the pth order linear differential operator L

$$L = D^p + x. \tag{20}$$

For an explicit basis for the state $|W\rangle$ one can take the wave-functions

$$v_n(z) = \sqrt{\frac{ipz^{p-1}}{2\pi\lambda}} e^{\frac{ip}{p+1}z^{p+1}/\lambda} \int_{-\infty}^{\infty} dy \cdot y^n \cdot e^{i(z^p y - \frac{y^{p+1}}{p+1})/\lambda} , \tag{21}$$

where the normalization is chosen such that we have the appropriate asymptotic expansion

$$v_n(z) = z^n(1 + O(z^{-1})) . \tag{22}$$

Since the wave-functions are moments in a Fourier transform, the fermionic determinant formula can be explicitly evaluated. This is due to the following identity, first given by Harish-Chandra, for the integral over the unitary group $U(N)$ [19]

$$\int dU\, e^{i\mathrm{Tr}[UXU^\dagger Y]} = c \cdot \frac{\det e^{ix_i y_j}}{\Delta(x)\Delta(y)}, \tag{23}$$

with dU the Haar measure and x_i, y_i the eigenvalues of the Hermitian matrices X and Y and c some normalization constant. As a special application of the above equation consider the matrix Fourier transform, *i.e.* the integral over a Hermitian matrix Y in an external field X, both $N \times N$ matrices, of the form

$$\tau = \int dY\, e^{i\,\mathrm{Tr}[XY + V(Y)]}. \tag{24}$$

By conjugation invariance, this is only a function of the eigenvalues $x_1, \ldots, x_N$. We can use (23) to integrate out the angular variables U in the decomposition

$$Y = U \cdot \mathrm{diag}(y_1, \ldots, y_N) \cdot U^\dagger, \tag{25}$$

which also introduces a Jacobian

$$dY = \Delta(y)^2 \cdot dU \cdot [dy]. \tag{26}$$

This leaves us with an integral over the eigenvalues y_i of the form

$$\tau = \int [dy]\, \Delta(y)\Delta(x)^{-1} \exp \sum_j i[x_j y_j + V(y_j)] \tag{27}$$

Now the Vandermonde determinant $\Delta(y)$ is a sum of terms of the form

$$\pm\, y_1^{i_1} \cdots y_N^{i_N}, \tag{28}$$

and for each of these terms the integral τ factorizes in separate integrals over the individual eigenvalues y_i. If we introduce the function

$$w(x) = \int dy \cdot e^{ixy+iV(y)} \tag{29}$$

and its derivatives

$$w_k(x) = \int dy \cdot y^k \cdot e^{ixy+iV(y)}, \tag{30}$$

the contribution of (28) is simply

$$\pm\, w_{i_1}(x_1)\cdots w_{i_N}(x_N). \tag{31}$$

So we can evaluate $\tau(X)$ straightforwardly as

$$\tau = \frac{\det(w_{j-1}(x_i))}{\Delta(x)}. \tag{32}$$

If we apply this method to the determinant (19) of the wave-functions (21), we find the matrix integral (29).

This result can be generalized to the 'generalized Kontsevich model' [13, 16] which features an arbitrary potential $V(z)$

$$\tau(t) = c \cdot \int DY \cdot e^{iTr(V'(Z)Y - V(Y))/\lambda}\,. \tag{33}$$

with

$$c = (2\pi i/\lambda)^{-\frac{N^2}{2}} \cdot \det V''(Z) \cdot \frac{\Delta(V'(z))}{\Delta(z)} \cdot e^{iTr(V(Z)-V'(Z)Z)/\lambda}\,. \tag{34}$$

It has been noticed by many authors that the case $p = -1$ (*i.e.* a logarithmic potential $V(z) = \log z$) is likely to be associated with the $c = 1$ model. In the next section we will proceed to show that this is indeed the case.

3. The $c = 1$ String

We will now return to the $c = 1$ string and discuss its solution within the matrix model framework.

3.1. The $c = 1$ matrix model

Matrix models have been introduced in two-dimensional quantum gravity to describe discretizations of smooth surfaces [1]. We will not explain this method here in any detail. The basic idea is that the path-integral over all metrics on the surface can be approximated by a summation over triangularizations. The combinatorics of these tessellations of surfaces by regular polyhedra is organized very efficiently by considering the perturbative expansions of Hermitian matrix integrals.

The coupling of matter fields to theories of quantum gravity, or equivalently random surface models, proceeds by making the matrix variables depend on other (discrete or continuous) variables. In this fashion the $c = 1$ matrix model describes the quantum mechanics of an $N \times N$ hermitian matrix $\Lambda(x)$. The real variable x should be considered as (Euclidean) time. In particular we can opt to compactify x with period $2\pi\beta$. This essentially means we consider matrix quantum mechanics at some finite temperature $1/\beta$. As our Lagrangian we choose a simple kinetic term plus a potential $\mathrm{Tr}\, U(\Lambda)$ with U some polynomial. The partition function is thus given by

$$Z(N,\beta) = \int d\Lambda \exp\left\{-N\int_0^{2\pi\beta} dx\, \mathrm{Tr}\left[\left(\frac{d\Lambda}{dx}\right)^2 + U(\Lambda)\right]\right\}. \tag{1}$$

This model has a nontrivial double scaling limit $N \to \infty$ if we fine-tune the potential U to a critical point [6]. The model can consistently be reduced to the eigenvalues λ_i of the matrix Λ. The wonderful simple result is that in the double scaling limit all the interesting dynamics is reproduced by considering *free* fermions $\psi(\lambda)$ moving in a potential [6]

$$U(\lambda) = -\lambda^2. \tag{2}$$

This is the famous inverted harmonic oscillator. The physical model is obtained by filling, say, the right-hand side of the Fermi sea to a level μ from the top of the potential (see *fig.* 2.). This parameter has naturally the interpretation of the cosmological constant. All tunneling effects (leaking through the barrier) are thus $e^{-\mu}$ effects. When we recall that for the two-dimensional string $\mu = 1/\lambda$, we see that all these effects are also non-perturbative in the string coupling constant. They can consequently not be given an interpretation in terms of surfaces.

Of course, the Hamiltonian

$$H = -\frac{d^2}{d\lambda^2} - \lambda^2 \tag{3}$$

has a continuous spectrum. Its eigenfunctions

$$H\psi_\omega = \omega\psi_\omega \tag{4}$$

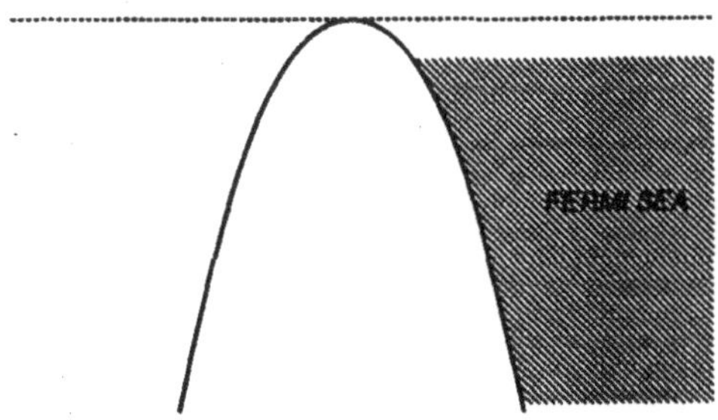

Fig. 2: The $c = 1$ matrix model reduces to a collection of non-interacting fermions in an inverted harmonic potential

can be expressed in parabolic cylinder functions [7]. These functions have the following interesting asymptotic behaviour. In the limit $\lambda \to \infty$ they behave as

$$\psi_\omega(\lambda) \sim \psi_0(\lambda) e^{i\omega(\pm\tau - x)} \tag{5}$$

where the new variable τ is defined as

$$\lambda = 2\sqrt{\mu} \cosh \tau. \tag{6}$$

That is, in (x, τ)-coordinates we find at spatial infinity (up to a factor) *relativistic* fermions.

Of course, the incoming and outgoing wave functions are not independent. They are related by a one-particle scattering process. That is, in general we will have

$$\psi_\omega^{out} = R_\omega \cdot \psi_\omega^{in} \tag{7}$$

where R_ω is the reflection factor that is fully determined by the shape of the potential. In this case the potential is the inverted quadratic, and the factor R_ω can be determined by a straightforward, but non-trivial, calculation. The outcome of the calculation certainly depends on how we treat the LHS of the potential. In the symmetric case we find [9]

$$R_\omega = (-i\mu)^{-\omega} \frac{\Gamma(\frac{1}{2} - i\mu + \omega)}{\Gamma(\frac{1}{2} - i\mu)}. \tag{8}$$

In some sense, this function is all one should know about the $c = 1$ matrix model in order to understand the scattering of tachyons. One verifies easily perturbative unitarity

$$R_\omega R^*_{-\omega} = 1 + O(e^{-\mu}). \tag{9}$$

In the Euclidean case we cannot really speak about scattering, since there are no separate past and future null infinities. We simply have a mixing of left and right-movers, given again by the factor R_ω. In the compactified case, to which we will restrict for the remainder, the energies will be furthermore quantized and are given by (recall the fermions are naturally anti-periodic)

$$\omega_n = \frac{1}{\beta}(n + \tfrac{1}{2}), \qquad n \in \mathbf{Z}. \tag{10}$$

The fermion S-matrix is now defined by

$$\psi^{in}(z) = S\psi^{out}(z)S^{-1}, \qquad \overline{\psi}^{in}(z) = S\overline{\psi}^{out}(z)S^{-1}, \tag{11}$$

with

$$S =: exp\Big[\sum_{m\in\mathbf{Z}} \log R_{\omega_m} \psi^{out}_{-(m+\frac{1}{2})} \overline{\psi}^{out}_{m+\frac{1}{2}}\Big] : \tag{12}$$

In order to calculate the tachyon scattering amplitudes we can now adopt the following simple strategy described in [9] (see *fig.* 3.). At spatial infinity the incoming and outgoing tachyons can be expressed in fermions through the standard formulas of relativistic fermionization and bosonization in two dimensions. The fermions have very simple scattering properties. Individual fermions just scatter with a reflection factor R_ω. In particular the number of fermions (and not only the fermion number) is conserved. After the scattering process the fermions propagate to infinity where they can be reassembled into bosons.

Thus we may write the full generating functional for connected Green's functions in terms of a single free boson

$$\partial\varphi(z) = \sum_n \alpha_n z^{-n-1} \,. \tag{13}$$

This scalar field should be thought of as the asymptotic limit of the tachyon field at τ, or equivalently $\phi, \to \infty$, where the Liouville interaction can be ignored. We now have a very simple representation for the generating function of all amplitudes [10]

$$\tau(t,\bar{t}) = \langle t|S|\bar{t}\rangle, \tag{14}$$

where the S-matrix is given by (12) and the coherent states now include an extra factor of μ

$$|\bar{t}\rangle = \exp\left(\sum_{n=1}^{\infty} i\mu\bar{t}_n\alpha_{-n}\right)|0\rangle, \qquad \langle t| = \langle 0|\exp\left(\sum_{n=1}^{\infty} -i\mu t_n\alpha_n\right). \tag{15}$$

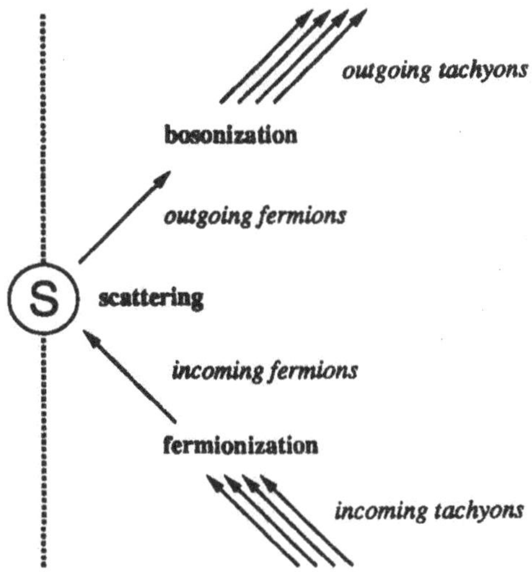

Fig. 3: The calculating of tachyon scattering amplitudes proceeds through fermionization and bosonization.

This formula is an enormous simplification over previous expressions for $c = 1$ amplitudes. It emphasizes the analogy with the $c < 1$ results which we recall were of the form

$$\tau(t) = \langle t|W\rangle = \langle t|S|0\rangle. \tag{16}$$

In particular we observe immediately from (14) that the $c = 1$ string forms a realization of the Toda Lattice hierarchy.

3.2. The Kontsevich-Penner model

As we have seen the partition function of the $c = 1$ string is a solution of the Toda Lattice hierarchy. For fixed values of the incoming times $\bar{t}_k$, the partition function $\tau(t, \bar{t})$ becomes a solution of the KP hierarchy. We want to determine in more detail the element $W_{\bar{t}}$ in the Grassmannian that parametrizes this particular orbit of the KP flows. To this end we have to consider the state

$$|W_{\bar{t}}\rangle = S \cdot U_{\bar{t}}|0\rangle, \qquad U_{\bar{t}} = \exp \sum_{n=1}^{\infty} i\mu\bar{t}_n\alpha_{-n}\ . \tag{17}$$

We will describe $|W_{\bar{t}}\rangle$ by giving a basis $v_k(z;\bar{t})$, $k \geq 0$, of one-particle wave-functions. First we observe that the operator $U_{\bar{t}}$ acts on the wave-functions z^n by simple multiplication

$$U_{\bar{t}}:\ z^n \to \exp\left(\sum i\mu\bar{t}_k z^{-k}\right)\cdot z^n\ . \tag{18}$$

Similarly we have for the action of S a multiplication

$$S:\ z^n \to R_n \cdot z^n\ . \tag{19}$$

We have already seen that the reflection factors R_n contain all the relevant information of the $c=1$ matrix model. At radius β they can be chosen to be

$$R_n = (-i\mu)^{-\frac{n+\frac{1}{2}}{\beta}} \frac{\Gamma(\frac{1}{2} - i\mu + \frac{n+\frac{1}{2}}{\beta})}{\Gamma(\frac{1}{2} - i\mu)}\ . \tag{20}$$

(Recall, we are only interested in the perturbative part in μ^{-1} of this expression.) The usual vacuum $|0\rangle$ is spanned by the non-negative powers z^k. Therefore the basis elements $v_k(z;\bar{t})$ of $W_{\bar{t}}$ are simply determined as

$$v_k(z;\bar{t}) = c_k \cdot S\ U_{\bar{t}} z^k\ , \tag{21}$$

with a normalization constant c_k such that $v_k(z;0) = z^k$. (This corresponds to the normal ordering of the S-matrix in (12).) Since the reflection factor is basically a gamma function, the result can be expressed as a Laplace transform

$$v_k(z;\bar{t}) = c'(z)\cdot\int_0^\infty dy\cdot y^k\cdot y^{-i\mu\beta+(\beta-1)/2} e^{i\mu(y/z)^\beta}\exp\left(\sum i\mu\bar{t}_k y^{-k}\right) \tag{22}$$

Here the constant $c'(z)$ is given by

$$c'(z) = \beta\frac{(-i\mu/z^\beta)^{\frac{1}{2}-i\mu}}{\sqrt{z}\Gamma(\frac{1}{2}-i\mu)}\ . \tag{23}$$

These integral representations are of Kontsevich type if and only if $\beta = 1$, that is, only at the self-dual radius. Indeed in that case we have

$$v_k(z;\bar{t}) = c'(z)\cdot\int_0^\infty dy\cdot y^k\cdot\exp i\mu\left(y/z - \log y + \sum\bar{t}_k y^{-k}\right)\ . \tag{24}$$

Therefore, following the procedure of §2.2 we can write the following matrix integral representation for the generating functional. Define the integral

$$\sigma(Z,\bar{t}) = \int dY e^{i\mu Tr[YZ^{-1}+V(Y)]}\ , \tag{25}$$

where

$$V(Y) = -\log Y + \sum \bar{t}_k Y^{-k} , \tag{26}$$

and we integrate over positive definite matrices Y. Then we have

$$\tau(t,\bar{t}) = \frac{\sigma(Z,\bar{t})}{\sigma(Z,0)} , \tag{27}$$

with the parameterization

$$t_n = \frac{\mu^{-1}}{n} Tr Z^{-n} . \tag{28}$$

Note that with this normalization $\tau(t,0) = 1$, which is appropriate since we consider normalized correlation functions. In order to write down the result we had to treat the incoming and outgoing tachyons very differently, parametrizing the outgoing states through whereas the coupling coefficients to the incoming states enter the matrix integral in a much more straightforward fashion.

3.3. Topological interpretation

Let us finish by mentioning some possible topological implications. The Hermitian one-matrix integral of logarithmic type has also been considered by Penner [20] in the special case $\bar{t} = 0$, $Z = 1$. In his work the matrix integral calculates the orbifold Euler characteristic of $\mathcal{M}_g$. A double-scaled version of this model was considered in [21] where it was conjectured to be related to the $c = 1$ string. Our analysis makes this relation much more explicit, and extends it to correlation functions. Actually, this result is good evidence of a conjectured relation between the $c = 1$ string at the self-dual radius $\beta = 1$ and topological field theory [22] that we will now briefly describe.

The correlation functions

$$\langle T_{n_1} \cdots T_{n_s} \rangle_g \tag{29}$$

should be related to the following modification of the Euler characteristic of moduli space. Consider the line bundle

$$L = K^2 \bigotimes_{i=1}^{s} \mathcal{O}(z_i)^{n_i+1} \tag{30}$$

on the surface Σ. We can now construct a vector bundle E over the moduli space $\mathcal{M}_{g,s}$ whose fiber over Σ is given by

$$E_\Sigma = H^0(\Sigma, L) \tag{31}$$

One easily verifies with Riemann-Roch that

$$\operatorname{rank} E = 3g - 3 + s. \tag{32}$$

In the case $n_i = 0$ this bundle E gives the cotangent bundle of $\mathcal{M}_{g,s}$. According to Witten we might have the following topological expression for the $\beta = 1$ tachyon correlators

$$\langle T_{n_1} \cdots T_{n_s} \rangle_g = \int_{\mathcal{M}_{g,s}} e(E). \tag{33}$$

Field theoretic evidence for this conjecture has been recently given in [23].

4. References

1. V. Kazakov, Phys. Lett. **159B** (1985) 303; F. David, Nucl. Phys. **B257** (1985) 45; V. Kazakov, I. Kostov, and A. Migdal, Phys. Lett. **157B** (1985) 295; J. Fröhlich, *The statistical mechanics of surfaces*, in *Applications of Field Theory to Statistical Mechanics*, L. Garrido ed. (Springer, 1985).
2. E. Brézin and V. Kazakov, *Exactly solvable field theories of closed strings*, Phys. Lett. **B236** (1990) 144.
   M. Douglas and S. Shenker, *Strings in less than one dimension*, Nucl. Phys. **B335** (1990) 635.
   D.J. Gross and A. Migdal, *Nonperturbative two dimensional quantum gravity*, Phys. Rev. Lett. **64** (1990) 127.
3. *Random Surfaces, Quantum Gravity and Strings*, Proceedings of the 1990 Cargèse workshop, O. Alvarez, E. Marinari, and P. Windey eds. (Plenum Press, 1991).
   Two Dimensional Quantum Gravity and Random Surfaces, Proceedings of the Jerusalem Winter School for Theoretical Physics, D. Gross, T. Piran, and S. Weinberg eds. (World Scientific, 1992)
   String Theory and Quantum Gravity '91, Proceedings of the Trieste Spring School and Workshop 1991, J. Harvey *et al* eds. (World Scientific, 1992).
4. T. Banks, M.R. Douglas, N. Seiberg, and S. Shenker, *Microscopic and macroscopic loops in nonperturbative two dimensional gravity*, Phys. Lett. **238B** (1990) 279.
   D.J. Gross and A. Migdal, *A nonperturbative treatment of two dimensional quantum gravity*, Nucl. Phys. **B340** (1990) 333.
5. M. Douglas, *Strings in less than one dimension and the generalized KdV hierarchies*, Phys. Lett. **238B** (1990) 176.
6. E. Brézin, V. Kazakov, and Al.B. Zamolodchikov, *Scaling violation in a field theory of closed strings in one physical dimension*, Nucl. Phys. **B338** (1990).
   D.J. Gross and N. Miljković, *A nonperturbative solution of $D = 1$ string theory*,

Phys. Lett. **238B** (1990) 217.
P. Ginsparg and J. Zinn-Justin, *2-d Gravity and 1-d matter,* Phys. Lett. **240B** (1990) 333.
D.J. Gross and I.R. Klebanov, *One-dimensional string theory on a circle,* Nucl. Phys. **B344** (1990) 475.

7. G. Moore, *Double-scaled field theory at* $c = 1$, Nucl. Phys. **B368** (1992) 557.
8. I.R. Klebanov and D. Lowe, *Correlation functions in two-dimensional gravity coupled to a compact scalar field,* Nucl. Phys. **B363** (1991) 543.
9. G. Moore, R. Plesser, and S. Ramgoolam, *Exact S-matrix for 2D string theory,* Nucl. Phys. **B377**(1992)143.
G. Moore and R. Plesser, *Classical scattering in 1+1 dimensional string theory,* Yale preprint YCTP-P7-92.
10. R. Dijkgraaf, G. Moore and R. Plesser, *The partition function of 2d string theory,* to be published in Nucl. Phys. B.
11. E. Witten, *Two dimensional gravity and intersection theory on moduli space,* Surveys In Diff. Geom. **1** (1991) 243-310.
12. M. Kontsevich, *Intersection theory on the moduli space of curves and the matrix Airy function,* Commun. Math. Phys. **147** (1992) 1–23.
13. R. Dijkgraaf, *Intersection theory, integrable hierarchies and topological field theory,* in Proceedings of the Cargèse Summer School on "New Symmetry Principles in Quantum Field Theory" (Plenum Press).
14. T. Curtright and C. Thorn, Phys. Rev. Lett. **48** (1982) 1309; Ann. Phys. **147** (1983) 365; Ann. Phys. **153** (1984) 147.
15. U. Danielsson and D.J. Gross, *On the correlation functions of the special operators in* $c = 1$ *quantum gravity,* Nucl. Phys. **B366** (1991) 3.
E. Witten, *Ground ring of two dimensional string theory,* Nucl. Phys. **B373** (1992) 187.
I.R. Klebanov and A.M. Polyakov, *Interaction of discrete states in two-dimensional string theory,* Mod. Phys. Lett. **A6** (1991) 3273.
D. Kutasov, E. Martinec, and N. Seiberg, *Ground rings and their modules in 2d gravity with* $c \leq 1$ *matter,* Phys. Lett. **276B** (1992) 437.
I.R. Klebanov, *Ward identities in two-dimensional string theory,* Mod. Phys. Lett. **A7** (1992) 723.
16. S. Kharchev, A. Marschakov, A. Mironov, A. Morozov, A. Zabrodin, *Unification of all string models with* $c \leq 1$, Phys. Lett. **275B** (1992) 311. *Towards unified theory of 2d gravity,* Lebedev Institute preprint FIAN/TD-10/91, hepth 9201013.
A. Marshakov, *On string field theory at* $c \leq 1$, Lebedev Institute preprint FIAN/TD-8/92, hepth 9208022.
17. G. Segal and G. Wilson, *Loop groups and equations of KdV type,* Publ. Math.

I.H.E.S. **61** (1985) 1.
E. Date, M. Jimbo, M. Kashiwara, T. Miwa, *Transformation groups for soliton equations,* RIMS Symp. *Nonlinear Integrable Systems—Classical Theory* and *Quantum Theory* (World scientific, Singapore, 1983).

18. K. Ueno and K. Takasaki, *Toda lattice hierarchy,* in *Group Representations and Systems of Differential Equations*, H. Morikawa Ed., Advanced Studies in Pure Mathematics 4.
19. Harish-Chandra, *Differential operators on a semisimple Lie algebra,* Am. J. Math. **79** (1987) 87.
C. Itzykson and J.-B. Zuber, *The planar approximation II,* J. Math. Phys. **21** (1980) 411.
M.L. Mehta, *A method of integration over matrix variables,* Commun. Math. Phys. **79** (1981) 327.
20. R.C. Penner, Commun. Math. Phys. **113** (1987) 299; J. Diff. Geom.**27** (1988) 35.
21. J. Distler and C. Vafa, Mod. Phys. Lett. **A6** (1991) 259; in *Random Surfaces and Quantum Gravity*, Eds: O. Alvarez *et al.*
22. E. Witten, *The N matrix model and gauged WZW models*, Nucl. Phys. **B371** (1992) 191–245.
23. S. Muhki and C. Vafa, *Two dimensional black-hole as a topological coset model of* $c = 1$ *string theory,* Harvard preprint HUPT-93/A002 hep-th/9301082.

MULTILOOPS FOR TWO-DIMENSIONAL GRAVITY COUPLED TO ONE-DIMENSIONAL MATTER

Paolo Di Vecchia
NORDITA, Blegdamsvej 17, DK-2100 Copenhagen Ø
Denmark

ABSTRACT

In this talk I review the results obtained by means of the sewing technique for the multiloop amplitudes in the closed bosonic string. The same technique is then used in order to compute multiloop amplitudes in non critical string theory where one has two-dimensional gravity coupled with a one-dimensional scalar field. Multiloop amplitudes in this theory are explicitly written.

1. Introduction

Several methods have been used for computing correlation functions of conformal field theories on arbitrary Riemann surfaces and multiloop amplitudes in string theories. A hopefully complete list can be found in my review talk given in Erice last summer [1] where also the status of multiloop calculations in superstring theories is discussed in some detail. In my Erice talk one can also find a short description of the sewing technique applied to both the bosonic and superstring theories.

This talk consists of two parts.

In the first part presented in sect. 1 I will summarize the explicit results obtained for the closed bosonic critical (D=26) string and I will discuss in some detail the factorization and modular invariance properties of the multiloop amplitudes.

In the second part presented in sect 2, after a brief discussion of two-dimensional gravity coupled to a one-dimensional scalar matter that after quantization results in a two-dimensional Weyl anomaly-free string theory in which the second dimension is provided by the Liouville field, I will apply the sewing technique obtaining explicit expressions for the multiloop amplitudes in this theory.

2. Critical Bosonic String

A complete expression for the N-string g-loop vertex in the closed bosonic string can be found at the beginning of sect 4 of Ref. [1]. Such an object is the 'generating functional' of all multiloop amplitudes of the theory. In particular in order to obtain the g loop scattering amplitude involving N particular states of the string one must saturate $V_{N;g}$ with those states.

If one wants to compute the g loop scattering amplitudes for N tachyons one must saturate the vertex with the states $|0; k_i>_i$ and one gets the following scattering amplitude:

$$\int \prod_{a=1}^{3g-3} d^2 m_a M(m) \int \prod_{i=1}^{N} d^2 z_i$$

$$\prod_{i<j} \left\{ |E(z_i, z_j)|^2 \exp\left[\frac{1}{\pi} Re \left(\int_{z_i}^{z_j} \omega \right) (Im\tau)^{-1} Re \left(\int_{z_i}^{z_j} \omega \right) \right] \right\}^{k_i \cdot k_j / 4} \tag{1}$$

In this section I have taken $\alpha' = 1/2$.

The function $M(m)$ is a function of the $3g-3$ moduli m_a that are used for describing the Riemann surface. In the sewing procedure outlined in Sect. 3 of Ref. [1] one gets the Riemann surface represented in terms of its Schottky parametrization. This parametrization is described in detail in Refs. [2, 3].

In the case of the Schottky parametrization of the Riemann surface the moduli are the fixed points η_μ and ξ_μ and the multipliers k_μ of the generators of the Schottky group. In this parametrization $M(\xi, \eta, k)$ is given by [4, 5, 6, 7]

$$\prod_a d^2 m_a M(m) = 1/dV_{abc}$$

$$\prod_{\mu=1}^{g} \left[\frac{d^2 k_\mu d^2 \eta_\mu d^2 \xi_\mu |1-k_\mu|^4}{|k_\mu|^4 |\xi_\mu - \eta_\mu|^4} \right] \prod_\alpha{}' \left| \prod_{n=1}^{\infty} (1-k_\alpha^n)^{-D} \prod_{n=2}^{\infty} (1-k_\alpha^n)^2 \right|^2 (\det(\pi Im\tau))^{-\frac{D}{2}} \tag{2}$$

In front of the measure we have included the volume of the projective group given by

$$dV_{abc} = \frac{d^2\rho_a d^2\rho_b d^2\rho_c}{|\rho_a - \rho_b|^2 |\rho_b - \rho_c|^2 |\rho_a - \rho_c|^2} \tag{3}$$

where ρ_a, ρ_b and ρ_c are three among the fixed points of the Schottky group, that must be fixed because of an overall projective invariance.

We have also not specified the value of the space-time dimension D that for a critical bosonic string must be $D = 26$.

Also the geometrical objects as the abelian differentials, the prime form and the period matrix come out expressed in the Schottky parametrization. They are respectively given by

$$\omega_\mu(z) = \sum_\alpha{}^{(\mu)} \left(\frac{1}{z - T_\alpha(\eta_\mu)} - \frac{1}{z - T_\alpha(\xi_\mu)} \right) dz \tag{4}$$

$$E(z,w) = (z-w) \prod_\alpha{}' \frac{[z - T_\alpha(w)]}{[z - T_\alpha(z)]} \frac{[w - T_\alpha(z)]}{[w - T_\alpha(w)]} \tag{5}$$

$$\tau_{\mu\nu} = \frac{1}{2\pi i} \left[\log k_\mu \, \delta_{\mu\nu} + {}^{(\mu)}\sum_\alpha{}'^{(\nu)} \log \frac{[\eta_\mu - T_\alpha(\eta_\nu)][\xi_\mu - T_\alpha(\xi_\nu)]}{[\eta_\mu - T_\alpha(\xi_\nu)][\xi_\mu - T_\alpha(\eta_\nu)]} \right] \tag{6}$$

Their detailed construction through the sewing procedure can be found in Ref. [3].

In order to make explicit the Belavin-Knizhnik singularities [8] it is convenient to fix $\eta_1 = 1$, $\eta_g = 0$ and $\xi_1 = \infty$ and to pass to the new variables [5]

$$\xi_\mu = (1 - H_\mu) A_2 A_3 \cdots A_\mu \qquad \eta_\mu = A_2 A_3 \cdots A_\mu \qquad 2 \leq \mu \leq g-1$$

$$\xi_g = A_2 A_3 \cdots A_g \tag{7}$$

In terms of the new variables the measure gets modified as follows

$$\prod_a d^2 m_a M(m) = \prod_{\mu=1}^{g} \left[\frac{d^2 k_\mu |1 - k_\mu|^4}{|k_\mu|^4} \right] \prod_{\mu=2}^{g} \frac{d^2 A_\mu}{|A_\mu|^4} \prod_{\mu=2}^{g-1} \frac{d^2 H_\mu}{|H_\mu|^4}$$

$$\prod_\alpha{}' \left| \prod_{n=1}^{\infty} (1 - k_\alpha^n)^{-D} \prod_{n=2}^{\infty} (1 - k_\alpha^n)^2 \right|^2 (\det(\pi Im\tau))^{-\frac{D}{2}} \tag{8}$$

The limit of small A_μ, H_μ or k_μ corresponds to the pinching limit in the appropriate channel as shown in figure 1 of the first paper of Ref. [9].

The singularity of the measure in these limits corresponds to the Belavin-Knizhnik singularities due to the tachyon and dilaton exchanges in the various channels. These singularities are the origin of the divergences of the multiloops in the bosonic string.

The divergence for $k_\mu \to 0$ can already be seen in the one loop case. In this case the N tachyon scattering amplitude can be written in the following form

$$A(k_1, k_2, \ldots k_N) = \int_F \frac{d^2\tau}{(Im\tau)^2} C(\tau) F(\tau) \tag{9}$$

where the integral should be performed in the fundamental region F specified by $-\frac{1}{2} \leq Re\tau \leq \frac{1}{2}; |\tau| > 0; Im\tau > 0$,

$$C(\tau) = (\frac{1}{2} Im\tau)^{-12} e^{4\pi Im\tau} \prod_{n=1}^{\infty} |1 - e^{2\pi i n\tau}|^{-48} \tag{10}$$

and

$$F(\tau) = 4(\pi\kappa)^N Im\tau \int \prod_{i=1}^{N-1} d^2\nu_i \prod_{i<j} [\chi_{ij}]^{k_i \cdot k_j/2} \tag{11}$$

where

$$\chi_{ij} = 2\pi \left| \frac{\Theta_1(\nu_i - \nu_j|\tau)}{\Theta_1'(0|\tau)} \right| e^{-\pi[Im(\nu_i - \nu_j)]^2/(Im\tau)} \tag{12}$$

In the case of one loop we have only one moduli $k = e^{2\pi i\tau}$ and the tachyon and dilaton singularities for $k \to 0$ correspond to the singular behaviour of the exponentially divergent term in $C(\tau)$ when $Im\tau \to \infty$. Unless the case of a pointlike particle there is no divergence for $Im\tau \to 0$ because this point is outside the fundamental region of integration F.

The Koba-Nielsen variables z_i are related to ν_i by the relation $z_i = e^{2\pi i\nu_i}$. Because of the overall projective invariance we have fixed the N-th Koba-Nielsen variable to zero ($\nu_N = 0$).

In order to see already at one loop the divergence arising from the pinching limit $A_\mu \to 0$ it is convenient to make the following change of variables

$$\nu_i = \epsilon\eta_i \qquad i = 1, 2, \ldots N-1 \qquad ; \qquad \nu_{N-1} = \epsilon e^{i\Phi} \tag{13}$$

where both ϵ and Φ are real variables.

One then obtains the following expression

$$\int \frac{d\epsilon}{\epsilon^3} d\Phi \int \prod_{i=1}^{N-2} d^2\eta_i \prod_{1\leq i<j\leq N-1} |\eta_i - \eta_j|^{k_i \cdot k_j/2} \int \frac{d^2\tau}{Im\tau} C(\tau) \tag{14}$$

that is divergent for $\epsilon \to 0$ exactly as $\frac{d|A|}{|A|^3}$ in expression (8). Remember that if we perform the same change of variables in the case of the four graviton amplitude in superstring theory we get a factor $d\epsilon\epsilon^5$ that is convergent for $\epsilon \to 0$[1].

In conclusion we see that the factors in the measure in eq. (8) that are divergent for $k_\mu \to 0$, $A_\mu \to 0$ or $H_\mu \to 0$ are responsible for the divergences in the bosonic string. Those divergences must disappear in superstring theories if we want them to provide finite theories of gravity.

Another interesting property of the measure in eq. (8) is its factorization. If we consider it for simplicity in the two loop case and we study its behaviour in the pinching limit $A \to 0$ we obtain that it factorizes into the product of two measures corresponding to the one loop case. This property follows from the following limiting expressions for the period matrix

$$2\pi i\tau_{\mu\mu} = \log k_\mu + 0(A^2) \qquad 2\pi i\tau_{12} = A + 0(A^2) \tag{15}$$

and for the rest of the measure

$$\prod_\alpha{}' \left| \prod_{n=1}^\infty (1 - k_\alpha^n)^{-26} \prod_{n=2}^\infty (1 - k_\alpha^n)^2 \right|^2 \prod_{\mu=1}^2 |1 - k_\mu|^4 = \prod_{\mu=1}^2 \prod_{n=1}^\infty |1 - k_\mu|^{-48} + 0(A^2) \tag{16}$$

The expression (8) for the measure has the advantage of being quite explicit, but has also the disadvantage of having a form in which it is very difficult to show that it is modular invariant.

In the case of two and three loops explicit expressions have been obtained using the elements of the period matrix as the $3g-3$ moduli.

One gets for $g = 2$ [11]

$$\prod_a [d^2 m_a] M(m) = d^2\tau_{11} d^2\tau_{12} d^2\tau_{22} \prod_{\alpha,\beta} |\theta \begin{pmatrix} \alpha \\ \beta \end{pmatrix} (0|\tau)|^{-4} \tag{17}$$

and for $g = 3$ [12]

$$\prod_a [d^2 m_a] M(m) = d^2\tau_{11} d^2\tau_{12} d^2\tau_{22} d^2\tau_{13} d^2\tau_{23} d^2\tau_{33} \prod_{\alpha,\beta} |\theta \begin{pmatrix} \alpha \\ \beta \end{pmatrix} (0|\tau)|^{-1} \tag{18}$$

[1]See for instance Ref. [10]

These expressions are not only very explicit, but also manifestly modular invariant.

By performing the transformation between the elements of the τ matrix and the moduli k, A and H it has been checked [9], by making an expansion for small values of A and H, that the expressions given in eqs. (17) and (18) and the corresponding expressions obtained with a parametrization of the moduli space given in terms of the Schottky parametrization do indeed coincide for the first few orders.

Finally also the part in eq. (1) depending on the punctures can be easily shown to be modular invariant by using the following transformation properties under the modular group of the period matrix

$$\tau \rightarrow (A\tau + B)(C\tau + D)^{-1} \qquad (Im\tau)^{-1} \rightarrow (C\tau + D)(Im\tau)^{-1}(\bar{\tau}C^t + D^t) \tag{19}$$

the abelian differentials

$$\omega \rightarrow \omega(C\tau + D)^{-1} \qquad \omega \rightarrow (\tau C^t + D^t)^{-1}\omega \tag{20}$$

and the prime form

$$E(z_i, z_j) \rightarrow \exp\left\{\frac{1}{4\pi i}\int_{z_i}^{z_j} \omega(C\tau + D)^{-1}C\int_{z_i}^{z_j}\omega\right\} E(z_i, z_j) \tag{21}$$

and the conservation of momentum $\sum_{i=1}^{N} k_i = 0$.

A, B, C, D are g-dimensional matrices satisfying the conditions

$$A^tC = C^tA, \qquad B^tD = D^tB \qquad A^tD - C^tB = 1 \tag{22}$$

Recently Mandelstam [13], choosing two other sets of moduli, has written two equivalent expressions for the measure in eq. (8) in terms of Θ-functions, prime forms and abelian differentials and has shown that they are equal to the expression given in eq. (8). His expressions are also manifestly modular invariant.

3. Noncritical String

Two-dimensional gravity coupled to $(D-1)$ scalar fields is described by the following action

$$Z = \int \frac{Dg_{\alpha\beta}Dx^{\mu}}{Vol.Diff.} \exp\left\{-\frac{1}{4\pi\alpha'}\int d^2\sigma\sqrt{g}\left[g^{\alpha\beta}\partial_\alpha x\cdot\partial_\beta x + \frac{1}{2}\alpha'\mu_0\right]\right\} \tag{23}$$

where in the case of a genus g topology the functional integral is over all metrics corresponding to a genus g topology. It is convenient to choose the conformal gauge where $g_{\alpha\beta} = e^{-\phi}\hat{g}_{\alpha\beta}$ and then it can be shown [14] that one can replace the covariant norm for the Liouville field with the gaussian norm by renormalizing the action. One finally arrives at the following expression

$$Z = \int [D\phi]_{\hat{g}}[DbDc]_{\hat{g}}[Dx]_{\hat{g}}e^{-S+S_{b,c}} \tag{24}$$

where $S_{b,c}$ is the usual ghost action and

$$S = \int d^2\sigma\sqrt{\hat{g}}\left[\frac{1}{4\pi\alpha'}\hat{g}^{\alpha\beta}\left(\partial_\alpha x \cdot \partial_\beta x + \partial_\alpha\phi\partial_\beta\phi\right) - \frac{1}{8\pi}QR_{\hat{g}} + \mu e^{\alpha\phi}\right] \tag{25}$$

The background charge for the Liouville field is fixed by requiring the theory to be Weyl invariant. This implies that the sum of the central charges of the matter fields, of the Liouville field and of the ghost field must be vanishing:

$$d + c_L - 26 = 0 \tag{26}$$

where $c_L = 1 + 3Q^2$ and $d = D - 1$.

The previous equations imply that

$$Q = \sqrt{\frac{25-d}{3}} \tag{27}$$

The choice of the sign is not important because it can be reabsorbed into the definition of ϕ.

The other parameter α can be determined by requiring that the Liouville exponential term be conformal invariant. This implies

$$1 = -\frac{1}{2}\alpha(Q+\alpha) \qquad \alpha_\pm = -\frac{1}{2\sqrt{3}}\left[\sqrt{25-d} \mp \sqrt{1-d}\right] \tag{28}$$

Notice that, in order α to be real, we need to restrict ourselves to the case $d \leq 1$.

If we limit ourselves to the case $d = 1$ we can see that two-dimensional gravity coupled to a scalar field reduces essentially to a two-dimensional string theory in which the role of the timelike coordinate is played by the Liouville field. Notice also that such a theory is not Lorentz-invariant because the two components are not treated symmetrically having the Liouville field a nonvanishing background charge together with an exponential Liouville term.

In a two-dimensional string one does not expect any transverse excitation. The only state of the spectrum is a tachyon, that in two dimensions is actually a massless state as follows from the zero energy argument of Brink and Nielsen [15].

This is actually not true because in addition to the "tachyon" there are also special states providing a very rich and unexpected structure.

The scattering of "tachyons" can also be studied in this theory. The integration over the zero modes of the matter and Liouville fields can be explicitly performed and one gets the following expression for the scattering of N tachyons on a genus g Riemann surface

$$2\pi\delta[\sum_{i=1}^{N} k_i]\frac{\Gamma(-s)}{\alpha}A_g(k_1, k_2, \ldots k_N) \tag{29}$$

where

$$A_g(k_1, k_2, \ldots k_N) = \prod_{a=1}^{3g} d^2m_a\hat{M}_g(m)\frac{1}{dV_{abc}}\prod_{i=1}^{N} d^2z_i$$

$$< \prod_{i=1}^{N} e^{ik_i\tilde{x}(z_i)} >_x < \left[\frac{\mu}{\pi}\int d^2w\sqrt{\hat{g}}e^{\alpha\phi(w)}\right]^s \prod_{i=1}^{N} e^{\beta_i\tilde{\phi}(z_i)} >_\phi \tag{30}$$

and

$$s \equiv -\frac{1}{\alpha}\left[Q(1-g)+\sum_{i=1}^{N}\beta_i\right] \tag{31}$$

$\alpha = -\sqrt{2}$, $Q = 2\sqrt{2}$. k_i is arbitrary and $\beta_i = -\sqrt{2}+\chi_i k_i$. χ_i is the chirality of the states that can be $\chi_i = \pm 1$ and $\alpha' = 2$. The fields $\tilde{x}$ and $\tilde{\phi}$ do not contain the zero modes of the matter and Liouville fields respectively.

The integral over the Liouville zero mode makes sense only if $s < 0$. On the other hand one is able to compute the correlation functions involving the Liouville field in eq. (30) only if $s > 0$. Then, in order to compute the amplitude in eq. (30) one must rely in some way of extending the expression obtained from positive to negative values of s. Notice, however, that this extension must be done only on the part depending on the punctures. In particular the measure of integration in moduli space is independent on the continuation from positive to negative s.

In the case of the sphere the calculation of the scattering amplitude has been explicitly performed obtaining [16]:

$$A_{g=0}(k_1, k_2, \ldots k_N) =$$

$$\int \prod_{i=1}^{N+s-3} d^2z_i \prod_{i<j} |z_i - z_j|^{2(k_ik_j - \beta_i\beta_j)} = \frac{\pi^{N-3}[\mu\Delta(\alpha^2/2)]^s}{\Gamma(N+s-2)} \prod_{i=1}^{N} \frac{\Gamma(1-\sqrt{2}k_i)}{\Gamma(\sqrt{2}k_i)} \tag{32}$$

where $\Delta(x) = \frac{\Gamma(x)}{\Gamma(1-x)}$. We have also taken $k_i = 0$ and $\beta_i = -\sqrt{2}$ for $i > N$.

Also the torus partition function has been computed in this theory in the case in which the scalar field is compactified on a circle of radius R. For $g = 1$ and no external particles one can take $s = 0$ and one gets [17]

$$Z_1 = \frac{T}{12}[r + 1/r] \tag{33}$$

where $r = R/\sqrt{\alpha'}$ and $T = 1/(2\pi\alpha')$ is the string tension.

The previous expression can be generalized to the case in which there are also external tachyonic states always keeping $s = 0$ and one gets

$$\int_F \frac{d^2\tau}{(Im\tau)^2} G(\tau) \tag{34}$$

where

$$G(\tau) = Im\tau \int \prod_{i=1}^{N-1} d^2\nu_i \prod_{i<j} |\chi_{ij}|^{2(k_ik_j - \beta_i\beta_j)} I(\nu_i, \tau) \tag{35}$$

and

$$I(\nu_i, \tau) = \sum_{n,m=-\infty}^{\infty} e^{-\pi r^2|n-m\tau|^2/Im\tau} e^{\sqrt{2}\pi r \sum_{i=1}^{N} k_i[\nu_i(m\bar{\tau}+n)-\bar{\nu}_i(m\tau+n)]} \tag{36}$$

It is easy to check that both the integration measure in eq. (34) and the functions $I(\nu_i, \tau)$ and $G(\tau)$ are modular invariant provided that the tachyons are on mass shell

$$\frac{1}{2}k_i^2 - \frac{1}{2}\beta_i^2 - \sqrt{2}\beta_i = 1 \tag{37}$$

and the momentum is conserved $\sum_{i=1}^N k_i = \sum_{i=1}^N \beta_i = 0$.

The fact that in this case the function $C(\tau)$ appearing in eq. (10) is equal to a constant implies that the contribution of the higher states cancels. However it does not seem easy anymore to perform the integral over τ because of the complicated dependence of χ_{ij} and $I(\nu_i|\tau)$ on τ.

If s is a positive integer then the Liouville correlation functions on a genus g Riemann surface can be computed and one gets the following g loop N tachyon amplitude

$$A_g(k_1, k_2, \ldots k_N) = \int \prod_{a=1}^{3g-3} d^2 m_a \hat{M}(m) \int \prod_{i=1}^{N+s} d^2 z_i \prod_{i<j} |E(z_i, z_j)|^{2(k_ik_j - \beta_i\beta_j)}$$

$$\exp\left\{-\frac{1}{2\pi}\left[\sum_{i=1}^N k_i Re\left(\int_{z_0}^{z_i} \omega_\mu\right)\right](Im\tau)^{-1}_{\mu\nu}\left[\sum_{j=1}^N k_j Re\left(\int_{z_0}^{z_j} \omega_\nu\right)\right]\right\}$$

$$\exp\left\{-\frac{1}{2\pi}L_\mu (Im\tau)^{-1}_{\mu\nu} L_\nu\right\} I(k_i) \tag{38}$$

where

$$L_\mu = \left[\sum_{i=1}^{N+s} \beta_i Re\left(\int_{z_0}^{z_i} \omega_\mu\right) + 4\pi\sqrt{2} Im\left(\Delta_\mu^{z_0}\right)\right] \tag{39}$$

and

$$I(k_i) = \sum_{n,m} e^{-\pi r^2 [m\tau - n](Im\tau)^{-1}[\bar{\tau}m - n]}$$

$$e^{-ir\left\{[m\tau - n](Im\tau)^{-1}\sum_{i=1}^N \frac{k_i}{\sqrt{2}}\int_{z_0}^{z_i}\omega + [m\bar{\tau} - n](Im\tau)^{-1}\sum_{i=1}^N \frac{k_i}{\sqrt{2}}\int_{z_0}^{z_i}\omega\right\}} \tag{40}$$

The measure is given by

$$\prod_{a=1}^{3g-3} d^2 m_a \hat{M}(m) = \prod_{\mu=1}^{g}\left[\frac{d^2 k_\mu}{|k_\mu|^4}\right]\prod_{\mu=2}^{g}\left[\frac{d^2 A_\mu}{|A_\mu|^4}\right]\prod_{\mu=2}^{g-1}\left[\frac{d^2 H_\mu}{|H_\mu|^4}\right]\frac{\prod_{\mu=1}^g |1-k_\mu|^4}{\prod'_\alpha |1-k_\alpha|^4}[\det(\pi Im\tau)]^{-1} \tag{41}$$

The previous measure differs from the measure (8) of the critical string ($D = 26$) by the factor

$$\hat{M}(m) = M(m)[\det(Im\tau)]^{12}|Z_1|^{24} \tag{42}$$

that is modular invariant. Z_1 is given in eq. (2.7) of Ref. [1] and is equal to

$$Z_1 = \prod_\alpha{}' \prod_{n=1}^{\infty}(1 - k_\alpha^n)^2 \tag{43}$$

Then from the modular invariance of the measure (8) we derive also the modular invariance of the measure (41). As shown in the previous section the measure satisfies also the required factorization properties. We have also taken $k_i = 0$ and $\beta_i = -\sqrt{2}$ for $i > N$.

In conclusion the measure $\hat{M}(m)$ provides an explicit, factorizable and modular invariant expression for the term $I_g(\tau)$ that appears in eq. (4.2) of the second paper of Ref. [17].

Finally by using the transformation properties under modular transformations of the Riemann constants

$$\Delta \to (\tau C^t + D^t)^{-1}\Delta \tag{44}$$

and those given in eqs. (19), (20) and (21) one can easily show that the part depending on the punctures is also modular invariant.

We conclude noticing that the factor $\Gamma(-s)\mu^s$ appearing in eqs. (29) and (30), is equal in the case with no external legs to $\Gamma(2g-2)\mu^{2-2g}$ (using eq. (31)), that reproduces the behaviour as a function of the cosmological constant of the free energy of matrix models [2]. The divergent factorial behaviour given by $\Gamma(2g-2)$ for $g \to \infty$ is also reproduced in matrix models [18]. In order to check the rest of the formula one needs to continue the part depending on the punctures from positive to negative values of s and then integrate over moduli space!

4. References

1. P. Di Vecchia, to be published in *International Journal of Modern Physics.* Nordita preprint 92/59 P.
2. S. Mandelstam, in *Unified String Theories*, eds. M. Green and D. Gross (World Scientific 1986) and in *Advances in Theoretical Physics*, ed. A.H. Luther (Pergamon Press)
3. P. Di Vecchia, F. Pezzella, M. Frau, K. Hornfeck, A. Lerda and S. Sciuto, **Nucl. Phys. B322** (1989) 317.
4. P. Di Vecchia, M. Frau, A. Lerda and S. Sciuto, **Phys. Lett.199B** (1987) 49.
   P. Di Vecchia, K. Hornfeck, M. Frau, A. Lerda and S. Sciuto, **Phys. Lett. 206B** (1988) 643.
   P. Di Vecchia, K. Hornfeck, M. Frau, A. Lerda and S. Sciuto, in *DST Workshop on Particle Physics-Superstring Theory*, eds. H. S. Mani and R. Ramachandran (World Scientific, 1988), p. 41.
5. J.L. Petersen and J. Sidenius, **Nucl. Phys. B301** (1988) 247.
   J.L. Petersen, J. Sidenius and A.K. Tollstén, **Nucl. Phys. B317** (1989) 109.
   J.L. Petersen, in *Physics and Mathematics of Strings*, eds. L. Brink, D. Friedan and A.M. Polyakov (World Scientific) p. 434.
6. G. Cristofano, F. Nicodemi and R. Pettorino, **Phys. Lett.200B** (1988) 292 and **Int. Jour. of Mod. Phys.A4** (1989) 857.

[2] See for instance eq. (39) in Ref. [18].

G. Cristofano, R. Musto, F. Nicodemi and R. Pettorino, **Phys. Lett.217B** (1989) 59.

7. H. Konno, **Phys. Lett. 212B** (1988) 165.
8. A.A. Belavin and V.G. Knizhnik. **Phys. Lett.168B** (1986) 201.
9. J.L. Petersen, K.O. Roland and J.R. Sidenius, **Phys. Lett.205B** (1988) 262.
   K.O. Roland, **Nucl. Phys. B313** (1989) 432.
10. M.B. Green, J.H. Schwarz and E. Witten, *Superstring Theory*, (Cambridge University Press, 1987), vol. 2 p.93.
11. A.A. Belavin, V.G. Knizhnik, A. Morozov and A. Perelomov, **Phys. Lett.177B** (1986) 324.
   G. Moore, **Phys. Lett.176B** (1986) 369.
   A. Kato, Y. Matsuo and S. Odake, **Phys. Lett.179B** (1986) 241.
12. A.A. Belavin, V.G. Knizhnik, A. Morozov and A. Perelomov, **Phys. Lett. 184B** (1987) 171.
13. S. Mandelstam, **Phys. Lett. 277B** (1992) 82.
14. F. David, **Mod. Phys. Lett. A3** (1988) 1651.
   J. Distler and H. Kawai, **Nucl. Phys. B321** (1989) 509.
   E. D'Hoker and P.S. Kurzepa, **Mod. Phys. Lett. A5** (1990) 1411.
15. L. Brink and H. B. Nielsen, **Phys. Lett. 45B** (1973) 332.
16. P. Di Francesco and D. Kutasov, **Phys. Lett. 261B** (1991) 385.
   V. Dotsenko, **Mod. Phys. Lett. A6** (1991) 3601.
   M. Goulian and M. Li, **Phys. Rev. Lett. 66** (1991) 2051.
   S. Gupta, A. Trivedi and M. Wise, **Nucl. Phys. B340** (1990) 475.
17. M. Bershadsky and I. R. Klebanov, **Phys. Rev. Lett. 65** (1990) 3088.
   N. Sakai and Y. Tanii, **Int. Jour. of Mod. Phys. A6** (1991) 2743.
18. I. Klebanov, in *String Theory and Quantum Gravity '91*, ed. J. Harvey el al. (World Scientific, 1992). p. 46.

THE OPERATOR ALGEBRA OF THE DISCRETE STATE OPERATORS IN 2D GRAVITY WITH NON-VANISHING COSMOLOGICAL CONSTANT

Vl.S.Dotsenko* †

LPTHE‡

Université Pierre et Marie Curie, PARIS VI

Tour 16, 1er étage

4 place Jussieu

75252 Paris CEDEX 05, FRANCE

ABSTRACT

Remarks are given to the structure of physical states in the 2D gravity coupled to $C \leq 1$ matter. The operator algebra of the discrete state operators is calculated for the theory with non-vanishing cosmological constant.

*E-mail address: dotsenko@lpthe.jussieu.fr

†Permanent address: Landau Institute for Theoretical Physics, Moscow.

‡Laboratoire associé No. 280 au CNRS

1 Introduction and results

The extra discrete physical states in the 2D gravity coupled to matter have been a subject of a number of recent studies. Their existance had been observed in the matrix model approach [1] for $C = 1$ theory, and they have been defined in the continuum theory and further analyzed in [2, 3, 4, 5, 6, 7], for more general class of theories.

The operator algebra of the operators corresponding to these extra discrete states have been defined in [8, 9], in the framework of $C = 1$ theory, for the vanishing value of the cosmological constant. It was found that the chiral current type operators form the w_∞ algebra, while the zero-form, zero ghost number operators form the ground ring algebra. In the matrix model approach the related results have been obtained in the papers [10].

Further consequences of these algebras, in the form of Ward identities for amplitudes of tachyonic operators have been analyzed recently in [11, 12, 13, 14, 15].

Here we shall calculate the deformation of the above mentioned algebras by the presence in the theory of a nonvanishing cosmological constant. Work in that direction have been done in a number of recent papers [16, 17, 18, 19].

But we shall repeat first some of the remarks made in [19] to the structure of the physical states in 2D gravity coupled to the minimal model as matter. This is to make further calculation of the algebras applicable to a wider class of theories.

2D gravity coupled to minimal matter could be represented, apart from ghosts, by two free fields with background charges, with the stress-energy tensor

$$T = T_M + T_L \tag{1.1}$$

$$T_M = -\frac{1}{2}(\partial\varphi_M)^2 + i\alpha_0\partial^2\varphi_M, \quad T_L = -\frac{1}{2}(\partial\varphi_L)^2 + \beta_0\partial^2\varphi_L \tag{1.2}$$

Here φ_M and φ_L are normalized as:

$$\langle\varphi_M\varphi_M\rangle = \langle\varphi_L\varphi_L\rangle = \log\frac{1}{|z-z'|^2} \tag{1.3}$$

The field φ_M is to represent, by the Coulomb gas technique, the conformal theory of matter, and φ_L that of Liouville. The relation for the central charges is:

$$C = C_M + C_L = 26 \tag{1.4}$$

$$C_M = 1 - 12\alpha_0^2, \quad C_L = 1 + 12\beta_0^2, \quad \beta_0^2 - \alpha_0^2 = 2 \tag{1.5}$$

The usual physical state operators will be of the form, by David and Distler-Kawai arguments [20]:

$$\Phi_{n'.n}^{(\mp)}(z,\bar{z}) = exp(i\alpha_{n'.n}\varphi_M(z,\bar{z}) + \beta_{n'.n}^{\mp}\varphi_L(z,\bar{z})) \tag{1.6}$$

$$\alpha_{n'.n} = \frac{1-n'}{2}\alpha_- + \frac{1-n}{2}\alpha_+ \tag{1.7}$$

$$\beta_{n'.n} = \frac{1-n'}{2}\beta_- + \frac{1-n}{2}\beta_+, \quad \beta_{n'.n}^{\mp} = \beta_{-n'.n}, \beta_{n'.-n} \tag{1.8}$$

$$\alpha_{\pm} = \alpha_0 \pm \sqrt{\alpha_0^2 + 2}, \quad \beta_{\pm} = \beta_0 \pm \sqrt{\beta_0^2 - 2} = \pm\alpha_{\pm} \tag{1.9}$$

with $(n'.n)$ taking their values inside the basic conformal grid:

$$1 \le n' \le p' - 1, \quad 1 \le n \le p - 1 \tag{1.10}$$

Here $p', p : \alpha_+^2 = 2p'/p, \quad p', p$ being relative prime numbers. The unusual physical operators, the extra ones as compared to the minimal midel without gravity, are the special states in the modules of the states $\alpha_{n'.n}$ in (1.6) (with $\beta_{n'.n}^{\mp}$ to be shifted appropriately), or the states (1.6) themselves with $(n'.n)$ outside the basic grid (1.10).

Under the linear transformation in the system of two fields φ_M and φ_L:

$$\begin{aligned}\varphi_M(z,\bar{z}) &= \frac{1}{2}(-i\alpha_0\varphi_1 + \beta_0\varphi_2)\\ \varphi_L(z,\bar{z}) &= \frac{1}{2}(-\beta_0\varphi_1 - i\alpha_0\varphi_2)\end{aligned} \tag{1.11}$$

the stress-energy tensor (1.1) takes the form

$$T = -\frac{1}{4}(\partial\varphi_2)^2 - \frac{1}{4}(\partial\varphi_1)^2 - \partial^2\varphi_1 = T_2 + T_1 \tag{1.12}$$

with fields φ_1 and φ_2 normalized as:

$$\langle\varphi_1\varphi_1\rangle = \langle\varphi_2\varphi_2\rangle = 2\log\frac{1}{|z-z'|^2} \tag{1.13}$$

By (12) we observe that we get an effective $C = 1$ theory: φ_2-'matter', with $\alpha_0 = 0$, $C_2 = 1$, and φ_1-'Liouville', with $\beta_0 = -1$, $C_1 = 25$.

The operators in (1.6) become:

$$\Phi^{(-)}_{n'.n}(z,\bar{z}) = exp(ij^{(-)}_{n'.n}\varphi_2 + (-1 - j^{(-)}_{n'.n})\varphi_1)$$

$$j^{(-)}_{n'.n} = \frac{n'}{2}\rho' - \frac{n}{2}, \quad \rho' = \frac{(\alpha)^2_-}{2} = \frac{p}{p'} \tag{1.14}$$

$$\Phi^{(+)}_{n'.n}(z,\bar{z}) = exp(ij^{(+)}_{n'.n}\varphi_2 + (-1 + j^{(+)}_{n'.n})\varphi_1)$$

$$j^{(+)}_{n'.n} = \frac{n'}{2} - \frac{n}{2}\rho, \quad \rho = \frac{(\alpha)^2_+}{2} = \frac{p'}{p} \tag{1.15}$$

Taking e.g. the branch (1.15) we observe that the states $(n'.0)$ are mapped, by the linear transformation (1.11), onto the $su(2)$ discrete tachyonic states of the $C = 1$ theory. Their modules involve $su(2)$ multiplets of extra physical states [2, 3, 5]. The usual physical states are mapped onto the tachyonic states with fractional values of the $su(2)$ momenta.

Notice that the extra physical states in the modules of the $\alpha_{n'.n}$ states (1.6) (with $\beta^{\mp}_{n'.n}$ to be shifted) get assembled in the modules of the $su(2)$ tachyonic states, $(n'.0)$ or $(0.n)$, after the linear transformation (1.11). We have to check for the first extra state in the module of $\alpha_{n'.n}$, as the inclusion repeats itself. The first extra state will have the form:

$$(\ldots)\exp(i\alpha_{n'.n}\varphi_M(z,\bar{z}) + \beta_{n'.n}\varphi_L(z,\bar{z})) \tag{1.16}$$

$\beta_{n',-n}$ is replaced with $\beta_{n'.n}$, and the conformal dimension of the exponential operator becomes $1 - n'n$; $(\ldots)$ stands for a polynomial of φ_M, φ_L oscillator states, of level $n'n$. The exponential in (1.16) transforms, under (1.11), into

$$(\ldots)\exp(i\frac{n'-n}{2}\varphi_2 + (-1 + \frac{n'+n}{2})\varphi_1) \tag{1.17}$$

This is the state in the module of $j = \frac{n'-n}{2}$ 'matter' state.

After these remarks it is clear that we may as well stay in the framework of the $C = 1$ theory with gravity, when calculating amplitudes (correlation functions). In this respect it is more a question of what particular set of operators we are interested in, than what is the C-matter of the theory.

We shall give next the results for the operator algebra of $su(2)$ physical state operators [8]:

$$\mathcal{T}_{j,m} = T_{j,m}\overline{T}_{j,m}, \quad \mathcal{O}_{j,m} = O_{j,m}\overline{O}_{j,m} \tag{1.18}$$

and

$$\mathcal{W}_{j,m} = T_{j+1,m}\overline{O}_{j,m} \tag{1.19}$$

$T_{j,m}$ are the conformal dimension one chiral operators:

$$T_{j,m} = (H^-)^{j-m}\exp(ijX + (-1+j)\phi) \tag{1.20}$$

$$H^- = \oint \frac{du}{2\pi i}\exp(-iX(u)) \tag{1.21}$$

They form $su(2)$ multiplets. The importance of the chiral (antichiral) operators $O_{j,m}$ ($\overline{O}_{j,m}$) have been stressed in [8]. The $j = 1/2$ states have the following explicit form:

$$O_{\frac{1}{2},\frac{1}{2}} =: (cb + \frac{i}{2}\partial X - \frac{1}{2}\partial\phi)\exp(\frac{i}{2}X + \frac{1}{2}\phi) : \tag{1.22}$$

$$O_{\frac{1}{2},-\frac{1}{2}} = H^- O_{\frac{1}{2},\frac{1}{2}} =: (cb - \frac{i}{2}\partial X - \frac{1}{2}\partial\phi)\exp(-\frac{i}{2}X + \frac{1}{2}\phi) : \tag{1.23}$$

We have assumed in (1.20–1.23) the notations and normalizations of [9, 11]. In particular

$$X = \sqrt{2}\varphi_M, \quad \phi = \sqrt{2}\varphi_L, \quad \alpha_0 = 0, \quad \beta_0 = -1$$

$$\langle XX\rangle = \langle\phi\phi\rangle = 2\log\frac{1}{|z - z'|^2} \tag{1.24}$$

b, c are covariant ghosts, dimension 2,–1 chiral fields. In the fully bosonized form the operator $O_{\frac{1}{2},\frac{1}{2}}$ in (1.22) becomes:

$$O_{\frac{1}{2},\frac{1}{2}} =: (i\partial\varphi + \frac{i}{2}\partial X - \frac{1}{2}\partial\phi)\exp(\frac{i}{2}X + \frac{1}{2}\phi) : \tag{1.25}$$

Here [21]

$$c = \exp(i\varphi), \quad b = \exp(-i\varphi), \quad : cb := i\varphi$$

$$\langle\varphi\varphi\rangle = \log\frac{1}{(z - z')} \tag{1.26}$$

The operator $\overline{O}_{j,m}$ in (1.19) balance the left-right momenta of physical chiral operators, the currents $\mathcal{W}_{j,m}$ [8].

Let us introduce the following operator:

$$\delta = -\oint \frac{du}{2\pi i} b(u) \exp(-\frac{i}{2}X + \frac{1}{2}\phi)$$
$$= -\oint \frac{du}{2\pi i} \exp(-i\varphi - \frac{i}{2}X + \frac{1}{2}\phi) \tag{1.27}$$

It is easy to check that

$$\delta(c(z)T_{1,1}(z))$$
$$= -\oint \frac{du}{2\pi i} \exp(-i\varphi - \frac{i}{2}X + \frac{1}{2}\phi) \times \exp(i\phi(z) + iX(z)) = O_{\frac{1}{2},\frac{1}{2}}(z) \tag{1.28}$$

Easy to see also that $\delta^2 = 0$, δ commutes with H^-, eq.(1.21), and it can be checked, with some algebra, that δ anticommutes with the BRST operator

$$d = \oint \frac{du}{2\pi i} : c(T_M + T_L + \frac{1}{2}T_{gh}) \tag{1.29}$$

Then, in the general case, we will have, by matching the quantum numbers:

$$O_{j,m} = \delta(cT_{j+\frac{1}{2},m+\frac{1}{2}}) \tag{1.30}$$

The calculation of each particular term of the operator algebra factorizes onto chiral and antichiral factor calculations, up to a small detail for the Liouville part which is remarked on in Section 2. The calculation of the operator algebra of the $T_{j,m}$ chiral factors in (1.18),(1.19) has already been done in [19], and is reproduced in Section 2, in a slightly different way. The calculation of the algebra of the chiral operators $O_{j,m}$ can be done using the representation (1.30). This is done in Section 3.

The general form of the algebras could be written as follows:

$$\mathcal{T}_{j_1,m_1}(z,\bar{z}) \times \mathcal{T}_{j_2,m_2}(z',\bar{z}') = \frac{1}{|z-z'|^2} \sum_{j_3} D^{(T)}_{(j_1,m_1),(j_2,m_2),(j_3,m_3)} \mathcal{T}_{j_3,m_3}(z',\bar{z}') \tag{1.31}$$

$$\mathcal{O}_{j_1,m_1}(z,\bar{z}) \times \mathcal{O}_{j_2,m_2}(z',\bar{z}') = \sum_{j_3} D^{(O)}_{(j_1,m_1),(j_2,m_2),(j_3,m_3)} \mathcal{O}_{j_3,m_3}(z',\bar{z}') \tag{1.32}$$

$$\mathcal{W}_{j_1,m_1}(z,\bar{z}) \times \mathcal{W}_{j_2,m_2}(z',\bar{z}') = \frac{1}{(z-z')} \sum_{j_3} D^{(W)}_{(j_1,m_1),(j_2,m_2),(j_3,m_3)} \mathcal{W}_{j_3,m_3}(z',\bar{z}') \tag{1.33}$$

with the operator algebra structure constants having the general form:

$$D^{(\mathcal{T})}_{(j_1,m_1),(j_2,m_2),(j_3,m_3)} = \left(A_{(j_1,m_1),(j_2,m_2),(j_3,m_3)}\right)^2 \tag{1.34}$$

$$D^{(\mathcal{O})}_{(j_1,m_1),(j_2,m_2),(j_3,m_3)} = \left(B_{(j_1,m_1),(j_2,m_2),(j_3,m_3)}\right)^2 \tag{1.35}$$

$$D^{(\mathcal{W})}_{(j_1,m_1),(j_2,m_2),(j_3,m_3)} = A_{(j_1+1,m_1),(j_2+1,m_2),(j_3+1,m_3)}B_{(j_1,m_1),(j_2,m_2),(j_3,m_3)} \tag{1.36}$$

In the r.h.s. of (1.31–1.33) just the BRST cohomology states are kept. The component constants A and B are found, in Section 2 and Section 3, to have the following form, for the appropriately normalized operators:

$$A_{(j_1,m_1),(j_2,m_2),(j_3,m_3)} = \begin{pmatrix} j_1 & j_2 & j_3 \\ m_1 & m_2 & -m_3 \end{pmatrix} d^{(A)}_{j_1,j_2,j_3} \tag{1.37}$$

$$d^{(A)}_{j_1,j_2,j_3} = \frac{(j_1+j_2-j_3)!(j_1-j_2+j_3)!(-j_1+j_2+j_3)!}{\Delta^{1/2}_{j_1,j_2,j_3}[(2j_1-1)!(2j_2-1)!(2j_3-1)!]^{1/2}} \tag{1.38}$$

$$B_{(j_1,m_1),(j_2,m_2),(j_3,m_3)} = \begin{pmatrix} j_1 & j_2 & j_3 \\ m_1 & m_2 & -m_3 \end{pmatrix} d^{(B)}_{j_1,j_2,j_3} \tag{1.39}$$

$$d^{(B)}_{j_1,j_2,j_3} = \frac{(j_1+j_2-j_3)!(j_1-j_2+j_3)!(-j_1+j_2+j_3)!}{\Delta^{1/2}_{j_1,j_2,j_3}[(2j_1)!(2j_2)!(2j_3)!]^{1/2}} \tag{1.40}$$

Here $\begin{pmatrix} j_1 & j_2 & j_3 \\ m_1 & m_2 & m_3 \end{pmatrix}$ is the $su(2)$ $3j$ symbol. The factor $\Delta^{1/2}_{j_1,j_2,j_3}$, which in fact cancels in the products of the $3j$ symbols and the d coefficients in (1.37),(1.39), is given by

$$\Delta^{1/2}_{j_1,j_2,j_3} = \left[\frac{(j_1+j_2-j_3)!(j_1-j_2+j_3)!(-j_1+j_2+j_3)!}{(j_1+j_2+j_3+1)!}\right]^{1/2} \tag{1.41}$$

The structure constants $D^{(\mathcal{W})}$ of the algebra of currents (1.33) can be written more explicitly as

$$D^{(\mathcal{W})}_{(j_1,m_1),(j_2,m_2),(j_3,m_3)} = \begin{pmatrix} j_1+1 & j_2+1 & j_3+1 \\ m_1 & m_2 & -m_3 \end{pmatrix}\begin{pmatrix} j_1 & j_2 & j_3 \\ m_1 & m_2 & -m_3 \end{pmatrix} d^{(A)}_{j_1+1,j_2+1,j_3+1} d^{(B)}_{j_1,j_2,j_3} \tag{1.42}$$

Under the permutation of e.g. (j_1, m_1) and (j_2, m_2) the $3j$ symbols produce the factors:

$$(-1)^{j_1+j_2+j_3+3} \times (-1)^{j_1+j_2+j_3} = -1 \tag{1.43}$$

So the constants (1.42) are properly antisymmetric. The current constants should satisfy also the Jacobi identities, which is harder to verify. This has not been done yet.

2 Calculation of the operator algebra of $T_{j,m}$

We shall look at the product of two particular operators and shall calculate the coefficient at a particular term (operator) in the r.h.s.:

$$T_{j_1,j_1} \times T_{j_2,j_3-j_1} \to \dots \quad T_{j_3,j_3} \tag{2.1}$$

The operators $T_{j,m}$ are defined in (1.20). It is easy to check that the coefficient at T_{j_3,j_3} is given by:

$$I = I_M I_L \tag{2.2}$$

$$I_M = \langle \exp(ij_1X(0) \times (\prod_1^k \oint \frac{dv_i}{2\pi i} \exp(-iX(v_i))) \times \exp(ij_2X(1)) \rangle$$

$$= \prod_1^k \oint \frac{dv_i}{2\pi i} (v_i)^{-2j_1} (v_i - 1)^{-2j_2} \prod_{i<j}^k (v_i - v_j)^2 \tag{2.3}$$

$$= k! \prod_1^k \frac{\Gamma(i)\Gamma(2j_3 + i)}{\Gamma(2j_1 + 1 - i)\Gamma(2j_2 + 1 - i)} \tag{2.4}$$

(double use of i in (2.3) should not be confusing). Here $k \equiv k_M = j_1 + j_2 - j_3$. The result (2.4) is obtained by transforming the contours in (2.3), which encircle the point $z = 1$, to the standart ones in [22], and then substituting the result for the integral. Some further slight transformations of the resulting products has been done to put the result in a more symmetric form.

$$I_L = \frac{(\mu)^{k_L}}{k_L!} \langle \exp((-1 + j_1)\phi(0) \times (\prod_1^{k_L} \int_{-\infty}^{+\infty} dw_i \exp(-\phi(w_i))) \times \exp((-1 + j_2)\phi(1)) \rangle$$

$$= \frac{(\mu)^{k_L}}{k_L!} \prod_1^{k_L} \int_{-\infty}^{+\infty} dw_i (w_i)^{2(-1+j_1)} (w_i - 1)^{2(-1+j_2)} \prod_{i<j}^{k_L} (w_i - w_j)^{-2} \tag{2.5}$$

$$= (\frac{\mu}{\Gamma(-1)})^{k_L} \prod_1^{k_L} \frac{\Gamma(2j_1 - i)\Gamma(2j_2 - i)}{\Gamma(1 + i)\Gamma(2j_3 + 1 + i)} \tag{2.6}$$

Here $k_L = j_1 + j_2 - j_3 - 1 = k_M - 1$. The contours in (2.5) extend from $-\infty$ to $+\infty$ going below $z = 0$ and above $z = 1$. They had been deformed to the standard ones, and the result for the integral in [22] had been used.

For the product of $I_M I_L$ in (2.2) we get

$$I = (\frac{\mu}{\Gamma(-1)})^{k_L}(j_1 + j_2 - j_3)! \frac{\Gamma(2j_3+1)}{\Gamma(2j_1)\Gamma(2j_2)} \tag{2.7}$$

The remark is in order. The Liouville integral in (2.5) is presumed to come from the expansion over the term

$$\mu \int d^2w \exp(-\phi(w, \overline{w})) \tag{2.8}$$

in the action for the field ϕ. The k_Lth order term is

$$\frac{(\mu)^{k_L}}{k_L!} \prod_1^{k_L} \int d^2w_i \exp(-\phi(w_i, \overline{w}_i)) \tag{2.9}$$

It is being picked up by the (anomalous) conservation of the ϕ field momenta (or 'charges', for the theory with the background charge $\beta_0 = -1$). The integrals in (2.9), which are over the 2D plane, are to be factorized onto chiral and antichiral contour integrals, as is explained e.g. in [23]. In the expansion over the exponential term (2.8) the chiral and the antichiral integrals will go in parallel, in equal numbers, so that the Liouville momenta have to match on both sides of the operator product algebra for the physical operators, i.e. in the relation (2.1) supplied by the antichiral partners:

$$(T_{j_1,j_1}\overline{T}_{j_1,j_1}) \times (T_{j_2,j_3-j_1}\overline{T}_{j_2,j_3-j_1}) \to \ldots \quad (T_{j_3,j_3}\overline{T}_{j_3,j_3}) \tag{2.10}$$

It should be noted also that, in undoing the 2D integrals over the plane into the product of the contour integrals, one gets, in a particular way of doing it, the contours extending from $-\infty$ to $+\infty$, one contour going below the other, and all going below the point $z = 0$ and above the point $z = 1$. This is on the chiral side. In the corresponding antichiral factor the integrations will be of the ordered type, one point after another, in the interval $0 < \overline{z} < 1$. This could be replaced also by the integral with all the contours going from $\overline{z} = 0$ to $\overline{z} = 1$, one below the other, which is to be devided then by $k_L!$. This replacement is useful to avoid the formal problem with the exponent -2 in (2.5). (Notice that j_1, j_2 could be thought of, in the calculation of the integral, as having general values, till the final result; so the rest of the exponents in the integrals pose no problem). Taken one way or another, the antichiral part will obviously not supply the extra factor of $(\mu)^{k_L}$; but also it will not produce the extra factor of $(1/\Gamma(-1))^{k_L}$, as can be checked. The factor

$$(\frac{\mu}{\Gamma(-1)})^{k_L} \equiv (\mu_{eff})^{k_L} \tag{2.11}$$

is in fact common for the chiral and the antichiral parts of the operator product algebra.

Since we want to consider the μ deformed algebra, the cosmological constant is being renormalized in a singular way, as in (2.11), which is the known point of the $C = 1$ theory. One also could think of it as of a singular renormalization of the corresponding screening operators, which would be the case for the $C < 1$ theory rotated to the effective theory with $C = 1$.

Performing the singular renormalization, or letting μ to vanish, appears in this respect as our own choice of the theory, both being well defined.

By $su(2)$ invariance the coefficient of (2.1) in (2.7) should have the form:

$$(\mu_{eff})^{k_L} \begin{pmatrix} j_1 & j_2 & j_3 \\ j_1 & j_3 - j_1 & -j_3 \end{pmatrix} d_{j_1,j_2,j_3} \tag{2.12}$$

where the first factor is the $su(2)$ $3j$ symbol, which in particular could be given as:

$$\begin{aligned} &\begin{pmatrix} j_1 & j_2 & j_3 \\ m_1 & m_2 & m_3 \end{pmatrix} \\ &= \Delta^{1/2}[(j_1+m_1)!(j_1-m_1)!(j_2+m_2)!(j_2-m_2)!(j_3+m_3)!(j_3-m_3)!]^{1/2} \\ &\times \sum_i (\frac{(-1)^{j_1-j_2-m_3+i}}{(i)!(j_1+j_2-j_3-i)!(j_1-m_1-i)!(j_2+m_2-i)!} \\ &\qquad \frac{1}{(j_3-j_2+m_1+i)!(j_3-j_1-m_2+i)!}) \end{aligned} \tag{2.13}$$

with $\Delta^{1/2}$ given in (1.41). Matching (2.13) with (2.7), to obtain the coefficients d_{j_1,j_2,j_3}, we have to remove first the extra $su(2)$ normalization of the operators $T_{j,m}$ in (1.20), which is

$$\langle T^+_{j,m} T_{j,m} \rangle \propto (2j)! \frac{(j-m)!}{(j+m)!} \tag{2.14}$$

Finally, for the coefficients of the operator product expansion

$$T_{j_1,m_1} \times T_{j_2,m_2} \to \sum_{j_3} T_{j_3,m_3} \tag{2.15}$$

we get

$$\binom{j_1 \quad j_2 \quad j_3}{m_1 \quad m_2 \quad -m_3} d_{j_1,j_2,j_3} \tag{2.16}$$

with

$$d_{j_1,j_2,j_3} = (\mu_{eff})^{j_1+j_2-j_3-1} \frac{(j_1+j_2-j_3)!(j_1-j_2+j_3)!(-j_1+j_2+j_3)!}{\Delta^{1/2}_{j_1,j_2,j_3}[(2j_1)!(2j_2)!(2j_3)!]^{1/2}}$$
$$\frac{(2j_3)!}{(2j_1-1)!(2j_2-1)!} \tag{2.17}$$

This could be given in a more symmetric form as:

$$(\mu_{eff})^{j_1+j_2-j_3-1} \times \frac{(j_1+j_2-j_3)!(j_1-j_2+j_3)!(-j_1+j_2+j_3)!}{\Delta^{1/2}_{j_1,j_2,j_3}[(2j_1-1)!(2j_2-1)!(2j_3-1)!]^{1/2}}$$
$$\times\left[\frac{(2j_3)}{(2j_1)(2j_2)}\right]^{1/2} \times \frac{(2j_3-1)!}{(2j_1-1)!(2j_2-1)!} \tag{2.18}$$

The above differs from the expression obtained in [19] by the normalization factors of individual operators. This is due to different representation of operators used in [19] and here.

The extra factor of $1/\Gamma(0)$ in [19] is that of the partition function normalization. It is absent here since we are doing directly the operator product expansion, instead of calculating the 3-point function.

The first, third, and fourth factors could be absorbed into the normalization of the operators, which leaves us with the symmetric coefficients $d^{(A)}_{j_1,j_2,j_3}$ in (1.38). We remark that we have not followed and kept the overall sign factors, as they cancel eventually in the products of the chiral and antichiral factors.

3 Calculation of the operator algebra of $O_{j,m}$

For the operators $O_{j,m}$, eq.(1.30), we shall again calculate the coefficient in

$$O_{j_1,j_1} \times O_{j_2,j_3-j_1} \rightarrow \ldots \quad O_{j_3,j_3} \tag{3.1}$$

which is easier, in a similar way as for the operators $T_{j,m}$ in Section 2. We have:

$$\begin{aligned}
&O_{j_1,j_1}(0) \times O_{j_2,j_3-j_1}(1) \\
&= \delta(c(0)T_{j_1+\frac{1}{2},j_1+\frac{1}{2}}(0)) \times \delta((H^-)^{j_2-j_3+j_1}c(1)T_{j_2+\frac{1}{2},j_2+\frac{1}{2}}(1)) \\
&\qquad \to \ldots \quad \delta(cT_{j_3+\frac{1}{2},j_3+\frac{1}{2}})
\end{aligned} \tag{3.2}$$

The contour of δ acting on the operator at $z = 0$ in l.h.s. can be tranformed to the contour which encircles both operators, at $z = 0$ and at $z = 1$, which is appropriate to obtain eventually the operator in r.h.s., plus the configuration where two δ contours encircle the operator at $z = 1$. The last vanishes because of $\delta^2 = 0$. Then it is sufficient to calculate a slightly reduced operator algebra, that of

$$cT_{j_1+\frac{1}{2},j_1+\frac{1}{2}} \times \delta((H^-)^{j_2-j_3+j_1}cT_{j_2+\frac{1}{2},j_2+\frac{1}{2}}) \to \ldots \quad cT_{j_3+\frac{1}{2},j_3+\frac{1}{2}} \tag{3.3}$$

It is easy to check that the coefficient at the operator in r.h.s. of (3.3) is given by the integral:

$$\begin{aligned}
I &= \langle \exp(i(j_1+\frac{1}{2})X(0) + (-\frac{1}{2}+j_1)\phi(0) + i\varphi(0)) \\
&\times \oint \frac{du}{2\pi i}\exp(-\frac{i}{2}X(u) + \frac{1}{2}\phi(u) - i\varphi(0)) \times \prod_1^{k_M}\oint \frac{dv_i}{2\pi i}\exp(-iX(v_i)) \\
&\times \frac{1}{k_L!}\prod_1^{k_L}\int_{-\infty}^{+\infty} dw_i \exp(-\phi(w_i)) \\
&\times \exp(i(j_2+\frac{1}{2})X(1) + (-\frac{1}{2}+j_2)\phi(1) + i\varphi(1))\rangle \\
&= \oint \frac{du}{2\pi i}(u)^{-(2j_1+1)}(u-1)^{-(2j_2+1)} \\
&\times \prod_1^{k_M}\oint \frac{dv_i}{2\pi i}(v_i)^{-(2j_1+1)}(v_i-1)^{-(2j_2+1)}(v_i-u)\prod_{i<j}^{k_M}(v_i-v_j)^2 \\
&\times \frac{1}{k_L!}\prod_1^{k_L}\int_{-\infty}^{+\infty} dw_i(w_i)^{2j_1-1}(w_i-1)^{2j_i-1}(w_i-u)\prod_{i<j}^{k_L}(w_i-w_j)^{-2}
\end{aligned} \tag{3.4}$$

This could be put into the form:

$$I = \oint \frac{du}{2\pi i}(u)^{-(2j_1+1)}(u-1)^{-(2j_2+1)}I_M(u)I_L(u) \tag{3.5}$$

with

$$I_M(u)=\prod_1^{k_M}\oint\frac{dv_i}{2\pi i}(v_i)^{-(2j_1+1)}(v_i-1)^{-(2j_2+1)}(v_i-u)\prod_{i<j}^{k_M}(v_i-v_j)^2 \tag{3.6}$$

$$I_L(u)=\frac{1}{k_L!}\prod_1^{k_L}\int_{-\infty}^{+\infty}dw_i(w_i)^{2j_1-1}(w_i-1)^{2j_i-1}(w_i-u)\prod_{i<j}^{k_L}(w_i-w_j)^{-2} \tag{3.7}$$

One checks that in this case (different from that in Sec.2)

$$k_M=k_L=j_1+j_2-j_3\equiv k \tag{3.8}$$

The u contour in (3.5) encircles $z=1$, like the v_i contours in (3.6). The w_i contours go below $z=0$ and above $z=1$,– see remarks on the Liouville contours in Section 2. We dropped here the factor $(\mu)^{k_L}$ in front of the Liouville integrals.

Some intermediate expressions are more symmetric if we present the integrals (3.6), (3.7) in the following forms:

$$I_M(u)\equiv(u)^kJ_M(t),\quad t=1/u$$

$$J_M(t)=\prod_1^{k}\oint\frac{dv_i}{2\pi i}(v_i)^{-(2j_1+1)}(v_i-1)^{-(2j_2+1)}(v_it-1)\prod_{i<j}^{k}(v_i-v_j)^2 \tag{3.9}$$

$$I_L(u)\equiv(u)^kJ_L(t),\quad t=1/u$$

$$J_L(t)=\frac{1}{k!}\prod_1^{k}\int_{-\infty}^{+\infty}dw_i(w_i)^{2j_1-1}(w_i-1)^{2j_2-1}(w_it-1)\prod_{i<j}^{k}(w_i-w_j)^{-2} \tag{3.10}$$

The integrals $J_M(t)$, $J_L(t)$ could be transformed to the following forms:

$$J_M(t)=(\frac{\sin\pi(-2j_2-1)}{\pi})^k$$

$$\times\prod_1^k\int_0^1dx_i(x_i)^{2j_3+1}(1-x_i)^{-2j_2-1}(t-x_i)\prod_{i<j}^k(x_i-x_j)^2 \tag{3.11}$$

$$J_L(t)=(\frac{\sin\pi(2j_2-1)}{\pi})^k$$

$$\times\frac{1}{k!}\prod_1^k\int_0^1dy_i(y_i)^{-3-2j_3}(1-y_i)^{2j_2-1}(t-y_i)\prod_{i<j}^k(y_i-y_j)^{-2} \tag{3.12}$$

The contours in (3.11),(3.12) go one below the other, all in the range $0<z<1$.

The following results have been found for these integrals:

$$J_M(t) = J_M(0)F_M(t) \tag{3.13}$$

$$J_M(0) = k!(-1)^k \prod_1^k \frac{\Gamma(i)\Gamma(2+2j_3+i)}{\Gamma(2+2j_1-i)\Gamma(2+2j_2-i)} \tag{3.14}$$

$$J_L(t) = J_L(0)F_L(t) \tag{3.15}$$

$$J_M(0) = \frac{(-1)^k}{\Gamma(-1)}{}^{k}\prod_1^k \frac{\Gamma(1+2j_1-i)\Gamma(1+2j_2-i)}{\Gamma(1+i)\Gamma(1+2j_3+i)} \tag{3.16}$$

$$F_M(t) = F_L(t) = \mathcal{F}(1+j_1-j_2+j_3, -k; 2+2j_3; t) \tag{3.17}$$

In (3.17) $\mathcal{F}(\alpha,\beta;\gamma;t)$ is the hypergeometric function – a polynomial, in fact, of order k, as $\beta = -k$, and k is always assumed to be integer. This form of $F_M(t)$, $F_L(t)$ is an assumption, verified by calculating the first two and the last coefficients of the expansions of $J_M(t)$, $J_L(t)$ in powers of t.

Using the parametrization

$$a = -2j_2 - 1, \quad b = 2j_3 \tag{3.18}$$

the function (3.17) could be presented in the following form:

$$\mathcal{F}(t) = (-1)^k k! \frac{\Gamma(2+b)}{\Gamma(2+k+b)} P_k^{(a,b+1)}(2t-1) \tag{3.19}$$

where $P_k^{(\alpha,\beta)}$ is the Jacobi polynomial.

On account of (3.9-3.19) the integral (3.5) could be transformed as

$$\begin{aligned} I &= \frac{\sin\pi(-2j_2-1)}{\pi} J_M(0)J_L(0)\int_1^\infty du(u)^{-(2j_1+1)}(u-1)^{-(2j_2+1)} \times (u)^{2k}(\mathcal{F}(\frac{1}{u}))^2 \\ &= \frac{\sin\pi(-2j_2-1)}{\pi} J_M(0)J_L(0)\int_0^1 du(u)^{2j_3}(1-u)^{-2j_2-1}(\mathcal{F}(u))^2 \\ &= \frac{\sin\pi(-2j_2-1)}{\pi} J_M(0)J_L(0)(k!\frac{\Gamma(2+b)}{\Gamma(2+k+b)})^2 \\ &\times \int_0^1 du(u)^b(1-u)^a(P_k^{(a,b+1)}(2u-1))^2 \end{aligned} \tag{3.20}$$

The parameters a, b are defined in (3.18). The last integral could be found in the tables and is given by

$$\frac{\Gamma(2+b+k)\Gamma(1+a+k)}{k!(b+1)\Gamma(2+a+b+k)} \tag{3.21}$$

Substituting the expressions for $J_M(0)$, $J_L(0)$ in (3.14), (3.16), and the expression (3.21) for the integral in (3.20), we find:

$$I = \frac{1}{(\Gamma(-1))^k}(j_1 + j_2 - j_3)!\frac{\Gamma(1+2j_3)}{\Gamma(1+2j_1)\Gamma(1+2j_2)} \qquad (3.22)$$

This is the value of the coefficient in the operator product expansion (3.3) and (3.1), the factor $(1/\Gamma(-1))^k$ is to be absorbed in the renormalization of $(\mu)^k$ which has been dropped in the above expressions. On general grounds this coefficient has to be of the form:

$$\begin{pmatrix} j_1 & j_2 & j_3 \\ j_1 & j_3 - j_1 & -j_3 \end{pmatrix} d'_{j_1,j_2,j_3} \qquad (3.23)$$

like that in Section 2. In a similar way, by matching (3.23) with (3.22), the coefficient d'_{j_1,j_2,j_3} is found to be:

$$d'_{j_1,j_2,j_3} = (\mu_{eff})^{j_1+j_2-j_3} \times \frac{(j_1+j_2-j_3)!(j_1-j_2+j_3)!(-j_1+j_2+j_3)!}{\Delta^{1/2}_{j_1,j_2,j_3}[(2j_1)!(2j_2)!(2j_3)!]^{1/2}} \times \frac{(2j_3)!}{(2j_1)!(2j_2)!} \qquad (3.24)$$

Here we have supplied back the factor $(\mu_{eff})^k$. The first and the last factors in (3.24) could be absorbed into the normalization of the operators, and we get the symmetric coefficients $d^{(B)}_{j_1,j_2,j_3}$ in (1.40).

4 Conclusions and discussion

One conclusion of the calculation is that the deformed operator algebra of currents is different from $\mathcal{T}(\mu)$, the enveloping algebra of $su(2)$ for a fixed value μ of the quadratic Casimir operator, which has been discussed in a number of recent papers.

The relation of the operators $O_{j,m}$ to $T_{j,m}$ by the operator δ in (1.30) might be of interest on its own, apart from the purpose of the calculation in this paper.

The relation by the linear transformation (1.11) of the minimal, $C < 1$, and $C = 1$ theories suggests the use of extra screening operators, of Liouville–matter mixed type. They would appear in particular in the representation of correlation functions of fractional

j operators (1.14),(1.15), in the context of $C = 1$ theory. The correlation function of this type are related to the unsolved yet problem of calculating the general multipoint functions in $C < 1$ theory,– to begin with the four–point ones, still on a sphere. Remarks to this problem are given in [24]. We hope to make some progress on this in the future.

Acknowledgements

I am grateful to the colleagues at CERN, at LPTHE, Jussieu, Paris, and to the organizers of the Workshop in Rome for their hospitality and stimulating discussions. I have benefited in particular from discussions with E. Kiritsis and with P. Bouwknegt.

References

[1] D. J. Gross, I. R. Klebanov and M. J. Newman, *Nucl.Phys.* **B350** (1991) 621; D. J. Gross, I. R. Klebanov, *Nucl.Phys.* **B352** (1991) 671

[2] B. Lian and G. Zuckerman, *Phys.Lett.* **254B** (1991) 417; *Phys.Lett.* **266B** (1991) 21

[3] A.M.Polyakov, *Mod.Phys.Lett.* **A6** (1991) 635

[4] C. Imbimbo, S. Mahapatra, and S. Mukhi, Preprint GEF-TH 8/91, TIFR/TH/91-27, May 1991

[5] P. Bouwknegt, J. McCarthy, and K. Pilch, Preprint CERN-TH.6162/91, July 1991

[6] S. Govindarajan, T. Jayaraman, V. John and P. Majumdar, Preprint IMSc -91/40, December 1991

[7] N. Chair, V. K. Dobrev and H. Kanno, Preprint IC/92/17, January 1992

[8] E. Witten, Preprint SLAC-PUB-IASSNS-HEP-91/51, August, 1991

[9] I. R. Klebanov and A. M. Polyakov, *Mod.Phys.Lett.* **A6** (1991) 3273

[10] J. Avan and A. Jevicki, *Phys.Lett.* **266B** (1991) 35; Preprints BROWN-HET-824 and BROWN-HET-839;
D. Mimic, J. Polchinski and Z. Yang, Preprint UTTG-16-91;
G. Moore and N. Seiberg, Preprint RU-91-29, YCTP-P19-91;
S. Das, A. Dhar, G. Mandal and S. Wadia, Preprints IASSNS-HEP-91/52 and 91/72

[11] I. R. Klebanov, *Mod.Phys.Lett.* **A7** (1992) 723

[12] E. Verlinde, Preprint IASSNS-HEP-92/5, February 1992

[13] E. Witten and B. Zwiebach, Preprint SLAC-PUB-IASSNS-HEP-92/2, MIT-CTP-2057, January 1992

[14] M. Bershadsky and D. Kutasov, Preprint PUPT-1315, HUTP-92/A016, April 1992

[15] I. R. Klebanov and A. Pasquinucci, Preprint PUPT-1313, April 1992

[16] M. Li, Preprint UCSBTH-91-47, October 1991

[17] S. Kachru, Preprint PUPT-1305, January 1992

[18] J. L. F. Barbón, Preprint CERN-TH.6379/92, FTUAM-92-02, January 1992

[19] Vl. S. Dotsenko, Preprint PAR-LPTHE 92-4, January 1992

[20] F. David, *Mod.Phys.Lett.* **A3** (1988) 1651
J. Distler and H. Kawai, *Nucl.Phys.* **B321** (1989) 509

[21] D. Friedan, E. Martinec and S. Shenker, *Nucl.Phys.* **B271** (1986) 93

[22] Vl. S. Dotsenko and V. A. Fateev, *Nucl.Phys.* **B251** (1985) 691

[23] Vl. S. Dotsenko, *Adv.Stud. in Pure Math.* **16** (1988)123

[24] Vl. S. Dotsenko, Preprint PAR-LPTHE 91-52, to be published in Proceedings of Cargèse Summer School, July 15 – July 27 1991

N=2 FIRST ORDER SYSTEMS: LANDAU–GINZBURG POTENTIALS AND TOPOLOGICAL TWISTS

P. FRÈ

SISSA - International School for Advanced Studies
Via Beirut 2, I-34100 Trieste, Italy, and I.N.F.N. sezione di Trieste

L. GIRARDELLO

Dipartimento di Fisica, Università di Milano
Via Celoria 16, I-20133, Milano, Italy and I.N.F.N. sezione di Milano

A. LERDA

Dipartimento di Fisica Teorica, Universitá di Torino,
Via P. Giuria 1, I-10125 Torino, Italy and I.N.F.N. sezione di Torino,
Institute for Theoretical Physics, S.U.N.Y. at Stony Brook
Stony Brook, N.Y. 11794, U.S.A.

P. SORIANI

SISSA - International School for Advanced Studies
Via Beirut 2, I-34100 Trieste, Italy, and I.N.F.N. sezione di Trieste

ABSTRACT

We review some recent results on the lagrangian formulation of N=2 superconformal theories with the use of first order "pseudo–ghost" fields. We discuss its connection with the Landau–Ginzburg formulation and implications on topological field theories.

1. Introduction

In the last few years the interest on N=2 superconformal theories with $c = 9$ has grown considerably, in connection with superstring compactifications on six-dimensional Calabi-Yau spaces [1, 2]. More generally, a $(2,2)$–superconformal theory with $c = 3d$ corresponds to the critical point of an N=2 σ-model on a target space with complex dimension d and vanishing first Chern class. These spaces are known as Calabi-Yau d-folds. It is well known [3] that to the deformations of Kähler and complex structures of the Calabi–Yau manifolds is associated a very peculiar geometrical structure: moduli spaces of CY manifolds are "special Kähler". This special geometry encode many relevant features of the low–energy effective supergravity action.

Recently it has become evident that there is a deep relation between this special Kähler geometry and the *flat* geometry of the topological field theories [4]–[11]

obtained by twisting and deforming N=2 superconformal models [12, 13, 14]. The best way to analyze these issues was based on the Landau–Ginzburg formulation of topological models, with deep relations on singularity theory and algebraic geometry [5]–[8], [15]–[18]. Indeed in the LG formulation of (2,2)-supersymmetric models, the superconformal theory is viewed as the infrared fixed point of a two-dimensional N=2 Wess-Zumino model with a polynomial superpotential W. When W is an analytic quasi-homogeneous function of the chiral superfields, we can assign a well defined $U(1)$ charge to these fields, which are in one to one correspondence with the chiral primary fields of N=2 superconformal theory. Furthermore, the polynomials W's can be identified with those used in the construction of Calabi-Yau d-folds. It can be shown [2, 17] that a superconformal model with $c = 3d$, corresponding to a LG potential W, is the same as that associated to a σ-model on the Calabi-Yau d-fold defined by the polynomial constraint $W(X_i) = 0$ in a suitable projective or weighted projective space [1].

In such a formulation the parameters entering the Landau-Ginzburg superpotential correspond either to "versal" deformations of the holomorphic (antiholomorphic) superpotential, or to "modal" deformations. From the point of view of quantum field theory this means that we are perturbing the theory, with some relevant (versal) or marginal (modal) deformations. The coupling constants parametrizing the deformations are interpreted as coordinates of some space (a moduli space in the case of marginal ones).

As it was shown in ref [5], once given the chiral ring of the N= 2 superconformal theory (i.e. the topological sector of the model), we can consider the perturbed three point functions:

$$c_{ijk}(t) = \langle \varphi_i \varphi_j \varphi_k exp(\sum_n t_n \int \phi_n^2) \rangle \tag{1}$$

where ϕ_n^2 are the two-form corresponding to the chiral superfield ϕ, with first component φ. In particular the two point function

$$c_{ij0}(t) \equiv \eta_{ij} \tag{2}$$

(where the index 0 refers to the identity operator) is independent of couplings and defines a flat metric in the space of couplings itself. This means that, when deforming the theory around the conformal point by two–form components of chiral primary fields, we define in the coupling space a flat constant geometry. This is not so in the Landau–Ginzburg formulation of the N=2 superconformal theories, just because the versal and modal deformations of the superpotential do not correspond directly to deformations around the conformal point; rather they are related to the latter by the solution of a uniformization problem, which in general involves higher transcendental functions. This is easily understood by the following consideration. The N=2

[1] In general $W(X_i)$ is actually the sum of several terms $W(X_i) = \sum_\alpha W_\alpha(X_i)$ and the Calabi-Yau d-fold is given by the complete intersection $W_1 = W_2 = \cdots = W_n = 0$ [2].

Landau–Ginzburg action can be put into correspondence with a N=2 superconformal theory only at its infrared fixed point. In particular the chiral primary ring of the conformal theory can be summarized as the quotient ring;

$$\mathcal{R} = \frac{C[\phi]}{dW} \tag{3}$$

of polynomials in ϕ modulo the vanishing relations $dW = 0$. When we turn on a coupling δs_i to one of the operators ϕ_i we change the potential as:

$$\delta W = \delta s_j \phi_j \tag{4}$$

This modifies also the chiral ring, which as a consequence will affect the potential W to the next order. This reflects the fact that the metric defined by residue pairing relations of the perturbed potential is no more constant in the couplings s, as shown in [5]–[11]. Moreover there exists a coordinate transformation $s_i = s_i(t_j)$ such that the new metric in t_i coordinates is constant. In the most interesting cases the search for such a transformation is the solution of the afore mentioned uniformization problem.

In a recent paper [19] we presented an alternative approach to such topological models where the relation to singularity theory is directly obtained in a natural system of flat coordinates. At the same time, these are the parameters of a Landau-Ginzburg superpotential as well as the deformations around the conformal point. The first step of our construction is the use of free first-order (b, c, β, γ)-systems to describe N=2 superconformal theories as proposed in [20]. We then show that an arbitrary interaction of the Landau-Ginzburg type – *i.e.* characterized by a polynomial potential V – can be added to the free Lagrangian without spoiling the superconformal invariance if V is a quasi-homogeneous function. The deformation parameters of the potential are the flat coordinates simply because they correspond to deformations around the conformal point. In this way we lose something: in the presence of a deformed potential we cannot use the residue pairing metric to define the perturbed correlation functions. We have to do any computation in the context of conformal field theory, using for example bosonization techniques.

2..Lagrangian Formulation of N=2 theories

In this section we review how it is possible to add a polynomial interaction V of the LG type to a collection of free first-order (b, c, β, γ)-systems in such a way that, if V is a quasi-homogeneous function, the theory possesses an N=2 superconformal symmetry at the classical level. We also show that the interaction potential unambiguously fixes the weights of the pseudo-ghost fields. In this formulation we can add in a natural way all the relevant perturbations and, whenever we use a quasi-homogeneous potential with modality different from zero, we can study marginal deformations and eventually Zamolodchikov's metric on the associated moduli space. Alternatively, we can consider topological models by "twisting" the generators of the superconformal algebra and compute topological correlation functions.

We start this program by defining our model. We consider a collection of pseudo-ghost fields $\{b_\ell, c_\ell, \beta_\ell, \gamma_\ell; \tilde{b}_r, \tilde{c}_r, \tilde{\beta}_r, \tilde{\gamma}_r\}$ where $\ell = 1, \ldots, N_L$ and $r = 1, \ldots, N_R$. β_ℓ and γ_ℓ form a bosonic first-order system with weights λ_ℓ and $1 - \lambda_\ell$ respectively whereas b_ℓ and c_ℓ form a fermionic first-order system with weights $\lambda_\ell + \frac{1}{2}$ and $\frac{1}{2} - \lambda_\ell$ respectively. The same can be said for the tilded fields with λ_ℓ replaced by $\tilde{\lambda}_r$. The action is

$$S = \int d^2z\, \mathcal{L} = \int d^2z\, (\mathcal{L}_0 + \Delta\mathcal{L}) \tag{5}$$

where

$$\begin{aligned}\mathcal{L}_0 &= \sum_\ell \left[-\lambda_\ell \beta_\ell \bar{\partial}\gamma_\ell + (1-\lambda_\ell)\gamma_\ell \bar{\partial}\beta_\ell - (\lambda_\ell + \tfrac{1}{2})\, b_\ell \bar{\partial} c_\ell + (\lambda_\ell - \tfrac{1}{2})\, c_\ell \bar{\partial} b_\ell\right] \\ &+ \sum_r \left[-\tilde{\lambda}_r \tilde{\beta}_r \partial\tilde{\gamma}_r + (1-\tilde{\lambda}_r)\tilde{\gamma}_r \partial\tilde{\beta}_r - (\tilde{\lambda}_r + \tfrac{1}{2})\tilde{b}_r \partial\tilde{c}_r + (\tilde{\lambda}_r - \tfrac{1}{2})\tilde{c}_r \partial\tilde{b}_r\right]\end{aligned} \tag{6}$$

and

$$\Delta\mathcal{L} = \sum_{\ell,r} b_\ell \tilde{b}_r\, \partial_\ell V(\beta)\, \tilde{\partial}_r \tilde{V}(\tilde{\beta}) \ . \tag{7}$$

Here and in the following we use the short-hand notations $\partial_\ell \equiv \partial/\partial\beta_\ell$, $\tilde{\partial}_r \equiv \partial/\partial\tilde{\beta}_r$ while $\partial, \bar{\partial}$ denotes the derivatives with respect the world–sheet coordinates. $\mathcal{L}_0$ in (6) represents the standard free Lagrangian for first-order systems [21] of the given weights and $\Delta\mathcal{L}$ in (7) defines an interaction of the LG type when V and $\tilde{V}$ are polynomial functions of β_ℓ and $\tilde{\beta}_r$ respectively. From (5) one can derive the following equations of motion

$$\begin{aligned}\bar{\partial}\beta_\ell &= 0 \ , \quad \bar{\partial}b_\ell = 0 \ , \\ \partial\tilde{\beta}_\ell &= 0 \ , \quad \partial\tilde{b}_\ell = 0 \ , \\ \bar{\partial}c_\ell &= \sum_r \tilde{b}_r \partial_\ell V(\beta)\tilde{\partial}_r\tilde{V}(\tilde{\beta}) \ , \\ \bar{\partial}\gamma_\ell &= \sum_{m,r} b_m \tilde{b}_r \partial_\ell\partial_m V(\beta)\tilde{\partial}_r\tilde{V}(\tilde{\beta}) \ , \\ \partial\tilde{c}_r &= -\sum_\ell b_\ell \partial_\ell V(\beta)\tilde{\partial}_r\tilde{V}(\tilde{\beta}) \ , \\ \partial\tilde{\gamma}_r &= \sum_{\ell,s} b_\ell \partial_\ell V(\beta)\tilde{b}_s\tilde{\partial}_r\tilde{\partial}_s\tilde{V}(\tilde{\beta}) \ .\end{aligned} \tag{8}$$

first two lines of (8) show that β_ℓ, b_ℓ, $\tilde{\beta}_r$ and $\tilde{b}_r$ satisfy the same classical equations of motion as in the free case, whereas c_ℓ, γ_ℓ, $\tilde{c}_r$ and $\tilde{\gamma}_r$ have no longer a definite holomorphic or anti-holomorphic character in the presence of the interaction.

It is not difficult to realize that there are no quantum corrections, due to the impossibility of generating closed loops, in the perturbative approach and hence the fundamental operator product expansions are the same as in the free case. Indeed, even in the presence of the interaction, we have

$$\beta_\ell(z)\,\gamma_m(w,\bar{w}) = -\frac{\delta_{\ell m}}{z-w} + \cdots \ , \quad b_\ell(z)\,c_m(w,\bar{w}) = \frac{\delta_{\ell m}}{z-w} + \cdots \ , \tag{9}$$

and similarly for the tilded fields. It is well-known that $\mathcal{L}_0$ in (6) describes a $(2,2)$–superconformal field theory with central charges $c_L = \sum_\ell(3-12\lambda_\ell)$ and $c_R = \sum_r(3-12\tilde{\lambda}_r)$ for the left and the right sectors respectively. We will now show that the addition of the interaction $\Delta\mathcal{L}$ does not destroy this (2,2)-superconformal invariance if V and $\tilde{V}$ are quasi-homogeneous functions, *i.e.* if for any $a \in \mathbf{R}^+$,

$$V(a^{\omega_\ell}\beta_\ell) = a\,V(\beta_\ell) \quad , \quad \tilde{V}(a^{\tilde{\omega}_r}\tilde{\beta}_r) = a\,\tilde{V}(\tilde{\beta}_r) \ . \tag{10}$$

The parameters ω_ℓ and $\tilde{\omega}_r$ are called the homogeneous weights of β_ℓ and $\tilde{\beta}_r$ respectively. By enforcing the requirement that the interaction Lagrangian $\Delta\mathcal{L}$ have the correct dimensions, one can see that

$$\omega_\ell = 2\lambda_\ell \quad , \quad \tilde{\omega}_r = 2\tilde{\lambda}_r \ ; \tag{11}$$

the parameters λ_ℓ and $\tilde{\lambda}_r$ of the free Lagrangian (6) are therefore *fixed* by the interaction terms. When (10) and (11) are satisfied, the action S in (5) is invariant under the following N=2 holomorphic supersymmetry transformations

$$\begin{aligned}
\delta\beta_\ell &= 2\sqrt{2}\epsilon^- b_\ell \quad , \quad \delta b_\ell = \frac{1}{\sqrt{2}}\epsilon^+\partial\beta_\ell + \sqrt{2}\lambda_\ell\,\partial\epsilon^+\beta_\ell \ , \\
\delta c_\ell &= 2\sqrt{2}\epsilon^-\gamma_\ell \quad , \quad \delta\gamma_\ell = \frac{1}{\sqrt{2}}\epsilon^+\partial c_\ell - \sqrt{2}\,(\lambda_\ell - \tfrac{1}{2})\,\partial\epsilon^+ c_\ell \ , \\
\delta\tilde{\beta}_r &= 0 \quad , \quad \delta\tilde{b}_r = 0 \ , \\
\delta\tilde{c}_r &= -\frac{1}{\sqrt{2}}\epsilon^+ V(\beta)\tilde{\partial}_r\tilde{V}(\tilde{\beta}) \quad , \quad \delta\tilde{\gamma}_r = \frac{1}{\sqrt{2}}\epsilon^+ V(\beta)\sum_s \tilde{\partial}_r\tilde{\partial}_s\tilde{V}(\tilde{\beta})\tilde{b}_s \ ,
\end{aligned} \tag{12}$$

where $\epsilon^\pm$ are arbitrary holomorphic functions ($\bar{\partial}\epsilon^\pm = 0$). The action S is also invariant under N=2 anti-holomorphic symmetries which are similar to the ones defined in (12), with the exchange of the tilded and untilded quantities, and the replacement of $\epsilon^\pm$ with arbitrary anti-holomorphic functions $\bar{\epsilon}^\pm$ ($\partial\bar{\epsilon}^\pm = 0$). Moreover, if we relax the hypothesis that V and $\tilde{V}$ are quasi-homogeneous, the transformations (12) and their $\bar{\epsilon}$-analogues remain symmetries of (5) provided $\epsilon^\pm$ and $\bar{\epsilon}^\pm$ are constant parameters. This means that our model has a global N=2 supersymmetry for any choice of V and $\tilde{V}$, and an N=2 superconformal invariance for quasi-homogeneous potentials.

The conserved Noether's currents associated to the symmetries (12) are [2]

$$\begin{aligned}
G^+_z &= \sqrt{2}\sum_\ell\left[(\tfrac{1}{2}-\lambda_\ell)\,c_\ell\partial\beta_\ell - \lambda_\ell\,\beta_\ell\partial c_\ell\right] \ , \\
G^+_{\bar{z}} &= \sqrt{2}\sum_\ell\left[\lambda_\ell\beta_\ell\bar{\partial}c_\ell + (\lambda_\ell - \tfrac{1}{2})\,\bar{\partial}\beta_\ell c_\ell\right] - \frac{1}{\sqrt{2}}\sum_r V(\beta)\tilde{b}_r\tilde{\partial}_r\tilde{V}(\tilde{\beta}) \ , \\
G^-_z &= 2\sqrt{2}\sum_\ell \gamma_\ell b_\ell \ , \\
G^-_{\bar{z}} &= 0 \ .
\end{aligned} \tag{13}$$

[2] From now on, to avoid repetitions we will discuss only the left sector and understand that similar considerations can be made in the right sector, with some obvious change of signs.

If we use the equations of motion (8) for quasi-homogeneous potentials, we see that $G^+_{\bar{z}}$ vanishes on-shell; thus from the conservation laws we deduce that G^+_z and G^-_z are holomorphic currents even if they contain the non-holomorphic fields c_ℓ and γ_ℓ. We denote these currents by $G^\pm(z)$.

The action (5) is also invariant under holomorphic conformal reparametrizations and $U(1)$-rescalings of the fields; the conserved Noether's currents associated to such symmetries are the stress-energy tensor $T_{\mu\nu}$ and the $U(1)$-current J_μ. For homogeneous potentials it is not difficult to see that the trace of $T_{\mu\nu}$ and the $\bar{z}$-component of J_μ are zero on-shell. Therefore, from the conservation laws, we deduce that

$$T_{zz} = \sum_\ell \left[-\lambda_\ell\beta_\ell\partial\gamma_\ell + (1-\lambda_\ell)\gamma_\ell\partial\beta_\ell - (\lambda_\ell + \tfrac{1}{2})\, b_\ell\partial c_\ell - (\tfrac{1}{2} - \lambda_\ell)\, c_\ell\partial b_\ell\right] \quad , \tag{14}$$

and

$$J_z = \sum_\ell \left[(2\lambda_\ell - 1)b_\ell c_\ell + 2\lambda_\ell\beta_\ell\gamma_\ell\right] \tag{15}$$

are holomorphic currents. We denote them by $T(z)$ and $J(z)$ respectively.

Using the OPE's in (9), it is straightforward to check that $T(z)$, $G^\pm(z)$ and $J(z)$ close an N=2 superconformal algebra with central charge $c_{\rm L} = \sum_\ell(3 - 12\lambda_\ell)$. Thus, the interaction $\Delta\mathcal{L}$ with homogeneous polynomials V and $\tilde{V}$ does not spoil the superconformal properties of $\mathcal{L}_0$. It is interesting to observe that the ring determined by the potential V, which contains all polynomials in β_ℓ's modulo the vanishing relations $\partial_\ell V = 0$, coincides with the ring of (left) chiral primary operators of the N=2 minimal model associated to V. Indeed, using (14) and (15), one can easily check that the $U(1)$-charge of $(\beta_\ell)^n$ is twice its conformal dimension.

In order to compare our formulation of N=2 supersymmetric models with the standard LG approach we shall consider $2d$–Lorentz invariant interactions of the form

$$\Delta\mathcal{L} = \sum_{i,j=1}^{N} b_i\tilde{b}_j\, \partial_i\tilde{\partial}_j W \tag{16}$$

($N_{\rm L} = N_{\rm R} = N$), where W is a quasi-homogeneous function of the variables

$$X_i = \beta_i\tilde{\beta}_i \quad , \quad i = 1, \ldots, N \ . \tag{17}$$

This is clearly a very special case of (7). It is well known that given an N=2 superconformal algebra generated by $T(z)$, $G^\pm(z)$ and $J(z)$, one obtains a topological conformal algebra by "twisting" the currents according to

$$\begin{aligned} \hat{T}_\pm(z) &= T(z) \pm \frac{1}{2}\partial J(z) \ , \\ \hat{J}_\pm(z) &= \pm J(z) \ , \\ Q_\pm(z) &= G^\pm(z) \ , \\ G_\pm(z) &= G^\mp(z) \ . \end{aligned} \tag{18}$$

In eq: (18) $Q_\pm(z)$ can be interpreted as a BRST current and the cohomology classes of the BRST charge

$$Q_\pm^{\mathrm{BRST}} = \oint dz\; Q_\pm(z) \tag{19}$$

are identified with the physical fields of the topological theory. The two choices of signs in (18) lead to two different sets of BRST invariant states: the chiral primary fields of the original N=2 superconformal algebra for the + sign, and the anti-chiral primary fields for the − sign.

It is now interesting to study the consequences of the twist (18) on our (b, c, β, γ) systems. We analyze the + case. From (14) and (15) we simply get

$$\hat{T}_+ = T + \frac{1}{2}\partial J = \sum_{i=1}^{N} (\partial c_i\, b_i + \gamma_i\, \partial\beta_i) \quad . \tag{20}$$

This is the canonical stress-energy tensor for a collection of N commuting (β,γ)-systems of weight $\lambda = 0$, and N anticommuting (b,c)-systems of weight $\lambda = 1$. $\hat{T}$ in (20) closes a Virasoro algebra with vanishing central charge. Indeed, the central charge of a first-order system of weight λ is $c_\lambda = \varepsilon\,(1 - 3Q^2)$, where

$$Q = \varepsilon\,(1 - 2\lambda) \tag{21}$$

is a "background charge" and $\varepsilon = 1$ or -1, depending on whether the system is anticommuting or commuting. In our case both the (β,γ)-systems and the (b,c)-systems have $Q = -1$, but since their statistics is different, their central charges exactly cancel.

The topological Lagrangian is:

$$\mathcal{L}_{\mathrm{top}} = \sum_{i=1}^{N} \left[\gamma_i\bar{\partial}\beta_i + \tilde{\gamma}_i\partial\tilde{\beta}_i - b_i\bar{\partial}c_i - \tilde{b}_i\partial\tilde{c}_i\right] + \sum_{i,j=1}^{N} \left[b_i\tilde{b}_j\partial_i\bar{\partial}_j W(X)\right] \tag{22}$$

The BRST transformations which leave (22) invariant, can be obtained from (12) and their analogue by identifying the BRST parameter θ with $\frac{\epsilon^+}{\sqrt{2}} = \frac{\bar{\epsilon}^+}{\sqrt{2}}$ (the factor of $1/\sqrt{2}$ is introduced for convenience). The action of the corresponding nilpotent Slavnov operator s on all fields read:

$$\begin{aligned}
s\,\beta_i &= 0 \quad , \qquad s\,\tilde{\beta}_i = 0 \; , \\
s\,b_i &= \partial\beta_i \quad , \qquad s\,\tilde{b}_i = \bar{\partial}\tilde{\beta}_i \; , \\
s\,\gamma_i &= \partial c_i - \sum_{j=1}^{N} b_j\,\partial_i\partial_j W \; , \quad s\,\tilde{\gamma}_i = \bar{\partial}\tilde{c}_i + \sum_{j=1}^{N} \tilde{b}_j\,\bar{\partial}_i\bar{\partial}_j W \; , \\
s\,c_i &= +\partial_i W \quad , \qquad s\,\tilde{c}_i = -\bar{\partial}_i W \; .
\end{aligned} \tag{23}$$

Using (23) it is quite easy to construct the representatives of the BRST-cohomology classes and the corresponding integrated invariants. According to the general theory, we have to consider multiplets composed by a 0-form Φ_P, a 1-form $\Phi_P^{(1)}$ and a 2-form $\Phi_P^{(2)}$ which satisfy the following descent equations

$$\begin{aligned} s\,\Phi_P &= 0 \quad , \quad s\,\Phi_P^{(1)} = -\,d\,\Phi_P \ , \\ s\,\Phi_P^{(2)} &= -\,d\,\Phi_P^{(1)} \quad , \quad d\,\Phi_P^{(2)} = 0 \ . \end{aligned} \tag{24}$$

Moreover the 0-form Φ_P must belong to a non-trivial BRST-cohomology class, *i.e.* it should not be BRST-exact. The solutions of the descent equations (23) provide the local physical observables Φ_P appearing in correlation functions as well as the integrated invariants $\Phi_P^{(2)}$ which can be used to deform the theory. Thus, the general form of a perturbed topological correlation function is

$$c_{P_1,\dots,P_m}(t_1,\dots,t_M) = \left\langle \Phi_{P_1}(z_1)\cdots\Phi_{P_m}(z_m)\ \exp\left[\sum_{k=1}^{M} t_k \int \Phi_{P_k}^{(2)}\right]\right\rangle_{\text{top}} \tag{25}$$

where $\langle\cdots\rangle_{\text{top}}$ means functional integration with the measure provided by the unperturbed Lagrangian $\mathcal{L}_{\text{top}}$ and t_k are coupling constants parametrizing its deformations $\int \Phi_{P_k}^{(2)}$.

In our (b,c,β,γ) formulation the general solution of the descent equations (24) is

$$\begin{aligned} \Phi_P &= P(X) \ , \\ \Phi_P^{(1)} &= -\sum_{i=1}^{N}\left[b_i\,\partial_i P\,dz + \tilde{b}_i\,\tilde{\partial}_i P\,d\bar{z}\right] \ , \\ \Phi_P^{(2)} &= \sum_{i,j=1}^{N}\left[b_i\tilde{b}_j\,\partial_i\tilde{\partial}_j P\right]\,dz\wedge d\bar{z} \ , \end{aligned} \tag{26}$$

where $P(X)$ is any polynomial in the variables $X_i = \beta_i\tilde{\beta}_i$ corresponding to a non trivial element of the local ring determined by the superpotential W of the Lagrangian (22). Indeed if the polynomial $P(X)$ is proportional to the vanishing relations (*i.e.* if $P(X) = \sum_i p^i(X)\frac{\partial W}{\partial X_i}$), then using the BRST transformations (23), we easily see that $P(X) = s\,K$ and so Φ_P would be exact. (For the proof it suffices to set $K = p^i(X)\,\frac{\partial\beta_j}{\partial X_i}\,c_j$.) Thus, the physical observables in the topological theory are simply *local* polynomials of β_i and $\tilde{\beta}_i$, which correspond to chiral primary fields of the original N=2 superconformal theory.

We can easily conclude that a deformation of the potential with some element $P(X)$ of the local ring, *i.e.*

$$W(X) \longrightarrow W(X) - t_P\,P(X) \ , \tag{27}$$

corresponds to a perturbation of the action with $\int \Phi_P^{(2)}$, *i.e.*

$$\int d^2z\ \mathcal{L}_{\text{top}} \longrightarrow \int d^2z\ \mathcal{L}_{\text{top}} - t_P \int \Phi_P^{(2)} \ . \tag{28}$$

Thus, the possible perturbations of the theory are in one-to-one correspondence with the possible deformations of the potential. Something similar happens also in the ordinary LG models, but only up to BRST-exact terms [19].

A few words concerning the other choice of sign in the topological twist: using the terminology of [4] "−" twist defines a topological field theory of the Witten-type, which corresponds to a BRST exact lagrangian. On the contrary, the "+" twist leads to the Lagrangian (22) which is not BRST commutator with respect to Q_+^{BRST}; thus the + twist defines a topological field theory of the Schwarz-type. As pointed out in [24], also the ordinary topological LG models are theories of the Schwarz-type.

All the properties of the model (5) we just discussed are at the classical level. Due to the very simple form of the interaction (7), we can easily extend it to a quantum level, as already mentioned. We do it in a qualitatively way by mean of the following consideration. From a conformal field theory point of view, once requiring that the interaction term has the right (1,1) dimension, we are perturbing the theory with a "candidate" marginal operator. The mere existence of a (1,1) operator is not sufficient to guarantee the existence of a fixed line. We have to require additional "integrability"conditions (see [22]), so that the perturbation generated by the marginal operator does not act to change its own conformal weight from (1,1). In our case, where we have collected the potential term as a single marginal operator $\mathcal{O}$ with coupling g, this reduce to one loop level to the requirement that in the operator product expansion of $\mathcal{O}$ with itself there are no term of the form:

$$\mathcal{O}(z,\bar{z})\mathcal{O}(w,\bar{w}) = c_{\mathcal{OOO}}(z-w)^{-1}(\bar{z}-\bar{w})^{-1}\mathcal{O} \tag{29}$$

If this condition is not satisfied. the conformal weight of $\mathcal{O}$ is shifted by a quantity proportional to the three point function $c_{\mathcal{OOO}}$: $\mathcal{O}$ would not remain marginal away from the fixed point. It is easily verified that our interaction satisfies this requirement, and the analogous one which is required to higher orders, leaving the theory conformally invariant. We can of course explicitly compute the stress–energy tensor and show that the interaction considered does not give any loop correction to it. We refer to [19] for details.

3. Bosonization of the (b,c,β,γ)-system

Let us specialize to the case of the A_n minimal models with potential:

$$W = \frac{1}{(n+1)^2}(\beta\tilde{\beta})^{n+1} \tag{30}$$

The result of our previous analysis is that for any value $\lambda = (2n+2)^{-1} = \frac{1}{2}\omega$ $(n = 2,3,\cdots)$, which fixes the marginal dimension of (30), we have a realization of the N=2 superconformal algebra with central charge

$$c = 3 - 12\lambda = \frac{3(n-1)}{n+1} \quad . \tag{31}$$

The Fock space generated by the modes of the almost-free fields b, c, β, γ and their spin fields, contains the irreducible representations of the N=2 minimal models. Such representations should be obtained from the Fock space through a suitable projection like in the case of the standard free-field realization of the minimal models as given for instance in [23]. In this section our aim is to make contact with this Coulomb gas formalism [25] , which, as we will see in the sequel, enables us to calculate explicitly (perturbed) topological correlation functions in the presence of a LG interaction.

Adopting the conventions of [23], an N=2 minimal model with central charge as in (31), can be described in terms of three scalar fields ϕ_0, ϕ_1 and ϕ_2. While ϕ_0 and ϕ_2 are really free fields, ϕ_1 is coupled to a background charge

$$Q_1^{(n)} = \sqrt{\frac{2}{n+1}} \ . \tag{32}$$

In this realization the holomorphic currents of the N=2 superconformal algebra are [3]

$$\begin{aligned} T &= \frac{1}{2}\left[(\partial\phi_0)^2 + (\partial\phi_1)^2 + (\partial\phi_2)^2\right] - \frac{1}{2}Q_1^{(n)}\,\partial^2\phi_1 \ , & (33)\\ J &= \sqrt{\frac{n-1}{n+1}}\,\partial\phi_0 \ , & (34)\\ G^\pm &= \sqrt{\frac{2n-2}{n+1}}\,\Psi^\pm \exp\left[\pm\sqrt{\frac{n+1}{n-1}}\,\phi_0\right] \ , & (35)\\ \Psi^\pm &= \frac{1}{\sqrt{2}}\exp\left[\pm \mathrm{i}\sqrt{\frac{2}{n-1}}\,\phi_2\right]\left(\sqrt{\frac{n+1}{n-1}}\partial\phi_1 \pm \mathrm{i}\,\partial\phi_2\right) \ . & (36)\end{aligned}$$

The field ϕ_0 bosonizes the U(1)-current and its exponentials realize the well known N=2 spectral flow [2]. The operators $\Psi^\pm$ in (36) are, instead, parafermionic currents and generate the non trivial part of the N=2 algebra. The complete Fock space which embeds the N=2 irreducible modules contains, in particular, N=2 primary fields, which are given by

$$\Lambda^{(n)}_{\ell,m;s} = \frac{1}{\sqrt{2}}\exp\left[\frac{m+sn-s}{\sqrt{n^2-1}}\phi_0 + \frac{\ell}{\sqrt{2(n+1)}}\phi_1 + \mathrm{i}\frac{m}{\sqrt{2(n-1)}}\phi_2\right] \tag{37}$$

where ℓ takes the integer values $0 \leq \ell \leq n-1$ and m takes the integer values $m = -l, -l+2, ..., l$. The quantum number s represents the sector and is 0 in the Neveu-Schwarz sector and $\pm 1/2$ in the Ramond sector. As we will see later, it is convenient to factor out the ϕ_0 contribution and rewrite the primary fields (37) as

$$\Lambda^{(n)}_{\ell,m;s} = \exp\left[\sqrt{\frac{3}{c}}q(m,s)\,\phi_0\right]\,\varphi^\ell_m \tag{38}$$

[3]Here and in the following, any exponential of free fields is understood as normal ordered.

where

$$\varphi_m^\ell = \exp\left[\frac{\ell}{\sqrt{2(n+1)}}\phi_1 + \mathrm{i}\frac{m}{\sqrt{2(n-1)}}\phi_2\right] .$$

The operators φ_m^ℓ are the principal primary fields of the $\mathbf{Z}_{n-1}$ parafermion algebra and must be identified according to

$$\varphi_m^\ell \sim \varphi_{m\pm(n-1)}^{n-1-\ell} . \tag{39}$$

In fact the Hilbert space created by φ_m^ℓ is isomorphic to the one created by $\varphi_{m\pm(n-1)}^{n-1-\ell}$ due to the existence of a map between the two that commutes with all generators of the algebra [23].

In order to relate this realization of the N=2 minimal models to the one provided by our (b, c, β, γ)-system, we bosonize the latter according to the standard rules and write[4]

$$b = \mathrm{e}^{-\pi_1} \quad , \quad \beta = \mathrm{e}^{\mathrm{i}\pi_2 - \pi_3} \quad , \quad c = \mathrm{e}^{\pi_1} \quad , \quad \gamma = \mathrm{e}^{-\mathrm{i}\pi_2 + \pi_3}\partial\pi_3 \tag{40}$$

where the π_i's are scalar fields coupled to the following background charges

$$\tilde{Q}_1^{(n)} = -\frac{1}{n+1} \quad , \quad \tilde{Q}_2^{(n)} = \mathrm{i}\frac{n}{n+1} \quad , \quad \tilde{Q}_3^{(n)} = -1 . \tag{41}$$

These numbers are explained as follows: π_1 bosonizes the anticommuting (b, c)-system whose weight is $\lambda + \frac{1}{2} = -\frac{1}{n+1}$. Insertion of this value in the general formula (21) yields $\tilde{Q}_1^{(n)}$ as listed in (41). The field $\mathrm{i}\pi_2$ bosonizes the commuting (β, γ)-system according to the rule

$$\gamma = \mathrm{e}^{-\mathrm{i}\pi_2}\,\partial\xi \quad , \quad \beta = \mathrm{e}^{\mathrm{i}\pi_2}\,\eta \tag{42}$$

where ξ and η form an anticommuting first-order system of weight $\lambda_{\xi\eta} = 1$. The background charge $\tilde{Q}_2^{(n)}$ of the field π_2 follows from (21) with $\lambda_{\beta\gamma} = \lambda = \frac{1}{2n+2}$. Finally π_3 is the scalar field that bosonizes the (ξ, η)-system and its backround charge $\tilde{Q}_3^{(n)}$ also follows from (21) upon use of the value $\lambda_{\xi\eta} = 1$.

Consequently, in terms of the fields π_i's , the stress-energy tensor of the N=2 model is

$$T = \frac{1}{2}\left[(\partial\pi_1)^2 + (\partial\pi_2)^2 + (\partial\pi_3)^2\right] - \frac{1}{2}\left(-\frac{1}{n+1}\,\partial^2\pi_1 + \mathrm{i}\frac{n}{n+1}\,\partial^2\pi_2 - \partial^2\pi_3\right) . \tag{43}$$

Similarly, using (40) in (13) and (15) we obtain

$$J = \frac{1}{n+1}(n\partial\pi_1 - \mathrm{i}\partial\pi_2) , \tag{44}$$

$$G^- = 2\sqrt{2}\exp\left[-\pi_1 - \mathrm{i}\,\pi_2 + \pi_3\right]\partial\pi_3 . \tag{45}$$

[4]Notice that the bosonization rules we are giving are actually true for the "free" holomorphic part of the c, γ fields. However, as we are going to see, we only need the bosonized b, β fields (which are correctly expressed by (40), (49) and (50) for the application to topological correlation functions

Comparing (43) and the last two equations with (33)–(36), we obtain the relation between the π's and the ϕ's, namely

$$\pi_1 = \frac{n}{\sqrt{n^2-1}}\,\phi_0 - \frac{1}{\sqrt{2(n+1)}}\,\phi_1 + \mathrm{i}\frac{1}{\sqrt{2(n-1)}}\,\phi_2 \ , \tag{46}$$

$$\mathrm{i}\,\pi_2 = \frac{1}{\sqrt{n^2-1}}\,\phi_0 - \frac{n}{\sqrt{2(n+1)}}\,\phi_1 + \mathrm{i}\frac{n}{\sqrt{2(n-1)}}\,\phi_2 \ , \tag{47}$$

$$\pi_3 = -\sqrt{\frac{n+1}{2}}\,\phi_1 + \mathrm{i}\sqrt{\frac{n-1}{2}}\,\phi_2 \ . \tag{48}$$

One can proceed even further and use (40) and (46)– (48) to identify the pseudo-ghost fields with the operators of the abstract N=2 superconformal model. Explicitly one finds

$$\beta = \exp\left[\frac{1}{\sqrt{n^2-1}}\phi_0 + \frac{1}{\sqrt{2(n+1)}}\phi_1 + \mathrm{i}\frac{1}{\sqrt{2(n-1)}}\phi_2\right] \ , \tag{49}$$

$$b = \exp\left[\frac{-n}{\sqrt{n^2-1}}\phi_0 + \frac{1}{\sqrt{2(n+1)}}\phi_1 + \mathrm{i}\frac{-1}{\sqrt{2(n-1)}}\phi_2\right] \ , \tag{50}$$

$$c = \exp\left[\frac{n}{\sqrt{n^2-1}}\phi_0 + \frac{-1}{\sqrt{2(n+1)}}\phi_1 + \mathrm{i}\frac{1}{\sqrt{2(n-1)}}\phi_2\right] \ , \tag{51}$$

$$\gamma = \exp\left[\frac{-1}{\sqrt{n^2-1}}\phi_0 + \frac{-1}{\sqrt{2(n+1)}}\phi_1 + \mathrm{i}\frac{-1}{\sqrt{2(n-1)}}\phi_2\right] \times$$
$$\times \sqrt{\frac{n-1}{2}}\left(-\sqrt{\frac{n+1}{n-1}}\partial\phi_1 + \mathrm{i}\partial\phi_2\right) \ . \tag{52}$$

From (49) one realizes that β is a chiral primary field and is given by $\beta = \Lambda^{(n)}_{1,1;0}$. More generally one can write

$$\beta^\ell = \Lambda^{(n)}_{\ell,\ell;0} \quad \text{for } \ell = 0,\ldots,n-1 \ , \tag{53}$$

which shows that at the quantum level the general left chiral primary field is simply the ℓ-th power of β and the vanishing relation is recovered by enforcing the bound $\ell \le n-1$. Moreover b is the first component of a chiral primary superfield and can be explicitly obtained by $\frac{1}{2\sqrt{2}}\oint_z G^-(w)\beta(w) = b(z)$; the same is true for fields of the form $b\beta^{l-1}$. On the contrary γ and c are in the Fock space of the three scalar fields, but not in the N=2 irreducible module.

If we now consider the case when the theory is topologically twisted with $Q_+(z) = G^+(z)$, it is not difficult to show that the relations (46)–(48) and the identifications (49)–(52) and (53) hold true also in the topological field theory, giving us a complete characterization of the fields b, c, β and γ at the quantum level. The net effect of the topological twist is simply to switch on a background charge for ϕ_0 given by

$$Q_0^{(n)} = \frac{1-n}{\sqrt{n^2-1}} \quad . \tag{54}$$

Moreover, using (49)–(52) and taking into account both left and right movers, we can easily check the descent equations (24) in the complete bosonized formalism. Therefore, even if the bosonized expressions for the topological b, c, β and γ are still given by (49)–(52), their conformal dimensions change with the twist. In particular the chiral primary fields β^ℓ loose their conformal weight and become dimensionless, as is appropriate for the physical operators of a topological field theory. Furthermore, the $U(1)$ current J acquires an anomaly proportional to $Q_0^{(n)}$.

The complete bosonization of the (b, c, β, γ)-system we have just presented is the technical tool which enables us to make explicit calculations of (perturbed) correlation functions for single minimal models as well as for tensor products thereof. To obtain a more adequate characterization of topological correlation functions in the bosonized formalism, it is more appropriate to use the fields which bosonize a twisted (b, c, β, γ)-system. As we have seen, only very few things change; most notably the field ϕ_0 acquires the background charge (54). Taking into account of this background contribution we can define the topological vacuum $|0\rangle_{\text{top}}$ as

$$|0\rangle_{\text{top}} \equiv |0,0,0\rangle \tag{55}$$

and its "topological" dual conjugate (see also [5]) ${}_{\text{top}}\langle 0|$ as

$${}_{\text{top}}\langle 0| \equiv \langle 0, -1-n, 1-n| \quad , \tag{56}$$

In the definition (56) we use the conjugate operator of the top chiral field β^{n-1}, as in [5] The topological correlator (or better to say, its holomorphic part) is expressed by:

$$\langle \prod_{i=1}^{N} \beta^{\ell_i}(z_i)\rangle_{\text{top}} \equiv {}_{\text{top}}\langle 0| \prod_{i=1}^{N} \beta_i^{\ell}(z_i)|0\rangle_{\text{top}} = \delta\left(\sum_i \ell_i - n + 1\right) \tag{57}$$

Now we are ready to test our formalism with some computation and to verify that the parameters of the deformed LG potential for the (b, c, β, γ)-system are the flat coordinates of the topological field theories. We stress that our techniques can be applied to single minimal models as well as to their tensor products, since there is practically no difference between the two cases. We give a sample computation for the torus case, which can be described by the tensor product of three minimal models A_2 deformed by its marginal operator. The potential we should consider is

$$W = \frac{1}{9}(\beta_x\tilde{\beta}_x)^3 + \frac{1}{9}(\beta_y\tilde{\beta}_y)^3 + \frac{1}{9}(\beta_z\tilde{\beta}_z)^3 - t\,(\beta_x\tilde{\beta}_x)(\beta_y\tilde{\beta}_y)(\beta_z\tilde{\beta}_z) \tag{58}$$

The dimensionless parameter t in (58) is (related to) the modulus of the torus. The chiral ring of the tensor product of three A_2 minimal models is generated by $\Phi_0 = \mathbf{1}$,

$(\Phi_1, \Phi_2, \Phi_3) = (\beta_x\tilde{\beta}_x, \beta_y\tilde{\beta}_y, \beta_z\tilde{\beta}_z)$, $\Phi_4 = \beta_x\tilde{\beta}_x\beta_y\tilde{\beta}_y$, $\Phi_5 = \beta_y\tilde{\beta}_y\beta_z\tilde{\beta}_z$, $\Phi_6 = \beta_x\tilde{\beta}_x\beta_z\tilde{\beta}_z$, and $\Phi_7 = \beta_x\tilde{\beta}_x\beta_y\tilde{\beta}_y\beta_z\tilde{\beta}_z$. Φ_7 is a marginal operator and is the one appearing in (58) as a deformation. The Lagrangian corresponding to (58) is

$$\mathcal{L} = \mathcal{L}_{\rm top} - t \int \Phi_7^{(2)} \tag{59}$$

where $\mathcal{L}_{\rm top}$ is the Lagrangian for three topological (b, c, β, γ)-systems as in (22), and $\Phi_7^{(2)}$ is the 2-form associated to Φ_7. According to (28), we have

$$\int \Phi_7^{(2)} = \int d^2w\, [(b_x\beta_y\beta_z + b_y\beta_x\beta_z + b_z\beta_x\beta_y)(w) \times (\text{c.c})] \quad . \tag{60}$$

The simplest way to check whether the parameter t in (59) is a flat coordinate or not, is to compute the topological metric

$$\eta_{ij}(t) = \left\langle \Phi_i\, \Phi_j\, e^{t \int \Phi_7^{(2)}} \right\rangle_{\rm top} \tag{61}$$

and see whether it is constant or not. To this end it is enough to consider one component, for example

$$\eta_{07}(t) = \left\langle \Phi_0\, \Phi_7\, e^{t \int \Phi_7^{(2)}} \right\rangle_{\rm top} \quad . \tag{62}$$

By looking at the $U(1)$-charges of the operators in (62), it is easy to realize that when expanding the exponential, only terms with $3n$ insertions of $\int \Phi_7^{(2)}$ will satisfy charge conservation. Thus, (62) can be rewritten as

$$\eta_{07}(t) = \sum_{n=0}^{\infty} \frac{1}{(3n)!} a_{3n}\, t^{3n} \tag{63}$$

where

$$a_{3n} = \left\langle \Phi_0\, \Phi_7 \left(\int \Phi_7^{(2)} \right)^{3n} \right\rangle_{\rm top} \quad . \tag{64}$$

It is immediate to see that $a_0 = 1$. The first non-trivial contribution is a_3 which, spelled out in detail, is

$$a_3 = \int d^2w_1 \int d^2w_2 \int d^2w_3\, f(u, w_1, w_2, w_3) \bar{f}(\bar{u}, \bar{w}_1, \bar{w}_2, \bar{w}_3) \tag{65}$$

where

$$f(u, w_1, w_2, w_3) = \left\langle (\beta_x\beta_y\beta_z)(u) \prod_{i=1}^{3} [(b_x\beta_y\beta_z + b_y\beta_x\beta_z + b_z\beta_x\beta_y)(w_i)] \right\rangle_{\rm top} \tag{66}$$

and $\bar{f}$ is its complex conjugate. To compute f, we split the r.h.s. of (66) into a sum of factorized correlation functions for each of the three minimal models and enforce on them (anomalous) conservation of the $U(1)$ charges. During this process the b fields in (66) must be suitably rearranged and proper minus signs arise from their anticommutation relations. After some straightforward algebra, it turns out that $f = 0$, which implies

$$a_3 = 0 \tag{67}$$

Actually it is very easy to generalize this result to all higher-order coefficients, and eventually conclude that

$$\eta_{07}(t) = 1 \tag{68}$$

The other entries of the metric $\eta_{ij}(t)$ can be computed in a similar way and all of them turn out to be constants. Thus the parameter t in (59) is a flat coordinate.

We want to emphasize that instead, in the standard LG formulation of the torus described by the potential

$$W = \frac{1}{3}X^3 + \frac{1}{3}Y^3 + \frac{1}{3}Z^3 - sXYZ \tag{69}$$

the topological correlation $\langle \Phi_0 \, \Phi_7 \rangle(s)$ *is* a non-trivial function of the LG parameter s, which therefore cannot be a flat coordinate [5].

It is interesting now to investigate the relation between s and t. On general grounds [7], it is possible to show that

$$s(t) = \frac{c_{111}(t)}{c_{123}(t)} \tag{70}$$

where

$$c_{ijk}(t) = \left\langle \Phi_i \, \Phi_j \, \Phi_k \, e^{t \int \Phi_7^{(2)}} \right\rangle_{\text{top}} \quad . \tag{71}$$

In our formulation it is easy to compute these perturbed 3-point functions as a power series in the flat coordinate t. Let us briefly see how $c_{123}(t)$ is evaluated. Expanding the exponential and looking for terms which satisfy charge conservation, we get

$$c_{123}(t) = \sum_{n=0}^{\infty} \frac{1}{(3n)!} \tilde{a}_{3n} \, t^{3n} \tag{72}$$

where

$$\tilde{a}_{3n} = \left\langle \Phi_1 \, \Phi_2 \, \Phi_3 \left(\int \Phi_7^{(2)} \right)^{3n} \right\rangle_{\text{top}} \quad . \tag{73}$$

It is immediate to see that $\tilde{a}_0 = 1$. The next contribution is

$$\tilde{a}_3 = \int d^2 w_1 \int d^2 w_2 \int d^2 w_3 \, g(x, y, z, w_1, w_2, w_3) \bar{g}(\bar{x}, \bar{y}, \bar{z}, \bar{w}_1, \bar{w}_2, \bar{w}_3) \tag{74}$$

---

[5] It turns out that $\langle \Phi_0 \, \Phi_7 \rangle(s) = \left[F(\frac{1}{3}, \frac{1}{3}; \frac{2}{3}; s^3) \right]^{-2}$, where F is the hypergeometric function.

where

$$g(x,y,z,w_1,w_2,w_3) = \left\langle \beta_x(x)\beta_y(y)\beta_z(z)\prod_{i=1}^{3}\left[(b_x\beta_y\beta_z + b_y\beta_x\beta_z + b_z\beta_x\beta_y)(w_i)\right]\right\rangle_{top} \quad . \tag{75}$$

By splitting the r.h.s. of (75) into a sum of factorized terms one can prove that the integrand of (74) gives just a numerical coefficient contribution that can be readsorbed in a rescaling of t, so that $\tilde{a}_3 = 1$, and

$$c_{123}(t) = 1 + \frac{1}{6}t^3 + O(t^6) \tag{76}$$

Similarly one can check that

$$c_{111}(t) = t + O(t^7) \tag{77}$$

so that from (70) it follows

$$s(t) = t - \frac{1}{6}t^4 + O(t^7) \tag{78}$$

These are precisely the first terms in the power series expansion of the solution of the schwarzian differential equation

$$\{s;t\} = -\frac{1}{2}\frac{(8+s^3)}{(1-s^3)^2}(s')^2 s \tag{79}$$

It has been recently shown [7, 8, 11] that (79) is equivalent to the requirement that t be a flat coordinate. Our methods provide automatically the solution to (79) as a power series.

References

1. M. Ademollo, L. Brink, A. D'Adda, R. D'Auria, E. Napolitano, S. Sciuto, E. del Giudice, P. di Vecchia, S. Ferrara, F. Gliozzi, R. Musto and R. Pettorino, *Phys. Lett.* **62B** (1976) 105.
   A. Sen, *Nucl. Phys.* **B278** (1986) 289 and *Nucl. Phys.* **B284** (1987) 423;
   T. Banks, L.J. Dixon, D. Friedan and E. Martinec, *Nucl. Phys.* **B299** (1988) 613 and *Nucl. Phys.* **B307** (88) 93.
   D. Gepner, *Nucl. Phys.* **B296** (1988) 757 and *Phys. Lett.* **199B** (1987) 380.
   T. Eguchi, H. Ouguri, A. Taormina and S. K. Yang, *Nucl. Phys.* **B315** (1989) 192.
   P. Candelas, C.T. Horowitz, A. Strominger and E. Witten, *Nucl. Phys.* **B298** (1988) 493.
   P. Candelas, A.M. Dale, C.A. Lutken and R. Schimmrick, *Nucl. Phys.* **B298** (1988) 493.
   M. Linker and R. Schimmrick, *Phys. Lett.* **208B** (1988) 216 and *Phys. Lett.* **215B** (1988) 681.
   C.A. Lutken and G.C. Ross, *Phys. Lett.* **213B** (1988) 152.
   P. Zoglin, *Phys. Lett.* **218B** (1989) 444.

2. W. Lerche, C. Vafa and N.P. Warner, *Nucl. Phys.* **B324** (1989) 427; B. Greene, C. Vafa and N.P. Warner, *Nucl. Phys.* **B324** (1989) 371.
3. B. de Wit and A. Van Proeyen, *Nucl. Phys.* **B245** (1984) 89.
   E. Cremmer, C. Kounnas, A. Van Proeyen, J.P. Derendinger, S. Ferrara, B. de Wit and L. Girardello, *Nucl. Phys.* **B250** (1985) 385.
   B. de Wit, P.G. Lauwers and A. Van Proeyen, *Nucl. Phys.* **B255** (1985) 560.
   L. Castellani, R. D'Auria and S. Ferrara, *Phys. Lett.* **241B** (1990) 57; *Class. and Quantum Grav.* **1** (1990) 163;
   R. D'Auria, S. Ferrara and P. Frè, *Nucl. Phys.* **B359** (1991) 705.
   A. Strominger, *Comm. Math. Phys.* **133** (1990) 163.
   S. Cecotti, S. Ferrara and L. Girardello, *Int. Jour. Mod. Phys* **A10** (1989) 2475 and *Phys. Lett.* **213B** (1988) 443.
   S. Ferrara and A. Strominger, Strings 89, Eds. R. Arnowitt, R. Bryan, M. Duff, D. Nanopulos and C. Pope, World Scientific, Singapore, 1989.
   P. Candelas and X. de la Ossa, *Nucl. Phys.* **B342** (1990) 246.
   L. J. Dixon, V. Kaplunowski and J. Louis, *Nucl. Phys.* **B329** (1990) 27.
4. For a review see D. Birmingham, M. Blau and M. Rakowski, *Phys. Rep.* **209** (1991) 129.
5. R. Dijkgraaf, E. Verlinde and H. Verlinde, *Nucl. Phys.* **B352** (1991) 59.
6. B. Block and A. Varchenko, Prepr. IASSNS-HEP-91/5.
7. E. Verlinde and N.P. Warner, *Phys. Lett.* **269B** (1991) 96.
8. A. Klemm, S. Theisen and M. Schmidt, Prepr. TUM-TP-129/91, KA-THEP-91-00, HD-THEP-91-32.
9. S. Ferrara, J. Louis, Prepr. CERN-TH-6334/91.
10. A. Ceresole, R. D'Auria, S. Ferrara, W. Lerche and J. Louis, CERN-TH-6441/92.
11. Z. Maassarani, Prepr. USC-91/023.
12. E. Witten, *Comm. Math. Phys.* **118** (1988) 411 and *Nucl. Phys.* **B340** (1990) 281;
    T. Eguchi and S.K. Yang *Mod Phys Lett* **A5** (1990) 1693
13. L. Baulieu and E. M. Singer, *Comm. Math. Phys.* **125** (1989) 125.
14. E. Witten, Prepr. IASSNS-HEP-91/83.
15. For a review see A. Pasquinucci, Ph.D. Thesis SISSA/EP 1990.
16. S. Cecotti, L. Girardello and A. Pasquinucci, *Nucl. Phys.* **B328** (1989) 701 and *Int. J. Mod. Phys.* **A6** (1991) 2427.
17. N.P. Warner, Lectures at Trieste Spring school 1988, World Scientific, Singapore;
    E. Martinec, *Phys. Lett.* **217B** (1989) 431;
    For a review see also "*Criticality, Catastrophe and Compactification*", V.G. Knizhnik memorial volume, 1989 .

18. V.I. Arnold, S.M. Gusein-Zade, A.N. Varchenko, "*Singularities of differentiable maps*", Vol I, II Birkäuser, Boston.

19. P. Frè, L. Girardello, A. Lerda and P. Soriani " *Topological first-order systems with Landau-Ginzburg interactions*" SISSA/ISAS 28/92/EP, ITP-SB-92-7, IFUM 416/FT, March 92, To appear in *Nucl. Phys. B.*

20. P. Frè, F. Gliozzi, R. Monteiro and A. Piras, *Class. and Quantum Grav* 8 (1991) 1455.

21. D. Friedan, E. Martinec and S. Shenker *Nucl. Phys.* **B271** (1986) 93

22. P. Ginsparg "*Applied conformal field theory*" Les Houches 1988

23. M. Frau, J.G. McCarthy, A. Lerda, S. Sciuto and J. Sidenius, *Phys. Lett.* **254B** (1991) 381 and *Phys. Lett.* **245B** (1990) 453.

24. C. Vafa, *Mod. Phys. Lett.* **A6** (1991) 337.

25. V.S. Dotsenko and V.A. Fateev, *Nucl. Phys.* **B240** (1984) 312.
G. Felder, *Nucl. Phys.* **B317** (1989) 215 and *Nucl. Phys.* **B324** (1989) 548E;

RECENT PROGRESS OF THE LIOUVILLE APPROACH TO 2D GRAVITY AND ITS TODA (W) GENERALIZATIONS

JEAN-LOUP GERVAIS
Laboratoire de Physique Théorique de l'École Normale Supérieure[1],
24 rue Lhomond, 75231 Paris CEDEX 05, France.

ABSTRACT

These lecture notes review current progress on the class of conformal theories which may be studied by quantising the conformal Toda dynamics. After summarising recent developments of the quantum Liouville theory, one recalls how two-dimensional black holes come out from the non-abelian Toda systems, and reviews the geometrical interpretation of the A_n-Toda theories, just put forward, that relate W geometries with the external geometry of particular (W) surfaces.

1 Introduction

The connection between conformally invariant field theories (CIFT) and integrable models has proven to be more and more useful in recent years. A particular role is played in this connection by the conformal Toda theories that generate, upon quantization, a whole class of the most important CIFT's. Two-dimensional gravity in the conformal gauge is notoriously equivalent to the Liouville theory which is the conformal Toda theory associated with the Lie algebra A_1. A step to generalize this situation was made in [3] where it was shown that the Toda theory associated[2] with any given simple Lie algebra gives two Noether realizations of the corresponding W-algebra. Thus, if the above A_1-Liouville scheme is repeated for the other Lie algebras, there should exist generalizations of two-dimensional gravity (called W-gravities) which are invariant by generalized diffeomorphisms, and coincide with the conformal Toda theories when a particular local coordinate frame is used.

These lecture notes cover recent developments.

In section 2, Liouville theory is discussed using the operator approach[6, 12, 14, 15, 17], where the quantum-group structure plays a fundamental role. After recalling some background material, one reviews recent results:
1) The complete derivation[20] of the holomorphic operator-algebra (braided category) where the role of the quantum group structure is fully established. The main

[1] *Unité Propre du Centre National de la Recherche Scientifique, associée à l'École Normale Supérieure et à l'Université de Paris-Sud.*

[2] Except in section 3, we only consider the so-called principal grading (see section 3)

new result here is that the fusion and braiding are not completely given by the quantum-group 3-j or 6-j symbols, as is usually assumed: There are coupling constants, which are not trigonometrical functions, and are not determined solely by the quantum-group symmetry.
2) The derivation[17] of the gravity-matter coupling in the weak-coupling regime, where the present method is found to agree with matrix-model calculations. One crucial point here is to show that the continuation in the number of screening operators which is made in the Coulomb gas picture is equivalent to the symmetry between quantum-group spin J and $-J-1$ put forward in my previous study[13, 14] of the strong-coupling regime.

The remaining sections cover recent progress about the more general Toda theories, considered at the purely classical level.

In section 3, it is first recalled how Toda theories are associated with any given embedding of $sl(2)$ into a given simple algebra. This defines a gradation. If the subgroub of gradation zero is non-abelian, the Toda theory is called non-abelian. The point of section 3 is to review the recent discovery[24] of the black-hole background metric of these non-abelian Toda theories.

The basic point of section 4 is the recent proposal[1, 2] that one can regard the W-geometries as the extrinsic geometry of particular two dimensional surfaces (W-surface) embedded into target spaces that are higher dimensional Kähler manifolds (We will restrict ourselves to the simplest particular situation, i.e. our target space is CP^n which corresponds to the A_n-type W-geometry). Instead of introducing higher-spin gauge generators, our approach makes use of the *extrinsic* curvatures of the embedded surface at its regular points, and relates it with the Toda dynamics mentioned above. The main virtue of our approach is that it is very simple to begin with. A W surface is characterized by the specific *chiral structure* of its embedding which we call chiral for short.

Section 5 deals with the geometry of the Toda hierachy. The dynamical variables of this hierarchy are shown to give particular coordinates of the higher dimensional Kähler manifolds. They are delt with by means of the free-fermion formalism, which is shown to be deeply connected with the concept of analytic continuation and with its W-generalizations. An important point here is to show that this allows us to extend the W transformations as reparametrizations of the target space, so that they become linear.

Finally, in section 6 we first reformulate our approach in terms of the intrinsic geometry of the family of associated surfaces in the Grassmannians $G_{n+1,k+1}$, $k = 1, \cdots, n$. This is needed to study singular points and global aspects of W-surfaces following the general scheme of refs. [1, 2], where these surfaces were shown to be intanton solutions of the associated non-linear σ models. The aim is to establish the generalization of the Gauss-Bonnet theorem to the W-surfaces discussed above. The instanton-number associated with each mapping are the global invariant of W-geometry. They are connected with the singularity indices of the W-surface.

2 Two dimensional gravity

2.1 The classical structure

let us first recall some feature of the classical Liouville dynamics. In the conformal gauge, it is governed by the action:

$$S = \frac{1}{4\pi}\int d_2x\sqrt{\hat{g}}\{\frac{1}{2}\hat{g}^{ab}\partial_a\Phi\partial_b\Phi + e^{2\sqrt{\gamma}\Phi} + \frac{1}{2\sqrt{\gamma}}R_0\Phi\} \tag{2.1}$$

$\hat{g}_{ab}$ is the fixed background metric. We work for fixed genus, and do not integrate over the moduli. As is well known, one can choose a local coordinate system such that $\hat{g}_{ab} = \delta_{ab}$. Thus we are reduced to the action

$$S = \frac{1}{4\pi}\int d\sigma d\tau\left(\frac{1}{2}(\frac{\partial\Phi}{\partial\sigma})^2 + \frac{1}{2}(\frac{\partial\Phi}{\partial\tau})^2 + e^{2\sqrt{\gamma}\Phi}\right) \tag{2.2}$$

where σ and τ are the local coordinates. The complex structure is assumed to be such that the curves with constant σ and τ are everywhere tangent to the local imaginary and real axis respectively. In a typical situation, one may work on the cylinder $0 \leq \sigma \leq 2\pi$, $-\infty \leq \tau \leq \infty$ obtained by an appropriate mapping from one of the handles of a general Riemann surface, and we shall do so in the present article. The action 2.2 corresponds to a conformal theory such that $\exp(2\sqrt{\gamma}\Phi)d\sigma d\tau$ is invariant. The classical equivalent of the chiral vertex operators may be obtained very simply[4, 5, 15] by using the fact that the field $\Phi(\sigma, \tau)$ satisfies the equation

$$\frac{\partial^2\Phi}{\partial\sigma^2} + \frac{\partial^2\Phi}{\partial\tau^2} = 2\sqrt{\gamma}\, e^{2\sqrt{\gamma}\Phi} \tag{2.3}$$

if and only if

$$e^{-\sqrt{\gamma}\Phi} = i\sqrt{\frac{\gamma}{2}}\sum_{j=1,2} f_j(x_+)g_j(x_-); \quad x_\pm = \sigma \mp i\tau \tag{2.4}$$

where f_j (resp.(g_j), which are functions of a single variable, are solutions of the same Schrödinger equation

$$- f_j'' + T(x_+)f_j = 0, \quad (\text{ resp. } - g_j'' + \overline{T}(x_-)g_j\,). \tag{2.5}$$

The solutions are normalized so that their Wronskians $f_1'f_2 - f_1f_2'$and $g_1'g_2 - g_1g_2'$ are equal to one. The proof of this basic fact is straightforward [4, 5, 15]. The potentials $T(x_+)$ and $\overline{T}(x_-)$ are the two components of the stress-energy tensor, and, after quantization, Eqs.2.5 become the Virasoro Ward-identities associated with the vanishing of the singular vector at the second level. As a result the Liouville theory also describes minimal models provided the coupling constant γ is taken to be negative. This is how, we shall treat the matter fields. For the dynamics associated with the action Eq.2.2, τ is the time variable, and the canonical Poisson brackets are

$$\{\Phi(\sigma_1,\tau), \frac{\partial}{\partial\tau}\Phi(\sigma_2,\tau)\}_{\text{P.B.}} = 4\pi\,\delta(\sigma_1 - \sigma_2), \quad \{\Phi(\sigma_1,\tau), \Phi(\sigma_2,\tau)\}_{\text{P.B.}} = 0 \tag{2.6}$$

The cylinder $0 \le \sigma \le 2\pi$, $-\infty \le \tau \le \infty$ may be mapped on the complex plane of $z = e^{\tau + i\sigma}$, and the above Poisson brackets lead to the usual radial quantization.

A priori, any two pairs f_j and g_j of linearly independent solutions of Eq.2.5 are suitable. In this connection, it is convenient to rename the functions g_j by letting $\overline{f}_1 = -g_2$, $\overline{f}_2 = g_1$. Then one easily sees that Eq.2.4 is left unchanged if f_j and $\overline{f}_j$ are replaced by $\sum_k M_{jk} f_k$ and $\sum_k M_{kj} \overline{f}_k$, respectively, where M_{jk} is an arbitrary constant matrix with determinant equal to one. Eq.2.4 is $sl(2,C)$-invariant with f_j transforming as a representation of spin 1/2. At the quantum level, the f_j's and $\overline{f}_j$'s become operators that do not commute, and the group $sl(2)$ is deformed to become the quantum group $U_q(sl(2))$. This structure plays a crucial role at the quantum level, and we now elaborate upon the classical $sl(2)$ structure where the calculations are simple.

At the classical level, it is trivial to take Eq.2.4 to any power. For positive integer powers $2J$, one gets (letting $\beta = i\sqrt{\frac{\gamma}{2}}$)

$$e^{-2J\sqrt{\gamma}\Phi} = \sum_{M=-J}^{J} \frac{\beta^{2J}(-1)^{J-M}(2J)!}{(J+M)!(J-M)!} \left(f_1(x_+)\overline{f}_2(x_-)\right)^{J-M} \left(f_2(x_+)\overline{f}_1(x_-)\right)^{J+M}. \tag{2.7}$$

It is convenient to put the result under the form

$$e^{-2J\sqrt{\gamma}\Phi} = \beta^{2J} \sum_{M=-J}^{J} (-1)^{J-M} f_M^{(J)}(x_+) \overline{f}_{-M}^{(J)}(x_-). \tag{2.8}$$

where $J \pm M$ run over integer. The $sl(2)$-structure has been made transparent by letting

$$f_M^{(J)} \equiv \sqrt{\binom{2J}{J+M}}\, (f_1)^{J-M} (f_2)^{J+M}, \quad \overline{f}_M^{(J)} \equiv \sqrt{\binom{2J}{J+M}}\, \left(\overline{f}_1\right)^{J+M} \left(\overline{f}_2\right)^{J-M}. \tag{2.9}$$

The notation anticipates that $f_M^{(J)}$ and $\overline{f}_M^{(J)}$ form representations of spin J. This is indeed true since f_1, f_2 and $\overline{f}_1$, $\overline{f}_2$ span spin 1/2 representations, by construction. Explicitely one finds

$$I_\pm f_M^{(J)} = \sqrt{(J \mp M)(J \pm M + 1)}\, f_{M\pm1}^{(J)}, \quad I_3 f_M^{(J)} = M f_M^{(J)}$$

$$\overline{I}_\pm \overline{f}_M^{(J)} = \sqrt{(J \mp M)(J \pm M + 1)}\, \overline{f}_{M\pm1}^{(J)}, \quad \overline{I}_3 f_M^{(J)} = M \overline{f}_M^{(J)}, \tag{2.10}$$

where I_ℓ and $\overline{I}_\ell$ are the infinitesimal generators of the x_+ and x_- components respectively. Moreover, one sees that

$$\left(I_\ell + \overline{I}_\ell\right) e^{-2J\sqrt{\gamma}\Phi} = 0 \tag{2.11}$$

so that the exponential of the Liouville field are group invariants.

2.2 The basic chiral operator-algebras

Denote by C the central charge of gravity. The standard screening charges $-\alpha_\pm$ of the Liouville theory[6, 18] are such that

$$\alpha_\pm = \frac{1}{2}\Big(\sqrt{\frac{C-1}{3}} \pm \sqrt{\frac{C-25}{3}}\Big),$$

$$\alpha_\pm = \frac{Q}{2} \pm \alpha_0, \quad Q = \sqrt{\frac{C-1}{3}}, \quad \alpha_0 = \frac{1}{2}\sqrt{\frac{C-25}{3}} \tag{2.12}$$

Q, and α_0 are introduced so that they agree with the standard notations. Kac's formula may be written as

$$\Delta_{Kac}(J,\widehat{J};C) = -\frac{1}{2}\beta(J,\widehat{J};C)\big(\beta(J,\widehat{J};C)+Q\big), \quad \beta(J,\widehat{J};C) = J\alpha_- + \widehat{J}\alpha_+, \tag{2.13}$$

where $2J$ and $2\widehat{J}$ are positive integers. We shall deal with generic values of C in order to avoid the complications of quantum group representations. As a result, we have to deal with h and $\widehat{h}$ on the same basis. This is in fact crucial for dealing with the strong-coupling regime[18, 13, 14]. There are two quantum-group parameters $h = \pi(\alpha_-)^2/2$, with $q = e^{ih}$, and $\widehat{h} = \pi(\alpha_+)^2/2$, with $\widehat{q} = e^{i\widehat{h}}$. Thus the number of fields doubles with respect with the classical case just recalled. The holomorphic operator will have a hat, or not, if they are related with $\widehat{h}$, or h, respectively; and the antiholomorphic operators are distinguished by an additional bar. According to Eq.2.13, the most general Liouville field is to be written as $\exp(-(J\alpha_- + \widehat{J}\alpha_+)\Phi)$. These fields have decompositions onto holomorphic and antiholomorphic operators to which we shall come below.

First we concentrate on holomorphic fields which are only functions of $x = \sigma - i\tau$. Then the whole structure may be described on the unit circle $\tau = 0$. The holomorphic operators associated with h form a subfamily which we next describe following refs.[12, 14, 20]. There are two useful basis of operators. First, the vertex operators $V_m^{(J)}$, with $-J \le m \le J$, and $2J$ a positive integer. They diagonalize the monodromy matrix

$$V_m^{(J)}(\sigma+2\pi) = e^{2ihm\varpi} e^{2ihm^2}\, V_m^{(J)}(\sigma), \tag{2.14}$$

where ϖ is the Liouville quasi-momentum (zero mode) around the cylinder. The meaning of the index m is to specify the shift of ϖ:

$$V_m^{(J)}(\sigma)\,\varpi = (\varpi + 2m)\, V_m^{(J)}(\sigma). \tag{2.15}$$

The Liouville momentum has been normalized so that this shift is integer. ϖ characterizes the Verma module. We shall work on the sphere, where the spectrum of ϖ is of the form $\varpi_J = \varpi_0 + 2J$. The momentum $\varpi_0 = 1 + 2\pi/h$ is one of the $sl(2, C)$

invariant state. There is a Verma modules $\mathcal{H}_J$ for each ϖ_J. The Moore Seiberg (MS) chiral vertex-operators connect three specified Verma modules and are thus of the form $\phi^{J_2}_{J_3,J_1}$. The operators $V^{(J)}_m$, on the contrary, act in the direct sum $\mathcal{H} = \oplus_J \mathcal{H}_J$. It is quite obvious, according to Eq.2.15, that the two are related by the projection operator $\mathcal{P}_J$:

$$\mathcal{P}_J \mathcal{H} = \mathcal{H}_J, \quad \mathcal{P}_{J_3} V^{(J_2)}_m \equiv \phi^{J_2}_{J_3, J_3+m} \tag{2.16}$$

The V fields are such that $< \varpi_2 | V^{(J)}_m | \varpi_1 >$ is equal to one of $\varpi_1 = \varpi_3 + 2m$, and is equal to zero otherwize. This normalization is required by the symmetry between three legs (sphere with three punctures). The complete fusion and braiding algebras take the form:

$$\mathcal{P}_K\, V^{(J_1)}_{m_1} V^{(J_2)}_{m_2} = \sum_{J=|J_1-J_2|}^{J_1+J_2} F_{K+m_1, J}\begin{bmatrix} J_1 & J_2 \\ K & K+m_1+m_2 \end{bmatrix}$$

$$\sum_{\{\nu\}} \mathcal{P}_K\, V^{(J,\{\nu\})}_{m_1+m_2} < \varpi_J, \{\nu\} | V^{(J_1)}_{J_2-J} | \varpi_{J_2} > \tag{2.17}$$

where $\mathcal{P}_K$ is the projector onto $\mathcal{H}_K$, and,

$$\mathcal{P}_K\, V^{(J_1)}_{m_1} V^{(J_2)}_{m_2} = \sum_{n_2} B_{K+m_1, K+n_2}\begin{bmatrix} J_1 & J_2 \\ K & K+m_1+m_2 \end{bmatrix} \mathcal{P}_K\, V^{(J_2,)}_{n_2} V^{(J_1)}_{m_1+m_2-n_2} \tag{2.18}$$

For simplicity of notation we omitted the dependence upon the world-sheet variables, which is standard. Using Eq.2.16, one may verify that these expressions have the general MS form. In ref.[20], it was shown that

$$F_{J_{23}, J_{12}}\begin{bmatrix} J_1 & J_2 \\ J_{123} & J_3 \end{bmatrix} = \frac{g^{J_{12}}_{J_1 J_2}\, g^{J_{123}}_{J_{12} J_3}}{g^{J_{23}}_{J_2 J_3}\, g^{J_{123}}_{J_1 J_{23}}} \left\{ \begin{matrix} J_1 & J_2 \\ J_3 & J_{123} \end{matrix} \middle| \begin{matrix} J_{12} \\ J_{23} \end{matrix} \right\} \tag{2.19}$$

where $\left\{ \begin{matrix} J_1 & J_2 \\ J_3 & J_{123} \end{matrix} \middle| \begin{matrix} J_{12} \\ J_{23} \end{matrix} \right\}$ denote the 6-j symbols of $U_q(sl(2))$, with $q = e^{ih}$. This term was of course expected, in view of the quantum-group structure previously exhibited, in particular, in refs.[12, 14]. However, there appear, in addition, coupling constants $g^{J_{12}}_{J_1 J_2}$, which are not trigonometrical functions of h. Their general expression is

$$g^J_{J_1 J_2} = \prod_{n=1}^{J_1+J_2-J} \sqrt{\frac{G(J_1 - n/2) G(J_2 - n/2) G(J + n/2)}{G((n-1)/2)}}, \tag{2.20}$$

where

$$G(J) \equiv \frac{\Gamma(1 + (1+2J)h/\pi)}{\Gamma(-(1+2J)h/\pi)}. \tag{2.21}$$

$\Gamma(x)$ is the usual (not q-deformed) Γ-function. The relation between braiding and fusion matrices is standard[19], and one may verify that the appearance of the coupling constant does not spoil the consistency conditions (pentagonal, and so on) satisfied by the 6-j symbols. In terms of the V fields the operator-algebra is entirely expressed in terms of quantum group invariant. Thus, it is clearly symmetric, but one does not see how the quantum group acts. This last feature is exhibited[11] by going to another basis of operators denoted $\xi_M^{(J)}$ (the quantum group basis) where M will really be the "third" component of the quantum-group spin. The change of basis is of the form[11, 12]

$$\mathcal{P}_K \xi_M^{(J)}(\sigma) := \sum_m C_{KK+2m}(J,M)\, \mathcal{P}_K V_m^{(J)}(\sigma). \tag{2.22}$$

where the coefficient $C_{KK+2m}(J,M)$ is proportional to a q-deformed hypergeometric function. The fusion properties become[20]

$$\xi_{M_1}^{(J_1)} \xi_{M_2}^{(J_2)} = \sum_{J=|J_1-J_2|}^{J_1+J_2} g_{J_1 J_2}^{J}\, (J_1, M_1; J_2, M_2|J) \sum_{\{\nu\}} \xi_{M_1+M_2}^{(J,\{\nu\})} < \varpi_J, \{\nu\}|V_{J_2-J}^{(J_1)}|\varpi_{J_2} >, \tag{2.23}$$

where $(J_1, M_1; J_2, M_2|J)$ denotes the q-deformed Clebsch-Gordan (CG) coefficients. The braiding relations are[3] as follows[12]

$$\xi_{M_1}^{(J_1)} \xi_{M_2}^{(J_2)} = \sum_{N_1 N_2} (J_1, J_2)_{M_1 M_2}^{N_2 N_1} \xi_{M_2}^{(J_2)} \xi_{M_1}^{(J_1)}. \tag{2.24}$$

The braiding matrix $(J_1, J_2)_{M_1 M_2}^{N_2 N_1}$ is given by the matrix element of the universal R-matrix of $U_q(sl(2))$:

$$(J_1, J_2)_{M_1 M_2}^{N_2 N_1} = \left(<< J_1, M_1| \otimes << J_2, M_2|\right) R \left(|J_1, N_1 >> \otimes |J_2, N_2 >>\right). \tag{2.25}$$

$|J, M >>$ are group theoretic states which span the representation of spin J of $U_q(sl(2))$, and R is the universal R-matrix:

$$R = e^{-2ihJ_3 \otimes J_3} \sum_{n=0}^{\infty} \frac{(1-e^{2ih})^n\, e^{ihn(n-1)/2}}{\lfloor n \rfloor !} e^{-ihnJ_3}(J_+)^n \otimes e^{ihnJ_3}(J_-)^n, \tag{2.26}$$

$J_\pm$, and J_3 are the quantum-group generators, which satisfy

$$\left[J_+, J_-\right] = \lfloor 2J_3 \rfloor, \quad \left[J_3, J_\pm\right] = \pm J_\pm. \tag{2.27}$$

We define $\lfloor r \rfloor \equiv \sin(hr)/\sinh$ in general. The fusion and braiding of the ξ fields are covariant under the quantum group action

$$J_3\, \xi_M^{(J)} = M \xi_M^{(J)}, \quad J_\pm\, \xi_M^{(J)} = \sqrt{\lfloor J \mp M \rfloor \lfloor J \pm M + 1 \rfloor}\, \xi_{M\pm 1}^{(J)}. \tag{2.28}$$

Indeed, first in the the operator-product Eq.2.23, the term $< \varpi_J, \{\nu\}|V_{J_2-J}^{(J_1)}|\varpi_{J_2} >$ is a quantum-group invariant since it does not involves the indices M_1, or M_2. Thus, by the basic property of the Clebsch-Gordan coefficients, it follows that $\xi_{M_1}^{(J_1)} \xi_{M_2}^{(J_2)}$

[3]We only coonsider one of the two possible orderings on the unit circle

also gives a representation of the quantum group algebra 2.27 with the co-product generators

$$\Lambda_\pm = J_\pm \otimes e^{ihJ_3} + e^{-ihJ_3} \otimes J_\pm, \quad \Lambda_3 = J_3 \otimes 1 + 1 \otimes J_3. \tag{2.29}$$

The tensor product is defined so that

$$(A \otimes B)\big(\xi_{M_1}^{(J_1)}(\sigma)\,\xi_{M_2}^{(J_2)}(\sigma')\big) := (A\xi_{M_1}^{(J_1)}(\sigma))\,(B\xi_{M_2}^{(J_2)}(\sigma')), \tag{2.30}$$

Roughly speaking the operator-product correspond to "adding" the quantum group spins. Since the result of the fusion should not depend upon the ordering, the non-commuativity of the ξ fields as quantum operators should precisely cancell the lack of symmetry of the co-product, and Clebsch-Gordan coefficients. This is why the braiding relations are given by the unversal R-matrix. Thus **the mathematical deformation of $sl(2)$, is precisely governed by the truly quantum mechnical effects of the theory**. Second, since the universal R-matrix which gives the braiding commutes with the co-product, the braiding relation Eq.2.24 is also covariant. This quantum-group symmetry, does not prevent invariant coupling constants $g_{J_1 J_2}^{J_3}$ to appear[4]. The value of these coupling constants makes the difference between Liouville theory and the $SU(2)$ WZNW model at this level.

An other important feature is that the fusion and braiding matrices do not depend upon the Verma module to which the fusion and braiding relations are applied. This is in contrast with the usual MS expressions. In the latter case, this means that the fusion and braiding matrices are not c-numbers since they depend upon the eigenvalue of the Liouville momentum. The corresponding solution of the Yang-Baxter equation thus takes a form that differs from the solutions obtained from the R-matrix of a quantum group. It is only when (and if) this dependence upon the Verma module may be removed that a direct connection with quantum-group representations is established. Note that, for A_n-W-algebras ($n \neq 2$), the same procedure works but gives[16] a deformation of $sl(n)$ that differs from the standard one.

2.3 The Liouville field, and matter-gravity couplings

We consider closed surfaces following ref.[17]. The above ξ fields together with their antiholomorphic counterparts $\bar{\xi}$ allow us to reconstruct the fields $\exp -J\alpha_-\Phi(\sigma,\tau)$. Imposing locality, and conservation of the winding number determines this operator completely. One finds

$$e^{-J\alpha_-\Phi(\sigma,\tau)} = \frac{1}{\sqrt{\varpi}} \sum_{M=-J}^{J} (-1)^{J-M}\, e^{ih(J-M)}\, \xi_M^{(J)}(x_+)\,\bar{\xi}_{-M}^{(J)}(x_-)\sqrt{\varpi} \tag{2.31}$$

[4] we simply see here an application of the well-known Wigner-Eckart theorem, which is so much used, for instance, in atomic physics

The quantum-group action on the $\overline{\xi}$ field is given by:

$$\overline{J}_3 \overline{\xi}_M^{(J)} = M\overline{\xi}_M^{(J)}, \quad \overline{J}_\pm \overline{\xi}_M^{(J)} = \sqrt{\lfloor J \mp M \rfloor \lfloor J \pm M + 1 \rfloor}\, \overline{\xi}_{M\pm 1}^{(J)}. \tag{2.32}$$

Thus if we define

$$\mathcal{J}_\pm = J_\pm e^{-ih\overline{J}_3} + e^{ihJ_3}\overline{J}_\pm, \quad \mathcal{J}_3 = J_3 + \overline{J}_3, \tag{2.33}$$

we obtain a representation of the quantum-group algebra Eq.2.28. Then one easily checks that

$$\mathcal{J}_\pm \exp(-J\alpha_- \Phi) = \mathcal{J}_3 \exp(-J\alpha_- \Phi) = 0, \tag{2.34}$$

so that the quantized Liouville field is a quantum-group invariant. This is the quantized version of Eq.2.11. Thus the $sl(2)$ symmetry recalled in section 2 has been deformed by the quantization in the mathematically standard way. The most general Liouville field $\exp(-(J\alpha_- + \widehat{J}\alpha_+)\Phi)$, is given by a similar expression in terms of the fields ξ, $\widehat{\xi}$, and $\overline{\xi}$, and $\overline{\widehat{\xi}}$ (see ref[17] for details).

Next we cansider the dressing of minimal models with central charge D by Liouville with central charge C. The balance of central charge requires

$$C + D = 26. \tag{2.35}$$

We shall be concerned with the case $D < 1$, where the Liouville theory is in its weakly coupled regime $C > 25$. The quantum structure of the Liouville theory just recalled is basically a consequence of operator differential equations which are equivalent to the Virasoro Ward-identities that describe the decoupling of null vectors. Thus the same operator algebra, with appropriate quantum deformation parameters also describes the matter with $D < 1$. We will thus have another copy of the quantum-group structure recalled above. It will be distinguished by primes. Thus we let

$$D = 1 + 6(\frac{h'}{\pi} + \frac{\pi}{h'} + 2) = 1 + 6(\frac{\widehat{h}'}{\pi} + \frac{\pi}{\widehat{h}'} + 2), \quad \text{with} \quad h'\widehat{h}' = \pi^2, \tag{2.36}$$

The appropriate dressing by gravity is such that one is concerned with matrix elements of operators of the type

$$\mathcal{V}_{J',\widehat{J}'}(\sigma,\tau) \equiv e^{((\widehat{J}'+1)\alpha_- - J'\alpha_+)\Phi(\sigma,\tau)}\, e^{-(J'\alpha'_- + \widehat{J}'\alpha'_+)X(\sigma,\tau)}, \tag{2.37}$$

where X is the matter field that is similar to the Liouville field. In ref.[17], the present approach was used to determine the three-point gravity-matter couplings:

$$\langle \prod_{\ell=1}^{3} \mathcal{V}_{J'_\ell,\widehat{J}'_\ell}(z_\ell, z_\ell^*) \rangle = \mathcal{C}_{1,2,3} \Big/ \Big(\prod_{k,l} |z_k - z_l|^2\Big). \tag{2.38}$$

For positiv J', Eq.2.37 involve negative quantum-group spins. This difficulty is equivalent with the need for negative number of dressing operator that is encountered in the Coulomb gas approach. Remarkably, this problem is precisely solved by symmetry $J \to -J-1$. which was found in refs.[13, 14]. Finally the result is very simple

$$C_{1,2,3} = \prod_l \frac{B_{J'_l,\widehat{J}'_l}}{\left[\Gamma\left(1+2\widehat{J}_l+(1+2J_l)h/\pi\right)\right]^2}. \tag{2.39}$$

It factorises, and thus gives back results of matrix models for ratios of correlators that do not depend upon the normalizations. The outcome of ref.[17] is that the symmetry between quantum-group spins J and $-J-1$ is the explanation of the continuation in the number of screening operators discovered by Goulian and Li. Moreover, and contrary to the previous discussions of this problem, the present approach clearly separates the emission operators for each leg. This clarifies the structure of the dressing by gravity. It is shown, in particular that the end points are not treated on the same footing as the mid point. Since the outcome is completely symmetric this suggests the possibility of a picture-changing mechanism.

Before leaving this topic we quote an older result[21] concerning open surfaces. The equivalent of Eq.2.31 is

$$e^{-\alpha_- J\Phi(\sigma)} = \left(\frac{h}{16\pi^3}\right)^J \sum_{M,N} < J,M|A|J,N > \xi_M^{(J)}(\sigma)\,\xi_N^{(J)}(2\pi-\sigma) \tag{2.40}$$

where

$$A = e^{-ihJ_3^2} \sum_{r,s=0}^{\infty} e^{ih(r+s)J_3} \frac{(J_+)^{r+s} e^{ihrs/2}}{\lfloor r \rfloor!\, \lfloor s \rfloor!}. \tag{2.41}$$

The matrix A is the universal reflection matrix associated with $U_q(sl(2))$, a concept that is gradually recognized as very important for integrable systems with boundaries.

3 BLACK HOLES FROM TODA THEORIES

In the present section, as well as in the coming ones, we restrict ourselves to classical Toda field theories, that describe the classical limits of conformal theories. Although this is a much simpler problem than the quantum one, it already teaches us a lot, since the appropriate operator quantization[3, 21] does preserve the key features of the classical Toda theories. The basic point of these theories is that their dynamical equations, no matter how complicated, are derivable from flatness conditions

$$\left[\partial_+ - \mathcal{A}_+, \partial_- - \mathcal{A}_-\right] \equiv \partial_- \mathcal{A}_+ - \partial_+ \mathcal{A}_- + \left[\mathcal{A}_+, \mathcal{A}_-\right] = 0, \tag{3.1}$$

which allow to obtain the general solutions in closed form[22][23]. We use ∂_+, and ∂_- as short hands for $\partial/\partial x_+$, and $\partial/\partial x_-$, respectively. The Lax pair $\mathcal{A}_\pm$ is systematically constructed once the Lie algebra, the gradation, and the grading spectrum of $\mathcal{A}_\pm$ are chosen[22][23]. Consider a finite-dimensional Lie group G, with Lie algebra $\mathcal{G}$, and a gradation $\mathcal{G} = \oplus_{m\in Z}\mathcal{G}_m$, with $[\mathcal{G}_m, \mathcal{G}_n] \in \mathcal{G}_{m+n}$. There is a grading operator H, such that $[H, \mathcal{G}_m] = 2m\mathcal{G}_m$. The basic idea is to consider the flatness condition Eq.3.1 taking $\mathcal{A}_\pm \in \mathcal{G}_0 \oplus \mathcal{G}_{\pm 1}$. The (abelian) A_N-Toda theories considered in the coming sections are associated with the principal grading of A_N where $\mathcal{G}_0$ is the Cartan subalgebra, and where $\mathcal{G}_{\pm 1}$ is generated by the simple (positive, resp. negative) roots. It is derivable from the Lagrangien

$$-\mathcal{L} = \frac{1}{2}\sum_{a,b=1}^{N} K_{ab}^{(A_N)}\partial_+\phi^a\partial_-\phi^b + \sum_{a=1}^{N}\prod_{b=1}^{N}\exp\left(K_{ab}^{(A_N)}\phi^b\right), \tag{3.2}$$

The matrix $K_{ab}^{(A_N)}$ is the Cartan matrix of A_N. The fields $\phi^a(x_+, x_-)$ are the bosonic Toda fields. We shall come back to this model in the next section. At the present time let us simply note that the fields ϕ^a may be regarded as components of a string-position in a N-dimensional target space with a metric $G_{ab} = K_{ab}^{(A_N)}$. It is somewhat trivial from this view point since this metric is constant. This feature is no more true for non-abelian Toda theories, namely those for which the subalgebra $\mathcal{G}_0$ is non-abelian. This was shown in ref[24] in the examples[5] of the Lie algebras B_{N-1}. Associated with a maximal subalgebra D_{N-1} (of maximal rank) in the Lie algebra B_{N-1}, we have the Lagrangian

$$\begin{aligned}-\mathcal{L} = \frac{1}{2}\Big[\sum_{j,k=1}^{N-1} K_{jk}^{(D_{N-1})}\partial_+\phi^j\partial_-\phi^k \quad - \quad \tanh^2\left(\frac{\phi^{N-1}-\phi^{N-2}}{2}\right)\frac{\partial\phi^N}{\partial z_+}\frac{\partial\phi^N}{\partial z_-}\Big] \\ + \sum_{i=1}^{N-1}\prod_{j=1}^{N-1}\exp\left(K_{ij}^{(D_{N-1})}\phi^j\right).\end{aligned} \tag{3.3}$$

In this case, the zero-grading part is given by $\mathcal{G}_0 = gl(1) \oplus \cdots \oplus gl(1) \oplus A_1$, where the one-dimensional linear algebra $gl(1)$ appears $N-1$ times. The ϕ-dependent part of the metric involves the hyperbolic-tangent-square function which is familiar in the 2D black-hole game. In particular if we choose $N = 3$, we find

$$-\mathcal{L} = \frac{1}{2}\left[2\sum_{i=1}^{2}\partial_+\phi^i\partial_-\phi^i - \tanh^2\left(\frac{\phi^1-\phi^2}{2}\right)\partial_+\phi^3\partial_-\phi^3\right] + \sum_{i=1}^{2}\exp(2\phi^i) \tag{3.4}$$

In order to clarify the black-hole aspect, it is convenient to change field variables. Let

$$\Phi = (\phi^1+\phi^2)/2, \quad r = (\phi^1-\phi^2)/2, \quad \theta = \phi^3/2. \tag{3.5}$$

One gets

$$-\mathcal{L} = \partial_+\Phi\partial_-\Phi + \partial_+ r\partial_- r - \tanh^2(r)\,\partial_+\theta\partial_-\theta + \cosh(2r)e^{2\phi}. \tag{3.6}$$

[5]Note that there is no useful non-abelian version of the Toda lattice for the series A_N

The target-space metric is

$$G = \begin{pmatrix} 1 & 0 & 0 \\ 0 & 1 & 0 \\ 0 & 0 & -\tanh^2 r \end{pmatrix} \tag{3.7}$$

One sees that the first component, which corresponds to the ϕ variable, defines a subspace that decouples from the rest from the viewpoint of Riemannian geometry. In the $(r,\, \theta)$ space we have $G = \begin{pmatrix} 1 & 0 \\ 0 & -\tanh^2 r \end{pmatrix}$, which coincides with Witten's black-hole metric[25] exactly.

The outcome of ref.[24] is that non-abelian Toda theories provide exactly solvable conformal systems in the presence of a black hole. They correspond to gauged WZNW models where the gauge group is nilpotent, and are thus basically different from the ones currently considered, following Witten. The non-abelian Toda potential gives a cosmological term which may be integrated exactly at the classical level.

4 GEOMETRY OF CHIRAL SURFACES

4.1 The background

The present discussion is concerned with the generalization of the situation. described in section 2 for 2D gravity. On the one hand, the above Liouville dynamics is a particular case of the Toda dynamics[6], which, as just recalled, exists for each simple Lie algebra. The Liouville case is associated with A_1. It was shown in ref.[3] that general Toda systems are related, through Noether theorem to the non-linear extensions of the Virasoro algebra called W-algebras. Thus going from Liouville to general Toda should correspond, in general cordinates to going from Einstein gravity to W gravity. As a first step towards formulating these fascinating theories, we next unravell the geometrical meaning of Toda systems. This is, at least a way to see the geometrical meaning of W algebras. We will show that it corresponds to the extrinsic geometry of embeddings of special (W) surfaces, a viewpoint which is natural since in general, we expect that conformal systems and their W generalizations are to be connected with string theories. The material covered untill the end is a summary of refs.[1, 2].

4.2 Chiral embeddings

The basic objects are two-dimensional surfaces embedded in Kähler manifolds. We shall only consider C^n here explicitly. The case of CP^n is treated by using homogeneous coordinates in C^{n+1}. We call z and $\bar{z}$ the surface parameters. One may think of z as an ordinary complex number, in which case the parametrization is Euclidian, and $\bar{z}$ is the complex conjugate of z; or take z, and $\bar{z}$ to be real, in which case

[6]from now on we only consider the principal grading

$x_0 \equiv (z+\bar{z})/2$, and $x_1 \equiv (z-\bar{z})/2$ are coordinates of the Minkowski type. The adjective chiral means function of a single variable z or $\bar{z}$ (if z is a complex variable this is equivalent to analytic or anti-analytic). We parametrize C^n by coordinates X^A, $\bar{X}^{\bar{A}}$, $1 \leq A, \bar{A} \leq n$. A chiral embedding is defined by equations of the form

$$X^A = f^A(z),\ A = 1, \cdots, n, \quad \bar{X}^{\bar{A}} = \bar{f}^{\bar{A}}(\bar{z}),\ \bar{A} = 1, \cdots, n. \tag{4.1}$$

where f and $\bar{f}$ are arbitrary f unctions. We call a W-surface the corresponding manifold Σ. We shall see that its extrinsic geometry is directly related to W transformations. It is convenient to introduce the matrix of inner products:

$$g_{i\bar{j}} \equiv \sum_A f^{(i)A}(z)\,\bar{f}^{(\bar{j})A}(\bar{z}), \quad 1 \leq i, \bar{j} \leq n. \tag{4.2}$$

We use ∂ and $\bar{\partial}$ as short hands for $\partial/\partial z$ and $\partial/\partial\bar{z}$ respectively. $f^{(i)A}$ stands for $(\partial)^i f^A$, and $\bar{f}^{(\bar{j})A}$ stands for $(\bar{\partial})^{\bar{j}}\bar{f}^A$. Later on we shall exhibit a particular parametrization of C^n, called W-parametrization, where the derivatives with i or $\bar{j}$ larger than one will be replaced by first order derivatives in other variables, so that the covariance properties of the present discussion will become more transparent. This section is concerned with generic regular points of Σ where the Taylor expansions of f^A and $\bar{f}^{\bar{A}}$ give linearly independent vectors. Then $f^{(a)}$, and $\bar{f}^{(a)}$, $a = 1, \cdots, n$, span the following

Definition 1 ***Moving frame in C^n.*** *Consider the vectors e_a, and $\bar{e}_a$, $a = 1, \cdots, n$, with components*

$$e_a^A = \frac{1}{\sqrt{\Delta_a \Delta_{a-1}}} \begin{vmatrix} g_{1\bar{1}} & \cdots & g_{a\bar{1}} \\ \vdots & & \vdots \\ g_{1\overline{a-1}} & \cdots & g_{a\overline{a-1}} \\ f^{(1)A} & \cdots & f^{(a)A} \end{vmatrix}, \qquad e_a^{\bar{A}} = 0, \tag{4.3}$$

$$\bar{e}_a^A = 0, \qquad \bar{e}_a^{\bar{A}} = \frac{1}{\sqrt{\Delta_a \Delta_{a-1}}} \begin{vmatrix} g_{\bar{1}1} & \cdots & g_{\bar{a}1} \\ \vdots & & \vdots \\ g_{\bar{1}a-1} & \cdots & g_{\bar{a}a-1} \\ \bar{f}^{(1)\bar{A}} & \cdots & \bar{f}^{(a)\bar{A}} \end{vmatrix}, \tag{4.4}$$

Δ_a is the determinant

$$\Delta_a \equiv \begin{vmatrix} g_{1\bar{1}} & \cdots & g_{a\bar{1}} \\ \vdots & & \vdots \\ g_{1\bar{a}} & \cdots & g_{a\bar{a}} \end{vmatrix}. \tag{4.5}$$

One may verify that the moving frame defined above is orthonormal, that is,

$$(e_a, e_b) = (\bar{e}_a, \bar{e}_b) = 0, \quad (e_a, \bar{e}_b) = \delta_{a,b}. \tag{4.6}$$

The vectors e_1 and $\bar{e}_1$ are tangents to the surface, while the other vectors are clearly normals. Thus the Gauss-Codazzi equations may be derived by studying their derivatives along the W-surface Σ. One derives equations of the form

$$\partial e_a = \sum_b \omega_{za}^{\ b} e_b, \quad \bar{\partial} e_a = \sum_b \omega_{\bar{z}a}^{\ b} e_b, \quad \bar{\partial} \bar{e}_a = \sum_b \bar{\omega}_{\bar{z}a}^{\ b} \bar{e}_b, \quad \partial \bar{e}_a = \sum_b \bar{\omega}_{za}^{\ b} \bar{e}_b. \tag{4.7}$$

which may be regarded as generalized Frenet-Serret formulae. Next we recall the

Definition 2 ***CP^n target space.*** *The complex projective space C^n is defined to be the quotient of the space C^{n+1} by the equivalence relation*

$$X \sim Y, \qquad if \;\; X^A = Y^A \rho(Y), \;\; and \;\; \bar{X}^{\bar{A}} = \bar{Y}^{\bar{A}} \bar{\rho}(\bar{Y}). \tag{4.8}$$

where ρ and $\bar{\rho}$ are arbitrary functions of $n+1$ variables.

Thus, the modification to go from C^n to CP^n is to use $n+1$ coordinates, so that now A, and $\bar{A}$ run from 0 to n, and to write formulae that are covariant under rescaling. This is achieved by letting the indices of the matrix $g_{i\bar{j}}$ run from 0 to n in the CP^n formulae for the moving frame. for this one includes derivatives of order 0. Our basic result is the following:

Theorem 1 ***Gauss-Codazzi equations.***
Define Toda fields by

$$\phi_a = -\ln(\tau_a), a = 1, \cdots, n; \; \tau_a \equiv \begin{vmatrix} g_{00} & \cdots & g_{a0} \\ \vdots & & \vdots \\ g_{0a} & \cdots & g_{aa} \end{vmatrix}. \tag{4.9}$$

The integrability conditions of the Frenet-Serret equations for the embedding in CP^n (analogous to Eqs.2.7) coincide with the Toda equations associated with A_n:

$$\partial\bar{\partial}\phi_i + \exp\left(\sum_{j=1}^{n} K_{ij}\phi_j\right) = 0. \tag{4.10}$$

K is the Cartan matrix associated with A_n. The functions τ_a relevant for CP^n are similar to the Δ_a of Eq.2.5. except that they include the derivatives of zeroth order.

In conclusion: The A_n Toda dynamics describes the extrinsic geometry of W surfaces.

4.3 Some basic facts about A_n Toda theories

Its general solution is of the form

$$e^{-\phi_k} = \sum_{i_1<\cdots<i_k} \begin{vmatrix} \chi^{i_1} & \cdots & \chi^{i_k} \\ \chi^{(1)i_1} & \cdots & \chi^{(1)i_k} \\ \vdots & \cdots & \vdots \\ \chi^{(k-1)i_1} & \cdots & \chi^{(k-1)i_k} \end{vmatrix} \begin{vmatrix} \bar{\chi}^{i_1} & \cdots & \bar{\chi}^{i_k} \\ \bar{\chi}^{(1)i_1} & \cdots & \bar{\chi}^{(1)i_k} \\ \vdots & \cdots & \vdots \\ \bar{\chi}^{(k-1)i_1} & \cdots & \bar{\chi}^{(k-1)i_k} \end{vmatrix}. \tag{4.11}$$

where k runs from 1 to n. It is expressed in terms of n functions of z $\chi^1, \cdots, \chi^n$, and n functions of $\bar{z}$ $\bar{\chi}^1, \cdots, \bar{\chi}^n$. Upper indices in between parenthesis denotes derivatives. These n functions χ^k (resp. $\bar{\chi}^k$) are restricted to be solution of the same differential equation $\chi^{(n+1)k} - \sum_{\ell=0}^{n-1} U_{n+1-\ell}\chi^{(\ell)k} = 0$ (resp. $\bar{\chi}^{(n+1)k} - \sum_{\ell=0}^{n-1} \bar{U}_{n+1-\ell}\bar{\chi}^{(\ell)k} = 0$). The set of potentials $\{U_\ell, \ell = 2, \cdots n+1\}$ and $\{\bar{U}_\ell, \ell = 2, \cdots n+1\}$, each generate a realization of the A_n W algebra by Poisson brackets. These are non-linear generalizations of the Virasoro algebra (conformal transformations). The Toda dynamics is non-chiral, and this is why the W algebra appears twice (for the holomorphic and anti-holomorphic components).

It follows that from the above geometrical derivation of the Toda eqautions, we may discuss the geometrical meaning of the W transformations.

4.4 Connection with the WZNW models

It is known, in general, that there is one W algebra associated with each simple[7] Lie algebra $\mathcal{G}$. This appears in several ways. First, as we have already seen, there is a Toda theory, and, hence, two PB realizations of the associated W-algebra for any given $\mathcal{G}$. On the other hand, consider the affine Lie algebra $\tilde{\mathcal{G}}$ associated with $\mathcal{G}$. The associated non-chiral theory is the WZNW model whose quantum solutions are given by representations of $\tilde{\mathcal{G}}$. It is possible to derive the Toda theory from the WZNW model by conformal reduction[9]. Here we have the

Theorem 2 ***Gauss decomposition from moving frame.*** *The moving-frame equations may be written as*

$$e_a = \sum_{b\leq a} C_{ab}(z,\bar{z})\sqrt{\frac{\tau_a}{\tau_{a+1}}}\, f^{(b)}(z), \quad \textit{with}\, C_{aa} = 1, \tag{4.12}$$

$$\bar{e}_a = \sum_{b\leq a} A_{ba}(z,\bar{z})\sqrt{\frac{\tau_a}{\tau_{a+1}}}\, \bar{f}^{(b)}(\bar{z}), \quad \textit{with}\, A_{aa} = 1. \tag{4.13}$$

The matrix $\theta = g^{-1}$ s such that
1) It has the Gauss decomposition

$$\theta = ABC, \tag{4.14}$$

where A, and C, which appear in eqs.2.11, and 2.12 are triangular with diagonal elements equal to one, and

$$B = \exp\left(\sum_i h_i\phi_{i+1}\right).$$

h_i are the Cartan generators. 2) It is a solution of the conformally reduced WZNW model associated with A_n.

[7]The non-simple Lie algebras are trivially reduced to the simple case by separating the invariant subalgebras

5 Geometry of Toda hierarchy

The Toda equations are a subsystem of the Toda hierarchy[7]. (This is the non-chiral version of the fact that the Virasoro algebra is identical with the second Poisson bracket of KdV, and that W algebras are obtained from KP hierarchies and Gelfand Dicki brackets). Introduce the additional variables as coordinates in our geometrical embedding problem. This is best done using the free-fermion formalism. Let

$$\begin{aligned} [\psi_n, \psi_m]_+ &= \left[\psi_n^+, \psi_m^+\right]_+ = 0, \\ \left[\psi_n, \psi_m^+\right]_+ &= \delta_{n,m}, \qquad (\, n,m \;=\; 0,\; 1,\; \cdots) \end{aligned} \tag{5.1}$$

$$\psi_n|\emptyset\rangle = 0, \qquad \langle\emptyset|\psi_n^+ = 0 \qquad \forall n. \tag{5.2}$$

We use the semi-infinite indices $n = 0, 1, 2, \cdots, \infty$ for the fermion-operators. The vacuum states $|\emptyset\rangle$ and $\langle\emptyset|$ correspond to the no-particle states. The n-particle ground state is created from them in the standard way:

$$|n\rangle = \psi_{n-1}^+ \psi_{n-2}^+ \cdots \psi_0^+ |\emptyset\rangle, \quad \langle n| = \langle\emptyset|\psi_0\psi_1 \cdots \psi_{n-1}. \tag{5.3}$$

The current operators,

$$J_n = \sum_{s=0}^{\infty} \psi_{n+s}^+ \psi_s, \qquad \bar{J}_n = \sum_{s=0}^{\infty} \psi_s^+ \psi_{n+s}, \tag{5.4}$$

will be taken as Hamiltonians as one does for the KP hierarchy. The rôle of these fermions may be understood as follows. Take the case where z is a complex variable. Then the embedding functions f^A are analytic, and each of them is entirely determined by its Taylor expansion around a single point of its analyticity domain. Its behaviour at any other point of its Riemann surface is fixed by analytic continuation. The following free-fermion formalism realizes this continuation automatically. Consider the Taylor expansions at the point z:

$$f^A(z+x) = \sum_{s=0}^{\infty} f^{(s)A}(z)\, \frac{x^s}{s!}, \quad \bar{f}^{\bar{A}}(\bar{z}+\bar{x}) = \sum_{s=0}^{\infty} \bar{f}^{(s)A}(\bar{z})\, \frac{\bar{x}^s}{s!}. \tag{5.5}$$

To these developements, we associate the free-fermion operators,

$$\psi_{f^A(z)} = \sum_{s=0}^{\infty} f^{(s)A}(z)\, \psi_s, \quad \psi^+_{\bar{f}^{\bar{A}}(\bar{z})} = \sum_{s=0}^{\infty} \bar{f}^{(s)A}(\bar{z})\, \psi_s^+. \tag{5.6}$$

The basic property of these operators are

Proposition 1 ***Fermionic representation of chiral functions.***
1) Any change of the Taylor-expansion point z, $\bar{z}$ can be absorbed by the action of the Hamiltonians J_1, and $\bar{J}_1$. In particular, one has

$$\psi_{f^A(z)} = e^{-J_1 z}\, \psi_{f^A(0)}\, e^{J_1 z} \quad \psi^+_{\bar{f}^{\bar{A}}(\bar{z})} = e^{\bar{J}_1 \bar{z}}\, \psi^+_{\bar{f}^{\bar{A}}(0)}\, e^{-\bar{J}_1 \bar{z}}. \tag{5.7}$$

2) The embedding functions are represented by the fermion expectation-values

$$f^A(z) = \langle\emptyset|\psi_{f^A(z_0)} e^{J_1(z-z_0)}|1\rangle, \qquad \bar{f}^{\bar{A}}(\bar{z}) = \langle 1|e^{\bar{J}_1(\bar{z}-\bar{z}_0)}\psi^+_{\bar{f}^{\bar{A}}(\bar{z}_0)}|\emptyset\rangle. \tag{5.8}$$

Definition 3 ***KP-parametrization of*** CP^n.

Given a chiral surface embedded into CP^n, *the associated KP-parameters of the target space are* $n+1$ *variables* $z^{(0)}$, $z^{(1)} = z$, $z^{(2)}$, $\cdots$, $z^{(n)}$, *noted* $[z]$, *and* $n+1$ *variables* $\bar{z}^{(0)}$, $\bar{z}^{(1)} = \bar{z}$, $\bar{z}^{(2)}$, $\cdots$, $\bar{z}^{(n)}$, *noted* $[\bar{z}]$. *The change of coordinates from* X^A, $\bar{X}^{\bar{A}}$ *to* $[z]$, $[\bar{z}]$ *is defined by*

$$X^A = f^A([z]), \qquad \bar{X}^{\bar{A}} = \bar{f}^{\bar{A}}([\bar{z}]) \tag{5.9}$$

where $f^A([z])$, *and* $\bar{f}^{\bar{A}}([\bar{z}])$, *are the solutions of the equations*

$$\frac{\partial}{\partial z^{(\ell)}} f^A([z]) = \frac{\partial^\ell}{\partial z^\ell} f^A([z]), \qquad \frac{\bar{\partial}}{\partial \bar{z}^{(\ell)}} \bar{f}^{\bar{A}}([\bar{z}]) = \frac{\bar{\partial}^\ell}{\partial \bar{z}^\ell} \bar{f}^{\bar{A}}([\bar{z}]) \tag{5.10}$$

with the initial conditions $f^A([z]) = f^A(z)$ *for* $z^{(0)}, z^{(2)}, \cdots, z^{(n)} = 0$, *and* $\bar{f}^{\bar{A}}([\bar{z}]) = \bar{f}^{\bar{A}}(\bar{z})$ *for* $\bar{z}^{(0)}, \bar{z}^{(2)}, \cdots, \bar{z}^{(n)} = 0$.

These coordinates coincide with the higher variables of the KP hierarchy. Indeed, their definition is most natural in the free-fermion language, where it is easy to see that

$$f^A([z]) = \langle\emptyset|\psi_{f^A} e^{\sum_0^n J_i z^{(i)}}|1\rangle, \quad \bar{f}^{\bar{A}}([\bar{z}]) = \langle 1|e^{\sum_0^n \bar{J}_i \bar{z}^{(i)}}\psi^+_{\bar{f}^{\bar{A}}}|\emptyset\rangle. \tag{5.11}$$

The dependence in $[z]$ and $[\bar{z}]$ is dictated by the action of the higher currents J, $\bar{J}$, defined by Eq.5.4,that is, $J_1 z \to \sum_{i=0}^n J_i z^{(i)}$, $\bar{J}_1\bar{z} \to \sum_{i=0}^n \bar{J}_i \bar{z}^{(i)}$ in Eq.5.8. The basic virtue of the KP coordinates is that they allow us to extend the previous discussion away from the W surface (they parametrize at least a neighborhood of it) in such a way that is becomes covariant. In particular, the metric g has the following extension

$$g_{i\bar{j}}([z],[\bar{z}]) = \sum_A \partial_i f^A([z])\, \bar{\partial}_j \bar{f}^A([z]), \quad \partial_i \equiv \frac{\partial}{\partial z^{(i)}}, \quad \bar{\partial}_i \equiv \frac{\bar{\partial}}{\bar{\partial} \bar{z}^{(i)}}. \tag{5.12}$$

Now, only first-order derivatives appear. This expression coincides with the *true* Riemannian metric with respect to the KP coordinates.

5.1 *W* transformations

A general infinitesimal W-transformation is a change of embedding functions which takes the form

$$\delta_W f^A(z) = \sum_{j=0}^n w^j(z)\partial^{(j)} f^A(z), \quad \delta_W \bar{f}^{\bar{A}}(\bar{z}) = \sum_{j=0}^n \bar{w}^j(\bar{z})\partial^{(j)} \bar{f}^{\bar{A}}(\bar{z}), \tag{5.13}$$

It is not difficult to show there exists a unique extension such that the differential equation 5.10 is left invariant. It is of the form

$$\begin{aligned} \delta_W f^A([z]) &= \sum_r W^r([z])\partial_r f^A([z]), \\ \delta_W \bar{f}^{\bar{A}}([\bar{z}]) &= \sum_r \bar{W}^r([\bar{z}])\partial_r \bar{f}^{\bar{A}}([\bar{z}]) \end{aligned} \tag{5.14}$$

where W^r and $\bar{W}^r$ are functionals of w^j and $\bar{w}^j$ respectively. Only first order derivative appear. Thus the W transformation become extended as reparametrizations

$$\delta_W z^{(r)} = W^r([z]), \quad \delta_W \bar{z}^{(r)} = \bar{W}^r([\bar{z}]). \tag{5.15}$$

They become particular diffeomorphisms of CP^n. Thus they are extended as linear transformations.

5.2 Dynamical equations

The followings topics are discussed in ref.[2]

The above functions τ_a when extended become tau-functions in the sense of Miwa-Jimbo-Sato[7].

The KP coordinates are related with a generalized moving frame, whose integrability condition is equivalent to the the well-known Zakharov-Shabat of the A_n Toda hierarchy.

The extension of the associated WZNW model gives solutions of a 2n dimensional generalization of the WZNW equations where the currents are replaced by the Christoffel symbols of the KP coordinates.

6 SINGULAR POINTS, GLOBAL STRUCTURE

At this point, it is useful to change the viewpoint, and make use of Grassmannians. This part is draws a lot of inspiration from ref.[10]

6.1 Associated mappings

Definition 4 ***Associated mappings.***

Consider the family of osculating hyperplanes with contact of order k denoted O_k ($k = 1, \cdots, n$) to the original W-surface. With CP^n as the target space, this family defines an embedding into the Grassmannian $G_{n+1,k+1}$, which we call the kth associated mappings to the original W-surface.

This formulation looks different, but is equivalent to the construction of the moving frame and only uses the intrinsic geometries of the induced metrics for $k = 1, \cdots, n$. In practice, what this means is that, instead of forming moving-frame vectors e_k out of

$f, \cdots, f^{(k)}$ ($k = 1, \cdots, n$), we consider the nested osculating planes $\mathcal{O}_1 \subset \mathcal{O}_2 \subset \cdots \subset \mathcal{O}_n$. It is obvious that those two have the same information. It is well-known that the Grassmannians are Kähler manifolds. The induced metric on the kth associated surface in $G_{n+1,k+1}$ is simply,

$$g_{z\bar{z}}^{(k)} = \partial\bar{\partial} \ln \tau_{k+1}(z, \bar{z}), \quad g_{zz}^{(k)} = g_{\bar{z}\bar{z}}^{(k)} = 0, \tag{6.1}$$

so that the Toda field $\phi_{k+1} \equiv -\ln(\tau_{k+1})$ appears naturally. Thus $-\phi_{k+1}$ is equal to the Kähler potential of the kth associated surface. At this point, it is very clear that by considering the associated surfaces, we can restrict ourselves to intrinsic geometries. In the discussion of section 4, the Toda equation came out from the Gauss-Codazzi equation. Here, it is equivalent to the local Plücker formula

$$R_{z\bar{z}}^{(k)} \sqrt{g_{z\bar{z}}^{(k)}} = -g_{z\bar{z}}^{(k+1)} + 2g_{z\bar{z}}^{(k)} - g_{z\bar{z}}^{(k-1)}. \tag{6.2}$$

where $R_{z\bar{z}}^{(k)}$ is the only non-vanishing component of the intrinsic Riemann tensor on the kth surface.

6.2 The instanton-numbers of a W-surface

A key point in the coming discussion is to use topological quantities that are instanton-numbers. W-surfaces are instantons of the associated non-linear σ-model. The general situation is as follows. W-surfaces are characterized by their chiral parametrizations which thus satisfy the Cauchy-Riemann relations. These are self-duality equations so that the coordinates of a W-surface define fields that are solutions of the associated non-linear σ model, with an action equal to the topological instanton-number. For a general Kähler manifold M with coordinates ξ^μ and $\bar{\xi}^{\bar{\mu}}$, and metric $h_{\mu\bar{\mu}}$, the action associated with 2D manifolds of M with equations $\xi^\mu = \varphi^\mu(z, \bar{z})$, and $\bar{\xi}^{\bar{\mu}} = \bar{\varphi}^{\bar{\mu}}(z, \bar{z})$ is

$$S = \frac{1}{2} \int d^2x \, h_{\mu\bar{\mu}} \, \partial_j \varphi^\mu \partial_j \bar{\varphi}^{\bar{\mu}}. \tag{6.3}$$

In this short digression we let $z = x_1 + ix_2$, and $\partial_j = \partial/\partial x_j$. The instanton-number is defined by

$$Q = \frac{i}{2\pi} \int d^2x \, \epsilon_{jk} \, h_{\mu\bar{\mu}} \, \partial_j \varphi^\mu \partial_k \bar{\varphi}^{\bar{\mu}}. \tag{6.4}$$

For W-surfaces and their associated surfaces, $\bar{\partial}\varphi^\mu = \partial\bar{\varphi}^{\bar{\mu}} = 0$, and one has $S = \pi Q$. Q is proportional to the integral of the determinant of the induced metric, that is $Q = \frac{i}{2\pi} \int d^2x \partial\bar{\partial} \ln \tau_1$.

Moreover we may also apply formula Eq.6.4 to the kth associated surface. This gives

Definition 5 ***Higher instanton-numbers of the W-surface.***

The kth instanton number of the W-surface Q_{k+1} is defined by,

$$Q_{k+1} \equiv \frac{i}{2\pi} \int_\Sigma dz d\bar{z} g_{z\bar{z}}^{(k)}, k = 1, \cdots, n-1. \tag{6.5}$$

Its topological nature is obvious from Eq.6.1 which shows that the integrand is indeed a total derivative. The collection of the (k-th) instanton-numbers together with the original one $Q \equiv Q_1$ gives a set of topological quantities which characterize the global properties of the original W-surface.

6.3 Global classification of W surfaces

In section 4, we have constructed the moving frames at the point where the tau-functions are regular. When those functions become irregular, we meet an obstruction to derive the moving frames. In the WZNW language, this signals that there appears a global obstruction to the Gauss decomposition. Toda equations should be modified at these points. In the following, we study the structure of such singularities.

6.4 Gauss-Bonnet Theorem for W-surfaces

For isolated singularities[8], the obstruction to constructing the moving frame may be reduced to the vanishing of certain terms in the Taylor expansion. The latter is characterized by the ramification indices β_k which are integers. Apply the Gauss-Bonnet theorem for each of the kth associated surfaces by computing $\int_{\Sigma_\epsilon} R^{(k)}_{z\bar z}\sqrt{g^{(k)}_{z\bar z}}$. The integral is first computed over Σ_ϵ where small neighborhoods of singularities are removed. The ramification indices at singularity was defined so that at a singular point the induced metric of the kth associated surface behaves as

$$g^{(k)}_{z\bar z} \sim (z-z_0)^{\beta_k(z_0)}(\bar z-\bar z_0)^{\bar\beta_k(\bar z_0)}\bar g^{(k)}_{z\bar z}, \tag{6.6}$$

where $\bar g^{(k)}_{z\bar z}$ is regular at z_0, $\bar z_0$. Since we do not assume that $\overline{f(z)} = \bar f(\bar z)$, $\beta_k(z_0)$ and $\bar\beta_k(\bar z_0)$ may be different. By letting $\epsilon \to 0$, one sees, that the contribution of the singularities to the Gauss-Bonnet formula is proportional to the k-th ramification index

$$\beta_k \equiv \frac{1}{2}\sum_{(z_0,\bar z_0)\in\Sigma}\left(\beta_k(z_0)+\bar\beta_k(\bar z_0)\right). \tag{6.7}$$

The contribution of the regular part does not depend upon k, since changing k is equivalent to using a different complex structure, while the result is equal to the Euler characteristic that does not depend upon it. The Gauss-Bonnet theorem for the kth associated surface finally gives

$$\frac{i}{2\pi}\int_\Sigma dz d\bar z R^{(k)}_{z\bar z}\sqrt{g^{(k)}_{z\bar z}} = 2-2g+\beta_k. \tag{6.8}$$

Combining these last relations with Eqs.6.2, one arrives at the

[8] if there is a cut with a finite number of sheets, one takes the appropriate covering

Theorem 3 ***Global Plücker formulae*** *The genus g of a W-surface is related to its instanton-numbers and ramification-indices by the relations*

$$2 - 2g + \beta_k = 2Q_k - Q_{k+1} - Q_{k-1}, \quad \left| \begin{array}{ll} k & = 1, \cdots, n \\ Q_{n+1} & \equiv 0, \; Q_0 \equiv 0 \end{array} \right. \tag{6.9}$$

6.5 Simple example

Consider the case of Liouville theory, for which one has $n = 1$. The simplest chiral surface corresponds to

$$f^0 = 1, \; f^1 = z; \quad \bar{f}^0 = 1, \; \bar{f}^1 = \bar{z}; \quad \tau_1 = 1 + z\bar{z}, \quad \tau_2 = 1.$$

The instanton number is thus $Q = (i/2\pi) \int dz d\bar{z}/(1 + z\bar{z})^2 = 1$. On the other hand, one has

$$R_{z\bar{z}}\sqrt{g_{z\bar{z}}} = -\partial\bar{\partial} \ln(g_{z\bar{z}}) = \frac{2}{(1 + z\bar{z})^2}$$

so that

$$\frac{i}{2\pi} \int dz d\bar{z} R_{z\bar{z}} \sqrt{g_{z\bar{z}}} = 2Q = 2,$$

and one verifies that the above formulae indeed hold with vanishing genus and ramification index. In this example, the Liouville solution coincides with the metric of the Lobatchevki half-plane. Upon quantization, it corresponds to the $SL(2, C)$ invariant vacuum of the Liouville theory.

1. J.-L. Gervais, Y. Matsuo, *Phys. Lett.* **B274** (1991) 309.
2. J.-L. Gervais, Y. Matsuo, "A_n–W-geometry" preprint LPTENS-95/35, hep-th 9201026, Comm. in Math. Phys. to be published.
3. A. Bilal and J.-L. Gervais, *Phys. Lett.* **B206** (1988) 412; *Nucl. Phys.* **B314** (1989) 646; *Nucl. Phys.* **B318** (1989) 579.
4. J.-L. Gervais, A. Neveu, *Nucl. Phys.* **B199** (59) 1982.
5. J.-L. Gervais, A. Neveu, *Nucl. Phys.* **B202** (125) 1982.
6. J.-L. Gervais, A. Neveu *Nucl. Phys.* **B238** (125) 1984; *Nucl. Phys.* **B238** (396) 1984.
7. M. Sato, *RIMS Kokyuroku* **439** (1981) 30; E. Date, M. Jimbo, M.Kashiwara and T.Miwa, "Transformation Groups for Soliton Equations" in *Proc. of RIMS Symposium on Non-linear Integrable Systems- Classical Theory and Quantum Theory* (Kyoto, Japan, May 1981) (World Scientific Publication Co. Singapore, 1983); G.Segal and G.Wilson, *Pub. Math. IHES* **61** (1985) 5.
8. V.G. Drinfeld and V.V. Sokolov, *Journ. Sov. Math.* **30** (1985) 1975.
9. P.Forgács, A.Wipf, J.Balog, L.Fehér, and L.O'Raifeartaigh, *Phys. Lett.* **227B** (1989) 214; J.Balog, L.Fehér, L.O'Raifeartaigh, P.Forgács and A.Wipf, *Ann.*

Phys. (N.Y.) **203** (1990) 76; *Phys. Lett.* **244B** (1990) 435; L.O'Raifeartaigh and A.Wipf, *Phys. Lett.* **251B** (1990) 361; L.O'Raifeartaigh, P.Ruelle, I. Tsutsui and A.Wipf, *W-Algebras for Generalized Toda Theories* Dublin preprint DIAS-STP-91-03, *Comm. Math. Phys.* to appear; L.Fehér, L.O'Raifeartaigh, P.Ruelle, I. Tsutsui and A.Wipf, *Generalized Toda theories and W-algebras associated with integer grading*, Dublin preprint DIAS-STP-91-17, *Ann. Phys.* to appear; L.O'Raifeartaigh, P.Ruelle and I. Tsutsui, *Phys. Lett.* **258B** (1990) 359.

10. P. Griffiths, S. Harris, "Principles of algebraic geometry" Wiley-Interscience (1978).
11. O. Babelon, *Phys. Lett.* **B215** (523) 1988.
12. J.-L. Gervais, *Comm. Math. Phys.* **130** (257) 1990.
13. J.-L. Gervais, *Phys. Lett.* **B243** (85) 1990.
14. J.-L. Gervais, *Comm. Math. Phys.* **138** (301) 1991.
15. J.-L. Gervais, "On the algebraic structure of Quantum gravity in two dimensions", Proceedings of the Trieste Conference on *Topological Methods in Quantum Field Theories*, June 1990, W. Nahm, S. Randjbar-Daemi, E. Sezgin, E. Witten editors, World Scientific.
16. E. Cremmer, J.-L. Gervais, *Comm. Math. Phys.* **134** (619) 1990.
17. J.-L. Gervais, "Quantum group derivation of 2D gravity-matter coupling" Invited talk at the Stony Brook meeting *String and Symmetry 1991*, LPTENS preprint 91/22, Nucl. Phys. B to be published.
18. J.-L. Gervais, A. Neveu, *Phys. Lett.* **151B** (271) 1985.
19. G. Moore, N. Seiberg, *Comm. Math. Phys.* **123** (177) 1989.
20. E. Cremmer, J.-L. Gervais, J.F. Roussel, LPTENS preprint to be published.
21. E. Cremmer, J.-L. Gervais, *Comm. Math. Phys.* **144** (279) 1992.
22. A.N.Leznov, M.V.Saveliev: Acta Appl. Math. **16**,1 (1989).
23. A.N.Leznov, M.V.Saveliev: Lett. Math. Phys. **6**,505 (1982); Comm. Math. Phys. **89**,59 (1983).
24. J.-L. Gervais, M. Saveliev, *Phys. Lett.* **274** (309) 1992.
25. E. Witten: Phys. Rev. **D44** (1991) 314, *On black holes in string theory*; Lecture at Strings '91, Stonybrook, June 1991.

SOME NEW/OLD APPROACHES TO QCD

DAVID J. GROSS[†]

Physics Division, Lawrence Berkeley Laboratory,
1 Cyclotron Road Berkeley, CA 94720

ABSTRACT

I discuss some recent attempts to revive two old ideas regarding an analytic approach to QCD–the development of a string representation of the theory and the large N limit of QCD.

⋆ Talk delivered at the Meeting on Integrable Quantum Field Theories, Villa Olmo and at STRINGS 1992, Rome, September 1992. This work was supported in part by the Director, Office of Energy Research, Office of High Energy and Nuclear Physics, Division of High Energy Physics of the U.S. Department of Energy under Contract DE-AC03-76SF00098 and in part by the National Science Foundation under grant PHY90-21984.

† On leave from Princeton University, Princeton, New Jersey.

1. Introduction

In this lecture I shall discuss some recent attempts to revive some old ideas to address the problem of solving QCD. I believe that it is timely to return to this problem which has been woefully neglected for the last decade. QCD is a permanent part of the theoretical landscape and eventually we will have to develop analytic tools for dealing with the theory in the infra-red. Lattice techniques are useful but they have not yet lived up to their promise. Even if one manages to derive the hadronic spectrum numerically, to an accuracy of 10% or even 1%, we will not be truly satisfied unless we have some analytic understanding of the results. Also, lattice Monte-Carlo methods can only be used to answer a small set of questions. Many issues of great conceptual and practical interest–in particular the calculation of scattering amplitudes, are thus far beyond lattice control. Any progress in controlling QCD in an explicit analytic fashion would be of great conceptual value. It would also be of great practical aid to experimentalists, who must use rather ad-hoc and primitive models of QCD scattering amplitudes to estimate the backgrounds to interesting new physics.

I will discuss an attempt to derive a string representation of QCD and a revival of the large N approach to QCD. Both of these ideas have a long history, many theorist-years have been devoted to their pursuit–so far with little success. I believe that it is time to try again. In part this is because of the progress in the last few years in string theory. Our increased understanding of string theory should make the attempt to discover a stringy representation of QCD easier, and the methods explored in matrix models might be employed to study the large N limit of QCD. For both political and intellectual reasons I fervently urge string theorists to try their hand at these tasks.

2. QCD as a String Theory

It is an old idea that QCD might be represented as a string theory. This notion dates back even before the development of QCD. Indeed, string theory itself was stumbled on in an attempt to guess simple mathematical representations of strong interaction scattering amplitudes which embodied some of the features gleamed from the experiments of the 1960's. Many of the properties of hadrons are understandable if we picture the hadrons as string-like flux tubes. This picture is consistent with linear confinement, with the remarkably linear Regge trajectories and with the approximate duality of hadronic scattering amplitudes.

Within QCD itself there is internal, theoretical support for this idea. First, the $\frac{1}{N}$ expansion of weak coupling perturbation theory can be interpreted as corresponding to an expansion of an equivalent string theory in which the string coupling is given by $\frac{1}{N}$. This is the famous result of 't-Hooft's analysis of the $\frac{1}{N}$ expansion of perturbative QCD [1]. The same is true for any *matrix model–i.e.* a model invariant under $SU(N)$ or $U(N)$, in which the basic dynamical variable is a matrix in the adjoint representation of the group. The Feynman graphs in such a theory can be represented as triangulations of a two dimensional surface. This is achieved by writing the gluon propagator as a double index line and tiling the graph with plaquettes that cover the closed index loops. 't-Hooft's principal result was that one can use $\frac{1}{N}$ to pick out the topology, *i.e.* the genus=number of handles, of the surface, since a diagram which corresponds to a genus G Riemann surface is weighted by $(\frac{1}{N})^{2G-2}$. The leading order in the expansion of the free energy in powers of $\frac{1}{N}$ is proportional to N^2 (reasonable since there are N^2 gluons, and is given by the *planar* graphs of the theory.

Another bit of evidence comes from the strong coupling lattice formulation of the theory. The strong coupling expansion of the free energy can indeed be represented as a sum over surfaces [5]. Again there is a natural large N expansion which picks out definite topologies for these surfaces. This result is an existence proof for a string formulation of QCD. However, the weights of the surfaces are extremely complicated and it is not at all clear how to take the continuum limit.*

From quite general considerations we expect that the large N limit of QCD is quite smooth, and should exhibit almost all of the qualitative features of theory. Thus an expansion in powers of $\frac{1}{3}$ or $(\frac{1}{3})^2$ might be quite good. The longstanding hope has been to find an equivalent (dual) description of QCD as some kind of string theory, which would be useful in to calculate properties of the theory in the infrared.

The problems with this approach are many. First, if QCD is describable as a string theory it is not as simple a theory as that employed for critical strings. It appears to be easier to guess the string theory of everything than to guess the string theory of QCD. Most likely the weights of the surfaces that one would have to sum over will depend on the *extrinsic geometry* of the surface and not only its intrinsic geometry. We know very little about such string theories. Also there are reasons to believe that a string formulation would require many (perhaps an infinite) new degrees of freedom in addition to the coordinates of the string. Finally, there is the important conceptual problem–how

---

* There is also the problem that for large N there is typically a phase transition between the strong and weak coupling regimes [2].

do strings manage to look like particles at short distances. The one thing we know for sure about QCD is that at large momentum transfer hadronic scattering amplitudes have canonical powerlike behavior in the momenta, up to calculable logarithmic corrections. String scattering, on the other hand, is remarkably soft. Critical string scattering amplitudes have, for large momentum transfer, Gaussian fall-off [3]. How do QCD strings avoid this?†

3. Two Dimensional QCD

Two dimensional QCD (QCD_2) is the perfect testing ground for the idea that gauge theories can be equivalent to string theory. First, many features of the theory are stringier in two than in four dimensions. For example, linear confinement is a perturbative feature which is exact at all distances. Most important is that the theory is exactly solvable. This is essentially because in two dimensions gluons have no physical, propagating degrees of freedom, there being no transverse dimensions. In fact QCD_2 is the next best thing to a topological field theory. The correlation functions in this theory will depend, as we shall see, only on the *topology* of the manifold on which formulate the theory and on its *area*. For this reason we will be able to solve the theory very easily and explicitly.

Consider for example the expectation value of the Wilson loop for pure QCD_2 , $\mathrm{Tr}_R\, Pe^{\oint_C A_\mu dx^\mu}$, for any contour, C, which does not intersect itself. Choose an axial gauge, say $A_1 = 0$, then the Lagrangian is quadratic, given by $\frac{1}{2}\,\mathrm{Tr}\, E^2$, where $E = \partial_1 A_0$ is the electric field. The Wilson loop describes a pair of charged particles propagating in time. This source produces, in two dimensions, a constant electric field. The Wilson loop is then given by the exponential of the constant energy of the pair integrated over space and time. This yields,

$$\mathrm{Tr}_R\, Pe^{\oint_C A_\mu dx^\mu} = e^{-g^2 C_2(R) A} \quad , \tag{3.1}$$

where g is the gauge coupling, $C_2(R)$ the quadratic Casimir operator for representation R and A the area enclosed by the loop. The expectation value of more complicated Wilson loops that do self intersect can also be calculated. Kazakov and Kostov worked out a set of rules for such loops

† Recently there have been some interesting speculations regarding this problem [4].

in the large N limit [6]. They are quite complicated. QCD_2 with quarks is also soluble, at least in the large N limit. The meson spectrum was solved for $N \to \infty$ by 't Hooft. It consists of an infinite set of confined mesons with masses m_n that increase as $m_n^2 \sim n$. This provides one with a quite realistic and very instructive model of quark confinement [1], [7].

Is QCD_2 describable as a string theory? The answer is not known, although there is much evidence that the answer is yes. I shall describe below a study that I have carried out to investigate this issue [8].

To simplify matters I shall discard the quarks and consider the pure gauge theory. This would correspond to a theory of closed strings only, quarks are attached to the ends of open strings. We shall consider the partition function for a $U(N)$ or $SU(N)$ gauge theory, on an arbitrary Euclidean manifold $\mathcal{M}$,

$$\mathcal{Z}_{\mathcal{M}} = \int [\mathcal{D}A^\mu] e^{-\frac{1}{4g^2}\int_{\mathcal{M}} d^2x\sqrt{g}\,\mathrm{Tr}\,F^{\mu\nu}F_{\mu\nu}} \quad . \tag{3.2}$$

One might think that in the absence of quarks the theory is totally trivial, since in two dimensions there are no physical gluon degrees of freedom. This is almost true, however the free energy of the gluons will depend non-trivially on the manifold on which they live. In fact, one cannot simply gauge the gluons away. If, for example, $\mathcal{M}$ contains a non-contractible loop C, then if $\mathrm{Tr}\,Pe^{\oint_C A_\mu dx^\mu} \neq 1$, one can not gauge A_μ to zero along C. Thus, the partition function will be sensitive to the *topology* of $\mathcal{M}$.

Although non-trivial the theory is extremely simple, almost as simple as a topological theory. It is easy to see that the partition function will only depend on the topology and on the area of the manifold $\mathcal{M}$. This is because the theory is invariant under all *area preserving diffeomorphisms.* To demonstrate this note that the two-dimensional field strength can be written as $F_{\mu\nu} = \epsilon_{\mu\nu} f$, where $\epsilon_{\mu\nu}$ is the anti-symmetric tensor and f a scalar field. Thus the action is $S = \int \mathrm{Tr}\, f^2 d\mu$, where $d\mu = \sqrt{g}d^2x$ is the volume form on the manifold. This action is independent of the metric, except insofar as it appears in the volume form. Therefore the theory is invariant under area preserving diffeomorphisms (W_∞). The partition function can thus only depend on the topology and on the area of the manifold $\mathcal{M}$,

$$\mathcal{Z}_{\mathcal{M}} = \mathcal{Z}[G, g^2, A, N] = \mathcal{Z}[G, g^2A, N] \quad , \tag{3.3}$$

where G is the genus of $\mathcal{M}$.

Now we can state the conjecture that the logarithm of this partition function, the free energy, is identical to the partition function on some string theory, with target space $\mathcal{M}$, where we would identify the string coupling with $\frac{1}{N}$ and the string tension with g^2N,

$$\ln(\mathcal{Z}[G, g^2A, N]) = \mathcal{Z}^{\mathrm{String}}_{\mathrm{TargetSpace}\ \mathcal{M}}[g_{\mathrm{st}} = \frac{1}{N}, \alpha' = g^2N] \ . \tag{3.4}$$

As a candidate for the type of string theory I am thinking of consider the Nambu action, wherein

$$\mathcal{Z}^{\mathrm{String}}_{\mathcal{M}} = \sum_{h=\mathrm{genus}} (g_{\mathrm{st}})^{2h-2} \int \mathcal{D}x^{\mu}(\xi) e^{\int d^2\xi\sqrt{g}} \ , \tag{3.5}$$

where g is the determinant of the induced metric

$$g = \det[g_{\alpha\beta}] = \det[\frac{\partial x^{\mu}}{\partial \xi_{\alpha}} \frac{\partial x^{\nu}}{\partial \xi_{\beta}} G_{\mu\nu}(x)] \ , \tag{3.6}$$

and $G_{\mu\nu}(x)$ is the metric on the manifold $\mathcal{M}$. This string theory, when the target space is two-dimensional, is indeed invariant under area preserving diffeomorphisms of the target space. To see this note that $\sqrt{g} = |\frac{\partial x^{\mu}}{\partial \xi^{\alpha}}|\sqrt{G}$, which is obviously unchanged by a map $x^{\mu} \rightarrow x'^{\mu}$ as long as $|\frac{\partial x^{\mu}}{\partial x'^{\nu}}| = 1$.* Unfortunately the only way we know to quantize this theory is to transform it into the Polyakov action, which upon quantization yields the standard non- critical string [10]. This is not what we want to do here, since the resulting theory is not even Lorentz invariant. Is there another quantization of the Nambu string that differs from the Polyakov quantization in two- dimensions? The answer is not known.

* Actually the Polyakov action with a two-dimensional target space also has a W_{∞} symmetry, although is is realized in a very nonlinear fashion. One might speculate that this is related to the well known W_{∞} symmetry of the $c = 1$ string theory [9].

3.1. Evaluation of the Partition Function

The partition function for QCD_2 can easily be evaluated by means of the following idea, originally due to Migdal [11]. The trick is to use a particular lattice regularization of the theory which is both exact and additive. For the lattice we take an arbitrary triangulation of the manifold and define the partition function as

$$\mathcal{Z}_{\mathcal{M}} = \int \prod_{L} dU_L \prod_{\text{plaq.}} Z_P[U_P] \quad , \tag{3.7}$$

where $U_P = \prod_{L \in \text{plaq.}} U_L$, and $Z_P[U_P]$ is some appropriate lattice action. Any action will do as long as it reduces in the continuum limit to the usual continuum action. Instead of the Wilson action, $Z_P(U) = e^{-\frac{1}{g^2}\text{Tr}(U+U^\dagger)}$, we shall choose the *heat kernel action*,

$$Z_P = \sum_R d_R \, \chi_R(U_P) e^{-g^2 C_2(R) A_P} \quad , \tag{3.8}$$

where the sum runs over representations R of $SU(N)$ (or $U(N)$), d_R is the dimension of R, $\chi_R(U_P)$ the character of U_P in this representation, $C_2(R)$ the quadratic Casimir operator of R and A_P the area of the plaquette.

It is easy to see, using the completeness of the characters to expand about $Z_P \overset{U_L \to 1 + iA_\mu dx^\mu}{\longrightarrow} \sum_R d_R \chi_R(U_P) = \delta(U_P - 1) + \ldots$, that in the continuum limit of this theory reduces to ordinary Yang-Mills theory. What is special about the heat kernel action is that it is additive. Namely, we can integrate over each link on the triangulation, say U_1, which appears in precisely two triangles, using the orthogonality of the characters, $\int dV \chi_a(XV)\chi_b(V^\dagger Y) = \frac{\delta_{ab}}{d_a}\chi_a(XY)$, and obtain,

$$\int dU_1 Z_{P_1}(U_2 U_3 U_1) Z_{P_2}(U_1^\dagger U_4 U_5) = Z_{P_1+P_2}(U_2 U_3 U_4 U_5) \quad . \tag{3.9}$$

This formula expresses the unitarity of the action, since in fact $Z_P(U) = \langle U | e^{-g^2 A \Delta} | 1 \rangle$, where Δ is the Laplacian on the group.

We can use this remarkable property of the heat kernel action to argue that the lattice representation is exact and *independent of the triangulation*. This is because we can use (3.9) in reverse to add as many triangles as desired, thus going to the continuum limit. On the other hand we can use (3.9) to reduce the number of triangles to the bare minimum necessary to capture the topology of $\mathcal{M}_G$. A two- dimensional manifold of genus G can be described by a $4G$-gon with identified sides: $a_1 b_1 a_1^{-1} b_1^{-1} \ldots a_G b_G a_G^{-1} b_G^{-1}$. The partition function can be written using this triangulation as,

$$\mathcal{Z}_{\mathcal{M}_G} = \sum_R d_R e^{-g^2 C_2(R) A} \int \prod \mathcal{D}U_i \mathcal{D}V_i \chi_R[U_1 V_1 U_1^\dagger V_1^\dagger \ldots U_G V_G U_G^\dagger V_G^\dagger] \, . \tag{3.10}$$

We can now evaluate the partition function using the orthogonality of the characters and the relation, $\int \mathcal{D}U \chi_a[AUBU^\dagger] = \frac{1}{d_a}\chi_a[A]\chi_a[B]$, to obtain [12], [13],

$$\mathcal{Z}_{\mathcal{M}_G} = \sum_R d_R^{2-2G} e^{-\frac{\lambda A}{N} C_2(R)} \quad , \tag{3.11}$$

where $\lambda \equiv g^2 N$ is kept fixed. Thus we have an explicit expression for the partition function. It depends, as expected, only on the genus and the area of the manifold.

3.2. The Large N expansion

The formula (3.11) for the partition function is quite complicated, being written as a sum over all representations of $SU(N)$. The representations of $SU(N)$ or $U(N)$ are labeled by the Young diagrams, with m boxes of length $n_1 \geq n_2 \geq n_3 \geq \ldots n_m \geq 0$. Such a representation has,

$$\begin{aligned} C_2(R) &= N \sum_{i=1}^m n_i + \sum_{i=1}^m (n_i + 1 - 2i); \\ d_R &= \frac{\Delta(h)}{\Delta(h^0)}, \quad h_i = N + n_i - 1, \quad h_i^0 = N - i \\ \Delta(h) &= \prod_{1 \leq i < j \leq N} (h_i - h_j) \quad . \end{aligned} \tag{3.12}$$

Thus we have a very explicit sum and one can, in principle, expand each term in powers of $\frac{1}{N}$ and evaluate the sum.

What do we expect if the string conjecture is correct? Consider the expansion in powers of $\frac{1}{N}$ of the free energy,

$$\ln[\mathcal{Z}_{\mathcal{M}_G}] = \sum_{g=0}^{\infty} \frac{1}{N^{2g-2}} f_g^G(\lambda A) \quad . \tag{3.13}$$

If this were given by a sum over maps of a two-dimensional surface of genus g onto a two-dimensional surface of dimension G we would expect that $f_g^G(\lambda A) \sim (\frac{1}{N})^{2g-2} e^{-\lambda A n}$, where n is the *winding number* of the map, *i.e.* the topological index that tells us how many times the map $x(\xi)$ covers $\mathcal{M}$. This is the integral of the Jacobian of the map $\xi \to x$, $\int d^2\xi \det[\frac{\partial x^\mu}{\partial \xi^i}]$, which differs from the Nambu area, $\int d^2\xi |\det[\frac{\partial x^\mu}{\partial \xi^i}]|$, since the surface can fold over itself.

Now there is a minimum value that G can take, given the genus G of the target space and the winding number n. Thus for example there are no smooth maps of a sphere onto a torus or a torus onto a genus two surface. Similarly there are no smooth maps of a genus g surface onto a genus g surface that wind around it more than once. To get an idea of the bound consider holomorphic maps, in which case the Riemann-Hurwitz theorem state states that $2(g-1) = 2n(G-1) + B$, where B is the total branching number. In the case of smooth maps there seems to be the following bound [14],

$$2(g-1) \geq 2n(G-1) \quad . \tag{3.14}$$

Thus if QCD_2 is described by a string theory we would expect that

$$f_g^G(\lambda A) = \sum_n \begin{cases} 0 & \text{if } (g-1) < n(G-1) \\ e^{-n\lambda A}\omega_g^n(A) & \text{otherwise} \end{cases} \quad . \tag{3.15}$$

We can use these inequalities as tests of whether our conjecture is correct. To do this we need to expand (3.11) in powers of $\frac{1}{N}$.

The hardest case is that of the sphere ($G = 0$), since the sum over representations blow up rapidly and it is not even evident that there exists a tamed large N expansion. We can break up the sum in (3.11) into a sum over representations with n boxes in the Young tableaux since, for large N, $C(R_n) \overset{N\to\infty}{\to} N\sum_i n_i = Nn$. Thus,

$$Z_{G=0} = \sum_R d_R^2 e^{-\frac{\lambda A}{N}C_2(R)} \overset{N\to\infty}{\to} \sum_n \sum_{R_n} d_{R_n}^2 e^{-n\lambda A}(1 + \ldots) \quad . \tag{3.16}$$

To evaluate this we need to evaluate the following sum, $\sum_{R_n} d_{R_n}^2$. This can be done using a method of discrete orthogonal polynomials [8], yielding,

$$\sum_{R_n} d_{R_n}^2 = \binom{N^2+n-1}{n} \ . \tag{3.17}$$

Then it follows that,

$$Z_{G=0} \to \exp[-N^2[\ln(1-e^{-\lambda A}) + \frac{2\lambda A e^{-2\lambda A}}{(1-e^{-\lambda A})^2} + \ldots] + O(N^0) + \ldots] \ . \tag{3.18}$$

Here there are no constraints implied by the inequality (3.15) , but the structure of the expansion is very interesting.

The case of the torus, $(G = 1)$, is some what simpler. One can easily derive that (for $SU(N)$) [8],

$$\begin{aligned} Z_{G=1} = \sum_R e^{-\frac{\lambda A}{N} C_2(R)} &\to \exp[-N^0 \ln \eta(-e^{-\lambda A}) + \\ \frac{\lambda A}{N^2} \sum_{n=1}^{\infty} &e^{-n\lambda A}[\sum_{ab=n} a^2 b + \sum_{ab+cd=n} ac] + \ldots] \ , \end{aligned} \tag{3.19}$$

where $\eta(x) = \prod_{n=1}^{\infty}(1-x^n)^{-1}$. This is totally consistent with the bound $g \geq 1$.

Most interesting is the case of $G > 1$, where the inequalities are quite stringent. In this case one can easily derive [8],

$$Z_G \to \sum_n (\frac{1}{N})^{2n(G-1)} e^{-n\lambda A} \sum_{r=\text{rep of } S_n} [\frac{n!}{d_r}]^{2(G-1)} \ , \tag{3.20}$$

where the sum is over representations of the symmetric group S_n and d_r is the dimension of the rth representation of S_n. Not only is this in total accord with our expectations, but one can also show that $\omega_g^n(A) =$, for $g = 1 + n(G-1)$, is precisely the number of topologically inequivalent maps on the genus g manifold onto the genus G manifold with winding number n[15].

So the large N expansion of QCD_2 looks precisely like what we would expect from string considerations. What remains to be understood are the all the rational numbers that appear as coefficients of the powers of $e^{-\lambda A}$ and of $\frac{1}{N}$ in terms of the counting of maps of $\mathcal{M}_g$ onto $\mathcal{M}_G$. Some of these are understood, but not all. Then it remains to construct a string action that reproduces these counting rules.

4. Induced QCD

4.1. The Large N Limit of QCD

QCD is hard to solve since it is a theory with no free, adjustable or small parameters. In pure QCD (no quarks) the only parameter we can adjust is the number of colors, N. Luckily, in the large N limit QCD simplifies enormously, and this limit remains the best hope to yield an exact or controllable treatment of the theory. We know that as $N = \infty$ only planar graphs survive. More generally we know that in terms of the appropriate variables the large N limit of gauge invariant observables is given, for $N = \infty$ by the *master field*, namely a solution of an appropriate classical equation of motion [16]. The large N limit is in the nature of a semi-classical expansion, with N playing the role of Planck's constant. Unlike the running coupling N does not vary with momentum and we expect the large N limit to be qualitatively correct for all momenta, to correctly capture the small distance asymptotic freedom of the theory as well as exhibit confinement at large distances. In the $N = \infty$ we should have an infinite spectrum of stable mesons and glueballs. Even baryons, bound states of N quarks, are describable, in this limit, as solitons of the effective Lagrangian for the master field [17]. Thus the hope has survived that we could find an exact solution of QCD for $N = \infty$, which would yield the hadronic spectrum, and would be the starting point for a systematic large N expansion which could allow us to calculate scattering amplitudes.

The standard method of solving a theory in the large N limit is to find an appropriate saddlepoint for the partition function. In the case of QCD this is difficult. Consider the standard (Wilson) lattice formulation of the theory,

$$\mathcal{Z}_{\mathrm{QCD}} = \int \prod_L \mathcal{D}U_L e^{-\sum_{\mathrm{plaq.}} \frac{N}{g^2(a)} \mathrm{Tr}\left[\prod_L U_L + \mathrm{h.c.}\right]} \quad . \tag{4.1}$$

The integrand behaves as the exponential of an action that is of order N^2, thus one might hope to evaluate it by saddlepoint techniques. However, the measure is also of order c^{N^2} and therefore one must somehow get rid of N^2 degrees of integration before this can be done. The reason QCD is not yet solved in the large N limit is that no one knows how to reduce the theory to N variables per site.

Another theory which is also insoluble in the large N limit is the non-critical string with $c > 1$. Following the recent success of the matrix model solutions of string theory [18], we can construct such strings if we could deal with the large N limit of a scalar matrix model in D dimensions, say

$$\mathcal{Z}_D^{\text{string}} = \int \prod_i \mathcal{D}\phi_i e^{-N\sum_i \operatorname{Tr} U(\phi_i) + N\sum_{i,\mu=1\ldots D} \operatorname{Tr}(\phi_i\phi_{i+\mu})} \quad , \tag{4.2}$$

which describes a scalar field on a D- dimensional lattice. The connection with string theory is made in the usual way, the Feynman diagrams of the perturbative expansion of (4.2) correspond, in an expansion in powers of $\frac{1}{N}$, to triangulations of two-dimensional surfaces. The scalar fields correspond to matter on this surface and thus, (4.2) , could yield, at the appropriate critical point where the mean number of triangles diverges, a $c = D$ string theory. The standard approach to the large N-limit of such a theory is to diagonalize the matrices ϕ, *i.e.* to pass to *radial* coordinates, $\phi_i = \Omega_i \lambda_i \Omega_i^\dagger$, where λ_i is diagonal. In terms of these variables,

$$\mathcal{Z}_D^{\text{string}} = \int \prod_i \mathcal{D}\lambda_i \mathcal{D}\Omega_i \Delta^2(\lambda_i) e^{-N\sum_i \operatorname{Tr} U(\lambda_i) + N\sum_{<ij>} \operatorname{Tr}\left(\lambda_i V_{ij} \lambda_j V_{ij}^\dagger\right)} \quad , \tag{4.3}$$

where $V_{ij} = \Omega_i \Omega_j^\dagger$ and $\Delta(\phi) = \prod_{i<j}(\phi_i - \phi_j)$.

The next step is to integrate out the diagonalization matrices, Ω_i. We can change variables from the Ω_i's,defined on the sites to the V_{ij}'s, defined on the links $\prod_i \mathcal{D}\Omega_i = \prod_{<ij>} \mathcal{D}V_{ij} \prod_{\text{plaq.}} \delta(1 - \prod_L V_{ij})$. The constraints arise since the V_{ij}'s are pure gauge fields. If not for the constraints we could perform the integral over the V_{ij}'s and reduce the integral to one over N variables per site that could be evaluated by saddlepoint techniques. It is these constraints that have prevented the construction of strings with $c > 1$.

Now let us combine these two models to consider QCD with adjoint scalar matter,

$$\mathcal{Z}_{\text{QCD}}^{\text{adj}} = \int \prod_L \mathcal{D}U_L \prod_i \mathcal{D}\phi_i e^{-N\sum_i \operatorname{Tr} U(\phi_i) + N\sum_{i,\mu=1\ldots D} \operatorname{Tr}(\phi_i U_\mu \phi_{i+\mu} U_\mu^\dagger)} \\ e^{-\frac{N}{g^2(a)}\sum_{\text{plaq.}}\left[\operatorname{Tr}(\prod_L U_L + \text{h.c.})\right]} \quad . \tag{4.4}$$

This theory is invariant under standard gauge transformations, $\phi_i \to V_i \phi_i V_i^\dagger$; $U_\mu \to V_i U_\mu V_{i+\mu}^\dagger$,which allow us to diagonalize the ϕ's. However the presence of the Wilson action prevents us from handling this theory for large N. If set the gauge coupling to zero, we recover the previous model, since in this limit we can drop the Wilson action term, as long as we enforce the constraints, $\mathrm{tr}[U_P] = 1$. However if we take the opposite limit, *i.e.* set $g = \infty$, then we can simply drop the Wilson action and the model will be soluble in the large N limit. This is *induced QCD* [19].

Induced QCD has the one great advantage of being soluble, or at least reducible to a well defined master field equation. This is because the integral over the link matrices can now be performed. This is the famous *Itzykson- Zuber* integral [20],

$$I(\phi,\chi) \equiv \int \mathcal{D}U e^{N\,\mathrm{Tr}\left[\phi U \chi U^{\dagger}\right]} = \frac{\det\left[e^{N\phi_i\chi_j}\right]}{\Delta(\phi)\Delta(\chi)} \quad . \tag{4.5}$$

This formula is very profound, underlies all the analysis of the $c = 1$, matrix model, and can be derived in many ways. One is the demonstration that the integral is given exactly by the WKB approximation, and the answer is simply the sum over the $N!$ saddlepoints, for which are the U are permutation matrices.

Although soluble this model appears to be very far from QCD, since asymptotic freedom instructs us to set the lattice coupling to zero, not infinity, in the continuum limit. However, Kazakov and Migdal argued that even though there is no kinetic term for the gauge field, it could be *induced* at large distances [19]. They argued that if one integrates out the scalar mesons (even in the case of noninteracting scalars with $U(\Phi) = \frac{1}{2}m^2\Phi^2$), then at distances large compared to a, one would induce in four dimensions an effective gauge interaction,

$$S_{\mathrm{eff}}(U) \sim \frac{N}{96\pi^2}\ln\left(\frac{1}{m^2a^2}\right)\mathrm{Tr}\,F_{\mu\nu}^2 + \text{finite as } a \to 0 \quad . \tag{4.6}$$

This is simply the one loop vacuum graph for the scalars in a background gauge field, which is logarithmically divergent in four dimension. Now this looks very much like the ordinary Yang-Mills action, $\frac{1}{g^2(a)}\mathrm{tr}\,F_{\mu\nu}F^{\mu\nu}$, if we recall that asymptotic freedom tell us that $\frac{1}{g^2(a)} = \frac{11N}{48\pi^2}\ln\left(\frac{1}{M_g^2a^2}\right)$, where M_g is a mass scale for QCD, say the glueball mass. We can therefore identify these two (the fact that there are N^2 scalars is crucial, as is the sign of the effective action which is due to the non-asymptotic freedom of the scalars.) If we do so then we find that, $M_g^2 = m^{\frac{23}{11}}a^{\frac{1}{22}}$. Thus in the continuum limit the adjoint scalars become infinitely massive and decouple, but not before they have drive $\frac{1}{g^2}$ up, from zero at distance a to the large QCD value atdistance $\frac{1}{m}$, where $\frac{1}{M_g} >> \frac{1}{m} >> a$.

The basic idea is that the infrared slavery of the scalars, at the size of the lattice spacing, produces an effective gauge theory at a larger scale (much larger than the inverse scalar mass), which then produces the usual asymptotically free fixed point theory.

There are many problems with this idea. For one the hard gluons are not absent and their contribution will overwhelm that of the scalars at short distances. Their asymptotic freedom is more powerful than the infrared slavery of scalars. Another issue is that the above theory possesses a much larger symmetry than the $SU(N)$ gauge symmetry of the usual lattice action. It is not difficult to see that, in D dimensions, it is invariant under $(D-1)\times(N-1)$ extra local $U(1)$-gauge symmetries. This is because the transformation $U_\mu(x) \to V_\mu^\dagger(x)U_\mu(x)V_\mu(x+\mu a)$, leaves the action invariant as long as $V_\mu(x)$ is a unitary matrix that commutes with $\Phi(x)$. If $V_\mu(x)$ were independent of μ then this would be the ordinary gauge invariance. Thus we have $D-1$ new gauge symmetries, which are of course isomorphic to the special unitary transformations that commute with Φ [21]. Thus $V_\mu(x) = D_\mu(x)\Omega(x)$, where $\Omega(x)$ is the unitary matrix that diagonalizes Φ and $D_\mu(x)$ is diagonal.

A subset of this symmetry is the, field independent, local Z_N symmetry, $U_\mu(x) \to Z_\mu U_\mu(x) Z_\mu^\dagger$, where Z_μ is an element of the center of the group. This symmetry alone prevents the Wilson loop from acquiring an expectation value. A Wilson loop contains different links, and thus $W(C) = \langle \prod_{L\in C} U_L\rangle \to (\prod_L Z_\mu) W(C) \Rightarrow W(C) = 0$. This symmetry must be broken if we are to recover the QCD fixed point from this formulation [22].

Finally, as we shall see, the simple Gaussian model is soluble and the answer is very simple and *not* equivalent to QCD [21]. However, there are interesting attempts to save the model and furthermore even if it does not yield a solution of QCD it might provide some interesting soluble matrix models which could yield new solutions of new string theories. Induced QCD is a matrix model and thus it corresponds to some kind of sum over surfaces. If we look at the Itzykson-Zuber integral we note that it could be expressed as,

$$I(\phi,\chi) = \exp\left[\tfrac{1}{2}\operatorname{Tr}\phi^2 \operatorname{Tr}\chi^2 + a \operatorname{Tr}\phi^4 \operatorname{Tr}\chi^4 + \frac{b}{N^2}(\operatorname{Tr}\phi^2)^2(\operatorname{Tr}\chi^2)^2 + \ldots\right] . \tag{4.7}$$

These terms will affect the structure of the large N expansion of the Feynman diagrams, and can be interpreted as yielding extra weights when the two-dimensional surfaces intersect [23]. Thus this model corresponds, perhaps, to some kind of string theory with weights that depend on the extrinsic geometry.

4.2. Solution of the Gaussian model

To try to solve the model of induced QCD we first integrate out the U_L's, then look for extrema of the effective action,

$$S[\phi_i] = N^2[\frac{1}{N}\,\mathrm{Tr}\sum_i U(\phi_i)+\frac{1}{N^2}\sum_{i,\mu}\ln I(\phi_i,\phi_{i+\mu})+\frac{1}{N^2}\sum_i \ln\Delta^2(\phi_i)] \quad . \tag{4.8}$$

In the large N limit the integral will be dominated by a translationally invariant saddlepoint for the density of eigenvalues of the matrices Φ_i, $\rho(x) \equiv \frac{1}{N}\sum_{a=1}^{N}\delta(x-\phi_a)$. Migdal has derived the master field equation for the saddlepoint, using the Schwinger-Dyson equations that are satisfied by $I(\phi,\chi)$ [24]. These are consequences of the fact that I satisfies $\mathrm{tr}[(\frac{1}{N}\frac{\partial}{\partial\phi})^k]I = \mathrm{tr}(\chi)^k I$. The net result is that one derives an equation for the function $F(z) \equiv \int dz \frac{\rho(\nu)}{z-\nu}$, whose imaginary part is $\mathrm{Im}F(\nu) = -\pi\rho(\nu)$,

$$\mathrm{Re}F(\lambda) = P\int \frac{d\nu}{2\pi i}\ln[\frac{\lambda - \frac{1}{2D}U'(\nu) - \frac{D-1}{D}\mathrm{Re}F(\nu) + i\pi\rho(\nu)}{\lambda - \frac{1}{2D}U'(\nu) - \frac{D-1}{D}\mathrm{Re}F(\nu) - i\pi\rho(\nu)}]. \tag{4.9}$$

This equation is much more complicated than the usual Riemann-Hilbert problem that one obtains for simple matrix models. It is sufficiently non-linear and complex that one might imagine that it describes QCD.

The master field equation simplifies dramatically for $D = 1$. This is because in one dimension the gauge field can be gauged away completely, thus the model is equivalent to a scalar field on a one-dimensional lattice. The large N limit of this model describes the $c = 1$ string on a discrete target space, a model which has been solved in the double- scaling limit for small lattice spacing [25]. It undergoes a phase transition at a finite lattice spacing and it might be very instructive to use (4.9) to explore this phenomenon.

In particular for the quadratic potential the path integral is Gaussian,

$$Z = \int\prod_n \mathcal{D}\Phi_n e^{-N\sum_n \mathrm{Tr}\{\frac{m^2}{2}\Phi_n^2 - \Phi_n\Phi_{n+1}\}} \ . \tag{4.10}$$

Thus the eigenvalues of Φ will be given by the semi-circular distribution, namely $\pi\rho(\nu) = \sqrt{\mu - \frac{\mu^2\nu^2}{4}}$, where μ is determined by the mean of the squares of the eigenvalues, $\langle\frac{1}{N}\mathrm{Tr}(\Phi^2)\rangle = \frac{1}{\mu}$. It is therefore sufficient to calculate the expectation value of $\frac{1}{N}\mathrm{Tr}(\Phi^2)$, which is given by the one loop integral,

$$\frac{1}{N}\mathrm{Tr}(\Phi^2) = \int_{-\pi}^{\pi} \frac{dp}{2\pi} \frac{1}{m^2 + 2\cos p} = \frac{1}{\sqrt{m^4 - 4}} \equiv \frac{1}{\mu} . \tag{4.11}$$

It is easy to verify that this solves (4.9), using the fact that

$$F(z) = \frac{\mu z}{2} - \sqrt{\frac{\mu^2 z^2}{4} - \mu}; \quad \mathrm{Re}F(\nu) = \tfrac{1}{2}\mu\nu, \tag{4.12}$$

However, if we return to (3.8), we see that the integral involved is of the same form for any D, as long as $\Re V'(\nu)$ is linear in ν. This suggests that we can find a solution of (4.9) with a semi-circular distribution of eigenvalues for a quadratic potential in any dimension [21].

Indeed, one can see that a semi-circular distribution of eigenvalues satisfies (4.9) for any D as long as,

$$\mu_{\pm}(D) = \frac{m^2(D-1) \pm D\sqrt{m^4 - 4(2D-1)}}{2D-1} . \tag{4.13}$$

This solution is much too trivial to describe QCD. In particular, for $D > 1$ there is no sensible continuum limit of the model.

Is the solution unique? To see that it is note that in the master field equation the dimension of space-time enters only via the number of nearest neighbors of a given site, the coordination number of the lattice.[⋆] The translationally invariant background scalar field is the same for any lattice with the same coordination number. The observables, say the scalar propagator, will of course depend on the full structure of the lattice, but not he background field. Therefore we can choose another simpler lattice with the same coordination number, say a Bethe lattice, which contains no closed loops. For such a lattice, as in the case of the $D = 1$ model, the gauge field can be eliminated completely, and the model is equivalent to,

$$Z_{\mathrm{BetheLattice}} = \int \mathcal{D}\phi_i e^{-N\sum_i \mathrm{Tr}\,\frac{m^2}{2}\phi_i^2 + N\sum_{<ij>} \mathrm{Tr}[\phi_i\phi_j]} . \tag{4.14}$$

This model is easily soluble. We define $Z(\phi)$ to be the partition function of a branch of the Bethe lattice with coordination number $2D$, so that $Z = \int \mathcal{D}\phi Z(\phi)^{2D} e^{-\frac{m^2}{2}N\,\mathrm{Tr}\,\phi^2}$. $Z(\phi)$ satisfies the equation,

⋆ I thank C. Bachas for emphasizing this point to me.

$$Z(\phi) = \int \mathcal{D}\phi' Z(\phi')^{2D-1} e^{-\frac{m^2}{2} N \operatorname{Tr} \phi'^2 + N \operatorname{Tr}[\phi\phi']} \quad . \qquad (4.15)$$

These equations are easily soluble. Take $Z(\phi)$ to have the form $Z(\phi) = ce^{-N\frac{\alpha}{2}\operatorname{Tr}\phi^2}$, then (4.15) determines α to equal $\alpha = \frac{-m^2 \pm \sqrt{m^4 - 4(2D-1)}}{2(2D-1)}$. Then (4.14) can be used to determine $\frac{1}{N}\operatorname{Tr}\phi^2 = \frac{1}{m^2+2\alpha D}$, which agrees precisely with $\frac{1}{\mu}$ as given by (4.13).

4.3. Prospects

The simplest Gaussian model fails, but all hope is not lost. It is certainly possible to induce QCD if one introduces enough flavors of matter. The problem is that one then loses solubility. It might be that the self interactions of the scalars could be adjusted to drive the theory towards the asymptotically free fixed point. This hope has been pursued with great vigor by Migdal, who has also considered adding fermions, not too many so that the model remains soluble, so as to break the Z_N symmetry [26]. Time will tell whether this will succeed. Even if it does not these model might yield a new class of interesting soluble matrix models which could teach us something about new classes of strings, perhaps strings that depend on extrinsic geometry. For this reason alone it is worth studying these models.

REFERENCES

1. G. 't Hooft, *Nucl. Phys.* **B72**, 461 (1974).
2. D. Gross and E. Witten, *Phys. Rev.* **D21**, 446(1980).
3. D. Gross and P. Mende,
4. J. Polchinski, *Phys. Rev. Lett* **68**, 1267 (1992); M. Green, QMW-91-24 (1991)
5. D. Weingarten, *Phys. Lett.* **B90**, 285 (1990); V. Kazakov and I. Kostov, *Phys. Lett.* **B128**, 316 (1983); I. Kostov *Phys. Lett.* **B138**, 191 (1984); *Nucl. Phys.* **B179**, 283 (1981); K. O'Brien and J. Zuber, *Nucl. Phys.* **B253**, 621 (1985)
6. V. Kazakov and I. Kostov, *Nucl. Phys.* **B220**, 167 (1983); I. Kostov *Nucl. Phys. Lett.* **B265**, 223 (1986).
7. C. Callan, N. Coote and D. Gross,
8. D. Gross, to be published

9. D. Gross and A. Polyakov, unpublished

10. A. Polyakov, *Nucl. Phys.* **B40**, 235 (1982).

11. A. Migdal, *Zh. Eksp. Teor. Fiz.* **69**, 810 (1975)(Sov. Phys. Jetp. **42**, 413).

12. B. Rusakov, *Mod. Phys. Lett.* **A5**, 693 (1990).

13. E. Witten, *Comm. Math. Phys*,**141**, 153(1991; D.Fine, *Comm. Math. Phys.* **134**, 273 (1990); M. Blau and G. Thompson, NIKHEF-H/91-09, MZ-TH/91-17.

14. S. Novikov and A. Schwarz, private communication.

15. D. Gross and W. Taylor, to be published

16. L. Yaffe, *Rev. Mod. Phys.***54** , 407 (1982).

17. E. Witten, *Nucl. Phys.* **B160**, 57 (1979); V. Kazakov and A. Migdal,*Phys. Lett.* **B103** , 214 (1981).

18. D. Gross and A.Migdal, *Phys. Rev. Lett.*, **64**, 717 (1990); M. Douglas and S. Shenker, *Nucl. Phys.* **B335**, 635 (1990); E. Brézin and V. Kazakov, *Phys. Lett.* **236B**, 144 (1990).

19. V. Kazakov, and A. Migdal, PUPT-1322, May 1992.

20. Harish-Chandra, *Amer. Jour. Math.* **79**, 87 (1957); C. Itzykson and J. Zuber, *Jour. Math. Phys.* **21**, 411 (1980); J. Duistermaat and G. Heckman, *Invent. Math.* **69**, 259 (1982).

21. D. Gross, PUPT 1335, (August 1992).

22. I. Kogan, G. Semenoff and N. Weiss, UBCTP 92-022, June 1992.

23. S. Das, A. Dhar, A. Sengupta and S. Wadia, *Mod. Phys. Lett.* **A5**, 1041 (1990); G. Korchemsky, UPRF-92-334 (1992).

24. A. Migdal, Princeton preprint PUPT-1323, June 1992.

25. D. Gross and I. Klebanov, *Nucl. Phys.* **B334**, 475 (1990).

26. A. Migdal, PUPT-1332, LPTENS-92/23 , PUPT-1343 ; S. Khoklachev and Yu. Makeenko, ITEP-YM-7-92 (August 1992);

Two-dimensional Models of Black Holes

JEFFREY A. HARVEY

Enrico Fermi Institute
University of Chicago
5640 Ellis Avenue, Chicago, IL 60637

Abstract

In this talk I review some of the recent progress in understanding black hole formation and evaporation through the use of $1+1$ dimensional models.

1. Introduction

Hawking's work of 15 years ago on the evaporation of black holes raises the possibility that pure states can evolve into mixed states, resulting in a net loss of information [1]. For example one could imagine starting with a pure state of infalling matter which collapses to form a black hole. The black hole will then slowly (at first) evaporate. According to Hawking's calculation, this process yields a purely thermal spectrum of outgoing radiation. This process continues until the black hole is Planck-sized, at which point quantum gravity effects are important and Hawking's semiclassical calculation is no longer expected to be a good approximation. What happens next is very controversial. Several possibilities that have been discussed are:

(I.) *The black hole disappears completely, carrying with it all the information carried in by the infalling particles.* This is the alternative advocated by Hawking [1]. While not logically inconsistent, it is a radical proposal, implying that the laws of quantum physics are not deterministic. Instead one can only assign a probability that the outcome of black hole formation/evaporation is a given quantum state and the rules for assigning this probability have yet to be fully understood.

(II.) *The evaporation process ceases, leaving a stable Planck-mass remnant. The final state, remnant + radiation is pure.* This has been advocated for example in [2]. Conservation of information requires an infinite number of "species" of stable remnants, one for each initial state which collapses to a black hole. This is potentially in conflict with low-energy physics. For example there should be a finite – although incredibly small – probability for producing a given species remnant via pair production in a non-static gravitational field. Assuming this is roughly the same for every species of Planck mass remnants then, after summing over species, the total remnant production rate would diverge. This argument assumes that the coupling of the remnant to gravity is dictated purely by its mass. For certain types of remnants this argument apparently breaks down, because the correct low-energy Lagrangian must take into account low-energy excitations of the remnant [3].

(III.) *The black hole disappears completely. Outgoing radiation is correlated with infalling matter and radiation in such a way that the final pure radiation state is pure.* This point of view has been advocated by Page [4], 'tHooft [5] and in [6] (to name a few). This is in fact how the final state of *e.g.* a burning lump of coal manages to be pure. While at each point in the process the emitted radiation appears thermal, late-time radiation has subtle correlations with early time radiation. The big difference for black holes (as stressed in [7]) is that until the final Planckian stage of the evaporation process they are surrounded by an apparent horizon which is very nearly null. The infalling particles therefore carry the information into a region causally shielded from that part of future null infinity which precedes (in retarded time) the final stage of evaporation. Thus the information cannot come back out without violating macroscopic causality until the black hole has evaporated down to the Planck size. It is conceivable that quantum coherence could be restored by radiation emitted in the final stage of evaporation which is governed by unknown laws of quantum gravity. However, since the total available energy is bounded and small (relative to the initial black hole mass), this is possible only if the radiation is emitted over an extremely long period [2]. One then has, for all practical purposes, a remnant and the objections to alternative (II) are applicable. A moving mirror model in which coherence is restored only over a long time period has recently been discussed in [8].

(IV.) *None of the above.*

Further discussion of these and other possibilities can be found in [7,9].

2. Extremal Black Hole–Particle Scattering

Despite intense efforts, it has not so far been possible to determine the final outcome of black hole evaporation. One reason is that quantum gravity appears to be important at late stages in the evaporation process, and one must therefore confront the problem of non-renormalizability of quantum gravity. In principle string theory could come to the rescue, but in practice string technology is not sufficiently developed to handle this problem. A second obstacle is the many degrees of freedom involved in the formation/evaporation of a macroscopic black hole. All of these degrees of freedom must be carefully kept track of in order to determine whether or not information is lost – a daunting task.

In fact it is possible to reduce the problem to a much simpler problem. In this simpler problem both of these obstacles are circumvented, yet the key conceptual puzzles remain. The first simplification is to consider extremal black holes. A key feature of extremal black holes (including those dilatonic variants to be discussed later) is that the Hawking temperature vanishes. Thus they are typically quantum mechanically as well as classically stable objects.[1] An $M > Q$ black hole will tend to Hawking radiate down to its extremal $M = Q$ state. Thus the extremal black holes are quantum ground states of the charge Q superselection sector of the Hilbert space. This view is reinforced by consideration of $N = 2$ supersymmetric versions of the theory, for which it can be shown that they are exactly stable, supersymmetric, quantum ground states.

Next consider throwing a low energy, neutral particle into an extremal black hole. This will excite it into a non-extremal state where, classically, it will remain forever. Quantum mechanically, however, one expects it to decay via Hawking emission back to its extremal ground state. The hope is that the entire process can be analyzed in a perturbation expansion about the extremal ground state. From the preceding discussion it should be clear that the problem of extremal black hole–particle scattering contains all the same puzzles as the problem of black hole formation/evaporation. Yet it is clearly much simpler, as there are few possible final states of the system, relative to the many possible outcomes of macroscopic black hole evaporation.

To make real progress, however, we must boil the problem down even more. We can do this by considering only low-energy scattering of states with zero angular momentum. In general one cannot simply truncate a quantum field theory to the S-wave sector and hope to obtain reliable results. One needs a small parameter to justify the approximation. In the next section we shall see that the S-wave approximation is justified for extremal dilatonic black holes if a) the black hole is large in Planck units and b) the Compton wavelength of the incident particle is large relative to the black hole. The problem can then be systematically reduced to a $1 + 1$ dimensional quantum field theory living in the (r, t) plane of the black hole. Clearly the number of degrees of freedom have been drastically reduced.

It is also important to note that quantum gravity is renormalizable in two dimensions. Thus the problems of the nonrenormalizability of four-dimensional quantum gravity are apparently avoided. The reason behind this is that particle-hole scattering is essentially a problem in low-energy physics, and the laws of physics at high energies should not be relevant.[2]

---

[1] This is not true for extremal electrically charged black holes in a world – such as the one we live in – with $m < e$ charged particles, which will lead to ordinary Schwinger pair production in the electric field of the black hole. In our world, magnetically charged black holes are possibly examples of the type of quantum mechanically stable objects we wish to discuss.

[2] Although it remains a logical possibility that instabilities inevitably drive one into the high energy regime.

3. Black Holes in Dilaton Gravity

The low-energy limit of string theory with unbroken supersymmetry generically includes a massless scalar field ϕ – termed the dilaton – which couples to other fields in a specific way. In "realistic" scenarios with broken supersymmetry it is expected (or hoped) that the dilaton will acquire a mass which makes its existence consistent with the standard post-newtonian tests of general relativity. Nonetheless, the theories with a massless dilaton should be relevant for the study of stringy black holes small relative to the Compton wavelength of the dilaton, and more generally provide a useful model for investigating modifications of black hole structure. The effects of a dilaton mass have recently been studied in [10].

The simplest four-dimensional example has the dilaton and a $U(1)$ gauge theory coupled to gravity with action

$$S = \int d^4x\sqrt{-g}e^{-2\phi}(R + 4g^{\mu\nu}\nabla_\mu\phi\nabla_\nu\phi - \frac{1}{2}g^{\mu\lambda}g^{\nu\rho}F_{\mu\nu}F_{\lambda\rho}). \tag{3.1}$$

In the form (3.1) the Einstein term in the action does not have its canonical form and the dilaton kinetic energy term has the "wrong" sign. This can be remedied by performing a conformal transformation

$$\tilde{g}_{\mu\nu} = e^{-2\phi}g_{\mu\nu}. \tag{3.2}$$

This results in the expression

$$S = \int d^4x\sqrt{\tilde{g}}(\tilde{R} - 2\tilde{g}^{\mu\nu}\nabla_\mu\phi\nabla_\nu\phi - \frac{1}{2}e^{-2\phi}\tilde{g}^{\mu\lambda}\tilde{g}^{\nu\rho}F_{\mu\nu}F_{\lambda\rho}). \tag{3.3}$$

The metric $\tilde{g}_{\mu\nu}$ is referred to as the canonical metric (because the Einstein action has the canonical prefactor) while $g_{\mu\nu}$ is often called the sigma-model or string metric.

Two important points follow trivially from inspection of the canonical form (3.3) of the action. The first is that for $F_{\mu\nu} = 0$ the solutions reduce to solutions of Einstein gravity coupled to a massless scalar field. The no-hair theorems then imply that the unique black hole solution is the Kerr solution characterized by (M, J). The second is that since there is no potential for ϕ, we are free to choose an arbitrary constant value ϕ_0 for ϕ at infinity. From (3.3) we then see that $e^{2\phi_0} \equiv g_s^2$ plays the role of the electromagnetic coupling constant squared at infinity.

Black holes with non-zero charge Q have been investigated in this theory in [11] and [12] and are reviewed in [13]. The general spherically symmetric solution with $J=0$ and magnetic charge Q is given in terms of $\tilde{g}_{\mu\nu}$ by

$$\begin{aligned} d\tilde{s}^2 &= \tilde{g}_{\mu\nu}dx^\mu dx^\nu = -(1-\frac{2M}{r})dt^2 + \frac{dr^2}{(1-2M/r)} + r(r-\frac{Q^2}{2M}e^{-2\phi_0})d\Omega_{II}^2 \\ e^{-2\phi} &= e^{-2\phi_0}(1-\frac{Q^2}{2Mr}e^{-2\phi_0}), \\ F &= Q\sin\theta d\theta\wedge d\phi. \end{aligned} \tag{3.4}$$

Comparing this solution to the usual Reissner-Nordstrøm solution one notices several differences. First of all, this solution has only one horizon at $r=2M$ and not two as for Reissner-Nordstrøm. Second, the curvature singularity in these variables occurs at $r=\frac{Q^2}{2M}e^{-2\phi_0}$ where the area of the two-spheres goes to zero. Finally, the extremal solution occurs at $Q^2=4M^2e^{2\phi_0}$, rather than at $Q^2=M^2$ as for Reissner-Nordstrøm black holes.

Let us look in more detail at the extremal limit of the dilatonic black hole solution (3.4). As $Q^2\to 4M^2e^{2\phi_0}$ it is clear that both the metric components and $e^{-2\phi}$ are becoming singular at the horizon $r=2M$. However the physical implications of this are not immediately obvious. In order to decide whether physical quantities become singular in theories with scalar fields coupled to gravity we must specify not only the coupling to the metric but also the coupling to the scalar fields. In the example at hand the most natural point of view is to ask whether or not string propagation becomes singular in this limit. To study this we want to look at the string metric $g_{\mu\nu}$ in the extremal limit. For $Q^2\to 4M^2e^{2\phi_0}$ the string metric behaves as

$$ds^2 \to e^{2\phi_0}\left(-dt^2 + \frac{dr^2}{(1-\frac{2M}{r})^2} + r^2 d\Omega_{II}^2\right). \tag{3.5}$$

(r and t may be rescaled by a constant factor of e^{ϕ_0} to restore the canonical asymptotic behavior of the metric.) As $r\to\infty$ this approaches flat space with a constant dilaton ϕ_0. As $r\to 2M$ we can introduce a new coordinate σ with

$$d\sigma = \frac{dr}{1-2M/r} = \frac{rdr}{r-2M}, \tag{3.6}$$

so that as $r\to 2M$ we have $\sigma\to 2M\ln(r-2M)$. We then find that as $\sigma\to-\infty$ ($r\to 2M$)

$$\begin{aligned} ds^2 &\to e^{2\phi_0}(-dt^2 + d\sigma^2 + (2M)^2 d\Omega_{II}^2), \\ e^{-2\phi} &\to \frac{e^{-2\phi_0}}{2M}e^{\sigma/2M}. \end{aligned} \tag{3.7}$$

We thus see that $\sigma = -\infty$ is an infinite proper distance away and that as we approach this limit the geometry approaches that of $1+1$ dimensional Minkowski space (σ, t) times a two-sphere of fixed radius $R = 2Me^{\phi_0} = |Q|$ with a dilaton field ϕ linear in the spatial coordinate σ. The extremal solution has two asymptotic regions, one "out at infinity" which approaches flat space, and the other "down the throat".

Note that in the extremal limit, the geometry is completely non-singular, and there are no horizons. The black hole is replaced by a narrow "bottomless hole" in space. This non-singular nature of extremal dilatonic black holes, in contrast to their Reissner-Nordstrøm cousins, simplifies the study of particle-hole scattering.

It is clear that low-energy physics deep into the throat region is effectively two-dimensional, and is confined to the (σ, t) plane. Excitations in the transverse directions have energies at least of order $1/R$. The two-dimensional effective field theory can be derived by viewing the throat region as a compactification from four to two dimensions for which the "internal" space is a two-sphere (of constant (σ, t)) threaded with magnetic flux. Standard Kaluza-Klein technology then leads to the effective two-dimensional action

$$S = \frac{1}{2\pi}\int d^2x\sqrt{-g}e^{-2\phi}(R + 4(\nabla\phi)^2 + 4\lambda^2 - \frac{1}{2}F^2), \tag{3.8}$$

where the cosmological constant $\lambda^2 = 1/4Q^2$ is a relic of the "internal" components of the scalar curvature and field strength tangent to the compactification two-sphere. Two-dimensional gauge fields have no local dynamics, and can and will be consistently set to zero if no charged particles are present.

If a low-energy particle is thrown into the non-singular extremal black hole, it produces a singularity and an event horizon. This behavior is directly reflected in the two-dimensional action (3.8). As will be seen in the next section, a particle incident on the two-dimensional vacuum (which corresponds to the extremal black hole) classically produces a singularity and an event horizon.

Before descending to two dimensions there are several more useful entries in the dictionary relating four and two-dimensional quantities which are useful. In a spherically symmetric four-dimensional spacetime, the area of the two-spheres is given by a function $4\pi\tilde{R}^2(\sigma^+, \sigma^-)$ where σ^+, σ^- are null coordinates. The two-sphere at σ^+, σ^-, will be trapped if $\tilde{R}$ is decreasing in both null directions, i.e. $\partial_\pm\tilde{R} < 0$. The area of the two spheres of constant σ^+, σ^- in the throat region of a near-extremal geometry as measured in the canonical $\tilde{g}$ metric is $4\pi Q^2 e^{-4\phi}$. Therefore it is natural to define a *trapped point* in the two-dimensional theory as a point at which

$$\partial_\pm \phi > 0. \tag{3.9}$$

An *apparent horizon* is then the outer boundary of such a region at which $\partial_+\phi = 0$ [14] (since asymptotically $\partial_+\phi < 0$ while $\partial_-\phi > 0$.) The phrase *apparent black hole* will also be used to refer to a region of trapped points. This is distinct from a real black hole, which is a region from which it is impossible to escape to future null infinity.

4. Classical 1+1 Dilaton Gravity

4.1. Eternal Black Holes

In this subsection we will be discussing a $1+1$ dimensional theory of gravity coupled to a dilaton field ϕ with action

$$S_D = \frac{1}{2\pi}\int d^2x\sqrt{-g}e^{-2\phi}\left[R + 4(\nabla\phi)^2 + 4\lambda^2\right]. \tag{4.1}$$

This action can be viewed as the dimensionally reduced low-energy effective action for the extremal dilatonic black hole as described in the previous section. It is also of interest in its own right as a "toy" model of quantum gravity in two dimensions which, as we shall see, contains black holes and Hawking radiation. This model arises in two-dimensional non-critical string theory and as such its black hole solutions were first discovered in [15], [16] and [17]. Previous work on two-dimensional black holes which is closely related can be found in [18], and on models of two-dimensional gravity with scalars in [19,20,21].

The classical equations of motion which follow from (4.1) are

$$2e^{-2\phi}\left[\nabla_\mu\nabla_\nu\phi + g_{\mu\nu}((\nabla\phi)^2 - \nabla^2\phi - \lambda^2)\right] = 0, \tag{4.2}$$

$$e^{-2\phi}\left[R + 4\lambda^2 + 4\nabla^2\phi - 4(\nabla\phi)^2\right] = 0, \tag{4.3}$$

where the first equation results from variation of the metric and the second is the dilaton equation of motion. We first note that there is a solution (often called the linear dilaton vacuum) characterized by

$$R = \nabla^2\phi = 0, \qquad (\nabla\phi)^2 = \lambda^2. \tag{4.4}$$

We shall refer to this simply as the vacuum. We can introduce coordinates (σ, τ) so that

$$g_{\mu\nu} = \eta_{\mu\nu}, \qquad \phi = -\lambda\sigma, \tag{4.5}$$

in the vacuum. Note that the vacuum is not translationally invariant – a feature which also occurred in the "throat" region of the extremal dilaton black hole. As was the case there, the natural coupling constant in this theory is $g_s = e^{\phi}$. Thus the vacuum can be divided into a strong coupling region ($\sigma \to -\infty$) and a weak coupling region ($\sigma \to +\infty$). It is sometimes useful to think of the strength of the coupling as providing a coordinate invariant notion of one's location in this one-dimensional world.

To introduce the black hole solution of this theory it is useful to introduce light-cone coordinates (the relation of these coordinates to the previous ones will be discussed momentarily)

$$x^{\pm} = x^0 \pm x^1, \tag{4.6}$$

and to choose conformal gauge $g_{\mu\nu} = e^{2\rho}\eta_{\mu\nu}$, or in light-cone coordinates

$$g_{+-} = -\frac{1}{2}e^{2\rho}, \qquad g_{++} = g_{--} = 0. \tag{4.7}$$

We then have $R = 8e^{-2\rho}\partial_+\partial_-\rho$ and the equations of motion become

$$\begin{aligned} \phi: \qquad & e^{-2(\phi+\rho)}\left[-4\partial_+\partial_-\phi + 4\partial_+\phi\partial_-\phi + 2\partial_+\partial_-\rho + \lambda^2 e^{2\rho}\right] = 0, \\ \rho: \qquad & e^{-2\phi}\left[2\partial_+\partial_-\phi - 4\partial_+\phi\partial_-\phi - \lambda^2 e^{2\rho}\right] = 0. \end{aligned} \tag{4.8}$$

Note that these two equations imply

$$\partial_+\partial_-(\rho - \phi) = 0, \tag{4.9}$$

so that $(\rho - \phi)$ is a free field. Since we have gauge fixed g_{++} and g_{--} to zero we must also impose their equations of motion as constraints. This gives

$$\begin{aligned} e^{-2\phi}(4\partial_+\rho\partial_+\phi - 2\partial_+{}^2\phi) = 0, \\ e^{-2\phi}(4\partial_-\rho\partial_-\phi - 2\partial_-{}^2\phi) = 0. \end{aligned} \tag{4.10}$$

Now (4.9) implies that ρ and ϕ are equal up to the sum of a function purely of x^+, $f_+(x^+)$ and a function purely of x^-, $f_-(x^-)$. But a coordinate transformation $x^{\pm} \to \tilde{x}^{\pm}(x^{\pm})$ preserves the conformal gauge (4.7) and can be used to set $f_{\pm} = 0$. Thus we can choose

$\rho = \phi$ in analyzing the equations of motion. With this choice the remaining equations and constraints reduce to

$$\begin{aligned} \partial_-\partial_+(e^{-2\rho}) &= -\lambda^2, \\ {\partial_+}^2(e^{-2\rho}) &= {\partial_-}^2(e^{-2\rho}) = 0, \end{aligned} \tag{4.11}$$

which has the general solution (up to constant shifts of $x^{\pm}$)

$$e^{-2\phi} = e^{-2\rho} = \frac{M}{\lambda} - \lambda^2 x^+ x^-, \tag{4.12}$$

where M is a free parameter which will turn out to be the mass of the black hole.

Calculating the curvature we find

$$R = 8e^{-2\rho}\partial_+\partial_-\rho = \frac{4M\lambda}{M/\lambda - \lambda^2 x^+ x^-}, \tag{4.13}$$

which is divergent at $x^+x^- = M/\lambda^3$. This solution has the same qualitative features as the (r, t) plane of the Schwarzschild black hole.

To see the relation to the flat space vacuum we can introduce coordinates

$$\begin{aligned} \lambda x^+ &= e^{\lambda\sigma^+}, \\ \lambda x^- &= -e^{-\lambda\sigma^-}. \end{aligned} \tag{4.14}$$

It is important to remember that in these coordinates ρ will no longer equal ϕ since ϕ transforms as a scalar under coordinate transformation while ρ does not. In these coordinates we find that as $\sigma = (\sigma^+ - \sigma^-)/2 \to \infty$

$$\begin{aligned} \phi &\to -\lambda\sigma - \frac{M}{2\lambda}e^{-2\lambda\sigma}, \\ \rho &\to 0 - \frac{M}{2\lambda}e^{-2\lambda\sigma}, \end{aligned} \tag{4.15}$$

and the solution approaches the vacuum up to exponentially small corrections. It is also important to note that $g_s = e^{\phi} \to 0$ as $\sigma \to \infty$ and that at the horizon $x^- = 0$, $g_s = \sqrt{\lambda/M}$. Thus we are in weak coupling outside the horizon for sufficiently massive black holes $(M >> \lambda)$.

4.2. Coupling to Conformal Matter

So far all we have constructed is an "eternal" black hole solution. To determine whether such solutions form from non-singular initial conditions and to study Hawking radiation we must couple in some dynamical matter degrees of freedom. To study this process in our 1 + 1 dimensional model we modify (4.1) by adding a matter term of the form

$$S_M = -\frac{1}{4\pi}\sum_{i=1}^{N}\int d^2x\sqrt{-g}(\nabla f_i)^2, \tag{4.16}$$

where the f_i are a set of N massless matter fields. For the moment we take $N = 1$ and will consider general N when we discuss Hawking radiation and back reaction. In conformal gauge the f equation of motion is simply

$$\partial_+\partial_- f = 0. \tag{4.17}$$

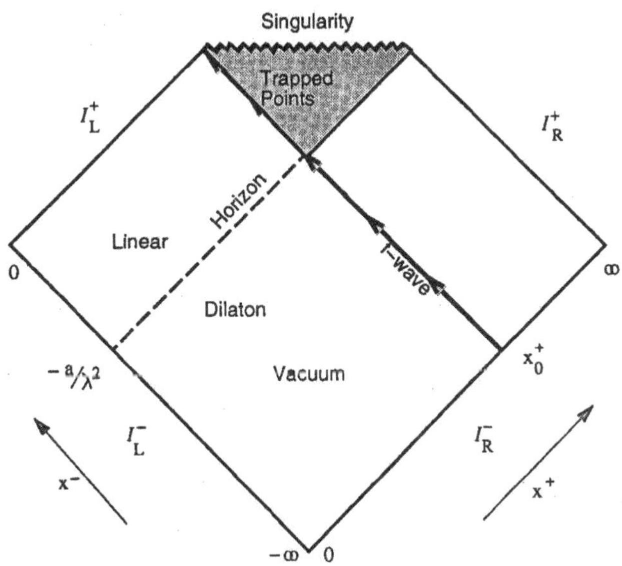

Fig. 1: Penrose diagram for formation of a black hole by an f shock-wave.

Let us consider sending in a pulse of energy from the right. Although we could consider taking f to be some function of x^+ with finite width, to simplify the calculation we take the f pulse to be a shock-wave traveling in the x^- direction with magnitude a described by the stress tensor

$$\frac{1}{2}\partial_+ f \partial_+ f = a\delta(x^+ - x_0^+)\ . \tag{4.18}$$

The only modification in the equations of motion and constraints due to the matter fields in this case is in the g_{++} constraint which becomes

$$e^{-2\phi}(4\partial_+\rho\partial_+\phi - 2\partial_+{}^2\phi) = -\frac{1}{2}\partial_+ f\partial_+ f. \tag{4.19}$$

For $x^+ < x_0^+$ we assume we are in the vacuum, while for $x^+ > x_0^+$ we know that the solution must be of the form (4.12). Matching the discontinuity across x_0^+ we obtain the solution

$$e^{-2\rho} = e^{-2\phi} = -a(x^+ - x_0^+)\Theta(x^+ - x_0^+) - \lambda^2 x^+ x^-. \tag{4.20}$$

For $x^+ > x_0^+$ this is identical to a black hole of mass $ax_0^+\lambda$ after shifting x^- by a/λ^2. The Penrose diagram for this spacetime is shown in fig. 1.

Both of these classical solutions have a straightforward interpretation from the four-dimensional viewpoint. The vacuum corresponds to an extremal $Q^2 = 1/(4\lambda^2)$ black hole, while the two-dimensional black hole solutions correspond to non-extremal four-dimensional black holes. The fact that an arbitrarily low-energy f-particle incident on the vacuum produces a two-dimensional black hole corresponds to the fact that an arbitrarily small particle thrown in to an extremal black hole turns it into a four-dimensional non-extremal black hole. Just as one expects this non-extremal black hole to revert to its extremal state via Hawking emission, one expects the two dimensional black hole to evaporate. Thus the four-dimensional process of particle-hole scattering corresponds to two-dimensional black hole formation/evaporation.

5. Hawking Radiation and the Trace Anomaly

So far we have achieved a satisfying description of the classical formation of a 1 + 1-dimensional black hole from collapsing matter. However the real motivation for studying this model is to understand quantum effects. We will do this in several parts. To begin with we will analyze the quantum effects of matter fields in the fixed classical background of a black hole formed by collapsing matter.

In two dimensions there is a beautiful relation between the trace anomaly and Hawking radiation discovered in [22]. For a massless scalar field the trace of the energy-momentum tensor is zero classically, $T \equiv T^{\mu}_{\mu} = 0$. Quantum mechanically there is a one-loop anomaly which relates the expectation value of the trace of the energy-momentum tensor to the Ricci scalar

$$\langle T \rangle = \frac{c}{24} R, \tag{5.1}$$

where $c = 1$ for a massless scalar and $c = 1/2$ for a Majorana fermion. In conformal gauge with $T = -4e^{-2\rho} T_{+-}$ this gives for N $c = 1$ scalars

$$\langle T^{f}_{+-} \rangle = -\frac{N}{12} \partial_+ \partial_- \rho. \tag{5.2}$$

Given the expectation value of T_{+-} as above we can use energy-momentum conservation to determine T_{++} and T_{--}. We have

$$\partial_+ T_{--} + \partial_- T_{+-} - \Gamma^{-}_{--} T_{+-} = 0, \tag{5.3}$$

and similarly for T_{++}. Using $\Gamma^{+}_{++} = 2\partial_+ \rho$, $\Gamma^{-}_{--} = 2\partial_- \rho$ the solution is found as

$$\begin{aligned} \langle T^{f}_{++} \rangle &= -\frac{N}{12} \left(\partial_+ \rho \partial_+ \rho - \partial_+^2 \rho + t_+(\sigma^+) \right) , \\ \langle T^{f}_{--} \rangle &= -\frac{N}{12} \left(\partial_- \rho \partial_- \rho - \partial_-^2 \rho + t_-(\sigma^-) \right) . \end{aligned} \tag{5.4}$$

The functions of integration $t_{\pm}$ are not determined purely by energy-momentum conservation and must be fixed by imposing physical boundary conditions. For the collapsing f-wave, T^f should vanish identically in the linear dilaton region, and there should be no incoming radiation along $\mathcal{I}^-_R$ except for the classical f-wave at σ_0^+.

We now turn to a calculation of Hawking radiation from a "physical" black hole formed by collapse of an infalling f shock-wave as in (4.18). The calculation and its physical interpretation is clearest in coordinates where the metric is asymptotically constant on $\mathcal{I}^{\pm}_R$. We thus set

$$\begin{aligned} e^{\lambda\sigma^+} &= \lambda x^+, \\ e^{-\lambda\sigma^-} &= -\lambda x^- - \frac{a}{\lambda}. \end{aligned} \tag{5.5}$$

This preserves the conformal gauge and gives for the new metric

$$-2g_{+-} = e^{2\rho} = \begin{cases} [1 + \frac{a}{\lambda} e^{\lambda\sigma^-}]^{-1}, & \text{if } \sigma^+ < \sigma_0^+; \\ [1 + \frac{a}{\lambda} e^{\lambda(\sigma^- - \sigma^+ + \sigma_0^+)}]^{-1} & \text{if } \sigma^+ > \sigma_0^+ \end{cases} \tag{5.6}$$

with $\lambda x_0^+ = e^{\lambda\sigma_0^+}$.

The formula for ρ, together with the boundary conditions on T^f at $\mathcal{I}^-_{L,R}$ then implies

$$t_+ = 0, \qquad t_- = \frac{-\lambda^2}{4}[1 - (1 + ae^{\lambda\sigma^-}/\lambda)^{-2}]. \tag{5.7}$$

The stress tensor is now completely determined, and one can read off its values on $\mathcal{I}^+_R$ by taking the limit $\sigma^+ \to \infty$:

$$\begin{aligned} &\langle T^f_{++}\rangle \to 0, \qquad \langle T^f_{+-}\rangle \to 0, \\ &\langle T^f_{--}\rangle \to \frac{N\lambda^2}{48}\left[1 - \frac{1}{\left(1 + ae^{\lambda\sigma^-}/\lambda\right)^2}\right] \quad . \end{aligned} \tag{5.8}$$

The limiting value of T^f_{--} is the flux of f-particle energy across $\mathcal{I}^+_R$. In the far past of $\mathcal{I}^+_R$ ($\sigma^- \to -\infty$) this flux vanishes exponentially while, as the horizon is approached, it approaches the constant value $N\lambda^2/48$. This is nothing but Hawking radiation. The surprising result that the Hawking radiation rate is asymptotically independent of mass has been found in other studies of two-dimensional gravity.

Although we have established that there is a net flux of energy which starts at zero and builds up to a constant value (ignoring backreaction) the skeptical reader might wonder whether this is in fact thermal Hawking radiation. There are two ways of arguing that the radiation is indeed thermal. The first involves a by now standard trick of rotating the black hole solution to Euclidean space. One then finds a non-singular solution only if the time coordinate is periodically identified. Once one does this, Green's functions constructed in this background will necessarily be thermal and it is easy to check from the periodicity that the temperature is $\lambda/2\pi$.

A more satisfying answer involves canonical quantization of the matter f fields in the black hole spacetime. Concentrating only on the right-moving modes which carry the Hawking radiation, one first considers the asymptotically flat "in" and "out" regions $\mathcal{I}^-_L$ and $\mathcal{I}^+_R$. For the "in" region one has a complete set of states for the mode expansion of the fields, but for the "out" region one must add a set of modes for the region internal to the black hole to obtain a complete set. At the end of the calculation these internal states are traced over since they are not observable. Although each asymptotic region has a natural timelike coordinate which allows one to define particles and anti-particles, the definitions do not agree so that positive frequency modes in one region will be a combination of positive and negative frequency modes in the other region. This has the interpretation of particle creation. In particular, the "in" vacuum corresponds to a thermal distribution of particles in the "out" region with temperature $\lambda/2\pi$. The details of this procedure in this model can be found in [23].

6. Including the Back-Reaction

If expression (5.8) is integrated along all of $\mathcal{I}_R^+$ to obtain the total energy emitted in Hawking radiation an infinite answer is obtained. This is obviously nonsense: the black hole can not radiate more energy than it owns.

The reason for this nonsensical result is simple: the backreaction of the Hawking radiation on the geometry has been neglected. While this should be unimportant at early times when the Hawking radiation is weak, ultimately it should be important enough to terminate the radiation process when the mass reaches zero.

6.1. The One Loop Action

The backreaction is easily included by simply letting the quantum stress tensor (5.2), (5.4) act as a source for the classical metric equations. For example the ρ equation (4.8) is modified to read

$$e^{-2\phi}(2\partial_+\partial_-\phi - 4\partial_+\phi\partial_-\phi - \lambda^2 e^{2\rho}) = \frac{N}{12}\partial_+\partial_-\rho, \tag{6.1}$$

while the constraint equations are modified by the addition of (5.4). These modified equations can be derived from the non-local action [24]

$$S_D - \frac{N}{96\pi}\int d^2x\sqrt{-g}R\Box^{-1}R, \tag{6.2}$$

where $\Box^{-1}$ is the scalar Greens function. Note that in conformal gauge $\Box^{-1}R = -2\rho$, so that (6.2) is local.

There is another, equivalent, method of deriving the extra term in (6.2). The quantum theory is defined by the functional integral in conformal gauge

$$Z = \int \mathcal{D}(b, c, \rho, \phi)\mathcal{D}f_i e^{i(S_D+S_{bc}+S_M)}, \tag{6.3}$$

where b and c are Fadeev-Popov ghosts arising from gauge fixing to conformal gauge, and S_{bc} is their action. In order to define the measures in Z one must introduce a short distance regulator. This should be done in a covariant manner, which implies that the measures will depend on ρ and so should be denoted e.g. $\mathcal{D}_\rho f_i$. This dependence of the measure on ρ is given by

$$\mathcal{D}_\rho f_i = \mathcal{D}_0 f_i e^{-\frac{iN}{12\pi}\int \partial_+\rho\partial_-\rho}, \tag{6.4}$$

where $\mathcal{D}_0$ is the measure with $\rho = 0$. The term in the exponent is precisely the conformal gauge version of the extra term in (6.2). Thus we see that this extra term arises from the metric dependence of the functional measure.

Of course, there is also metric dependence hidden in the (b, c, ρ, ϕ) measure. One might suspect that this simply leads to a shift of N in (6.2) but this is not necessarily the case [25]. To see why not, note that the functional measure is canonically derived from the metric governing small fluctuations of the fields. This is determined from the kinetic part of the action as [26,27] :

$$dS^2 = \frac{1}{4\pi}\int d^2\sigma\sqrt{-g}\left(\sum_{i=1}^{N}\delta_i f\delta_i f - 8e^{-2\phi}\delta\phi\delta\phi + 8e^{-2\phi}\delta\phi\delta\rho\right). \tag{6.5}$$

Thus in computing the ρ, ϕ (as well as b, c) measures it is natural to replace g by $e^{-2\phi}g$, or equivalently ρ by $\rho - \phi$. One then finds [28,25]

$$Z = \int \mathcal{D}_0(b, c, \rho, \phi, f_i)e^{iS_1}, \tag{6.6}$$

where

$$\begin{aligned} S_1 = \frac{1}{\pi}\int d^2\sigma\Big[&e^{-2\phi}(2\partial_+\partial_-\rho - 4\partial_+\phi\partial_-\phi + \lambda^2 e^{2\rho}) \\ &-\frac{N}{12}\partial_+\rho\partial_-\rho + \frac{1}{2}\sum_{i=1}^{N}\partial_+ f_i\partial_- f_i + 2\partial_+(\rho-\phi)\partial_-(\rho-\phi)\Big], \end{aligned} \tag{6.7}$$

and now all ρ dependence has been explicitly exhibited. The corresponding modifications of the constraints will be given below.

The difference in the ghost-gravity and matter measures has a physically sensible consequence: there is no Hawking radiation of ghosts or gravity modes for large black holes. This can be seen by going to a gauge in which $\rho - \phi$ vanishes to leading order. This is as expected because the dilaton and metric are non-dynamical, while ghosts are not real particles.

6.2. The Large N Approximation

The quantum theory defined by (6.6) is still rather complicated. The theory simplifies in the limit $N \to \infty$, with $Ne^{2\phi}$ held fixed [28]. The first three terms in (6.7) are then of order N, while the last ghost-gravity term is order N^0 and therefore (formally) negligible. Furthermore, since the entire action is large the stationary phase approximation is valid, so we need merely solve the semiclassical equations. The semiclassical ρ, ϕ equations can be cast in the form

$$8P\partial_+\partial_-\phi = -P'(4\partial_+\phi\partial_-\phi + \lambda^2 e^{2\rho}), \tag{6.8}$$

$$2P\partial_+\partial_-\rho = e^{-4\phi}(4\partial_+\phi\partial_-\phi + \lambda^2 e^{2\rho}), \tag{6.9}$$

where

$$P = e^{-4\phi} - \frac{N}{12}e^{-2\phi}, \tag{6.10}$$

$$P' = -4e^{-4\phi} + \frac{N}{6}e^{-2\phi}. \tag{6.11}$$

The $++$ constraint equation is

$$\begin{aligned} T_{++} = e^{-2\phi}(4\partial_+\phi\partial_+\rho - 2\partial_+^2\phi) + \frac{1}{2}\sum_{i=1}^{N}\partial_+ f_i\partial_+ f_i \\ -\frac{N}{12}(\partial_+\rho\partial_+\rho - \partial_+^2\rho) + t_+ = 0, \end{aligned} \tag{6.12}$$

while a similar equation holds for T_{--}.

An immediately obvious feature of (6.8) and (6.9) is [14,29] that the prefactor P on the left hand side vanishes at the critical value of the dilaton field:

$$\phi_{cr} = \frac{1}{2}\ell n\frac{12}{N}. \tag{6.13}$$

Unless the right hand sides of (6.8) and (6.9) vanish when ϕ reaches ϕ_{cr} the second derivatives of these fields will have to diverge. While the RHS of (6.8) and (6.9) do vanish for the vacuum, this will not be the case for perturbations of the vacuum, and singularities will occur.

These singularities can be viewed as a quantum version of the classical black hole singularities [14]. Consider a matter shock wave at x_0^+ as given by equation (4.18). Beneath the shock wave ($x^+ < x_0^+$), the geometry is the vacuum. The equations imply that ρ and ϕ, but not their first derivatives $\partial_+\rho$ and $\partial_+\phi$, are continuous across the shock wave. The geometry above the shock wave can then be perturbatively computed in a Taylor expansion about the shock wave. One finds that just above the shock wave [14,29]

$$\partial_+\phi(x_0^+, x^-) = \frac{1}{2x_0^+}\left(\frac{M/\lambda}{\sqrt{P(\phi(x_0^+, x^-))}} - 1\right), \tag{6.14}$$

where by continuity $\phi(x_0^+, x^-)$ is given by its vacuum value.

There are two notable features of this expression. The first is that $\partial_+\phi$ diverges when the shock wave crosses the timelike line in the vacuum where $\phi = \phi_{cr}$ (and P vanishes). Before diverging, however, it must cross zero at an earlier value x_H^- of x^-. This point marks the beginning of an apparent horizon. Behind this horizon and above the shock wave there is a region of trapped points, or an apparent black hole. The singularity at $\phi = \phi_{cr}$ is thus inside an apparent black hole.

In a region of trapped points lines of constant ϕ are spacelike. Therefore the singularity at $\phi = \phi_{cr}$ leaves the shock wave on a spacelike trajectory. It can also be seen [14] that the apparent horizon leaves the shock wave on a timelike trajectory, corresponding to the fact that the black hole is radiating and shrinking.

Unfortunately it is difficult to ascertain from the equations the final fate of the apparent black hole. It was initially believed [30] that the singularity and apparent horizon were both asymptotic to a null line (the event horizon) which meets $\mathcal{I}_R^+$ at future timelike infinity. This is similar to the "remnant" alternative (II). Precisely what does happen is a matter of controversy at the moment. See [31] for more detail.

It is important to stress that the large$-N$ approximation can not be trusted in regions where the fields themselves grow to be of order N. In particular the semiclassical equations must break down before the singularity is reached, and one cannot reliably conclude that a singularity does indeed exist. To probe the region near the "singularity" apparently requires a non-perturbative treatment of the quantum theory.

7. Beyond One Loop

While much was learned in the semiclassical analysis of the previous sections, it was also clear that the answers to some questions must await a more complete analysis of the quantum theory. Recent attempts in this direction can be found in [32,33,34,35,27,36,37]. In this section we will discuss some of the issues which arise in this regard.

The first point is that there is not a unique quantization of dilaton gravity. If the quantum theory is defined as an expansion in $e^{2\phi}$, there are new finite, renormalizable, counterterms at every order in perturbation theory. For example at nth order there is the term $e^{2(n-1)\phi}(\nabla\phi)^2$. While some constraints on these terms will be discussed, they are far from being completely fixed.

One elementary constraint is that the theory should have a stable ground state. In fact it is quite easy to destabilize the ground state in the process of adding terms to the action. General criteria for the existence of a positive energy theorem are discussed in [38].

Further properties of the quantum theory follow from the connection between two-dimensional gravity and conformal field theory [39]. This connection is best understood by quantizing the theory in conformal gauge:

$$\begin{aligned} g_{+-} &= -\frac{1}{2}e^{2\rho}, \\ g_{++} &= g_{--} = 0. \end{aligned} \tag{7.1}$$

This gauge leaves unfixed a group of residual diffeomorphisms for which

$$\begin{aligned} \delta g_{++} &= \nabla_+\zeta_+ = g_{+-}\partial_+\zeta^- = 0, \\ \partial g_{--} &= \nabla_-\zeta_- = g_{+-}\partial_-\zeta^+ = 0. \end{aligned} \tag{7.2}$$

These equations imply

$$\zeta^{\pm} = \zeta^{\pm}(\sigma^{\pm}), \tag{7.3}$$

and that the residual diffeomorphisms generate the conformal group. Correspondingly the moments of T_{++} and T_{--} generate Virasoro algebras.

Invariance of the quantum theory under the residual symmetry group can be insured, for example, by constructing a BRST charge Q which obeys $Q^2 = 0$ and identifying physical states as Q - cohomology classes.

At this point it should be clear that – although a slightly different set of words is being used – what is being constructed here is a $c = 26$ conformally invariant sigma model with ρ, ϕ and f_i as fields living in an $N + 2$ dimensional target space. If one demands that the matter fields f_i constitute a free $c = N$ conformal field theory, then the ρ, ϕ sigma model must be conformally invariant with $c = 26 - N$.

Letting $X^\mu = (\rho, \phi)$, the ρ, ϕ sigma model can be written in the form:

$$S = -\frac{1}{2\pi}\int d^2x\sqrt{-\hat{g}}[\mathcal{G}_{\mu\nu}\nabla X^\mu\nabla X^\nu + \frac{1}{2}\Phi\hat{R} + T], \tag{7.4}$$

$\hat{g}$ here is a fiducial metric and $\mathcal{G}$, Φ and T are functions of X^μ. The couplings $\mathcal{G}, \Phi$ and T are severely restricted by conformal invariance. Namely, the beta functions must vanish:

$$\begin{aligned} 0 = \beta^{\mathcal{G}}_{\mu\nu} &= \mathcal{R}_{\mu\nu} + 2\nabla_\mu\nabla_\nu\Phi + \cdots, \\ 0 = \beta^{\Phi} &= \frac{N-24}{3} - \frac{\mathcal{R}}{4} + (\nabla\Phi)^2 - \nabla^2\Phi + \cdots, \\ 0 = \beta^T &= \nabla^2 T - 2\nabla\Phi\cdot\nabla T + 8T + \cdots \end{aligned} \tag{7.5}$$

where $\mathcal{R}$ is the curvature of $\mathcal{G}$. These equations are indeed obeyed, to leading order, by the $\mathcal{G}, \Phi$ and T implicit in section eight. While conformal invariance severely constrains the quantum theory, there are still an infinite number of solutions. This may be viewed as an initial data problem in which initial data is specified as a function of ϕ at fixed ρ, and the beta function equations are then used to solve for $\mathcal{G}, \Phi$ and T at every value of ρ.

In order to correspond to the theory of dilaton gravity that we are interested in, the values of $\mathcal{G}$, Φ and T at weak coupling ($\phi \to -\infty$) should agree with those appearing in our one-loop expressions in section eight.

It is natural to ask whether $\mathcal{G}$, Φ and T can be specified in such a way so as to correspond to an exactly solvable conformal field theory. In [34,35,33] it was shown that if $\mathcal{G}$ and Φ are given exactly by their one-loop expressions, T can be adjusted at higher orders in such a way as to obtain a soluble Liouville-like conformal field theory. This construction manifestly gives a theory which agrees with dilaton gravity at weak coupling. Unfortunately, the required adjustments of T destabilize the vacuum [27].

Another approach to finding an underlying conformal field theory is described in the talk by Martellini in these proceedings [40]. It is also possible that making the model supersymmetric will lead to better control of the quantum corrections. For recent progress in this direction see [38].

8. Conclusion

The models described above have in some ways only strengthened the paradox first raised by Hawking. They allow much more detailed calculation of the process of Hawking evaporation from a black hole formed by collapse, and show no signs that the information is contained in the outgoing radiation. Of the various scenarios for resolving the puzzle of Hawking radiation, this probably gives most support to the remnant scenario. A clear resolution of the problem in these toy models will clearly require a more sophisticated analysis, but there is reason to hope that our improved understanding of two-dimensional gravity will eventually lead to such a resolution. We will still have to tackle the more challenging problem in four dimensions, but a solution even in these toy models would be an important milestone in the study of quantum gravity.

Acknowledgments

I would first of all like to acknowledge my collaborators C. Callan, S. Giddings, and A. Strominger for their contributions to the work described here. I would also like to thank the organizers of the Lincei meeting for their hospitality and the invitation to speak. This work was supported in part by NSF Grants PHY90-00386 and PYI Grant No. PHY-9196117.

References

[1] S. W. Hawking, *Phys. Rev.* **D14** (1976) 2460.

[2] Y. Aharonov, A. Casher and S. Nussinov, *Phys. Lett.* **191B** (1987) 51.

[3] T. Banks, M. O'Loughlin and A. Strominger, "Black Hole Remnants and the Information Puzzle," hep-th/9211030. ;
A. Strominger and S. P. Trivedi, "Information Consumption by Reissner-Nordstrom Black Holes," hep-th/9302080.

[4] D. Page, *Phys. Rev. Lett.* **44** (1980) 301.

[5] G. 't Hooft, *Nucl. Phys.* **B335** (1990) 138.

[6] A. Peet, L. Susskind and L. Thorlacius, Stanford preprint SU-ITP-92-16 (1992).

[7] R. M. Wald, "Black holes, singularities, and predictability," in Quantum Theory of Gravity, S. M. Christensen ed. (Adam Hilger, Bristol U.K. 1984).

[8] C. Holzhey, Princeton University Thesis, (1992), unpublished ;
F. Wilczek, "Quantum Purity at a Small Price: Easing a Black Hole Paradox," hep-th/9302096.

[9] S. Giddings, "Black Holes and Massive Remnants," *Phys. Rev.* **D46** (1992) 1347.

[10] R. Gregory and J. Harvey, "Black Holes with a Massive Dilaton," hep-th/9209070 ;
J. Horne and G. Horowitz, "Black Holes Coupled to a Massive Dilaton," hep-th/9210012.

[11] G. Gibbons and K. Maeda, *Nucl. Phys.* **B298** (1988) 741.

[12] D. Garfinkle, G. Horowitz, and A. Strominger, *Phys. Rev.* **D43** (1991) 3140; Erratum: *Phys. Rev.* **D45** (1992) 3888.

[13] G. Horowitz, in *String theory and quantum gravity '91*, Proceedings of the Trieste Spring School 1991, eds. J. Harvey, R. Iengo, K. Narain, S. Randjbar-Daemi and H. Verlinde, World Scientific (1991).

[14] J.G. Russo, L. Susskind, and L. Thorlacius, "Black hole evaporation in 1+1 dimensions," *Phys. Lett.* **B292** (1992) 13.

[15] E. Witten, "On string theory and black holes," *Phys. Rev* **D44** (1991) 314.

[16] G. Mandal, A Sengupta, and S. Wadia, "Classical solutions of two-dimensional string theory," *Mod. Phys. Lett.* **A6** (1991) 1685.

[17] S. Elitzur, A. Forge, and E. Rabinovici, *Nucl. Phys.* **B359** (1991) 581.

[18] R. B. Mann, "Lower dimensional gravity," in Proc. of 4th Canadian Conf. on General Relativity and Relativistic Astrophysics, Winnipeg Canada and references therein.

[19] R. Jackiw, "Liouville field theory: a two-dimensional model for gravity?," in *Quantum theory of gravity*, S. Christensen, ed. (Hilger, Bristol U.K. 1984); D. Cangemi and R. Jackiw "Gauge Invariant Formulations of Lineal Gravity" MIT preprint CTP-2085 (1992).

[20] C. Teitelboim " The hamiltonian structure of spacetime and its relation with the conformal anomaly," in *Quantum theory of gravity*, S. Christensen, ed. (Hilger, Bristol U.K. 1984)

[21] A. Chamseddine, *Nucl. Phys.* **B368** (1992) 98.

[22] S. M. Christensen and S. A. Fulling, "Trace anomalies and the Hawking effect," *Phys. Rev.* **D15** (1977) 2088.

[23] S.B. Giddings and W.M. Nelson, "Quantum emission from two-dimensional black holes," *Phys. Rev.* **D46** (1992) 2486.

[24] A. M. Polyakov, "Quantum geometry of bosonic strings," *Phys. Lett.* **103B** (1981) 207.

[25] A. Strominger, "Fadeev-Popov ghosts and 1+1 dimensional black hole evaporation," *Phys. Rev. D*, to appear, hep-th/9205028.

[26] H. Verlinde, "Black holes and strings in two dimensions," in the proceedings of the Sixth Marcel Grossman Meeting, World Scientific, Singapore (1992).

[27] S. Giddings and A. Strominger, "Quantum theories of dilaton gravity," preprint UCSBTH-92-28, hep-th/9207034.

[28] C.G. Callan, S.B. Giddings, J.A. Harvey, and A. Strominger, "Evanescent black holes," *Phys. Rev.* **D45** (1992) R1005.

[29] T. Banks, A. Dabholkar, M.R. Douglas, and M O'Loughlin, "Are horned particles the climax of Hawking evaporation?" *Phys. Rev.* **D45** (1992) 3607.

[30] L. Susskind and L. Thorlacious, "Hawking radiation and back-reaction," Stanford preprint SU-ITP-92-12, hep-th/9203054.

[31] S. W. Hawking and J. M. Stewart, "Naked and thunderbolt singularities in black hole evaporation," hep-th/9207105 ;
D. Lowe, "Semiclassical Approach to Black Hole Evaporation," hep-th/9209008.

[32] T. Banks and M. O'Loughlin, "Two-dimensional quantum gravity in Minkowski space," *Nucl. Phys.* **B362** (1991) 649.

[33] J. G. Russo and A. A. Tseytlin, "Scalar-Tensor Quantum Gravity in Two Dimensions," *Nucl. Phys.* **B382** (1992) 259.

[34] A. Bilal and C. Callan, "Liouville models of black hole evaporation," Princeton preprint PUPT-1320, hep-th/9205089.

[35] S.P. deAlwis, "Quantization of a theory of 2d dilaton gravity," *Phys. Lett.* **B289** (1992) 278 ;
"Black hole physics from Liouville theory," Boulder preprint COLO-HEP-284, hep-th/9206020 ;
Quantum Black Holes in Two Dimensions," *Phys. Rev.* **D46** (1992) 5429.

[36] K. Hamada, "Quantum theory of dilaton gravity in 1+1 dimensions," *Phys. Lett.* **B300** (1993) 322.

[37] A. Mikovic, "Exactly solvable models of 2d dilaton quantum gravity," *Phys. Lett.* **B291** (1992) 19.

[38] Y. Park and A. Strominger, "Supersymmetry and Positive Energy in Classical and Quantum Two-Dimensional Dilaton Gravity," hep-th/9210017.

[39] S. R. Das, S. Naik and S. R. Wadia, "Quantization of the Liouville mode and string theory," *Mod. Phys. Lett.* **A4** (1989) 1033 ;
J. Polchinski, "A two-dimensional model for quantum gravity," *Nucl. Phys.* **B324** (1989) 123 ;
T. Banks and J. Lykken, "String theory and two-dimensional quantum gravity," *Nucl. Phys.* **B331** (1990) 173.

[40] F. Belgiorno, A. S. Cattaneo, F. Fucito, and M. Martellini, "Quantum Models of Black Hole Evaporation," hep-th/9211136.

THE W ALGEBRA STRUCTURE OF $N = 2$ CP_n COSET MODELS

KATSUSHI ITO
Niels Bohr Institute, Blegdamsvej 17, DK-2100 Copenhagen Ø, Denmark

ABSTRACT

We discuss the $N = 2$ super W algebras from the hamiltonian reduction of affine Lie superalgebras $A(n|n-1)^{(1)}$ and $A(n|n)^{(1)}$. From the quantum hamiltonian reduction of $A(n|n-1)^{(1)}$ we get the free field realisation of $N = 2$ CP_n super coset models. In the case of the affine Lie superalgebras $A(n|n)^{(1)}$, the corresponding conformal field theories do not have $N = 2$ superconformal symmetry. However we show that these models are twisted $N = 2$ CP_n models and may be regarded as topological conformal field theories.

1. Introduction

The W-algebra symmetry[1] in two dimensions plays a fundamental role in various integrable systems such as generalized KdV hierarchies and Toda field theories. In conformal field theories the W algebras provide an important class of chiral algebras and have been studied extensively[2]. A systematic construction of the W-algebra associated with a simple Lie algebra is given by the hamiltonian reduction of the affine Lie algebra[3,2]. One of the recent developments in this direction is that superconformal symmetries and their W-extensions are also obtained by considering the hamiltonian reduction of a certain class of affine Lie superalgebras[4]. In particular, the W-extension[5-11] of the $N = 2$ superconformal symmetry is an interesting subject in view of its relation to topological field theories and the compactifications of superstrings.

In previous papers[5,6], the author has studied the quantum hamiltonian reduction of the affine Lie superalgebras $A(n|n-1)^{(1)}$ and the Feigin-Fuchs representations of the $N = 2$ super W algebras, which characterize the $N = 2$ super CP_n coset models constructed by Kazama and Suzuki[12] (see also ref. 7). Inami and Kanno[9] observed that the classical $N = 2$ super W algebra appears also in the $N = 2$ super KdV hierarchies associated with the affine Lie superalgebras $A(n|n)^{(1)}$.

The purpose of the present talk is to explain why these different Lie superalgebras correspond to the same $N = 2$ super W algebra. We shall discuss the classical and quantum hamiltonian reduction of the affine Lie superalgebras $A(n|n-1)^{(1)}$ and $A(n|n)^{(1)}$. We will show that the conformal field theories associated with the affine

Lie superalgebras $A(n|n)^{(1)}$, are the twisted $N = 2$ CP_n models.

2. Lie superalgebras and their hamiltonian reductions

2.1. Notations

Denote by $\mathbf{g}$ a basic classical Lie superalgebra of rank r [13]. $\mathbf{g} = \mathbf{h} \oplus (\oplus_{\alpha\in\Delta}\mathbf{g}_\alpha)$ is a root space decomposition of $\mathbf{g}$. $\mathbf{h}$ is the Cartan subalgebra. The root system Δ of $\mathbf{g}$ is $\Delta^0 \cup \Delta^1$, where Δ^0 (Δ^1) is the set of even (odd) roots. $\Delta_+ = \Delta_+^0 \cup \Delta_+^1$ is the set of positive roots of $\mathbf{g}$, where Δ_+^0 (Δ_+^1) is the set of even (odd) positive roots. $\mathbf{g}_0 = \mathbf{h} \oplus (\oplus_{\alpha\in\Delta^0}\mathbf{g}_\alpha)$ is an even subalgebra of $\mathbf{g}$. $\mathbf{g}_1 = \oplus_{\alpha\in\Delta^1}\mathbf{g}_\alpha$ is an odd subspace. ρ_0 (ρ_1) is half the sum of positive even (odd) roots and ρ is defined as $\rho_0 - \rho_1$. $h^\vee$ is the dual Coxeter number of $\mathbf{g}$. For an affine Lie superalgebra $\hat{\mathbf{g}}$ at level k associated with $\mathbf{g}$, we define a constant $\alpha_+ = \sqrt{k + h^\vee}$. In the following we shall discuss the Lie superalgebras $A(m|n)$ ($h^\vee = m - n$) in detail.

2.2. The Lie superalgebras $A(m|n)$

A Lie superalgebra $sl(n+1|m+1)$ may be represented by matrices

$$X = \left(\begin{array}{c|c} A & a \\ \hline b & B \end{array}\right), \tag{1}$$

satisfying $\mathrm{str}X = \mathrm{tr}A - \mathrm{tr}B = 0$, where A and B are $(n+1)\times(n+1)$ and $(m+1)\times(m+1)$ matrices with grassmann even elements and a and b are $(n+1)\times(m+1)$ and $(m+1)\times(n+1)$ matrices with grassmann odd elements. The commutation relation for two elements $X_i = \left(\begin{array}{c|c} A_i & a_i \\ \hline b_i & B_i \end{array}\right)$ $(i = 1, 2)$ is given by

$$[X_1, X_2] = \left(\begin{array}{c|c} [A_1, A_2] + a_1b_2 - a_2b_1 & a_1B_2 - a_2B_1 + A_1a_2 - A_2a_1 \\ \hline B_1b_2 - b_1B_2 + b_1A_2 - b_2A_1 & [B_1, B_2] + b_1a_2 - b_2a_1 \end{array}\right). \tag{2}$$

In the case of $n = m$, the identity matrix $\mathbf{1}_{2n+2}$ spans an ideal of $sl(n+1|n+1)$ and the Lie superalgebra $A(n|n)$ is defined as $sl(n+1|n+1)/ < \mathbf{1}_{2n+2} >$. Even subalgebras of $A(n|m)$ for $n \neq m$ ($n = m$) are $A_n \oplus A_m \oplus u(1)$ ($A_n \oplus A_n$). For $A(n|n)$ it is convenient to use a pseudo-representation[14]. Namely we take a representative of elements of $A(n|n)$ in (1) such that $\mathrm{tr}A = \mathrm{tr}B = 0$, and modify the commutation relation like

$$[\![X_1, X_2]\!] = [X_1, X_2] - \frac{1}{n+1}\mathrm{tr}(a_1b_2 - a_2b_1)\mathbf{1}_{2n+2}. \tag{3}$$

In contrast to the simple Lie algebras, there is a variety of choices of the simple root system of Lie superalgebras, which correspond to different Dynkin diagrams. In the case of $m = n - 1$ and n, we may take the simple roots as purely odd roots. For $A(n|n-1)$ they are given by

$$\alpha_{2i-1} = e_i - \delta_i, \quad \alpha_{2i} = \delta_i - e_{i+1}, \quad i = 1, \ldots, n, \tag{4}$$

where e_i $(i = 1, 2, \ldots, n+1)$ and δ_i $(i = 1, \ldots, m+1)$ are orthonormal bases with positive and negative metric. Similarly the simple roots of $A(n|n)$ are given by $\alpha_1, \ldots, \alpha_{2n}, \alpha_{2n+1} = e_{n+1} - \delta_{n+1}$. The even positive roots of $A(n|m)$ $(m = n, n-1)$ are $e_i - e_j$ $(1 \leq i < j \leq n+1)$ and $\delta_i - \delta_j$ $(1 \leq i < j \leq m+1)$. The odd positive roots are $e_i - \delta_j$ $(1 \leq i \leq j \leq m+1)$ and $\delta_i - e_j$ $(1 \leq i < j \leq n+1)$. In the matrix representation of the type (1), we have the fundamental representation of $\mathbf{g}$:

$$
\begin{aligned}
E_{e_i - e_j} &= E_{i,j+1}, \quad E_{\delta_i - \delta_j} = E_{n+1+i,n+2+j}, \\
E_{e_i - \delta_j} &= E_{i,n+1+j}, \quad E_{\delta_i - e_j} = E_{n+1+i,j},
\end{aligned} \tag{5}
$$

for positive roots. We define $E_{-\alpha} = {}^tE_\alpha$ for negative roots $-\alpha$ $(\alpha \in \Delta_+)$. The Cartan elements are defined as $\alpha \cdot H = [E_\alpha, E_{-\alpha}]$. Instead of using (5), we can take a different representation[5,6], in which the odd simple root structure is manifest:

$$
E_{e_i - e_j} = E_{2i-1,2j-1}, \quad E_{\delta_i - \delta_j} = E_{2i,2j}, \quad E_{e_i - \delta_j} = E_{2i-1,2j}, \quad E_{\delta_i - e_j} = E_{2i,2j-1}. \tag{6}
$$

Note that in these expressions, $A(n|n-1)$ and $A(n|n)$ have a structure similar to that of the simple Lie algebras A_{2n+1} and A_{2n}, respectively.

The superalgebra $A(n|m)$ has rank $n+m+1$, but the rank of $A(n|n)$ reduces to $2n$ due to the existence of the ideal. Moreover the root vectors are not linearly independent. In fact, from the relation $\sum_{i=1}^{n} [\![E_{\alpha_{2i+1}}, E_{-\alpha_{2i+1}}]\!] = 0$, we have

$$
\alpha_1 + \alpha_3 + \cdots + \alpha_{2n+1} = 0. \tag{7}
$$

For both Lie superalgebras $A(n|n-1)$ and $A(n|n)$ half the sum of positive roots ρ may be shown to be zero.

2.3. The classical hamiltonian reduction

Let us consider the hamiltonian reduction of the affine Lie superalgebras $\hat{\mathbf{g}} = A(n|n-1)^{(1)}$ and $A(n|n)^{(1)}$. On the dual space $\hat{\mathbf{g}}^*$ of $\hat{\mathbf{g}}$ one may introduce the hamiltonian structure which is generated by the coadjoint action or the gauge transformation:

$$
\delta_\Lambda J(z) = [\Lambda(z),\ J(z)] + \partial\Lambda(z). \tag{8}
$$

Let $J_\alpha(z)$ $(\alpha \in \Delta)$ and $H^i(z)$ $(i = 1, \ldots, r)$ be the currents in the canonical basis. We consider the constraint space $\mathcal{M}$ in $\hat{\mathbf{g}}^*$:

$$
J_{\alpha_i + \alpha_{i+1}}(z) = 1, \quad J_{\alpha_i + \cdots + \alpha_j}(z) = 0, \quad \text{for } |i-j| > 1. \tag{9}
$$

In the present case we treat a dynamical system with second class constraints[15] since the fermionic currents J_{α_i} must satisfy $J_{\alpha_i}(z) J_{\alpha_{i+1}}(w) \sim 1/(z-w)$. By introducing auxiliary fermionic fields one can convert these second class constraints into first class constraints[16]. In this extended phase space, one may use the ordinary gauge fixing procedure. The first class constrains generate the gauge symmetries in the extended phase space. One finds that these gauge symmetries are generated by the subalgebra

$\hat{\mathbf{n}}_-$, the affine extension of the nilpotent subalgebra $\mathbf{n}_-$ generated by the negative roots. The hamiltonian structure on the reduced phase space $\mathcal{M}/\hat{\mathbf{n}}_-$ is introduced by projecting the original gauge symmetries onto the reduced phase space[17]. This method is also known as Polyakov's soldering procedure[18]. One may take the Drinfeld-Sokolov type gauge for $A(n|n-1)^{(1)}$:

$$\begin{aligned} J_{DS}(z) &= \sum_{i=1}^{n+1} U_{n+2-i}(z)E_{n+1,i} + \sum_{i=1}^{n} V_{n+1-i}(z)E_{2n+1,n+1+i} \\ &\quad + \sum_{i=1}^{n}(G_{n+1-i}(z)E_{n+1,n+1+i} + \bar{G}_{n+1-i}(z)E_{2n+2,i}) + \Lambda^{A(n|n-1)} \end{aligned} \tag{10}$$

where

$$\Lambda^{A(n|n-1)} = \sum_{i=1}^{n} E_{i,i+1} + \sum_{i=1}^{n-1} E_{n+1+i,n+2+i} \tag{11}$$

and $V_1 = U_1$. For $A(n|n)^{(1)}$, we can choose a similar type of gauge fixing. It is obtained simply by replacing $\Lambda^{A(n|n-1)}$ by

$$\Lambda^{A(n|n)} = \sum_{i=1}^{n} E_{i,i+1} + \sum_{i=1}^{n} E_{n+1+i,n+2+i}. \tag{12}$$

3. The classical $N = 2$ Super W_3 Algebra

In this section we shall give a non-trivial example of the $N = 2$ super W-algebra from the classical hamiltonian reduction of the affine Lie superalgebra $A(2|1)^{(1)}$. Let us take the Drinfeld-Sokolov gauge

$$J_{DS}(z) = \left(\begin{array}{ccc|cc} 0 & 1 & 0 & 0 & 0 \\ 0 & 0 & 1 & 0 & 0 \\ U_3(z) & U_2(z) & U_1(z) & G_2(z) & G_1(z) \\ \hline 0 & 0 & 0 & 0 & 1 \\ \bar{G}_2(z) & \bar{G}_1(z) & 0 & V_2(z) & U_1(z) \end{array}\right). \tag{13}$$

The gauge transformation (8) has the gauge parameter

$$\Lambda(z) = \left(\begin{array}{ccc|cc} x_{11}(z) & x_{12}(z) & x_{13}(z) & \xi_{11}(z) & \xi_{12}(z) \\ x_{21}(z) & x_{22}(z) & x_{23}(z) & \xi_{21}(z) & \xi_{22}(z) \\ x_{31}(z) & x_{32}(z) & x_{33}(z) & \xi_{31}(z) & \xi_{32}(z) \\ \hline \eta_{11}(z) & \eta_{12}(z) & \eta_{13}(z) & y_{11}(z) & y_{12}(z) \\ \eta_{21}(z) & \eta_{22}(z) & \eta_{23}(z) & y_{21}(z) & y_{22}(z) \end{array}\right), \tag{14}$$

where the diagonal elements are parametrized as $x_{11} = \varepsilon_1 + 2\varepsilon$, $x_{22} = \varepsilon_2 - \varepsilon_1 + 2\varepsilon$, $x_{33} = -\varepsilon_2 + 2\varepsilon$ and $y_{11} = \varepsilon_3 + 3\varepsilon$ and $y_{22} = -\varepsilon_3 + 3\varepsilon$. By imposing the conditions that the gauge transformations preserve the Drinfeld-Sokolov gauge (13), one can reduce the number of independent gauge parameters and find that all gauge parameters are expressed in terms of $x_1 \equiv x_{13}$, x_{23}, y_{12}, ε, $\eta_1 \equiv \eta_{13}$, $\eta_2 \equiv \eta_{23}$, $\xi_1 \equiv \xi_{12}$ and $\xi_2 \equiv \xi_{22}$.

Note that we may regard the bosonic part of this system as the coupled one of $gl(3)$ and $gl(2)$ W algebras which share the same $u(1)$ current U_1. But if one requires the $N = 2$ superconformal symmetry which is generated by ξ_2 and η_2, one finds that the $N = 2$ supermultiplet of the $u(1)$ current U_1 is $(U_1, G_1, \bar{G}_1, U_2 - V_2)$. Hence $T \equiv U_2 - V_2$ becomes the energy-momentum tensor, and V_2 turns out to be a spin two field. Corresponding to this change of physical variables, we redefine the gauge parameters as $x = x_{23} + y_{12}$ and $y = y_{12}$ instead of x_{23} and y_{12}. One may add suitable differential polynomials of fields in order to get well defined primary fields. This can be done by replacing a gauge parameter ε by

$$\varepsilon + \frac{1}{6}\{U_1 x + 3\partial x - 3U_1 y - 3\partial y + 2U_2 x_1 - 2\partial(U_1 x_1) - 2\partial^2 x_1 + 3G_1\eta_1 - 2\bar{G}_1\xi_1\}. \quad (15)$$

After this change of variables, one gets the gauge transformations on the reduced phase space of the form:

$$\delta X_i = \mathcal{D}_{ij} Y_j, \quad (16)$$

where $X =^t (U_1, T, V_2, U_3, G_1, G_2, \bar{G}_1, \bar{G}_2)$, $Y =^t (\varepsilon, x, y, x_1, \xi_1, \xi_2, \eta_1, \eta_2)$ and $\mathcal{D}$ is a 8×8 matrix valued differential operator which satisfies $\mathcal{D}_{ji} = \mathcal{D}^*_{ij}$, where for a differential operator $\sum_i a_i(z)\partial^i$, a formal adjoint $*$ is defined as $(\sum_i a_i(z)\partial^i)^* = \sum_i (-\partial)^i a_i(z)$.

Let us introduce the Poisson bracket structure by expressing δ_Λ as

$$\begin{aligned}
\delta_\Lambda &= \int \frac{dz}{2\pi i} \mathrm{str}\{ \left(\begin{array}{ccc|cc} 2\varepsilon & 0 & x_1 & 0 & \xi_1 \\ 0 & 2\varepsilon & x - y & 0 & \xi_2 \\ 0 & 0 & 2\varepsilon & 0 & 0 \\ \hline 0 & 0 & \eta_1 & 3\varepsilon & y \\ 0 & 0 & \eta_2 & 0 & 3\varepsilon \end{array}\right) \left(\begin{array}{ccc|cc} 0 & 0 & 0 & 0 & 0 \\ 0 & 0 & 0 & 0 & 0 \\ U_3 & T + V_2 & U_1 & G_2 & G_1 \\ \hline 0 & 0 & 0 & 0 & 0 \\ \bar{G}_2 & \bar{G}_1 & 0 & V_2 & U_1 \end{array}\right) \} \\
&= \int \frac{dz}{2\pi i} (U_3 x_1 + Tx - yV_2 - \varepsilon U_1 + \xi_1 \bar{G}_2 + \xi_2 \bar{G}_1 - \eta_1 G_2 - \eta_2 G_1). \quad (17)
\end{aligned}$$

From Eqs. (16) and (17), we get the Poisson bracket structure on the reduced phase space in the form of operator product expansions. First we write down the $N = 2$ superconformal algebra:

$$\begin{aligned}
T(z)T(w) &\sim \frac{T(z) + T(w)}{(z-w)^2}, \quad T(z)U_1(w) \sim \frac{6}{(z-w)^3} + \frac{U_1(z)}{(z-w)^2}, \\
\bar{G}_1(z)T(w) &\sim \frac{\bar{G}_1(z) + \bar{G}_1(w)}{(z-w)^2}, \quad G_1(z)T(w) \sim \frac{G_1(w)}{(z-w)^2}, \\
U_1(z)U_1(w) &\sim \frac{-6}{(z-w)^2}, \\
U_1(z)G_1(w) &\sim \frac{G_1(w)}{(z-w)}, \quad U_1(z)\bar{G}_1(w) \sim \frac{-\bar{G}_1(w)}{(z-w)},
\end{aligned}$$

$$G_1(z)\bar{G}_1(w) \sim \frac{-6}{(z-w)^3} + \frac{U_1(w)}{(z-w)^2} + \frac{T(w)}{(z-w)}. \tag{18}$$

Note that in this expression $T(z)$ is the twisted $N = 2$ energy-momentum tensor. Second the $N = 2$ supermultiplet structures for $(V_2, G_2, \bar{G}_2, U_3)$ are given by

$$\begin{aligned}
T(z)V_2(w) &\sim \frac{6}{(z-w)^4} + \frac{-U_1(w)}{(z-w)^3} + \frac{V_2(w)+V_2(z)}{(z-w)^2}, \\
T(z)U_3(w) &\sim \frac{U_3(z)+2U_3(w)}{(z-w)^2}, \\
T(z)\bar{G}_2(w) &\sim \frac{2\bar{G}_2(w)+\bar{G}_2(z)}{(z-w)^2}, \quad T(z)G_2(w) \sim \frac{-2G_1(w)}{(z-w)^3} + \frac{G_2(w)+G_2(z)}{(z-w)^2}, \\
U_1(z)V_2(w) &\sim \frac{-6}{(z-w)^3} + \frac{3U_1(w)}{(z-w)^2}, \\
U_1(z)U_3(w) &\sim \frac{-12}{(z-w)^4} + \frac{4U_1(w)}{(z-w)^3} + \frac{2(T+V_2)(w)}{(z-w)^2}, \\
U_1(z)G_2(w) &\sim \frac{3G_1(w)}{(z-w)^2} + \frac{G_2(w)}{(z-w)}, \quad U_1(z)\bar{G}_2(w) \sim \frac{2\bar{G}_1(w)}{(z-w)^2} + \frac{-\bar{G}_2(w)}{(z-w)}, \\
\bar{G}_1(z)V_2(w) &\sim \frac{\bar{G}_1(w)}{(z-w)^2} + \frac{\bar{G}_2(w)}{(z-w)}, \quad G_1(z)V_2(w) \sim \frac{-G_2(w)}{(z-w)}, \\
\bar{G}_1(z)U_3(w) &\sim \frac{2\bar{G}_2(w)+\bar{G}_2(z)}{(z-w)^2}, \quad G_1(z)U_3(w) \sim \frac{4G_1(w)}{(z-w)^3} + \frac{2G_2(w)}{(z-w)^2}, \\
\bar{G}_2(z)G_1(w) &\sim \frac{-12}{(z-w)^4} + \frac{-4U_1(z)}{(z-w)^3} + \frac{2V_2(z)}{(z-w)^2} + \frac{-U_3(w)}{(z-w)}, \\
\bar{G}_1(z)G_2(w) &\sim \frac{6}{(z-w)^4} + \frac{-2U_1(w)}{(z-w)^3} + \frac{V_2(w)+V_2(z)+T(w)}{(z-w)^2} + \frac{-U_3(w)}{(z-w)}.
\end{aligned} \tag{19}$$

Finally the remaining nontrivial operator product expansions take the forms:

$$\begin{aligned}
V_2(z)V_2(w) &\sim \frac{-12}{(z-w)^4} + \frac{4[U_1(w)-U_1(z)]}{(z-w)^3} + \frac{V_2(z)+V_2(w)}{(z-w)^2} + \frac{2U_1(z)U_1(w)}{(z-w)}, \\
U_3(z)V_2(w) &\sim \frac{-24}{(z-w)^5} + \frac{6[U_1(w)-U_1(z)]}{(z-w)^4} + \frac{2[U_2(z)+U_1(z)U_1(w)]}{(z-w)^3} \\
&\quad + \frac{-U_1(w)U_2(z)+\bar{G}_1(w)G_1(z)}{(z-w)^2} + \frac{[\bar{G}_2G_1-\bar{G}_1G_2](w)}{(z-w)}, \\
U_3(z)U_3(w) &\sim \frac{2[U_3(z)-U_3(w)]}{(z-w)^3} \\
&\quad + \frac{-U_3(w)U_1(z)-U_1(w)U_3(z)+\bar{G}_2(w)G_1(z)-G_1(w)\bar{G}_2(z)}{(z-w)^2}, \\
\bar{G}_2(z)V_2(w) &\sim \frac{2[\bar{G}_1(z)-\bar{G}_1(w)]}{(z-w)^3}
\end{aligned}$$

$$+\frac{-\bar{G}_2(w)-\bar{G}_1(w)U_1(z)-U_1(w)G_1(z)}{(z-w)^2}+\frac{[\bar{G}_1V_2-\bar{G}_2U_1](w)}{(z-w)},$$

$$G_2(z)V_2(w)\sim\frac{4G_1(z)}{(z-w)^3}+\frac{-2U_1(w)G_1(z)-G_2(z)}{(z-w)^2}+\frac{[U_1G_2-V_2G_1](w)}{(z-w)},$$

$$\bar{G}_2(z)U_3(w)\sim\frac{2[\bar{G}_2(z)-\bar{G}_2(w)]}{(z-w)^3}-\frac{U_1(w)\bar{G}_2(z)+\bar{G}_2(w)U_1(z)}{(z-w)^2}+\frac{[U_3\bar{G}_1-T\bar{G}_2](w)}{(z-w)},$$

$$G_2(z)U_3(w)\sim\frac{6[G_1(z)-G_1(w)]}{(z-w)^4}+\frac{-2G_2(w)-2G_1(w)U_1(z)-2U_1(w)G_1(z)}{(z-w)^3}$$
$$+\frac{-G_2(w)U_1(z)-G_1(w)T(z)-U_2(w)G_1(z)}{(z-w)^2}+\frac{-[G_2T+G_1U_3](w)}{(z-w)},$$

$$\bar{G}_2(z)G_2(w)\sim\frac{-24}{(z-w)^5}+\frac{6[U_1(w)-U_1(z)]}{(z-w)^4}+\frac{2[U_1(w)U_1(z)+V_2(z)+T(w)]}{(z-w)^3}$$
$$+\frac{U_3(w)+T(w)U_1(z)-U_1(w)V_2(z)+G_1(w)\bar{G}_1(z)}{(z-w)^2}+\frac{[U_3U_1-V_2T](w)}{(z-w)},$$

$$G_2(z)G_2(w)\sim\frac{[2G_1\partial G_1+2G_2G_1](w)}{(z-w)},\quad \bar{G}_2(z)\bar{G}_2(w)\sim\frac{2\bar{G}_1\bar{G}_2(w)}{(z-w)}. \tag{20}$$

The present Poisson bracket structure is the same as that obtained from the super Gel'fand-Dickii algebra[9,10] but different from the results in ref. 8 due to the different choice of the gauges.

4. The quantum hamiltonian reduction and $N=2$ CP_n coset models

So far we have discussed the classical hamiltonian reduction. In the quantum case, we use the BRST gauge fixing procedure by introducing ghost systems for the constraints[3,2]. In order to impose the constraints at the quantum level, we must improve the energy-momentum tensor T_{WZNW} of the Wess-Zumino-Novikov-Witten model corresponding to the affine Lie superalgebra $\hat{\mathbf{g}}$ by the Cartan currents $H^i(z)$,

$$T_{improved}(z)=T_{WZNW}(z)+\mu\cdot\partial H(z). \tag{21}$$

After this improvement, the conformal dimensions of the currents $J_\alpha(z)$ become $1-\mu\cdot\alpha$.

First we discuss the $A(n|n-1)^{(1)}$ case. The improvement vector $\mu=\mu_{A(n|n-1)}$ is defined by the conditions:

$$\alpha_i\cdot\mu_{A(n|n-1)}=\frac{1}{2},\quad i=1,\ldots,2n. \tag{22}$$

After this improvement, the conformal dimensions of the currents for the odd simple roots $J_{\alpha_i}(z)$ becomes $\frac{1}{2}$ and zero for the the currents for the simple roots $\alpha_i+\alpha_{i+1}$ of the even subalgebras. Hence we may introduce the "diagonal" gauge:

$$J_{diag}(z)=\partial\varphi(z)\cdot H+\sum_{i=1}^{n}(\alpha_{2i-1}\cdot\chi(z)E_{\alpha_{2i-1}}+\alpha_{2i}\cdot\chi(z)E_{\alpha_{2i}})+\Lambda^{A(n|n-1)}, \tag{23}$$

where $\chi^i(z)$ $(i=1,\ldots,2n)$ are $2n$ real fermions, satisfying $\chi^i(z)\chi^j(w) \sim \delta_{ij}/(z-w)$. The improvement vector $\mu_{A(n|n-1)}$ is uniquely determined by (22) and is expressed as

$$\mu_{A(n|n-1)} = \frac{1}{2}\sum_{i=1}^{n}[(n+1-i)\alpha_{2i-1} + i\alpha_{2i}] = \frac{1}{2}\sum_{i=1}^{2n}\lambda_i, \tag{24}$$

where λ_i are fundamental weights of $A(n|n-1)$ satisfying $\lambda_i \cdot \alpha_j = \delta_{ij}$:

$$\lambda_{2i} = \alpha_1 + \alpha_3 + \cdots + \alpha_{2i-1}, \quad \lambda_{2i-1} = \alpha_{2i} + \alpha_{2i+2} + \cdots + \alpha_{2n}, \tag{25}$$

for $i = 1,\ldots,n$. By using the Wakimoto realization of affine Lie superalgebras, one finds that the total energy-momentum tensor $T_{total} = T_{improved} + T_{ghosts} + T_\chi$ is BRST-equivalent to that of the $N=2$ CP_n coset model[5,7]:

$$T_{total}(z) = T_{CP_n}(z) + \{Q_{BRST}, *\}, \tag{26}$$

where $T_\chi = -\frac{1}{2}\sum_{i=1}^{2n}\chi^i\partial\chi^i$ and

$$T_{CP_n}(z) = -\frac{1}{2}(\partial\varphi)^2 - \mathrm{i}\alpha_+\mu_{A(n|n-1)}\cdot\partial^2\varphi + T_\chi. \tag{27}$$

The remaining $N=2$ generators are expressed as

$$\begin{aligned} U_1(z) &= \sum_{i=1}^{n}\lambda_{2i}\cdot\chi\alpha_{2i}\cdot\chi + \mathrm{i}\alpha_+\nu\cdot\partial\varphi, \\ G_1(z) &= \sum_{j=1}^{n}(\mathrm{i}\alpha_{2j}\cdot\partial\varphi\lambda_{2j}\cdot\chi - \alpha_+\lambda_{2j}\cdot\partial\chi), \\ \bar{G}_1(z) &= \sum_{j=1}^{n}(\mathrm{i}\alpha_{2j-1}\cdot\partial\varphi\lambda_{2j-1}\cdot\chi - \alpha_+\lambda_{2j-1}\cdot\partial\chi), \end{aligned} \tag{28}$$

where $\nu = \sum_{i=1}^{n}(\lambda_{2i} - \lambda_{2i-1})$. The other W currents can be obtained by the quantum Miura transformation, which connects the diagonal gauge (23) with the Drinfeld-Sokolov gauge (10). The free field representation of the $N=2$ CP_n models has been studied in ref. 6 in detail.

Now we proceed to the affine Lie superalgebra $A(n|n)^{(1)}$. In this case one cannot impose the spin $\frac{1}{2}$ constraints (22) for the simple roots α_i $(i=1,\ldots,2n+1)$ due to the relation (7). Instead we should require the conditions:

$$\alpha_{2i-1}\cdot\mu = 0, (i=1,\ldots,n+1), \quad \alpha_{2j}\cdot\mu = 1, (j=1,\ldots,n), \tag{29}$$

that are consistent with (7). This means that the conformal weights are 1 for $J_{\alpha_{2i-1}}$ but 0 for $J_{\alpha_{2i}}$ after the improvement. The vector μ is determined uniquely up to $\sum_{i=1}^{n+1}\alpha_{2i-1}(=0)$:

$$\mu = \mu_{A(n|n)} = \sum_{i=1}^{n+1}(n+1-i)\alpha_{2i-1}. \tag{30}$$

The diagonal gauge for $A(n|n)^{(1)}$ is

$$J_{diag}(z) = \partial\varphi(z)\cdot H + \sum_{i=1}^{n+1}\eta_{2i-1}(z)E_{\alpha_{2i-1}} + \sum_{i=1}^{n}\xi_{2i}(z)E_{\alpha_{2i}} + \Lambda^{A(n|n)}, \tag{31}$$

where $\eta_{2i-1} = \eta_{2i} - \eta_{2i-2}$ ($\eta_{2n+2} = \eta_0 \equiv 0$) and (η_{2i}, ξ_{2i}) are fermionic ghosts with conformal weight $(1,0)$. The energy-momentum tensor of the reduced theory becomes

$$\tilde{T}_{CP_n}(z) = -\frac{1}{2}(\partial\varphi)^2 - i\alpha_+\mu_{A(n|n)}\cdot\partial^2\varphi - \sum_{i=1}^{n}\eta_{2i}\partial\xi_{2i}, \tag{32}$$

which is equal to $T_{CP_n} + \frac{1}{2}\partial U_1$. This is nothing but the energy-momentum tensor of the twisted[19] $N=2$ CP_n model. Hence the conformal field theory corresponding to the affine Lie superalgebra $A(n|n)^{(1)}$ should be regarded as a topological conformal field theory rather than an $N=2$ superconformal field theory. This implies that the $A(n|n)$ Toda field theory has the *twisted* $N=2$ super W-algebra symmetry instead of the $N=2$ superconformal symmetry. Recently Evans and Hollowood[20] also pointed out that the $A(n|n)$ Toda field theory does not have $N=2$ superconformal symmetry.

In this sense the $A(n|n)^{(1)}$ affine Toda field theory can be also regarded as a topological field theory rather than an $N=2$ theory. It is a quite interesting problem to study this affine Toda field theory as a topological field theory since this model gives a different class of topological solvable models which are not classified by the $N=2$ Landau-Ginzburg type models.

References

1. A.B. Zamolodchikov, *Theor. Math. Phys.* **63** (1985) 1205.
2. For review, see P. Bouwknegt and K. Schoutens, preprint CERN-TH.6538/92, ITP-SB-92-23 July 1992 and references therein.
3. M. Bershadsky and H. Ooguri, *Commun. Math. Phys.* **126** (1989) 719,
   J. Balog, L. Fehér, L. O'Raifeartaigh, P. Forgács and A. Wipf, *Ann. Phys.* **203** (1990) 76.
4. J. Evans and T. Hollowood, *Nucl. Phys.* **B352** (1991) 723;
   S. Komata, K. Mohri and H. Nohara, *Nucl. Phys.* **B359** (1991) 168;
   F. Delduc, E. Ragoucy and P. Sorba, *Commun. Math. Phys.* **146** (1992) 403;
   L.A. Ferreira, J.F. Gomes, R.M. Ricotta and A.H. Zimmerman, preprint IFT-P-45/91;
   K. Ito, J.O. Madsen and J.L. Petersen, NBI preprint NBI-HE-92-42, to be published in Nucl. Phys. **B**.
5. K. Ito, *Phys. Lett.* **B259** (1991) 73.
6. K. Ito, *Nucl. Phys.* **B370** (1992) 123.
7. D. Nemeschansky and S. Yankielowicz, preprint USC-91/005.

8. H. Lu, C.N. Pope, L.J. Romans, X. Shen and X.-J. Wang, *Phys. Lett.* **B264** (1991) 91.
9. T. Inami and H. Kanno, *J. Phys.* **A 25** (1992) 3729; in *Infinite Analysis*, eds. A. Tsuchiya, T. Eguchi and M. Jimbo (World Scientific, Singapore, 1992).
10. J.M. Figueroa-O'Farrill and E. Ramos, *Nucl. Phys.* **B368** (1992) 361; K. Hiutu and D. Nemeschansky, *Mod. Phys. Lett.* **A 6** (1991) 3179.
11. L.J. Romans, *Nucl. Phys.* **B369** (1991) 403.
12. Y. Kazama and H. Suzuki, *Nucl. Phys.* **B321** (1989) 232.
13. V.G. Kac, *Adv. Math.* **26** (1977) 8.
14. B. DeWitt, *Supermanifolds* (Cambridge Univ. Press, Cambridge, 1984), p. 185.
15. M. Bershadsky and H. Ooguri, *Phys. Lett.* **B229** (1989) 374.
16. I.A. Batalin and E.S. Fradkin, *Phys. Lett.* **B180** (1986) 157.
17. V.G. Drinfeld and V.V. Sokolov, *Sov. J. Math* **30** (1985) 1975.
18. A.M. Polyakov, *Int. J. Mod. Phys.* **A5** (1990) 833.
19. T. Eguchi and S.-K. Yang, *Mod. Phys. Lett.* **A5** (1990) 1693.
20. J. Evans and T. Hollowood, Oxford preprint OUTP-92-12P.

EXTENDED SUPERCONFORMAL ALGEBRAS FROM CLASSICAL AND QUANTUM HAMILTONIAN REDUCTION

KATSUSHI ITO, JENS OLE MADSEN

and

JENS LYNG PETERSEN

Niels Bohr Institute, University of Copenhagen, Blegdamsvej 17, DK 2100 Copenhagen Ø, Denmark

ABSTRACT

The classification of extended superconformal algebras of the Knizhnik-Bershadsky type with W-algebra like composite operators occurring in the commutation relations, but with generators only of conformal dimension $1, \frac{3}{2}$ and 2, has recently been reconsidered by various authors from various points of view. We argue that a particularly natural classification seems to arise on the basis of hamiltonian reduction of affine Lie superalgebras with even subalgebras $G \oplus sl(2)$. Based on generic formulas for the Poisson bracket structure of the classical Gel'fand-Dickey algebra, we introduce similar generic formulations for all these algebras at the quantum level. Similarly, we rewrite the free field (Feigin-Fuchs) representations of all these algebras by using the BRST formalism and the free field realization of the affine Lie superalgebra. Again we emphasize the unifying aspects partly based on previously known results, partly on our own work when everything is formulated in the language of reduced affine Lie superalgebras. We also discuss the screening operators of these algebras. Although completely explicit results for all of those are known only in a limited number of cases, enough may be said about the general structure, to in fact provide what appears to be the complete pattern of singular vectors in the free field realisation.

1. Introduction

The role of extended superconformal algebras in formulations of such related subjects as string theory, topological field theory and conformal field theory has been remarkable. This has been true both as far as the world sheet supersymmetry and the target space geometry is concerned. It seems not unlikely that new and unforeseen applications will turn up. Thus it appears to be of general interest to have these extended superconformal algebras reasonably classified.

One must distinguish several different ways in which the supersymmetries have been implemented. The first way studied considered only what may be termed extended superconformal *Lie* (or perhaps, super-Lie) algebras: those for which commutators (anticommutators) of generators are themselves members of the algebra, i.e., the expression for the commutators involve only a *linear* combination of the generators themselves. This class of algebras appears to have been adequately classified for some time. For these algebras it seems reasonable to restrict the number, N of supersymmetries to be less than 4. For example, the linear $o(N)$ supersymmetric extension of the Virasoro algebra, which was introduced by Ademollo et al.,[1] was realized as the superconformal transformation on extended superspace with internal $o(N)$ symmetry. However, this formulation leads to negative conformal weight generators and the absence of a central extension unless $N \leq 4$,[1,2]

In a slightly different approach one regards linear $N = 2,4$ extended superconformal algebras as a symmetry of supersymmetric non-linear sigma models on group manifolds or on coset spaces such as hermitian symmetric spaces[3] or Wolf spaces.[4]

Knizhnik and Bershadsky,[5] however, have introduced the *non-linear* extension of superconformal algebras with $u(N)$ and $so(N)$ affine symmetries. The precise form of the algebra follows from closure and associativity of the operator product expansions (the OPE method). This non-linearity means that these extended superconformal algebras are closely related to Zamolodchikov's W-algebras.[6] W-algebras are known to play a very useful role in the classification of chiral algebras in conformal field theories, so this non-linear extension deserves further study. The representation theory has been studied in the case of $so(N)$-extended algebras[7] and in special cases such as the so-called doubly extended $N = 4$ algebra[4,8] for which a fair amount by now is known.[9]

For actual applications in conformal field theory or string theory, it must be expected that the extended superconformal algebra will be a sub-algebra of an even larger, perhaps W-extended (or super W-extended) chiral algebra. However, it seems of some interest to first address the classification problem for the extended superconformal algebras themselves, especially since that problem appears so much simpler. The classification presented here has independently been found by Bowcock and by Fradkin and Linetsky.[10,11] Our emphasis here is on the unifying aspect provided by the idea of hamiltonian reduction of Lie superalgebras.

As is by now well known, there is a beautiful way of obtaining W-algebras from the hamiltonian reduction[12–15] of suitable affine Lie algebras by considering the constraints on the phase space of currents using a gauge fixing procedure, (see also the recent review by Bouwknegt and Schoutens[16]). Recently various types of further extensions of the W-algebras[17–21] including their supersymmetric generalization[22–24] have been considered, also by using the hamiltonian reduction technique.

There is a simple rule for finding at least the conformal dimension of the W-generators in any one particular case. In fact the labelling of the W-algebra as proposed by the review of Bouwknegt and Schoutens[16] makes use of that directly. The reduction is characterized by a particular $sl(2)$ subalgebra of the original Lie algebra. The generators of the original Lie algebra occur as spin multiplets of that $sl(2)$ subalgebra with spins $\{s_i\}$, say. Then the conformal dimensions of the generators of the ensuing W-algebra are simply the set of numbers, $\{s_i + 1\}$, the constant 1,

Table 1: Lie Superalgebras with an even subalgebra A_1

$\mathbf{g}$	$\mathbf{g}$	$\mathbf{g}_0$	G	G (alternatively)
$A(n\|1)$	$sl(n+1,2)$	$A_n \oplus A_1 \oplus u(1)$	$A_n \oplus u(1)$	$sl(n+1) \oplus u(1)$
$B(n\|1)$	$osp(2n+1\|2)$	$B_n \oplus A_1$	B_n	$so(2n+1)$
$D(n\|1)$	$osp(2n\|2)$	$D_n \oplus A_1$	D_n	$so(2n)$
$D(2\|n)$	$osp(4\|2n)$	$A_1 \oplus A_1 \oplus C_n$	$A_1 \oplus C_n$	$sp(2n) \oplus sl(2)$
$D(2\|1;\alpha)$		$A_1 \oplus A_1 \oplus A_1$	$A_1 \oplus A_1$	$sl(2) \oplus sl(2)$
$F(4)$		$B_3 \oplus A_1$	B_3	$so(7)$ (spin(7))
$G(3)$		$G_2 \oplus A_1$	G_2	G_2
$B(1\|n)$	$osp(3\|2n)$	$A_1 \oplus C_n$	C_n	$sp(2n)$

being the conformal dimension of the WZNW currents of the unconstrained affine lie-algebra.

In our case we want to consider the set of extended superconformal algebras, which may be characterized as follows: (i) There is an energy-momentum tensor of conformal dimension 2, (ii) there is a number of supercurrents, G_γ, primary conformal fields of dimension $\frac{3}{2}$ and primary affine fields of some affine algebra $\hat{G}$, transforming according to some completely reducible representation of G, and where the label γ, is a weight label of that representation, or indeed, as we shall see, an odd root of the Lie superalgebra; (iii) there are the generators of the affine Lie algebra $\hat{G}$, with conformal dimension 1; (iv) all other generators are composites of the above.

It follows that within the framework of hamiltonian reduction of Lie superalgebras which we are going to adopt, we need consider situations where a certain $sl(2)$ subalgebra is embedded so that the generators of the subalgebra, G, have spin 0. That immediately implies that we are looking for situations where there is a subalgebra of the form $G \oplus sl(2)$. The generators of the $sl(2)$ itself have spin 1 and indeed give rise as usual to the energy momentum tensor after reduction. We see that we are looking for situations where the remaining generators have spin $\frac{1}{2}$.

It is easy from the known theory of Lie superalgebras[25] to look up the ones having even subalgebras of the form $G \oplus sl(2)$. The result is given in table (1). Previously we[26] have considered these in some detail after two of us considered the classical reduced algebras.[27] The well known superconformal algebras occur as special cases. Thus $N = 1$ corresponds to $osp(1|2)$ and $N = 2$ to $osp(2|2)$ as is very well known. For $N = 4$, there are several cases with $D(2|1)$ and $A(1|1)$ forming extremes between which $D(2|1;\alpha)$ interpolates. It is amusing and perhaps significant to notice that this classification corresponds to that of the reduced holonomy groups of non-symmetric Riemannian manifolds.[28] In nearly all of the cases in table (1) the odd generators do indeed carry spin $\frac{1}{2}$ of the $sl(2)$ subalgebra. There is only one exception: in the case of $B(1|n)$ the spin is 1, so that the supercurrents acquire conformal dimension 2 after reduction. Hence this kind of algebra falls outside the scope of what we consider here, but may deserve separate investigation. In the appendix we provide details on the root systems of these algebras.

We expect that these extended superconformal algebras might correspond to non-linear sigma models with rich geometrical structures such as quaternionic or octonionic structures. In the similar classification based mostly on the OPE method[10] the role of an underlying Lie superalgebra was also hinted at in various ways. Our treatment presented here, however, provides a complete account also of the free field realizations.

To avoid confusion, is should be emphasized that the affine Lie superalgebras we are going to consider differ from the supersymmetrization of WZNW models[29] giving rise to super affine Lie algebras. However, it may be of interest to consider the two ideas put together, as has recently been done.[30] The results have a slightly different perspective and seems of particular value when further W type extensions need be considered. Thus one considers the embedding, not of $sl(2)$, but of the "supersymmetric extension", $osp(1|2)$. But that does not seem necessary for our purpose of understanding the extended superalgebras per se, without further extensions.

The purpose of the present paper is to present the hamiltonian reduction in the classical and quantum cases and to provide a unifying account of all these extended superconformal algebras. Similarly we study the quantum algebras further using free field representations, emphasizing again the unifying aspects offered by the formalism, so that generic formulas may be written down. For practical calculations the free field representations are useful for the computation of correlation functions once the pattern of singular vectors have been understood.[31] To that end we analyze the structure of screening operators.

This paper is organized as follows: In sect. 2, we first review some basic properties of affine Lie superalgebras, and then we discuss the classical hamiltonian reduction for an affine Lie superalgebra $\hat{\mathbf{g}}$ associated with a Lie superalgebra $\mathbf{g}$ which has the even subalgebra $G \oplus sl(2)$, and we derive the classical G extended superconformal Gel'fand-Dickey algebra. In sect. 3 we use the BRST gauge fixing procedure and the Wakimoto realization[32] of the affine Lie superalgebra to derive the free field realization of the extended superconformal algebra at the level of BRST charges and energy momentum tensors. In sect 4, we explain how the classical results for the algebras generalize in all cases and we complete the free field construction with generic formulas pertaining to all cases. In sect. 5, we investigate the structure of screening operators and the null field structure of degenerate representations of the G extended superconformal algebras.

2. The Classical Hamiltonian Reduction

2.1 Lie superalgebras

We start with explaining some definitions of basic classical Lie superalgebras[25] and their affine extensions. Let $\mathbf{g} = \mathbf{g}_0 \oplus \mathbf{g}_1$ be a rank n basic classical Lie superalgebra with an even subalgebra $\mathbf{g}_0$ and an odd subspace $\mathbf{g}_1$. $\Delta = \Delta^0 \cup \Delta^1$ is the set of roots of $\mathbf{g}$, where Δ^0 (Δ^1) is the set of even (odd) roots. Denote the set of positive even (odd) roots by Δ^0_+ (Δ^1_+). The superalgebra $\mathbf{g}$ has a canonical basis $\{E_\alpha, e_\gamma, h^i\}$ ($\alpha \in \Delta^0$, $\gamma \in \Delta^1$, $i = 1, \ldots, n$), which satisfies (anti-)commutation relations

$$\begin{aligned}
[E_\alpha, E_\beta] &= \begin{cases} N_{\alpha,\beta} E_{\alpha+\beta}, & \text{for } \alpha,\ \beta,\ \alpha+\beta \in \Delta^0, \\ \dfrac{2\alpha \cdot h}{\alpha^2}, & \text{for } \alpha + \beta = 0, \end{cases} \\
\{e_\gamma, e_{\gamma'}\} &= N_{\gamma,\gamma'} E_{\gamma+\gamma'}, \quad \text{for } \gamma, \gamma' \in \Delta^1,\ \gamma + \gamma' \in \Delta^0, \\
\{e_\gamma, e_{-\gamma}\} &= \gamma \cdot h, \quad \text{for } \gamma \in \Delta^1_+, \\
[e_\gamma, E_\alpha] &= N_{\gamma,\alpha} e_{\gamma+\alpha}, \quad \text{for } \alpha \in \Delta^0,\ \gamma, \gamma + \alpha \in \Delta^1, \\
[h^i, E_\alpha] &= \alpha^i E_\alpha, \quad [h^i, e_\gamma] = \gamma^i e_\gamma.
\end{aligned} \tag{1}$$

The even subalgebra $\mathbf{g}_0$ is generated by $\{E_\alpha, h^i\}$. The odd subspace $\mathbf{g}_1$ is spanned by $\{e_\gamma\}$. $\mathbf{g}_0$ acts on $\mathbf{g}_1$ as a faithful, completely reducible representation. The Killing form (,) on $\mathbf{g}$ is defined by

$$(E_\alpha, E_\beta) = \frac{2}{\alpha^2}\delta_{\alpha+\beta,0}, \quad (e_\gamma, e_{-\gamma'}) = -(e_{-\gamma}, e_{\gamma'}) = \delta_{\gamma,\gamma'}, \quad (h^i, h^j) = \delta_{ij}, \tag{2}$$

for $\alpha, \beta \in \Delta^0$, $\gamma, \gamma' \in \Delta^1_+$, $i, j = 1, \ldots, n$.

An affine Lie superalgebra $\hat{\mathbf{g}}$ at level k is the (untwisted) central extension of $\mathbf{g}$ and consists of the elements of the form $(X(z), x_0)$, where $X(z)$ is a $\mathbf{g}$-valued Laurent polynomial of $z \in \mathbf{C}$ and x_0 is a number.[33] The commutation relation for two elements $(X(z), x_0)$ and $(Y(z), y_0)$ is given by

$$[\ (X(z), x_0),\ (Y(z), y_0)\ ] = (\ [\ X(z),\ Y(z)\ ],\ k \oint \frac{dz}{2\pi i}(X(z), \partial Y(z))\ ). \tag{3}$$

The dual space $\hat{\mathbf{g}}^*$ of $\hat{\mathbf{g}}$ is generated by the current $(J(z), a_0)$. Using the Killing form (,) on $\mathbf{g}$, we may identify $\hat{\mathbf{g}}^*$ with $\hat{\mathbf{g}}$. The inner product $\langle\ ,\ \rangle$ of $(J(z), a_0) \in \hat{\mathbf{g}}^*$ and $(X(z), x_0) \in \hat{\mathbf{g}}$ is given by

$$\langle\ (J, a_0),\ (X, x_0)\ \rangle = \oint \frac{dz}{2\pi i}(\ J(z),\ X(z)\ ) + a_0 x_0. \tag{4}$$

One defines the coadjoint action ad* of $\hat{\mathbf{g}}$ on $\hat{\mathbf{g}}^*$ by

$$\langle\ \mathrm{ad}^*(X, x_0)\ (J, a_0),\ (Y, y_0)\ \rangle = -\langle\ (J, a_0),\ [\ (X, x_0),\ (Y, y_0)\ ]\ \rangle. \tag{5}$$

Using Eq.(3), one gets

$$\mathrm{ad}^*(X, x_0)(J(z), a_0) = (\ [\ X(z), J(z)\ ] + k a_0 \partial X(z),\ 0\ ). \tag{6}$$

Namely, the coadjoint action simply takes the form of an infinitesimal gauge transformation of the current $J(z)$. Denote this gauge transformation with the gauge parameter $\Lambda(z)$ as δ_Λ:

$$\delta_\Lambda J(z) = [\ \Lambda(z),\ J(z)\ ] + ka_0\partial\Lambda(z), \tag{7}$$

In the following we take the number a_0 to be 1. In terms of the canonical basis

$$J(z) = \sum_{\alpha\in\Delta^0} \frac{\alpha^2}{2} J_\alpha(z)E_\alpha + \sum_{\gamma\in\Delta^1} j_\gamma(z)e_\gamma + \sum_{i=1}^{n} H^i(z)h^i. \tag{8}$$

and

$$\Lambda(z) = \sum_{\alpha\in\Delta^0} \varepsilon_\alpha(z)E_\alpha + \sum_{\gamma\in\Delta^1} \xi_\gamma(z)e_\gamma + \sum_{i=1}^{n} \varepsilon^i(z)h^i, \tag{9}$$

we can express the gauge transformation Eq.(7) in terms of components:

$$\begin{aligned}
\delta_\Lambda J_\alpha = & \sum_{\beta,\alpha-\beta\in\Delta^0} \frac{(\alpha-\beta)^2}{\alpha^2} N_{\beta,\alpha-\beta}\varepsilon_\beta J_{\alpha-\beta} - \frac{2(\alpha\cdot H)}{\alpha^2}\varepsilon_\alpha + \frac{2k}{\alpha^2}\partial\varepsilon_\alpha + (\alpha\cdot\varepsilon)J_\alpha \\
& + \sum_{\gamma,\alpha-\gamma\in\Delta^1} \frac{2}{\alpha^2} N_{\gamma,\alpha-\gamma}\xi_\gamma j_{\alpha-\gamma}, \\
\delta_\Lambda j_\gamma = & \sum_{\substack{\alpha\in\Delta^0 \\ \gamma-\alpha\in\Delta^1}} N_{\alpha,\gamma-\alpha}\varepsilon_\alpha j_{\gamma-\alpha} + (\gamma\cdot\varepsilon)j_\gamma + \sum_{\substack{\alpha\in\Delta^0 \\ \gamma-\alpha\in\Delta^1}} \frac{\alpha^2}{2} N_{\gamma-\alpha,\alpha}\xi_{\gamma-\alpha}J_\alpha - \xi_\gamma\gamma\cdot H + k\partial\xi_\gamma, \\
\delta_\Lambda H^i = & \sum_{\alpha\in\Delta^0} \alpha^i\varepsilon_\alpha J_{-\alpha} + \sum_{\gamma\in\Delta^1_+} \gamma^i(\xi_\gamma j_{-\gamma} + \xi_{-\gamma}j_\gamma) + k\partial\varepsilon^i.
\end{aligned} \tag{10}$$

Writing the gauge transformations δ_Λ as

$$\begin{aligned}
\delta_\Lambda &= \oint \frac{dz}{2\pi i}(\Lambda(z), J(z)) \\
&= \oint \frac{dz}{2\pi i} \sum_{\alpha\in\Delta^0} \varepsilon_\alpha J_{-\alpha} + \sum_{\gamma\in\Delta^1_+} (j_\gamma\xi_{-\gamma} + \xi_\gamma j_{-\gamma}) + \sum_{i=1}^{n} \varepsilon^i H^i,
\end{aligned} \tag{11}$$

one can introduce a canonical Poisson structure on the dual space $\hat{\mathbf{g}}^*$. This Poisson structure is conveniently summarized in the form of "the operator product expansions": Poisson brackets between modes are obtained from these exactly as commutators from OPE's in conformal field theory. The result is:

$$J_\alpha(z)J_\beta(w) = \begin{cases} \dfrac{N_{\alpha,\beta}J_{\alpha+\beta}(w)}{z-w} + \cdots, & \text{for } \alpha,\ \beta,\ \alpha+\beta\in\Delta^0, \\ \dfrac{\frac{2k}{\alpha^2}}{(z-w)^2} + \dfrac{\frac{2\alpha\cdot H(w)}{\alpha^2}}{z-w} + \cdots, & \text{for } \alpha+\beta=0, \end{cases}$$

$$\begin{aligned}
j_{\pm\gamma}(z)j_{\pm\gamma'}(w) &= +\frac{N_{\mp\gamma,\mp\gamma'}J_{\pm(\gamma+\gamma')}(w)}{z-w}+\cdots, \quad \text{for } \gamma,\gamma'\in\Delta^1_+,\ \gamma+\gamma'\in\Delta^0,\\
j_{\pm\gamma}(z)j_{\mp\gamma'}(w) &= -\frac{N_{\mp\gamma,\pm\gamma'}J_{\pm(\gamma-\gamma')}(w)}{z-w}+\cdots, \quad \text{for } \gamma,\gamma'\in\Delta^1_+,\ \gamma-\gamma'\in\Delta^0,\\
j_{\gamma}(z)j_{-\gamma}(w) &= -\frac{k}{(z-w)^2}-\frac{\gamma\cdot H(w)}{z-w}+\cdots, \quad \text{for } \gamma\in\Delta^1_+,\\
J_\alpha(z)j_\gamma(w) &= \frac{N_{-\alpha,\gamma+\alpha}j_{\gamma+\alpha}(w)}{z-w}+\cdots, \quad \text{for } \gamma,\gamma+\alpha\in\Delta^1,\ \alpha\in\Delta^0,,\\
H^i(z)J_\alpha(w) &= \frac{\alpha^i J_\alpha(w)}{z-w}+\cdots, \quad H^i(z)j_\gamma(w)=\frac{\gamma^i j_\gamma(w)}{z-w}+\cdots.\\
H^i(z)H^j(w) &= \frac{k\delta^{ij}}{(z-w)^2}+\cdots.
\end{aligned} \tag{12}$$

Here we have used some identities for the structure constants coming from the Jacobi identities:

$$\begin{aligned}
\frac{2N_{\alpha,\beta}}{(\alpha+\beta)^2} &= \frac{2N_{\beta,-\alpha-\beta}}{\alpha^2}=\frac{2N_{-\alpha-\beta,\alpha}}{\beta^2}, \quad \text{for } \alpha,\beta\in\Delta^0,\\
\frac{2N_{\gamma,\gamma'}}{(\gamma+\gamma')^2} &= N_{\gamma',-\gamma-\gamma'}=-N_{-\gamma-\gamma',\gamma}, \quad \text{for } \gamma,\gamma'\in\Delta^1_+\\
\frac{2N_{\gamma,-\gamma'}}{(\gamma-\gamma')^2} &= N_{-\gamma',-\gamma+\gamma'}=N_{-\gamma+\gamma',\gamma}, \quad \text{for } \gamma,\gamma'\in\Delta^1_+\\
\frac{2N_{-\gamma,-\gamma'}}{(\gamma+\gamma')^2} &= -N_{-\gamma',\gamma+\gamma'}=N_{\gamma+\gamma',-\gamma}, \quad \text{for } \gamma,\gamma'\in\Delta^1_+ .
\end{aligned} \tag{13}$$

In the following, as argued in the introduction, we will study the class of Lie superalgebras, with even algebras of the form $G\oplus A_1$. Using Kac's notation, these algebras are classified as follows $A(n|1)$ $(n\geq 1)$, $A(1,0)=C(2)$, $B(n|1)$ $(n\geq 0)$, $B(1|n)$ $(n\geq 1)$, $D(n|1)$ $(n\geq 2)$, $D(2|n)$ $(n\geq 1)$, $D(2|1;\alpha)$, $F(4)$ and $G(3)$ (see table 1). The embedding of A_1 in $\mathbf{g}$ carried by $\mathbf{g}_1$ has spin $\frac{1}{2}$, except for $B(1|n)$. In the case of $B(1|n)$ the embedding has spin 1. In the spin $\frac{1}{2}$ embedding case, which is the only case we will be studying in this paper for the reasons given in the introduction, the odd subspace $\mathbf{g}_1$ belongs to the spin $\frac{1}{2}$ representation with respect to the even subalgebra A_1 and therefore splits into two parts:

$$\mathbf{g}_1=(\mathbf{g}_1)_{+\frac{1}{2}}\oplus(\mathbf{g}_1)_{-\frac{1}{2}}. \tag{14}$$

Likewise, the odd root space Δ^1 may be divided into two parts

$$\Delta^1=\Delta^1_{+\frac{1}{2}}\cup\Delta^1_{-\frac{1}{2}}, \tag{15}$$

where $(\mathbf{g}_1)_{\pm\frac{1}{2}}$ is spanned by the generators whose roots belong to $\Delta^1_{\pm\frac{1}{2}}$. More explicitly, the sets $\Delta^1_{\pm\frac{1}{2}}$ consist of the roots $\gamma\in\Delta^1$ satisfying

$$\frac{\gamma\cdot\theta}{\theta^2}=\pm\frac{1}{2}, \tag{16}$$

where θ is the simple root of A_1. By choosing appropriate simple roots for **g**, we can take the root space $\Delta^1_{+\frac{1}{2}}$ as the space of odd positive roots;

$$\Delta^1_+ = \Delta^1_{+\frac{1}{2}}, \tag{17}$$

and then

$$\Delta^1_{+\frac{1}{2}} = \Delta^1_{-\frac{1}{2}} + \theta, \tag{18}$$

holds. We note that each odd space $(\mathbf{g}_1)_{\pm\frac{1}{2}}$ belongs to a fundamental representation of G of dimension $|\Delta^1_+|$, this being the number of positive odd roots. These will be related to the supercurrents as discussed below. The corresponding representation of G (possibly of the form $G^{(1)} \oplus G^{(2)}$)has weights simply equal to

$$\gamma_\perp(G^{(i)}) \equiv P_{G^{(i)}}(\gamma)$$

namely the (positive in this case) odd root projected onto the root space of $G^{(i)}$. If we denote the representation matrices in the canonical basis by

$$t^\alpha_{\gamma,\gamma'}, t^i_{\gamma,\gamma'}$$

then we have (using the Jacobi identities, see also below) for $\gamma \in \Delta^1_+$

$$\begin{aligned} t^\alpha_{\gamma,\gamma'} &= -N_{-\alpha,-\gamma}\delta_{\gamma+\alpha,\gamma'} \\ t^i_{\gamma,\gamma'} &= (\gamma_\perp)^i\delta_{\gamma,\gamma'} \\ tr(t^i t^j) &= c_F\delta_{ij} \\ tr(t^\alpha t^\beta) &= c_F\frac{2}{\alpha^2}\delta_{\alpha+\beta,0} \\ c_F &= \sum_{\gamma\in\Delta^1_+}\gamma^2_\perp/r(G) \end{aligned} \tag{19}$$

where

$$r(G) = \mathrm{rank}(G).$$

Table (2) gives several parameters relevant for the different Lie super algebras we are considering. One may show that

$$t^\alpha_{\gamma,\gamma'} = -t^\alpha_{\theta-\gamma',\theta-\gamma}.$$

The Casimir operator in that representation takes the form

$$C_{\gamma,\gamma'} = (\sum_{\alpha\in\Delta^0(G)}\frac{\alpha^2}{2}t^\alpha t^{-\alpha} + \sum_{i=1}^{r(G)} t^i t^i)_{\gamma,\gamma'} \tag{20}$$

Then it is convenient to introduce the currents corresponding to $\widehat{G}$ in the following basis

$$J_{\gamma,\gamma'}(z) \equiv \sum_{\alpha\in\Delta^0(G)}\frac{\alpha^2}{2}t^\alpha_{\gamma,\gamma'}J_{-\alpha}(z) + \sum_{i=1}^{r(G)} t^i_{\gamma,\gamma'}H_i(z) \tag{21}$$

Table 2: Properties of Lie superalgebras giving rise to ESA's

g	sdim	rank	$h^\vee$	$\lvert\Delta_+^1\rvert$	$c_F^{(1)}$	$c_F^{(2)}$
$A(n\|1)$	n^2-2n (-2 for $n=1$)	$n+2$	$n-1$	$2(n+1)$	2	$\frac{(n-1)^2}{2(n+1)}$
$B(n\|1)$	$(n-1)(2n-1)$	$n+1$	$3-2n$	$2n+1$	-2	
$D(n\|1)$	$(n-1)(2n-3)$	$n+1$	$4-2n$	$2n$	-2	
$D(2\|n)$	$(n-2)(2n-3)$	$n+2$	$2n-2$	$4n$	4	$-2n$
$D(2\|1;\alpha)$	1	3	0	4	-2γ	$-2(1-\gamma)$
$F(4)$	8	4	-3	8	-2	
$G(3)$	3	3	2	7	2	

Notice that for given γ, γ', $J_{\gamma,\gamma'}$ contains at most a single term, which may be written in several ways, for example as:

$$J_{\gamma,\gamma'}(z) = \begin{cases} N_{\gamma',-\gamma} J_{\gamma-\gamma'}(z) & \text{for} \quad \gamma-\gamma' \in \Delta^0(G) \\ \gamma \cdot H(z) & \text{for} \quad \gamma = \gamma' \\ 0 & \text{otherwise} \end{cases} \tag{22}$$

It follows that the root system of the even subalgebra of G is expressed in the form of $\gamma - \gamma'$, where γ and γ' are some positive odd roots. This fact turns out to be essential for the study of the structure of general G extended superconformal algebras.

2.2 Classical hamiltonian reduction

We now consider the hamiltonian reduction of the phase space of the currents $\hat{\mathbf{g}}^*$ associated with any one of the set of Lie superalgebras $\mathbf{g}$ with an even subalgebra $G \oplus A_1$. We want to impose the constraint on the even subalgebra A_1 while keeping the affine G algebra symmetry. We start from the phase space $\mathcal{F}$ of the currents with the constraint

$$J_{-\theta}(z) = 1. \tag{23}$$

Denote by $\mathcal{N}$ the gauge group which preserves the constraint Eq.(23). Putting $-\theta$ for α in the first formula of eqs. Eq.(10), we get

$$\delta_\Lambda J_{-\theta} = \frac{2(\theta \cdot H)}{\theta^2}\varepsilon_{-\theta} + \frac{2k}{\theta^2}\partial\varepsilon_{-\theta} - (\theta\cdot\varepsilon)J_{-\theta} + \sum_{\gamma\in\Delta_+^1} \frac{2}{\theta^2} N_{-\gamma,-\theta+\gamma}\xi_{-\gamma} j_{-\theta+\gamma}. \tag{24}$$

Here we use the relation Eq.(18) and $N_{\alpha,-\theta-\alpha} = 0$ for $\alpha \in \Delta(G)$. Thus $\mathcal{N}$ is characterized by the algebra conditions

$$\varepsilon_{-\theta} = \theta\cdot\varepsilon = 0, \quad \xi_{-\gamma} = 0, \quad \text{for } \gamma \in \Delta_+^1. \tag{25}$$

So the Lie algebra $\mathbf{n}$ of the gauge group $\mathcal{N}$ is equal to $\widehat{G}\oplus\mathbf{n}_1$, where $\widehat{G}$ is the affine Lie algebra corresponding to the even subalgebra G and $\mathbf{n}_1$ is generated by the elements:

$$\Lambda(z)=\varepsilon_\theta(z)E_\theta+\sum_{\gamma\in\Delta^1_+}\xi_\gamma(z)e_\gamma. \tag{26}$$

Let $\mathcal{N}_1$ be the gauge group corresponding to $\mathbf{n}_1$. The reduced phase space $\mathcal{M}$ is defined as the quotient space $\mathcal{M}=\mathcal{F}/\mathcal{N}_1$. By the standard gauge fixing procedure, one chooses one or the other gauge slice as a representative of the reduced phase space. In the so-called "Drinfeld-Sokolov (DS)"-gauge the generators of the classical extended algebra are rather easily read off. It is defined by:[12]

$$J_\theta(z)=T(z),\quad \theta\cdot H(z)=0,\quad j_\gamma(z)=G_\gamma(z),\quad j_{-\gamma}(z)=0, \tag{27}$$

for $\gamma\in\Delta^1_+$. And the generic gauge transformation projected on the DS gauge slice becomes the transformation corresponding to the extended superconformal algebra[12-15,17-23] in our case. Hence the final step is to consider the generic gauge transformation preserving the gauge condition Eq.(27). In particular

$$\begin{aligned}
\delta_\Lambda J_{-\theta} &= -(\theta\cdot\varepsilon)+\frac{2k}{\theta^2}\partial\varepsilon_{-\theta},\\
\delta_\Lambda(\theta\cdot H) &= \theta^2\varepsilon_\theta-\theta^2\varepsilon_{-\theta}T+k\partial(\theta\cdot\varepsilon)+\sum_{\gamma\in\Delta^1_+}\theta\cdot\gamma\xi_{-\gamma}G_\gamma,\\
\delta_\Lambda j_{-\gamma} &= N_{-\theta,\theta-\gamma}\varepsilon_{-\theta}G_{\theta-\gamma}+\sum_{\substack{\gamma'\in\Delta^1_+\\ \gamma'-\gamma\in\Delta(G)}}N_{\gamma,-\gamma'}\xi_{-\gamma'}J_{\gamma'-\gamma}\\
&+ N_{\theta-\gamma,\gamma}\xi_{\theta-\gamma}+\xi_{-\gamma}(\gamma\cdot H)+k\partial\xi_{-\gamma},
\end{aligned} \tag{28}$$

for $\gamma\in\Delta^1_+$. By solving the conditions which preserve the DS-gauge $\delta J_{-\theta}=0$, $\delta(\theta\cdot H)=0$ and $\delta j_{-\gamma}=0$, we can express the parameters ε_θ, $\theta\cdot\varepsilon$ and ξ_γ ($\gamma\in\Delta^1_+$) in terms of the other parameters, $\varepsilon_{-\theta}$, $\xi_{-\gamma}$ ($\gamma\in\Delta^1_+$), ε_α and $\alpha\cdot\varepsilon$ ($\alpha\in\Delta(G)$):

$$\theta\cdot\varepsilon=\frac{2k}{\theta^2}\partial\varepsilon_{-\theta},\qquad \varepsilon_\theta=\varepsilon_{-\theta}T-\frac{2k^2}{(\theta^2)^2}\partial^2\varepsilon_{-\theta}-\frac{1}{2}\sum_{\gamma\in\Delta^1_+}\xi_{-\gamma}G_\gamma, \tag{29}$$

$$\xi_{\theta-\gamma}=\frac{-1}{N_{\theta-\gamma,\gamma}}\{N_{-\theta,\theta-\gamma}G_{\theta-\gamma}\varepsilon_{-\theta}+\sum_{\substack{\gamma'\in\Delta^1_+\\ \gamma-\gamma'\in\Delta(G)}}N_{\gamma,-\gamma'}\xi_{-\gamma'}J_{\gamma'-\gamma}+\xi_{-\gamma}(\gamma\cdot H)+k\partial\xi_{-\gamma}\},$$

In the DS gauge, the gauge transformation of $J_\theta(z)=T(z), J_\beta(z),\beta\cdot H(z)$ and $G_\gamma(z)$ may then be worked out in terms of these. We refer to our papers[26] for the straightforward but slightly messy details.

We then convert this result into formal "operator product expansions". We denote the gauge transformation in the DS-gauge as

$$\delta=\oint\frac{dz}{2\pi i}(\Lambda(z)\ ,J_{DS}(z)), \tag{30}$$

where

$$J_{DS}(z) = \frac{\theta^2}{2}T(z)E_\theta + \sum_{\gamma\in\Delta^1_+} G_\gamma(z)e_\gamma + \sum_{\alpha\in\Delta(G)} \frac{\alpha^2}{2}J_\alpha(z)E_\alpha + \sum_{i=1}^{\mathrm{rank}(G)} H^i h^i, \tag{31}$$

This conveniently generates the Poisson bracket structure on the reduced phase space. Here we take h^i ($i = 1, \ldots, \mathrm{rank}(G)$) as the generators of the Cartan subalgebra of G. We also introduce the rescalings $\tilde{T} = \frac{\theta^2}{2k}T$ and $\tilde{G}_\gamma = \sqrt{\frac{N_{\theta-\gamma,\gamma}}{k}}G_\gamma$ for $\gamma \in \Delta^1_+$. We further use several identities for the structure constants, like the Jacobi identities Eq.(13), as well as

$$\frac{N_{\theta-\gamma,\gamma'-\theta}}{N_{\gamma,\theta-\gamma}} = -\frac{N_{\gamma',-\gamma}}{N_{\gamma',\theta-\gamma'}}, \tag{32}$$

for $\gamma, \gamma' \in \Delta^1_+$, as may be checked using explicit matrix representations of $\mathbf{g}$[34,35] in table 1. Similarly one may derive

$$N_{\gamma,\theta-\gamma} = -N_{\gamma-\theta,-\gamma} = \frac{\theta^2}{2} \tag{33}$$

from the fact that $\mathbf{g} = G \oplus A_1$. Using all these properties one may achieve a major simplification of the complicated, generic expression we have previously presented.[26] To this end it is very useful to make use of the affine currents of G in the basis of Eq.(21). The result finally is:

$$\begin{aligned}
\tilde{T}(z)\tilde{T}(w) &= \frac{\frac{-6k}{\theta^2}}{(z-w)^4} + \frac{2\tilde{T}(w)}{(z-w)^2} + \frac{\partial\tilde{T}(w)}{z-w} + \cdots, \\
\tilde{T}(z)\tilde{G}_\gamma(w) &= \frac{\frac{3}{2}\tilde{G}_\gamma(w)}{(z-w)^2} + \frac{\partial\tilde{G}_\gamma(w)}{z-w} - \frac{1}{k}\frac{\sum_{\gamma'\in\Delta^1_+} J_{\gamma,\gamma'}G_{\gamma'}(w)}{z-w} \\
J_\beta(z)\tilde{G}_\gamma(w) &= \frac{-N_{-\beta,-\gamma}\tilde{G}_{\beta+\gamma}(w)}{z-w} + \cdots, \quad H^i(z)\tilde{G}_\gamma(w) = \frac{\gamma^i\tilde{G}_\gamma(w)}{z-w} + \cdots, \\
\tilde{G}_{\theta-\gamma'}(z)\tilde{G}_\gamma(w) &= \frac{2k\delta_{\gamma',\gamma}}{(z-w)^3} - 2\frac{J_{\gamma,\gamma'}(w)}{(z-w)^2} \\
&+ \frac{\delta_{\gamma,\gamma'}\frac{\theta^2}{2}\tilde{T} - \partial J_{\gamma,\gamma'}(w) + \frac{1}{k}(J^2)_{\gamma,\gamma'}(w)}{z-w} + \cdots,
\end{aligned} \tag{34}$$

Here we imply the notation:

$$(J^2)_{\gamma,\gamma'} \equiv \sum_{\gamma''\in\Delta^1_+} J_{\gamma,\gamma''}J_{\gamma'',\gamma} \tag{35}$$

Notice that this expression is nonvanishing only if either $\gamma - \gamma' \in \Delta(G)$, $\gamma = \gamma'$ or $\gamma + \gamma' = \theta$. At this point, the supercurrents $\tilde{G}_\gamma$ are not primary fields with respect to

$\tilde{T}(z)$. However, if we define the total energy-momentum tensor T_{ESA} by adding the Sugawara form $T_{Sugawara}$ of the $\widehat{G}$ affine Lie algebra: $T_{ESA} = \tilde{T} + T_{Sugawara}$ where

$$
\begin{aligned}
T_{Sugawara} &= \frac{1}{2k}\left\{\sum_{\alpha\in\Delta(G)} \frac{\alpha^2}{2} J_\alpha J_{-\alpha} + \sum_{i=1}^{\mathrm{rank}(G)} H^i H^i\right\} \\
&= \frac{1}{2kc_F} tr(J^2),
\end{aligned} \tag{36}
$$

we may check that the supercurrents have conformal weight 3/2 with respect to T_{ESA}. The result is the classical or Gel'fand-Dickey extended superconformal algebra. In sect. 4 we shall give the explicit form generalized to the arbitrary quantum case. The classical value of the central charge c_{ESA} is $-12k/\theta^2$. Similarly, the classical central extension of the surviving affine Lie algebra $\widehat{G}$ (or $\widehat{G} = \oplus_i \widehat{G}^{(1=i)}$ is given by $K_{cl}^{(i)} = \frac{2k}{\alpha_{Li}^2}$,where α_L denotes the "long root" so that this definition is the one pertaining to a normalization where the longest root has length squared equal to 2. The relative minus sign between these expressions is significant. In fact the metrics in the root space of the A_1 to be reduced and that of G have opposite signs in all cases except for $D(2|n)$, cf. Appendix A. Thus in most cases $c_{ESA} \geq 0 \Leftrightarrow K_{cl} \geq 0$. This property turns out to hold in the quantum case as well. In the case of $D(2|n)$, $\widehat{G}$ splits into two commuting algebras having opposite sign central extensions.[10]

3. The Quantum Hamiltonian Reduction

3.1 BRST formalism

In this section we discuss the quantum hamiltonian reduction[14] for an affine Lie superalgebra $\hat{\mathbf{g}}$ at level k, generated by $J_\alpha(z)$ ($\alpha \in \Delta^0$), $j_\gamma(z)$ ($\gamma \in \Delta^1$) and $H^i(z)$ ($i = 1, \ldots, n$). Let $T_{WZNW}(z)$ be the energy-momentum tensor of an affine Lie superalgebra $\hat{\mathbf{g}}$, defined by the Sugawara form:

$$
\begin{aligned}
T_{WZNW} &= \frac{1}{2(k+h^\vee)}\Bigg(: \sum_{\alpha\in\Delta_+^0} \frac{\alpha^2}{2}(J_\alpha J_{-\alpha} + J_{-\alpha}J_\alpha) + \sum_{i=1}^{n} H^i H^i \\
&+ \sum_{\gamma\in\Delta_+^1} (j_\gamma j_{-\gamma} - j_{-\gamma} j_\gamma) : \Bigg)
\end{aligned} \tag{37}
$$

where $h^\vee$ is the dual Coxeter number of $\mathbf{g}$ and : : denotes the normal ordering. In order to impose the constraint for the currents at the quantum level, we have to "improve" the energy-momentum tensor by a contribution from the Cartan currents $H^i(z)$:

$$T_{improved}(z) = T_{WZNW}(z) - \mu \cdot \partial H(z). \tag{38}$$

Here μ is an n-vector. With respect to the improved energy-momentum tensor $T_{improved}(z)$, the currents corresponding to the roots α have conformal weights $1 + \mu \cdot \alpha$.

We are concerned with a class of Lie superalgebras whose even subalgebras are $G \oplus A_1$, where G is a semisimple Lie algebra. In the previous section we have considered the constraint $J_{-\theta}(z) = 1$ but the currents have conformal dimension one, so in order for the constraint to make sense, $J_{-\theta}(z)$ must have improved conformal dimension, 0, whereas the currents of $\widehat{G}$ should continue to have conformal dimension 1. This means that the vector μ should satisfy

$$\mu \cdot \theta = 1, \quad \mu \cdot \alpha = 0, \quad \text{for } \alpha \in \Delta(G). \tag{39}$$

These equations Eq.(39) determine the vector μ uniquely:

$$\mu = \frac{\theta}{\theta^2}. \tag{40}$$

From Eq.(16), we find that the fermionic currents $j_\gamma(z)$ ($j_{-\gamma}(z)$) for the positive roots $\gamma \in \Delta_+^1$ have conformal weight 3/2 (1/2). The current $J_\theta(z)$ has conformal weight 2.

Now we use the BRST-gauge fixing procedure. In the previous section we took the Drinfeld-Sokolov gauge Eq.(27) and derived the Poisson bracket structure for the currents. In order to study the representation of the algebra, it is very useful to consider the free field representation. This is realized in the classical case by taking the "diagonal" gauge

$$\begin{aligned} J_\theta(z) &= 0, \quad \theta \cdot H(z) = a_0 \partial\phi(z), \\ j_{-\gamma}(z) &= \sqrt{N_{-\gamma,-\theta+\gamma}}\chi_\gamma(z), \quad \text{for } \gamma \in \Delta_+^1. \end{aligned} \tag{41}$$

in Eq.(8). In the quantum case we similarly introduce a free boson $\phi(z)$, coupled to the world sheet curvature, and free fermions $\chi_\gamma(z)$, satisfying operator product expansions

$$\chi_\gamma(z)\chi_{\theta-\gamma'}(w) = \frac{\delta_{\gamma,\gamma'}}{z-w} + \cdots, \quad \text{for } \gamma, \gamma' \in \Delta_+^1. \tag{42}$$

This is consistent with the OPE's of $\hat{\mathfrak{g}}$ together with the constraint. The role of the free fermions is to act as auxiliary fields that convert the constraints into first class ones.

We further introduce fermionic ghosts ($b_\theta(z)$,$c_\theta(z)$) with conformal weights (0,1) and bosonic ghosts ($\tilde{b}_\gamma(z)$, $\tilde{c}_\gamma(z)$) of weight ($\frac{1}{2}$, $\frac{1}{2}$) for $\gamma \in \Delta_+^1$. The BRST current $J_{BRST}(z)$ is defined as[23]

$$J_{BRST}(z) = c_\theta(J_{-\theta} - 1) + \sum_{\gamma\in\Delta_+^1} \tilde{c}_\gamma(j_{-\gamma} - \sqrt{N_{-\gamma,-\theta+\gamma}}\chi_\gamma) + \frac{1}{2}\sum_{\gamma\in\Delta_+^1} N_{\gamma,\theta-\gamma} : \tilde{c}_\gamma\tilde{c}_{\theta-\gamma}b_\theta : . \tag{43}$$

We can easily show that the BRST charge $Q_{BRST} = \oint \frac{dz}{2\pi i} J_{BRST}(z)$ satisfies the nilpotency condition $Q_{BRST}^2 = 0$. The total energy-momentum tensor is expressed as

$$T_{total}(z) = T_{improved}(z) + T_\chi + T_{ghost}(z), \tag{44}$$

Table 3: Central charges for G extended superconformal algebras

$\mathbf{g}$	c_{total}
$A(n\|1)$	$[6k^2 + k(n^2 + 3n - 9) - n^2 - 2n + 3]/[k + n - 1]$
$B(n\|1)$	$[(k+1)(4n^2 + 4n - 15 - 6k)]/[2(k + 3 - 2n)]$
$D(n\|1)$	$[(k+1)(4n^2 - 16 - 6k)]/[2(k + 4 - 2n)]$
$D(2\|n)$	$[6k^2 + k(2n^2 + 3n - 8) + 4 - 4n^2]/[k + 2n - 2]$
$D(2\|1;\alpha)$	$-3 - 6k$
$F(4)$	$[2(-2k^2 + 7k + 9)]/[k - 3]$
$G(3)$	$[9k^2 + 13k - 22]/[2(k+2)]$

where

$$\begin{aligned} T_{ghost}(z) &= :(\partial b_\theta)c_\theta + \frac{1}{2}\sum_{\gamma\in\Delta^1_+}(\tilde{b}_\gamma\partial\tilde{c}_\gamma - (\partial\tilde{b}_\gamma)\tilde{c}_\gamma):, \\ T_\chi(z) &= -\frac{1}{2}\sum_{\gamma\in\Delta^1_+} :\chi_{\theta-\gamma}\partial\chi_\gamma: . \end{aligned} \tag{45}$$

The central charge c of the total system is computed to be

$$c_{total} = c_{WZNW} - 12k\mu^2 + \frac{1}{2}|\Delta^1_+| - 2 - |\Delta^1_+|, \tag{46}$$

(where $|\Delta^1_+|$ is still the number of positive odd roots). The last two terms are contributions from ghost fields. The central charge c_{WZNW} of the WZNW models on the Lie supergroup at level k is given by the formula:

$$c_{WZNW} = \frac{k\ \mathrm{sdim}\mathbf{g}}{k + h^\vee}, \tag{47}$$

where the super dimension $\mathrm{sdim}\mathbf{g}$ of a Lie superalgebra $\mathbf{g}$ is defined as $\dim\mathbf{g}_0 - \dim\mathbf{g}_1$. The list for the corresponding central charges is shown in table 3. The results here are in complete agreement with previous calculations using a variety of different techniques. Thus, the results for $A(n|1), B(n|1), D(n|1)$ were obtained by Knizhnik and Bershadsky;[5] the one for $D(2|1;\alpha)$ agrees with original literature on the doubly extended $N = 4$ algebra[4,9] (see also our recent free field representation[26]), the result for $D(2|n)$ and $G(3)$ agrees with Bowcock and Fradkin and Linetsky,[10] and the result for $F(4)$ with that of Fradkin and Linetsky.[11] Our treatment here, however, provides a unifying framework.

In the classical limit $k \to \infty$, the expression Eq.(53) becomes $-12k/\theta^2$, which agrees with the result (34) obtained in the previous section.

3.2 The free field representation

We consider the free field representations of G extended superconformal algebra based on the Wakimoto construction[32] of the affine Lie superalgebra $\hat{\mathbf{g}}$.[36] Let us introduce bosonic ghosts $(\beta_\alpha(z),\gamma_\alpha(z))$ for even positive roots $\alpha\in\Delta^0_+$ with conformal dimensions (1,0), fermionic ghosts $(\eta_\gamma(z),\xi_\gamma(z))$ for odd positive roots $\gamma\in\Delta^1_+$ with (unimproved) conformal dimensions (1,0) and n free bosons $\varphi(z)=(\varphi_1(z),\ldots,\varphi_n(z))$ coupled to the world-sheet curvature, satisfying the operator product expansions:

$$\begin{aligned}\beta_\alpha(z)\gamma_{\alpha'}(w) &= \frac{\delta_{\alpha,\alpha'}}{z-w}+\cdots,\quad \eta_\gamma(z)\xi_{\gamma'}(w)=\frac{\delta_{\gamma,\gamma'}}{z-w}+\cdots,\\ \varphi_i(z)\varphi_j(w) &= -\delta_{ij}\ln(z-w)+\cdots.\end{aligned}\tag{48}$$

Using these free fields the energy-momentum tensor is expressed as

$$T_{WZNW}(z)=\sum_{\alpha\in\Delta^0_+}:\beta_\alpha\partial\gamma_\alpha:-\sum_{\gamma\in\Delta^1_+}:\eta_\gamma\partial\xi_\gamma:-\frac{1}{2}:(\partial\varphi)^2:-\frac{\mathrm{i}\rho\cdot\partial^2\varphi}{\alpha_+},\tag{49}$$

where $\alpha_+=\sqrt{k+h^\vee}$ and $\rho=\rho_0-\rho_1$. ρ_0 (ρ_1) is half the sum of positive even (odd) roots:

$$\rho_0=\frac{1}{2}\sum_{\alpha\in\Delta^0_+}\alpha,\quad \rho_1=\frac{1}{2}\sum_{\gamma\in\Delta^1_+}\gamma.\tag{50}$$

The Cartan part currents $\hat{H}^i(z)$ are given by

$$\hat{H}^i(z)=-\mathrm{i}\alpha_+\partial\varphi^i+\sum_{\alpha\in\Delta^0_+}\alpha^i:\gamma_\alpha\beta_\alpha:+\sum_{\gamma\in\Delta^1_+}\gamma^i:\xi_\gamma\eta_\gamma:.\tag{51}$$

The currents indicated by a "hat" will differ by the final full ones by certain fermionic contributions to be described. After improvement of the energy momentum tensor, the ghost systems acquire conformal dimensions, (0,1) for $(\beta_\theta,\gamma_\theta)$ and $(\frac{1}{2},\frac{1}{2})$ for (η_γ,ξ_γ). In the BRST gauge fixing procedure it is natural to assume that the multiplets $(\beta_\theta,\gamma_\theta,b_\theta,c_\theta)$ and $(\eta_\gamma,\xi_\gamma,\tilde{b}_\gamma,\tilde{c}_\gamma)$ for $\gamma\in\Delta^1_+$ form Kugo-Ojima quartets,[37] and that the corresponding energy-momentum tensor is BRST-exact. In the case of the $\widehat{sl(N)}$ affine Lie algebra, this has been proven both by homological techniques,[38] and by a more direct method.[39] This ansatz enables us to write the total energy-momentum tensor in the form:

$$T_{total}(z)=T_{ESA}(z)+\{Q_{BRST},*\},\tag{52}$$

where T_{ESA} is the energy-momentum tensor of G extended superconformal algebra:

$$\begin{aligned}T_{ESA}(z) &= -\frac{1}{2}:(\partial\varphi)^2:-\mathrm{i}\left(\frac{\rho}{\alpha_+}-\alpha_+\mu\right)\cdot\partial^2\varphi\\ &\quad+\sum_{\alpha\in\Delta_+(G)}:\beta_\alpha\partial\gamma_\alpha:+\frac{1}{2}\sum_{\gamma\in\Delta^1_+}:(\partial\chi_\gamma)\chi_{\theta-\gamma}:.\end{aligned}\tag{53}$$

The central charge of T_{ESA} is given by the formula

$$c=n-12\left(\frac{\rho}{\alpha_+}-\alpha_+\mu\right)^2+2|\Delta_+(G)|+\frac{1}{2}|\Delta^1_+|.\tag{54}$$

This may be shown to be equal to c_{total} Eq.(46).

In the case that $\widehat{G}$ is expressed as a direct sum of simple affine Lie algebras $\oplus_i \widehat{G}_i$, the relation between the level k of the affine Lie superalgebra $\hat{\mathfrak{g}}$ and that, K_i, of the i'th affine Lie algebra $\widehat{G_i}$, is given by considering the decomposition of the vector $\rho/\alpha_+ - \alpha_+\mu$ into the roots of the even subalgebras $(\oplus_i G_i) \oplus A_1$:

$$\frac{\rho}{\alpha_+} - \alpha_+\mu = \sum_i \frac{\rho_{G_i}}{\alpha_+} + \left(\frac{\frac{1}{2} - \frac{|\Delta_+^1|}{4}}{\alpha_+} - \frac{\alpha_+}{\theta^2}\right)\theta. \tag{55}$$

Here ρ_{G_i} is half the sum of positive roots of the subalgebra G_i. The above decomposition means that

$$k + h^\vee = \frac{\alpha_{Li}^2}{2}(K_i + H_i^\vee), \tag{56}$$

where $H_i^\vee$ is the dual Coxeter number of the even subalgebra G_i in a notation appropriate to the long root being normalized to length squared, 2, whereas α_{Li} is an actual long long root of the subalgebra G_i. Notice that the central extensions of the affine sub algebras receive additional contributions from the free fermions, see next section.

The n bosons $\varphi(z) = (\varphi_1(z), \ldots, \varphi_n(z))$ coupled to the world sheet curvature, can be divided into two classes, due to the decomposition of the Cartan subalgebra $\mathrm{h} = H_G \oplus H_{A_1}$, where H_G and H_{A_1} are the Cartan subalgebras of the even subalgebras G and A_1, respectively. A boson $\theta \cdot \varphi$ in the θ direction of the root space of the even subalgebra A_1, commutes with the bosons lying along the root space of the even subalgebra G. The remaining $n-1$ free bosons are used for the free field representation of the $\hat{G}$ affine Lie algebra, combined with $(\beta_\alpha, \gamma_\alpha)$-systems $(\alpha \in \Delta_+(G))$. If we define a Feigin-Fuchs boson ϕ:

$$\phi(z) = \frac{\theta \cdot \varphi(z)}{\sqrt{\theta^2}}, \tag{57}$$

the energy-momentum tensor Eq.(53) becomes $T_{ESA}(z) = T_\phi(z) + T_G(z) + T_\chi(z)$, where

$$T_\phi = -\frac{1}{2}(\partial\phi)^2 - \frac{iQ}{2}\partial^2\phi, \quad Q = \sqrt{\theta^2}\left(\frac{2 - |\Delta_+^1|}{2\alpha_+} - \frac{2\alpha_+}{\theta^2}\right). \tag{58}$$

$T_G(z)$ is the Sugawara energy-momentum tensor of the affine Lie algebra $\widehat{G}$.

4. Generic Free Field Representations and Explicit Form of all Algebras

Having provided a free field representation for the energy momentum tensor, we still have to do the same for the affine G-currents and in particular for the $|\Delta_+^1|$ supercurrents, G_γ. We have at our disposal already the free (Feigin Fuchs) scalar, $\phi(z)$ and the $|\Delta_+^1|$ free fermions, $\chi_\gamma(z)$. We then build the affine currents $J_\alpha(z), H_i(z)$ as

$$\begin{aligned} J_\alpha(z) &= \hat{J}_\alpha(z) + J_\alpha^f(z) \\ H_i(z) &= \hat{H}_i(z) + H_i^f(z) \end{aligned} \tag{59}$$

where the fermionic parts are given by

$$
\begin{aligned}
J_\alpha^f(z) &= \frac{1}{2}\sum_{\gamma,\gamma'\in\Delta_+^1} t^\alpha_{\gamma',\gamma} :\chi_\gamma(z)\chi_{\theta-\gamma'}(z): \\
H_i^f(z) &= \frac{1}{2}\sum_{\gamma\in\Delta_+^1} (\gamma_\perp)_i :\chi_\gamma(z)\chi_{\theta-\gamma}(z):
\end{aligned} \tag{60}
$$

The "hatted" currents commute with the fermionic ones but are also affine G currents with levels such that the total match up to what is required. More precisely, for $G = \oplus_i G^{(i)}$, the central extensions of the affine subalgebras are denoted by K_i^{tot} in a notation pertaining to the length squared of the maximal root being 2. We then have

$$
\begin{aligned}
K_i^{tot} &= K_i + K_i^f \\
K_i^f &= \frac{c_F^{(i)}}{(\alpha_L^2)_i}
\end{aligned} \tag{61}
$$

where K_i is the contribution from the hat algebra tabulated in table (5). These may be treated by the standard Wakimoto like free field realisations[32] as above.

So far we have used the parameter, k, the central extension of the Lie superalgebra given by the "hat'ed" currents, for expressing everything. It may be convenient for various purposes and comparisons to express the central charges as functions of the central extensions of the $\widehat{G}$ algebra in a standard normalization pertaining to the longest root having length squared equal to 2. This we do in table 4. For $D(2|n)$ we have written the central charge in terms of the level of the affine $sp(2n)$, and for $D(2|1;\alpha)$ we have written it in terms of the level of $su(2)^+$, one of two commuting $su(2)$'s. The parameter, γ (not to be confused with the odd roots!) in the tables is related to the continuous parameter, α, occurring in $D(2|1;\alpha)$ as

$$\gamma \equiv \frac{\alpha}{1+\alpha}.$$

Introducing the previously defined combinations Eq.(21), we find for the supercurrents in all cases the following remarkably simple general expression

$$
\begin{aligned}
G_\gamma(z) &= \sum_{\gamma'\in\Delta_+^1} \{\delta_{\gamma,\gamma'}[\frac{(k+\theta^2/2)}{\alpha_+}\partial + \frac{i}{\sqrt{2}}\sqrt{\frac{\theta^2}{2}}\partial\phi(z)] \\
&- \frac{1}{\alpha_+}\sum_i [\hat{J}^{(i)}_{\gamma,\gamma'}(z) + c_2^{(i)} J^{f,(i)}_{\gamma,\gamma'}(z)]\}\chi_{\gamma'}(z)
\end{aligned} \tag{62}
$$

The only difference between the various different algebras is that the constants, c_1, c_2 take different values. These are provided in table (5). In the cases $A(n|1), D(2|n)$ and $D(2,1;\alpha)$ the surviving affine algebra, G, is nonsimple: $G = G^{(1)} \oplus G^{(2)}$, and the constants similarly break up in two. These are listed separately. It is worth while

Table 4: Central charges for G extended superconformal algebras, in terms of the level K^{tot} of the full affine algebra.

g	k	c_{total}
$A(n\|1)$	$K^{tot}+1$	$\frac{6(K^{tot})^2+K^{tot}(n^2+3n+3)+n}{K^{tot}+n}$
$B(n\|1)$	$-K^{tot}-1$	$\frac{K^{tot}(6K^{tot}+4n^2+4n-9)}{2(K^{tot}+2n-2)}$
$D(n\|1)$	$-K^{tot}-1$	$\frac{K^{tot}(6K^{tot}+4n^2-10)}{2(K^{tot}+2n-3)}$
$D(2\|n)$	$2K_1^{tot}$	$\frac{12(K_1^{tot})^2+K_1^{tot}(2n^2+3n-8)+2-2n^2}{K_1^{tot}+n-1}$
$D(2\|1;\alpha)$	$-\gamma(K_1^{tot}+1)$	$-3+6\gamma(K_1^{tot}+1)$
$F(4)$	$-K^{tot}-1$	$\frac{2(2(K^{tot})^2+11K^{tot})}{K^{tot}+4}$
$G(3)$	$K^{tot}+1$	$\frac{9(K^{tot})^2+31K^{tot}}{2(K^{tot}+3)}$

Table 5: Constants occurring in the free field representations.

g	$\theta^2/2$	$c_2^{(1)}$	$c_2^{(2)}$	$k+h^\vee$	$\frac{\alpha_L^2}{2}(K+H^\vee)^{(1)}$	$\frac{\alpha_L^2}{2}(K+H^\vee)^{(2)}$
$A(n\|1)$	-1	1	$\frac{4(n+1)}{(n-1)^2}$	$k+n-1$	$K+n+1$	
$B(n\|1)$	-2	0		$k-2n+3$	$-(K+2n-1)$	
$D(n\|1)$	-2	0		$k-2n+4$	$-(K+2n-2)$	
$D(2\|n)$	-1	1	0	$k+2n-2$	$2(K^{(1)}+n+1)$	$-(K^{(2)}+2)$
$D(2\|1;\alpha)$	-1	$\frac{2}{3}$	$\frac{2}{3}$	k	$-\gamma(K^{(1)}+2)$	$-(1-\gamma)(K^{(2)}+2)$
$F(4)$	$\frac{3}{2}$	$\frac{2}{3}$		$k-3$	$-(K+5)$	
$G(3)$	$-\frac{4}{3}$	$\frac{2}{3}$		$k+2$	$K+4$	

Table 6: Coefficient constants in the general Extended Superconformal Algebra.

g	$f_1/f_2^{(1)}$	$f_2^{(1)}$	$f_2^{(2)}$
$A(n\|1)$	$k-1$	$\frac{1}{2}(k+h^\vee+k)$	$k-1$
$B(n\|1)$	$k+1$	$\frac{1}{2}(k+h^\vee+k+2)$	
$D(n\|1)$	$k+1$	$\frac{1}{2}(k+h^\vee+k+2)$	
$D(2\|n)$	$k+1$	$k+h^\vee+1$	$k+1$
$D(2\|1;\alpha)$	$k+\gamma$	$k+(1-\gamma)$	$k+\gamma$
$F(4)$	$k+1$	$k+h^\vee+2$	
$G(3)$	$k-1$	$k+h^\vee-\frac{4}{3}$	

noticing that these constants are independent of the affine levels, expressed variously in terms of the affine Lie super algebra level, k, or in terms of the ordinary affine $G^{(i)}$ levels.

The above exceedingly compact form of the generic free field realisation specializes to (much more complicated looking) results in the special cases of $osp(N|2)$ and $A(n|1)$, $n > 1$,[40] $A(1|1)$,[41,8] and $D(2|1;\alpha)$,[8] as well as our results for $D(2|n)$ and $D(2|1;\alpha)$.[26] The results for $F(4)$ and $G(3)$ are new.

The full quantum algebra may now be constructed directly from the free field expressions. Indeed the constants are partly determined by demanding closure. Most of the OPE's of the quantum ESA are rather trivial. Thus the OPE's between the energy momentum tensor and the other generators merely indicate that the conformal dimensions of the affine currents is 1, whereas that of the supercurrents is 3/2. Finally the energy momentum tensor with itself requires knowledge of the central charges already provided in table (3). Similarly the OPE's with the affine currents partly just indicate that the energy momentum tensor is an affine scalar whereas the supercurrents are primaries transforming according to suitable representations as already discussed at length. The affine currents with themselves generate the central extensions of the affine algebras as wee have also discussed.

There remains the highly non trivial OPE's between the supercurrents themselves. A glance at the several partly covered examples already presented in the literature will convince anyone that previous formulations were unsatisfactory. Now however, we are able to provide a very simple general expression:

$$\begin{aligned} G_{\theta-\gamma'}(z)G_\gamma(w) &= \frac{f_1(k)}{k+h^\vee}\frac{\delta_{\gamma',\gamma}}{(z-w)^3} - \frac{1}{k+h^\vee}\sum_i f_2^{(i)}(k)\left(2\frac{J^{(i)}_{\gamma,\gamma'}(w)}{(z-w)^2} + \frac{\partial J^{(i)}_{\gamma,\gamma'}(w)}{z-w}\right) \\ &- \frac{2}{\theta^2}\delta_{\gamma,\gamma'}\frac{T(w)-T_G(w)}{z-w} + \frac{1}{k+h^\vee}\frac{(J^2)_{\gamma,\gamma'}(w)}{z-w} \end{aligned} \tag{63}$$

Here

$$T_G(w) \equiv \frac{1}{2(k+h^\vee)}\{\sum_{\alpha\in\Delta^0(G)}\frac{\alpha^2}{2}J_\alpha J_{-\alpha}(w) + \sum_i H_iH_i(w)\}$$

is similar to the sugawara energy momentum tensor, but in fact carries the incorrect normalization, namely the normalization that would be needed for the "hat" part whereas here we are only dealing with the full generators. Also, the trace part of $(J^2)_{\gamma,\gamma'}(w)$ is of a similar form but (in general) with yet another normalization. In each concrete case one may wish to combine these two contributions. The great similarity between this expression and the classical one given previously (and corresponding to the limit $k\to\infty$) is only obtained when the composite operator J^2 is "normal ordered" according to the *symmetrized* prescription

$$: A(z)B(z) :\equiv \frac{1}{2}\oint_z \frac{dw}{2\pi i}\frac{A(z)B(w)+A(w)B(z)}{z-w}.$$

It should be emphasized that the striking simplicity of these expressions is perhaps slightly deceptive. This point is particularly relevant for the cases where the algebra G is non-simple. In these cases, namely, there is no à priori natural relative normalization between the representation matrices and indeed between the roots of the two commuting algebras, had it not been for the fact that the structure arose from a Lie superalgebra reduction. But the fact that these commuting algebras are embedded in a *simple* Lie super algebra implies a precise choice for that relative normalization. It is only with that very choice, which allows us to introduce the currents $J_{\gamma,\gamma'}$ in a particular normalization, that the simplicity obtains. So even though the final expressions appear superficially to make no reference to the Lie super algebra, in fact that underlying structure remains of importance.

5. Degenerate Representation of G Extended Superconformal Algebras

The free field realizations, which we have summarized for all extended superconformal algebras above, can give rise to efficient calculational methods. To begin with, however, they give rise to very highly reducible representations. Thus, in order to discuss representation theory based on them we must introduce appropriate screening charges and understand how reducible representations can be viewed as cohomologies of those. We here take a number of steps in this direction.

The Fock space of the G extended superconformal algebras is a tensor product of ones for $|\Delta_+^1|$ fermions χ_γ, free fields for the affine Lie algebra $\widehat{G}$, and a free boson $\phi(z)$ coupled to the world sheet curvature.

The free field representations of affine Lie algebras $\widehat{G}$ follows from our discussion above.[32] Let $\bar{\alpha}_i$ $(i = 1,\ldots,r)$ be simple roots of the even subalgebra G. $\bar{\lambda}_i$ $(i = 1,\ldots,r)$ the fundamental weights of G satisfying $\frac{2\bar{\lambda}_i\cdot\bar{\alpha}_j}{\bar{\alpha}_j^2} = \delta_{ij}$. $\Phi_{\bar{\lambda}}^{\bar{\Lambda}}(z)$ is a primary field of the affine Lie algebra $\widehat{G}$ at level K, with weight $\bar{\lambda}$ in the highest weight module with highest weight $\bar{\Lambda}$. In the free field representation, this field can be expressed as $p_{\bar{\lambda}}^{\bar{\Lambda}}(z)\exp(\frac{i\bar{\Lambda}\cdot\varphi(z)}{\alpha_+})$, where $p_{\bar{\lambda}}^{\bar{\Lambda}}$ is a polynomial consisting of terms, $\gamma_{\alpha_1}\cdots\gamma_{\alpha_k}(z)$ $(\alpha_i \in \Delta_+(G))$ such that $\bar{\lambda} = -\bar{\Lambda} + \alpha_1 + \cdots + \alpha_k$. (Note that in the present prescription, the vertex operator $\exp(\frac{i\bar{\Lambda}\cdot\varphi(z)}{\alpha_+})$ represents the lowest weight state $\Phi_{-\bar{\Lambda}}^{\bar{\Lambda}}(z)$.)

Denote the total Fock space as $F_{\chi,\bar{\Lambda},p} = F^{\chi} \otimes F_{\bar{\Lambda}}^{G} \otimes F_p^{\phi}$, where F^{χ} is a fermionic Fock space constructed from χ_γ $(\gamma \in \Delta_+^1)$. $F_{\bar{\Lambda}}^{G}$ is a Fock space of the algebra $\widehat{G}$ built on a primary field $\Phi_{-\bar{\Lambda}}^{\bar{\Lambda}} = e^{\frac{i\bar{\Lambda}\cdot\varphi(z)}{\alpha_+}}$. F_p^{ϕ} is a Fock space built on a vertex operator $V_p(z) = e^{ip\sqrt{\theta^2}\phi(z)}$. The dual spaces $(F_{\bar{\Lambda}}^{G})^*$ and $(F_p^{\phi})^*$ are isomorphic to $F_{-2\rho_G-\bar{\Lambda}}^{G}$ and F_{-Q-p}^{ϕ}, respectively.

A primary field of a G extended superconformal algebra is expressed as the products of three fields:

$$V_{\gamma_1,\ldots,\gamma_l\bar{\lambda}p}^{\bar{\Lambda}}(z) = \chi_{\gamma_1}(z)\cdots\chi_{\gamma_l}(z)\Phi_{\bar{\lambda}}^{\bar{\Lambda}}(z)e^{ip\sqrt{\theta^2}\phi(z)}, \tag{64}$$

where γ_i are positive odd roots. The conformal weight of Eq.(64) is given by

$$\Delta = \frac{l}{2} + \frac{\Lambda(\Lambda + 2\rho_G)}{2\alpha_+^2} + \frac{1}{2}(p^2 + Qp)\theta^2. \tag{65}$$

5.1 Screening operators

In order to study the representation of the algebra using free fields, we must specify the screening operators which commute with the generators of the extended superconformal algebra. We consider screening operators which correspond to the simple roots of the Lie superalgebra $\mathbf{g}$. These screening operators are BRST-equivalent to those of the affine Lie superalgebra $\hat{\mathbf{g}}$.[14] In the present choice of the simple root system of the Lie superalgebra in table 1, the simple roots of the even subalgebra G are a subset of those of $\mathbf{g}$ (see Appendix A). Thus, first of all we shall get the screening operators corresponding to the simple roots of the affine Lie algebra $\widehat{G}$. Since the remaining simple roots are odd, they will correspond to fermionic type screening operators. As discussed for example by Kato and Matsuda[42] fermionic screening operators may only occur in single contour integrals without producing ambiguous short distance behaviours. Hence it is convenient to introduce yet another screening operator, which characterizes the A_1 even subalgebra corresponding to the root θ, even though that root does not belong to our choice of simple roots.

5.1.1. Affine screening operators

First we can take the standard screening operators $S_{\bar{\alpha}_i}(z)$ of the affine Lie algebra $\widehat{G}$ as those of G extended superconformal algebra:

$$S_{\bar{\alpha}_i}(z) = s_{\bar{\alpha}_i}(z) e^{-\frac{i\alpha_i \cdot \varphi(z)}{\alpha_+}}, \tag{66}$$

where $s_{\bar{\alpha}_i}$ consists of terms like $\beta_{\bar{\alpha}_i}$ and $\gamma_{\alpha_1} \cdots \gamma_{\alpha_k} \beta_{\alpha_1 + \cdots + \alpha_k + \bar{\alpha}_i}$ with $\alpha_1, \ldots, \alpha_k \in \Delta_+(G)$. These screening operators are used for the characterization of singular vectors in the Fock modules of the affine Lie algebra $\widehat{G}$.

5.1.2. Fermionic screening operators

Next, we consider the screening operators, which corresponds to the odd simple roots. In the case of $A(n|1)$ there are two of those, in all other cases there is only one, cf. Appendix A. We have a fermion for every (negative) odd root (our labeling pertains to the positive odd roots), corresponding to the fundamental representation of the even subalgebra G of dimensions $|\Delta_+^1|$. Denote the highest weight of these representations as $\bar{\Lambda}^*$. As we have seen, the odd roots themselves (suitably projected to the root subspaces of the appropriate even sub algebras) may be taken as the weights. In order that the screening operator commutes with the $\widehat{G}$ currents, it should be a $\widehat{G}$ singlet operator *i.e.* the operator product expansion with the $\widehat{G}$ currents should be regular. These observations lead to fermionic screening operators of the following type:

$$S_f(z) = \sum_\gamma \chi_\gamma \Phi_{\theta-\gamma}^{\bar{\Lambda}^*} \exp\left(-\frac{i\sqrt{\theta^2}\phi(z)}{2\alpha_+}\right), \tag{67}$$

where γ runs over the roots corresponding to the weights of the representation (with the highest weight $\bar{\Lambda}^*$).

The case of $A(n|1)$, where the algebra G is $sl(n+1) \oplus u(1)$, is slightly more complicated than the other cases due to the fact that the representation of the fermions (or the supercurrents) is reducible: Roots γ and $\theta - \gamma$ belong to different irreducible representations. Correspondingly we get two odd roots and two fermionic screening operators by restricting the sum above to weights of one or the other irreducible representation. Further, it may be convenient to explicitly bosonize the ("hat part" of the) $u(1)$ algebra. We refer to our paper[26] for more details.

One may verify that these fermionic screening operators indeed either commute with or produce total derivatives in OPE's with the generators. For the supercurrents this holds by virtue of the Knizhnik-Zamolodchikov equations.[43]

This treatment generalizes previous ones in special cases[40,41,44,26] and again provides a unified account.

5.1.3. Screening operators corresponding to the θ-direction

Finally we need a screening operator to characterize the θ direction. Denote this screening operator by $S_\theta(z)$. Based on several worked out examples we expect that this operator takes the form:

$$S_\theta(z) = s_\theta(z) \exp\left(\frac{2i\alpha_+ \phi(z)}{\sqrt{\theta^2}}\right), \tag{68}$$

where $s_\theta(z)$ is a G singlet operator containing (among many others) a term $\prod_{\gamma \in \Delta_+^1} \chi_\gamma$, and hence having conformal dimension $|\Delta_+^1|/2$. The requirement that $S_\theta(z)$ has conformal dimension 1, then fixes the "momentum" in the vertex operator part. This assumption is justified in part by the following list of results:

When we consider the $osp(N|2)$ case, for $N = 1$ and 2, the screening operators are the well known ones of $N = 1$ and 2 minimal models, and they are of this form: In fact they may be given as $S_\theta = \psi_1 \exp(-\alpha_+\phi(z))$ and $S_\theta = (\psi_1\psi_2 - \frac{1}{K} J_{12}) \exp(-\alpha_+\phi(z))$, respectively.

For the $N = 3$ case, this kind of screening operator has been found by Miki.[44]

In the case of $N = 4$ $sl(2)$ superconformal algebra this screening operator has been obtained by Matsuda,[41] and it is again of the above form. Also we[26] have recently provided the explicit form for the so-called doubly extended $N = 4$ algebra[4,9] corresponding to the case of $D(2|1;\alpha)$ which generalizes the two $N = 4$ cases corresponding to affine $so(4)$ and $su(2)$.

One must expect that this procedure can be generalized to any N, but so far we have not found completely general expressions. From the cases studied it seems clear that we must expect all kinds of terms with pairs of fermions from $\prod_{\gamma \in \Delta_+^1} \chi_\gamma$ being replaced by appropriate affine current operators, in such a way that the combined group theory quantum numbers work out correctly.

In the following we assume the existence of this kind of screening operator for any G. Even though we do not have the explicit form of $s_\theta(z)$ in all cases, the knowledge of the quantum numbers suffices for the determination of singular vectors as we shall see.

5.2. Structure of null fields

Based on the above observations on the screening operators, we discuss the structure of singular vectors of G extended superconformal algebras. We consider the Neveu-Schwarz sector for simplicity.

Firstly we consider the singular fields corresponding to the affine Lie algebra $\widehat{G}$.[32] These are given by the following screened vertex operators:

$$\Psi^{\bar{\Lambda}}_{\beta_1,\ldots,\beta_m}(z) = \oint du_1 \cdots du_m S_{\beta_1}(u_1)\cdots S_{\beta_m}(u_m)\Phi^{\bar{\Lambda}}_{-\bar{\Lambda}}(z), \tag{69}$$

where β_a $(a = 1,\ldots,m)$ are simple roots of the algebra G: i.e. $\beta_a = \bar{\alpha}_{i_a}$ for some $i_a = 1,\ldots,r$. The contours of integrations are taken for example as in the prescription by Kato and Matsuda.[42] The "mass-shell-condition" that the above contour integral represents a singular field is derived either by considerations of pole singularities,[42] or equivalently by noticing that the quantum numbers of the contour integral are the same as those of the field $\Phi^{\bar{\Lambda}}_{-\bar{\Lambda}}(z)$ whereas, if the result is a descendant it must be of a primary pertaining to the weight

$$\bar{\Lambda} - \sum_{a=1}^{m} \beta_a.$$

It is then a simple matter to work out the difference in conformal dimension of these two fields, using Eq.(65) and demand that it be an integer:

$$\frac{1}{\alpha_+^2}\sum_{a<b}\beta_a\cdot\beta_b - \frac{1}{\alpha_+^2}\sum_{a=1}^{m}\beta_a\cdot\bar{\Lambda} = -M, \tag{70}$$

with a positive integer M. If $\sum_{a=1}^{m}\beta_a = n'\bar{\alpha}$ and $M = nn'$ for a positive root $\bar{\alpha} \in \Delta_+(G)$ and positive integers n and n', we get the Kac-Kazhdan formula:[45]

$$(\bar{\Lambda} + \rho_G)\cdot\bar{\alpha} = +n\alpha_+^2 + n'\frac{\bar{\alpha}^2}{2}. \tag{71}$$

This formula and its dual, which is obtained by replacing $\bar{\Lambda}$ by $-2\rho_G - \bar{\Lambda}$, characterize the singular vectors of the Fock module of the affine Lie algebra $\widehat{G}$.

The fermionic singular vectors are given by considering a screened vertex operator of the form:

$$\Psi^{\bar{\Lambda},p}(z) = \oint du S_f(u) V_{\bar{\Lambda},p}(z), \tag{72}$$

where $V_{\bar{\Lambda},p}(z) = \Phi^{\bar{\Lambda}}_{\bar{\Lambda}} \exp(ip\sqrt{\theta^2}\phi(z))$. Again the contour integral has the quantum numbers of $V_{\bar{\Lambda},p}(z)$, but (if singular) must be a descendant of a primary of the Fock module $F^G_{\bar\Lambda+\bar\Lambda^*} \otimes F^\phi_{p-\frac{1}{2\alpha_+}}$ since S_f carries momentum $-1/2\alpha_+$. Equivalently the non-zero existence of the above contour integral requires the condition:

$$\frac{\bar\Lambda \cdot \bar\Lambda^*}{\alpha_+^2} - \frac{p\theta^2}{2\alpha_+} = -M, \tag{73}$$

where M is a positive integer. In this case $\Psi^{\Lambda,p}(z)$ is a singular vector at level $M - \frac{1}{2}$.

The null fields in the θ direction can be obtained from the screened vertex operators:

$$\Psi_r^p(z) = \oint du_1 \cdots du_r S_\theta(u_1) \cdots S_\theta(u_r) V_p(z), \tag{74}$$

where $V_p(z) = e^{ip\sqrt{\theta^2}\phi(z)}$. The on-shell condition becomes

$$\frac{r(r-1)}{2}\left(\frac{2\alpha_+}{\theta^2}\right)^2\theta^2 + 2rp\alpha_+ = -M. \tag{75}$$

with an positive integer M. Writing M as $\frac{r(s-|\Delta_+^1|+2)}{2}$, we find that p is given by

$$p_{r,-s} = -\frac{r-1}{\theta^2}\alpha_+ - \frac{s-|\Delta_+^1|+2}{4\alpha_+}. \tag{76}$$

In this case $\Psi_r^p(z)$ is a null field in the Fock module $F^\phi_{-Q-p_{r,s}}$ at level $\frac{rs}{2}$, where

$$p_{r,s} = \frac{1-r}{\theta^2}\alpha_+ + \frac{s+|\Delta_+^1|-2}{4\alpha_+}. \tag{77}$$

The precise range for s cannot be identified unambiguously until the multiplicities of zeros of the Kac determinant (or something equivalent) has been dealt with.

The formulas Eq.(71), Eq.(73) and Eq.(77) characterize the whole singular vector structure of the G extended superconformal algebras.

6. Conclusions and Discussion

In the present paper we have studied G extended superconformal algebras from the viewpoint of classical and quantum hamiltonian reductions of an affine Lie superalgebra $\hat{\mathbf{g}}$, with even Lie subalgebras $\widehat{G} \oplus \widehat{sl(2)}$. That framework seems to provide an attractive classification scheme for extended superconformal algebras of the type with generators consisting of just the energy-momentum tensor, the (dimension 3/2) supercurrents and some affine currents, the commutation relations, however, being allowed to contain quadratic composites of these generators.

We have derived very compact generic expressions for all these algebras and their free field realizations in the classical case, and we have demonstrated that these expressions also hold in the quantum case modulo simple "renormalizations" of certain constants. We have provided all these constants, thereby completing previous works with partial results.[5,14,18,23,26,10,40,41,44]

We have introduced a set of screening operators for all these G extended superconformal algebras. Using the null field construction, we have identified the primary fields corresponding to degenerate representations of these algebras.

For future work, it would be of interest to work out in detail the cohomologies of the BRST operators we have constructed in order to understand the origin of the free field realizations more completely from that point of view. In general the quantum hamiltonian reduction which we have employed works "algorithmically" only at the level of the energy momentum tensor. It would be very useful to have a better understanding of how the remaining generators are built in terms of free fields as well. The fact that we have found expressions of such generic nature for the supercurrents, may perhaps yield some clues towards such an understanding.

For the screening operators our understanding seems to be enough for constructing singular vectors. Thus it would seem possible to find the Kac-determinants in all cases. For detailed calculation of correlation functions there still remains the task of constructing the completely explicit version of one of the screening charges, which is so far known only in a finite number of cases. With this knowledge it is in principle possible to calculate correlation functions and characters for these models. They will then be expressed as products of those of $\hat{G}$ affine Lie algebras, Virasoro minimal models and free fermions.

Compared to the linearly extended superconformal algebras, a geometrical interpretation of the present non-linear algebras is unclear. It seems an interesting problem to try to find a way of interpreting these non-linear symmetries in terms of non-linear σ-models on non-symmetric Riemannian manifolds. This problem is important in order to clarify the geometrical meaning of the W-algebras.

The present construction of the extended superconformal algebra is based on the hamiltonian reduction of the affine Lie superalgebras. It is well understood that in the case of W-algebras associated with simple Lie algebras, the hamiltonian reduction[12–15] provides a connection to various integrable systems such as Toda field theory[46] and the generalized KdV hierarchy.[47] In the present case it is natural to expect the super Liouville model coupled to Wess-Zumino-Novikov-Witten (WZNW) models or the KdV hierarchy coupled to affine Lie algebras to arise. In the bosonic case, this kind of integrable system has been partially studied.[20]

One might further generalize the hamiltonian reduction procedure to any Lie superalgebra. This would then give rise to a super W-algebra coupled to WZNW models. Interesting work in this direction has been undertaken in particular in the classical case.[30]

Appendix. Root Systems of Lie Superalgebras

In this appendix, we describe the root systems of the Lie superalgebras with the even subalgebra $G \oplus A_1$ as given in table 1 (with the exception of $B(1|n)$, cf. the discussion in the text). θ is the simple root of A_1. We use the orthonormal basis e_i ($i \geq 1$) with positive metric and δ_j ($j \geq 1$) with negative metric:

$$e_i \cdot e_j = \delta_{ij}, \quad \delta_i \cdot \delta_j = -\delta_{ij}, \quad e_i \cdot \delta_j = 0. \tag{78}$$

1. $A(n|1)$ $(n \geq 1)$, (rank $n+2$, the dual Coxeter number $h^\vee = n-1$)
   Simple roots:

   $\alpha_1 = \delta_1 - e_1, \quad \alpha_i = e_{i-1} - e_i, \ (i = 2, \ldots, n+1), \quad \alpha_{n+2} = e_{n+1} - \delta_2.$
   Positive even roots: $e_i - e_j, \ (1 \leq i < j \leq n+1), \quad \theta = \delta_1 - \delta_2.$
   Positive odd roots: $\delta_1 - e_j, \quad e_j - \delta_2, \ (j = 1, \ldots, n+1).$

2. $B(n|1)$ $(n \geq 0)$, (rank $n+1$, $h^\vee = 3-2n$)
   Simple roots:

   $\alpha_1 = e_1 - \delta_1, \quad \alpha_{i+1} = \delta_i - \delta_{i+1}, \ (i = 1, \ldots, n-1), \quad \alpha_{n+1} = \delta_n.$
   Positive even roots: $\theta = 2e_1, \quad \delta_i \pm \delta_j, \ (1 \leq i < j \leq n), \ \delta_i, \ (i = 1, \ldots, n)$
   Positive odd roots: $e_1, \quad e_1 \pm \delta_j, \ (j = 1, \ldots, n).$

3. $D(n|1)$ $(n \geq 2)$, (rank $n+1$, $h^\vee = 4-2n$)
   Simple roots:

   $\alpha_1 = e_1 - \delta_1, \quad \alpha_{i+1} = \delta_i - \delta_{i+1}, \ (i = 1, \ldots, n-1), \quad \alpha_{n+1} = \delta_{n-1} + \delta_n.$
   Positive even roots: $\theta = 2e_1, \quad \delta_i \pm \delta_j, \ (1 \leq i < j \leq n).$
   Positive odd roots: $e_1 \pm \delta_j, \quad (j = 1, \ldots, n).$

4. $D(2|n)$ $(n \geq 1)$ (rank $n+2$, $h^\vee = 2n-2$
   Simple roots:

   $\alpha_1 = -\delta_2 - \delta_1, \quad \alpha_2 = \delta_1 - e_1, \quad \alpha_{i+2} = e_i - e_{i+1}, \ (i = 1, \ldots, n-1), \quad \alpha_{n+2} = 2e_n.$
   Positive even roots: $-\delta_2 - \delta_1, \ \theta = \delta_1 - \delta_2, \quad e_i + e_j, \ (1 \leq i \leq j \leq n), \ e_i - e_j,$ $(1 \leq i < j \leq n).$
   Positive odd roots: $\delta_1 \pm e_j, \ -\delta_2 \pm e_j, \ (j = 1, \ldots, n).$

5. $D(2|1;\alpha)$ $(\alpha \neq 0, -1, \infty)$, (rank 3, $h^\vee = 0$)
   Simple roots:

   $\alpha_1 = \frac{1}{2}(\sqrt{2\gamma}\delta_1 + \sqrt{2(1-\gamma)}\delta_2 + \sqrt{2}e_3), \quad \alpha_2 = -\sqrt{2\gamma}\delta_1, \quad \alpha_3 = -\sqrt{2(1-\gamma)}\delta_2,$
   where $\alpha = \gamma/(1-\gamma)$.
   Positive even roots: $\alpha_2, \quad \alpha_3, \quad \theta = 2\alpha_1 + \alpha_2 + \alpha_3.$
   Positive odd roots: $\alpha_1, \quad \alpha_1 + \alpha_2, \quad \alpha_1 + \alpha_3, \quad \alpha_1 + \alpha_2 + \alpha_3.$

6. $F(4)$ (rank:4, $h^\vee = -3$)
   Simple roots:

   $\alpha_1 = \frac{1}{2}(\sqrt{3}e_1 + \delta_1 + \delta_2 + \delta_3), \quad \alpha_2 = -\delta_1, \quad \alpha_3 = \delta_1 - \delta_2, \quad \alpha_4 = \delta_2 - \delta_3.$
   Positive even roots: $\alpha_2, \quad \alpha_3, \quad \alpha_4, \quad \alpha_2 + \alpha_3, \quad \alpha_3 + \alpha_4, \quad 2\alpha_2 + \alpha_3, \quad \alpha_2 + \alpha_3 + \alpha_4, \ 2\alpha_2 + \alpha_3 + \alpha_4, \quad 2\alpha_2 + 2\alpha_3 + \alpha_4, \quad \theta = 2\alpha_1 + 3\alpha_2 + 2\alpha_3 + \alpha_4.$
   Positive odd roots: $\alpha_1, \quad \alpha_1 + \alpha_2, \quad \alpha_1 + \alpha_2 + \alpha_3, \quad \alpha_1 + 2\alpha_2 + \alpha_3, \quad \alpha_1 + \alpha_2 + \alpha_3 + \alpha_4, \quad \alpha_1 + 2\alpha_2 + \alpha_3 + \alpha_4, \quad \alpha_1 + 2\alpha_2 + 2\alpha_3 + \alpha_4, \quad \alpha_1 + 3\alpha_2 + 2\alpha_3 + \alpha_4.$

7. $G(3)$ (rank:3, $h^\vee = 2$)
   Simple roots:

$\alpha_1 = \sqrt{\frac{2}{3}}\delta_1 + \frac{2e_1-e_2-e_3}{3}, \quad \alpha_2 = \frac{-e_1+2e_2-e_3}{3}, \quad \alpha_3 = -e_2 + e_3.$
Positive even roots: $\alpha_2, \quad \alpha_3, \quad \alpha_2 + \alpha_3, \quad 2\alpha_2 + \alpha_3, \quad 3\alpha_2 + \alpha_3, \quad 3\alpha_2 + 2\alpha_3,$ $\theta = 2\alpha_1 + 4\alpha_2 + 2\alpha_3$.
Positive odd roots: $\alpha_1, \quad \alpha_1+\alpha_2, \quad \alpha_1+\alpha_2+\alpha_3, \quad \alpha_1+2\alpha_2+\alpha_3, \quad \alpha_1+3\alpha_2+\alpha_3,$ $\alpha_1 + 3\alpha_2 + 2\alpha_3, \quad \alpha_1 + 4\alpha_2 + 2\alpha_3$.

References

1. M. Ademollo et al., *Phys. Lett.* **B62** (1976) 105; *Nucl. Phys.* **B114** (1976) 297.
2. K. Schoutens, *Nucl. Phys.* **B295** (1988) 634.
3. Y. Kazama and H. Suzuki, *Nucl. Phys.* **B321** (1989) 232.
4. A. Sevrin, W. Troost and A. Van Proeyen, *Phys. Lett.* **B208** (1988) 447; A. Sevrin, Ph. Spindel, W. Troost and A. Van Proeyen, *Nucl. Phys.* **B308** (1988) 662; **B311** (1988) 465; A. Van Proeyen, Class. Quantum Grav. 6 (1989) 1501.
5. V.G. Knizhnik, Theor. Math. Phys. **66** (1986) 68; M. Bershadsky, *Phys. Lett.* **B174** (1986) 285.
6. A.B. Zamolodchikov, *Theor. Math. Phys.* **63** (1985) 1205; V.A. Fateev and A.B. Zamolodchikov, *Nucl. Phys.* **B280** (1987) 644; V.A. Fateev and S.L. Lukyanov, *Int. J. Mod. Phys.* **A3** (1988) 507.
7. K. Schoutens, *Nucl. Phys.* **B314** (1989) 519.
8. E. Ivanov, S. Krivonos and V. Leviant, *Phys. Lett.* **B215** (1989) 689; *Phys. Lett.* **B221** (1989) 432E; *Int. J. Mod. Phys.*A7 (1992) 287.
9. P. Goddard and A. Schwimmer, *Phys. Lett.* **B214** (1988) 209;
M. Günaydin, J.L. Petersen, A. Taormina and A. Van Proeyen, *Nucl. Phys.* **B322** (1989) 402,
J.L. Petersen and A. Taormina, *Nucl. Phys.* **B331** (1990) 556; *Nucl. Phys.* **B333** (1990) 833; H. Ooguri, J.L. Petersen and A. Taormina, *Nucl. Phys.* **B368** (1992) 611.
10. P. Bowcock, *Chicago preprint EFI 92-09;* E.S. Fradkin and V.Ya. Linetsky, *preprint ITP-SB-92-34, preprint ETH-TH/92-6.*
11. E.S. Fradkin and V.Ya. Linetsky, *Phys. Lett.* **B275** (1992) 345.
12. V.G. Drinfeld and V.V. Sokolov, J. Sov. Math. **30** (1984) 1975.
13. A.A. Belavin, in *"Quantum String Theory" Proceedings in Physics vol.* **31** (Springer Verlag, Berlin ,1989); A. Alekseev and S. Shatashvili, *Nucl. Phys.* **B323** (1989) 719.
14. M. Bershadsky and H. Ooguri, *Commun. Math. Phys.*.**126** (1989) 429.
15. J. Balog, L. Fehér, P. Forgács, L. O'Raifeartaigh and A. Wipf, *Ann. Phys.* **203** (1990) 76.

16. P. Bouwknegt adn K. Schoutens, *preprint CERN-TH.6583/92, ITP-SB-92-23, Physics Reports,* to be published.
17. A.M. Polyakov, *Int. J. Mod. Phys.* **A5** (1990) 833.
18. M. Bershadsky, *Commun. Math. Phys.***139** (1991)71; I. Bakas and D.A. Deprieux, *Mod. Phys. Lett.* **A6** (1991) 1561; M.F.de Groot, T.J. Hollowood and L.J. Miramontes, *preprint IASSNS-HEP-91/19.*
19. L. O'Raifeartaigh, P. Ruelle, I. Tsutsui and A. Wipf, *Commun. Math. Phys.*(1992) 333; L. Fehér, L. O'Raifeartaigh, P. Ruelle and I. Tsutsui and A. Wipf, *Ann. Phys.* **213** (1992) 1; *preprint DIAS-STP-91-29;* P. Bowcock and G.M.T. Watts,*preprint EFI 91-63.*
20. F.A. Bais, T.Tjin and P. van Driel, *Nucl. Phys.* **B357** (1991) 632; T. Tjin and P.van Driel, *preprint ITFA-91-04;*
21. L. Romans, *Nucl. Phys.* **B357** (1991) 549; J. Fuchs, *Phys. Lett.* **B262** (1991) 249.
22. T.G. Khovanova, *Theor. Math. Phys.* **72** (1987) 899;
23. M. Bershadsky and H. Ooguri, *Phys. Lett.* **B229** (1989) 374.
24. J. Evans and T. Hollowood, *Nucl. Phys.* **B352** (1991) 723; S. Komata, K. Mohri and H. Nohara, *Nucl. Phys.* **B359** (1991) 168;
25. V.G. Kac, Adv. Math. **26** (1977) 8; *Lect. Notes in Math. vol.* **676**, p.597 (Springer-Verlag, Berlin, 1978); *Commun. Math. Phys.*53 (1977) 31.
26. K. Ito, J.O. Madsen and J.L. Petersen, *NBI preprint NBI-HE-92-42; Nucl. Phys.***B** to be published; K. Ito, J.O. Madsen and J.L. Petersen, *Phys. Letters* **B292** (1992) 298.
27. K. Ito and J.O. Madsen, *Phys. Lett.* **B283** (1992) 223.
28. M. Berger, *Bull. Soc. Math. France,* **83** (1955) 279.
29. P. Di Vecchia, V.G. Knizhnik, J.L. Petersen and P. Rossi, *Nucl. Phys.* **B253** (1985) 701; E. Abdalla and M.C.B. Abdalla, *Phys. Lett.* **B152** (1985) 59.
30. L. Frappat, E. Ragoucy and P. Sorba, *preprint ENSLAPP-AL-391/92*; F. Delduc, E. Ragoucy and P. Sorba, *Commun. Math. Phys.***146** (1992) 403.
31. B. Feigin and D.B. Fuchs, *Representation of the Virasoro algebra, Seminar on supermanifolds no 5,* ed. D. Leites; Vl.S. Dotsenko and V.A. Fateev, *Nucl. Phys.* **B240** (1984) 232; *Nucl. Phys.* **B251** (1985) 691.
32. M. Wakimoto, *Commun. Math. Phys.***104** (1986) 605; D. Bernard and G. Felder, *Commun. Math. Phys.***127** (1990) 145; B. Feigin and E. Frenkel, in *"Physics and Mathematics of Strings",* eds. L. Brink et al., (World Scientific, 1990 Singapore).
33. A.G. Reĭman and M.A. Semenov-Tjan-Šanskiĭ, *Soviet Math. Dokl.* **21** (1980) 630.
34. J.F. Cornwell, *" Group Theory in Physics" vol. III* (Academic Press, London, 1990)

35. B.S. DeWitt and P.van Nieuwenhuizen, *J. Math. Phys.* **23** (1982) 1953.
36. K. Ito, *Int. J. Mod. Phys.* **A7** (1992) 4885.
37. T. Kugo and I. Ojima, *Prog. Theor. Phys. Suppl.* **66** (1979) 1.
38. J.M. Figueroa-O'Farrill, *Nucl. Phys.* **B343** (1990) 450.
39. N. Hayashi, *Nucl. Phys.* **B363** (1991) 681.
40. P. Mathieu, *Phys. Lett.* **B218** (1989) 185.
41. S. Matsuda, *Phys. Lett.* **B282** (1992) 56.
42. M. Kato and S. Matsuda, *Adv. Stud. Pure. Math.* **16** (1988) 205.
43. V.G. Knizhnik and A.B. Zamolodchikov, *Nucl. Phys.* **B247** (1984) 83.
44. K. Miki, *Int. J. Mod. Phys.* **A5** (1990) 1293.
45. V.G. Kac and D.A. Kazhdan, *Adv. Math.* **34** (1979) 97.
46. A. Bilal and J.L. Gervais, *Nucl. Phys.* **B318** (1989) 579.
47. T.G. Khovanova, *Funct. Anal. Appl.* **20** (1986) 332. K. Yamagishi, *Phys. Lett.* **B205** (1988) 466; P. Mathieu, *Phys. Lett.* **B208** (1988) 101; I. Bakas, *Commun. Math. Phys.*(1989) 627.

TOPOLOGICAL VERSUS NON–TOPOLOGICAL THEORIES AND $p-q$ DUALITY IN $c \leq 1$ 2D GRAVITY MODELS.

S.KHARCHEV, A.MARSHAKOV [1]
P.N.Lebedev Physics Institute, Leninsky pr. 53,
Moscow, 117924, Russia [2]

ABSTRACT

We discuss the non–perturbative formulation for $c \leq 1$ string theory. The field theory like formulation of topological and non–topological models is presented. The integral representation for arbitrary (p, q) solutions is derived which explicitly obeys $p-q$ duality of these theories. The exact solutions to string equation and various examples are also discussed.

1. Introduction

Recent years brought us to a great progress in understanding of non-perturbative string theory. The key idea, established at least for the most simple set of $c \leq 1$ conformal theories interacting with two-dimensional gravity, is the appearance of the structure of integrable hierarchy in the description of generating function for physical correlators in these models [1, 2].

Fortunately, the particular solutions to non-perturbative string theory can be singled from the whole set of solutions to the Kadomtsev-Petviashvili (KP) or rather Toda lattice hierarchy by an additional requirement usually known in the form of string equation, which allows to present these particular solutions in a conventional "field-theory-like" form. Indeed, in series of papers [3, 4, 5, 6, 7] it was shown, that there exists a certain matrix model, describing the particular subset of solutions to (reduced) KP-hierarchy satisfying at the same time the string equation. The proposed matrix theory can be considered as unifying theory for $c \leq 1$ coupled to 2d gravity string models, allowing one to interpolate among them [3, 4, 6], thus being a sort of effective string field theory [7].

Below, we are going to investigate solutions to various (p, q)-models (with central charges $c_{p,q} = 1 - 6\frac{(p-q)^2}{pq}$) coupled to $2d$ gravity in more details. Moreover, we would stress the advantages of matrix integral (or better multiple integral [3]) representation for the particular "stringy" solutions to KP hierarchy.

---

[1] talk given by A.M. at the International Workshop on String Theory, Quantum Gravity and the Unification of Fundamental Interactions, Rome, 21 - 26 September 1992.

[2] E-mail address: tdparticle@glas.apc.org

[3] Indeed, the naive very simple matrix form for such solutions might not be really their basic feature. For example, in the case of ordinary discrete matrix models the generalization to "multimatrix

The basic feature of these matrix integrals is that they give an explicit solution to string equation around a *topological* point. By definition, for the simplest case of $(p = 2, q = 1)$ theory such integral was derived by Kontsevich [8, 9] when studying topological characteristics of 2d gravity module space. In its basic form it obeys all the properties of topological theory – trivial continuum limit (the size of matrix $N \to \infty$) means actually N-independence while N can be interpreted as a cutoff parameter, very simple form of particular solutions (Airy function etc), deep interrelation with the Landau-Ginzburg topological theories and good quasiclassical properties [6]. However, the problem of higher critical points is much more complicated question.

Below, we are going to argue, that it also gets much better understanding along this line. Shortly, higher critical points can be described using the same "action" principle, based on study of the quasiclassical limit [10, 11], but the exact answer has much more complicated form. It means that the higher critical points are no more *topological* theories in the naive sense we used above. [4] The integral representation for these solutions besides the "action" functional has very complicated structure of integration measure. Nevertheless, this integral representation obeys the basic property of $p - q$ duality in the spirit of [12] and might turn to be useful for studying the exact solutions in various cases.

In sect.2 we are going to repeat the main results of [6] on topological solutions and speculate on naive "analytic continuation" to higher critical points. In sect.3 we will formulate the "action" principle and derive an integral representation for arbitrary (p, q)-solutions. In sect.4 we consider $p - q$ symmetry and present several simple and useful examples. Sect.5 contains several examples of $c < 1$ exact (p, q)-solutions and sect.6 – some comments on what is supposed to be a particular example of $c = 1$ situation. In sect.7 we give several concluding remarks.

2. Topological $(p, 1)$ models.

First, we remind that the partition function is defined [3, 4] as a matrix integral

$$Z^{(N)}[V|M] \equiv C^{(N)}[V|M]e^{TrV(M)-TrMV'(M)} \int DX\ e^{-TrV(X)+TrV'(M)X} \tag{1}$$

over $N \times N$ "Hermitean" matrices, with the normalizing factor given by Gaussian integral

$$C^{(N)}[V|M]^{-1} \equiv \int \mathrm{DY}\ e^{-TrV_2[M,Y]},$$

$$V_2 \equiv \lim_{\epsilon\to 0}\frac{1}{\epsilon^2}Tr[V(M+\epsilon Y) - V(M) - \epsilon YV'(M)] \tag{2}$$

---

case" is better done via *conformal multimatrix models* which do not have an elegant representation by *matrix* integral, but are *multiple* integral solutions to the extended (discrete) Virasoro-W constraints [13].

[4]Of course, they could still be reminisents of some more complicated topological theories (and their module spases) like W-gravity etc (see for example [26] and J.-L.Gervais's contribution to this volume).

and Z actually depends on M only through the invariant variables

$$T_k = \frac{1}{k} Tr\ M^{-k},\ k \geq 1\ ; \tag{3}$$

moreover, if rewritten in terms of T_k, $Z[V|T] = Z^{(N)}[V|M]$ is actually independent of the size N of the matrices.

As a function of T_k $Z[V|T_k]$ is a τ-function of KP-hierarchy, $Z[V|T_k] = \tau_V[T_k]$, while the potential V specifies (up to certain invariance) the relevant point of the infinite-dimensional Grassmannian.

For various choices of the potential $V(X)$ the model (1) formally reproduces various (p,q)-series: the potential $V(X) = \frac{X^{p+1}}{p+1}$ can be associated with the entire set of (p,q)-minimal string models with all possible q's. In order to specify q one needs to make a special choice of T-variables: all $T_k = 0$, except for T_1 and T_{p+q} (the symmetry between p and q is implicit in this formulation).

However, this is only a formal consideration. For the potential $V(X) = \frac{X^{p+1}}{p+1}$ the partition function $Z[V|T_k] = \tau_V[T_k] \equiv \tau_p[T_k]$ satisfies the string equation which looks like

$$\sum_{k=1}^{p-1} k(p-k)T_k T_{p-k} + \sum_{k=1}^{\infty}(p+k)(T_{p+k} - \frac{p}{p+1}\delta_{k,1})\frac{\partial}{\partial T_k}\log\ \tau_p[T] = 0 \tag{4}$$

i.e. τ-function is defined with all Miwa times (3) around zero values (in $1/M$ decomposition like in original Kontsevich model) with the only exception - T_{p+1} is shifted what corresponds obviously to $(p,1)$ model. Thus, we see that the matrix integral gives an explicit solution to $(p,1)$ string models which should have mentioned above topological properties and must be nothing but particular topological matter coupled to topological gravity.

Of course, we still have an opportunity for analytic continuation in string equation, using the definition of Miwa's times (3). We have to satisfy the following conditions:

$$\begin{aligned} T_1 &= x \\ T_2 &= 0 \\ &\dots \\ T_{p+1} - \frac{p}{p+1} &= 0 \\ T_{p+q} = t_{p+q} &= fixed \\ T_{p+q+1} &= 0 \\ &\dots \end{aligned} \tag{5}$$

which is a system of equations on the Miwa parameters $\{\mu_i\}$, $i = 1, ..., N$. So, to do this analytical continuation one has to decompose the whole set

$$\{\mu_i\} = \{\xi_a\} \oplus \{\mu'_s\}$$

$$T_k = \frac{1}{k} TrM^{-k} = \frac{1}{k}\sum_{j=1}^{N}\mu_j^{-k} = \frac{1}{k}\sum\xi_a^{-k} + \frac{1}{k}\sum_{j=1}^{N'}\mu'^{-k}_j \equiv T_k^{(cl)} + T'_k \tag{6}$$

into "classical" and "quantum" parts respectively. In principle it is clear that we have now to solve the equations

$$T_k^{(cl)} = \frac{1}{k}\sum\xi_a^{-k} = t_{p+q}\delta_{k,p+q} - \frac{p}{p+1}\delta_{k,p+1} \tag{7}$$

and this can be done adjusting a certain block form of the matrix M [4, 7]. However, in such a way we can only vanish several first times, and the rest ones can be vanished only adjusting correct behaviour in the limit $N \to \infty$. The most elegant way to do this [5] is to look at the formula

$$\exp(-\sum_{k=1}^{\infty}\lambda^k T_k^{(cl)}) = \lim_{K\to\infty}(1 - \frac{1}{K}\sum_{k=1}^{\infty}\lambda^k T_k^{(cl)})^K = \prod_a(1-\frac{\lambda}{\xi_a}) \tag{8}$$

and then the solution to (7) will be given by K sets of roots of the equation

$$\sum_{k=1}^{\infty}\lambda^k T_k^{(cl)} - K = t_{p+q}\lambda^{p+q} - \frac{p}{p+1}\lambda^{p+1} - K = 0 \tag{9}$$

Obviously, the eigenvalues ξ_a will now depend on the size of the matrix $N = (p+q)K + N'$ through explicit K-dependence ($\xi_a \sim K^{1/(p+q)}$) and we lose one of the main features of $(p,1)$ theories mentioned above — trivial dependence of the size of the matrix. Now we can consider only matrices of *infinite* size and deal only with the infinite determinant formulas.

That is why we call such way to get higher critical points as a formal one. Below we will try to understand an alternative way of thinking, connected with so-called p-times. Indeed, it was noticed in [6] that there exists *a priori* another integrable structure in the model (1), connected with time variables, related to the non-trivial coefficients of the potential V. As a results, the cases of monomial potential $V_p(X) = \frac{X^{p+1}}{p+1}$ and arbitrary polynomial of the same degree $(p+1)$ are closely connected with each other. The direct calculation shows (see [6] for details)

$$Z[V|T_k] = \tau_V[T_k] =$$

$$= \exp\left(-\frac{1}{2}\sum A_{ij}(t)(\tilde{T}_i - t_i)(\tilde{T}_j - t_j)\right)\tau_p[\tilde{T}_k - t_k]\ , \tag{10}$$

where

$$V(x) = \sum_{i=0}^{p}\frac{v_i}{i}x^i$$

$$\tilde{T}_k = \frac{1}{k}Tr\tilde{M}^{-k}\ ,$$

[5] suggested by A.Zabrodin

$$\tilde{M}^p = V'(M) \equiv W(M) \ ,$$
$$A_{ij} = Res_\mu W^{i/p} dW_+^{j/p} \ , \tag{11}$$

where $f(\mu)_+$ denotes the positive part of the Laurent series $f(\mu) = \Sigma\ f_i\mu^i$ and

$$\tau_p[T] \equiv \tau_{V_p}[T] \tag{12}$$

– is the τ-function of p-reduction. The parameters $\{t_k\}$ are certain linear combinations of the coefficients $\{v_k\}$ of the potential [17, 10]

$$t_k = -\frac{p}{k(p-k)} Res\ W^{1-k/p}(\mu) d\mu \tag{13}$$

Formula (10) means that "shifted" by flows along p-times (13) τ-function is easily expressed through the τ-function of p-reduction, depending only on the difference of the time-variables $\tilde{T}_k$ and t_k. The change of the spectral parameter in (5) $M \to \tilde{M}$ (and corresponding transformation of times $T_k \to \tilde{T}_k$) is a natural step from the point of view of equivalent hierarchies.

The τ-functions in (10) are defined by formulas

$$\tau_V[T] = \frac{\det\ \phi_i(\mu_j)}{\Delta(\mu)} \tag{14}$$

and

$$\frac{\tau_p[\tilde{T} - t]}{\tau_p[t]} = \frac{\det \hat{\phi}_i(\tilde{\mu}_j)}{\Delta(\tilde{\mu})} \tag{15}$$

with the corresponding points of the Grassmannian determined by the basic vectors

$$\phi_i(\mu) = [W'(\mu)]^{1/2} \exp\left(V(\mu) - \mu W(\mu)\right) \int x^{i-1} e^{-V(x) + xW(\mu)} dx \tag{16}$$

and

$$\hat{\phi}_i(\tilde{\mu}) = [p\tilde{\mu}^{p-1}]^{1/2} \exp\left(-\sum_{j=1}^{p+1} t_j \tilde{\mu}_j\right) \int x^{i-1} e^{-V(x) + x\tilde{\mu}^p} dx \tag{17}$$

respectively. Then it is easy to show that $\hat{\tau}_p(T)$ satisfies the L_{-1}- constraint with *shifted* KP-times in the following way

$$\sum_{k=1}^{p-1} k(p-k)(\tilde{T}_k - t_k)(\tilde{T}_{p-k} - t_{p-k}) + \sum_{k=1}^{\infty} (p+k)(\tilde{T}_{p+k} - t_{p+k}) \frac{\partial}{\partial \tilde{T}_k} \log \hat{\tau}_p[\tilde{T} - t] = 0 \tag{18}$$

where t_i defined by (13) are *identically* equal to zero for $i \geq p+2$.

The formulas (10,18) demonstrate at least two things. First, the partition function in the case of deformed monomial potential ($\equiv$ *polynomial* of the same degree) is expressed through the equivalent solution (in the sense [14, 15]) of the same p-reduced KP hierarchy, second – not only t_{p+1} but all t_k with $k \leq p+1$ are not equal to zero in the deformed situation. We will call such theories as *topologically deformed* $(p,1)$ models (in contrast to *pure* $(p,1)$ models given by monomial potentials $V_p(X)$), the deformation is "topological" in the sense that it preserves all the features of topological models we discussed above. Moreover, this "topological" deformation preserves almost all features of $2d$ Landau-Ginzburg theories and from the point of view of continuum theory they should be identified with the twisted Landau-Ginzburg topological matter interacting with gravity.

From the point of view of KP hierarchy we deal now again with p-reduction. Indeed, from eq.(15) in the limit when all of the Miwa variables $\tilde{\mu}_i$ go to infinity except of the first eigenvalue ($\tilde{\mu}_i \equiv \tilde{\mu}$) one can obtain the expression for the Baker-Akhiezer function which is almost equal to the first basic vector:

$$\psi(\tilde{\mu},t) = \exp\left(\sum_{j=1}^{p+1} t_j \tilde{\mu}_j\right)\hat{\phi}_1(\tilde{\mu}) = [p\tilde{\mu}^{p-1}]^{1/2}\int e^{-V(x)+x\tilde{\mu}^p}dx \tag{19}$$

where potential $V(x)$ is parameterized by p-times $\{t_k\}$ due to eq.(13). It is evident that $\psi(\tilde{\mu},t)$ has the usual asymptotic

$$\psi(\tilde{\mu},t) \underset{\mu\to\infty}{\longrightarrow} \exp\left(\sum_{j=1}^{p+1} t_j\tilde{\mu}_j\right)\left(1+O(\tilde{\mu}^{-1})\right) \tag{20}$$

Using equations of motion for *quasiclassical* KP hierarchy [17, 10]

$$\frac{\partial V}{\partial t_i} = -W_+^{i/p} \tag{21}$$

(this is the consequence of parameterization (13)) one can easy to show that the Baker-Akhiezer function (19) satisfies the usual equations of the p-reduced KP hierarchy:

$$[W(\frac{\partial}{\partial t_1}) + t_1]\psi(\tilde{\mu},t) = \tilde{\mu}^p\psi(\tilde{\mu},t)\ ,$$

$$\frac{\partial\psi}{\partial t_i} = W_+^{i/p}(\frac{\partial}{\partial t_1})\psi\ . \tag{22}$$

where polynomials $W_+^{i/p}(\mu)$ are functions of p-times (13). It is important that $W_+^{i/p}(\mu)$ does not depend on t_1 for $i<p$ and, therefore, in the corresponding Zakharov-Shabat equations we can treat $\partial/\partial t_1$ as a formal *parameter*, not an operator. Thus, we see that topologically deformed $(p,1)$ models which are quasiclassical limit of the KP hierarchy in the sense of [10] are simultaneously the *exact* solutions of the full p-KP hierarchy restricted on the "small phase space" [16]. The Baker-Akhiezer function (19) represent the explicit solution of evolution equations along first p flows and all basic vectors of the deformed $(p,1)$ model can be obtained from $\psi(\tilde{\mu},t)$ with the help of the formula

$$\phi_i(\tilde{\mu},t) = \exp\left(-\sum_{j=1}^{p+1} t_j\tilde{\mu}_j\right)\frac{\partial^{i-1}\psi(\tilde{\mu},t)}{\partial t_1^{i-1}} \tag{23}$$

(Of course, the basic vectors of the pure $(p,1)$ model corresponding to monomial potential V_p can be obtained by setting $t_1 = ... = t_p = 0$, $t_{p+1} = \frac{p}{p+1}$). As a solution to string equation this deformed case differs only in analytic continuation along first p times.

These topologically deformed $(p,1)$ models as we already said preserve all topological properties of $(p,1)$ models. Indeed, according to [2] shifting of first times $t_1, ..., t_{p+1}$ is certainly not enough to get higher critical points. To do this one has to obtain $t_{p+q} \neq 0$, but this cannot be done using above formulas naively, because it is easily seen from definition (13) of p-times, that $t_k \equiv 0$ for $k \geq p+2$. To do this we have to modify the above procedure and we are going to this in next section.

3. Action principle.

The above scheme has a natural quasiclassical interpretation. Indeed, the solution to $(p,1)$ theories given by the partition function (1) can be considered as a "path integral" representation of the solution to Douglas equations [1]

$$[\hat{P}, \hat{Q}] = 1 \tag{24}$$

where $\hat{P}$ and $\hat{Q}$ are certain differential operators (of order p and q) respectively and obviously $p-th$ order of $\hat{P}$ dictates p-reduction, while q stands for $q-th$ critical point. Quasiclassically, (24) turns into Poisson brackets relation [10, 11]

$$\{P, Q\} = 1 \tag{25}$$

where $P(x)$ and $Q(x)$ are now certain (polynomial) functions. It is easily seen that the above case corresponds to the first order polynomial $Q(x) \equiv x$ and the p-th order polynomial $P(x)$ should be identified with $W(x) \equiv V'(x)$ [10]. Thus, the exponentials in (1), (16) and (17) acquire an obvious sense of action functionals

$$S_{p,1}(x,\mu) = -V(x) + xW(\mu) = -\int_0^x dy\ W(y)Q'(y) + Q(x)W(\mu)$$

$$W(x) = V'(x) = x^p + \sum_{k=1}^{p} v_k x^{k-1}$$

$$Q(x) = x \tag{26}$$

and we claim that the generalization to arbitrary (p,q) case must be

$$S_{W,Q} = -\int_0^x dy\ W(y)Q'(y) + Q(x)W(\mu)$$

$$W(x) = V'(x) = x^p + \sum_{k=1}^{p} v_k x^{k-1}$$

$$Q(x) = x^q + \sum_{k=1}^{q} \bar{v}_k x^{k-1} \tag{27}$$

Now the "true" co-ordinate is Q, therefore the extreme condition of action (27) is still

$$W(x) = W(\mu) \tag{28}$$

having $x = \mu$ as a solution, and for extreme value of the action one gets

$$S_{W,Q}\big|_{x=\mu} = \int_0^\mu dy\ W'(y)Q(y) =$$

$$= \sum_{k=-\infty}^{p+q} t_k \tilde{\mu}^k \tag{29}$$

where $\tilde{\mu}^p = W(\mu)$ and

$$t_k \equiv t_k^{(W,Q)} = -\frac{p}{k(p-k)} Res\ W^{1-k/p} dQ\ . \tag{30}$$

We should stress that the extreme value of the action (27), represented in the form (29), determines the quasiclassical (or dispersionless) limit of the p-reduced KP hierarchy [10, 11] with $p+q-1$ independent flows. We have seen that in the case of topologically deformed $(p,1)$ models the quasiclassical hierarchy is exact in the strict sense: topological solutions satisfy the full KP equations and the first basic vector is just the Baker-Akhiezer function of our model (1) restricted to the small phase space. Unfortunately, this is not the case for the general (p,q) models: now the quasiclassics is not exact and in order to find the basic vectors in the explicit form one should solve the original problem and find the exact solutions of the full KP hierarchy along first $p+q-1$ flows. Nevertheless, we argue that the presence of the "quasiclassical component" in the whole integrable structure of the given models is of importance and it can give, in principle, some useful information, for example, we can make a conjecture that the coefficients of the basic vectors are determined by the derivatives of the corresponding *quasiclassical* τ-function [6].

Returning to eq.(30) we immediately see, that now only for $k \geq p+q+1$ p-times are identically zero, while

$$t_{p+q} \equiv t_{p+q}^{(W,Q)} = \frac{p}{p+q} \tag{31}$$

and we should get a correct critical point adjusting all $\{t_k\}$ with $k < p+q$ to be zero. The exact formula for the Grassmannian basis vectors in general case acquires the form

[6] The difference between generic (p,q) and $(p,1)$ cases is also crucial from the point of view of topological nature. We can see here again a distinction between what we call topological and naively non-topological models. The complications in general (p,q) case might be connected with the fact that we use not the most convenient representation for these theories (see also footnote on the second page)

$$\phi_i(\mu) = [W'(\mu)]^{1/2} \exp(- S_{W,Q}|_{x=\mu}) \int d\mathcal{M}_Q(x) f_i(x) \exp\, S_{W,Q}(x,\mu) \tag{32}$$

where $d\mathcal{M}_Q(x)$ is the integration measure. We are going to explain, that the integration measure for generic theory determined by two arbitrary polynomials W and Q has the form

$$d\mathcal{M}_Q(z) = [Q'(z)]^{1/2} dz \tag{33}$$

by checking the string equation. For the choice (33) to insure the correct asymptotics of basis vectors $\phi_i(\mu)$ we have to take $f_i(x)$ being functions (not necessarily polynomials) with the asymptotics

$$f_i(x) \sim x^{i-1}(1 + O(1/x)) \tag{34}$$

4. p-reduction and the Kac-Schwarz operator

To satisfy the string equation, one has to fulfill two requirements: the reduction condition

$$W(\mu)\phi_i(\mu) = \sum_j C_{ij}\phi_j(\mu) \tag{35}$$

and the Kac-Schwarz [19, 20] operator action

$$A^{(W,Q)}\phi_i(\mu) = \sum A_{ij}\phi_j(\mu) \tag{36}$$

with

$$\begin{aligned} A^{(W,Q)} &\equiv N^{(W,Q)}(\mu)\frac{1}{W'(\mu)}\frac{\partial}{\partial\mu}[N^{(W,Q)}(\mu)]^{-1} \\ &= \frac{1}{W'(\mu)}\frac{\partial}{\partial\mu} - \frac{1}{2}\frac{W''(\mu)}{W'(\mu)^2} + Q(\mu) \\ N^{(W,Q)}(\mu) &\equiv [W'(\mu)]^{1/2}\exp(- S_{W,Q}|_{x=\mu}) \end{aligned} \tag{37}$$

These two requirements are enough to prove string equation (see [4] for details). The structure of action immediately gives us that

$$A^{(W,Q)}\phi_i(\mu) = N^{(W,Q)}(\mu)\int d\mathcal{M}_Q(z)Q(z)f_i(z)\exp\, S_{W,Q}(z,\mu) \tag{38}$$

and the condition (36) can be reformulated as a Q-reduction property of basis $\{f_i(z)\}$

$$Q(z)f_i(z) = \sum A_{ij}f_i(z) \tag{39}$$

Let us check now the reduction condition. Multiplying $\phi_i(\mu)$ by $W(\mu)$ and integrating by parts we obtain

$$W(\mu)\phi_i(\mu)$$

$$= N^{(W,Q)}(\mu) \int d\mathcal{M}_Q(z) f_i(z) \frac{1}{Q'(z)} \frac{\partial}{\partial z} [\exp\ Q(z)W(\mu)] \exp[-\int_0^z dy\ W(y)Q'(y)]$$

$$= -N^{(W,Q)}(\mu) \int d\mathcal{M}_Q(z) \exp[S_{W,Q}(z,\mu)] \left(\frac{1}{Q'(z)} \frac{\partial}{\partial z} - \frac{1}{2} \frac{Q''(z)}{Q'(z)^2} - W(z) \right) f_i(z)$$

$$\equiv -N^{(W,Q)}(\mu) \int d\mathcal{M}_Q(z) \exp[S_{W,Q}(z,\mu)] A^{(Q,W)} f_i(z) \tag{40}$$

Therefore, in the "dual" basis $\{f_i(z)\}$ the condition (31) turns to be

$$A^{(Q,W)} f_i(z) = -\sum C_{ij} f_j(z) \tag{41}$$

with $A^{(Q,W)} (\neq A^{(W,Q)})$ being the "dual" Kac-Schwarz operator

$$A^{(Q,W)} = \frac{1}{Q'(z)} \frac{\partial}{\partial z} - \frac{1}{2} \frac{Q''(z)}{Q'(z)^2} - W(z) \tag{42}$$

The representation (32), (33) is an exact integral formula for basis vectors solving the (p,q) string model. It has manifest property of $p-q$ duality (in general $W-Q$), turning the (p,q)-string equation into the equivalent (q,p)-string equation.

Now let us transform (32), (33) into a little bit more explicit $p-q$ form. As before for $(p,1)$ models we have to make substitutions, leading to equivalent KP solutions:

$$\tilde{\mu}^p = W(\mu),\ \tilde{z}^q = Q(z) \tag{43}$$

Then we can rewrite (32) as

$$\hat{\phi}_i(\tilde{\mu}) = [p\tilde{\mu}^{p-1}]^{1/2} \exp\left(-\sum_{k=1}^{p+q} t_k \tilde{\mu}^k \right) \int d\tilde{z} [q\tilde{z}^{q-1}]^{1/2} \hat{f}_i(\tilde{z}) \exp\ S_{W,Q}(\tilde{z},\tilde{\mu}) \tag{44}$$

where action is given now by

$$S_{W,Q}(\tilde{z},\tilde{\mu}) = -\left[\int_0^{\tilde{z}} d\tilde{y} q \tilde{y}^{q-1} W(y(\tilde{y})) \right]_+ + \tilde{z}^q \tilde{\mu}^p$$

$$= \sum_{k=1}^{p+q} \tilde{t}_k \tilde{z}^k + \tilde{z}^q \tilde{\mu}^p \tag{45}$$

In new coordinates the reduction conditions are

$$\tilde{\mu}^p \hat{\phi}_i(\tilde{\mu}) = \sum_j \tilde{C}_{ij} \hat{\phi}_j(\tilde{\mu})$$

$$\tilde{z}^q \hat{f}_i(\tilde{z}) = \sum_j \tilde{A}_{ij} \hat{f}_j(\tilde{z}) \tag{46}$$

and for the Kac-Schwarz operators one gets conventional formulas [19, 20, 4]

$$\tilde{A}^{(p,q)} = \frac{1}{p\tilde{\mu}^{p-1}} \frac{\partial}{\partial \tilde{\mu}} - \frac{p-1}{2p} \frac{1}{\tilde{\mu}^p} + \frac{1}{p} \sum_{k=1}^{p+q} k t_k \tilde{\mu}^{k-p}$$

$$\tilde{A}^{(q,p)} = \frac{1}{q\tilde{z}^{q-1}}\frac{\partial}{\partial\tilde{z}} - \frac{q-1}{2q}\frac{1}{\tilde{z}^q} + \frac{1}{q}\sum_{k=1}^{p+q} k\bar{t}_k\tilde{z}^{k-q} \tag{47}$$

where for (q,p) models we have introduced the "dual" times:

$$\bar{t}_k \equiv t_k^{(Q,W)} = \frac{q}{k(q-k)} Res\ Q^{1-k/q}dW \tag{48}$$

in particularly, $\bar{t}_{p+q} = -\frac{q}{p}t_{p+q} = -\frac{q}{p+q}$. Now string equations give correspondingly

$$\tilde{A}^{(p,q)}\hat{\phi}_i(\tilde{\mu}) = \sum \tilde{A}_{ij}\hat{\phi}_j(\tilde{\mu})$$

$$\tilde{A}^{(q,p)}\hat{f}_i(\tilde{z}) = -\sum \tilde{C}_{ij}\hat{f}_j(\tilde{z}) \tag{49}$$

By these formulas we get a manifestation of $p-q$ duality if solutions to $2d$ gravity.

5. Examples: topological and non-topological theories

Now, let us consider briefly several explicit examples. First, for monomials $W(x) = x^p$ and $Q(x) = x^q$, $\tilde{\mu} \equiv \mu$, $\tilde{z} \equiv z$, $\hat{\phi}_i \equiv \phi_i$ and $\hat{f}_i \equiv f_i$, thus, the formulas of the previous section will be

$$\phi_i(\mu) = [p\mu^{p-1}]^{1/2}\exp\left(-\frac{p}{p+q}\mu^{p+q}\right) \times$$

$$\times \int dz[qz^{q-1}]^{1/2}f_i(z)\exp\left(-\frac{q}{p+q}z^{p+q} + z^q\mu^p\right) \tag{50}$$

and the Kac-Schwarz operators acquire the most simple form

$$A^{(p,q)} = \frac{1}{p\mu^{p-1}}\frac{\partial}{\partial\mu} - \frac{p-1}{2p}\frac{1}{\mu^p} + \mu^q$$

$$A^{(q,p)} = \frac{1}{qz^{q-1}}\frac{\partial}{\partial z} - \frac{q-1}{2q}\frac{1}{z^q} - z^p \tag{51}$$

For any (p,q) theory with $q > p$ the formula (50) maps it onto the corresponding "dual" theory with $q < p$ and vice versa.

In such way one can easily consider the $(p,1)$ topological theories as dual to the higher critical points of the $(1,p)$ theory with the potential $V_2(x) = \frac{1}{2}x^2$, $W_2 = x$. For this theory the "topological" solution is trivial (for example, the partition function is given by a Gaussian integral and equals to unity) so the basis vectors are

$$f_i^{(1,p)}(z) = z^{i-1} \tag{52}$$

and the Kac-Schwarz operator

$$A^{(1,p)} = \frac{\partial}{\partial z} - z^p \tag{53}$$

preserves reduction of the corresponding $(p,1)$ model in a trivial way

$$A^{(1,p)}f_i^{(1,p)}(z) = [\frac{\partial}{\partial z} - z^p]z^{i-1}$$

$$= -z^{i+p-1} + (i-1)z^{i-2} = -f_{i+p-1}^{(1,p)}(z) + (i-1)f_{i-1}^{(1,p)}(z) \tag{54}$$

In this particular case we see how the duality formula turns the problem of finding nontrivial basis of [19, 3, 4, 9] to the trivial basis in the Grassmannian (52), corresponding to sphere.

In general case, we have no more the situation when a non-trivial problem reduces to a trivial one. Moreover, it can be easily proven that for a generic (p,q) model the string equation reduces to a sort of higher hypergeometrical equation giving rise to (linear combinations of) generalized hypergeometric functions [21, 22] (for integral formulas and connection to free-field representation see also [23] and references therein).

Indeed, we can obtain some particular solutions of the conditions (36) concerning only the shift $T_{p+q} \to T_{p+q} - t_{p+q}$ as follows. Let us consider the (p,q) model with $q = pn + \alpha$, $\alpha = 1, \ldots, p-1$; $n = 1, 2, \ldots$ Using condition of p-reduction we can choose the whole basis in the form

$$\phi_{i+pk} = \mu^{pk}\phi_i \ , \ i = 1, \ \ldots \ , p \tag{55}$$

and therefore eq.(36) give the system of equations for first p vectors:

$$A^{(p,q)}\phi_i = \phi_{i+pn+\alpha} = \mu^{pn}\phi_{i+\alpha} \ , \ i = 1, \ \ldots \ , p \tag{56}$$

where in the case under consideration

$$A^{(p,q)} = \frac{1}{p\mu^{p-1}}\frac{\partial}{\partial\mu} - \frac{p-1}{2p}\frac{1}{\mu^p} + \mu^{pn+\alpha} \equiv$$

$$\equiv N^{(p,q)}(\mu)\frac{1}{p\mu^{p-1}}\frac{\partial}{\partial\mu}[N^{(p,q)}(\mu)]^{-1} \tag{57}$$

and

$$N^{(p,q)}(\mu) = [p\mu^{p-1}]^{1/2}\exp\left(-\frac{p}{p+q}\mu^{p+q}\right)$$

$$q = pn + \alpha \tag{58}$$

After the substitution

$$\phi_i = \mu^{i-1}N^{(p,q)}(\mu)u_i(\mu) \tag{59}$$

and changing the spectral parameter

$$z = \frac{p}{p+q}\mu^{p+q} \tag{60}$$

the system (57) acquires the remarkably simple form

$$A_i u_i = u_{i+\alpha}\ ,\ i = 1,\ \dots\ ,\ p;\ u_{j+p} \equiv u_j \tag{61}$$

with

$$A_i = \frac{d}{dz} + \frac{i-1}{p+q}\frac{1}{z}\ . \tag{62}$$

Equations (61) can be easily solved, say, by series expansion

$$u_i(z) = \sum_{j=0}^{\infty} u_{ij} z^j \tag{63}$$

giving

$$u_{1j} = \prod_{k=0}^{p-1} [j + (\frac{1}{p+q} - 1)k]^{-1} u_{1,j-p} \tag{64}$$

which determines function $u_1(z)$ (up to p arbitrary constants) and others can be obtained by action of (62). Up to these constants a particular solution will be determined by the following formula

$$u_{1,np} = \prod_{l=0}^{n} \prod_{k=0}^{p-1} (pl + (\frac{1}{p+q} - 1)k)^{-1} \tag{65}$$

from which follows that the solutions can be expressed through generic hypergeometric functions ${}_rF_s(z)$. We are going to return to this problem in a separate publication [25].

As a concrete example of the relation to the Hypergeometric equation, we present here the solution to $(2, 2k-1)$ model

$$u_1(\mu) = exp(-\frac{2}{2k+1}\mu^{2k+1})\,{}_1F_1(\frac{1}{4k+2}, \frac{1}{2k+1}; \frac{4}{2k+1}\mu^{2k+1}) \tag{66}$$

For the particular case of $k = 1$ this reduces to the well-known solution of the Kontsevich model [8] being a linear combination of the Airy functions [3, 4, 9].

Now, let us only finish with a remark, that formulas (61) are practically equivalent for all $(p, q = np + \alpha)$ series with different n and α. They give a manifestation for a certain cyclic $\mathbf{Z}_p$-symmetry. The only difference in solution to different (p, q)-string equations is the exact order established within the multiplets of order p. The exact sense of this symmetry deserves further investigation.

6. Remarks on $c \to 1$ limit

Let us now make some comments on $c = 1$ situation. From basic point of view we need in generic situation to get the most general (unreduced) KP or Toda-lattice tau-function satisfying some (unreduced) string equation. In a sense this is not a limiting case for $c < 1$ situation but rather a sort of "direct sum" for all (p,q) models. This reflects that in conformal theory coupled to $2d$ gravity there is no more a big difference between $c < 1$ and $c = 1$ situations - quite different before this coupling.

However, there are several particular cases when one can construct a sort of direct $c \to 1$ limit and which should correspond to certain highly "degenerate" $c = 1$ theories. From the general point of view presented above these are nothing but very specific cases of (p, q) string equations, and they could correspond only to a certain very reduced subsector of $c = 1$ theory.

Indeed, it is easy to see, that for two special cases $p = \pm q$ the equations (61) can be simplified drastically, actually giving rise to a single equation instead of a system of them. Of course, these two cases don't correspond to minimal series where one needs (p, q) being coprime numbers. However, we still can fulfill both reduction and Kac-Schwarz condition and these solutions to our equations using naively the formula for the central charge, one mightidentify with $c = 1$ for $p = q$ and $c = 25$ for $p = -q$.

Now, the simplest "topological" theories should be again with $q = 1$. For such case "$c = 1$" turns to be equivalent to a discrete matrix model [5] while "$c = 25$" is exactly what one would expect from generalization of topological Kontsevich-Penner approach [7, 24] (see also R.Dijkgraaf's contribution to this volume). Indeed, taking *non-polynomial* finctions, like

$$
\begin{aligned}
W(x) &= x^{-\beta} \\
Q(x) &= x^{\beta}
\end{aligned}
\tag{67}
$$

the action would acquire a logariphmic term

$$
S_{-\beta,\beta} = -\beta log x + \frac{x^{\beta}}{\mu^{\beta}} \tag{68}
$$

while equations (61) give rise just to rational solutions. It is very easy to see that $\beta = 1$ immediately gives "Kontsevich-Penner" result, which rather corresponds to "dual" to $c = 1$ situation with matter central charge being $c_{matter} = 25$ with a highly non-unitary realization of conformal matter.

On the other hand, $p = q = 1$ solution is nothing but a trivial theory, which however becomes a nontrivial discrete matrix model for unfrozen zero-time. Moreover, these particilar $p = \pm q$ solutions become nontrivial only if one considers the Toda-lattice picture with negative times being involved into dynamics of the effective theory. On the contrary, we know that $c < 1$ (p, q)-solutions in a sense trivially depends on negative times with the last ones playing the role of symmetry of string equation [5]. It means, that we don't yet understand enough the role of zero and negative times in the Toda-lattice formulation.

7. Conclusion

Let us make some conclusive remarks. We tried to present in the paper the exact mechanism of transitions among different (p, q) solutions of non-perturbative

$2d$ gravity in the framework of general scheme proposed in papers [3, 4, 5, 6, 7]. We demonstrated that a naive analytic continuation in the space of Miwa parameters though correct formally leads to certain practical difficulties in explicit description of higher critical points even in trivial topological situation. Instead, we demonstrated a concrete scheme, which allow one to shift "classical" counterparts of the KP times, determined by the coefficients of the potential and by the choice of right variable.

The corresponding integral representation is a direct consequence of the action principle and can be interpreted as a certain field theory integral with a highly non-trivial measure. It obeys manifest $p - q$ symmetry which is evident and restores equivalence in motion along naively two different p- and q- directions. Moreover, the appearance of higher degrees of polynomials can be obtained by transformation from the higher critical points of lower p models.

There is still a lot of open questions. Even in a dual to topological $(p, 1)$ series model there exists nontriviality after $\alpha log X$ term (and negative times terms) are added to the potential. For the $p = 1$ model this gives rise to a separate interesting problem – the discrete Hermitean matrix model [5] and the question is about interpretation of such generalizations of nontrivial theories.

The other question is more deep understanding of generic $c = 1$ situation (which is not reduced to particular "degenerate" cases considered in sect.6) and the role of negative times: symmetry between positive and negative times, the "dissappearing" of negative times in $c < 1$ case etc. Naively, the duality formula leads to a Fredholm equation on basis vector, which can be solved by using the Hermit polynomials, giving rise to the trivialized situation of a discrete 1-matrix model. It is also quite interesting to study the quasiclassical limit of general (p, q) solutions and to compare them with topological theories. This might shed light to the underlying topological structure of naively non-topological theories.

All these problems deserves further investigation and we are going to return to them elsewhere.

8. Acknowledgements

We are deeply indebted to R.Dijkgraaf, L.Girardello, A.Lossev, A.Mironov and A.Zabrodin for illuminating discussions. A.M. is grateful to the organizers of the International Workshop on String Theory, Quantum Gravity and the Unification of Fundamental Interactions for warm hospitality in Rome.

9. References

1. M.Douglas *Phys.Lett.*, **238B** (1990) 176
2. M.Fukuma, H.Kawai, and R.Nakayama *Int.J.Mod.Phys.*, bf A6 (1991) 1385
3. S.Kharchev et al. *Phys.Lett.*, **275B** (1992) 311.
4. S.Kharchev et al. *Nucl.Phys.*, **B380** (1992) 181.

5. S.Kharchev et al. *Generalized Kontsevich model versus Toda hierarchy and discrete matrix models, preprint FIAN/TD/3-92, ITEP-M-3/92* (February, 1992).

6. S.Kharchev et al. *Landau-Ginzburg topological theories in the framework of GKM and equvalent hierarchies, preprint FIAN/TD/7-92, ITEP-M-5/92* (July, 1992).

7. A.Marshakov, *On string field theory for* $c \leq 1$, *preprint FIAN/TD/08-92* (June, 1992).

8. M.Kontsevich sl Funk.Anal.&Prilozh, **25** (1991) 50 (in Russian)

9. M.Kontsevich, *preprint Max-Planck Inst.*, 1991

10. I.Krichever, *Comm.Math.Phys.*, **143** (1992) 415; *The tau-function of the universal Whitham hierarchy, matrix models and topological field theories, preprint LPTENS-92/18.*

11. K.Takasaki and T.Takebe, *Sdiff(2) KP hierarchy, preprint RIMS-814* (October, 1991).

12. M.Fukuma, H.Kawai, and R.Nakayama *Explicit solution for* $p-q$ *duality in two-dimensional quantum gravity, preprint UT-582, KEK-TH-289* (May 1991)

13. S.Kharchev at al. *Conformal multimatrix models*, sl preprint FIAN/TD/9-92, ITEP-M-4/92 (July, 1992).

14. T.Shiota, *Invent.Math.*, **83** (1986) 333.

15. T.Takebe, *From general Zakharov-Shabat equations to the KP and the Toda lattice hierarchies, preprint RIMS-779* (1991).

16. R.Dijkgraaf and E.Witten, *Nucl.Phys.*, **B342** (1990) 486.

17. R.Dijkgraaf, H.Verlinde, and E.Verlinde, *Nucl.Phys.*, **B352** (1991) 59.

18. A.Lossev, *Descendans constructed from matter fields and K.Saito higher residue pairing in Landau-Ginzburg theories coupled to topological gravity, preprint ITEP/TPI-MINN* (May, 1992).

19. V.Kac, A.Schwarz, *Phys.Lett.* **B257** (1991) 329

20. A.Schwarz, *On some mathematical problems of 2d-gravity and* W_h*-gravity, preprint UCD*, 1990

21. W.Bailey *Generalized hypergeometric series.* London, Cambridge Univ. press, 1935

22. H.Bateman, A.Erdelyi *Higher trancendental functions, vol.1, NY*, 1953

23. A.Morozov, L.Vinet, *q-hypergeometric functions in the formalism of free fields, preprint ITEP/Univ. of Montreal*, 1992

24. R.Dijkgraaf, G.Moore, R.Plessner *The partition function 2d string theory, preprint IASSNS-HEP-92/48* (1992).

25. S.Kharchev et al. *preprint FIAN/TD/13-92* (1992), *to be published.*

26. A.Gerasimov et al. *Nucl.Phys.* **B360** (1991) 537

Exact Duality Symmetries in CFT and String Theory*

Elias Kiritsis[†‡]

Laboratoire de Physique Théorique
de l'Ecole Normale Supérieure
24 rue Lhomond
Paris, CEDEX 05, F-75231, FRANCE

Abstract

The duality symmetries of WZW and coset models are discussed. The exact underlying symmetry responsible for semiclassical duality is identified with the symmetry under affine Weyl transformations. This identification unifies the treatement of duality symmetries and shows that in the compact and unitary case they are exact symmetries of string theory to all orders in α' and in the string coupling constant. Non-compact WZW models and cosets are also discussed. A toy model is analyzed suggesting that duality will not generically be a symmetry.

---

*Invited talk presented at the International Workshop on String Theory, Quantum Gravity and the Unification of Fundamental Interactions, Roma, 21-26 Sept. 1992.

[†]Present address: Theory Division, CERN, CH-1211, Geneva 23, Switzerland

[‡]email: KIRITSIS@SURYA1.CERN.CH

1 Introduction, Results and Conclusions

Strings, being extended objects, sense the target space, into which they are embended, in a different way than point particles. The difference comes because, strings, embended in a compact space, except from their local excitations, that mimic point particle behaviour ("momentum" modes), have "winding" excitations where the string wraps around non-contractible cycles of the manifold. The masses of momentum modes are inversely proportional to the volume of the manifold, whereas those of the winding modes are proportional to the volume, since it costs energy in order to stretch the string. In the simplest possible example, that of a string moving on a circle, it was observed that the spectrum of the theory with radius R and that with radius $1/R$ are identical, [1]. This duality symmetry is the same as the electric-magnetic duality symmetry of the underlying 2-d gausian model. Such duality symetries persist in all flat compact backgrounds, [2] and imply the existence of discrete symmetries for the effective theory of string theory around such backgrounds. These discrete symmetries are local, in the sense that they can be considerent as remnants of broken gauge symmetries, present at special points in the space of such flat backgrounds, [3].

The existence of such symmetries poses important questions about the background interpretation of such string ground states (CFTs). Obviously, the string senses the geometry of the target space in a rather "confusing" way. For example, when the string moves on a circle of radius R, just looking at the scattering data, we cannot tell if the radius is R or $1/R$. When R is large or small, then the distinction of the momentum and winding modes makes sense (although which is which depends on whether R is large or small). For $R \sim \mathcal{O}(1)$ however, such a distinction does not make sense any more.

To be more specific, we will discuss here the $R \to 1/R$ duality of a free scalar field in order to set the notation and to derive the formula that will be of use for all semiclassical σ-model duality symmetries, [4]. Consider a scalar field ϕ taking values on a circle of radius R. We will use the convention that the R dependence is explicit and $\phi \in [0, 2\pi)$. Let's consider the partition function in the presence of an external current J_μ

$$Z_R(J) = \int_0^{2\pi} [RD\phi] exp\left[-\frac{R^2}{4\pi}\int \partial_\mu\phi\partial^\mu\phi + \int \partial_\mu\phi J^\mu\right]. \tag{1.1}$$

In order to perform the duality transformation, we will use an infinite dimensional version of the gaussian integration formula,

$$e^{-ab^2} = \frac{1}{2\sqrt{\pi a}}\int_{-\infty}^{+\infty} dx e^{-\frac{x^2}{4a}+ibx} \tag{1.2}$$

in order to make the exponent in (1.1) linear in ϕ. Thus, we obtain

$$Z_R(J) = \int_0^{2\pi} [RD\phi] \int \left[\frac{DB_\mu}{R^2}\right] exp\left[-\frac{\pi}{R^2}\int B_\mu B^\mu + i\int B_\mu\left(\partial^\mu\phi - \frac{2\pi}{R^2}J^\mu\right) + \frac{\pi}{R^2}\int J_\mu J^\mu\right]. \tag{1.3}$$

The crucial step is to go from the (dummy) vector field B_μ to its dual, $B_\mu = \varepsilon_{\mu\nu}A^\nu$. By integrating out ϕ we obtain

$$Z_R(J) = \int \left[\frac{2\pi}{R^2}DA_\mu\right]\delta(F(A))exp\left[-\frac{\pi}{R^2}\int A_\mu A^\mu - \frac{2\pi i}{R^2}\int \varepsilon^{\mu\nu}J_\mu A_\nu + \frac{\pi}{R^2}\int J_\mu J^\mu\right], \tag{1.4}$$

where $F(A) = \varepsilon^{\mu\nu}\partial_\mu A_\nu$. The original theory was invariant under translations of ϕ by a constant. This implies that $\int F(A) = 0$. We will subsequently solve the δ-function constraint by $A_\mu = \partial_\mu\phi/2\pi$ (the jacobian for this is 1) to finally obtain

$$Z_R(J) = \int_0^{2\pi}\left[\frac{D\phi}{R}\right]exp\left[-\frac{1}{4\pi R^2}\int \partial_\mu\phi\partial^\mu\phi - \frac{i}{R^2}\int \varepsilon^{\mu\nu}J_\mu\partial_\nu\phi + \frac{\pi}{R^2}\int J_\mu J^\mu\right]. \tag{1.5}$$

Eq. (1.5) will be enough to derive all σ-model duality transformations. In particular, setting $J = 0$, we obtain the usual duality symmetry $Z_R = Z_{1/R}$.*

The discussion above generalizes to strings propagating on an d-dimensional torus, where there are d generating duality transformations, each for every coordinate.

In order to apply (1.5) to a general σ-model, the presence of a Killing symmetry is needed. In the appropriate coordinates, one can write the action of such a σ-model as

$$S = \frac{1}{4\pi}\int \left[G_{ij}\partial_\mu x^i\partial^\mu x^j + iB_{ij}\varepsilon^{\mu\nu}\partial_\mu x^i\partial_\nu x^j\right], \tag{1.6}$$

where we assume that G_{ij}, B_{ij} do not depend on the coordinate x^0. In terms of x^0 the action (1.6) has the same form as in (1.1) (we will assume here that G_{00} is a constant although this is not necessary†). The identifications are $R^2 \to G_{00}$ and

$$J_\mu \to -\frac{1}{2\pi}\left(G_{0i}\partial_\mu x^i + B_{0i}\varepsilon_{\mu\nu}\partial^\nu x^i\right). \tag{1.7}$$

Then application of (1.5) gives a dual action with

$$\tilde{G}_{00} = \frac{1}{G_{00}}\quad,\quad \tilde{G}_{0i} = \frac{B_{0i}}{G_{00}}\quad,\quad \tilde{B}_{0i} = \frac{G_{0i}}{G_{00}} \tag{1.8a}$$

$$\tilde{G}_{ij} = G_{ij} + \frac{B_{0i}B_{0j} - G_{0i}G_{0j}}{G_{00}}\quad,\quad \tilde{B}_{ij} = B_{ij} + \frac{B_{0i}G_{0j} - B_{0j}G_{0i}}{G_{00}}. \tag{1.8b}$$

There is a change also in the measure, as in (1.5), which can be interpeted as a shift of the dilaton, (see for example [7, 4, 8]).

In a σ-model with d Killing symmetries, the structure of the group of duality transformations is as follows. There are d generating duality transformations D_i, corresponding

---

*Similar results can be obtained for correlation functions. In the path integral framework, the general conformal operator (affine U(1) primary) with $(\Delta, \bar{\Delta}) = ((mR + nR^{-1})^2/4, (mR - nR^{-1})^2/4)$ is represented by the insertion of the field $e^{in\phi(z_0,\bar{z}_0)}$ and the instruction to do the integration over maps $\phi(z,\bar{z})$ which wind m times around the point z_0.

†Non-constant, but x^0-independent G_{00} can be handled by the quotient method, [5, 6].

to doing the transformation (1.8) in the i-th Killing direction. These transformations are commutative

$$D_iD_j = D_jD_i \tag{1.9a}$$

and each one generates a Z_2 group,

$$D_iD_i = 1 \ , \ \ \forall i = 1, 2, \cdots, d \tag{1.9b}$$

The transformation $\prod_{i=1}^{d} D_i$ inverts the generalized metric $G + B \to (G + B)^{-1}$.

When the target space is a d-torus, the σ-model is described by (1.6) with G, B constants. The partition function can be calculated directly via instanton sums

$$Z = \left(\frac{\sqrt{Im\tau}}{|\eta|^2}\right)^d \sum_{\vec{m},\vec{n}} e^{-\frac{\pi}{Im\tau}(\tau\vec{m}+\vec{n})_i(G+B)_{ij}(\bar{\tau}\vec{m}+\vec{n})_j} \ . \tag{1.10}$$

Modular invariance is obvious in (1.10). Let us introduce the $2d \times 2d$ matrices

$$M = \begin{pmatrix} G^{-1} & -G^{-1}B \\ BG^{-1} & G - BG^{-1}B \end{pmatrix} \ , \ H = \begin{pmatrix} 0 & 1 \\ 1 & 0 \end{pmatrix} \ . \tag{1.11}$$

They satisfy

$$M = M^T \ , \ M^{-1} = HMH \ , \ M^THM = H \ , \tag{1.12}$$

which imply that $M \in O(d, d, R)$. $O(d, d, R)$ transformations act implicitly on G, B via $M \to \Omega M \Omega^T$. Upon Poisson resumming (1.10), it can be cast in character form

$$Z = \frac{det(G + B)^{-d/2}}{|\eta|^{2d}} \sum_{\vec{m},\vec{n}} q^{\Delta_{\vec{m},\vec{n}}} \bar{q}^{\bar{\Delta}_{\vec{m},\vec{n}}} \ . \tag{1.13}$$

Introducing a $2d$ vector $\vec{N} \sim \begin{pmatrix} \vec{n} \\ \vec{m} \end{pmatrix}$, we can write the conformal weights as

$$\Delta_{\vec{m},\vec{n}} = Q_i(G^{-1})_{ij}Q_j = N_i(M + H)_{ij}N_j \ , \tag{1.14a}$$

$$\bar{\Delta}_{\vec{m},\vec{n}} = \bar{Q}_i(G^{-1})_{ij}\bar{Q}_j = N_i(M - H)_{ij}N_j \ , \tag{1.14b}$$

$$Q_i = n_i + (G - B)_{ij}m_j \ , \ \bar{Q}_i = n_i - (G + B)_{ij}m_j \ . \tag{1.14c}$$

The generating duality transformations can be represented as $O(d, d)$ transformations

$$D_i \to \begin{pmatrix} 1 - e_i & e_i \\ e_i & 1 - e_i \end{pmatrix} \ , \tag{1.15}$$

where e_i is a $d \times d$ matrix with all elements zero except the diagonal ii element being 1. It is obvious from (1.14) that D_i interchanges $m_i \leftrightarrow n_i$ and thus the invariance of the partition function is obvious. In the basis for the currents in which the metric is unity, this amounts to the transformation $\bar{J}^i \to -\bar{J}^i$.

Provided that J_μ is a classical source, eq. (1.5) is exact. However, subtleties can (and do) arise when J_μ depends on other quantum fields, as is the case in (1.7). Of course our semiclassical considerations will still be valid, but, in general, we expect discrepancies in higher loops. One of our tasks in this paper is to investigate when semiclassical duality is exact (in the sense that it can be corrected beyond 1-loop to yield a genuine symmetry). There are two points of view relevant here. One is the σ-model point of view, which has the advantage that the background interpretation is manifest. The other is the CFT point of view, where, although the background interpretation is not always obvious, it has the advantage that one can get exact results easier.

In this paper we will discuss all the duality symmetries of well-understood CFTs, that is WZW models [9] and their cosets [10, 11].[‡] This class is quite large and contains (modulo a mild assumption) all the CFTs which describe string propagation in a target space with d Killing symmetries. It was argued in [6] that any such σ-model can be obtained by gauging d abelian currents in a WZW model.

The first step is to understand duality in the WZW model. From the σ-model point of view, there are many semiclassical duality transformations, of the type (1.8). By analyzing their effect on the affine primaries, we will be able to identify them with Weyl transformations acting on the current algebra representations. The question of exact duality invariance then translates into invariance under the affine Weyl group. In the case of compact current algebra and unitary (integrable) representations the affine Weyl group is a genuine symmetry. This is not the case in general (where it relates inequivalent representations). We will also see explicitly that the action of the exact duality transformation on the fields is, in general, more complicated than the semiclassical duality transformation.

Once we understand how duality works in the WZW model, we can proceed to the coset models. When we gauge a semisimple subgroup, then the duality symmetry of the coset theory is inherited from that of the original WZW model, and the different dual actions are obtained by gauging the different dual actions of the WZW model. The non-trivial duality transformations are those that leave the subgroup structure invariant. The generic coset G/H model with H semisimple, has extra Killing symmetries, which can be used to generate duality transformations. However these transformations are included in the ones mentioned above. When H is maximal, then the σ-model describing the G/H coset has no Killing symmetries. However, according to our previous discussion it still posseses duality symmetries.

More interesting things happen when H is abelian. In this case, we have the option to gauge an axial or a vector abelian current. Semiclassically, it can be shown that, these

[‡]There is a more general class of CFTs whose structure is much less understood, namely the affine-Virasoro constructions, [12].

two theories are dual to each other, [4, 13]. We will see that the original affine Weyl symmetry of the WZW model guarantees that this extra axial-vector duality is an exact symmetry (although in the σ-model language it needs corrections beyond one-loop). This type of duality is a generalization of the order-disorder (Kramers-Wannier) duality of the critical Ising model. Using axial-vector duality, one can generate new conformal σ-models using $O(d,d,R)$ transformations, [14, 15, 16]. The $O(d,d,R)$ transformations need to be corrected beyong one-loop, however, in the compact case, this can always be done. One implication of this result is that there are marginal $J\bar{J}$ perturbations in σ models with Killing symmetries. If the currents are abelian and chiral this already known. However marginality persists for some combinations of non-chiral abelian currents.

The presence of duality symmetries in compact targets complicates the background interpretation of the σ model. When σ-model couplings are strong, it is difficult to have a geometric notion of a target manifold (even the notion of dimensionality can break down, and many such instances are known, for example $SU(2)_{k=1} \sim U(1)_{R=1}$ etc.). The only case where one has an (almost) unabiguous notion of a manifold is when all couplings are weak, ($\alpha' \to 0$). In curved backgrounds, the dual versions obtained for example by (1.8) are not trustworthy guides of geometry since the dual background describes strong coupling regions.

When one considers string propagation in non-compact backgrounds, the situation is quite different. For Euclidean non-compact cosets, generically, duality is not expected to be a symmetry, since the underlying affine Weyl group relates, in general, inequivalent representations. If one considers a model where the spectrum can be classified into complete orbits of the affine Weyl group, then duality will be restored.[§] There are two potential problems with this procedure. The first is that the required orbits contain representations that are not positive. However, this might not be lethal for the associated string model, but positivity of the string Hilbert space needs to be addressed. [¶] The second is that the background interpretation of such theories is obscure.

In order to investigate whether the semiclassical duality is exact in the non-compact case we will analyse the simplest possible model, where axial-vector duality relates the 2-d Euclidean plane (free field theory), to a certain singular manifold. Although we cannot compute the latter partition function exactly, we will compute it in the "minisuperspace" approximation where it will turn out to be different than that of the plane. We will show, however, that at weak coupling the two coincide. Although, this computation does not settle the issue of exactness of non-compact duality symmetries it does give some useful indications.

[§] This has been effectively done for the SL(2,R)/U(1) coset in [17].

[¶] In the case of $SL(2,R)$ this orbit method works for the discrete series but it is not at all obvious how it could be implemented in the continuous series.

For non-compact cosets with Minkowskian signature, the meaning of the duality transformation is different. Instead of relating two different manifolds, it interchanges various regions of spacetime, [18, 13]. The same remarks apply here as in the Euclidean case. Duality here, although it might not be a symmetry, provides a map that can give meaning to regions of spacetime that one otherwise would traditionally neglect.

The structure of the rest of this paper is as follows. In section 2 we will analyze semiclassically and exactly the duality symmetries of compact WZW models. The same will be done for compact cosets in section 3. The extension of duality symmetries to $O(d,d)$ symmetries will be discussed in section 4. Section 5 contains some remarks on marginal current-current perturbations implied by $O(d,d)$ covariance. Finally, in section 6 we will discuss non-compact cosets.

2 Duality in the WZW model.

In this section we will analyze in detail the duality symmetries of WZW model, both from the σ-model and the CFT (affine current algebra) point of view.

We will consider for simplicity a compact group G which is simple and simply laced. It will turn out that understanding the simplest such group, SU(2), will suffice. In the case of non-simply laced simple groups there are some minor changes due to the short roots that will be dealt with latter on. The case of non-simple groups has further complications that we will not consider here.

The action of the WZW model is

$$I(g) = \frac{k}{4\pi} I_{NS}(g) + \frac{ik}{6\pi}\Gamma_{WZ}(g) \tag{2.1}$$

$$I_{NS}(g) = \int d^2x Tr[U_\mu U^\mu] \ , \ \Gamma_{WZ}(g) = \int_{\substack{B \\ \partial B = S^2}} d^3y \varepsilon^{\mu\nu\rho} Tr[U_\mu U_\nu U_\rho] \tag{2.2}$$

where

$$U_\mu = g^{-1}\partial_\mu g \ , \ V_\mu = \partial_\mu g g^{-1} \tag{2.3}$$

g is a matrix in the fundamental representation of G and Tr is a properly normalized trace such that

$$\frac{1}{12\pi^2}\int_{S^3} Tr[U \wedge U \wedge U] \in Z \ . \tag{2.4}$$

The action $I(g)$ is invariant under the group $G_R \otimes G_L$, generated by left and right group transformations, $g \to h_1 g h_2$, with associated conserved currents

$$J^\mu_R = \frac{k}{2\pi} P^{\mu\nu}_- U_\nu \ , \ J^\mu_L = \frac{k}{2\pi} P^{\mu\nu}_+ V_\nu \tag{2.5}$$

with $P_\pm^{\mu\nu} \equiv \delta^{\mu\nu} \pm i\epsilon^{\mu\nu}$. These currents are conserved and chirally conserved and they generate two copies of the affine $\hat{G}$ current algebra. An important property of the WZW action is that it satisfies the Polyakov-Wiegman formula

$$I(gh) = I(g) + I(h) - \frac{k}{2\pi}\int d^2x P_+^{\mu\nu} Tr[U_\mu(g)V_\nu(h)] \tag{2.6}$$

To generate duality transformations in the WZW model, we pick a generator of the Lie algebra of G, T^0, normalized as $Tr[(T^0)^2] = 1$. We can then parametrize $g = e^{i\phi T^0}h$. Using (1.6), the action $I(g)$ takes the form

$$I(g) = I(h) + \frac{k}{4\pi}\int \partial_\mu\phi\partial^\mu\phi - \frac{ik}{2\pi}\int P_+^{\mu\nu}\partial_\mu\phi V_\nu^0(h) \tag{2.7}$$

where $V_\mu^0(h) = Tr[T^0 V_\mu(h)]$. We can now apply the duality map (1.5) $\to$ (2.7) to obtain[‖]

$$I^{\text{dual}}(g) = I(h) + \frac{1}{4\pi k}\int \partial_\mu\phi\partial^\mu\phi - \frac{i}{2\pi}\int P_+^{\mu\nu}\partial_\mu V_\nu^0(h)\,. \tag{2.8}$$

The angle ϕ was originally normalized to take values in $[0, 2\pi]$. It is obvious from (2.8) that the effect of the duality transformation is to change the range of values to $[0, 2\pi/k]$. To see how many independent duality transformations exist, we have to explicitly parametrize the Cartan torus dependence of the WZW model. Pick a basis in the Cartan algebra, T^i, $i = 1, 2, \cdots, r$, $[T^i, T^j] = 0$, $Tr[T^iT^j] = \delta^{ij}$ and parametrize,

$$g = e^{i\sum_{i=1}^r \alpha^i T^i}\, h\, e^{i\sum_{i=1}^r \gamma^i T^i}\,. \tag{2.9}$$

Then using (2.6) the WZW action becomes

$$I(g) = I(h) + \frac{k}{4\pi}\int(\partial_\mu\alpha^i\partial^\mu\alpha^i + \partial_\mu\gamma^i\partial^\mu\gamma^i) - \frac{ik}{2\pi}\int(P_+^{\mu\nu}\partial_\mu\alpha^i V_\nu^i(h) + P_-^{\mu\nu}\partial_\mu\gamma^i U_\nu^i(h)) +$$
$$+\frac{k}{2\pi}\int P_+^{\mu\nu}\partial_\mu\alpha^i\partial_\nu\gamma^j M^{ij}(h) \tag{2.10}$$

where

$$U_\mu^i(h) = Tr[T^iU_\mu(h)]\,,\ V_\mu^i(h) = Tr[T^iV_\mu(h)]\,,\ M^{ij}(h) = Tr[T^ihT^jh^{-1}]\,. \tag{2.11}$$

It is obvious from (2.10) that we can apply the duality transformation using any of the α^i, γ^i. Thus, there are $2^{2r} - 1$ non-trivial duality transformations. A duality transformation on α^i effectively makes the substitution $\alpha^i \to \alpha^i/k$ in the action whereas a duality transformation on γ^i makes the substitution $\gamma^i \to -\gamma^i/k$.

In order to identify the underlying property of the WZW model, responsible for the invariance under these duality transformations, we have delve a bit into such elements of the the representation theory of the affine Lie algebras as the affine Weyl group and

[‖]The measure also changes by a finite computable piece, see [4]

external automorphisms. Here, I will just state some properties that we need. More information can be obtained in [19] and references therein.

The affine Weyl group $\hat{W}$ is a semidirect product of the Lie algebra Weyl group W times a translation group, $\hat{W} = W \triangleright T$. Appart from the action of finite Weyl group elements, there are Weyl transformations associated to roots which have a component in the direction of the imaginary simple root. The action of such an element $\hat{W}_{\vec{\alpha}}$ on a finite Lie algebra weight $\vec{\lambda}$ and on the grade n is

$$\hat{W}_{\vec{\alpha}}(\vec{\lambda}) = W_{\vec{\alpha}}(\vec{\lambda}) - k\vec{\beta} \tag{2.12a}$$

$$\hat{W}_{\vec{\alpha}}(n) = n - \vec{\lambda}\cdot\vec{\beta} - \frac{k}{2}\vec{\beta}\cdot\vec{\beta} \tag{2.12b}$$

where $\vec{\beta} = 2\vec{\alpha}/\vec{\alpha}\cdot\vec{\alpha}$ is the coroot associated to the finite Lie algebra root $\vec{\alpha}$, the grade n is basically the mode number* and $W_{\vec{\alpha}}(\vec{\lambda}) = \vec{\lambda} - \vec{\alpha}(\vec{\lambda}\cdot\vec{\beta})$ is a finite Weyl transformation. It is important to note that affine Weyl transformations, in general, map states inside a representation at different levels.

There are also external automorphisms of the affine algebra which are essentially associated to symmetries of the affine Dynkin diagram. For the $SU(n)$ case, the affine Dynkin diagram consists of n nodes connected around a circle. The external automorphisms are generated by a basic rotation, and a reflection which corresponds to the finite Lie algebra external automorphism (that maps a representation to its complex conjugate). When we write a highest weight $\vec{\Lambda} = \sum_{i=1}^{n-1} m_i\vec{\Lambda}_i$ in terms of the fundamental weights $\vec{\Lambda}_i$, (m_i are non-negative integers), the action of the generating rotation of the affine Dynkin diagram is as follows

$$\sigma(\vec{\Lambda}) = (k - \sum_{i=1}^{n-1} m_i)\vec{\Lambda}_1 + m_1\vec{\Lambda}_2 + \cdots + m_{n-2}\vec{\Lambda}_{n-1}\ . \tag{2.13}$$

σ generates a Z_n group† where $\sigma^n = 1$ on the heighest weights, but acts as an affine Weyl transformation in the representation. Specializing to SU(2), let $m \in Z/2$ be the weight, and $j \in Z/2$ the highest weight (spin of a representation). Then the finite Weyl group acts as $m \to -m$, and combined with the affine translation $m \to m + k$ they generate the affine Weyl group. The only nontrivial outer automorphism σ acts as $j \to k - j$ and σ^2 is a Weyl translation.

The non-trivial statement now is: For compact groups, integer level and integrable heighest weight representations, both the affine Weyl group and the external automorphisms are symmetries. In particular, in a WZW model the Hilbert space is constructed by tying together (in a modular invariant way) two copies of representations of the affine algebra. Thus, we have invariance under independent affine Weyl transformations acting

*In a highest weight representation where the affine primaries have L_0 eigenvalue Δ, the grade n of a state is the eigenvalue of $L_0 - \Delta$ on that state.

†In general this group is isomorphic to the center of the finite Lie group

on left or right representations. Moreover, since the modular transformation properties of the affine characters reflect the external automorphism symmetries, the theory is invariant under external automorphisms that act at the same time on left and right representations. These invariance properties can be verified for correlation functions on the sphere and the torus. This then implies that they hold on an arbitrary Riemann surface since the sphere and torus data are sufficient in order to construct the correlators at higher genus.

As an example, we will present the SU(2) case and focus on the spectrum. We introduce the (affine) $SU(2)_k$ characters

$$\chi_l(q=e^{2\pi i\tau},w)=Tr_l\left[q^{L_0}e^{2\pi i w J_0^3}\right]=\sum_{m=-k+1}^{k}c^l_m(q)\vartheta_{m,k}(q,w) \tag{2.14}$$

where l is twice the spin (a non-negative integer) and m is twice the J_0^3 eigenvalue. The trace is in the affine hw representation of spin l,

$$\vartheta_{m,k}(q,w)=\sum_{n\in Z}q^{k(n+\frac{m}{2k})^2}e^{2\pi i w(kn+\frac{m}{2})} \tag{2.15}$$

and c^l_m are the standard string functions [20] which satisfy $c^l_m=0$ when $l-m=1(mod\ 2)$ (which means that the spin is increased or decreased in units of 1). For integrable representations (k is a positive integer and $0\le l\le k$), invariance under the affine Weyl group is equivalent to

$$c^l_m=c^l_{-m}\ ,\ \ c^l_m=c^l_{m+2k} \tag{2.16}$$

The first relation is due to the Weyl group of $SU(2)$ while the second is the generating translation in the affine Weyl group. There is another important relation

$$c^l_m=c^{k-l}_{k-m} \tag{2.17}$$

which is a consequence of the external affine automorphism, [20].

The duality tranformation on α^i amounts to replacing $\bar{J}^i\to-\bar{J}^i$, where $\bar{J}^i$ is the right Cartan current in the T^i basis of the Cartan subalgebra. Similarly the duality transformation on γ^i amounts to the replacement $J^i\to-J^i$ at the level of the Cartan subalgebra. This is not the whole story however. With a bit more effort one can see that they act as Weyl transformations on the left or right SU(2) currents. This identification can be seen clearly by coupling the WZW action to external gauge fields and monitoring the effect of the duality transformation on the currents. It can also be recovered from the twisted partition function via the action of the duality transformation on the gauge field moduli (for the Cartan). Let us now check that the duality transformations

$$D_i\ :\ J^i\to-J^i \tag{2.18a}$$

$$\bar{D}_i\ :\ \bar{J}^i\to-\bar{J}^i \tag{2.18b}$$

are exact symmetries of the model. We will consider again for simplicity SU(2) and then generalize to an arbitrary group.

The partition function is

$$Z(q,\bar{q}) = \sum_{l,\bar{l}=0}^{k} N^{l,\bar{l}} \chi_l(q, w=0) \bar{\chi}_{\bar{l}}(\bar{q}, \bar{w}=0) \tag{2.19}$$

where $N^{l,\bar{l}}$ is one of the CIZ modular invariants. The diagonal one $N^{l,\bar{l}} = \delta_{l,\bar{l}}$ corresponds to the usual WZW model. Putting everything together we obtain

$$Z(q,\bar{q}) = \sum_{l,\bar{l}=0}^{k} \sum_{m=-k+1}^{k} \sum_{\bar{m}=-k+1}^{k} N^{l,\bar{l}} c^l_m(q) \bar{c}^{\bar{l}}_{\bar{m}}(\bar{q}) \cdot$$
$$\cdot \sum_{n,\bar{n} \in Z} exp\left[2\pi i \left(\tau k(n + \frac{m}{2k})^2 - \bar{\tau} k(\bar{n} + \frac{\bar{m}}{2k})^2\right)\right] \tag{2.20}$$

The two generating duality transformations here correspond to $n \to -n$, $m \to -m$, and $\bar{n} \to -\bar{n}$, $\bar{m} \to -\bar{m}$. They are symmetries of (2.20) if we use the invariance of the string functions under the affine Weyl group, (2.16).

This invariance is similar, but qualitatively different than that present in flat backgrounds. There, one has a family of theories parametrized by G, B and duality is the statement that two theories are equivalent for different values of the parameters. Here, there is no parameter present and, in this sense, this is what we could call self-duality. This becomes more transparent if we consider the one parameter family of theories, parametrized by the radius of the cartan torus of SU(2). The partition function is known, [21]

$$Z(R) = \sum_{l,\bar{l}=0}^{k} \sum_{m=-k+1}^{k} \sum_{r=0}^{k-1} N^{l,\bar{l}} c^l_m(q) \bar{c}^{\bar{l}}_{m-2r}(\bar{q}) \sum_{M,N \in Z} q^{\Delta_{M,N}} \bar{q}^{\bar{\Delta}_{M,N}} \tag{2.21}$$

with

$$\Delta_{M,N} = \frac{1}{4k}\left(\frac{kM+m-r}{R} + R(kN+r)\right)^2 \quad , \quad \bar{\Delta}_{M,N} = \frac{1}{4k}\left(\frac{kM+m-r}{R} - R(kN+r)\right)^2 \tag{2.22}$$

In (2.21) there is a duality symmetry $R \to 1/R$, which becomes self-duality at the point $R = 1$, that corresponds to the WZW model.

Now we are in a position to discuss the general WZW model for a simple group G. Let M be the root lattice, M_L the long root lattice and M^* the weight lattice. The character of a hw representation of $\hat{G}$ with hw $\vec{\Lambda}$ is defined as

$$\chi_{\vec{\Lambda}}(q, \vec{w}) = Tr[q^{L_0} e^{2\pi i \vec{w} \cdot \vec{J}_0}] \tag{2.23}$$

where $\vec{J}_0$ generates the cartan subalgebra of G. The character admits the string function decomposition, [20]

$$\chi_{\vec{\Lambda}} = \sum_{\vec{\lambda} \in M^*/kM_L} c^{\vec{\Lambda}}_{\vec{\lambda}}(q) \Theta_{\vec{\lambda}}(\vec{w}, q) \tag{2.24}$$

with $\Theta_{\vec{\lambda}}$ being the classical ϑ-function of level k of the Lie algebra of G

$$\Theta_{\vec{\lambda}}(\vec{w},q)=\sum_{\vec{\gamma}\in M_L} q^{\frac{k}{2}\left(\vec{\gamma}+\frac{\vec{\lambda}}{k}\right)^2} e^{2\pi i\vec{w}\cdot(k\vec{\gamma}+\vec{\lambda})}\,. \tag{2.25}$$

The string functions are invariant under the Weyl group and Weyl translations

$$c^{\vec{\Lambda}}_{w(\vec{\lambda})}=c^{\vec{\Lambda}}_{\vec{\lambda}}\ ,\quad c^{\vec{\Lambda}}_{\vec{\lambda}+k\vec{\beta}}=c^{\vec{\Lambda}}_{\vec{\lambda}} \tag{2.26}$$

where w is a Weyl transformation and $\vec{\beta}\in M_L$.

The (left) generating duality transformations D_i correspond to Weyl reflections generated by the simple roots $\vec{\alpha}_i$ which implement the transformations (2.18a). The invariance of the spectrum (and partition function) is encoded in the fact, obvious from (2.25,26), that $\chi_{\vec{\Lambda}}$ is invariant under $w_i\to -w_i$. Although $w_{\vec{\alpha}_i}$ do not commute, they do so when applied to the character, thus at the level of the partition function they generate a group isomorphic to (1.9). However, at the level of correlation functions the (left) duality group is larger and in fact isomorphic to the finite Weyl group of G, W_G. Thus the full duality group of the WZW model is $W_G\times W_G$ the first acting on the left current modules while the second acting on the right current modules.[‡] The structure of the (self)-duality group is different than the one present in flat backgrounds.

3 Compact Cosets

A host of CFTs can be obtained from the coset costruction [10]. In Langrangian form it amounts to gauging a subgroup H of G in a conformally invariant way, [11]. We will assume G to be simple, and H regularly embedded[*].

Let us first consider H to be semi-simple. Then, there is one possible gauging , the vectorial one, [4]. The gauged WZW action is

$$S_V(g,A)=I(g)+\frac{k}{2\pi}\int d^2x Tr[(J^\mu_R-J^\mu_L)A_\mu+P^{\mu\nu}_+A_\mu gA_\nu g^{-1}-A_\mu A^\mu] \tag{3.1}$$

where A_μ belongs to the Lie algebra of H. The action (3.1) is invariant under

$$g\to hgh^{-1}\quad ,\quad A_\mu\to h^{-1}A_\mu h+h^{-1}\partial_\mu h\,. \tag{3.2}$$

Since the gauge field is quadratic in the action (3.1) one can integrate it out, and fix a physical gauge in order to obtain a σ-model desription of the coset theory.

[‡] In [22] a non-abelian form of duality transformations was introduced. It is not clear if these are related to the extended duality group introduced above.

[*] The analysis can be extended to non-simple G and/or irregularly embedded H, but it is certainely more involved.

There are two possible ways to generate duality transformations for the non-abelian coset theory. The first is to gauge different dual versions of the original WZW model. The group of the left duality transformations obtained this way is equivalent to the original Weyl group W_G with the restriction that its subgroup W_H acts trivially. The action of W_H can be absorbed in a redefinition of the gauge fields, and in the σ-model form (where the gauge fields have be integrated out) is trivial.

The other possibility is that the action (3.1) has Killing symmetries which can be exploited in order to generate duality transformations. The constant vector gauge transformations are symmetries of (3.1) but will not survive the passage (gauge fixing) to the σ-model. We can directly hunt for such Killing symmetries. The result is that whenever there exists a subgroup H' of G, such that $[H, H'] = 0$, then, there are extra conserved currents which can be calculated from (3.1),

$$J^\mu_R = P^{\mu\nu}_+ g^{-1}(\partial_\nu g + \{g, A_\nu\})|_{H'} \tag{3.3a}$$

$$J^\mu_L = P^{\mu\nu}_-(\partial_\nu g - \{g, A_\nu\})g^{-1}|_{H'} \tag{3.3b}$$

where $\{,\}$ stands for anti-commutator, and $|_{H'}$ implies a projection onto the Lie algebra of H'. The currents (3.3) transform covariantly under H gauge transformations and they are conserved: $\partial_\mu J^\mu_{R,L} = 0$. A less obvious, but verifiable statement is that these currents are also chirally conserved: $\epsilon_{\mu\nu}\partial^\mu J^\nu_{L,R} = 0$. Thus, they generate a H' current algebra, and this implies that, locally the G/H model can be factorized into a $G/(H \times H')$ model times a H' WZW model. This guarantees the presence of the Killing symmetries associated to the Cartan of H' and the duality transformations they imply have been discussed in the previous section.

A more interesting case is when H is abelian. We will assume without much loss of generality that $H = U(1)$. The situation with more $U(1)$'s will become obvious. In this case, there are two possible ways to gauge. The vector as in the non-abelian case with action given in (3.1) and the axial with action

$$S_A(g, A) = I(g) + \frac{k}{2\pi}\int d^2x Tr[(J^\mu_R + J^\mu_L)A_\mu - P^{\mu\nu}_+ A_\mu g A_\nu g^{-1} - A_\mu A^\mu] \,. \tag{3.4}$$

The axial and vector actions are related by a duality transformation, [13, 4]. In order to show this, we have to parametrize the group element g as in (2.7) where T^0 is the generator of the U(1) subgroup. In order to write the gauged action (3.1), we need the expressions for the left and right U(1) currents

$$J^\mu_R = \frac{k}{2\pi}P^{\mu\nu}_-(iX(h)\partial_\nu\phi + U^0_\nu(h)) \;\;,\;\; J^\mu_L = \frac{k}{2\pi}P^{\mu\nu}_+(i\partial_\nu\phi + V^0_\nu(h)) \tag{3.5}$$

where $X(h) = Tr[T^0 h T^0 h^{-1}]$, as well as (2.7) for the WZW action. Then,

$$S_V(g, A) = I(h) + \frac{k}{4\pi}\int \partial_\mu\phi\partial^\mu\phi - \frac{ik}{2\pi}\int P^{\mu\nu}_+ \partial_\mu\phi V^0_\nu(h) +$$

$$+\frac{ik}{2\pi}\int A_\mu\left(P_-^{\mu\nu}(iX(h)\partial_\nu\phi+U_\nu^0(h))-P_+^{\mu\nu}(i\partial_\nu\phi+V_\nu^0(h))\right)-\frac{k}{2\pi}\int(1-X(h))A_\mu A^\mu. \tag{3.6}$$

Applying the duality transformation (1.5), we obtain

$$S_V\to S_V^{dual}=I^{dual}(g)+\frac{ik}{2\pi}\int A_\mu(\tilde{J}_R^\mu+\tilde{J}_L^\mu)-\frac{k}{2\pi}\int(1+M)A_\mu A^\mu \tag{3.7}$$

where $\tilde{J}^\mu_{R,L}$ are the respective currents of the dual theory

$$\tilde{J}_R^\mu=\frac{k}{2\pi}P_-^{\mu\nu}\left(\frac{i}{k}X(h)\partial_\nu\phi+U_\nu^0(h)\right)\ ,\ \tilde{J}_L^\mu=\frac{k}{2\pi}P_+^{\mu\nu}\left(\frac{i}{k}\partial_\nu\phi+V_\nu^0(h)\right) \tag{3.8}$$

Inspection of (3.7) shows that it is the axially gauged dual WZW model action. Of course, this is not unexpected, since, at the naive level, the vector coset is the WZW model with the constraint $J_L-J_R=0$. As we have seen, a duality transformation of the WZW model changes the sign of one of the currents, and this gives the axial constraint $J_L+J_R=0$.

In order to investigate to what extend this semiclassical axial-vector duality is exact, we will analyze the partition function. In particular we will need a method to compute exactly the partition function for both the axial and the vector gauge theory. The easiest way is the operator method.*

We will start by considering the $SU(2)_k/U(1)$ coset which captures the relevant effects. Once we understand it, the generalization will be simple.

Since we are concerned with the partition function, we will be working on the torus, in the standard flat metric. It is well known, [11] that, in the gauge $\partial_\mu A^\mu=0$, the gauged WZW action factorizes (up to gauge field moduli) to that of the original WZW plus the quadratic action for the gauge field (and the FP determinant, det$'\Box$). The effect of the gauge field moduli is to introduce twisted boundary conditions for the field g of the WZW model around the two non-contactible cycles of the torus. The strategy will be to compute the WZW partition function in the presence of the gauge-field moduli (twists), then integrate over them as specified by the gauge field measure, and then add the contribution of the (decoupled) local part of the gauge field. Consider first the twist in the "space" direction, (vector gauging is considered here). Its effect is to impose a boundary condition on g which is a global $U(1)_V$ transformation ,

$$g(\sigma+1)=e^{i\pi\alpha\sigma_3}g(\sigma)e^{-i\pi\alpha\sigma_3}\ . \tag{3.9}$$

The left and right currents are defined in the standard fashion

$$J=\frac{k}{2\pi}g^{-1}\partial g\ ,\ \bar{J}=\frac{k}{2\pi}\bar{\partial}gg^{-1}\ . \tag{3.10}$$

*More precisely it is a hybrid of operator methods and the path integral approach of Gawedski, [11].

If we introduce cylinder coordinates

$$z = e^{2\pi(t+i\sigma)} \quad , \quad \bar{z} = e^{2\pi(t-i\sigma)} \tag{3.11}$$

(3.9) amounts to

$$J^{\pm}(ze^{2\pi i}) = e^{\pm 2\pi i\alpha} J^{\pm}(z) \Rightarrow J^{\pm}(z) = \sum_{m\in Z\mp\alpha} \frac{J^{\pm}_m}{z^{m+1}} \tag{3.12a}$$

$$\bar{J}^{\pm}(\bar{z}e^{-2\pi i}) = e^{\pm 2\pi i\alpha} \bar{J}^{\pm}(\bar{z}) \Rightarrow \bar{J}^{\pm}(\bar{z}) = \sum_{m\in Z\mp\alpha} \frac{\bar{J}^{\pm}_m}{\bar{z}^{m+1}} \tag{3.12b}$$

while it leaves $J^3, \bar{J}^3$ almost invariant. In fact, due to the central term in the current algebra, both $J^3, \bar{J}^3$ are shifted by the same constant, to be determined below. The vector current $J^3 - \bar{J}^3$ is invariant, as it should be.

The twisted currents satisfy an algebra that is isomorphic to the untwisted SU(2) current algebra. In particular, the Cartan currents are shifted,

$$J^3_m(\alpha) = J^3_m + \frac{k\alpha}{2}\delta_{m,0} \tag{3.13}$$

and similarly for $\bar{J}^3$. Then,

$$[J^{+}_{m-\alpha}, J^{-}_{n+\alpha}] = \frac{k}{2} m\delta_{m+n,0} + J^3_{m+n}(\alpha) \tag{3.14a}$$

$$[J^3_m(\alpha), J^{\pm}_{n\mp\alpha}] = \pm J^{\pm}_{m+n\mp\alpha} \tag{3.14b}$$

$$[J^3_m(\alpha), J^3_n(\alpha)] = \frac{k}{2} m\delta_{m+n,0} \tag{3.14c}$$

and similarly for the left sector.

The Virasoro operator also get shifted. This is standard, we can see it by either doing the Sugawara construction using the twisted currents or checking that the following expression has the proper commutation relations

$$L_m(\alpha) = L_m + \alpha J^3_m + \frac{\alpha^2 k}{4}\delta_{m,0} \,. \tag{3.15}$$

Now we need to twist in the "time" direction. This is achieved in the standard way by inserting a factor

$$e^{2\pi i\beta(J^3_0(\alpha) - \bar{J}^3_0(\alpha))} \tag{3.16}$$

which will eventually project onto invariant states (this is similar to the orbifold case). Thus, collecting everything together, we obtain the (vectorially) twisted WZW partition function

$$Z^{\beta}_{\alpha}(q,\bar{q};V) = Tr_H\left[q^{L_0(\alpha)}\bar{q}^{\bar{L}_0(\alpha)} e^{2\pi i\beta(J^3_0(\alpha) - \bar{J}^3_0(\alpha))}\right] \tag{3.17}$$

where H is the Hilbert space of the WZW theory. Using (2.14,15,19) and (3.13,15) we can explicitly evaluate (3.17),

$$Z_{\alpha}^{\beta}(q,\bar{q})=\sum_{l,\bar{l}=0}^{k}\sum_{m=-k+1}^{k}\sum_{\bar{m}=-k+1}^{k}N^{l,\bar{l}}c_{m}^{l}(q)\bar{c}_{\bar{m}}^{\bar{l}}(\bar{q})\cdot$$

$$\cdot\sum_{n,\bar{n}\in Z}exp\left[2\pi i\left(\tau k(n+\frac{m}{2k}+\frac{\alpha}{2})^{2}-\bar{\tau}k(\bar{n}+\frac{\bar{m}}{2k}+\frac{\alpha}{2})^{2}+\beta(k(n-\bar{n})+\frac{m-\bar{m}}{2})\right)\right]. \tag{3.18}$$

The twisted partition function satisfies

$$Z_{\alpha}^{\beta}=Z_{\alpha+1}^{\beta}=Z_{\alpha}^{\beta+1}=Z_{-\alpha}^{-\beta} \tag{3.19}$$

which can be shown, using the invariance under the affine Weyl group, (2.16) *and* the invariance under the proper outer automorphism, (2.17). Eq. (3.19) specifies the fundamental domain for the gauge field moduli, and agrees with the periodicity implied by the $U(1)$ transformations (3.9,16). Under modular transformations it transforms as

$$Z_{\alpha}^{\beta}(\tau+1,\bar{\tau}+1)=Z_{\alpha}^{\alpha+\beta}(\tau,\bar{\tau}) \tag{3.20a}$$

$$Z_{\alpha}^{\beta}(-\frac{1}{\tau},-\frac{1}{\bar{\tau}})=Z_{\beta}^{-\alpha}(\tau,\bar{\tau}) \tag{3.20b}$$

Eqs. (3.15) imply that, under a modular transformation

$$\tau\to\frac{a\tau+b}{c\tau+d}\;,\quad\begin{pmatrix}a&b\\c&d\end{pmatrix}\in SL(2,Z) \tag{3.21}$$

the gauge field moduli transform linearly

$$\begin{pmatrix}\alpha\\\beta\end{pmatrix}\to\begin{pmatrix}a&b\\c&d\end{pmatrix}\begin{pmatrix}\alpha\\\beta\end{pmatrix} \tag{3.22}$$

The meaning of (3.20-22) becomes more transparent if we introduce complex coordinates in the gauge field moduli space, $u=\alpha\tau+\beta$. Then, using (3.19) we can see that the twisted partition function $Z(u,\bar{u},\tau,\bar{\tau})$ is invariant under the mapping class group of a torus with coordinate u and modulus τ,

$$u\to u+1\;,\quad u\to u+\tau \tag{3.23a}$$

$$\tau\to\tau+1\;,\quad u\to u \tag{3.23b}$$

$$\tau\to-\frac{1}{\tau}\;,\quad u\to\frac{u}{\tau}\,. \tag{3.23c}$$

It remains to calculate the integral over the fundamental region of the moduli

$$\int_{0}^{1}d\alpha\int_{0}^{1}d\beta\, Z_{\alpha}^{\beta}(q,\bar{q};V)\,.$$

We can do the integral over β first. The only terms in the sum (3.18) that contribute are those that satisfy $k(n-\bar{n})+\frac{m-\bar{m}}{2}=0$ Taking into account the ranges of $m,\bar{m}$ and the fact that they are both even or both odd, the only solution is $n=\bar{n}$ and $m=\bar{m}$. Using

$$\int_0^1 d\alpha \sum_{n\in Z} F(n+\alpha)=\int_{-\infty}^{\infty} d\alpha F(\alpha) \tag{3.24}$$

we finally obtain

$$\int_0^1 d\alpha \int_0^1 d\beta Z_\alpha^\beta(q,\bar{q};V)=\frac{\pi}{\sqrt{kIm\tau}}\sum_{l,\bar{l}=0}^{k}\sum_{m=-k+1}^{k} N^{l,\bar{l}}c_m^l(q)\bar{c}_m^{\bar{l}}(\bar{q}) \tag{3.25}$$

To obtain the full partition function for the coset we have to multiply (3.25) with the contribution from the local part of the gauge field , $(det'\Box)^{-1/2}$ and the FP determinant, $(det'\Box)$ giving a net contribution $(det'\Box)^{1/2}=|\eta(q)|^2$ and an extra factor of $\sqrt{Im\tau}$ coming from the measure of the twists (This factor is standard and can be read from the norm of the gauge field $|\delta A|^2=\int\sqrt{g}g^{\mu\nu}\delta A_\mu\delta A_\nu$ using the proper flat metric for a torus parametrized by τ). Putting everything together we obtain (up to constants)

$$Z_{SU(2)/U(1)}^{V}=|\eta(q)|^2\sum_{l,\bar{l}=0}^{k}\sum_{m=-k+1}^{k} N^{l,\bar{l}}c_m^l(q)\bar{c}_m^{\bar{l}}(\bar{q}) \tag{3.26}$$

which is the correct parafermionic partition function, [23].

Let us now consider the axial case. The boundary condition (3.9) is replaced by

$$g(\sigma+1)=e^{i\pi\alpha\sigma_3}g(\sigma)e^{i\pi\alpha\sigma_3} \tag{3.27}$$

From (3.10) we can verify that $\bar{J}^{\pm}$ is twisted with the oposite sign of α compared to $J^{\pm}$. Thus, the axial partition function is proportional to

$$\int_0^1 d\alpha \int_0^1 d\beta Tr_H\left[q^{L_0(\alpha)}\bar{q}^{\bar{L}_0(-\alpha)}e^{2\pi i\beta(J_0^3(\alpha)+\bar{J}_0^3(-\alpha))}\right]. \tag{3.28}$$

Doing the integrals over the moduli, we obtain in this case

$$Z^A=\frac{1}{2}|\eta(q)|^2\sum_{l,\bar{l}=0}^{k}\sum_{m=-k+1}^{k} N^{l,\bar{l}}c_m^l(q)(\bar{c}_{-m}^{\bar{l}}(\bar{q})+\bar{c}_{-m-2k}^{\bar{l}}(\bar{q})) \tag{3.29}$$

Using again the symmetry under the affine Weyl group (2.26) we obtain that

$$Z^A=Z^V \tag{3.30}$$

One final comment is in order here, concerning the SU(2)/U(1) case: We can also compute the parafermionic partition function $Z_{SU(2)/U(1)}(r,s)$ twisted around the two cycles of the torus by two elements of its parafermionic symmetry $Z_k\times\bar{Z}_k$, ($e^{2\pi ir/k}$, $e^{2\pi is/k}$). The way to do this is to allow a general twist

$$g(\sigma+1)=e^{i\pi\alpha\sigma_3}g(\sigma)e^{i\pi\bar{\alpha}\sigma_3} \tag{3.31}$$

with $k(\alpha-\bar{\alpha})/2 = r \bmod k$. We must also project in the time direction on $J-\bar{J} = s \bmod k$. Since the twist is now neither axial nor vector there is the standard modular anomaly that can be cancelled by multiplying the twisted partition function by $exp(\pi k|\tau|^2(\alpha-\bar{\alpha})^2/4Im\tau)$.

The procedure described above for the $SU(2)/U(1)$ coset easily generalizes. Consider a simple group G. We will gauge the maximal abelian subgroup, namely the Cartan subalgebra. Thus we will be looking at the theory of generalized parafermions, [24]. A convenient basis to work with is the Chevaley basis. Let J^i be a basis of the Cartan, and $\vec{\alpha}, \vec{\alpha}_i \in M$ denote the roots and simple roots respectively. The zero modes of the currents in this basis satisfy

$$[J^i, J^j] = 0 \ , \ [J^i, J^{\vec{\alpha}}] = \frac{2\vec{\alpha}\cdot\vec{\alpha}_i}{\vec{\alpha}^2} J^{\vec{\alpha}} \ , \ [J^{\vec{\alpha}}, J^{-\vec{\alpha}}] = \sum_i m_i J^i \tag{3.32a}$$

$$\frac{2\vec{\alpha}}{\vec{\alpha}^2} = \sum_i m_i \frac{2\vec{\alpha}_i}{\vec{\alpha}_i^2} \ , \ [J^{\vec{\alpha}}, J^{\vec{\beta}}] = \varepsilon_{\vec{\alpha},\vec{\beta}} r_{\vec{\alpha},\vec{\beta}} J^{\vec{\alpha}+\vec{\beta}} \ for \ \vec{\alpha}+\vec{\beta} \in M \tag{3.32b}$$

$$[J^{\vec{\alpha}}, J^{\vec{\beta}}] = 0 \ for \ \vec{\alpha}+\vec{\beta} \notin M \tag{3.32c}$$

where $r_{\vec{\alpha},\vec{\beta}}$ is the smallest integer such that $\vec{\beta} - r\vec{\alpha} \notin M$ and $\varepsilon_{\vec{\alpha},\vec{\beta}} = \pm 1$. The non-zero components of the Killing form in this basis are given by

$$\kappa(J^i, J^j) \equiv \kappa_{ij} = \frac{4\vec{\alpha}_i\cdot\vec{\alpha}_j}{\vec{\alpha}_i^2\vec{\alpha}_j^2} \ , \ \kappa(J^{\vec{\alpha}}, J^{-\vec{\alpha}}) = \frac{2}{\vec{\alpha}^2} \tag{3.33}$$

Finally, the central term in the current algebra is given by k times the Killing form.

We can now impose the twisted boundary conditions similar to (3.9)

$$g(\sigma+1) = e^{2\pi i z_i T^i} g(\sigma) e^{-2\pi i z_i T^i} \tag{3.34}$$

Their effect is to twist the algebra in the following way

$$J^{\vec{\alpha}}_m \to J^{\vec{\alpha}}_{m-s(\vec{\alpha})} \ , \ s(\vec{\alpha}) = \sum_i z_i \frac{2\vec{\alpha}\cdot\vec{\alpha}_i}{\vec{\alpha}^2} \tag{3.35a}$$

$$J^i_m \to J^i_m + k(\sum_j \kappa_{ij} z_j)\delta_{m,0} \tag{3.35b}$$

$$L_m \to L_m + \sum_i z_i J^i_m + \frac{k}{2}\left(\sum_{i,j} \kappa_{ij} z_i z_j\right)\delta_{m,0} \, . \tag{3.35c}$$

The projection factor now becomes

$$e^{2\pi i \sum_{i,j} w_i \kappa_{ij} (J^j_0 - \bar{J}^j_0)} \, . \tag{3.36}$$

Using the string decomposition formulae, (2.24,25) (and properly accounting for the change in the metric) we obtain

$$Z_G = \sum_{\Lambda,\bar{\Lambda}} \sum_{\vec{\lambda}_{1,2} \in \frac{M^*}{kM_L}} N^{\Lambda,\bar{\Lambda}} c^{\Lambda}_{\vec{\lambda}_1}(q) \bar{c}^{\bar{\Lambda}}_{\vec{\lambda}_2}(\bar{q}) \sum_{\vec{\alpha},\vec{\beta} \in M_L} q^{\frac{k}{2}(\vec{\alpha}+\vec{z}+\frac{\vec{\lambda}_1}{k})^2} \bar{q}^{\frac{k}{2}(\vec{\beta}+\vec{z}+\frac{\vec{\lambda}_2}{k})^2} e^{2\pi i \vec{w}\cdot(k(\vec{\alpha}-\vec{\beta})+\vec{\lambda}_1-\vec{\lambda}_2)}. \tag{3.37}$$

At this stage, inspection of (3.37) reveals that all we have found in the SU(2) case goes through here. In particular, a Weyl reflection generated by the simple root $\vec{\alpha}_i$ is generating from (3.37) the partition function with the $U(1)$ subgroup in that direction axially gauged.

An interesting point is that the (axial-vector) duality transformation in the abelian coset can be effected also via an orbifold construction , (this has been observed for $SU(2)_k/U(1)$ in [23]). Let us consider the parafermionic partition function on the torus with boundary conditions around the two cycles twisted by elements $e^{2\pi i r/k}$, $e^{2\pi i s/k}$ of the Z_k parafermionic symmetry. This can be evaluated to be

$$Z(r,s) = \frac{1}{2}|\eta|^2 e^{-2\pi i \frac{s}{k}} \sum_{l=0}^{k} \sum_{m=-k+1}^{k} c^l_m \bar{c}^l_{m-2r} \tag{3.38}$$

and the usual vector partition function is $Z(0,0)$. If we construct the orbifold of the original theory with repect to the Z_k symmetry, (which amounts to summing over r, s), we obtain the axial partition function. In this respect the orbifold projection throws out the order operators and adds as twisted sectors the disorder operators. This is precisely the generalization of what is known to happen in the Ising model. This automatically generalizes to arbitrary abelian cosets where the parafermionic symmetry group is isomorphic to M^*/kM_L. Thus we have the following sequence. We gauge the vector U(1), and thus obtain a model with a Killing symmetry associated with axial U(1). This U(1) symmetry is broken to a discrete group (the parafermionic symmetry). Doing an orbifold on that symmetry (which amounts to a flat gauging) we obtain the axially gauged theory.

4 $O(d,d)$ symmetries

Combining duality transformations with antisymmetric tensor shifts and linear transformations on the Cartan angles, we can generate a bigger "duality" group which at the level of the partition function acts as $O(d,d,Z)$, [6]. In the case of toroidal backgrounds it is easy to see how this works. Invariance under $B_{ij} \to B_{ij} + N_{ij}$ is obvious from (1.14c), where N_{ij} is an antisymmetric matrix with integer entries. Also, invariance under $G + B \to U(G+B)U^T$ is obvious from (1.10), where U is an arbitrary matrix with integer entries. The duality group (1.9) and the transformations above generate the $O(d,d,Z)$ group.

In the non-flat case, similar arguments apply, [6], with one difference: the duality group acting on the full operator content of the theory is more complicated than its reduction on the partition function. This is the reflection of our observation that the full duality group in the non-abelian case is isomorphic to the finite Weyl group (or its reductions by subgroups). Keeping this in mind, we can reproduce easily the argument for the partition function. The most general σ-model action with d chiral currents is of the form (2.10). We will rewrite it in chiral form, and we will explicitly parametrize the $I(h)$. We will be a bit more general than [6] by allowing arbitrary radii for the Cartan angles. The general action takes then the form (up to total derivatives)

$$S=\frac{1}{4\pi}\int\Big[\kappa_{ij}(\partial\alpha^i\bar\partial\alpha^j+\partial\gamma^i\bar\partial\gamma^j)+2\Sigma_{ij}(x)\partial\alpha^i\bar\partial\gamma^j+\Gamma^1_{ai}(x)\partial x^a\bar\partial\gamma^i+\Gamma^2_{ia}(x)\bar\partial x^a\partial\alpha^i+$$
$$+\Gamma_{ab}(x)\partial x^a\bar\partial x^b\Big]-\frac{1}{8\pi}\int R^{(2)}\Phi(x) \tag{4.1}$$

where α^i,γ^i take values in $[0,2\pi]$. The action (4.1) is invariant under $\alpha^i\to\alpha^i+\varepsilon^i(\bar z)$ and $\gamma^i\to\gamma^i+\zeta^i(z)$ with associated chiral (abelian) currents,

$$J^i=\partial\gamma^j\kappa_{ji}+\partial\alpha^j\Sigma_{ji}+\frac{1}{2}\partial x^a\Gamma^1_{ai} \tag{4.2a}$$

$$\bar J^i=\kappa_{ij}\bar\partial\alpha^j+\Sigma_{ij}\bar\partial\gamma^j+\frac{1}{2}\Gamma^2_{ia}\bar\partial x^a\,. \tag{4.2b}$$

This automatically implies (assuming conformal invariance) that S describes a (not direct, in general) tensor product of a WZW model and some arbitrary decoupled CFT. The currents (4.2) generate the Cartan subalgebra of the full current algebra of the WZW model. We can gauge vectorially the Cartan subalgebra,

$$S_V=S+\frac{1}{4\pi}\int\Big[A^i\bar J^i-\bar A^iJ^i+\frac{1}{2}A^i(\kappa-\Sigma)_{ij}\bar A^j\Big]\ . \tag{4.3}$$

Integrating out the gauge fields and gauge fixing $\alpha^i=\gamma^i$ we obtain

$$S_c=\frac{1}{4\pi}\int\Big[E_{ij}(x)\partial\alpha^i\bar\partial\alpha^j+F^1_{ai}(x)\partial x^a\bar\partial\alpha^i+F^2_{ia}(x)\partial\alpha^i\bar\partial x^a+F_{ab}(x)\partial x^a\bar\partial x^b\Big]-\frac{1}{8\pi}\int R^{(2)}\phi(x) \tag{4.4}$$

where

$$E_{ij}(x)=4\kappa(1+\kappa^{-1}\Sigma)(1-\kappa^{-1}\Sigma)^{-1} \tag{4.5a}$$

$$F^2(x)=2\kappa(\kappa-\Sigma)^{-1}\Gamma^2\ ,\quad F^1(x)=2\Gamma^1(\kappa-\Sigma)^{-1}\kappa\ ,\quad F=\Gamma-\frac{1}{2}\Gamma^1(\kappa-\Sigma)^{-1}\Gamma^2 \tag{4.5b}$$

$$\phi(x)=\Phi+\log(\det(\kappa-\Sigma))\,. \tag{4.5c}$$

The interesting observation is that, given a σ-model (4.4) with d Killing symmetries we can always construct it as an abelian coset of a WZW model (4.1). The reason is that relations (4.5) are generically invertible. There is an underlying assumption in this, that should be kept in mind, namely that conformal invariance of (4.4) implies conformal invariance of (4.1).

The duality generators D_i correspond to switching from vector to axial in the i-th component of the gauging. There are also the following obvious symmetries, integer shifts of the antisymmetric tensor $E_{ij}(x) \to E_{ij}(x) + N_{ij}$ with N an antisymmetric integer matrix, and integer linear transformations of the angles α^i which act as $E \to UEU^T$, $F^2 \to UF^2$, $F^1 \to F^1U^T$. The full group of invariance of the partition function is the $O(d,d,Z)$ group acting as

$$\begin{pmatrix} E & F^2 \\ F^1 & F \end{pmatrix} \to \begin{pmatrix} (aE+b)(cE+d)^{-1} & (ac^{-1}d-b)(cE+d)^{-1}cF^2 \\ F^1(cE+d)^{-1} & F-F^1(cE+d)^{-1}cF^2 \end{pmatrix} \tag{4.6}$$

where

$$\begin{pmatrix} a & b \\ c & d \end{pmatrix} \in O(d,d,Z) \tag{4.7}$$

The exact underlying picture of the symmetries above is as follows. The antisymmetric tensor shift in the action corresponds to combined affine Weyl translations on the left and right parts of the theory. The duality transformations, as we argued in the previous section are isomorphic to Weyl transformations. Finally the $GL(d)$ group acts a linear integer transformations of the weight lattice. Again, these are exact symmetries at least when the coset is compact.

The reasoning above can be extended to derive the $O(d,d,R)$ action on conformal backgrounds. This was first observed as an invariance of the one-loop string effective action, with backgrounds having d Killing symmetries, [14]. The observation is the following. Starting from a background (CFT) with d Killing symmetries, an arbitrary constant shift of the antisymetric tensor, as well as an arbitrary linear combination of the coordinates corresponding to the Killing directions provide another theory which is also conformally invariant. If these transformations are intertwined with the duality transformations D_i it can be shown that the full group of transformations is isomorphic to $O(d,d,R)$ which acts as in (4.6). The $O(d,d,Z)$ subgroup generates the same string theory. It is obvious that linear transformations and antisymmetric tensor shifts are exact to all orders in a' and the string loop expansion. In the compact case we have shown in the previous sections that, although the action of the duality transformations D_i has to be modified beyond one-loop, there is such a modification, that is exact again non-perturbatively in α' and perturbatively in the string loop expansion.†

One further comment applicable to the compact case: $O(d,d,R)$ transformations do not in general preserve the positivity (unitarity in Minkowski space) of the appropriate conformal field theory. This is obvious for antisymmetric tensor shifts, since they correspond to arbitrary shifts of the weight lattice and thus map integrable to non-integrable reps. It remains to be seen if the string theory constructed from such CFTs remains unitary.

---

†Arguments, to the extend that $O(d,d,R)$ transformations can be made exact symmetries to all orders in α' where also given in [15] from a string field theory point of view.

5 On marginal current-current perturbations

In a CFT with chiral abelian currents like (4.1) it is well known [25] that the perturbation

$$S_I = \lambda \int g_{ij} J^i \bar{J}^j \tag{5.1}$$

is marginal. This corresponds to the deformation of the Cartan torus, and generalizes the SU(2) case which we explicitly discussed in section 2. It is also well known from CFT that such perturbations break the non-abelian symmetry while leaving the (abelian) Cartan symmetries intact. However a simple calculation shows that the action $S + S_I$ has (deformed) chiral symmetries only to order $\mathcal{O}(\lambda)$. The way to improve this situation is via (in this case) a special $O(2d, 2d)$ transformation, or equivalently by considering the tensor product of this theory with d free scalar fields and gauging an arbitrary linear combination of the two $U(1)^d$ symmetries.[‡]

To see that we can get the current-current perturbation from an $O(2d, 2d)$ transformation we can study first infinitesimal perturbations. Let us consider the infinitesimal form of the transformations in (4.6).

$$a \sim 1 + \lambda A + \mathcal{O}(\lambda^{\epsilon}) \ , \ \lceil \sim \infty - \lambda \mathcal{A}^T + \mathcal{O}(\lambda^{\epsilon}) \tag{5.2a}$$

$$b \sim \lambda B + \mathcal{O}(\lambda^{\epsilon}) \ , \ \rfloor \sim \lambda C + \mathcal{O}(\lambda^{\epsilon}) \ , \ B^T = -B \ , \ C^T = -C \tag{5.2b}$$

It is not difficult to see that the following infinitesimal $O(2d, 2d)$ transformation

$$\begin{pmatrix} 1 & 0 & 0 & -\lambda g^T \\ 0 & 1 & \lambda g & 0 \\ 0 & -\lambda g^T & 1 & 0 \\ \lambda g & 0 & 0 & 1 \end{pmatrix} + \mathcal{O}(\lambda^{\epsilon}) \tag{5.3}$$

generates the perturbation (5.1) Of course there will be also a non-trivial dilaton that can be calculated from (4.5c), which will ensure conformal invariance at the one-loop level.

For the $SU(2)$ case there is only one marginal perturbation and (5.3) can be integrated automatically to obtain the finite transformation. The advantage of the finite transformation is that the theory with the fully transformed action will have chiral abelian currents for all values of the parameters. However conformal invariance will still have to corrected beyong one loop.

What we remarked so far is hardly surprising. When we look however at the action of $O(d, d)$ transformations on the abelian coset theory, which generically has no chiral currents, we can observe that it still implies that certain current-current perturbations are marginal.

---

‡ Similar observations were made independently in [26].

Let us first compute the conserved currents in (4,4) associated to the Killing symmetries $\alpha^i \to \alpha^i + \varepsilon^i$

$$J^i = \partial\alpha^j E_{ji}(x) + \partial x^a F^1_{ai}(x) \tag{5.4a}$$

$$\bar{J}^i = E_{ij}(x)\bar{\partial}\alpha^j + F^2_{ia}(x)\bar{\partial}x^a \,. \tag{5.4b}$$

These currents are conserved

$$\bar{\partial}J^i + \partial\bar{J}^i = 0 \tag{5.5}$$

but not chirally conserved.

The infinitesimal transformations corresponding to (4.6) become

$$\delta E = \lambda(AE + EA^T + B - ECE) \;\;,\;\; \delta F = -\lambda F^1 C F^2 \tag{5.6a}$$

$$\delta F^1 = \lambda F^1(A^T - CE) \;\;,\;\; \delta F^2 = \lambda(A - EC)F^2 \,. \tag{5.6b}$$

We can now observe that the infinitesimal change in the action (4.4) under the special transformation $A = B = 0$ has the form

$$\delta S = -\lambda \int J^i C_{ij} \bar{J}^j \tag{5.7}$$

which is a specific current-current perturbation (because the matrix C is forced by (5.2b) to be antisymmetric). Of course, there are corrections again to the dilaton via (4.5c).

In CFT, whenever there are conserved but not chirally conserved currents, they are bad conformal fields. This can be proven in general in 2-d by showing that normal conservation of a current and conformal invariance (which fixes the form of the two-point function) implies chiral conservation. Moving a bit off criticality we can see that conserved but not chirally conserved currents have severe IR divergences and decouple from the spectrum as one approaches the critical point. It is thus surprising that a perturbation of the form (5.7) is a marginal perturbation of such models.

6 Non-compact cosets

Non compact cosets attracted attention recently, [27, 18, 13, 4] as CFTs that provide curved backgrounds for consistent string propagation. They also generically exhibit (semi-classically) spacetime singularities.

The prototype theory (and apparently the simplest) is the $SL(2,R)/U(1)$ model describing a two dimensional target manifold. We will consider Euclidean targets, which means that the U(1) we are going to gauge will be compact. If we parametrize the $SL(2,R)$ matrix using Euler angles as

$$g = e^{i\frac{\phi}{2}\sigma_2} e^{\frac{r}{2}\sigma_1} e^{i\frac{\psi}{2}\sigma_2} \tag{6.1}$$

then, upon integrating out the gauge fields, we arrive at the following partition functions for the axial and vector theory

$$Z_A = \int_0^\infty \frac{\sinh r dr}{1+\cosh r} \int_0^{2\pi} d\phi e^{-\frac{k}{4\pi}\int[\partial_\mu r\partial^\mu r + 4\tanh^2\frac{r}{2}\partial_\mu\phi\partial^\mu\phi]} \tag{6.2}$$

$$Z_V = \int_0^\infty \frac{\sinh r dr}{1-\cosh r} \int_0^{2\pi} d\phi e^{-\frac{k}{4\pi}\int[\partial_\mu r\partial^\mu r + 4\coth^2\frac{r}{2}\partial_\mu\phi\partial^\mu\phi]} \tag{6.3}$$

where we have incorporated the dilaton into the measure. In the axial theory, the string propagates on a manifold with the shape of a cigar, which becomes a cylinder asymptotically ($r \to \infty$). However the manifold of the vector theory, although similar when $r \to \infty$, has a different, and in fact singular behaviour as $r \to 0$. The line element and scalar curvature behave as follows, in this region

$$ds^2 \sim dr^2 + \frac{1}{r^2}d\phi^2 \quad ; \quad R \sim \frac{1}{r^2} \,. \tag{6.4}$$

In the Minkowskian (2-d black hole) model the analytic continuation of the axial Euclidean model (6.2) generates region I of spacetime (the asymptotically flat region till the horizon). Region III (from the horizon to the sigularity) corresponds to the self-dual SU(2)/U(1) model. Finally region V (behide the singularity) corresponds to the vector Euclidean model (6.3).

The pertinent question here is: are the two models (6.2,3) equivalent, like in the compact case? Our semiclassical derivation of the axial to vector duality is still valid here. However there are reasons to make us distrustful of such a semiclassical reasoning in the non-compact case. One is that the two targets are radically different , unlike the compact case where the target manifold of the axial theory is a reparametrization of that of the vector theory (even when higher loop corrections are included, [13]). The other reason is that the semiclassical spectrum of the two theories in the non-compact case is quite different. In the axial theory, only the continuous series of the SL(2,R) representations contribute, while in the vector theory there are extra contributions from the discrete series. The partition function of the axial model has been computed by Gawedski [28], however that of the vector model remains a mystery.

Trying to understand the situation, we will analyse a simpler (but not trivial) case of (potential) axial-vector duality in a non-compact model. Let us consider the conformal field theory on a 2-d Euclidean plane

$$Z_E = \int d^2x e^{-\frac{1}{4\pi}\int[(\partial x^1)^2 + (\partial x^2)^2]} \tag{6.5}$$

This is a free field theory that we know everything about, in particular, its exact torus partition function is (up to constants)

$$Z_E^t = \frac{1}{Im\tau}\frac{1}{|\eta(\tau)|^4} \tag{6.6}$$

where the $1/Im\tau$ factor comes from the integration of the zero modes. We can write this theory in polar coordinates (r,θ),

$$Z_E = \int_0^\infty r dr \int_0^{2\pi} d\theta e^{-\frac{1}{4\pi}\int[(\partial r)^2+r^2(\partial\theta)^2]}. \tag{6.7}$$

We can now apply the O(1,1) duality transformation corresponding to the Killing symmetry associated with translations of θ to obtain the "dual" theory

$$Z_{\tilde{E}} = \int_0^\infty \frac{dr}{r} \int_0^{2\pi} d\theta e^{-\frac{1}{4\pi}\int[(\partial r)^2+r^{-2}(\partial\theta)^2]} \tag{6.8}$$

where the effects of the dilaton transformation were also taken into account by the change in the measure. A naive extrapolation of our results from the compact case would imply that these theories are equivalent, and in particular that the theory (6.8) is a free field theory. However, as in the SL(2,R)/U(1) example, the manifold corresponding to (6.8) is radically different from the flat Euclidean plane of (6.6); it is a curved manifold with a curvature singularity at the origin. It concides with a region close to the origin, of the vector SL(2,R)/U(1) model as can be seen from (6.4).

It is also interesting to note that the two theories (6.6,8) can be viewed as axial and vector gauged models of the following σ-model, with 3-d target

$$Z_3 = \int_0^\infty g(r)dr \int_0^{2\pi} d\theta d\varphi \;\; e^{-\frac{1}{4\pi}\int[(\partial\theta)^2+(\partial\varphi)^2+(\partial r)^2+2f(r)\partial\theta\bar{\partial}\varphi]} \tag{6.9}$$

Choosing

$$g(r) = \frac{r}{1+r^2} \quad , \quad f(r) = \frac{1-r^2}{1+r^2} \tag{6.10}$$

we can verify with a simple computation that by gauging the axial symmetry $\theta \to \theta + \varepsilon$, $\varphi \to \varphi + \varepsilon$ we obtain the free model (6.7) while gauging the vector symmetry $\theta \to \theta + \varepsilon$, $\varphi \to \varphi - \varepsilon$ we obtain model (6.8). It is an interesting question whether the the model (6.9,10) is conformally invariant. There are abelian chiral currents in this model associated with the symmetries $\theta \to \theta + \varepsilon(\bar{z})$ and $\varphi \to \varphi + \zeta(z)$

$$J = \partial\varphi + f(r)\partial\theta \;\; , \;\; \bar{J} = \bar{\partial}\theta + f(r)\bar{\partial}\varphi \;\; ; \;\; \partial\bar{J} = \bar{\partial}J = 0 \tag{6.11}$$

We can easily check that unless $f(r)$ is the one which corresponds to the SL(2,R) model there are no other chiral currents in the theory (this might seem trivial, but it is possible in principle that the model can be mapped to that of SL(2,R) through a complicated reparametrization). Thus if (6.9) is conformally invariant it describes the product (certainely not direct) of a U(1) theory and some other CFT.

We will now proceed to tackle the question posed above: is the free model (6.6) and (6.8) equivalent? Unfortunately it seems extremely difficult to compute exactly the torus partition function of the $\tilde{E}$ model. Thus we will resort to the so called "minisuperspace" approximation. This amounts essentially to a dimensional reduction to 1-d, that is, neglecting the σ dependence. Thus, we will have to deal with a quantum mechanical model with Langrangian given by

$$L_{\tilde{E}} = \frac{1}{4\pi}[\dot{r}^2 + r^{-2}\dot{\theta}^2] \tag{6.12}$$

The minisuperspace approximation of a CFT is not an approximation in the usual sense of the word. However, it is well understood that, since it describes the quantum mechanics of zero modes, it does not "see" the oscilator part of the spectrum, finite renormalizations of couplings and unitary truncations of the Hilbert space. For example in the "minisuperspace" approximation to the $SU(2)_k$ WZW model all representations of $SU(2)$ contribute whereas in the 2-d theory their range is restricted to $0 \le j \le k/2$. However, if two CFTs are different in this approximation they are certainely different as 2-d theories, whereas the converse is not necessarily true.

The Euclidean time quantum mechanical partition function of the E theory is given by dropping the η-function contributions from (6.6)

$$\Omega_E \sim \int d^2x \int d^2p e^{-\tau\vec{p}^2} \sim \frac{1}{\tau}\int d^2x \tag{6.13}$$

In general, the quantum mechanical partition function will be given by the trace of $exp[-\tau\hat{H}]$

$$\Omega_H = \int \langle x|e^{-\tau\hat{H}}|x\rangle \tag{6.14}$$

and is proportional to the volume of the manifold.

The Hamiltonian of the $\tilde{E}$ theory is minus the Laplacian on the manifold

$$H_{\tilde{E}} = -\frac{\partial^2}{\partial r^2} + \frac{1}{r}\frac{\partial}{\partial r} - r^2\frac{\partial^2}{\partial\theta^2} \tag{6.15}$$

$\Omega_{\tilde{E}}$ is then the integrated trace of the heat kernel and in order to compute it we need to know the spectrum of this Laplacian. This is done in a straightforward manner. The wave functions are labeled by the energy E and the eigenvalues m of angular momentum $(\frac{\partial}{\partial\theta})$, which are integers. When $m \neq 0$, the energy spectrum is discrete, $E_{m,n} = 4|m|(n+1)$, $n = 0, 1, 2, \cdots$, and the energy eigenvalues are doubly degenerate. Their respective eigenfunctions are

$$\Psi_{m,n}(r,\theta) = \frac{mr^2}{\sqrt{\pi(n+1)}}e^{-\frac{|m|}{2}r^2}L_n^1(|m|r^2)e^{im\theta} \tag{6.16}$$

properly normalized

$$\int_0^\infty \frac{dr}{r}\int_0^{2\pi} d\theta\ \Psi^*_{m,n}(r,\theta)\Psi_{m',n'}(r,\theta) = \delta_{m,m'}\delta_{n,n'} \tag{6.17}$$

where L_n^1 is a Laguerre polynomial.

When $m = 0$, the energy is continuous and non-negative, the corresponding eigenfunctions being

$$\Psi_E(r,\theta) = rJ_1(\sqrt{E}r) \quad , \quad E \geq 0 \tag{6.18}$$

normalized as

$$\int_0^\infty \frac{dr}{r} \int_0^{2\pi} d\theta \; \Psi_E^*(r,\theta)\Psi_{E'}(r,\theta) = 4\pi\delta(E - E') \tag{6.19}$$

and J_1 is the standard Bessel function. Already, at the level of the spectrum, the two theories E and $\tilde{E}$ look quite different. The E theory has positive continuous spectrum of infinite multiplicity. The $\tilde{E}$ theory has both continuous and discrete spectrum, both of finite multiplicity. These features are also common in the two versions of the SL(2,R)/U(1) model, presented above, namely (6.2,3).

The completeness condition can be verified explicitly using standard formulae of special functions

$$\frac{1}{4\pi}\int_0^\infty dE\Psi_E^*(r,\theta)\Psi_E(r',\theta') + \sum_{m\neq 0}\sum_{n=0}^{\infty} \Psi_{m,n}^*(r,\theta)\Psi_{m,n}(r',\theta') = r\delta(r-r')\delta(\theta-\theta') \tag{6.20}$$

We can now write the trace of the heat kernel

$$\langle r,\theta|e^{-\tau\hat{H}_{\tilde{E}}}|r,\theta\rangle = \frac{r^2}{4\pi\tau}e^{r^2/2\tau}I_1(\frac{r^2}{2\tau}) +$$

$$+\frac{r^2}{2\pi}\sum_{m=1}^{\infty}\frac{m}{sinh(2m\tau)}e^{-mr^2 coth(2m\tau)}I_1\left(\frac{mr^2}{sinh(2m\tau)}\right) \tag{6.21}$$

where the first term comes from the continuous part of the spectrum whereas the second from the discrete. I_1 is the standard Bessel function. The partition function is given by

$$\Omega_{\tilde{E}}(\tau) = \int_0^\infty \frac{dr}{r}\int_0^{2\pi} \langle r,\theta|e^{-\tau\hat{H}_{\tilde{E}}}|r,\theta\rangle \tag{6.22}$$

Performing this integral we find that the continuous spectrum contributes a τ-independent divergent piece (a linear divergence) which moreover does not even scale with the volume of the manifold that is logarithmically divergent. The discrete part of the spectrum gives a finite contribution

$$\Omega_{\tilde{E}}^{finite}(\tau) = \sum_{m=1}^{\infty}\frac{1}{e^{4m\tau}-1} \tag{6.23}$$

Thus, not only the two quantum mechanical partition functions differ but model (6.8) has the pathological behaviour that the free energy per unit volume is infinite.

In order to rederive our semiclassical expectations it is instructive to go back to model (6.9) (where α' was set to one) and re-introduce it explicitly. The free model is of course insensitive to this, since the α' dependence can be scaled away. However in the "dual" model (6.8) α' can be scaled away from the action at the expense of changing the range of θ from $[0,2\pi]$ to $[0,2\pi\alpha']$. Thus we see that at weak coupling, we have to go to the universal cover of the manifold, the eigenvalues of the angular momentum become continuous and the partition function is given by

$$\Omega^c_{\hat{E}} \sim \int_0^{infty} dx \frac{1}{e^{4x\tau}-1} \sim \frac{1}{\tau} \times log - divergence \tag{6.24}$$

which maps properly to the free model.

What we have seen so far is that semiclassically we have recovered the duality of the compact case. However the example above raises serious doubts about its validity beyond weak coupling. One could of course contemplate modifications that could bypass the discussion above (like alternative quantization of the dual theory)[§]. Our point here is that, unlike the compact case, duality if present is certainly not manifest. We should also note that the standard toroidal duality cannot be seen by a naive dimensional reduction. One has also to sum up the winding sectors. Here however there are no stable winding sectors, although there is still the possibility that a full 2-d consideration might restore what was left out in the dimensional reduction. This is currently under study.

Similar remarks apply to the more "realistic" SL(2,R)/U(1) model, (6.2,3). As we mentioned earlier the two versions (6.2) and (6.3) differ substantially only in a neighbourhood of $r = 0$, and there they are approximated (possibly crudely) by our toy models (6.7) and (6.8). An analysis similar to the above is underway for (6.2,3) in order to settle this question.

At the full 2-d level, duality might require, as in the compact case, invariance of the theory under affine translations. This was manifestly true in compact unitary cosets but it is not difficult to see that n the non-compact case, affine Weyl translations map in general a representation to a different one. This can be seen at the level of the non-compact string functions, [29]. One way to proceed is to consider orbits under the translation group, but in that case one has always to cope with non-positive representations and the spacetime interpretation is not manifest.

In many issues associated with black-holes one usually invokes some analytic continuation from Minkowski to Euclidean space. As we have seen it plausible that there are two inequivalent such continuations depending on the region of spacetime. This might imply a different behaviour (and maybe interpretation) for such issues as Hawking radiation etc.

NOTE ADDED

After the completion of this work I have found that the connection between the duality transformations in k=1 (toroidal) WZW models for simply laced groups and the finite Weyl group (a special case of what has been discribed here) has been noticed before, [30]. I would like to thank A. Schwimmer for bringing this to my attention.

[§]A little analysis shows that the freedom in quantizing the theory is so wide that can reproduce any possible Hamiltonian, that commutes with the angular momentum.

References

[1] K. Kikkawa, M Yamazaki, Phys. Lett. B149 (1984) 357; N. Sakai, I. Senda, Prog. theor. Phys. 75 (1986) 692; V. P. Nair, A. Shapere, A. Strominger, F. Wilczek, Nucl. Phys. B287 (1987) 402.

[2] A. Giveon, E. Rabinovici, G. Veneziano, Nucl. Phys. B322 (1989) 167;

[3] A. Giveon, N. Malkin, E. Rabinovici, Phys. Lett. B238 (1990) 57.

[4] E. Kiritsis, Mod. Phys. Lett. A6 (1991) 2871.

[5] E. Verlinde, M. Rocek, Nucl. Phys. B373 (1992) 630.

[6] A. Giveon, M. Rocek, "*Generalized Duality in Curved String Backgrounds*", Princeton preprint, hep-th/9112070.

[7] T. H. Buscher, Phys. Lett. B201 (1988) 466.

[8] A. Schwarz, A. Tseytlin, "*Dilaton Shift under Duality and Torsion of the Elliptic Complex*", Imperial preprint, hep-th/9210015.

[9] E. Witten, Comm. Math. Phys. 92 (1984) 455.

[10] K. Bardakci, M. Halpern, Phys. rev. D3 (1971) 2493; M. Halpern, Phys. Rev. D4 (1971) 2398; P. Goddard, A. Kent, D. Olive, Phys. Lett. B152 (1985) 88; M. Douglas, PhD Thesis, Caltech 1988, unpublished.

[11] K. Bardakci, E. Rabinovici, B. Säring, Nucl. Phys. B299 (1988) 157; D. Altschuler, K. Bardakci, E. Rabinovici, Comm. math. Phys. 118 (1988) 241; K. Gawedski, A. Kupiainen, Phys. Lett. B215 (1988) 119; Nucl. Phys. B320 (1989) 625; D. Karabali, Q.-H. Park, H. Schnitzer, Z. Yang, Phys. Lett. B216 (1989) 307; D. Karabali, H. Schnitzer, Nucl. Phys. B329 (1990) 649.

[12] M. Halpern, E. Kiritsis, Mod. Phys. Lett. A4 (1989) 1373; ibid. A4 1797 (E).

[13] R. Dijkgraaf, E. Verlinde, H. Verlinde, Nucl. Phys. B371 (1992) 269.

[14] K. Meissner, G. Veneziano, Phys. Lett. B267 (1991) 33; Mod. Phys. Lett. A6 (1991) 3397; M. Gasperini, G. Veneziano, Phys. Lett. B277 (1992) 256; M. Gasperini, J. Maharana, G. Veneziano, "*Boosting away Singularities from Conformal String Backgrounds*", CERN preprint, hep-th/9209052.

[15] A. Sen, Phys. Lett. B271 (1991) 295; ibid. B274 (1992) 34; Phys. Rev. Lett. 69 (1992) 1006; S. Hassan, A. Sen, Nucl. Phys. B375 (1992) 103.

[16] J. Maharana, J. Schwarz, "*Non-compact Symmetries in String Theory*", Caltech preprint, CALT-68-1790.

[17] M. Henninngson, S. Hwang, P. Roberts, B. Sundborg, Phys. Lett. B267 (1991) 350.

[18] A. Giveon, Mod. Phys. Lett. A6 (1991) 2843.

[19] D. Gepner, E. Witten, Nucl. Phys. B278 (1986) 493.

[20] V. Kac, D. Peterson, Adv. Math. 53 (1984) 125.

[21] S. K. Yang, Phys. Lett. B209 (1988) 242.

[22] X. De La Ossa, F. Quevedo, "*Duality Symmetries from Non-Abelian Isometries in String Theory*", Neuchâtel preprint, hepth/9210021.

[23] D. Gepner, Z. Qiu, Nucl. Phys. B285 [FS19] (1987) 423.

[24] D. Gepner, Nucl. Phys. B290 [FS20] (1987) 10.

[25] S. Chaudhuri, J. A. Schwarz, Phys. Lett. B219 (1989) 291.

[26] S. Hassan, A. Sen, "*Marginal Deformations of WZNW and Coset Models from O(d,d) Transformations*", Tata preprint, hepth/9210121.

[27] E. Witten, Phys. Rev. D44 (1991) 314; E. Martinec, S. Shatasvili, Nucl. Phys. B368 (1992) 338; D. Kutasov, A. Bershadsky, Phys. Lett. B266 (1991) 345; J. Horne, G. Horowitz, A. Steif, Phys. Rev. Lett. 68 (1992) 568; N. Ishibashi, M. Li, A. Steif, Phys. Rev. Lett. 67 (1992) 3336; P. Horava, Phys. Lett. B278 (1992) 101; C. Nappi, E. Witten, Phys. Lett. B293 (1992) 309; C. Kounnas, D. Lüst, Phys. Lett. B289 (1992) 56; D. Gershon, "*Exact Solutions of 4-D Black Holes in String Theory*", Tel Aviv preprint, TAUP-1937-91; P. Ginsparg, F. Quevedo, "*Strings in Curved Spacetimes, Black Holes, Torsion and Duality*", Los Alamos preprint, hepth/9202092; S. Chaudhuri, J. Lykken, "*String Theory, Black Holes and SL(2,R) Current Algebra*", Fermilab preprint, hepth/9206107; J. L. Gervais, M. Saveliev, Phys. Lett. B286 (1992) 271; J. Ellis, N. Mavromatos, D. Nanopoulos, Phys. Lett. B289 (1992) 25, and references therein; I. Bars, "*Superstrings in Curved Spacetimes*", USC preprint, hepth/9210079 and references therein;

[28] K. Gawedski, Lectures given at the Cargese Summer Inst. "New Symmetry Principles in QFT", July 1991.

[29] I. Bakas, E. Kiritsis, Proceedings of the RIMS "Infinite Analysis" project, Kyoto, 1991, published in Int. J. Mod. Phys. A7 [Sup. 1A] (1992) 55.

[30] L. Caneschi, Phys. Lett. B228 (1989) 332.

CONSTRUCTION OF STRING SOLUTIONS AROUND NON-TRIVIAL BACKGROUNDS

Costas KOUNNAS[†)]
Theoretical Physics Division, CERN
CH - 1211 Geneva 23

ABSTRACT

We present a way of constructing string solutions around non-trivial gravitational backgrounds. The proposed solutions are constructed using $N = 4$ superconformal building blocks with $\hat{c} = 4$. We give two different and inequivalent realizations of non-trivial four-dimensional subspaces, and we show the emergence of the $N = 4$ globally defined superconformal symmetry. The existence of $N = 4$ world-sheet symmetry stabilizes our solutions and implies in target space a number of covariantized supersymmetries around space-time dependent gravitational and dilaton backgrounds.

A way to better understand string theories and their induced low-energy field theories coupled to Einstein gravity is to study classical string solutions in the presence of non-trivial space-time backgrounds.

There are two ways to proceed in this direction:

(i) The first is to use a two-dimensional σ-model where the non-trivial backgrounds correspond to some two-dimensional field-dependent coupling constants. The vanishing of the corresponding β-functions is then identified with the background field equation of motion in the target space-time [1].
(ii) The second approach consists of constructing directly some underlying non-trivial conformal field theory, and then trying to interpret the obtained string vacuum in terms of the target space-time backgrounds.

The two methods are useful and complementary. The σ-model approach provides a clear geometric interpretation, but it has the disadvantage of the α'-expansion which is valid only when all curvatures and derivatives on space-time background fields are small. Via the σ-model approach one can easily obtain approximate solutions which are in fact identical to the classical equation of gravity in the presence of a dilaton, antisymmetric field and some gauged minimally coupled matter. However, the possible extension of these approximate solutions to exact string solutions is in general difficult and an unsolved problem at the present time.

The two-dimensional conformal field theory approach takes into account all orders in α' automatically and has the main advantage of providing exact string vacua. The background interpretation of a given exact string solution is a notion which is ill-defined

†) On leave from Lab. de Physique Théorique de l' ENS, F-75231, Paris, FRANCE.

in general. Indeed, the notion of space-time dimensionality and topology breaks down for a solution which involves highly curved backgrounds, namely when the metric and/or gauge field curvatures are of the order of the string scale. A typical example of the space-time dimensional and topological confusion is that of the $SU(2)$ level k group manifold compactification. For large k (small curvature) the target space is a three-dimensional sphere S^3, $[\frac{SO(4)}{SO(3)} \simeq SO(3) \simeq SU(2)]$. For small k (high curvature) this background interpretation fails. It is in fact well known that the $SU(2)_{k=1}$ WZW model is equivalent to a $c = 1$ conformal system defined by one free bosonic co-ordinate compactified on a cycle with a radius $R = \frac{2}{R} = \sqrt{2}$ (self-dual point).

Naively one may interpret $T_{(R=\sqrt{2})}$ compactification as one-dimensional space with S^1 topology, which is in contradiction to the three-dimensional interpretation with S^3 topology of the $SU(2)_{k=1}$. This shows that both the dimensionality as well as the topology of the target space are not well-defined concepts in string theory. In general, a background interpretation of a given string solution exists only when the lower Kaluza-Klein excitations have masses much smaller than the typical string scale ($M_{st} = \alpha'^{-1/2}$).

In this talk, I will present a class of string solutions which is constructed by using some $N = 4$ superconformal systems with $\hat{c} = 4$ as building blocks. In the limit of small curvatures, these solutions have a non-trivial ten-dimensional background interpretation. Furthermore, the globally defined underlying $N = 4$ worldsheet symmetry stabilizes these solutions under string-loop perturbation, and implies some covariantized space-time sypersymmetry around a non-trivial gravitational and dilaton background on the target space.

More explicitly, we arrange the degrees of freedom of the ten supercoordinates in three superconformal systems [2]-[5]:

$$\hat{c} = 10 = \{\hat{c} = 2\} + \{\hat{c} = 4\}_1 + \{\hat{c} = 4\}_2 \tag{1}$$

The $\hat{c} = 2$ subsystem is saturated by two free superfields. In one variation of our solutions, one of the two free superfields is chosen to be the time-like supercoordinate and the other to be one of the nine space-like supercoordinates. In other variations, both supercoordinates are Euclidean or even compactified on a one- or two-dimensional torus.

The remaining eight supercoordinates appear in a group of four in $\{\hat{c} = 4\}_1$ and $\{\hat{c} = 4\}_2$. Both $\{\hat{c} = 4\}_A$ subsystems show an $N = 4$ superconformal symmetry of the Ademollo et al. type [6]. The non-triviality of our solutions follows from the fact that some realizations of the $\hat{c} = 4, N = 4$ superconformal systems exist which are based on geometrical and topological non-trivial spaces other than the T^4/Z_2 orbifold and the K_3 compact Calabi-Yau space. I will now present two different realizations.

(A) - $W_k^{(4)}$, semi-wormhole realization [7],[8] $\hat{c}[W_k^{(4)}] = 4$

It is based on a supersymmetric version of the $\{U(1) \times SU(2)_k\}$ WZW model. The three bosonic coordinates parametrize the $SU(2)_k$ group manifold while the fourth one is a free field with background charge Q. The four fermionic coordinates are free. In order to obtain a $\hat{c}[U(1)\times SU(2)_k]=4$ for any value of k, it is necessary to balance the central charge deficit of the $SU(2)_k$ by a central charge benefit of the $U(1)$ background charge Q

$$\begin{aligned}\hat{c}[SU(2)_k] &= \frac{2}{3}[3-\frac{6}{k+2}+\frac{3}{2}] = 3-\frac{4}{k+2}\\ \hat{c}[U(1)_Q] &= \frac{2}{3}[1+3Q^2+\frac{1}{2}] = 1+2Q^2\end{aligned} \tag{2}$$

(The contributions $\frac{3}{2}$ and $\frac{1}{2}$ inside [...] in the first and second line are those of the 3+1 free fermions).

From Eq. (2) one has

$$\hat{c}[SU(2)_k]+\hat{c}[U(1)_Q] = 4+2(Q^2-\frac{2}{k+2})\,,$$

and so $\hat{c}[W_k^{(4)}]=4$ only if

$$Q=\sqrt{\frac{2}{k+2}} \tag{3}$$

The existence of $N=4, \hat{c}=4$ superconformal symmetry with this value of Q is found in Ref. [7]. What is extremely interesting is the background interpretation of the $W_k^{(4)}$ space in terms of a four-dimensional (semi)-wormhole space given by Callan, Harvey and Strominger in Ref. [8]. Indeed, for large k, the three $SU(2)_k$ coordinates define a three-dimensional subspace with a non-trivial topology S^3, while the fourth coordinate with a background charge corresponds to the scale factor of the S^3 sphere.

Another interesting interpretation of the $W_k^{(4)}$ space is its connection with four-dimensional dilaton-axion instantons [8]-[10]. Here, one assumes a string solution based on a six-dimensional compact manifold K^6 with $\hat{c}=6$; our four-dimensional space-time (after Euclidean rotation) is saturated by $W_k^{(4)}$ with $\hat{c}=4$.

The explicit realization of the $N=4$ symmetry of $W_k^{(4)}$, as well as some other properties of string solutions based on it, will be discussed later after I present a new and inequivalent realization of a $\hat{c}=4, N=4$ superconformal system.

B - $\Delta_k^{(4)}$ Cigar-Bell and Trumpet-Bell realizations [5] $\hat{c}[\Delta_k^{(4)}]=4$

For large values of k (small curvatures), the $\Delta_k^{(4)}$ has a four-dimensional interpretation which is different from $W_k^{(4)}$. The underlying two-dimensional action associated to $\Delta_k^{(4)}$ is an exact superconformal theory based on a supersymmetric version of a gauged WZW model, namely,

$$\Delta_k^{(4)} \equiv \left\{\left[\frac{SU(2)}{U(1)}\right]_k \times \left[\frac{SL(2,R)}{U(1)}\right]_{k+4}\right\}_{\text{SUSY}} \tag{4}$$

The relation among the levels k and $k'=k+4$ ensures $\hat{c}[\Delta_k^{(4)}]=4$ for any k

$$\hat{c}[\frac{SU(2)}{U(1)}]_{ss} = \frac{2}{3}[\frac{3k}{k+2}-1+1] = 2-\frac{4}{k+2}$$

$$\hat{c}[\frac{SL(2,R)}{U(1)}]_{ss} = \frac{2}{3}[\frac{3k'}{k'-2} - 1 + 1] = 2 + \frac{4}{k'-2} \tag{5}$$

and so $\hat{c}[\Delta_k^{(4)}] = 4$ when $k' = k + 4$. As we will show later, the above relation among the levels is necessary for the existence of an $N = 4$ superconformal symmetry of Ademollo type [6].

For large k, the $\Delta_k^{(4)}$ space has a dimensional target space interpretation based on a non-trivial background metric G_{ij} and dilaton field Φ given by [11]-[17],[2]-[5]

$$ds^2 = k\Big\{(d\alpha)^2 + \tan^2\alpha d\theta^2\Big\} + k'\Big\{(d\beta)^2 + \tanh^2\beta d\varphi^2\Big\} \tag{6a}$$

$$2\Phi = \log\cos^2\alpha + \log\cosh^2\beta + \text{const.} \tag{6b}$$

with

$$\alpha \in [0, \frac{\pi}{2}]\bigcup[\pi, \frac{3}{2}\pi] \ , \quad \beta \in [0, \infty] \ , \quad \theta, \varphi \in [0, 2\pi]$$

The term proportional to k in Eq. (6a) parametrizes the two-dimensional subspace defined by the $\left[\frac{SU(2)}{U(1)}\right]_k$ parafermionic theory [18] and the one proportional to k' is the two-dimensional subspace which is defined by the non-compact parafermionic theory [19],[20],[12] based on the $\left[\frac{SL(2,R)}{U(1)}\right]_{k'}$ axial gauged WZW model. It is well known that a different metric $\tilde{G}_{ij}$ and dilaton function $\tilde{\Phi}$ are obtained if one chooses a vector gauging instead of the axial one [15],[13]:

$$\begin{aligned}(U(1)^V \times U(1)^V) \rightarrow d\tilde{s}^2 = \quad & k\Big\{(d\alpha)^2 + \tfrac{1}{\tan^2\alpha} d\theta^2\Big\} \\ & + k'\Big\{(d\beta)^2 + \tfrac{1}{\tanh^2\beta} d\varphi^2\Big\}\end{aligned} \tag{7a}$$

$$2\tilde{\Phi} = \quad \log\sin^2\alpha + \log\sinh^2\beta + \text{const.} \tag{7b}$$

In both versions of gaugings, $\Delta_k^{(4)}$ has always one non-compact coordinate (β) and three compact ones (α, θ, φ). (G_{ij}, Φ) and $(\tilde{G}_{ij}, \tilde{\Phi})$ are related by a generalized duality transformation [14]:

$$R(t) \rightarrow \frac{1}{R(t)} \ , \quad \Phi \rightarrow \Phi + \log R(t) \tag{8}$$

For later purposes, it is more convenient to use the complex notation:

$$z = (\sin\alpha)e^{i\theta} \ , \quad \omega = (\sinh\beta)e^{i\varphi} \ , \quad \text{(axial case)}$$

and (9)

$$\tilde{z} = (\cos\alpha)e^{i\theta} \ , \quad y = (\cosh\beta)e^{i\varphi} \ , \quad \text{(vector case)}$$

In terms of z and w, G_{ij} and Φ are given by

$$(ds)^2 = k\,\frac{dzd\bar{z}}{1 - z\bar{z}} + k'\,\frac{dwd\bar{w}}{w\bar{w} + 1} \ , \tag{10}$$

and

$$2\Phi = \log(1 - z\bar{z}) + \log(w\bar{w} + 1) + \text{const.}$$

The dual metrics $\tilde{G}_{ij}$ and $\tilde{\Phi}$ are given in terms of $\tilde{z}$ and y as:

$$(d\tilde{s})^2 = k\,\frac{d\tilde{z}d\tilde{z}*}{1-\tilde{z}\tilde{z}*} + k'\,\frac{dyd\bar{y}}{y\bar{y}-1}\,,$$

and (11)

$$2\tilde{\Phi} = \log(1 - \tilde{z}\tilde{z}*) + \log(y\bar{y} - 1)\,.$$

From Eqs. (10) and (11), one observes that the $\left(\frac{SU(2)}{U(1)}\right)_k$ metric and dilaton part are self-dual, (z and $\tilde{z}$), while the $\left(\frac{SU(2,R)}{U(1)}\right)_{k'}$ is not. The w subspace is regular while the y subspace is singular).

The z subspace (or $\tilde{z}$) defines a two-dimensional Bell. Its metric $G_{z\bar{z}}$, the Ricci tensor $R_{z\bar{z}}$ and its scalar curvature $R^{(z)}$ are singular at the boundary of the Bell ($z = 1$):

$$G_{z\bar{z}} = \frac{k}{1-z\bar{z}}\,,\quad R_{z\bar{z}} = \frac{-1}{(1-z\bar{z})^2}\,,\quad R^{(z)} = \frac{-1}{k(1-z\bar{z})}\,, \tag{12}$$

The z-Bell is negatively curved for any value of $|z| < 1$.

The w subspace is regular everywhere with finite positive curvature and is asymptotically flat for $|w| \to \infty$

$$G_{w\bar{w}} = \frac{k'}{w\bar{w}+1}\,,\quad R_{w\bar{w}} = \frac{+1}{(w\bar{w}+1)^2}\,,\quad R^{(w)} = \frac{+1}{k'(w\bar{w}+1)}\,, \tag{13}$$

It has a cigar shape with maximal curvature at $w = 0$.

The y subspace has a different shape from its w-dual. It looks like a two-dimensional trumpet with infinite curvature at the boundary ($y = 1$). It is positively curved everywhere ($|y| > 1$) and is asymptotically flat for $|y| \to \infty$:

$$G_{y\bar{y}} = \frac{k'}{y\bar{y}-1}\,,\quad R_{y\bar{y}} = \frac{+1}{(y\bar{y}-1)^2}\,,\quad R^{(y)} = \frac{+1}{k'(y\bar{y}-1)}\,, \tag{14}$$

We have therefore two different versions of the $\Delta_k^{(4)}$ space. The first one is that of the (z, w)-BELL-CIGAR four-dimensional space and the second one is that of the $(\tilde{z}, y)$ BELL-TRUMPET four-dimensional space.

Up to now we have discussed the geometrical structure of the bosonic coordinates of the $\Delta_k^{(4)}$ space. In order to complete the description of our solution, we must include the fermionic superpartners of (z, w) coordinates, e.g., four Weyl Majorana left-handed two-dimensional fermions ($\psi_+^a, a = 1,2,3,4$) as well as four Weyl-Majorana right-handed ones ($\psi_-^a, a = 1,2,3,4$).

Because of the $N = 1$ local supersymmetry, the interactions among fermions are fixed in terms of the two-dimensional σ-model backgrounds G_{ij}, B_{ij} and Φ). In the superconformal gauge, one has the following generic form for the $N = 1$ σ-model action (in the absence of $B_{ij} = 0$):

$$S = -\frac{1}{2\pi}\int d\xi d\bar{\xi}\Big\{V_+^a V_-^a - \frac{1}{2}(\psi_+^a \nabla_- \psi_+^a - \psi_-^a \nabla_+ \psi_-^a) - \frac{1}{2}R_{ab,cd}\psi_+^a\psi_+^b\psi_-^c\psi_-^d + \Phi R^{(2)}\Big\} \quad (15)$$

where $a = 1, 2, 3, 4$ are local flat indices and

$$V_+^a = E_i^a \partial X^i \ , \quad V_-^a = E_i^a \bar{\partial} X^i \ , \quad \text{with} \ \ G_{ij} = E_i^a E_j^a \quad (16)$$

The ∇_- and ∇_+ denote the left- and right-handed covariant derivatives acting on left- and right-handed fermions ψ_+^a, ψ_-^a:

$$\begin{aligned} \nabla_- \psi_+^a &= \bar{\partial}\psi_+^a + \Gamma^a{}_{bc}\psi_+^c V_-^b \\ \nabla_+ \psi_-^a &= \partial\psi_-^a + \Gamma^a{}_{bc}\psi_-^c V_+^b \end{aligned} \quad (17)$$

Since $\Delta_K^{(4)}$ is defined as a direct product of two two-dimensional subspaces, the σ-model action can be written as:

$$S(\Delta_K^{(4)}) = S\Big[\frac{SU(2)}{U(1)}\Big]_k + S\Big[\frac{SL(2,R)}{U(1)}\Big]_{k'} \quad (18)$$

Because of the identity

$$R_{ab,cd}\psi_+^a\psi_+^b\psi_-^c\psi_-^d = 2R\psi_+^1\psi_+^2\psi_-^1\psi_-^2 \ , \quad (19)$$

valid in both subspaces and thanks to the relations [5]

$$-\left(R^{(z)} + \Gamma_z G^{z\bar{z}}\Gamma_{\bar{z}}\right) = \tfrac{1}{k}\ ; \ \left(\tfrac{SU(2)}{U(1)}\right)_k \quad (20a)$$

$$-\left(R^{(w)} + \Gamma_w G^{z\bar{z}}\Gamma_{\bar{w}}\right) = \tfrac{-1}{k'}\ ; \ \left(\tfrac{SL(2,R)}{U(1)}\right)_{k'} \quad (20b)$$

it is possible to rewrite the σ-model action in a more convenient form which shows in particular (at least at the classical level) that the fermions can be described in terms of two free bosonic fields compactified in a special radius. Indeed, using Eqs. (19) and (20), one finds [5]:

$$\begin{aligned} S\Big[\frac{SU(2)}{U(1)}\Big]_k = \frac{-1}{4\pi}\int d\xi d\bar{\xi} \quad & \Big\{\partial A\bar{\partial}A + t(A)^2(\partial\theta_A - \sqrt{\tfrac{2}{k}}\,\psi_+^1\psi_+^2)\,(\bar{\partial}\theta_A - \sqrt{\tfrac{2}{k}}\,\psi_-^1\psi_-^2) \\ & -(\psi_+^1\bar{\partial}\psi_+^2 + \psi_+^2\bar{\partial}\psi_+^2 + \psi_-^1\partial\psi_-^1 + \psi_-^2\partial\psi_-^2) \\ & +\frac{2}{k}(\psi_+^1\psi_+^2)\,(\psi_-^1\psi_-^2) + \log C^2(A)R^{(2)}\Big\} \end{aligned} \quad (21)$$

where A, θ_A are rescaled fields so that, in the large k limit, real bosonic fields are conventionally normalized [see Eq. (9)]:

$$\sin^2 \frac{A}{\sqrt{2k}} = \sin^2 \alpha = z\bar{z}$$
$$-i\,\frac{1}{2}\,\log\,\frac{z}{\bar{z}} = \theta = \frac{\theta_A}{\sqrt{2k}}$$
$$t(A)^2 = \frac{z\bar{z}}{1 - z\bar{z}} = \tan^2 \frac{A}{\sqrt{2k}}$$
$$C(A)^2 = 1 - z\bar{z} = \cos^2 \frac{A}{\sqrt{2k}} \tag{22}$$

In a similar way, the

$$S\left[\frac{SL(2,R)}{U(1)}\right]_{k'} = \frac{-1}{4\pi}\int d\xi d\bar{\xi} \quad \Big\{\partial B\bar{\partial}B + T(B)^2(\partial\theta_B - \sqrt{\frac{2}{k'}}\,\psi^3_+\psi^4_+)\,(\bar{\partial}\theta_B - \sqrt{\frac{2}{k'}}\,\psi^3_-\psi^4_-)$$
$$-(\psi^3_+\bar{\partial}\psi^4_+ + \psi^4_+\bar{\partial}\psi^4_+ + \psi^3_-\partial\psi^3_- + \psi^4_-\partial\psi^4_-)$$
$$-\frac{2}{k'}(\psi^3_+\psi^4_+)\,(\psi^3_-\psi^4_-) + \log C(B)^2 R^{(2)}\Big\} \tag{23}$$

where B and θ_B are now defined in terms of w fields:

$$\sinh^2 \frac{B}{\sqrt{2k'}} = \sin^2 \beta = w\bar{w}$$
$$-i\,\frac{1}{2}\,\log\,\frac{w}{\bar{w}} = \varphi = \frac{\theta_B}{\sqrt{2k}}$$
$$T(B)^2 = \frac{w\bar{w}}{1 + w\bar{w}} = \tanh^2 \frac{B}{\sqrt{2k'}}$$
$$C(B)^2 = 1 + w\bar{w} = \cosh^2 \frac{B}{\sqrt{2k'}} \tag{24}$$

In both Eqs. (21) and (23), the pure fermionic part of the action is given by the free-fermion kinetic terms together with some current-current interaction. This fact permits us to describe the four left and the four right fermions in terms of free bosonic fields ϕ_A, ϕ_B compactified in the shifted radii [5]:

$$R_A^2 \;=\; 1 + \tfrac{2}{k} = \tfrac{k+2}{k}\;; \quad \left(\tfrac{SU(2)}{U(1)}\right)_k \tag{25a}$$
$$R_B^2 \;=\; 1 - \tfrac{2}{k'} = \tfrac{k'-2}{k'}\;; \quad \left(\tfrac{SL(2,R)}{U(1)}\right)_{k'} \tag{25b}$$

The deviation from the value $R_A = R_B = 1$ is due to current-current interactions. The decoupling of ϕ_A and ϕ_B fields can be seen via the bosonization of fermions and the redefinition of the $\overset{(-)}{\partial}\,\theta_A$ and $\overset{(-)}{\partial}\,\theta_B$ bosonic currents order by order in a $\frac{1}{k}$ or $\frac{1}{k'}$ expansion. This statement is indeed exact and it follows from the fact that both $\left(\frac{SU(2)}{U(1)}\right)_k$ and $\left(\frac{SL(2,R)}{U(1)}\right)_{k'}$ are exact $N = (2,2)$ superconformal models. This fact is well known in both $\frac{SU(2)}{U(1)}$ and $\frac{SL(2,R)}{U(1)}$ supersymmetric coset models in the compact and non-compact parafermionic representations. The $N = 2$ generators, $J(\xi), G(\xi), \bar{G}(\xi)$ and $T(\xi)$ are given in terms of free-scalar fields (ϕ_A, ϕ_B) and in terms of (non-local) parafermionic currents [18]-[21], $(P_k, \Pi_{k'})$,

(i) - $\left(\frac{SU(2)}{U(1)}\right)_k$

$$
\begin{aligned}
J(\xi) &= \sqrt{\frac{k}{k+2}}\partial\phi_A \\
\overset{(-)}{G}(\xi) &= \overset{(-)}{P}_k \; e^{\pm i\sqrt{\frac{k+2}{k}}\phi_A} \\
T(\xi) &= -\frac{1}{2}(\partial\phi_A)^2 + T_P(\xi)
\end{aligned}
\tag{26}
$$

(ii) - $\left(\frac{SL(2,R)}{U(1)}\right)_{k'}$

$$
\begin{aligned}
J(\xi) &= \sqrt{\frac{k'}{k'-2}}\partial\phi_B \\
\overset{(-)}{G}(\xi) &= \overset{(-)}{\Pi}_{k'} \; e^{\pm i\sqrt{\frac{k'-2}{k'}}\phi_B} \\
T(\xi) &= -\frac{1}{2}(\partial\phi_B)^2 + T_\Pi(\xi)
\end{aligned}
\tag{27}
$$

ϕ_A and ϕ_B are free bosons compactified on radii $R_A = \sqrt{\frac{k+2}{k}}$ and $R_B = \sqrt{\frac{k'-2}{k'}}$ and parametrize the fermions $\psi^a_\pm, a = 1,2,3,4$ which appear in the σ-model actions in Eqs. (21) and (23) according to our previous discussion. For large k, k', the operators $e^{i\sqrt{\frac{k+2}{k}}\phi_A}$ and $e^{i\sqrt{\frac{k'-2}{k'}}\phi_B}$ have conformal dimensions of almost $\frac{1}{2}$ $(h_A = \frac{1}{2}+\frac{1}{k},\ h_B = \frac{1}{2}-\frac{1}{k'})$ P_k and $\Pi_{k'}$ are the parafermionic currents with dimensions $h_P = 1-\frac{1}{k}$ and $h_\Pi = 1+\frac{1}{k'}$, so that $G(\xi)$ in Eqs. (26) and (27) are the $N=2$ supercurrents of dimension $\frac{3}{2}$.

As usual, the compact and non-compact parafermions satisfy the algebra which follows from their O.P.E. [18]-[20]:

$$
\begin{aligned}
P_k(\xi)\bar{P}_k(\xi') &= 2\left[\left(\frac{k}{k+2}\right)\frac{1}{(\xi-\xi')^2} + T_P(\xi')\right](\xi-\xi')^{\frac{2}{k}} \\
\Pi_{k'}(\xi)\bar{\Pi}(\xi') &= 2\left[\left(\frac{k'}{k'-2}\right)\frac{1}{(\xi-\xi')^2} + T_\Pi(\xi')\right](\xi-\xi')^{-\frac{2}{k'}}
\end{aligned}
\tag{28}
$$

and the dimension-2 operators $T_P(\xi)$ and $T_\Pi(\xi')$ are their corresponding stress tensors. $T_P(\xi)$ and $T_\Pi(\xi')$ satisfy the Virasoro algebra

$$
T(\xi)T(\xi') = \frac{c}{2(\xi-\xi')^4} + \frac{2T(\xi')}{(\xi-\xi')^2} + \frac{\partial T(\xi')}{(\xi-\xi')}
\tag{29}
$$

with central charge $c_P = \frac{3k}{k+2} - 1, c_\Pi = \frac{3k'}{k'-2} - 1$. Using the above O.P.E. [Eqs. (28), (29)] and those of the free fields

$$
\partial\phi_A\,\partial\phi_A = -\frac{1}{(\xi-\xi')^2}\,,\quad \partial\phi_B\,\partial\phi_B = -\frac{1}{(\xi-\xi')^2}\,,
\tag{30}
$$

it is easy to show the closure of the $N=2$ algebra for both $\left(\frac{SU(2)}{U(1)}\right)_k$ and $\left(\frac{SL(2,R)}{U(1)}\right)_{k'}$ supersymmetric coset models.

What is interesting is the interpretation of the almost dimension-one, non-local [22] currents P_k and $\Pi_{k'}$ for k and k' large [5]. P_k are the conjugate momenta of the two-dimensional subspace z while $\Pi_{k'}$ are those of the w subspace. As we mention in the introduction, this interpretation breaks down for small values of k and k', e.g., when the target backgrounds are strongly curved.

For large values of k and k', P_k generalizes two out of the four flat space currents while $\Pi_{k'}$ generalizes the two remaining ones [5], [23]

$$\begin{array}{lcl} \overset{(-)}{P}_k & \overset{k\to\infty}{\longrightarrow} & \pm\partial x_1 + i\partial x_2 \\ \overset{(-)}{\Pi}_{k'} & \overset{k'\to\infty}{\longrightarrow} & \pm\partial x_4 + i\partial x_3 \end{array} \tag{31}$$

We would like to show that for $k' = k+4$ the $N=2$ superconformal symmetry is extended to $N=4$ with $\hat{c} = 4(c=6)$ for any value of k. For this purpose it is more convenient to redefine the fields ϕ_A and ϕ_B in terms of two other bosonic fields H_+ and H_- as follows:

$$\begin{aligned} \sqrt{2}H_+ &= \sqrt{\frac{k}{k+2}}\,\phi_A + \sqrt{\frac{k+4}{k+2}}\,\phi_B \\ \sqrt{2}H_- &= +\sqrt{\frac{k+4}{k+2}}\,\phi_A - \sqrt{\frac{k}{k+2}}\,\phi_B \end{aligned} \tag{32}$$

The $N=2$ current of the $\Delta_k^{(4)}$ is then given uniquely in terms of the ∂H_+ current. It is then possible (due to the $N=2$ algebra) to factor the H_+ dependence from the (complex) supercurrent. One finds [5]

$$G = \left[P_k\, e^{i\frac{\alpha}{\sqrt{2}}H_-} + \Pi_{k'}\, e^{-i\frac{1}{\alpha\sqrt{2}}H_-}\right] e^{i\frac{1}{\sqrt{2}}H_+} \tag{33}$$

with $\alpha = \sqrt{\frac{k+4}{k}}$.

It is suggestive to compare the above expression of the supercurrent with that of the free-field realization [6] $(k \to \infty, \alpha = 1)$

$$G_{free} = \left[(\partial x_1 + i\partial x_2)e^{i\frac{1}{\sqrt{2}}H_-} + (\partial x_4 + i\partial x_3)e^{-i\frac{1}{\sqrt{2}}H_-}\right] e^{i\frac{1}{\sqrt{2}}H_+} \tag{34}$$

Here H_+ and H_- are given in terms of free fermions via the conventional bosonization

$$\begin{aligned} \sqrt{2}H_+ &= \psi_1\psi_2 + \psi_4\psi_3 \\ \sqrt{2}H_- &= \psi_1\psi_2 - \psi_4\psi_3 \end{aligned} \tag{35}$$

There are some similarities and some basic differences between the two realizations. First, the rôle of the H_+ field is the same and, as is expected from the $N=2$ algebra, is a free field compactified on a radius $R_{H^+} = \sqrt{\frac{c}{3}} = \sqrt{2}$. This value of the radius is special and extends in both cases the $U(1)$ to $SU(2)_1$ current algebra. In free field realization, however, the $N = 2$ superconformal symmetry is extended to an $N = 4$, due to the existence of an additional supercurrent, namely:

$$\tilde{G}_{free} = \left[(\partial x_1 - i\partial x_2)e^{-i\frac{1}{\sqrt{2}}H_-} - (\partial x_4 - i\partial x_3)e^{i\frac{1}{\sqrt{2}}H_-}\right] e^{i\frac{1}{\sqrt{2}}H_+}$$

In a similar way, an additional supercurrent exists also in the $\Delta_k^{(4)}$ realization [5]:

$$\tilde{G} = \left[\bar{P}_k \; e^{-i\frac{a}{\sqrt{2}}H^-} - \bar{\Pi}_{k'} \; e^{i\frac{1}{a\sqrt{2}}H^-}\right] e^{i\frac{1}{\sqrt{2}}H^+} \tag{36}$$

The $N = 4$ superconformal algebra closes in both realizations among $[T_B, G, \tilde{G}$ and $S_i]$ with central charge $\hat{c} = 4$. This shows that the $\Delta_k^{(4)}$ space shares the same global superconformal symmetries as the T^4/Z_2 orbifold model, the four-dimensional Calabi-Yau space K_3 and the previously discussed $W_k^{(4)}$ wormhole space. $\Delta_k^{(4)}$ and $W_k^{(4)}$ are not only non-trivial spaces but are also non-compact spaces (K_3 and T^4/Z_2 are compact).

For completeness, we will present below the basic operators and fields of the $W_k^{(4)}$ realization [7],[8],[2]-[5],[10]. As we already mentioned, $W_k^{(4)}$ is based on a supersymmetric $SU(2)_k \times U(1)_Q$ WZW model with a background term $Q = \sqrt{\frac{2}{k+2}}$ in the $U(1)_Q$ current. The four fermions of the model are free and are parametrized by the H^+ and H^- fields (via bosonization) as in the free-field representation [Eq. (35)]. The four coordinate currents are the three $SU(2)_k$ $(J^i, i = 1,2,3)$ currents and the $U(1)_Q$ J_4 current

$$\begin{aligned} J^i(\xi)J^i(\xi') &= -\frac{k}{2}\frac{\delta^{ij}}{(\xi-\xi')^2} + \epsilon^{ij\ell}\frac{J^\ell}{(\xi-\xi')^2}, \quad i = 1,2,3 \\ J^4(\xi)J^4(\xi') &= \frac{-1}{(\xi-\xi')^2}\cdot\frac{1}{Q^2} \end{aligned} \tag{37}$$

(the Q rescaling of J^4 is for convenience).

The $T_B, B, \tilde{G}$ and S_i associated to $W_k^{(4)}$ are

$$\begin{aligned} T_B &= -\frac{1}{2}\left[(\partial H^+)^2 + (\partial H^-)^2 + Q^2 \;\; (J_1^2 + J_2^2 + J_3^2 + J_4^2 + \partial J_4)\right] \\ G &= Q\left[(J_1 + iJ_2)\; e^{i\frac{1}{\sqrt{2}}H^-} + (J_4 + i(J_3 + \sqrt{2}\partial H^-))\; \bar{e}^{i\frac{1}{\sqrt{2}}H^-}\right] e^{i\frac{1}{\sqrt{2}}H^+} \\ \tilde{G} &= Q\left[(J_1 - iJ_2)\; \bar{e}^{i\frac{1}{\sqrt{2}}H^-} - (J_4 - i(J_3 + \sqrt{2}\partial H^-))\; e^{i\frac{1}{\sqrt{2}}H^-}\right] e^{i\frac{1}{\sqrt{2}}H^+} \\ S_0 &= \frac{1}{\sqrt{2}}\partial H^+, \quad S_\pm = e^{\pm i\sqrt{2}H^+} \end{aligned} \tag{38}$$

In the $W_k^{(4)}$ realization, both H^+ and H^- are compactified on a torus with radius $R_{H^+} = R_{H^-} = \sqrt{2}$ as in the free field case. There are in total three underlying $SU(2)$ Kac-Moody currents: (i) the $SU(2)_k$ defined by the coordinate currents, (ii) the $SU(2)_1^+$ defined by the H^+ field $R_{H^+} = \sqrt{2}$, (iii) the $SU(2)_1^-$ defined by the H^- field $R_{H^-} = \sqrt{2}$.

The background term in $T_B, (Q\partial J^4)$, comes from a non-trivial dilaton background: $\Phi = QX^4$. The term $Q\sqrt{2}\partial H^- \; exp[\pm i\frac{1}{\sqrt{2}}H^- + i\frac{1}{\sqrt{2}}H^+]$ in G and $\tilde{G}$ describes, at the same time, the standard fermionic torsion term $\pm Q\psi^i\psi^j\psi^k$ $(i = 1,2,3,4)$ as well as the fermionic background term $\pm Q\partial\psi^i$. They are arranged together in $\sqrt{2}\partial H^-$ which shifts the J_3 current. Contrary to the $\Delta_k^{(4)}$ space in which the torsion terms were absent, in $W_k^{(4)}$ space there is a non-vanishing torsion due to the WZ term proportional to the $SU(2)$ structure constant. So, in $W_k^{(4)}$ the antisymmetric field background B_{ij} is non-zero with non-trivial field strength H_{ijk}

$$H_{ijk} = e^{-2\phi}\epsilon_{ijk}{}^{\ell}\partial_\ell\Phi \simeq e^{-2\phi}Q\cdot\epsilon_{ijk}{}^{4} \tag{39}$$

Summarizing our results, we propose a special class of exact string solutions [2]-[5] with $N = (1,0)$ or $N = (1,1)$ local worldsheet supersymmetry. In the type II case, we arrange the degrees of freedom of the ten supercoordinates in three superconformal systems:

$$\hat{c} = 10 = \{\hat{c} = 2\}_0 + \{\hat{c} = 4\}_1 + \{\hat{c} = 4\}_2$$

The $\hat{c} = 2$ system is saturated by two free superfields (compact or non-compact), and so the background metric is flat:

$$ds^2\{F^2\} = dxd\bar{x}\ , \quad (x = x_1 + ix_2)\ .$$

The remaining eight supercoordinates appear in a group of four. In both $\{\hat{c} = 4\}_{1,2}$ subsystems an $N = 4$ globally defined superconformal symmetry is assumed. We propose three exact $N = 4$ superconformal systems with $\hat{c} = 4$.

(i) - <u>Flat realization : $\hat{c} = 4$</u>

$$ds^2(F) = dzd\bar{z} + dwd\bar{w}$$

$$\Phi = \text{constant}\ , \quad B_{ij} = \text{constant}$$

This realization includes the flat space as well as the toroidal T^4 and orbifold T^4/Z_2 models.

(ii) - <u>Semi – wormhole space $W_k^{(4)}$ with $\hat{c} = 4$</u>

$$ds^2(W) = k\ \frac{dzd\bar{z} + dwd\bar{w}}{z\bar{z} + w\bar{w}}$$

$$2\Phi = \log(z\bar{z} + w\bar{w})$$

$$H_{ijk} = \epsilon_{ijk\ell}\partial^\ell\Phi$$

(iii) - <u>The $\Delta_k^{(4)}$realization with $\hat{c} = 4$</u>

$$ds^2(\Delta) = k\ \frac{dzd\bar{z}}{1 - z\bar{z}} + (k+4)\ \frac{dwd\bar{w}}{w\bar{w} + \epsilon}$$

$$2\Phi(\Delta) = \log(1 - z\bar{z}) + \log(w\bar{w} + \epsilon)\ , \quad \epsilon = \pm 1$$

$$H_{ijk} = 0$$

$$\epsilon = 1\ , \quad \text{Bell – Cigar space}$$

$$\epsilon = -1\ , \quad \text{Bell – Trumpet space}$$

Using as building blocks the $F^2, F^4, \Delta^4_\epsilon$ or W^4 superconformal systems, one obtains several string solutions based on non-trivial ten-dimensional backgrounds. The advantage of our solutions compared to the subclass of models ($F^6 \times W^{(4)}$) studied in the literature [8] lies in the complete knowledge of the $\{\hat{c}=2\}_0$ and $\{\hat{c}=4\}_A$ superconformal theories. Due to this we are able to study not only the background solutions for large k_A but also the full string spectrum and derive the partition functions of the models in a modular invariant way respecting the $N=4$ superconformal symmetry [2]-[5]. Indeed, the full spectrum is given in terms of several characters of known conformal theories. In the W^4_k realization [2]-[4] one uses a character combination of $SU(2)_k, SU(2)^+_1$ and $SU(2)^-_1$ together with the $U(1)_Q$ Liouville-type characters. In $\Delta^{(4)}_k$, one [5] uses the compact [24] and non-compact [25] parafermionic characters (string functions) together with the $SU(2)^+_1$ and $U(1)^-_{R_H}$ $(R_H = \sqrt{2}\sqrt{\frac{k+4}{k}})$. Finally in F^2 and F^4, one uses non-compact, toroidal or orbifold characters.

It is important to stress here that the character combinations above, are not arbitrary but are dictated by the global existence of $N=4$ superconformal symmetry as well as modular invariance.

These requirements define some generalized GSO projections [2]-[5], similar to that of the fermionic construction [26] and that of the conformal block construction [27] of Gepner. One of these projections is fundamental and guarantees the existence of some space-time supersymmetries via the $N=4$ spectral flows. It can be expressed in terms of the two $N=4$ $(\hat{c}=4)_A, A=1,2$ $\{SU(2)^+_1\}_A$ spins, $[j(s_1), j(s_2)]$ [2]-[5]:

$$2(j(s_1)+j(s_2)) = \text{odd integer} . \tag{40}$$

Equation (40) guarantees the existence of some covariantized target space symmetries around the proposed non-trivial backgrounds. This stabilizes our solutions (at least perturbatively) and projects out all kinds of tachyonic or complex conformal weight states. This projection phenomenon is similar to the one observed by Kutasov and Seiberg in the framework of non-critical superstrings with an $N=2$ globally defined superconformal symmetry [28].

Heterotic solutions are simply obtained via a generalized [27],[29] heterotic map. Here also the solutions are stable but the number of covariantized space-time supersymmetries is reduced by a factor of two.

We hope that our explicit construction of a family of consistent and stable solutions will give a better understanding of some fundamental string properties, especially in the case of strongly curved backgrounds (small k_A) where the notion of the space-time dimensionality and topology breaks down.

ACKNOWLEDGEMENTS

I would like to thank I. Antoniadis and S. Ferrara. Some results about the $W_k^{(4)}$ space of this work are products of our collaboration. I am also grateful to C. Callan, D. Gross, E. Kiritsis and D. Lüst for fruitful discussions.

REFERENCES

[1] C. Callan, D. Friedman, E. Martinec and M. Perry, *Nucl.Phys.* **B262** (1985) 593; E. Fradkin and A. Tseytlin, *Nucl.Phys.* **B261** (1985) 1.
[2] I. Antoniadis, S. Ferrara and C. Kounnas, LPTENS-CERN preprint in preparation.
[3] I. Antoniadis and C. Kounnas, LPTENS-CERN preprint in preparation.
[4] C. Kounnas, Proceedings of the Geneva International Conference, on "Lepton-photon Symposium and Europhysics Conference on High Energy Physics", 25 July - 1 August 1991, Vol. 1, pp. 302-306.
[5] C. Kounnas, CERN preprint, CERN-TH.6799/93.
[6] M. Ademollo et al., *Nucl.Phys.* **B114** (1976) 297; T. Eguchi and A. Taormina, *Phys.Lett.* **B200** (1988) 634; A. Sevrin, W. Troost and A. Van Proeyen, *Phys.Lett.* **B208** (1988) 447.
[7] C. Kounnas, M. Porrati and B. Rostand, *Phys.Lett.* **B258** (1991) 61.
[8] C. Callan, J. Harvey and A. Strominger, *Nucl.Phys.* **B359** (1991) 611; C. Callan, Lectures at Sixth J.A. Swieca Summer School, PUPT-1278 (1991).
[9] S.J. Rey, *Phys.Rev.* **D43** (1991) 526 and SLAC-PUB-5659 (1991).
[10] M. Billó, P. Fré, L. Girardello and A. Zaffaroni, preprint SISSA 159/92/EP or IFUM/431/FT.
[11] E. Witten, *Phys.Rev.* **44** (1991) 314.
[12] I. Bars, *Nucl.Phys.* **B334** (1990) 125; I. Bars and D. Nemeschansky, *Nucl.Phys.* **B348** (1991) 89.
[13] R. Dijgkraaf, E. Verlinde and H. Verlinde,*Nucl.Phys.* **B371** (1992) 269.
[14] A. Giveon, *Mod.Phys.Lett.* **A6** (1991) 2843.
[15] E. Kiritsis, *Mod.Phys.Lett.* **A6** (1991) 2871; CERN preprint, CERN-TH.6797/93.
[16] P. Ginsparg and F. Quevedo, Nucl. Phys. **B385** (1992) 527.
[17] C. Kounnas and D. Lüst, *Phys.Lett.* **B289** (1992) 56.
[18] A.B. Zamolodchikov and V.A. Fateev, *Sov.Phys.-JETP* **62** (1985) 215.
[19] L. Dixon, J. Lykken and M. Peskin, *Nucl.Phys.* **B325** (1989) 329.
[20] J. Lykken, *Nucl.Phys.* **B313** (1989) 473.
[21] A.B. Zamolodchikov and V.A. Fateev, *Sov.Phys.-JETP* **63** (1986) 913; P. Di Vecchia, J. Petersen, M.Yu and H. Zheng, *Phys.Lett.* **B174** (1986) 280; Z. Qiu, *Phys.Lett.* **B188** (1987) 207.
[22] I. Bakas and E. Kiritsis, ENS preprint, LPTENS-92-30 (1992).
[23] I. Bakas, E. Kiritsis, Nucl. Phys. **B343** (1990) 185; Mod. Phys. Lett. **A5** (1990) 2039.

[24] V. Kac and D. Peterson, *Adv.Math.* **53** (1984) 125.i
[25] I. Bakas and E. Kiritsis, *Int.J.Mod.Phys.* **A7** [Suppl. 1A] (1992) 339.
[26] I. Antoniadis, C. Bachas and C. Kounnas, *Nucl.Phys.* **B289** (1987) 87; H. Kawai, D.C. Lewellen and S.H.-H. Tye, *Nucl.Phys.* **B288** (1987) 1.
[27] D. Gepner, *Phys.Lett.* **B199** (1987) 370; *Nucl.Phys.* **B296** (1988) 757.
[28] D. Kutasov and N. Seiberg, *Phys.Lett.* **B251** (1990) 67.
[29] W. Lerche, D. Lüst and A.N. Schellekens, *Phys.Lett.* **B181** (1986) 71; W. Lerche, D. Lüst and A.N. Schellekens, *Nucl.Phys.* **B287** (1987) 477.

A tour through $N = 2$ strings.

NEIL MARCUS*
School of Physics and Astronomy
Raymond and Beverly Sackler Faculty of Exact Sciences
Tel-Aviv University
Ramat Aviv, Tel-Aviv 69978, ISRAEL

ABSTRACT

I give an overview of open, closed and heterotic N=2 strings. At the tree level I derive the effective field theories of all the strings, and discuss the group theory of the N=2 open string and the interaction between its open and closed sectors. The two-dimensional effective field theory of the open N=2 string is a sigma model, while the four-dimensional theory gives self-dual Yang-Mills (SDYM) in a self-dual gravity (SDG) background. The theory can have any gauge group, unlike the usual Chan-Paton *ansatz*. The four-dimensional closed string gives SDG, and the heterotic string is related to SDYM. At one loop N=2 string loop amplitudes and partition functions have incurable infra-red divergences, and show puzzling disagreements on the dimension of spacetime when compared to their effective field theories. I show that the known closed-string three-point amplitude can be written directly in terms of a Schwinger parameter, so explicitly exhibiting the inconsistency. I finally discuss the possibility that the puzzles posed by the loop amplitudes could be solved if the N=2 theories were Lorentz invariant and supersymmetric, and I speculate on possible modifications of the string calculations.

1. The basics of $N = 2$ strings

1.1. Introduction

I would like to thank the organizers of this conference for giving me the opportunity to talk about $N = 2$ strings in Rome. It would be a pleasure to give any talk here, but it is particularly appropriate to talk about $N = 2$ strings, since they were first studied in the famous papers by Ademollo *et al.*[1,2], and so have 11 Italian (and 2 foreign) fathers. Now, although $N = 2$ strings naturally live in four dimensions, it is a four-dimensional spacetime with signature $(2,2)$ or $(4,0)$. Thus, unlike the subjects of the first two talks here, $N = 2$ strings can not be directly relevant to nature. So, before jumping into the physics, I would like to give you some reasons for talking to you about them, aside from the aptness of the location.

First, there are interesting mathematical aspects to these theories. As we shall see, the field theories describing these strings are self-dual gravity (SDG) for the closed string and self-dual Yang-Mills (SDYM) for the open and heterotic strings.

*Work supported in part by the US-Israel Binational Science Foundation, and the Israel Academy of Science. E-Mail: NEIL@HALO.TAU.AC.IL.

SDYM in Euclidean space is of course related to instantons, which have a whole host of interesting physical and mathematical properties. In addition, the dimensional reductions of SDYM to two dimensions gives rise to many integrable systems[3], and it has even been conjectured that all two-dimensional integrable systems are reductions of SDYM[4]. $N = 2$ strings thus may give us some notion of what a quantum theory of instantons should be.

What is more interesting from my point of view, and perhaps also from the point of view of the fathers of the string, are the "stringy" reasons for examining these theories. Although $N = 2$ strings are unrealistic, studying them may provide us with better insight into string theory in general because, in some ways, they are remarkably simple theories. An example of this will be our ability to directly compare the loop amplitudes of the $N = 2$ strings to those of their effective field theories. Since $N = 2$ strings do not have Regge trajectories, and contain only massless particles, they give rise to field theories containing only a finite number of fields. In this way they are similar to the $c = 1$ theories, but are even more simple since, being Lorentz invariant, they can not contain any discrete states. Also, $N = 2$ amplitudes are local and many of them vanish[5-7], possibly indicating some kind of topological structure of the theories. It will turn out that, despite their simplicity, our understanding of $N = 2$ theories is still immature with many puzzles remaining. I would like to view this as a challenge, rather than a problem, and I hope that other people will be encouraged to enter into the subject.

The original work that I will present in this talk consists of the reconsideration of the $N = 2$ open string in $2_{\mathbb{C}}$ dimensions[7]. However, instead of concentrating on this particular case, I shall rather give a general tour through various $N = 2$ strings, telling you their status and pointing out their problems. This means that I shall be discussing the work of many other people, with additional commentary of my own in various places, and I shall of course be biased to those aspects of the theories that are particularly interesting to me. For example, I shall not discuss non-critical $N = 2$ strings at all, nor the $N = 2$ strings with background charges[8]. Also I shall not go into too much detail on aspects of any particular theory; you can find more details in the various original references.

The outline of my talk is the follows: I first give a basic review of $N = 2$ strings. Then I calculate various tree-level string amplitudes, to find the effective field theory actions describing the strings. I then present some one-loop amplitudes to illustrate the problems intrinsic in the amplitudes, and in comparing the amplitudes to those of the effective field theory. I conclude with a discussion of the status of the $N = 2$ strings, and of possible solutions to their various problems.

1.2. $N = 2$ supergravity in two dimensions

At this stage, I should explain to those who do not know that the "N" of $N = 2$ refers to the number of local world-sheet supersymmetries of the string. Thus the $N = 0$ string is the bosonic string, which is basically a theory of matter coupled to 2-dimensional world-sheet gravity, and the $N = 1$ string is a theory of supermatter coupled to world-sheet $N = 1$ supergravity. (In heterotic strings, the numbers of left- and right-handed supersymmetries are different.) $N = 2$ supergravity was first considered by Brink and Schwarz[9]. The $N = 2$ supergravity multiplet consists of a vielbein, a *complex* gravitino and a $U(1)$ gauge field denoted, respectively:

$$e_\alpha^a \, , \quad \begin{pmatrix} \chi_\alpha^{(1)} \\ \chi_\alpha^{*(-1)} \end{pmatrix} \quad \text{and} \quad A_\alpha \, . \tag{1}$$

Here the numbers in parenthesis show the $U(1)$ charges of the gravitini. As is usual in string theory, all the supergravity fields can be locally gauged away using the gauge symmetries of the theory, and there is no action for them: The vielbein is removed by general coordinate invariance, local Lorentz invariance and local Weyl transformations; the gravitini by the complex supersymmetry and complex super-Weyl transformations, and the two components of the $U(1)$ gauge field by vector and chiral $U(1)$ gauge symmetries on the world sheet[†]. These $U(1)$ symmetries have no counterpart in the $N = 0$ and $N = 1$ theories, and their existence leads to many of the special features of $N = 2$ strings.

To get a string theory, one must couple the supergravity to some ($N = 2$) matter. The simplest such matter is an $N = 2$ chiral superfield X^i, with i some internal (space-time!) index. The component fields are seen in the θ expansion of these superfields given, somewhat schematically, by:

$$X^i \sim x^i + \theta^{(-1)}\psi^{i(1)} \quad \text{and} \quad \bar{X}^{\bar{i}} \sim \bar{x}^{\bar{i}} + \theta^{*(1)}\psi^{*\bar{i}(-1)} \, . \tag{2}$$

The string world-sheet action is[9]:

$$\begin{aligned} S = \int d^2z\sqrt{g}\,\Big(\frac{1}{2} g^{\alpha\beta}\partial_\alpha x_i \partial_\beta \bar{x}^{\bar{i}} + i\,\bar{\psi}^{\bar{i}} \not{D}\, \psi_i + A_\alpha \bar{\psi}^{\bar{i}}\gamma^\alpha \psi_i \\ + (\partial_\alpha \bar{x}^{\bar{i}} + \bar{\psi}^{\bar{i}}\chi)\,\chi_\beta \gamma^\alpha \gamma^\beta \psi_i \; + \text{c.c.} \Big) \, , \end{aligned} \tag{3}$$

where $\not{D}$ denotes a gravitationally covariant derivative of the spinor, containing a spin-connection piece. Note that, although the scalar field x^i is complex, it does not have a $U(1)$ charge and does not couple to the gauge field A_α. On the other hand, the spinors ψ^i are charged, and I have explicitly shown their minimal coupling in the action.

1.3. The critical dimension of $N = 2$ strings

The first question to ask in any string theory is what is the critical dimension of the theory? Ademollo *et al.* found that in $N = 2$ strings $D = 2$, where D is the number of chiral superfields X^i. The modern derivation of this result goes as follows[10]: gauge fixing the $N = 2$ supergravity algebra gives rise to independent left and right-handed $N = 2$ constraint algebras:

$$\begin{pmatrix} & T & \\ G & & G^* \\ & J & \end{pmatrix}_L \quad \text{and} \quad \begin{pmatrix} & \bar{T} & \\ \bar{G} & & \bar{G}^* \\ & \bar{J} & \end{pmatrix}_R \, . \tag{4}$$

[†] The existence of the chiral invariance was pointed out by Fradkin and Tseytlin in ref. 10.

In both the left and right sectors one has the usual (b,c) ghosts, *complex* (β,γ) supersymmetry ghosts and a (b',c') system of $U(1)$ ghosts with conformal spins $(1,0)$. Canceling the conformal anomaly means

$$c = -26 + 2\cdot 11 - 2 + 2D\cdot(1+\frac{1}{2}) = 0\ , \tag{5}$$

resulting in $D = 2$. An even easier calculation is to cancel the axial-vector $U(1)$ anomaly between the ψ's and the (β,γ) ghosts:

$$c = 0 = -2 + D\ . \tag{6}$$

This gives the same result, since all the anomalies of the theory fall into a single supermultiplet.

Since we have 2 *complex* scalars x^i, it now appears to be obvious that the theory lives in a $2_{\mathbb{C}}$-spacetime; in terms of real dimensions this is a $4_{\mathbb{R}}$-dimensional spacetime that must have a signature $(2,2)$ or $(4,0)$. However, here the history of the subject is a little peculiar. In their original work, Ademollo *et al.* defined a real superfield $Y = X + \bar{X}$, instead of working with the chiral superfield X. This superfield contains the same information as X, except that the imaginary part of the zero-mode x appears only as a derivative. Based on this, they decided not to excite positions and momenta in these "imaginary" coordinates. This means that they dimensionally reduced the theory by brute force to $2_{\mathbb{R}}$ dimensions. (Note that the anomaly cancellation still works in the truncated theory, since only the zero-modes of the fields have been changed, and the theory still contains 4 real bosonic and fermionic fields.) It was only much later, following the work of D'Adda and Lizzi[11], that the theory was returned to $2_{\mathbb{C}}$ dimensions.

1.4. The spectrum of $N = 2$ strings

To see the spectrum of the theory one needs to calculate the theory on the cylinder. This means that we must discuss the boundary conditions of the fields. As usual, since they are world-sheet tensor fields, the x^i fields and the anticommuting ghosts are periodic (although other possibilities were considered in ref. 2). However, a new feature of the $N = 2$ string is that the fermions ψ^i and the commuting ghosts are charged under the $U(1)$ symmetry. Since one can have a constant gauge field on the cylinder (more precisely one can have non-trivial Wilson lines around the cylinder), one can continuously change the boundary conditions of the fermions from NS to R by turning on this gauge field. Because of this, one usually considers just the purely NS part of the theory, and argues that all other sectors are equivalent to it by the $N = 2$ spectral flow[12]. I shall generally do this throughout the talk, although I shall discuss other possibilities in the conclusion.

Now naively, since we are in $D = 2$, one would expect that one could go to a light-cone gauge with $D - 2 = 0$ transverse dimensions! This suggests that $N = 2$ theories should have no oscillator excitations. The light-cone argument also suggests that the mass squared of this state is $(2 - D)/24 = 0$. The decoupling of the massive states of the string was indeed seen by Ademollo *et al.*, and the same result has been confirmed by a BRST analysis[13]. But the simplest way to find the spectrum is to calculate the partition function of the theory: Aside from zero modes, the four x oscillators are canceled by the (b, c) and (b', c') ghosts, while, for any boundary conditions, the two complex charged ψ^i's are canceled by the charged (β, γ) ghost system. This leaves only massless states in the spectrum. Summarizing:

a) $N = 2$ strings live in $2_{\mathbb{C}}$ dimensions.

b) All their oscillator excitations vanish.

c) In each sector of the theory, only a massless "scalar" state propagates.

and, thus:

d) $N = 2$ string amplitudes must satisfy duality with no infinite sums over massive states!

e) $N = 2$ string field theory is an ordinary field theory with a finite number of particles.

1.5. The Lorentz symmetry of the $N = 2$ strings

We have now seen where the $N = 2$ strings live and have calculated their spectrum. One remaining question that we should ask is what is the "Lorentz" invariance of these theories. Note that in the gauge-fixed theory, where one turns off the supergravity fields A_α and χ_α in the string action of eq. (3), there is no difference between matter fields and their complex conjugates in the action. Based on this, D'Adda and Lizzi argued that the the theory has an $SO(2,2)$ Lorentz invariance (chosing the "Minkowski" rather than the Euclidean signature). However, the ψ^i's and $\psi^{*\bar{i}}$'s couple differently to the gauge field A_α, since they have opposite charges. This difference is then fed to the x^i and $\bar{x}^{\bar{i}}$ fields by the gravitini χ_α. Another way of looking at this is that the constraint multiplets of eq. (4) distinguish the different fields. In the works of Ooguri and Vafa that sparked the renewed interest in $N = 2$ strings[5], spacetime is thus considered to be intrinsically complex (actually Kähler), with a $U(1,1)$ symmetry group.

Recently, however, Siegel has argued[14] that the $N = 2$ string is the same as the "$N = 4$" string, which was also discovered by Ademollo *et al.*[2]. This string has an $SU(2)$ local symmetry, is hard to quantize covariantly, and apparently has $D = -2$! If Siegel is correct, the manifest $U(1,1)$ invariance of the $N = 2$ string should be extendible to the full $SO(2,2)$ Lorentz invariance. This is known to be the case in the tree-level amplitudes of refs. 5–7, as I will discuss. A bigger Lorentz symmetry would be important, since the $U(1,1)$ spaces have no little group, so one cannot define the "spins" of particles, while the $SO(2,2)$ Poincaré group can have representations of continuous spin. Using the larger Lorentz group, Siegel has also argued that the Ramond sectors of the $N = 2$ strings should describe fermions[14], and that the effective field theories of the $N = 2$ strings should be maximally spacetime supersymmetric[15].

Such results necessitate a modification of the string amplitudes. I shall return to these issues in the discussion section at the end of my talk.

2. Tree amplitudes and effective actions

2.1. $N = 2$ strings in two real dimensions

We have seen that $N = 2$ strings do not have any oscillator excitations, and so can be described by effective field theories with finite numbers of fields. The most straightforward approach to finding these field theories is simply to calculate the tree-level amplitudes of the string theory, and to equate them to those of the field theory. Here I shall follow an approximately historical approach, and first consider the strings in two *real* dimensions, *à la* Ademollo *et al.* It may seem a bit strange to the post-Polyakov physicist that the $2_{\mathbb{R}}$-dimensional theory makes any sense at all. However, we have already argued that (local) anomalies are still canceled in this truncated theory, and tree amplitudes are consistent even in dimensions which do not give anomaly cancellation. We would expect this truncation to make troubles at some stage, but since the rules for writing down $N = 2$ strings are so ill-defined, it is even possible that these theories are consistent as is. It turns out that these truncated theories illustrate many of the general properties of $N = 2$ strings, and have interesting effective field theories in their own right.

I now turn to the case of the open string in $2_{\mathbb{R}}$ dimensions[2]. As I suggested above, the spectrum of this theory should be a single massless scalar. However, as is usual in open strings, Ademollo *et al.* added ($SU(2)$) Chan-Paton factors[16] to the string, ending up with a theory of three scalars in the adjoint of SU(2). Since the $2_{\mathbb{R}}$-dimensional theory is the truncation of the $2_{\mathbb{C}}$ theory, which I shall consider in detail later, I shall simply present its amplitudes without derivation: The first nonvanishing amplitude in the theory is the 4-point function, since all odd-point functions vanish by the G-parity of the theory. It is given by

$$A_4 = \frac{g^2}{2}\, u \,\mathrm{Tr}\,(\lambda_1\lambda_2\lambda_3\lambda_4) + \text{perms}\;, \tag{7}$$

where u is the usual Mandelstam variable, and the λ's are the Chan-Paton factors. Note that this amplitude would vanish without the Chan-Paton factors, since it would be proportional to $s + t + u = 0$. Also, recall that in the scattering of two massless particles in $2_{\mathbb{R}}$-dimensions one has the kinematic relation

$$stu = 0\;. \tag{8}$$

One can get the equation of motion corresponding to eq. (7) from an $SU(2)$ sigma model‡, with action

$$S = \int d^2x\, \mathrm{Tr}\left(g^{-1}\partial_i g\; g^{-1}\partial^i g\right)\;, \tag{9}$$

‡Because the sign of the amplitude has not been fixed, the correct theory may be the coset theory $SU(2)_{\mathbb{C}}/SU(2)$[6].

and equation of motion

$$\partial_i \left(g^{-1}\partial^i g\right) = 0 \ . \tag{10}$$

Ademollo *et al.* argued that this equivalence should be true to all orders, using uniqueness arguments based on Regge behaviour and Adler zeroes, which I do not claim to follow.

We can already see several intriguing features of this theory, which will generalize to all $N = 2$ theories. Unlike other string theories, and about any other field theories, the amplitude of eq. (7) is a local function of the momenta. That this had to happen is pretty much forced on us by the opposing requirements of the duality of the amplitudes and the fact that the theory contains only massless particles. This should therefore be true for all nonvanishing amplitudes. Ademollo *et al.* argued that all $4N + 2$ point functions vanish. (Thus the next possibly nonvanishing amplitude is the eight-point function—not surprisingly, it has never been calculated.) It will turn out that there are many vanishing amplitudes in $N = 2$ string calculations, for reasons that are not completely clear. Finally, as one might hope for a theory based on strings, the effective field theory has a nice geometrical nature.

I shall now briefly turn to the case of closed and heterotic strings in $2_{\mathbb{R}}$ dimensions. These were both studied in a paper by Green in 1987[17]. Using the rules for obtaining closed-string amplitudes from open ones[18], he found that the closed $2_{\mathbb{R}}$-dimensional $N = 2$ string has a vanishing four-point function, and is presumably free. On the other hand, the heterotic string has a four-point amplitude given by

$$A_4^{\rm het} = A_4^{\rm open} + \alpha' \, tu \, \delta_{12}\delta_{34} \ , \tag{11}$$

which can be interpreted as a sigma-model amplitude with a correction from a graph with an internal "graviton".

2.2. Closed $N = 2$ strings in $2_{\mathbb{C}}$-dimensions

Continuing on the historical path, I now turn to a discussion of the closed $N = 2$ string in $2_{\mathbb{C}}$ dimensions, following Ooguri and Vafa[5]. I will consider this case in somewhat more detail, since I will take over many of the conventions and results for the open case. The world-sheet action of the theory, with all the supergravity fields gauge fixed, can be written as

$$S = \int \frac{d^2z}{\pi} \, d^2\theta \, d^2\bar{\theta} \, K_0(X^i, \bar{X}^{\bar{i}}) \ , \tag{12}$$

where X^i is the $N = 2$ chiral superfield. Here i runs from 0 to 1, corresponding to a real (2,2)-dimensional spacetime, and K_0 is the flat Kähler potential $\eta_{i\bar{j}} X^i \bar{X}^{\bar{j}}$. The only state of the theory is a single massless scalar, corresponding to a perturbation around the flat Kähler potential. The superspace-vertex operator for emitting this scalar with (complex) momentum k is

$$V_c = \frac{\kappa}{\pi} \, e^{i(k\cdot\bar{X} + \bar{k}\cdot X)} \ . \tag{13}$$

The first amplitude that one can calculate is the three-point function. Fixing the super-Möbius transformations, this is given by

$$\begin{aligned} A_{ccc} &= \left\langle V_c|_{\theta=0}(0) \cdot \int d^2\theta d^2\bar{\theta}\, V_c(1) \cdot V_c|_{\theta=0}(\infty) \right\rangle \\ &= \kappa\, c_{12}^2 \,, \end{aligned} \tag{14}$$

where

$$c_{12} \equiv (k_1 \cdot \bar{k}_2 - \bar{k}_1 \cdot k_2) \tag{15}$$

is the extra invariant product of the momenta (other than the usual dot product which has a plus sign) that exists in $2_{\mathbb{C}}$ dimensions, when the Lorentz group $SO(2,2)$ is reduced to $U(1,1)$. Note that c_{ij} is antisymmetric with respect to its two indices, and is additive in the sense that $c_{i,j} + c_{i,k} = c_{i,j+k}$. Using momentum conservation, one sees that A_{ccc} is totally symmetric, as it should be. It is important to note that in $(2,2)$ dimensions, unlike the familiar $(3,1)$-dimensional case, there is sufficient phase space to describe one massless particle splitting into two others. The three-point function of the $2_{\mathbb{C}}$-dimensional closed string therefore implies a truly nontrivial S-matrix element, and not just some unphysical vertex.

The four-point function can be calculated similarly, and results in an apparently standard string amplitude:

$$\begin{aligned} A_{cccc} &\sim \int d^2z \left\langle V_c|_{\theta=0}(0) \cdot \int d^2\theta d^2\bar{\theta}\, V_c(z) \cdot \int d^2\theta d^2\bar{\theta}\, V_c(1) \cdot V_c|_{\theta=0}(\infty) \right\rangle \\ &= \frac{\kappa^2}{\pi} F^2 \frac{\Gamma(1-s/2)\,\Gamma(1-t/2)\,\Gamma(1-u/2)}{\Gamma(s/2)\,\Gamma(t/2)\,\Gamma(u/2)} \,. \end{aligned} \tag{16}$$

However, Ooguri and Vafa noted that, like the $2_{\mathbb{R}}$-dimensional kinematic identity $stu = 0$, in the scattering of massless particles in $(2,2)$ dimensions there is again a relation:

$$F \equiv 1 - \frac{c_{12}c_{34}}{su} - \frac{c_{23}c_{41}}{tu} = 0 \,. \tag{17}$$

This means that the four-point function of eq. (16) vanishes on shell. Ooguri and Vafa proposed that this vanishing comes from some kind of topological nature of the theory, but this is still somewhat of a mystery. They also conjectured that all higher-point functions in the $2_{\mathbb{C}}$-dimensional case vanish; this is, as yet, unchecked. (As I argued before, duality arguments just show that the amplitudes should be local.) It is simple to see that if the string is reduced to $2_{\mathbb{R}}$ dimensions, as in the work of Green[17], the theory becomes trivial: In $2_{\mathbb{R}}$ dimensions c_{ij} vanishes identically, so the three-point function now vanishes. Also, while in $2_{\mathbb{R}}$ dimensions $F \to 1$, the four-point function still vanishes because of the identity $stu = 0$.

Returning to $2_{\mathbb{C}}$ dimensions, one sees that the local three-point function and vanishing four-point function can be obtained from the action[5]

$$\mathcal{L}_c = \int d^4x \left(\frac{1}{2} \partial^i \phi\, \bar{\partial}_i \phi + \frac{2\kappa}{3} \phi\, \partial\bar{\partial}\phi \wedge \partial\bar{\partial}\phi + O(\phi^5) \right) \,, \tag{18}$$

where the $O(\phi^5)$ term needs to be determined from the five-point function. If all higher-order terms are absent the equation of motion of this action has a name—the Plebanski equation[19]—and has a nice geometrical meaning. It is therefore very reasonable that the action is exact. (Note that this conjecture and the conjecture that higher-point amplitudes vanish are logically independent.) The Plebanski equation is

$$\partial^i \bar{\partial}_{\bar{i}} \phi - 2\kappa\, \partial\bar{\partial}\phi \wedge \partial\bar{\partial}\phi = 0 \ . \tag{19}$$

Its geometrical meaning is seen by considering ϕ to be a perturbation of a Kähler potential around the flat $2_{\mathbb{C}}$ space, resulting in a spacetime metric

$$g_{i\bar{j}} = \partial_i \bar{\partial}_{\bar{j}} \left(x_k \bar{x}^{\bar{k}} + 4\kappa\phi \right) \ . \tag{20}$$

The Plebanski equation of motion then becomes simply

$$\det g_{i\bar{j}} = -1 \ . \tag{21}$$

Now since in a Kähler space the Ricci tensor $R_{i\bar{j}}$ is given by

$$R_{i\bar{j}} = \partial_i \bar{\partial}_{\bar{j}} \log \det g \ , \tag{22}$$

the Plebanski equation is the condition for the Ricci flatness of the $2_{\mathbb{C}}$-dimensional Kähler space. An important theorem of Atiyah, Hitchin and Singer[20], which is quite nontrivial in the forward direction, is that a $4_{\mathbb{R}}$-dimensional Riemann space is self dual *iff* it is Kähler and Ricci flat. Thus the equation of motion of the $N = 2$ closed string can be written elegantly as

$$R = \tilde{R} \ , \tag{23}$$

and the $N = 2$ string describes self-dual gravity. It is intriguing that although the entire formulation of the string has been carried out in a $2_{\mathbb{C}}$-dimensional Kähler space, the final equation of motion can be written in a completely $(2,2)$ Lorentz-invariant way. I shall discuss this further at the end of the talk.

2.3. Purely open $N = 2$ strings in $2_{\mathbb{C}}$-dimensions

Since, despite apparently being a scalar theory, the closed $N = 2$ string turns out to describe self-dual gravity, it is reasonable to expect that the open and heterotic strings describe self-dual Yang-Mills. In the heterotic string[6], which was studied by Ooguri and Vafa soon after the closed case, this is basically true, but there are some complications. Since the geometrical structure of the heterotic string is not that well understood, I shall leave the historical path and turn rather to the open string, returning to make a few comments on the heterotic string later.

I shall thus now turn to my work; the reconsideration of the open string in $2_{\mathbb{C}}$ dimensions[7§]. The open string sweeps out a world sheet that is a super-Riemann

[§]There were previous attempts to calculate open string amplitudes in $2_{\mathbb{C}}$ dimensions in refs. 11 and 21. However in the first paper it was not realised that the three-point amplitude is non-zero, and in the second that the four-point amplitude is zero. The nature of the theory was therefore misunderstood.

surface, but now with boundaries. For example, in the case of the super upper-half plane which gives the tree-level amplitudes of the theory, the boundary is given by $z = \bar{z} \equiv \sigma$, $\theta = \bar{\theta} \equiv \theta$. The action of the string is the same as that of the closed string in eq. (12). To calculate with it, one also needs the boundary conditions for the fields, which are given by $\partial x = \bar{\partial} x|_{z=\bar{z}}$ and $\psi_R = \psi_L \equiv \psi|_{z=\bar{z}}$.

As in the closed string, the spectrum of the open string is a single massless scalar. However, as is usually the case in open strings, we want to append "Chan-Paton" group theory factors to the string amplitudes, and so end up with a multiplet of scalars φ^a. The superspace vertex operator to emit these scalars is the same as that of the closed scalars:

$$V_o = g\, e^{i(k\cdot\bar{X}+\bar{k}\cdot X)}\,, \tag{24}$$

but it has a different interpretation, since it is inserted on the boundary of the super-Riemann surface. After integrating out the fermionic coordinates one obtains

$$V_o^{int} = \int d^2\theta\, V_o = \frac{g}{2}\,(ik\cdot\partial_\sigma\bar{x} - i\bar{k}\cdot\partial_\sigma x - 4k\cdot\bar{\psi}\,\bar{k}\cdot\psi)\, e^{i(k\cdot\bar{x}+\bar{k}\cdot x)}\,. \tag{25}$$

(The strange factors of 2 are because I am using conventions appropriate to the closed string.)

The open-string three-point amplitude is then given by

$$\begin{aligned} A_{ooo} &= \Big\langle V_o|_{\theta=0}(0)\cdot\int d^2\theta\, V_o(1)\cdot V_o|_{\theta=0}(\infty)\Big\rangle \\ &= g\, c_{12}\times(-i\, f^{abc})\,, \end{aligned} \tag{26}$$

where I have inserted the "group-theory" factor $-if^{abc}$ by hand. Without this factor the amplitude would be totally antisymmetric with respect to the three scalars, and so vanish. In the Chan-Paton *ansatz* one would have $f^{abc} = \mathrm{Tr}\,(\Lambda^a[\Lambda^b\,,\,\Lambda^c])$, but since we have not yet established the principles for constructing $N=2$ strings, I shall here consider the most general possible *ansatz* consistent with principles to be given later. Thus, at this stage, f^{abc} is a general unspecified totally antisymmetric tensor. (The notation may be a little suggestive, however.) Note that, as usual[18], the open amplitude is the square root of the closed-string amplitude of eq. (14). It is therefore the same as that of the heterotic string, and can be derived from the same field-theory action[6]:

$$\mathcal{L}_{3o} = \int d^4x\,\left(\frac{1}{2}\,\partial^i\varphi^a\bar{\partial}_{\bar{i}}\varphi^a - i\,\frac{g}{3}\,f^{abc}\varphi^a\partial^i\varphi^b\bar{\partial}_{\bar{i}}\varphi^c\right) + O(\varphi^4)\,. \tag{27}$$

The four-point amplitude of the string is given by

$$\begin{aligned} A_{oooo} &\sim \int_0^1 dx\,\Big\langle V_o|_{\theta=0}(0)\cdot\int d^2\theta\, V_o(x)\cdot\int d^2\theta\, V_o(1)\cdot V_o|_{\theta=0}(\infty)\Big\rangle \\ &= \frac{g^2}{4}\,F\,\frac{\Gamma(1-2s)\,\Gamma(1-2t)}{\Gamma(2u)}\,, \end{aligned} \tag{28}$$

and, as in the closed string, vanishes because of the F factor. Because this amplitude vanishes, the usual unitarity constraints[22] on the group-theory factors do not apply.

However, one can still find a constraint on them by calculating the four-point function of the field theory action of eq. (27) (supplemented by a quartic term $\mathcal{L}_{4o}$), and demanding that it vanishes in agreement with the string result of eq. (28). Using some kinematical identities, the field-theory result can be written as

$$\begin{aligned} A_{oooo\,\mathrm{FT}} = -g^2\Big\{ & \frac{c_{12}c_{34}}{s}\left(f^{abx}f^{xcd} - f^{acx}f^{xbd} - f^{bcx}f^{xda}\right) \\ & + u f^{bcx}f^{xda} + t f^{cax}f^{xbd}\Big\} - V_{4o} \,. \end{aligned} \tag{29}$$

Since V_{4o} must be a local vertex, the factor in the parentheses must vanish for the amplitude to be zero. This factor is simply the Jacobi identity, so we see that the f^{abc}'s must be the structure constants of some group, as one might have been expected. The group is unspecified, and can be any semisimple group times a product of $U(1)$'s. This is unlike the case of the bosonic and $N = 1$ open strings, where classically the group can only be $SO(N)$, $USp(N)$ or $U(N)$[22¶]. We thus have an *ansatz* for the group theory that is more general than (but also even more ugly and *ad hoc* than) that of Chan and Paton[16]. The resulting quartic interaction determined by eq. (29) is now[‖]:

$$\mathcal{L}_{4o} = \int d^4x \left(-\frac{g^2}{6} f^{adx}f^{xbc}\, \partial^i\varphi^a\varphi^b\bar{\partial}_{\bar{i}}\varphi^c\varphi^d\right) . \tag{30}$$

The action of eqs. (27) and (30) does not appear to have any clear meaning. However, its resulting equation of motion can be written in a nice compact way: Defining the *anti*-hermitian matrix φ, the equation of motion becomes

$$\bar{\partial}_{\bar{i}}\left(e^{-2ig\varphi}\partial^i e^{2ig\varphi}\right) = 0 \,, \tag{31}$$

which is known as Yang's equation[24]. As expected, since we again have an equation with a name, it has a nice meaning and is indeed related to self-dual Yang-Mills (SDYM). This can be seen by noting that in a Kähler space the SDYM equation $F = \tilde{F}$ breaks into three pieces. Defining the holomorphic $(2,0)$ form ω and the Kähler form k, one first has

$$\begin{aligned} F_{ij} = 0 \quad (F\wedge\bar{\omega} = 0) \quad &\Longleftrightarrow \quad A_i \equiv e^{-ig\varphi}\partial_i e^{ig\varphi} \\ F_{\bar{i}\bar{j}} = 0 \quad (F\wedge\omega = 0) \quad &\Longleftrightarrow \quad \bar{A}_{\bar{i}} \equiv e^{ig\varphi}\bar{\partial}_{\bar{i}} e^{-ig\varphi} \,. \end{aligned} \tag{32}$$

(Here, choosing the φ's in A_i and $\bar{A}_{\bar{i}}$ to be the same means that we have fixed the gauge-invariance of the theory.) The third self-duality equation is then

$$\begin{aligned} F_i^{\bar{i}} = 0 \quad (F\wedge k = 0) \quad &\Longleftrightarrow \quad \left[\mathcal{D}_i\,,\,\bar{\mathcal{D}}^{\bar{i}}\right] = 0 \\ &\Longleftrightarrow \quad g^{i\bar{j}}\bar{\partial}_{\bar{j}}\left(e^{-2ig\varphi}\partial_i e^{2ig\varphi}\right) = 0 \,. \end{aligned} \tag{33}$$

¶Of course, in the case of the superstring, we know that the cancellation of anomalies uniquely picks out the group $SO(32)$[23].

‖Since the heterotic string also has a vanishing four-point amplitude, this result is again in agreement with that of the pure Yang-Mills sector of 6. However, the heterotic string also gets contribution from graphs with an intermediate graviton, modifying $\mathcal{L}_{4o}$. In an open theory such graphs have the topology of an annulus, and are not part of the classical theory.

Thus the purely open string equation, eq. (31), describes SDYM in a flat (2,2)-dimensional spacetime. The action giving Yang's equation can be written order by order in φ, as we have started to do in eqs. (28) and (30). Since the equation is a generalization of the Wess-Zumino-Witten equation in two dimensions, it may not be surprising that one can also write a more geometrical action in terms of coset elements in a *five*-dimensional space[25].

Summarizing, once the dust has settled, the equations of motion of the string can again be written in a completely Lorentz-invariant way:

$$F = \tilde{F} \; ! \tag{34}$$

2.4. Coupled open and closed $N = 2$ strings in $2_{\mathbb{C}}$-dimensions

We have seen that, as we expected, the open string describes SDYM. However, we know that all strings should contain gravity, so there should also be a gravitational sector of the theory. In the heterotic string the gravitational sector indeed exists and, as in the $2_{\mathbb{R}}$ case, it changes the equations of motion of the string from being exactly the self-duality of the Yang-Mills. In open strings, gravity exists because open strings can always join together to form closed ones. Open theories thus always have closed sectors, and one should consider interactions between the sectors. Actually, not many calculations of this type have been carried out, at least in the Polyakov formalism, so such calculations have several relatively unfamiliar aspects to them. For example, you should bear in mind that in open string theories the tree-level equations of the closed sector are found on the sphere, at genus 0, so the classical gravitational equations of motion are the same as those of the closed string. However, the tree level of the open sector comes from the disk, or upper-half plane (UHP), at "genus 1/2", so the very notion of the classical limit of the theory is not so well defined. For example, the closed and open-string couplings are related by $\kappa \sim \sqrt{\hbar} g^2$.

I shall not go into the details of the mixed open-closed amplitude calculations[7], but shall just give the results. First, all amplitudes with only one open string and an arbitrary number of closed ones vanish, because of the twist symmetry of the theory. This is fortunate, since the open sector has a group theory index and such amplitudes imply a breaking of the group. The only mixed three-point amplitude therefore involves the scattering of two open strings with one closed one on the UHP. It is given by

$$\begin{aligned} A_{ooc} &\sim \int_{-\infty}^{\infty} dx \left\langle V_o|_{\theta=0}(x) \cdot \int d^2\theta d^2\bar{\theta}\, V_c(z=i) \cdot V_o|_{\theta=0}(\infty) \right\rangle \\ &= \kappa\, \delta^{ab} c_{12}^2 \,. \end{aligned} \tag{35}$$

(Note that this amplitude involves an integration over the position of one of the vertices, even though it is a three-point function!) Since the amplitude gives the coupling of the gravitational sector to the quadratic term of the open scalars, I have chosen to append the group-theory factor δ^{ab}, as in the kinetic term. This vertex is similar to that of the gravitational self-interaction of eq. (14), showing some kind of universality in the couplings of the various fields to gravity. The existence of the vertex means that one has to add the interaction

$$\mathcal{L}_{ooc} = \int d^4x \left(2\kappa\, \phi\, \partial\bar{\partial}\varphi^a \wedge \partial\bar{\partial}\varphi^a\right) \tag{36}$$

to our open-closed action.

The first interesting four-point function involves the scattering of three open strings with one closed one on the UHP. It is given by

$$A_{oooc} = \frac{i}{2}\kappa g f^{abc}\, F\, \frac{\Gamma(s)\,\Gamma(t)\,\Gamma(u)}{\Gamma(-s)\,\Gamma(-t)\,\Gamma(-u)}\,(c_{12}t + c_{23}s)\ , \tag{37}$$

and it again vanishes on shell because of the F factor. Calculating the same amplitude in the field theory, one sees that one has to add yet another term to the action:

$$\mathcal{L}_{oooc} = \int d^4x \left(-\frac{4}{3}\, ig\kappa\, f^{abc}\, \partial\bar{\partial}\phi \wedge \varphi^a \partial\varphi^b \bar{\partial}\varphi^c\right)\ . \tag{38}$$

In principle, one should also consider the amplitude with two closed and two open scalars. Since it is really messy to calculate, I shall assume that it vanishes because of an F factor, like all the other four-point amplitudes. This vanishing is found from the field theory without having to add any new terms to the Lagrangian.

There is one final type of amplitude that I have not yet discussed, because it has a different structure and interpretation than all the other amplitudes. This is the scattering of three closed strings on the UHP and on the projective plane $\mathbb{R}P^2$, which should be combined with the UHP graph, since it is also of Euler number 1. Unlike graphs with open vertices, whose tree-level amplitudes are defined on the UHP (since $\mathbb{R}P^2$ has no boundaries, it does not appear in open amplitudes), the closed three-point tree-level amplitude of the theory comes from the sphere. These genus 1/2 graphs thus give a type of quantum corrections to the theory. Their amplitudes are difficult to calculate, but one can see that they both have the form:

$$A'_{ccc} \propto \frac{\kappa^3}{g^2}\, c_{12}^4\ , \tag{39}$$

with some finite coefficients. Since the amplitudes contains no open-string vertices, their group-theory factors are not fixed and one can consider them with arbitrary coefficients. (In the Chan-Paton scheme the overall factor would be proportional to $N-2$ for the group $SO(N)$, *etc.*) The interpretation of eq. (39) is unclear and, in the following, I will generally choose the coefficients to make the contribution vanish.

At this stage, we have enough of the action to see what the full equations of motion of the theory should be. The open-sector equation is:

$$g_{i\bar{j}}(\phi)\,\bar{\partial}^{\bar{j}}\left(e^{-2ig\varphi}\partial^i e^{2ig\varphi}\right) = 0 \iff F \wedge k = 0\ , \tag{40}$$

where $k(\phi)$ and $g(\phi)$ are now the full Kähler form and metric, defined in terms of ϕ as in eq. (20). This gives the self-duality of the Yang-Mills field strength in the background of the curved Kähler space of the closed sector. The closed equation of motion is now modified to

$$\det g_{i\bar{j}} = -1 - \frac{2\kappa^2}{g^2} \operatorname{Tr}\left(F_{i\bar{j}} F^{i\bar{j}}\right) , \tag{41}$$

or

$$\partial\bar{\partial}K \wedge \partial\bar{\partial}K = 2\omega \wedge \bar{\omega} - \frac{4\kappa^2}{g^2} \operatorname{Tr}\left(F \wedge F\right) . \tag{42}$$

Since the Ricci tensor is defined in terms of derivatives of $\det g$ (eq. (22)), one sees that the Ricci-flatness condition is modified by a source term from the open sector, in the same way that the Einstein tensor gets a contribution from the matter stress tensor in usual gravity. Recall that, since there is relation $\kappa \sim \sqrt{\hbar} g^2$, the source term is some kind of quantum mechanical correction to the equation of motion. (If we had included the quantum amplitude of eq. (39), there would also have been a gravitational source term on the right of eqs. (41) and (42).) The spacetime is thus no longer Ricci flat, and is no longer self-dual. I do not know of any Lorentz-invariant way of writing this gravitational equation with the source.

2.5. Heterotic $N = 2$ strings in $2_{\mathbb{C}}$ (?) dimensions

For completeness, I will now briefly discuss the heterotic $N = 2$ strings[6]. The major problem with this string is that the left-hand side of the string lives in $(2,2)$ dimensions, while the right-hand side lives in either $(25,1)$ or $(9,1)$ dimensions. One therefore needs to get rid of one of the LHS time coordinates. This is done, essentially, by compactifying the theory to either $(2,1)$ or $(1,1)$ dimensions, somewhat messing up the geometrical interpretation of the theories. As in the open case, the pure gauge sector of the theory is described by Yang's equation for SDYM, now reduced to three (or two) dimensions[6]. Somewhat surprisingly, the resulting Yang-Mills particles are actually tachyonic scalars! In addition, the theory also contains massless vector-like particles in the gravitational sector, whose couplings are rather poorly understood. As in the $(1,1)$-dimensional case, the intermediate vector particles also induce an $O(\alpha')$ modification to the equation of motion of the scalars, so the gauge sector of the theory is not simply SDYM. Since our understanding of these theories is so confused, I shall not say anything more about them in this talk.

3. Loops

3.1. The closed string partition function

Thus far, aside from the heterotic string, we have had a nice interpretation of all of the amplitudes in the theory. However once we continue to loop amplitudes we are going to see all kinds of problems. I will give the results for several amplitudes here, and discuss various possible ways around the problems in the conclusion of the talk.

The first loop calculation done in an $N = 2$ string—the partition function of the closed string—was carried out by Mathur and Mukhi[27]. Unlike other strings, this calculation includes an integration over the possible Wilson lines on the torus, which is equivalent to an integration over the boundary conditions of the world-sheet fermions. However, since each field in the theory is accompanied by a ghost field of the same charge but different spin, and spin is irrelevant on the torus, this means that *the partition function is independent of the boundary conditions.* After doing a careful integration over all the zero modes in the theory, one obtains[27]

$$Z_{\text{string}} = \frac{1}{4\pi} \int_{\mathcal{M}} \frac{d\tau d\bar{\tau}}{\tau_2^{D/2}} , \tag{43}$$

where $\mathcal{M}$ is the usual "keyhole" of the moduli space of the torus. Recall that in the $2_{\mathbb{R}}$-dimensional version of the theory one still has all four scalars and only the x zero-mode integration is changed, so this result is valid both in $2_{\mathbb{R}}$ and $2_{\mathbb{C}}$ dimensions. The partition function is only modular invariant in the $2_{\mathbb{C}}$ case, giving the first evidence that this is where the theory should actually be defined.

However, *the string partition is not that which one would expect of a* $(2,2)$-*dimensional scalar, such as that of our Plebanski lagrangian of eq. (18)!* That would be given by

$$\begin{aligned} Z_{part} &= \sum \frac{1}{2} \hbar\, \omega \\ &= \frac{1}{2} \operatorname{Tr} \log(p^2 + m^2) \\ &= \frac{1}{2} \frac{1}{(4\pi)^{D/2}} \int_0^\infty \frac{ds}{s^{1+D/2}} e^{-sm^2} , \end{aligned} \tag{44}$$

where $s = \pi\alpha'\tau_2$ is the Schwinger parameter. As usual, the range of integration in the string case is different from that of the particle[26]. What is really peculiar is that the string result agrees with that of a massless particle in $2_{\mathbb{R}}$ and *not* $2_{\mathbb{C}}$ dimensions. Technically, this is because the zero modes of the extra $U(1)$ ghosts of the string contribute an extra factor of τ_2, but this does not explain the physical discrepancy between eqs. (43) and (44).

The discrepancy also gives rise to another problem in $N = 2$ loop amplitudes: Instead of considering the string in a $(2,2)$-dimensional Minkowski space, one can compactify one of the complex dimensions to a complex torus. This was done by Ooguri and Vafa[5], using the results of Dixon *et al.*[28]. The partition function then becomes

$$\begin{aligned} Z_{\text{torus}} &\sim \int_{\mathcal{M}} \frac{d\tau d\bar{\tau}}{\tau_2} \cdot \sum q^{p_L^2/2} \bar{q}^{p_R^2/2} \\ &\xrightarrow{\tau_2 \to \infty} \int^\infty \frac{d\tau_2}{\tau_2} \left(1 + e^{-2\pi M^2 \tau_2} + \cdots\right) , \end{aligned} \tag{45}$$

and so develops a IR divergence as $\tau_2 \to \infty$. Such divergences will turn out to plague the interpretation of $N = 2$ loop amplitudes.

3.2. The open string partition function

The open-string partition function shows many of the same problems: In an open theory the partition function is found by calculating the path integral on all the genus zero graphs: the torus, the Klein bottle, the Möbius band and the annulus. Defining the proper time t, the total partition function should be given by

$$Z = \frac{1}{2} Z_{torus} + \frac{1}{4\pi} \int_0^\infty \frac{dt}{t^2} \left(\frac{1}{2} + \frac{1}{2} c_{annulus} - \frac{1}{2} c_{Möbius} \right) , \tag{46}$$

where the coefficients c are group-theory factors. (I have put $c_{Klein} = 1$, since I want the sum of the contributions of the torus and Klein bottle to give the one scalar of the closed sector. The factors of 1/2 are due to the nonorientability of the theory, and should be dropped if nonorientable graphs are discarded.) In our general group-theory *ansatz*, we do not have any *a priori* knowledge of the coefficients. In order to get the spectrum right, one needs the relation

$$c_{annulus} + c_{Möbius} = 2 \dim G , \tag{47}$$

which is a new, not very obvious, constraint on the *ansatz*. In the Chan-Paton case this constraint is natural, since $c_{annulus} = N^2$ and $c_{Möbius} = \pm N$ for the groups $SO(N)$ and $USp(N)$.

One can isolate the divergences of the partition function by doing appropriate "modular-like" transformations on eq. (46)[29]. The result is

$$Z = \frac{1}{2} Z_{torus} + \frac{1}{16\pi} \left[\int_0^1 \frac{dq}{q} \left(2 + \frac{1}{2} c_{annulus} \right) + \int_{-1}^0 \frac{dq}{q} 2 c_{Möbius} \right] . \tag{48}$$

This has an IR divergence at $q = 0$, which in the Chan-Paton case can be regulated only for the (very uninteresting) group $SO(2)$. This is the analogue of the groups $SO(32)$ for the superstring[30] and $SO(8192)$ for the bosonic string[29]. However, since there are no known anomalies for the bosonic and $N = 2$ strings, the argument for these special groups is not very compelling in these cases.

3.3. One-loop three-point functions

So far we have seen problems in the calculation of partition functions. One could argue that these are not very relevant physically, although this would not be the case for free energy calculations, but actual scattering amplitude calculations also show peculiar behaviour. This was first seen by Bonini, Gava and Iengo[31] in the one-loop three-point scattering of the closed string. They found the result

$$A_3^{(1)} = 2 (2\pi)^6 g^3 c_{12}^6 \times \int_{\mathcal{M}} \frac{d\tau d\bar{\tau}}{\tau_2^2} \frac{3}{32\pi^6} \sum_{n,m}{}' \frac{\tau_2^3}{|n + m\tau|^6} , \tag{49}$$

where c_{12} is the usual kinematic factor of eq. (15). This amplitude presents several surprising features: Although it is a one-loop amplitude, it is completely local! (This is also the case for the similar genus 1/2 term of the open-closed theory in eq. (39), but that is more expected since it is not a true quantum term.) Also, the terms in the sum with $m = 0$ again give a nasty IR divergence. Note that this divergence can not be cured in the usual way: Normally, one would calculate the interference terms between this graph and the tree-level graph in the first-order correction to the three-point cross section. The IR divergence would then be expected to be canceled by the singular contribution of the tree-level four-point function. However, in this theory the four-point function vanishes identically, so this cancellation can not work (even if one knew how to define a cross-section in a spacetime with two times)!

A good thing about this amplitude is that, because of the simplicity of the integrand, the integration over modular space can (almost) be carried out explicitly[32], using the techniques of Dixon *et al.*[28]. These allow one to convert the sum over m and n in the amplitude to a sum over different copies of the moduli space, giving:

$$A_3^{(1)} = 2\,(2\pi)^6 g^3 c_{12}^6 \times \int_0^\infty d\tau_2\, \frac{\tau_2}{7!}\;. \tag{50}$$

The "almost" in carrying out the calculation is that we are still stuck with the IR divergence. Nevertheless, it is very interesting that the string amplitude can be rewritten as an integral with a field-theory Schwinger parameter, rather than as an integral over some complicated region in moduli space. This means that the amplitude can be directly compared to the corresponding amplitude of the effective field theory, in the same way as can the free-energy of string theories[26]. It is a reasonable conjecture that, except for partition functions, string theory amplitudes can always be rewritten in this way. This means that common statements about "strings naturally providing a UV cut-off", and so being intrinsically different from field theories are not true, and string theories can really be studied in terms of their effective field theories.

Unfortunately, when we attempt to compare this particular amplitude to that of its field theory we find the same problem that arose in the partition function calculations. The amplitude calculated from the Plebanski action of eq. (18) is

$$A_3^{(1)}{}_{FT} = \frac{1}{(4\pi)^{D/2}}\, g^3 c_{12}^6 \times \int_0^\infty ds\, \frac{s^{2-D/2}}{7!}\;. \tag{51}$$

This almost agrees with the string result (even up to the 7!, if not up to 2's and π's!) except that, one again, the field theory wants to live in $2_{\mathbb{R}}$ dimensions!

4. Conclusions

4.1. Tree level theory

Having told you all that I know about the amplitudes of the $N = 2$ theories, I would now like to stand back and to summarize the status of the strings. At the classical level, they are elegant and well understood. They are naturally defined in $2_{\mathbb{C}}$ dimensions, although one can consider them truncated to $2_{\mathbb{R}}$ dimensions. This implies that they live in a spacetime with signature $(2,2)$, and are theories with two times! Their spectrum contains only massless scalar particles, so their space-time string field theories are simply field theories of the usual sort. These theories all turn out to have nice geometrical meanings: In $2_{\mathbb{R}}$ dimensions, the closed string is trivial[17], while the open string describes a sigma model theory, either on a group manifold[2] G, or on the coset space[6] $G_{\mathbb{C}}/G$. The heterotic string gives a modified sigma model[17].

In $2_{\mathbb{C}}$ dimensions, the closed theory describes self-dual gravity (SDG) in a (2, 2)-dimensional spacetime, with the scalar of the string describing the Kähler potential of the spacetime[5]. The open string describes self-dual Yang-Mills (SDYM) propagating in the Kähler background of the closed sector[7], and its scalars parameterize the gauge field as described by Yang[24]. The $N = 2$ open theory can have any gauge group, since the usual unitarity constraints[22] on the Chan-Paton factors no longer apply, but the insertion of the group theory into the string is then very *ad hoc*. The heterotic string is less understood. Its gauge sector also describes SDYM, with a gauge group coming from the compactification of an internal 24-dimensional space, but the theory must be compactified to (2, 1) dimensions[6]. This means that the scalars becomes tachyonic! In addition, the geometry of the gravity sector of the heterotic string is poorly understood, and it in turn induces interactions modifying the SDYM gauge structure.

The amplitudes of $N = 2$ strings are very unusual since, although the strings only have massless particles, they must still have consistent dual amplitudes. This means that all the amplitudes of the theory—even the loops!—are local in momenta. In $2_{\mathbb{R}}$ dimensions, the (open and heterotic) theories have nontrivial local four-point amplitudes; the next possibly nonvanishing amplitude is the eight-point function, which has never been calculated. In $2_{\mathbb{C}}$ dimensions, all the theories have nonvanishing three-point functions, which imply a "physical" decay process in the (2, 2)-dimensional spacetime. All the four-point functions that have been calculated vanish, and it is reasonable to conjecture that the higher-point functions also vanish. This conjecture is related to the folk-theorem that there is no classical scattering in SDYM[33], and it has been suggested that one may be able to prove it using the much-studied twistor formalism[34] for instantons. However, although there is a light-cone field theory calculation to support it[35], the conjecture remains unproven. The vanishing of these amplitudes suggests that there is some kind of topological structure to the $N = 2$ string theories, but it is not of the usual form.

4.2. Loops and their problems

In contrast to the situation at the tree level, $N = 2$ loops are a bit of a mess! They present two problems: First, they have nasty infra-red divergences. These are not intrinsically stringy, but can also be seen in the effective field theory. They may thus not seem to be so surprising, since one is dealing with massless particles, but the divergences cannot be cured in the usual ways. In addition, when one tries to compare the partition functions of the strings to those of their effective field theories, the field-theory loop integrations must be carried out in $2_{\mathbb{R}}$ dimensions to obtain the $2_{\mathbb{C}}$-dimensional result of the strings[27]! A very nice feature of the $N = 2$ string is that, because of the simplicity of its spectrum, one can take a (closed three-point) one-loop amplitude[31], and rewrite it so that it is explicitly an integration over a Schwinger parameter instead of an integration over moduli space[32]. I conjecture that this should be possible for all amplitudes in all string theories, except for vacuum amplitudes. However, this particular $N = 2$ string amplitude also shows both the IR divergence and the incompatibility of dimension with that of its field theory.

4.3. Lorentz invariance and the other spin structures

An important issue that may have a bearing on the problems of the loop amplitudes, is "what is the Lorentz invariance of the $N=2$ strings in $(2,2)$ dimensions?". As we have seen, the $N=2$ theories are naturally defined in a Kähler space, and so have a $U(1,1)$ invariance group. Since the $U(1,1)$ Poincaré group has no little group, this means that one can not define the "spin" of the particles of the various sectors of the theories. However, despite the Kähler nature of the theory (seen, for example, in the ubiquitous c_{12} factors in the amplitudes), the equations of motion of the closed and open $N=2$ strings can be written simply as $R=\tilde{R}$ and $F=\tilde{F}$, respectively, and have a manifest $SO(2,2)$ Lorentz invariance! The reason for this is unclear. While these equations of motion are Lorentz invariant, it is well known that one can not write a Lorentz invariant action for them without including new anti-self-dual fields[36], unless one uses nonlinear Lagrange multiplier fields[37]. This makes it difficult to calculate quantum corrections in the field theory in a Lorentz invariant way. There has been a recent attempt to write a Lorentz-invariant action for SDYM in harmonic space[38], which seems very reasonable since the harmonic approach is deeply related to twistors. However, one can show that this action does not give a correct description of the quantum SDYM theory[39].

I should now like to amplify a bit more on the propositions of Siegel concerning the spectrum and invariances of the $N=2$ strings[14,15]. These are rather controversial, and I shall not try to pronounce a final verdict on them. He first argues that the $N=2$ strings are the same as the "$N=4$" strings of Ademollo *et al.*[2], which have an $SU(2)$ world sheet symmetry[14]. This would mean that the x oscillators of the theory can be taken to be $SO(2,2)$ vectors, while the ψ's would be $SO(2,2)$ spinors. The constraint system of the $N=4$ theories is complicated to quantize, but the equivalence immediately implies that $N=2$ strings *must* be completely Lorentz invariant. This also has consequences for the non-NS sectors of the $N=2$ string, which I have generally ignored in the talk. These sectors are normally regarded as duplicates of the NS sector[2,5], but if the theory is Lorentz invariant the R sectors of the $N=2$ strings must describe fermions[14]. Since no amplitudes of non-NS sectors have ever been calculated, there is as yet no stringy evidence for or against this proposition. The NS tree amplitudes are unaffected by the existence of other sectors, and it is hard to draw conclusions from the loop amplitudes, since they are so confusing.

Siegel has also argued that the $N=2$ strings should actually be *maximally* spacetime supersymmetric[15]! This conjecture is based on the fact that (without getting into nonlinear Lagrange multipliers) one can write a Lorentz-invariant superspace action for SDYM and SDG only if they have $N=4$ and $N=8$ spacetime supersymmetries, respectively. I feel that the arguments for this conjecture can be evaded, since there are supersymmetric theories which do not have full-superspace actions[40], and the ordinary type II string does not have a usual Lorentz-invariant action, since it has a self-dual field with no anti-dual partner. If this proposition is nevertheless true, it has very far-reaching consequences for the $N=2$ theories: First, all the other fields of the extended supermultiplet must show up in the theories. For example, the other sectors of the open string must give rise to *four* spinors and *six* scalars, in addition to the aforementioned anti-dual fields. These numbers have to be included by hand, and appear to be somewhat *ad hoc*. In the case of the heterotic string, the gauge groups must also be far smaller than those derived by Green[17] and Ooguri and Vafa[6]. In the open string, it should not be possible for the open and closed sectors to interact, since the two sectors have different $N = 4$ and $N = 8$ supersymmetries. This contradicts to the various mixed amplitudes that I have presented here[7]. Somehow forcing these amplitudes to vanish would have the good feature of removing the non-Lorentz invariant "1/2-loop" corrections to the SDG equations in eqs. (41) and (42), but it also means that the SDYM of the open string could live only in flat space!

The true loop amplitudes of the various strings would also have to be changed: supersymmetry means that the partition functions of all the strings must vanish, with cancellations between the bosonic and fermionic sectors of the theories. This contradicts the standard result of ref. 27, in which all sectors contribute with the same phase. In addition, the IR-divergent non-Lorentz-invariant one loop closed-string amplitude of eq. (50)[31], which gave us so much trouble, should also vanish. In fact, the proposition of maximal supersymmetry solves all the loop problems of the $N = 2$ theories rather dramatically: All loops must vanish identically!

4.4. Modifying string amplitudes?

All these results follow from calculations in the (supposed) effective field theories of the $N = 2$ strings. If they are to be relevant to the strings themselves, one has to implement them at the string level. I would like to conclude my talk by considering the possibility that the $N = 2$ string amplitudes that I have talked about at such length should be modified, and if so, how.

The mixed open-closed amplitudes can only be made to vanish by declaring that they should not exist. Since I inserted group-theory factors by hand into these amplitudes, this is possible to do. It is, however, somewhat unsatisfying. If one does not wish to do this, the gravity sector of the $N = 2$ open string could have at most an $N = 4$ supersymmetry, as in the usual superstring, and not the maximal $N = 8$ supersymmetry.

In the case of the true one-loop amplitudes of eqs. (43) and (50), one has more freedom to modify the string. These amplitudes involve the adding together of different spin structures, all of which give the same contribution. Normally, one would add them with fixed phases, but the only convincing argument for this is that the final result should be modular invariant and factorizable. In the case of $N = 2$ strings all the sectors are independently modular invariant, and unitarity and factorizability are not understood in a spacetime with two times. It thus may be reasonable to make the total amplitudes vanish. In fact, since the spacetime constraints on $N = 2$ strings are so unclear, one could be even more extreme and drop the possibility that the theories need be modular invariant: The 2d-dimensional truncations of the theories could then be consistent. While this may seem to be a remarkably ugly thing to do, the possibility can not be ruled out by space-time arguments.

Finally, I should also point out that all the calculations that I have presented here are in the zero-instanton sector of the theory. The existence of the $U(1)$ gauge field of the string means that one should sum over world-sheet configurations with different instanton number, possibly weighted with a θ-term: $\theta \int d^2zF$. The necessity of such calculations has been mentioned by various groups but they have never been carried out. One such calculation that we can do very simply is the partition function on the torus: Since the contribution of the charged ψ fields is canceled by that of the (β,γ) ghosts, the partition function is independent of the instanton number, giving:

$$\begin{aligned} Z &\to Z_0 \times \sum_n e^{in\theta} \\ &= 2\pi\delta(\theta)\, Z_0 \ . \end{aligned} \tag{52}$$

This is a rather odd result (although it explains the strong-CP problem!), and it may need to be modified—possibly to zero—as suggested above. The contribution of the sectors with nonvanishing instanton number to scattering amplitudes should be further investigated.

As a final summing up: our investigations of the $N = 2$ strings have led to many unexpected results. The theories are still not understood deeply, and we still need to find the basic rules for constructing and calculating with these theories. It should be interesting!

References

1. M. Ademollo, L. Brink, A. D'Adda, R. D'Auria, E. Napolitano, S. Sciuto, E. Dél Guido, P. Di Vecchia, S. Ferrara, F. Gliozzi, R. Musto and R. Pettorini *Phys. Lett.* **62B** (1976) 105.
2. M. Ademollo, L. Brink, A. D'Adda, R. D'Auria, E. Napolitano, S. Sciuto, E. Del Guido, P. Di Vecchia, S. Ferrara, F. Gliozzi, R. Musto, R. Pettorini and J.H. Schwarz, *Nucl. Phys.* **B111** (1976) 77, *Nucl. Phys.* **B114** (1976) 297.
3. E. Witten, *Phys. Rev. Lett.* **38** (1977) 121;
   A.N. Leznov and M.V. Saveliev, *Comm. Math. Phys.* **74** (1980) 111;
   L. Mason and G. Sparling, *Phys. Lett.* **137A** (1989) 29;
   I. Bakas and D.A. Depireux, *Mod. Phys. Lett.* **A6** (1991) 399; *Int. J. Mod. Phys.* **A7** (1992) 1767; *Mod. Phys. Lett.* **A6** (1991) 1561, erratum-*ibid.* **A6** (1991) 2351.
4. R.S. Ward, *Phil. Trans. Roy. Soc. Lond.* **A315** (1985) 451;
   M.F. Atiyah, unpublished;
   N.J. Hitchin, *Proc. Lond. Math. Soc.* **5** (1987) 59.
5. H. Ooguri and C. Vafa, *Mod. Phys. Lett.* **A5** (1990) 1389, *Nucl. Phys.* **B361** (1991) 469.
6. H. Ooguri and C. Vafa, *Nucl. Phys.* **B367** (1991) 83.
7. N. Marcus, *"The N = 2 open string"*, Tel-Aviv preprint **TAUP–1929–91**, (HEP–TH/9207024), to appear in Nuclear Physics B.

8. I. Antoniadis, C. Bachas and C. Kounnas, *Phys. Lett.* **242B** (1990) 185; J. Distler, Z. Hlousek and H. Kawai, *Int. J. Mod. Phys.* **A5** (1990) 391; H. Lu, C.N. Pope, X.J. Wang and K.W. Xu, *Phys. Lett.* **284B** (1992) 268.
9. L. Brink and J.H. Schwarz, *Nucl. Phys.* **B121** (1977) 285.
10. E.S. Fradkin and A.A. Tseytlin, *Phys. Lett.* **106B** (1981) 63.
11. A. D'Adda and F. Lizzi, *Phys. Lett.* **191B** (1987) 85.
12. A. Schwimmer and N. Seiberg, *Phys. Lett.* **184B** (1987) 191.
13. J. Bieńkowska, *Phys. Lett.* **281B** (1992) 59.
14. W. Siegel, *Phys. Rev. Lett.* **69** (1992) 1493.
15. W. Siegel, *The $N = 2(4)$ string is self-dual $N = 4$ Yang-Mills*, ITP–SB–92–24, hep-th/9205075; *Self-dual $N = 8$ supergravity as closed $N = 2$ ($N = 4$) strings*, ITP–SB–92–31, hep-th/9207043.
16. H.M. Chan and J. Paton, *Nucl. Phys.* **B10** (1969) 519.
17. M.B. Green, *Nucl. Phys.* **B293** (1987) 593.
18. H. Kawai, D.C. Lewellen and S.H. Tye, *Nucl. Phys.* **B269** (1986) 1.
19. J.F. Plebanski, *J. Math. Phys.* **16** (1975) 2395.
20. M.F. Atiyah, N.J. Hitchin and I.M. Singer, *Proc. R. Soc. London* **A362** (1978) 425.
21. M. Corvi, *Phys. Lett.* **231B** (1989) 240.
22. N. Marcus and A. Sagnotti, *Phys. Lett.* **119B** (1982) 97.
23. M.B. Green and J.H. Schwarz, *Phys. Lett.* **149B** (1984) 117.
24. C.N. Yang *Phys. Rev. Lett.* **38** (1977) 1377.
25. V.P. Nair and J. Schiff, *Phys. Lett.* **246B** (1990) 423; *Nucl. Phys.* **B371** (1992) 329.
26. J. Polchinski, *Comm. Math. Phys.* **104** (1986) 37.
27. S.D. Mathur and S. Mukhi, *Nucl. Phys.* **B302** (1988) 130.
28. L.J. Dixon, V. Kaplunovsky and J. Louis, *Nucl. Phys.* **B355** (1991) 649.
29. N. Marcus and A. Sagnotti, *Phys. Lett.* **188B** (1987) 58.
30. M.B. Green and J.H. Schwarz, *Phys. Lett.* **151B** (1985) 21.
31. M. Bonini, E. Gava and R. Iengo, *Mod. Phys. Lett.* **A6** (1991) 795.
32. N. Marcus, unpublished.
33. R.S. Ward, *"The twistor approach to differential equations"*, in *Quantum Gravity 2,* (Oxford 1980).
34. R.S. Ward, *Phys. Lett.* **61A** (1977) 81. M.F. Atiyah and R.S. Ward, *Comm. Math. Phys.* **55** (1977) 117; E.F. Corrigan, D.B. Fairlie, R.G. Yates and P. Goddard, *Comm. Math. Phys.* **58** (1978) 223.
35. A. Parkes, *Phys. Lett.* **286B** (1992) 265.

36. N. Marcus and J.H. Schwarz, *Phys. Lett.* **115B** (1982) 111.
37. W. Siegel, *Nucl. Phys.* **B238** (1984) 307.
38. S. Kalitzin and E. Sokatchev, *Phys. Lett.* **257B** (1991) 151.
39. N. Marcus, Y. Oz and S. Yankielowicz, *Nucl. Phys.* **B379** (1992) 121.
40. W. Siegel and M. Roček, *Phys. Lett.* **105B** (1981) 275.

A CONTINUUM APPROACH TO 2D-QUANTUM GRAVITY FOR C>1

M. MARTELLINI*
I.N.F.N., Sezione di Roma "La Sapienza", Roma, Italy

M. SPREAFICO
Dipartimento di Fisica, Università di Milano, Milano, Italy

and

K. YOSHIDA
Dipartimento di Fisica, Università di Roma, Roma, Italy and I. N. F. N., sezione di Roma, Italy

ABSTRACT

A Lagrangian continuum model for a string theory with central charge $c>1$ is formulated by incorporating Weyl and diffeomorphism gauge fixing. In particular the tachyon scattering amplitudes are deduced generalizing the standard $c\leq 1$ computation.

1. Lagrangian formulation

Despite the remarkable progress of last years in understanding the quantum theory of gravity in 2 dimensional space-time, a serious problem still remains, connected with the inability of going beyond the matter central charge at value $c = 1$.

A possible way to overcome this difficulty lies in a slight generalization of the David, Distler and Kawai[1] (DDK) model in the conformal gauge, obtained by introducing a new scalar field coupled to the "Liouville" one. This generalization can be motivated by on observation on the procedure used to deduce the partition function for 2DQG by summing on smooth 2-dimensional random surfaces, from which one is induced to introduce in the path integral a new Jacobian term to fix all the gauge freedom in the right way. Unfortunately by this way we define a infinite set of model, characterized by the value of a new free parameter.

In this note, after resuming the meaningful result of this approach, we proceed to compute the tachyon correlation functions of the model. In the simplest case the kinematical and regularity constraints allow to gain more information on the free parameter of the theory.

⋆ On leave of absence from Dipartimento di Fisica, Università di Milano, Milano, Italy and I.N.F.N., Sezione di Pavia, Italy

Taking care of both diffeomorphisms and Weyl symmetries in quantizing the non critical string, one is lead to a simple extension of the DDK[1] ansatz where the total gauge fixing jacobian give rise to the following combined Liouville-Weyl conformal field theory[2,3]

$$\begin{aligned} S_{WL} &= \frac{1}{8\pi}\int d^2\xi\sqrt{\hat{g}}[-M_{a,b}\Phi^a\Delta_{\hat{g}}\Phi^b - Q_a\Phi^a R_{\hat{g}}] \\ M_{a,b} &\equiv \begin{pmatrix} 1 & B \\ B & -1 \end{pmatrix} \end{aligned} \tag{1}$$

Here indices are summed when repeated and take values 1 and 2, Φ^1 is the Liouville field, Φ^2 is the bosonized Weyl-ghost field and B, Q_1 and Q_2 are general free parameters. In the following, we shall take the attitude to understand (1) as the defining theory, no matter the way followed to get it.

Holomorphic energy-momentum tensor associated with (1) reads locally as

$$T = -\frac{1}{2}[M_{a,b}\partial\Phi^a\partial\Phi^b + Q_a(\partial^2)\Phi^a]$$

The corresponding central charge is

$$\begin{aligned} c_{W+L} &= 2 + 3M^{a,b}Q_aQ_b \\ M^{a,c}M_{c,b} &= \delta^a_b \end{aligned}$$

and Weyl invariance of the whole system requires

$$c_{W+L} + c_M - 26 = 0$$

which gives

$$\begin{aligned} Q_1^2 + 2BQ_1Q_2 - Q_2^2 + \frac{M}{3}(c_M - 24) &= 0 \\ M = -det(M_{a,b}) &= 1 + B^2 \end{aligned} \tag{2}$$

Eq. (2) represents the conformal invariance of the whole system as a relation between the central charges of the two fields Φ^1 and Φ^2. One can choose the special class of solutions of (2) setting up the ansatz

$$\begin{aligned} Q_2(B, c_M) &= \sqrt{\frac{1 + c_M B}{3}} \\ Q_1(B, c_M) &= -\frac{1}{\sqrt{3}}\left[B\sqrt{1 + Bc_M} - \sqrt{1 + B^2}\sqrt{25 + (B-1)c_M}\right] \end{aligned}$$

To gain more information on the behaviour of the model and to test its validity we proceeded to calculate string susceptibility.[2] To do that we follow the method of DDK who used a constant rescaling argument of the Liouville field, that is we evaluate the variation of the partition function for a variation of the area of the riemann surface. Our ansatz is that this area is expressed by the usual relation: $A = \int d^2x\sqrt{\hat{g}}e^{\alpha_1\Phi^1}$, where the Weyl field does not appear. This is because that if Weyl symmetry has been fixed then that is equivalent to fix the area of the riemann surface and so this area can not vary any more, vice versa if the area varies then the Weyl symmetry has not been fixed and so path integral must remain ambiguous.

We can then calculate string susceptibility by using

$$\Gamma(h) = \chi(h)\frac{Q_1}{2\alpha_1} + 2 \tag{3}$$

where α_1 is the momentum of the Liouville field, and its value can be obtained by considering that the conformal dimension of the gravitational dressing $e^{\alpha_1\Phi^1}$ for the area operator, where again the Weyl field does not appear, must be one:

$$\Delta_0[e^{\alpha_1\phi_1}] = -\frac{1}{2}\alpha_1 M^{1,b}(\alpha_b + Q_b) = 1 \tag{4}$$

from which we get (we chose the solution $\alpha_1 = \alpha_{1+}$ for the reason understood in ref. 2)

$$\begin{aligned}\alpha_1 &= -\frac{1}{2}\left[Q_1 + BQ_2 - \sqrt{(Q_1+BQ_2)^2 - 8M}\right] \\ &= -\frac{1}{2}\left[\sqrt{M(Q_2^2 + 8 - \frac{c_M}{3})} - \sqrt{M(Q_2^2 - \frac{c_M}{3})}\right]\end{aligned} \tag{5}$$

Inserting the above solutions the final expressions for α_1 and Γ are

$$\begin{aligned}\alpha_1(B, c_M) &= -\frac{\sqrt{1+B^2}}{2\sqrt{3}}\left[\sqrt{25+(B-1)c_M} - \sqrt{1+(B-1)c_M}\right] \\ \Gamma(B, c_M, h) &= \frac{2(1-h)}{\sqrt{1+B^2}}\frac{B\sqrt{1+Bc_M} - \sqrt{1+B^2}\sqrt{25+(B-1)c_M}}{\sqrt{25+(B-1)c_M} - \sqrt{1+(B-1)c_M}} + 2\end{aligned} \tag{6}$$

Studying Γ as a function of B, we find[2] that it is a real quantity for $B > 1 - \frac{1}{c_M}$ if $c_M > 0$.

To proceed to calculate the correlation functions of our model we couple it to a conformal matter represented by the following action

$$S_M = \frac{1}{8\pi}\int d^2\xi\sqrt{\hat{g}}[-X^a\Delta_{\hat{g}}X_a]$$

where X^a are d bosonic field and the central charge of the matter system is

$$c_M = d$$

Matter vertex operators $V_{k^a} = e^{ik_aX^a}$ will be dressed by the Liouville field alone

$$T_{k^a} = e^{ik_a X^a} e^{\beta_1 \Phi^1}$$

and again conformal invariance requires (we set $k_a = k \ \forall a$)

$$\begin{aligned}
&\Delta[T_{k^a}] = 1 \\
&\frac{1}{2}k_a k^a - \frac{1}{2}\beta_a M^{a,b}(\beta_b + Q_b) = 1 \\
&\frac{d}{2}k^2 - \frac{1}{2}\beta_1 M^{1,b}(\beta_b + Q_b) = 1 \quad (\beta_2 = 0) \\
&\beta^2 + (Q_1 + BQ_2)\beta + M[2 - dk^2)] = 0 \quad (\beta := \beta_1)
\end{aligned}$$

with solution

$$\begin{aligned}
&\beta(B,k) = \\
&= -\frac{\sqrt{1+B^2}}{2\sqrt{3}}\left[\sqrt{25 + d(B-1)} - \sqrt{1 + d(B-1) + 12dk^2}\right]
\end{aligned}$$

2. Tachyon scattering amplitudes

To compute tachyon scattering amplitudes we follow the procedure defined firstly by Gupta et al.[4] and then stated in a more complete way by Di Franceso and Kutasov.[5] This procedure is really introduced for a $c \leq 1$ Liouville theory, anyway we can generalize it to our model and we get the result below (where the vector index a in k_i^a has been omitted)

$$\begin{aligned}
A(k_1,...,k_n) &= < T_{k_1},...,T_{k_n} > \\
&= \frac{2\sqrt{\pi}}{\alpha_1} A^{-1-s} \int D\Phi' DX' e^{-S_{WL}[\Phi'] - S_M[X']} \left(\int d^2 z e^{\alpha_1 \Phi_1'(z)}\right)^s T_{k_1}...T_{k_n}
\end{aligned} \quad (7)$$

here A is the area of the riemann surface, Φ' and X' represent the non-zero modes of the fields and s is given by

$$\sum_{i=1}^{n} \beta(k_i) + s\alpha_1 = -Q_1 \quad (8)$$

A second condition to take in account is given by the charge neutrality of the matter theory; in the actual case, where the background charge of the matter sector is zero, this traduces simply in the following requirement (that is there are not screening charges)

$$\sum_{i=1}^{n} k_i = 0 \tag{9}$$

It is clear that we can compute (7) when s is a positive integer since the amplitude reduces to a free field one, in fact we obtain

$$A(k_1, ..., k_n) = \frac{2\sqrt{\pi}}{\alpha_1} A^{-1-s}$$
$$\prod_{i=1}^{n}\prod_{j=1}^{s} \int d^2 t_j |x_i - t_j|^{-\frac{2}{1+B^2}\alpha_1\beta_i} \prod_{i'<i} |x_{i'} - x_i|^{-\frac{2}{1+B^2}\beta_{i'}\beta_i} \prod_{j'<j} |t_{j'} - t_j|^{-\frac{2}{1+B^2}\alpha_1^2}$$

Otherwise it becomes unclear how to perform this calculation when s is an arbitrary number. To overcome this difficulty the idea of Kutasov and Di Francesco is based on the theorem of Carlson which states that it is possible, under favorable conditions, to continue a function calculated for an infinite set of positive integer values to the whole complex positive half-plane.

Following this idea we start by computing (7) for $n = 3$ and s integer, we obtain

$$\begin{aligned} A(k_1, k_2, k_3) &= < T_{k_1}, T_{k_2}, T_{k_3} > \\ &= \frac{2\sqrt{\pi}}{\alpha_1} A^{-s-1} \prod_{i=1}^{s} \int_C d^2 z_i |z_i|^{2a} |1 - z_i|^{2b} \prod_{j<i} |z_j - z_i|^{4c} \end{aligned} \tag{10}$$

where

$$a := -\frac{1}{1+B^2}\alpha_1\beta(k_1), \qquad b := -\frac{1}{1+B^2}\alpha_1\beta(k_2)$$
$$c := -\frac{1}{2(1+B^2)}\alpha_1^2$$

This integral presents a deep analogy with the famous Selberg integral, which is also computed by using the Carlson theorem, and in fact the result, indeed obtained by different method,[6] show very suggestive similarities. It is clear that this result must underlie some conditions on the three parameters a, b, c to ensure the convergence of the integral. We examine these conditions in the simple cases of

$s = 1$ and $s = 2$. We set:

$$I(a,b,c,s) = \prod_{i=1}^{s} \int_C d^2 z_i |z_i|^{2a} |1 - z_i|^{2b} \prod_{j<i} |z_j - z_i|^{4c}$$

We can evaluate directly (appendix B) the case $s = 1$

$$I(a,b,c,1) = \int_C d^2 z |z|^{2a} |1 - z|^{2b} = \pi \frac{\Gamma(-a-b-1)}{\Gamma(-a)\Gamma(-b)} \frac{\Gamma(a+1)\Gamma(b+1)}{\Gamma(a+b+2)}$$

under the conditions

$$Re\ a > -1, \quad Re\ b > -1, \quad Re(a+b) < -1 \tag{11}$$

For $s = 2$

$$I(a,b,c,2) = \int_C d^2 z_1 d^2 z_2 |z_1|^{2a} |z_2|^{2a} |1 - z_1|^{2b} |1 - z_2|^{2b} |z_1 - z_2|^{4c}$$

even if a direct evaluation is not still known, we can deduce the following convergence conditions (appendix A)

$$\begin{gathered} Re\ a > -1, \quad Re\ b > -1 \\ -Min\left[\frac{1}{2}, Re(a+1), Re(b+1)\right] < Re\ c < -Max\left[\frac{1}{2}Re(a+b+1), Re(a+b+1)\right] \end{gathered} \tag{12}$$

It appears that the case of $s = 1$ is fundamentally different from the general case $s > 1$, so we shall consider condition (12) rather than condition (11) as representing the generic case (in the sense that from it we can guess the conditions for the generic case).

3. Explicit calculation: d=4 target space time case

So concentrating on the convergence conditions in the simplest case (12), we can look for a range for the free parameters which would ensure it. In the actual case just three of parameters B, d, k_1, k_2, k_3 are still free, k_2 being fixed by condition (9) ($k_2 = -k_1 - k_3$), and B, d, k_1, k_3 related by (8). Moreover we must remember the condition $B > 1 - \frac{1}{c_M}$ which ensures the reality of Γ. Since we are interested in the case of a physical matter, for simplicity we set $c = d = 4$ (as the dimensions for the space time). We then get the following expression for the quantity of interest:

$$\alpha_1(B) = -\frac{\sqrt{1+B^2}}{2\sqrt{3}}\left(\sqrt{21+4B}-\sqrt{4B-3}\right)$$

$$Q_1(B) = -\frac{1}{\sqrt{3}}\left[B\sqrt{1+4B}-\sqrt{(1+B^2)(21+4B)}\right]$$

$$\Gamma(B,h) = \frac{2(1-h)}{\sqrt{1+B^2}}\frac{B\sqrt{1+4B}-\sqrt{(1+B^2)(21+4B)}}{\sqrt{21+4B}-\sqrt{4B-3}}+2$$

$$\beta(B,k) = -\frac{\sqrt{1+B^2}}{2\sqrt{3}}\left[\sqrt{21+4B)}-\sqrt{4B-3+48k^2}\right]$$

and

$$a = f(B,k_1)$$

$$b = f(B,k_3)$$

$$f(B,k) = -\frac{1}{1+B^2}\alpha_1(B)\beta(B,k)$$

$$= -\frac{1}{12}\left(\sqrt{21+4B}-\sqrt{4B-3}\right)\left(\sqrt{21+4B}-\sqrt{4B-3+48k^2}\right)$$

$$c = -\frac{1}{2(1+B^2)}[\alpha_1(B)]^2$$

$$= -\frac{1}{24}\left(\sqrt{21+4B}-\sqrt{4B-3}\right)^2$$

For the from eq. (8) for $s = 2$ we get

$$2\alpha_1 + Q_1 + \beta(k_1) + \beta(k_3) + \beta(-k_1-k_3) = 0$$

which, after the substitution of above quantities, becomes

$$2\left(B\sqrt{\frac{1+4B}{1+B^2}}+3\sqrt{21+4B}\right) - f(B,k_1) - f(B,k_3) - f(B,-k_1-k_3) = 0$$

Analytic solution of this equation is rather hard, and even harder is the solution of the set of disequations following from condition (12), the only way to handle these constraints being by numerical evaluation. On this way we immediately find that to satisfy the necessary condition $c > -\frac{1}{2}$, B must be greater than $\frac{3}{2}$. We then start by the minimum value for B getting the correspondent value for c:

$$B = B_0 = \frac{3}{2} + \epsilon = 1.50001$$

$$c = -0.499998$$

Keeping fixed the value of $B = B_0$, we can now find a minimum value also for the momenta k_i. In fact from (12) we see that $-a-1$ $(-b-1)$ must be less than

c, so evaluating the variation of $-a-1$ $(-b-1)$ in function of k_1 (k_3) for $B = B_0$ we find that $k_i > k_{min} = 0.433010$. Then fixed $k_1 = k_{min} + \epsilon = 0.433011$ we can look for a value for k_3 by solving numerically (8) , we get $k_3 \sim 0.690571$, which satisfy the condition stated above for k_i. At this point we analyze the complete condition (12) for these values, and we find that the condition is verified and the final values of the parameters are:

$$a = -0.500001$$
$$b = -0.0311387$$
$$c = -0.499998$$
$$\Gamma = 0.270978$$

We can now try to vary the value of k_1 (and then the one of k_3) by keeping $B = B_0$ fixed, to define the whole range for the momenta. The stronger condition we must keep in mind is now $c < -a - b - 1$. What we find is that increasing the value of k_1 the one of k_3 decreases, so we can do it until k_3 reach the minimum value allowed for the momenta, k_{min}. We find that (as was readily obvious from the symmetry of condition (8) when $k_2 = -k_1 - k_3$) the maximum value for the momenta is $k_{max} \sim 0.690571$. We note also that the condition $c < -a-b-1$ is always satisfied by the couple of values for (k_1, k_3) in the range $[k_{min}, k_{max}] \times [k_{min}, k_{max}]$, so it does not introduce more constraints. More for all these couples of values the whole condition (12) is satisfied, and so they represents a good set of solutions of our problem. Here below some sets of allowed values.

$$B = B_0 = 1.50001$$
$$c = -0.499998$$

k_1	k_3	a	b	$\lvert k_1 - k_3\rvert$
0.433011	0.690571	−0.500001	−0.0311387	0.25756
0.5	0.625464	−0.381965	−0.152847	0.125464
0.563012	0.563012	−0.267956	−0.267956	0

We can now try to vary the value of B from the value B_0, looking for new sets of solutions. We immediately see that the value of c increases by increasing B, with the limit $\lim_{B\to\infty} c = 0^-$, so the first condition $c > -\frac{1}{2}$ is always satisfied. By fixing successively different increasing values of $B = B_n, n = 1, 2, ...$ we look for the minimum values for the k_i by considering again condition $-a - 1 < c$ $(-b - 1 < c)$. We find that $k_{min,n}$ decreases respect to the above value set for $B = B_0$. Then fixing $k_1 = k_{min,n}$ the correspondent value for k_3, as deduced from (8), increases. We finally must consider the last condition $c < -a-b-1$, differently from the case of $B = B_0$, now this condition must be kept in account. In fact while for the case of $B = B_0$, all the allowed values of (k_1, k_3) satisfied this condition, now we find that as the value of c increases not all the couples (k_1, k_3) give values of a and b that satisfy it: the range for the momenta reduces increasing the value of B. What happens increasing B is that once one have fixed an acceptable value for k_1 then the value deduced using eq. (8) for k_3 is often too big to satisfy the condition $c < -1 - a - b$.

This traduces in the result that a maximum value for B exists. More over, by numerical evaluation we find that the maximum value for B is really close to the minimum one, in fact $B = 1.6$ is still acceptable while $B = 1.7$ is not, so the set of allowed values for B is now almost reduced to a point as we need. To summarize for $B = 1.6$ we have, for instance:

$$B = B_1 = 1.6$$
$$c = -0.479005$$

k_1	k_3	a	b	$\|k_1 - k_3\|$
0.53	0.594627	−0.318033	−0.203711	0.064627
0.55	0.574767	−0.282916	−0.239095	0.024767
0.562395	0.562395	−0.261029	−0.261029	0

4. Acknowledgements

K. Yoshida wishes to thank for stimulating discussions H. Kawai and K. Fujikawa, M. Spreafico is grateful to Vl. S. Dotsenko and E. Montaldi for helpful conversations, and M. Martellini acknowledges partial support from MPI-40%.

5. References

1. J. Distler and H. Kawai, *Nucl. Phys.* **B321** (1989) 509.
2. M. Martellini, M. Spreafico and K. Yoshida, *Mod. Phys. Lett.* **A7** (1992) 1281.
3. J. Cohn and V. Periwal, *Phys. Lett.* **B270** (1991) 18.
4. A. Gupta, S. P. Trevedi and M. B. Wise, *Nucl. Phys.* **B340** (1990) 475.
5. P. Di Francesco and D. Kutasov, *Phys. Lett.* **B261** (1991) 385.
6. Vl. S. Dotsenko and V. A. Fateev, *Nucl. Phys.* **B240** (1984) 312.

6. Appendix A

We consider the integral:

$$I[a,b,c,s] := \prod_{i=i}^{s} \int_{C_i} d^2 z_i |z_i|^{2a} |1-z_i|^{2b} \prod_{j<i} |z_j - z_i|^{4c} \tag{13}$$

that can be written in the following alternative ways:

cartesian coordinates

$$I[a,b,c,s] := \prod_{i=i}^{s} \int_{-\infty}^{+\infty} dx_i dy_i (x_i^2 + y_i^2)^a [(1-x_i)^2 + y_i^2]^b \prod_{j<i} [(x_j - x_i)^2 + (y_j - y_i)^2]^{2c}$$

polar coordinates

$$I[a,b,c,s] := \prod_{i=i}^{s} \int_0^{2\pi} d\theta_i \int_0^{+\infty} dr_i r_i^{2a+1} (1 + r_i^2 + 2 r_i cos\theta_i)^b \prod_{j<i} [r_j^2 + r_i^2 + 2 r_j r_i cos(\theta_j - \theta_i)]^{2c}$$

Convergence of integral (13) depends on the behaviour of the integrand in function of the parameters a, b, c, s in the singular points, which can be identified in $z_i = 0 \neq z_k \ \forall k \neq i$; $z_i = 1 \neq z_k \ \forall k \neq i$; $z_i = +\infty \neq z_k \ \forall k \neq i$. Then also the cases when one or more different variables assume the same value, as for example: $z_i = z_k \neq 0, 1, \infty \ \forall k$; $z_i = 0 \ \forall i$; $z_i = 1 \ \forall i$; $z_i = \infty \ \forall i$.

We start by simplicity to consider the case of $s = 1$ and $s = 2$.

For $s = 1$ integral (13) becomes

$$\begin{aligned} I[a,b,c,1] &= \int_C d^2 z |z|^{2a} |1-z|^{2b} \\ &= \int_{-\infty}^{+\infty} dx \ dy (x^2 + y^2)^a [(1-x)^2 + y^2]^b \end{aligned}$$

and can be calculated (appendix B) under the conditions:

$$Re\ a > -1, \ Re\ b > -1, \ Re(a+b) < -1$$

We can obtain the same conditions by considering the behaviour in $z = 0$

and $z = 1$

$$z = 0 \quad I \sim z^{2a+2}$$

where the second term (2) in the exponent comes from the fact that we integrate over two variables. From this behaviour we get the condition

$$Re\ a > -1 \tag{14}$$

Then by using the symmetry on the variable change $z \to 1 - z$, we can evaluate the behaviour in $z = 1$ getting the condition for b,

$$Re\ b > -1 \tag{15}$$

In $z = \infty$ we have

$$z = \infty, \quad I \sim z^{2a+2b+2}$$

from which we get the last condition $Re(a + b) < -1$.

For $s = 2$ integral (13) becomes

$$\begin{aligned} I[a,b,c,2] &= \int_C d^2 z_1 d^2 z_2 |z_1|^{2a} |z_2|^{2a} |1 - z_1|^{2b} |1 - z_2|^{2b} |z_1 - z_2|^{4c} \\ &= \int_{-\infty}^{+\infty} dx_1 dy_1 dx_1 dy_1 (x_1^2 + y_1^2)^a (x_2^2 + y_2^2)^a [(1 - x_1)^2 + y_1^2]^b [(1 - x_2)^2 + y_2^2]^b \\ &\quad [(x_1 - x_2)^2 + (y_1 - y_2)^2]^{2c} \end{aligned}$$

Let proceed to evaluate the behaviour in the singular points. If just one of the variable is zero, integral behaviour is the same as in the case above from which the condition $Re\ a > -1$, in the same way for $z_i = 1$ we can deduce the condition $Re\ b > -1$. If just one of the variable goes to infinity we have

$$z_1 = \infty, \quad z_2 = k, \quad I \sim (z_1 z_2)^{2a+2b+4c+2}$$

(from now on k represents a finite constant different from zero and one) from which the condition

$$Re\ c < -\frac{1}{2} Re(a + b + 1) \tag{16}$$

Let now evaluate what happens when both the variables have singular values.

If both the variables are zero, integral behaviour is the following

$$z_1 = 0, \ z_2 = 0 \quad I \sim (z_1 z_2)^{2a+2a+4c+2+2}$$

from which we deduce the condition

$$-Re(a+1) < Re\ c \tag{17}$$

and as above, for $z_1 = z_2 = 1$ we get the condition

$$-Re(b+1) < Re\ c \tag{18}$$

If both the variables go to infinity, we have

$$z_1 = z_2 = \infty, \quad I \sim (z_1 z_2)^{2a+2a+2b+2b+4c+4}$$

from which the condition

$$Re\ c < -Re(a+b+1).$$

If, finally, both the variables have the same not singular value, we get

$$z_1 = z_2 = k, \quad I \sim (z_1 - z_2)^{4c+2}$$

in this case the term 2 comes from the fact that singularities occurs from values not of the variables but of their differences, that is there are two, not four degrees of freedom. From this we get the following condition

$$Re\ c > -\frac{1}{2}$$

We can resume all the conditions as follows

$$-Min\left[\frac{1}{2}, Re(a+1), Re(b+1)\right] < Re\ c < -Max\left[\frac{1}{2}Re(a+b+1), Re(a+b+1)\right].$$

7. Appendix B

We can calculate directly the integral in the case $s = 1$,

$$I[a,b,c,1] = \int_C d^2z|z|^{2a}|1-z|^{2b}$$
$$= \int_{-\infty}^{+\infty} dx\ dy(x^2+y^2)^a[(1-x)^2+y^2]^b$$

We start by rewriting the integral by using the following propriety of function

$\Gamma(x)$

$$\Gamma(x) = s^x \int_0^{+\infty} dt\ e^{-sx} t^{x-1} \qquad Re\ s > 0, \quad Re\ t > 0$$

we get under the conditions: $Re\ a < 0, \quad Re\ b < 0,$

$$= \frac{1}{\Gamma(-a)\Gamma(-b)} \int_0^{+\infty} du\ dv\ u^{-a-1} v^{-b-1} \int_{-\infty}^{+\infty} dx\ dy\ e^{-u(x^2+y^2)} e^{-v[(x-1)^2+y^2]}$$

$$= \frac{1}{\Gamma(-a)\Gamma(-b)} \int_0^{+\infty} du\ dv\ u^{-a-1} v^{-b-1} e^{-v} \int_{-\infty}^{+\infty} dx\ dy\ e^{-(u+v)x^2+2xv} e^{-(u+v)y^2}$$

$$= \frac{1}{\Gamma(-a)\Gamma(-b)} \int_0^{+\infty} du\ dv\ u^{-a-1} v^{-b-1} e^{-v} \sqrt{\frac{\pi}{u+v}} e^{\frac{v^2}{u+v}} \sqrt{\frac{\pi}{u+v}}$$

$$= \frac{\pi}{\Gamma(-a)\Gamma(-b)} \int_0^{+\infty} du\ dv\ u^{-a-1} v^{-b-1} \frac{1}{u+v} e^{-\frac{uv}{u+v}}$$

if now we consider the variable change $u = tv, \quad du = v\ dt$

$$= \frac{\pi}{\Gamma(-a)\Gamma(-b)} \int_0^{+\infty} dt\ dv\ v(tv)^{-a-1} v^{-b-1} \frac{1}{tv+v} e^{-\frac{tv}{tv+v}}$$

$$= \frac{\pi}{\Gamma(-a)\Gamma(-b)} \int_0^{+\infty} dt\ t^{-a-1} \frac{1}{1+t} \int_0^{+\infty} dv\ v^{-a-b-1} e^{-\frac{t}{t+1}v}$$

using again the above propriety of $\Gamma(x)$ function we get, under the condition $Re(a+b) < -1$,

$$= \frac{\pi}{\Gamma(-a)\Gamma(-b)} \int_0^{+\infty} dt\ t^{-a-1} \frac{1}{1+t} \Gamma(-a-b-1) \left(\frac{t}{1+t}\right)^{a+b+1}$$

$$= \pi \frac{\Gamma(-a-b-1)}{\Gamma(-a)\Gamma(-b)} \int_0^{+\infty} dt\ t^b (t+1)^{-a-b-2}$$

here we can use the definition of function $B(x, y)$ and its expression in function of

$\Gamma(x)$

$$B(x,y) := \int_0^{+\infty} dt\ t^{x-1}(1+t)^{-x-y} = \frac{\Gamma(x)\Gamma(y)}{\Gamma(x+y)}, \qquad Re\ x > 0, \quad Re\ y > 0$$

and we get, under the conditions $Re\ a > -1, \quad Re\ b > -1$, the final result

$$I(a,b,c,1) = \pi \frac{\Gamma(-a-b-1)}{\Gamma(-a)\Gamma(-b)} \frac{\Gamma(a+1)\Gamma(b+1)}{\Gamma(a+b+2)}$$

that holds under the conditions

$$Re\ a > -1, \quad Re\ b > -1, \quad Re(a+b) < -1$$

which obviously comprehend also the conditions $a < 0$ and $b < 0$ since

$$a+b < -1, \quad a > -1 \quad \rightarrow \quad b < -1-a < -1+1 = 0.$$

CHIRAL NON-CRITICAL STRINGS

ROBERT C. MYERS
Physics Department, McGill University, Ernest Rutherford Building, Montréal, Québec, H3A 2T8, Canada

and

VIPUL PERIWAL
The Institute for Advanced Study, Princeton, New Jersey, 08540-4920, USA

ABSTRACT

It is shown that conformal matter with $c_L \neq c_R$ can be consistently coupled to two-dimensional 'frame' gravity. The theory is quantized in conformal gauge, following David, and Distler and Kawai. There is no analogue of the $c = 1$ barrier found in nonchiral non-critical strings. A non-critical heterotic string is constructed—it has 744 states in its spectrum, transforming in the adjoint representation of $(E_8)^3$. Correlation functions are calculated in this example, revealing the existence of extra discrete states.

Conformal matter in two dimensions couples to quantum gravity via the conformal anomaly. The coupling is characterized by the central charge of the conformal matter. The central charges for holomorphic and anti-holomorphic fields in a conformal field theory may differ. Hence, it is natural to ask how such theories, with $c_L \neq c_R$, interact with quantum gravity. Since these theories have a Lorentz anomaly, in addition to the conformal anomaly, there must be degrees of freedom other than the conformal factor that become dynamical. Under combined scale (ρ) and Lorentz (χ) transformations, the zweibein transforms

$$\mathrm{e}^{\pm}_{\mu} \rightarrow \exp(\rho \pm i\chi)\,\mathrm{e}^{\pm}_{\mu}\,.$$

Just as the conformal anomaly provides dynamics for the scale factor ρ, the Lorentz anomaly provides dynamics for local Lorentz field χ. It is the action that governs these fields that we study here.

This paper is organized as follows: Sect. 1 briefly reviews results from the Liouville field treatment of nonchiral non-critical strings. In sect. 2, we determine an analogous conformal field theory representation of chiral gravity. We consider the additional Lorentz moduli, and the gravitational dressings in this new theory. Finally, we derive some critical exponents. In sect. 3 we give an example, the $\jmath$-string, and compute the partition function and correlation functions in this example. Sect. 4 contains some concluding remarks.

The present paper is primarily a review of the work appearing in Ref. [1], and we refer the reader there for a more detailed account of this work. Sect. 3 presents some new results about scattering amplitudes for the $\jmath$–string. We note that the coupling of chiral matter to two-dimensional theories of quantum gravity has been previously considered by several authors.[2–4] Of particular relevance to the present work are Ref. [3], which studied chiral non-critical strings in light-cone gauge, and Ref. [4], which also derived the conformal field theory presented in sect. 2.

1. Review of non-critical strings

The initial success in producing an analytic understanding of non-critical strings made use of light-cone gauge.[5] The present discussion follows the David–Distler–Kawai (DDK) approach[6] in conformal gauge. Scattering amplitudes in a (non-supersymmetric) string theory are calculated with a Polyakov path integral over metrics $g_{\mu\nu}$ and matter fields ψ on the two-dimensional string world-sheet

$$\int \frac{\mathrm{D}g\ \mathrm{D}\psi}{\text{vol.(symmetries)}}\ \exp(-S_o[g,\psi])\ \mathcal{O}_1 \ldots \mathcal{O}_n \ . \tag{1}$$

The classical symmetries include both diffeomorphisms and local Weyl rescalings (*i.e.*, $g \to e^{2\rho} g$). In the quantum theory, the path integral measure introduces an anomaly for the latter transformations. Polyakov[7] showed that this anomaly is proportional to $c_m - 26$, where c_m is the central charge for the matter fields.

In critical string theories, one chooses the matter fields with $c_m = 26$ so that the anomaly vanishes. In this case, diffeomorphisms and Weyl rescalings are both symmetries of the quantum theory, and all of the local degrees of freedom of the metric decouple. A non-critical string theory is one with $c_m \neq 26$, and so the only symmetry volume divided out in Eq. 1 is that of diffeomorphisms. Hence the path integral still includes a path integral over the scale factor of the metric. Choosing conformal gauge, $g_{\mu\nu} = \mathrm{e}^{2\rho}\hat{g}_{\mu\nu}(m)$, and fixing diffeomorphisms à la Faddeev-Popov, the path integral includes a factor of the form

$$\int \mathrm{d}m\ \mathrm{D}\rho\ \exp\left[\frac{c_m - 26}{24\pi}\int \mathrm{d}^2x\,\sqrt{\hat{g}}\left\{(\hat{\nabla}\rho)^2 + \hat{R}\,\rho\right\}\right] \tag{2}$$

where $\mathrm{d}m$ stands for an integration over the moduli space of the world-sheet, and $\hat{R}$ is the curvature scalar for the background metric. The scale factor action can be regarded as the Jacobian, which arises when the original functional measures for the matter and ghost fields, defined in terms of the full metric g, are replaced by measures based on the background metric $\hat{g}$.

The problem is now to understand the measure $\mathrm{D}\rho$, which is the Riemannian measure induced by

$$(\delta\rho,\delta\rho) \equiv \int \mathrm{d}^2x\ \sqrt{g}\ (\delta\rho)^2 = \int \mathrm{d}^2x\ e^{2\rho}\sqrt{\hat{g}}\ (\delta\rho)^2 .$$

It is much more convenient to use the translation-invariant measure $\mathrm{D}_o\rho$, induced by

$$(\delta\rho,\delta\rho)_o \equiv \int \mathrm{d}^2x\sqrt{\hat{g}}\ (\delta\rho)^2 ,$$

which allows one to treat the functional integral over the scale factor as a standard quantum field theory. The DDK ansatz[6] is that Eq. 2 is replaced by

$$\int dm\ \mathrm{D}_0\rho\ \exp\left[\frac{c_m - 25}{24\pi}\int d^2x\,\sqrt{\hat{g}}\left\{(\hat{\nabla}\rho)^2 + \hat{R}\,\rho\right\}\right] \quad . \tag{3}$$

This result was later rigorously derived in Ref. [8]. Note that for the purposes of the present discussion, we have assumed that a local counterterm has been introduced to produce a vanishing cosmological constant on the world-sheet.

Given this field theory, one can perform several interesting calculations. First, the off-diagonal components of the stress tensor are

$$\begin{aligned} T_{zz} &= -\frac{c_m - 25}{6}\left(\partial\rho\partial\rho - \partial^2\rho\right) , \\ \bar{T}_{\bar{z}\bar{z}} &= -\frac{c_m - 25}{6}\left(\bar{\partial}\rho\bar{\partial}\rho - \bar{\partial}^2\rho\right) . \end{aligned} \tag{4}$$

The central charge computed from Eq. 4 is $c_\rho = 26 - c_m$. Therefore summing the contributions from the scale factor, the ghosts and the matter fields, one finds that the total central charge vanishes. This result verifies that Eq. 3 correctly represents the integral over the world-sheet metrics, since $c_{tot} = 0$ ensures that the total path integral is independent of the choice of the background metric $\hat{g}$.

Another aspect of these theories is that when a spinless primary field Φ_m in the matter theory is inserted on the world-sheet, it acquires a gravitational dressing $e^{\beta\rho}$. This combination produces an operator of conformal dimension (1,1), whose position is integrated over the surface, $\mathcal{O} = \int d^2x\,\sqrt{\hat{g}}\;e^{\beta\rho}\Phi_m$.

Finally, one may calculate various critical exponents for these theories.[6] For instance, the string susceptibility Γ is defined in terms of the fixed area partition function

$$Z(A) = \left\langle \delta\Big(\int d^2x\,\sqrt{\hat{g}}\,e^{\alpha\rho} - A\Big) \right\rangle \propto A^{\Gamma-3} \quad .$$

One finds the following result

$$\Gamma = 2 + \frac{h-1}{12}\left[25 - c_m + \sqrt{(25 - c_m)(1 - c_m)}\right] \tag{5}$$

where h is the genus of the world-sheet. Note that Γ has a simple linear dependence on the genus h, and also that it becomes complex for $c_m > 1$ (and $c_m < 25$). Therefore these calculations seem to give nonsensical results in this regime. This is the well-known $c_m = 1$ barrier. In the following sections, we will investigate these three aspects for chiral non-critical strings: background independence, gravitational dressings, and critical exponents.

2. Chiral non-critical strings

In this section, consider chiral non-critical string theories in which the central charges of the holomorphic and anti-holomorphic matter fields are different. To introduce chiral matter fields (*e.g.*, a Weyl fermion) on a curved world-sheet, one must introduce a zweibein or frame field. The classical symmetries are then diffeomorphisms, Weyl scalings and local Lorentz transformations, but anomalies arise in the quantum theory. Using a diffeomorphism-invariant regularization, these anomalies may be described by the non-local effective action[9]

$$\frac{1}{96\pi}\int R(x)G(x,y)[(c_+-26)R(y)-ic_-U(y)]$$

where G is the inverse of the scalar Laplacian, R is the curvature scalar, and $U\equiv 2\nabla\cdot\omega$ is given by the divergence of the spin connection ω_μ. We also define $c_\pm\equiv(c_R\pm c_L)/2$. An important new feature of the chiral theories is that there exists a local counterterm, $\int\omega^\mu\omega_\mu$, which may be added to this anomaly action. Therefore the coefficient of this term will appear as a free parameter, ξ, in these theories, which reflects the ambiguity in the choice of the regularization scheme.

Generically then, diffeomorphisms are the only symmetry which survive in the quantum theory. Choosing a background gauge $\mathrm{e}^{\pm}{}_{\mu}=\exp(\rho\pm i\chi)\,\hat{\mathrm{e}}^{\pm}{}_{\mu}$, the integration over all zweibein reduces to

$$\frac{\mathrm{De}}{\mathrm{vol.(diffeo.)}}=\mathrm{d}m\ \mathrm{D}\rho\ \mathrm{d}n\ \mathrm{D}\chi\ \times\ (Faddeev-Popov\ determinant) \tag{6}$$

where $\mathrm{d}n$ is a integration over additional Lorentz moduli, discussed in sect. 2.1. The functional measures for ρ and χ above are defined in terms of the full zweibein $\mathrm{e}^{\pm}$, but as in the nonchiral case they are exchanged for measures based on the background zweibein $\hat{\mathrm{e}}^{\pm}$. This change of the two measures introduces Jacobian factors, which are in fact identical to the factor found in nonchiral gravity. The final functional integral over zweibein may be written as $\int \mathrm{d}m\,\mathrm{d}n\,\mathrm{D}_0\rho\,\mathrm{D}_0\theta\,\mathrm{e}^{-S_{\mathrm{cft}}}$, where

$$S_{\mathrm{cft}}=\int\frac{\mathrm{d}^2x\sqrt{\hat{g}}}{24\pi}\left[X\left\{(\hat{\nabla}\rho)^2+\hat{R}\,\rho\right\}+\xi\left\{(\hat{\nabla}\theta)^2-(\hat{U}+\frac{ic_-}{2\xi}\hat{R})\theta\right\}\right], \tag{7}$$

and we defined $\theta\equiv\chi-\frac{ic_-}{2\xi}\rho$, so that the two fields appearing in Eq. 7 decouple. The part of the action for ρ resembles that for the scale factor in Eq. 3 except that the overall factor of (c_m-25) is replaced by

$$X=24-c_++\xi+\frac{c_-^2}{4\xi}. \tag{8}$$

In Eq. 7, θ couples to both $\hat{R}$ and $\hat{U}$. The latter coupling is not invariant under parity inversion on the world-sheet, and therefore the holomorphic and anti-holomorphic components of the stress tensor differ

$$\begin{aligned} T_{zz} &= -\frac{X}{6}\left(\partial\rho\partial\rho-\partial^2\rho\right)-\frac{\xi}{6}\left(\partial\theta\partial\theta+i\left(\frac{c_-}{2\xi}+1\right)\partial^2\theta\right), \\ \bar{T}_{\bar{z}\bar{z}} &= -\frac{X}{6}\left(\bar\partial\rho\bar\partial\rho-\bar\partial^2\rho\right)-\frac{\xi}{6}\left(\bar\partial\theta\bar\partial\theta+i\left(\frac{c_-}{2\xi}-1\right)\bar\partial^2\theta\right). \end{aligned}$$

As a result the left and right central charges differ, but this produces precisely the desired contributions, $c_{e,R} = 26 - c_R$ and $c_{e,L} = 26 - c_L$. Thus the total central charge vanishes for both the holomorphic and anti-holomorphic sectors, ensuring the theory is independent of the background zweibein. Note that the free parameter ξ associated with the regularization ambiguity does not appear in the expressions for the central charge of the combined conformal and Lorentz induced action, but we will find that it does affect the physical exponents.

Given a matter operator of weight (Δ_L, Δ_R), one may expect that it acquires an exponential dressing $\exp[\alpha\rho + ik\theta]$ to make it a (1,1) operator, as in nonchiral gravity.[6] Note that this is possible even when $\Delta_L \neq \Delta_R$, because $\exp[ik\theta]$ has different left and right weights due to the form of the stress tensor. Explicitly, one finds $k = \Delta_L - \Delta_R$ for such a dressing. In fact, this description is incomplete as we will find in the next subsection.

2.1. Lorentz moduli

In Eq. 6, the integration over all frames includes an ordinary integral over Lorentz moduli dn. These moduli are extra global phases that must be integrated over, above and beyond the moduli associated with the integration over conformal equivalence classes of metrics. One can realize these phases by shifting the spin connection, $\omega \to \omega + \sum_{i=1}^{2h} \lambda_i \beta^i$, where β^i is a basis for the harmonic differentials on the genus h surface. Since β^i are closed and divergenceless, this shift leaves R and U everywhere unchanged. Given any closed contour around a particular nontrivial cycle though, one acquires an additional phase: $\int_a \omega \to \int_a \omega + \lambda_a$. These global phases are most conveniently incorporated into the Lorentz field θ (rather than the background geometry). To be precise, one lets

$$\mathrm{d}\theta = \mathrm{d}\tilde{\theta} + \sum_{i=1}^{2h} \lambda_i \beta^i .$$

where $\tilde{\theta}$ is a (single-valued) function on the surface. Since the β^i are not exact forms, θ must be multivalued on the surface (or alternatively, θ contains discontinuities). The measure for the Lorentz moduli, $dn = \prod \mathrm{d}\lambda_i$, would then be included as a part of the functional measure $\mathrm{D}_0\theta$.

These global phases are associated with the nontrivial cycles on higher genus surfaces. Additional nontrivial cycles occur in correlation functions surrounding the operator insertions. Therefore we should allow θ to have discontinuities around such cycles. Such cuts would be produced by dressing the matter operators with exponentials of the form: $\exp[\alpha\rho + ik_R\theta_R + ik_L\theta_L]$. Now given a matter operator of weight (Δ_L, Δ_R), one may fix α and $k_R + k_L$ in the gravitational dressing to produce a (1,1) operator. This leaves one free parameter, namely $k_R - k_L$. Hence in chiral gravity, each matter operator acquires a family of gravitational dressings, corresponding to a continuum of Lorentz phases on the contours enclosing the operator.

These dressing operators are nonlocal, and hence may seem unnatural. In fact, one is forced to introduce such operators because of another aspect of the theory. Because ρ and θ couple to background charges, their momentum conservation rules are nontrivial. One finds the following superselection rules

$$\begin{aligned}\sum_a \alpha^{(a)} &= \frac{X}{3}(1-h),\\ \sum_a k_R^{(a)} &= -\frac{\xi}{3}\left[\frac{c_-}{2\xi}+1\right](1-h),\\ \sum_a k_L^{(a)} &= -\frac{\xi}{3}\left[\frac{c_-}{2\xi}-1\right](1-h).\end{aligned} \tag{9}$$

These results may be derived by demanding SL(2,C) invariance of scattering amplitudes on the sphere. It is the novel coupling of θ to the divergence of the background spin connection $\hat{U}$, which leads to $\sum k_R \neq \sum k_L$. As a result, nonvanishing amplitudes must include nonlocal operators of the form discussed above.

2.2. String susceptibility

In this section, we examine critical exponents to address the question of the critical 'dimension' for chiral gravity. One might begin by demanding the reality of the string susceptibility.[6] Define the area operator, $\int d^2x\sqrt{\hat{g}}\, e^{\alpha\rho}$ with $\alpha = [X - \sqrt{X(X-24)}]/6$. This definition is chosen to involve only a local dressing operator, and to make no explicit reference to the Lorentz field, which should be irrelevant to defining the area in analogy to the classical geometry. Then one considers the fixed area partition function

$$Z(A) = \left\langle \delta\Big(\int d^2x\sqrt{\hat{g}}\, e^{\alpha\rho} - A\Big)\right\rangle = 0 \ .$$

The result vanishes, except for genus one surfaces,* because the super-selection rules for θ in Eq. 9 are not satisfied. To properly fix the *two* θ zero mode integrals, one can consider inserting punctures with dressings which absorb the appropriate θ-momenta. If one introduces a single puncture, there is a unique dressing which yields a nonvanishing result for $g = 0, 1$. $P_0 = \int d^2x\sqrt{\hat{g}}\exp[\beta\rho + ik_R\theta_R + ik_L\theta_L]$ with $\beta = \alpha$ precisely as in the area operator, and $k_R = (h-1)(c_-/6 + \xi/3)$ and $k_L = (h-1)(c_-/6 - \xi/3)$. Now one has

$$Z'(A) = \left\langle P_0\, \delta\Big(\int d^2x\sqrt{\hat{g}}\, e^{\alpha\rho} - A\Big)\right\rangle \propto A^{\Gamma-2}$$

where

$$\Gamma = 2 + \frac{h-1}{12}\left[X + \sqrt{X(X-24)}\right] \tag{10}$$

for $h = 0, 1$. This is precisely analogous to the result for nonchiral gravity[6] with the replacement $(25 - c_m) \to X$. If we make the assumption that X is positive so that the ρ-action Eq. 7 has a positive coefficient, then from Eq. 8 requiring that Γ be real imposes the restriction

$$X - 24 = \xi + \frac{c_-^2}{4\xi} - c_+ > 0 \ . \tag{11}$$

*For $h = 1$, one finds $Z(A) \propto A^{\Gamma-3}$ with $\Gamma = 2$, exactly as in nonchiral gravity.[6]

For any fixed values of $c_\pm$, there will exist values of ξ for which this inequality is satisfied. Therefore chiral gravity has no barriers for the allowed values of central charges in the matter sector.

Of course, this conclusion relies on results for only $h = 0, 1$, since $Z'(A)$ always vanishes for $h > 1$. Introducing two punctures yields nontrivial results for higher genera as well. In this case, one finds that $Z''(A) \propto A^{\tilde{\Gamma}-1}$ where $\tilde{\Gamma} = \Gamma + (\alpha_+ + \alpha_- - 2\alpha)/\alpha$. Γ is given by Eq. 10, while $\alpha_\pm = \left[X + \sqrt{XY_\pm}\right]/6$ with

$$Y_\pm = \left[2h^2 - 1 \pm 2h\sqrt{h^2-1}\right]\xi + \left[2h^2 - 1 \mp 2h\sqrt{h^2-1}\right]\frac{c_-^2}{4\xi} - c_+$$

for $h \geq 1$. Thus one finds that the genus dependence of the string susceptibility is far more complicated than the simple linear dependence found in Eq. 5 for nonchiral gravity. Further requiring real exponents by imposing $Y_\pm > 0$, only produces constraints which are less restrictive than Eq. 11. Finally note that we have ignored ghost zero modes in this entire discussion, but that a more rigorous account can be given completely equivalent to that appearing in Ref. [6].

3. The $\jmath$–string

We now consider as a concrete example the heterotic non-critical string theory constructed from the holomorphic conformal field theory associated with the $E_8 \times E_8 \times E_8$ root lattice. One may view the world-sheet matter as 24 chiral bosons, or as 48 chiral fermions with appropriate sums over spin structures. We designate this theory as the $\jmath$–string for the amusing result that the partition function is the $\jmath$–invariant[10] on the torus.

The calculation of the torus partition function for this theory is straightforward.[11] The only novel feature is the θ factor which incorporates the Lorentz moduli. Covering the torus with a fixed coordinate patch, $0 \leq \sigma^1, \sigma^2 \leq 1$, the world-sheet metric may be written $ds^2 = |d\sigma^1 + \tau\, d\sigma^2|^2$. We introduce arbitrary phases in θ around the σ^i cycles by setting

$$\theta(\sigma^1,\sigma^2) = \tilde{\theta}(\sigma^1,\sigma^2) + 2\pi\lambda_i\sigma^i.$$

Here $\tilde{\theta}$ is a smooth function on the torus, and the second term incorporates the multi-valued contribution of the Lorentz moduli. The contribution to the partition function is then found to be

$$\begin{aligned} Z_\theta &= (\mathrm{Im}\tau)^{-1/2}|\eta(q)|^{-2}\int_{-\infty}^{\infty} d\lambda_1 d\lambda_2\, \exp\left(-\frac{\pi\xi|\lambda_2 - \lambda_1\tau|^2}{6\,\mathrm{Im}\tau}\right) \\ &= (\mathrm{Im}\tau)^{-1/2}|\eta(q)|^{-2}\, 6/\xi \end{aligned} \tag{12}$$

where $q \equiv \exp(2\pi i\tau)$, and we have included the phase integral, which one may note is invariant under modular transformations. After the phase integral is performed, the final θ factor is simply that of a free scalar field (up to an arbitrary normalization).

Combining Eq. 12 with the contributions of the other fields yields

$$Z_{\text{torus}} = \int \frac{d^2\tau}{(\mathrm{Im}\tau)^2}\,\jmath(\tau) = \int \frac{d^2\tau}{(\mathrm{Im}\tau)^2}\left(\frac{\Theta_2(q)^2 + \Theta_3(q)^2 + \Theta_4(q)^2}{\eta(q)^8}\right)^3.$$

To understand the spectrum of the theory, we investigate the region of large Imτ, where

$$\jmath(\tau) \sim \frac{1}{q} + 744 + \dots \quad . \tag{13}$$

The integration over Reτ projects out any term in Eq. 13 with a non-zero power of q. The only term left is the constant 744, indicating that there are 744 states in this string theory. Of these, 720 correspond to winding number states in the 24-dimensional lattice, and 24 come from the maximal torus of the group, the so-called oscillator states. If one counted the states before performing the phase integral in Eq. 12, each of the 744 states corresponds to a continuum with arbitrary phases around the σ^1 cycle.

3.1. Correlation functions

We now consider some sample correlation functions in this theory. In the partition function, we found that the physical states correspond to (0,1) operators in the matter sector. Thus we construct exponential (1,0) dressings, as discussed in sect. 2.1. First though in the present theory with $c_{\pm} = 12$, computations are greatly simplified if one rescales the fields to a new basis ($\hat{\rho} = (x + 2/x)\rho$, $\hat{\theta} = x\theta$) where $x = \sqrt{\xi/3}$. There are two possible (1,0) operators of the form $\exp[\alpha\hat{\rho} + ik\hat{\theta}_R + i\tilde{k}\hat{\theta}_L]$, with good semiclassical limits. For fixed k, they correspond to

i. $\tilde{k} = k + x,\ \alpha = -k$

ii. $\tilde{k} = -2/x - k,\ \alpha = -k$.

The first introduces a fixed phase in the θ field with $k - \tilde{k} = -x$, while the phase of the second dressing varies continuously with k, $k - \tilde{k} = 2/x + 2k$. The momentum super-selection rules for correlation functions are: $\sum \alpha^{(a)} = x + 2/x = -\sum k^{(a)}$, and $\sum \tilde{k}^{(a)} = x - 2/x$.

One can begin by considering three-point amplitudes of some given set of matter operators. For a fixed set, there would be eight possible amplitudes corresponding to all of the different combinations of dressings. Some of these amplitudes vanish since they combine dressings which are incompatible with the momentum conservation rules. The amplitudes which survive are those which involve either two (i) and one (ii) dressings, or one (i) and two (ii). In those cases where the amplitude is non-vanishing, one finds that as expected the results are SL(2,C) invariant, but also *independent* of how the various dressings are combined with the matter operators. There are two classes of nonvanishing amplitudes: those with three winding number states, $\exp(i\gamma^{(a)}.X_R)$, which yield simply $\delta(\gamma^{(1)} + \gamma^{(2)} + \gamma^{(3)})$, and those with two winding number states and one oscillator state, $i\beta\cdot\partial X_R$, which yield $\beta \cdot \gamma^{(1)}\ \delta(\gamma^{(1)} + \gamma^{(2)})$.†

The four-point amplitudes produce more interesting results, as we illustrate here. Consider the following correlation function

$$\mathcal{A} = \int d^2z^{(3)} \left\langle c\bar{c}\, V_{\rm i}\, e^{i\gamma^{(1)}.X_R}(z^{(1)})\, c\bar{c}\, V_{\rm i}\, e^{i\gamma^{(2)}.X_R}(z^{(2)})\, V_{\rm ii}\, e^{i\gamma^{(3)}.X_R}(z^{(3)})\, c\bar{c}\, V_{\rm i}\, e^{i\gamma^{(4)}.X_R}(z^{(4)}) \right\rangle , \tag{14}$$

†These results use Kronecker δ-functions, since the winding number vectors are discrete.

where $c\bar{c}$ are ghost dressings for the fixed operators, $V_{\mathrm{i(ii)}}$ are gravitational dressings of type i (ii) given above, and $\exp[i\gamma^{(a)} \cdot X_R]$ are winding state operators in the matter sector with $\gamma^{(a)} \cdot \gamma^{(a)} = 2$. Momentum conservation requires $\sum \gamma^{(a)} = 0$ in the matter sector, while the super-selection rules, for the gravity sector given above, restrict the momenta to be

$$z^{(1)}: \; k = q,\; \tilde{k} = q + x,\; \alpha = -q$$
$$z^{(2)}: \; k = p,\; \tilde{k} = p + x,\; \alpha = -p$$
$$z^{(3)}: \; k = x/2 - 1/x,\; \tilde{k} = -1/x - x/2,\; \alpha = -x/2 + 1/x$$
$$z^{(4)}: \; k = -3x/2 - 1/x - p - q,\; \tilde{k} = -x/2 - 1/x - p - q,\; \alpha = 3x/2 + 1/x + p + q\,.$$

One can explicitly verify that the result is SL(2,C) invariant, and fixing $z^{(a)} = (\infty, 1, z, 0)$ yields

$$\mathcal{A} = \int d^2z \; (1-z)^{\gamma^{(2)}\cdot\gamma^{(3)}} \; z^{\gamma^{(4)}\cdot\gamma^{(3)}} \; (1-\bar{z})^{-1-x^2/2-px} \; \bar{z}^{x^2+x(p+q)} \quad . \tag{15}$$

On the holomorphic side of this integral, one always has integral exponents since $\gamma^{(a)} \cdot \gamma^{(b)} = -2, -1, 0, 1, 2$, while on the anti-holomorphic side, arbitrary exponents arise as p and q are varied.

Thus we would like to evaluate integrals of the form

$$\mathcal{A}(a,b,c,d) = \int d^2z \; (1-z)^a \; z^b \; (1-\bar{z})^c \; \bar{z}^d \quad . \tag{16}$$

At present, we do not know how to perform this integral for arbitrary exponents, but it is straightforward to evaluate in the special case when several of the exponents are integers. Using repeated integration by parts and the identity, $\bar{\partial}(z-w)^{-1} = \pi\delta^2(z-w)$, one finds

$$\mathcal{A}(a,b,c,d) = \pi \frac{\Gamma(-a-b-1)}{\Gamma(-a)\Gamma(-b)} \frac{\Gamma(c+1)\Gamma(d+1)}{\Gamma(c+d+2)} \tag{17}$$

which is valid when $a+b+c+d < -2$, a and b are negative integers, and c is a positive integer while $d > -1$ or d is a positive integer while $c > -1$. Eq. 17 is an enticing formula since setting $a = c$ and $b = d$ precisely reproduces the Virasoro-Shapiro amplitude,[12] as $\mathcal{A}(a,b,a,b)$ should. If we allow the exponents c and d to take arbitrary values in this result (as they would in Eq. 15), we find that the amplitude has an infinite set of poles for negative integer values of c and d. This would be a curious result since the partition function indicated that the theory only contains a finite number of physical states. One can disregard these poles though since we have not justified Eq. 17 as the correct analytic continuation for Eq. 16 with arbitrary exponents.

Even within the restrictions for the derivation for Eq. 17, we do in fact find that $\mathcal{A}$ has unexpected poles. We illustrate this by focusing on one particular choice of exponents: $a = \gamma^{(2)} \cdot \gamma^{(3)} = -2$, $b = \gamma^{(4)} \cdot \gamma^{(3)} = -2$, and $d = x^2 + x(p+q) = 2$. In this case, Eq. 17 reduces to

$$\mathcal{A} = \frac{4\pi}{(c+1)(c+2)(c+3)} \tag{18}$$

where $c = -1 - x^2/2 - px$. While the original derivation applies for $-1 < c < 0$, Eq. 18 provides a unique analytic continuation to the entire complex plane, with poles at $c = -1, -2, -3$. In terms of the θ_R-momenta appearing in the external states in Eq. 14, these poles occur at $p = -\frac{x}{2} + \frac{n}{x}$ and $q = -\frac{x}{2} + \frac{2-n}{x}$ where $n = 0, 1, 2$.

Now the momentum super-selection rules may be used to determine the character of the intermediate states producing these poles. For instance, $\alpha^{(a)} = -k^{(a)}$ is satisfied by all of the external states, and through the momentum conservation rules, must also be satisfied by the intermediate states. As a result, the dressing for the latter states has $\Delta_R = 0$, and so they will correspond to (0,1) matter operators, as expected.[‡] Since a, b and d are fixed, one must first determine in what channel the poles occur. Note that since $\gamma^{(a)} \cdot \gamma^{(a)} = 2$ and $\sum \gamma^{(a)} = 0$ with $\gamma^{(2)} \cdot \gamma^{(3)} = -2 = \gamma^{(4)} \cdot \gamma^{(3)}$, one must have $\gamma^{(1)} = -\gamma^{(2)} = \gamma^{(3)} = -\gamma^{(4)}$. Therefore $\mathcal{A}$ cannot factorize in the (3,1) particle channel but only in the (3,2) or (3,4) channels where the external winding number states can couple to an oscillator state in the intermediate channel. Examining the (3,2) channel, one finds that the exponential dressing would have a fixed weight $\Delta_L = 4$ at all of the poles. Since this is not one(, zero or a negative integer—see below), the amplitude cannot factorize in this channel as two SL(2,C) invariant three-point amplitudes.

Hence one is lead to conclude that the poles must occur in the (3,4) channel, for which one finds the exponential dressings have $\Delta_L = 1-n$. For $n = 0$ (*i.e.*, $c = -1$), the intermediate state is a conventional state combining a (0,1) matter operator, $i\beta\cdot\partial X_R$, with an exponential ($\hat{\rho}$, $\hat{\theta}$) dressing with weight (1,0). However for $n = 1, 2$ (*i.e.*, $c = -2, -3$), the exponential gravity dressings have $\Delta_L = 0, -1$. Thus the complete dressings cannot be simply exponentials, but must also have oscillator contributions to produce $\Delta_L = 1$. Hence, the intermediate states must involve gravity dressings of the form: $(A\,\bar\partial\hat\rho + B\,\bar\partial\hat\theta)\exp[\alpha\hat\rho + ik\hat\theta_R + i\tilde{k}\hat\theta_L]$ or $(A\,\bar\partial^2\hat\rho + B\,\bar\partial^2\hat\theta + C(\bar\partial\hat\rho)^2 + D(\bar\partial\hat\theta)^2 + E\,\bar\partial\hat\rho\,\bar\partial\hat\theta)\exp[\alpha\hat\rho + ik\hat\theta_R + i\tilde{k}\hat\theta_L]$. Note that it is possible to construct dressings with oscillator contributions in chiral gravity, because the gravity sector includes two scalar fields. This contrasts with the nonchiral case, which only involves the Liouville field, and hence only allows for exponential dressings.

In fact the above description is still incomplete. Explicit construction of a (1,0) primary field of the form $(A\,\bar\partial\hat\rho + B\,\bar\partial\hat\theta)\exp[\alpha\hat\rho + ik\hat\theta_R + i\tilde{k}\hat\theta_L]$ produces a unique solution up to an overall normalization: $\bar\partial(\exp[\alpha\hat\rho + ik\hat\theta_R + i\tilde{k}\hat\theta_L])$, where the momenta are chosen such that the exponential is a (0,0) operator. It is easy to verify that states with such a total derivative dressing decouple. Therefore the correct dressing for the intermediate state at the $c = -2$ pole must incorporate the ghost number current as well, $(A\bar\partial\rho + B\bar\partial\theta + C\bar{c}\bar{b})\exp[\alpha\hat\rho + ik\hat\theta_R + i\tilde{k}\hat\theta_L]$. In this respect, the states producing these extra poles resemble the discrete states appearing in the $D = 1$ non-critical string.[13] Another feature in common with the latter is that these new states make no appearance in the analysis of the partition function presented above.[14]

---

‡The conclusions, here and below, with regard to the dressings are unaffected if the arguments are extended for the possible inclusion of unconventional ghost dressings.

4. Concluding remarks

We have examined two-dimensional quantum gravity coupled to chiral matter, and found that for fixed values of $c_\pm$, there is in fact a family of theories labelled by the free parameter, ξ. This freedom reflects an ambiguity in the choice of a diffeomorphism invariant regularization scheme used to define the theory. The effects of ξ are quite intricate: It does not appear in the central charge of the combined conformal and Lorentz induced action, nor in the partition function. However, ξ does affect critical exponents, the spectrum of discrete states, and the positions of poles in amplitudes. One might think of the freedom to specify ξ in analogy with the cosmological constant, which arises as a free parameter in nonchiral gravity. In fact since chiral gravity contains two scalar fields, it is possible to construct a large number of new (1,1) primary fields, beyond the cosmological constant operator, which may be included as terms in the action. Then, ξ should be counted as only one of the undetermined couplings which arise in association with these terms.

A physically consistent analysis required some restrictions on ξ in the form of the inequality Eq. 11, but no barriers appeared for the matter theories. Note as well that one can easily produce theories with space-time tachyons, which do not yield unphysical critical exponents. Thus at this level, the properties of space-time tachyons and physical consistency on the world-sheet, appear to be divorced in the present theory, in contrast to nonchiral gravity.[15] Clearly, the appearance of the Lorentz field has drastic effects on the quantum theory of the world-sheet geometry. The most pressing question would appear to be to understand the complete space of physical states.[16]

R.C.M. was supported by NSERC of Canada, and Fonds FCAR du Québec. V.P. was supported by D.O.E. grant DE-FG02-90ER40542.

References

1. R.C. Myers and V. Periwal, "Chiral gravity in two dimensions," to appear in *Nuclear Physics B*.
2. K. Li, *Phys. Rev.* **D34** (1986) 2292; T. Fukuyama and K. Kamimura, *Phys. Lett.* **200B** (1988) 75; T. Berger and I. Tsutsui, *Nucl. Phys.* **B335** (1990)245; *Z. Phys.* **C49** (1991) 337; S. Hwang and R. Marnelius, *Nucl. Phys.* **B271** (1986) 369.
3. Y. Oz, J. Pawelczyk and S. Yankielowicz, *Phys. Lett.* **B249** (1990) 417; *Nucl. Phys.* **B363** (1991) 555.
4. T. Lee, 'Two-Dimensional Chiral Quantum Gravity,' preprint SNUTP-91-54.
5. A.M. Polyakov, *Mod. Phys. Lett.* **A2** (1987) 893; V.G. Knizhnik, A.M. Polyakov and A.B. Zamolodchikov, *Mod. Phys. Lett.* **A3** (1988) 819.
6. F. David, *Mod. Phys. Lett.* **A3** (1988) 1651; J. Distler and H. Kawai, *Nucl. Phys.* **B321** (1989) 509.
7. A.M. Polyakov, *Phys. Rev. Lett.* **103B** (1981) 207.
8. N.E. Mavromatos and J.L. Miramontes, *Mod. Phys. Lett.* **A4** (1989) 1847; E. D'Hoker and P.S. Kurzepa, *Mod. Phys. Lett.* **A5** (1990) 1411; E. D'Hoker, *Mod. Phys. Lett.* **A6** (1991) 745.
9. H. Leutwyler, *Phys. Lett.* **153B** (1985) 65.
10. J.-P. Serre, *A Course in Arithmetic*, (Springer, New York, 1973).
11. See, for example, E. D'Hoker and D.H. Phong, *Rev. Mod. Phys.* **60** (1988) 917.

12. M.A. Virasoro, *Phys. Rev.* **177** (1969) 2309; J. Shapiro, *Phys. Rev.* **179** (1969) 1345.
13. D.J. Gross, I.R. Klebanov and M. Newman, *Nucl. Phys.* **B350** (1991) 621; A.M. Polyakov, *Mod. Phys. Lett.* **A6** (1991) 635; "Singular States in 2D Quantum Gravity," preprint PUPT-1289 (1991).
14. M. Bershadsky and I.R. Klebanov, *Phys. Rev. Lett.* **65** (1990) 3088; *Nucl. Phys.* **B360** (1991) 559.
15. N. Seiberg, *Prog. Theor. Phys. Suppl.* **102** (1990) 319.
16. In preparation

OFF-SHELL STRING PHYSICS*

ROBERT C. MYERS
Physics Department, McGill University, Ernest Rutherford Building, Montréal, Québec, H3A 2T8, Canada

and

VIPUL PERIWAL
The Institute for Advanced Study Princeton, New Jersey 08540-4920, U.S.A.

ABSTRACT

Recent advances in non-critical string theory allow a unique continuation, preserving conformal invariance, of critical Polyakov string amplitudes to off-shell momenta. These continuations possess unusual, apparently stringy, characteristics, which are unlikely to be reproduced in a string field theory. Thus our results may be an indication that some fundamentally new formulation, other than string field theory, will be required to extend our understanding of critical strings beyond the Polyakov path integral. Three-point functions are explicitly calculated. The tree-level effective potential is computed for the tachyon.

1. Introduction

Off-shell amplitudes are of great physical interest in string theory, just as in field theory. They are essential for the derivation of effective actions, *e.g.*, the derivation of effective potentials for particles such as the tachyon and the dilaton, free from the ambiguities of on-shell trivial field redefinitions. They can be used to derive measures for integrating over moduli of space-time instantons in string theory, and for the calculation of hadronic form-factors when one attempts to interpret certain aspects of quantum chromodynamics in terms of effective string theories. Naturally, given the physics involved, off-shell continuations of string amplitudes have been studied a great deal in the past, beginning in the early days of dual models. Despite this effort, off-shell string amplitudes have proven to possess a remarkable intransigence. A more extensive discussion of previous investigations will be presented elsewhere[1], but we mention three of these approaches here to put our work in perspective.

Every string field theory naturally defines off-shell amplitudes[2]. These amplitudes have been known for a long time to possess spurious singularities that can be traced to the fact that s.f.t.'s construct surfaces from building blocks (propagators and vertices) of fixed geometries. One only obtains results invariant under

* Talk by V.P.

conformal mappings, or Weyl rescalings of the geometry, when all the external legs are on-shell. Thus off-shell s.f.t. amplitudes are not geometric, in the sense that they cannot be associated to a surface independent of the representation of the conformal structure on the surface. Naturally enough, the singularities in off-shell amplitudes of different s.f.t.'s differ. Furthermore, duality is lost in this approach.

Attempts have also been made to compute the Polyakov path integral on surfaces with finite boundaries. The idea here is that specifying the matter configurations on the boundaries of the surface defines off-shell amplitudes. As the boundaries shrink to punctures, one expects to see singularities for boundary conditions corresponding to on-shell states. While this approach appears promising, it remains to properly understand the rôle of reparametrizations of the boundaries. These reparametrizations are not treated in the calculations that use the theory of doubled Riemann surfaces[3], while the remaining calculations have not taken into account the anomalous Weyl dependence which such reparametrizations introduce[4].

In the dual model literature, Bardakçi[5] investigated spontaneous symmetry breaking by summing over tachyon emissions at zero space-time momentum. Faced with the problem that such tachyons are not vertex operators of dimension (1,1), he introduced a fictitious dimension and let the momentum in this direction be ± 1. These 'charged' tachyon vertex operators were then (1,1) operators. It will turn out that this approach is the closest to the approach based on the Polyakov path integral that we shall pursue.

2. The Weyl measure in the Polyakov functional integral

Polyakov's derivation of the connection between conformal anomalies and the critical dimensions of string theories[6] elucidated a multitude of features of string physics, gleaned piece-meal in pioneering work. Space-time scattering amplitudes of string excitations are calculated as correlation functions of vertex operators in a functional integral over the metric on the string world-sheet, and the space-time string configurations:

$$\left\langle \prod_i \int \mathrm{d}^2 z_i \sqrt{g}\, V_i(z_i) \right\rangle \equiv \int \frac{\mathrm{D}g\, \mathrm{D}X}{\text{vol.(Diff)vol.(Weyl)}} \exp(-S[g,X]) \prod_i \int \mathrm{d}^2 z_i \sqrt{g}\, V_i(z_i)\,. \tag{1}$$

The measure is divided by the 'volume' of the symmetries of the classical action $S \equiv (8\pi)^{-1} \int \mathrm{d}^2 z \sqrt{g} g^{ab} \partial_a X^\mu \partial_b X_\mu$, with $\mu = 1, \ldots, D$—namely, diffeomorphisms and local Weyl rescalings on the world-sheet. Choosing conformal gauge, $g_{ab} \equiv \mathrm{e}^{2\phi} \hat{g}_{ab}(m)$, and fixing diffeomorphisms à la Faddeev-Popov, these functional integrals reduce to

$$\int \mathrm{d}m \, \frac{\mathrm{D}\phi}{\text{vol.(Weyl)}} \frac{\mathrm{D}X \;\, \mathrm{Det}'_{\mathrm{FP}}}{\text{vol.(c.k.v.)}} \exp(-S[\hat{g},X]) \prod_i \int \mathrm{d}^2 z_i \sqrt{\hat{g}(m)} V_i(z_i)\,. \tag{2}$$

Here, c.k.v. stands for the conformal Killing vectors that must be taken into account if the world-sheet is a sphere or a torus, and $\mathrm{d}m$ denotes the measure for integrating over moduli labelling distinct conformal equivalence classes of metrics on surfaces with one or more handles. In Eq. (2), the integration over the Weyl factor should cancel against the volume of the group of Weyl rescalings in the denominator, and hence all of the local degrees of freedom in the world-sheet metric would completely decouple from the theory. This decoupling is only actually achieved if Weyl rescaling survives as a symmetry of the quantum path integral. This requires[6] that $D = 26$ in order to cancel the anomalous dependences on the Weyl field, ϕ, in the measure factor, $DX\ \mathrm{Det}'_{\mathrm{PP}}/\mathrm{vol.(c.k.v.)}$. Also, one must impose various *space-time* mass-shell and polarization/gauge conditions on the external string states to avoid any anomalous Weyl dependences from normal-ordering the vertex operators. Combined these restrictions ensure that ϕ is decoupled from on-shell correlation functions in critical string theory, and the Weyl factor simply disappears from the functional integral (*i.e.*, $\int \mathrm{D}\phi/\mathrm{vol.(Weyl)} \equiv 1$). The presence of the Weyl volume in the denominator is, of course, justified only by the fact that we are considering critical (*i.e.*, Weyl-invariant) string theories. For non-critical string theories no such factor occurs.

Therefore the mass-shell conditions can be obtained from requiring Weyl invariance. It follows, in the Polyakov approach, that the calculation of amplitudes for *off-shell* string states requires the ability to compute correlation functions of vertex operators with an anomalous Weyl dependence, in the normalized measure $\mathrm{D}\phi/\mathrm{vol.(Weyl)}$. Why are such computations difficult? The problem resides in the non-linearity of the Riemannian metric that defines $\mathrm{D}\phi$. The norm on infinitesimal changes of the conformal factor is constructed with the full world-sheet metric g_{ab}

$$(\delta\phi, \delta\phi) = \int \mathrm{d}^2x \sqrt{g} (\delta\phi)^2 = \int \mathrm{d}^2x \sqrt{\hat{g}}\, \mathrm{e}^{2\phi} (\delta\phi)^2, \tag{3}$$

which then explicitly depends on ϕ. The functional integral over ϕ would be a standard quantum field theory with the measure, $\mathrm{D}_0\phi$, defined by the translation invariant norm, $(\delta\phi, \delta\phi)_0 = \int \mathrm{d}^2x \sqrt{\hat{g}} (\delta\phi)^2$. The crucial insight that we shall use is due to Mavromatos and Miramontes, and, independently, D'Hoker and Kurzepa[7]. These authors computed the relation between these two measures, and found the remarkably simple result

$$\mathrm{D}\phi = \mathrm{D}_0\phi \, \exp\left(S_L - \frac{\mu}{\pi} \int \mathrm{d}^2z\, \mathrm{e}^{\alpha\phi} \right), \tag{4}$$

where $S_L \equiv \int \frac{\mathrm{d}^2z}{6\pi} \left[\partial\phi\bar{\partial}\phi + \frac{1}{4}\sqrt{\hat{g}}\hat{R}\phi \right]$. The 'cosmological constant' μ is the coefficient of a local counterterm, and remains undetermined in this computation. The constant α in this interaction is explicitly fixed (see below). This relation was conjectured originally by David, and Distler and Kawai[8], in their study of two-dimensional gravity coupled to conformal matter in conformal gauge. It is important to note that the derivation of Eq. (4) is mathematically entirely independent of the rest of the functional integrals involved. It is valid in non-critical string theory, and equally valid in the context of critical string theory. It follows, therefore, that any insight into non-critical string physics, or into quantum Liouville theory, directly translates into insights into off-shell critical string physics.

3. Correlation functions

The only assumption in our work is in treating the correlation functions of interest using the methods of conformal field theory. For non-critical strings, this approach has been verified by comparison with results determined by matrix model techniques. The stress tensor deduced from S_L is

$$T_L = \frac{1}{6}\left[(\partial\phi)^2 - \partial^2\phi\right], \tag{5}$$

and it is easily checked that the central charge $c_L = 0$. Thus the total central charge for the matter fields, the ghosts and now, the Liouville field remains zero. The weight of an exponential operator $e^{\beta\phi}$ is $\frac{3}{2}\beta(\beta+\frac{1}{3})$. An off-shell vertex operator V_i of weight (Δ,Δ), is dressed in the same way as matter operators in non-critical string theory to produce a (1,1) operator $\exp(\beta_\Delta\phi)V_i$ with

$$\beta_\Delta = \frac{1}{6}\left[\sqrt{25-24\Delta} - 1\right]. \tag{6}$$

This is the unique solution of $\frac{3}{2}\beta_\Delta(\beta_\Delta+\frac{1}{3}) = 1-\Delta$ such that $\Delta = 1 \Leftrightarrow \beta_\Delta = 0$, which insures that in the on-shell limit, these off-shell amplitudes reduce precisely to the usual on-shell amplitudes. Rather puzzling is the non-analyticity in this prescription at $\Delta = \frac{25}{24}$. While one expects cuts in loop amplitudes in field theories, it seems difficult to interpret this non-analyticity as arising from similar physics. For the present time, we will limit our attention to $\Delta \leq \frac{25}{24}$. Certainly, there is no obvious physical reason for such a restriction, and we will address this question further in the concluding remarks.

The presence of the cosmological constant in Eq. (4) is important for defining the integration over ϕ. Insertions of cosmological constant interaction 'cancel' Liouville momentum carried by the off-shell vertex operators, and the background charge term in S_L. However, the treatment of the complete action is rather subtle[9,10]. Here, treating the cosmological constant term as a perturbatively defined interaction, we determine $\alpha = \beta_{\Delta=0} = \frac{2}{3}$. One could consider the other branch of the square root, which gives $\alpha = -1$, but $\alpha = \frac{2}{3}$ may be preferred since then this interaction can be interpreted as a zero-momentum tachyon, hence as obtained from the off-shell continuation of a physical state. Also, if used as the area operator of the quantum theory, a vanishing area results in the limit $\phi \to -\infty$, in accord with classical expectations.

Explicit computations can be performed on the two-sphere, using the idea of Goulian and Li[9] to perform the integral over the constant zero-mode, ϕ_0. The classic calculation of Dotsenko and Fateev[11] can then be used to compute the resulting correlation function, with appropriate analytic continuations along the way[9]. The zero-mode integral is

$$\int d\phi_0 \exp\left(\frac{1}{6}\phi_0 - Ce^{\frac{2}{3}\phi_0}\right)\exp(\gamma\phi_0) = \frac{3}{2}\Gamma\left(\frac{1}{2}(3\gamma+1)\right)C^{-\frac{1}{2}(3\gamma+1)}, \tag{7}$$

where $\gamma \equiv \sum\beta_i \equiv \sum\beta(\Delta_i)$, and $C \equiv (\mu/\pi)\int d^2z \exp(\frac{2}{3}\tilde{\phi})$, with $\int d^2z\,\tilde{\phi} = 0$. The amplitude is now

$$\left\langle \prod_i \int d^2z_i e^{\beta_i\phi} V_i(z_i)\right\rangle = \tfrac{3}{2}\Gamma(-s)\prod_i \int d^2z_i \left\langle C^s \prod_j e^{\beta_j\tilde{\phi}}(z_j)\right\rangle_L \left\langle \prod_k V_k(z_k)\right\rangle_m, \tag{8}$$

where $s \equiv -\frac{1}{2}(3\gamma+1)$, and the subscript $L(m)$ stands for Liouville (matter) expectation values. For three-point functions with positive integer values of s, these correlations were treated by Dotsenko and Fateev[11]. Choosing three tachyon operators, $V_j = \exp(ik_j^\mu X_\mu)$, and fixing their positions $\{z_1, z_2, z_3\}$ at $\{0, \infty, 1\}$, yields

$$\mathcal{A} = \tfrac{3}{2}\mu^s\Gamma(-s)\Gamma(s+1)\Delta(\frac{1}{3})^s \prod_{k=0}^{s-1}\prod_{i=0}^{3}\Delta(1+2\beta_i+\tfrac{2}{3}k)\,. \tag{9}$$

Here $\Delta(z) \equiv \Gamma(z)/\Gamma(1-z)$, and we have defined $\beta_0 \equiv -\frac{1}{6}$, but $\gamma = \sum_{i=1}^3\beta_i$. As it stands this formula is sensible only for positive integer values of s, and hence negative γ. Using the ideas of Ref. 9, the above formula can be continued to expressions which are valid for positive values of γ

$$\begin{aligned}
\mathcal{A} = {} & \left[\mu\Delta\left(\tfrac{1}{3}\right)\right]^{-\frac{3\gamma+1}{2}}\Gamma\left(\tfrac{1+3\gamma}{2}\right)\Gamma\left(\tfrac{1-3\gamma}{2}\right) \\
& \times\left(\tfrac{2}{3}\right)^{\gamma-\frac{2}{3}}\prod_{i=0}^{3}\prod_{p=0}^{\gamma-\frac{2}{3}}\Delta(1-3\beta_i+\tfrac{3}{2}p) \\
\text{or}\quad & \times\left(\tfrac{2}{3}\right)^{\chi_+}\prod_{i=0}^{3}\prod_{p=0}^{\gamma-\frac{2}{3}(N+1)}\Delta(1-3\beta_i+\tfrac{3}{2}p)\prod_{m=1}^{N}\Delta(1+\gamma-2\beta_i-\tfrac{2}{3}m) \\
\text{or}\quad & \times\left(\tfrac{2}{3}\right)^{\chi_-}\prod_{i=0}^{3}\prod_{p=0}^{\gamma+\frac{2}{3}(N-1)}\Delta(1-3\beta_i+\tfrac{3}{2}p)\prod_{m=0}^{N-1}\Delta(2\beta_i-\gamma-\tfrac{2}{3}m),
\end{aligned} \tag{10}$$

where $\chi_+ = \frac{2}{3}(4N+1)(N-1)-(4N-1)\gamma$ and $\chi_- = \frac{2}{3}(4N-1)(N+1)+(4N+1)\gamma$. In these formulæ, N is a positive integer, and γ must be such that the upper limits of the products are integers. This collection of expressions is clearly redundant, but any two representations valid for the same value of γ can be shown to be equivalent. We illustrate the full set, retaining the auxiliary parameter N, since different formats may prove most useful in examining different problems. Note that from the first expression and from the last, with $N = 1, 2$, one has results which are valid for $\gamma = n/3$ where n is a positive integer or zero. A more extensive description of the analytic continuations above will appear elsewhere[1].

The amplitudes given above must still be normalized by the division by the Weyl volume. At tree-level it is possible to evade a direct computation of the Weyl volume by considering ratios of amplitudes. It is then interesting to investigate the analytic structure of the amplitudes when N and γ are held fixed. Considering the ratio of two amplitudes (with the same values of N and γ), one finds that the interesting dependence on β_i resides, *e.g.*, for $N = 1$ and positive integer values of γ, in

$$\prod_{i=1}^{3}\left\{\Delta(2\beta_i - \gamma)\prod_{p=0}^{\gamma}\Delta(1 - 3\beta_i + \tfrac{3}{2}p)\right\} . \tag{11}$$

This expression is a product of three factors with poles and *zeroes* depending on the value of β_i for each individual particle, and γ as well. Note that the restriction which arose in the discussion of the dressings, $\Delta \le \frac{25}{24}$, also constrains $\beta_i \ge -\frac{1}{6}$. For a fixed γ, this restricts the number of poles and zeroes which actually occur. A case of interest because the particles can all go on-shell is $\gamma = 0$, where we find $\prod_{i=1}^{3}\Delta(1 - 3\beta_i)\Delta(2\beta_i)$. This expression has poles where $\beta_i \to 1/3$ (*i.e.*, $k_i^2 \to \frac{4}{3}$), and no zeroes—in particular, it remains finite as $\beta_i \to 0$.

Explicit computations are possible for other amplitudes. For example, four point functions are calculable, with the restriction that one of the particles is either on-shell, and hence decoupled from the Weyl functional integral, or when one of the particles is a tachyon at zero momentum, when the amplitude reduces essentially to the computations above. Amplitudes with any number of zero-momentum tachyons are readily calculated, as we consider in the next section.

4. Tachyon Potential

It is relatively straightforward to calculate an effective tree-level potential for the tachyon within our approach. As commented above, the vertex operator for a zero-momentum tachyon, $\int d^2z \exp(\frac{2}{3}\phi)$, is identical to the cosmological constant interaction. A generating function for connected tree-level amplitudes for n zero-momentum tachyons is

$$W(t) = \sum_{n=1}^{\infty}\frac{t^n}{n!}\left\langle\left[\int d^2z\, e^{\frac{2}{3}\phi}\right]^n\right\rangle = \left\langle \exp\left[t\int d^2z\, e^{\frac{2}{3}\phi}\right] - 1\right\rangle . \tag{12}$$

In these amplitudes, the operators decouple from the matter part of the functional integral, and hence the contribution of the latter reduces to $\int d^{26}X_0$. We have set an arbitrary normalization constant to one, but the zero-mode integral is explicitly retained, as usual. The exponential of vertex operators shifts the cosmological constant, $\mu \to \mu - \pi t$. In the absence of any vertex operators, Eq. (7) shows that the unnormalized amplitude is proportional to $\mu^{-1/2}$, and hence the generating function is

$$W(t) = \int d^{26}X_0 \left(1 - \frac{\pi t}{\mu}\right)^{-\frac{1}{2}} , \tag{13}$$

where we have dropped the irrelevant constant term in Eq. (12). This result has a non-analytic singularity at $t = \mu/\pi$ because on the two-sphere, the ϕ_0 integral only converges with a positive cosmological constant. Hence Eq. (13) is valid for a source $t < \mu/\pi$.

A Legendre transformation $\Gamma(T) = -W(t) + \int d^{26}X_0\, Tt$ produces an effective tree-level potential

$$\Gamma(T) = \int d^{26}X_0 \left[3\left(\frac{T}{T_c}\right)^{\frac{1}{3}} - \frac{T}{T_c}\right] , \tag{14}$$

where we have introduced an undetermined scale T_c. This uncertainty arises because the correct normalization of t, which would allow for the precise identification of the Polyakov amplitudes with space-time Green functions, is unknown. It also accounts for a constant factor which must multiply the sources in Eq. (13) to correct for the fact that the Polyakov amplitudes are truncated, while the Legendre transformation requires a generating function for connected amplitudes with propagators on the external legs.

Further the above Legendre transformation produces a valid effective potential if all other string excitations decouple, as follows. Eq. (13) is the generating function with arbitrary t, but all other sources set to zero, $W = W(t, J^{(m)} = 0)$. We have only made a Legendre transformation with respect to a single source (*i.e.*, t), and so Eq. (14) is the correct effective potential with all other fields set to zero if $\partial W(t, J^{(m)})/\partial J^{(n)}|_{J^{(p)}=0} = 0$ for all currents $J^{(n)}$. This condition is equivalent to the vanishing of any amplitudes with any number of zero-momentum tachyons and a single vertex operator for any other string excitation. For a vertex corresponding to any state with nonvanishing momentum, such amplitudes vanish as a result of momentum conservation (*i.e.*, the X_0 integrals). Similarly any zero-momentum vertex operator at higher mass levels produces vanishing amplitudes because of the matter or ghost oscillator contributions. Some states may also have ϕ oscillator contributions—we have not yet proved that these all decouple.

Let us consider the potential in Eq. (14). One of the most striking features is that it is non-analytic at $T = 0$. This non-analyticity appears as a result of the singularity in the generating function, discussed above. Given the restriction $t < \mu/\pi$, the potential is valid for positive T. In this range, there is a single extremum at $T = T_c$. Hence one arrives at the conclusion that the bosonic string cannot be stabilized by a constant shift of the tachyon.

Perhaps an even more interesting fact made apparent by these calculations is that the zero-momentum dilaton D does not couple as expected to the tachyon. Previous investigations of critical strings give the expectation that there is a trilinear DT^2 interaction. Such a coupling would lead to singularities in the amplitudes in

Eq. (12), which are in fact not observed. In discussing the Legendre transformation, we actually showed that there are no interactions DT^n for any n. Of course, the original expectations are based on calculations with on-shell tachyons, and so it is perhaps not too surprising that they are not fulfilled.

5. Conclusions and prospects

It has been our aim here to show that the effort expended on the study of non-critical strings in somewhat unphysical contexts has important physical consequences in critical string theories. Any future progess in non-critical string physics, or in quantum Liouville theory, will be of use in understanding off-shell critical string physics. There are a great many physical questions that become accessible in our approach to off-shell string physics. The computational limitations of the conformal field theory treatment of the Liouville correlators obviously leave much to be desired, a problem that appears in non-critical string theory as well. Another technical question, with what may be very interesting physics lurking underneath, is that of the analytic continuation of our discrete product formulæ as a function of γ.

The restriction of being able to dress operators with dimension $\Delta \leq \frac{25}{24}$ may be a computational limitation. On the other hand, such a limitation is precisely what prevents us from attempting to sew our amplitudes together as we would if these amplitudes were off-shell amplitudes of a field theory. Note that modular invariance dictates that if we were able to sew amplitudes together, we would end up with infinite answers. Thus, it is tempting to speculate that these non-analyticities are a reflection of modular invariance, especially since as a function of space-time momentum, they vary with the mass level of the vertex operators being dressed. Since our computations preserved conformal invariance, any amplitude we compute should be well-defined on moduli space, *i.e.*, should be independent of the coordinates on moduli space. This is, of course, not true of s.f.t. off-shell amplitudes, which describe a particular cell decomposition of moduli space.

There is no conceptual barrier to the extension of our results to supersymmetric strings, or to open string theories[1]. Above we have only considered simple exponential dressings, but one can also find many new (1,1) primary fields with Liouville oscillator contributions (*e.g.*, $\partial\phi$) which will couple in amplitudes. Some of these may account for longitudal polarizations which only couple off-shell[1].

More subtle is the computation of the Weyl volume. The presence of the factor vol.(Weyl) in the denominator of Eq. (2) is an important feature which distinguishes our off-shell amplitudes from those of non-critical string. At tree-level though, one can avoid a direct computation of this factor by considering ratios of amplitudes. One could follow the prescription of Ref. 9 which uses Eq. (11) with $\gamma = 2$ and $\beta_i = 2/3$ to compute a result for the two-sphere (which actually vanishes). On higher genus surfaces, this factor in the denominator ensures that the Weyl field does not show up in any counting of states via degenerations. In particular, the dependence on the moduli in $D\phi$ is precisely cancelled by the denominator, unless there are off-shell vertex operators present. Note then that in Eq. (2), dm and $D\phi$/vol.(Weyl) must be explicitly ordered as given. This crucial cancellation at higher genus shows that our off-shell continuation is not merely defined by an additional $c = 0$ conformal field theory tacked to a critical string theory.

A striking feature of the amplitudes is the presence of poles that are not accounted for by excitations in the matter sector (even if combined with the ghost sector). They may indicate the presence of excitations that are entirely stringy in nature. Independent of the existence of new poles, the fact that the amplitudes have products which have upper limits determined by γ is something entirely unlike the amplitudes one obtains from a field theory. This prevents them from factorizing into separate terms depending only on each individual β_i. In field theories, the off-shell character of the amplitude is a function of individual external states. Here, one can obtain the value $\gamma = 0$ when all external states are on-shell, *or* if they are off-shell. It is difficult to imagine how this γ dependence could be reproduced in a string field theory. Thus our results may indicate that some fundamentally new framework, other than string field theory, will be required to extend our understanding of critical string theory beyond the Polyakov path integral.

Note: Since this talk was given, E. Witten has computed some off-shell quantities in his background independent approach to open-string field theory[12].

It is a pleasure to thank the organizers of the Rome Workshop, especially Liù Catena, for their hospitality. R.C.M. was supported by NSERC of Canada, and Fonds FCAR du Québec. V.P. was supported by D.O.E. grant DE-FG02-90ER40542.

References

1. R.C. Myers and V. Periwal, Inst. for Adv. Study/McGill preprints IASSNS-HEP-92-20/McGill/92-48 (1992); IASSNS-HEP-92-36 (to appear)
2. J.H. Sloan, *Nucl. Phys.* **B302** (1988) 349 ; S. Samuel, *Nucl. Phys.* **B308** (1988) 285, 317; O. Lechtenfeld and S. Samuel, *Nucl. Phys.* **B308** (1988) 361, **B310** (1988) 254; V.A. Kostelecky and S. Samuel, *Phys. Lett.* **207B** (1988) 169; R. Bluhm and S. Samuel, *Nucl. Phys.* **B323** (1989) 337, **B325** (1989) 275; B. Sathiapalan, *Phys. Lett.* **206B** (1988) 211; D.Z. Freedman, S.B. Giddings, J. Shapiro and C.B. Thorn, *Nucl. Phys.* **B298** (1988) 253; A. Ukegawa, *Phys. Lett.* **261B** (1991) 391; J. Feng, *Nucl. Phys.* **B338** (1990) 459; K. Sakai, *Prog. Theor. Phys.* **80** (1988) 294
3. M. Bershadsky, *Sov. J. Nucl. Phys.* **45** (1987) 925; S.K. Blau, M. Clements, S. Della Pietra, S. Carlip and V. Della Pietra, *Nucl. Phys.* **B301** (1988) 285; J. Bolte and F. Steiner, *Nucl. Phys.* **B361** (1991) 451
4. A. Cohen, G. Moore, P. Nelson and J. Polchinski, *Nucl. Phys.* **B267** (1986) 143, **B281** (1987) 127, *Phys. Lett.* **169B** (1986) 47; F. Fucito, *Phys. Lett.* **193B** (1987) 233; W.I. Weisberger, *Nucl. Phys.* **B294** (1987) 113; A.N.

Redlich, *Phys. Lett.* **205B** (1988) 295; C. Varughese, SUNY (Stony Brook) thesis (1989), UMI-90-11440-mc; Z. Jaskolski, *Comm. Math. Phys.* **139** (1991) 353; M.A. Martin-Delgado and J. Ramirez Mittelbrunn, *Int. J. Mod. Phys.* **A6** (1991) 1719

5. K. Bardakçi, *Nucl. Phys.* **B68** (1974) 331, **B70** (1974) 397, **B133** (1978) 297
6. A.M. Polyakov, *Phys. Lett.* **103B** (1981) 207, 211
7. N.E. Mavromatos and J.L. Miramontes, *Mod. Phys. Lett.* **A4** (1989) 1847; E. D'Hoker and P.S. Kurzepa, *Mod. Phys. Lett.* **A5** (1990) 1411; E. D'Hoker, *Mod. Phys. Lett.* **A6** (1991) 745
8. F. David, *Mod. Phys. Lett.* **A3** (1988) 1651; J. Distler and H. Kawai, *Nucl. Phys.* **B321** (1989) 509
9. M. Goulian and M. Li, *Phys. Rev. Lett.* **66** (1991) 2051; see also A. Gupta, S.P. Trivedi and M.B. Wise, *Nucl. Phys.* **B340** (1990) 475
10. A.M. Polyakov, *Mod. Phys. Lett.* **A6** (1991) 635
11. Vl.S. Dotsenko and V. Fateev, *Nucl. Phys.* **B240** (1984) 312, **B251** (1985) 691
12. E. Witten, Inst. for Adv. Study preprints IASSNS-HEP-92-53 and -63 (1992)

QUANTUM SYMMETRIES IN SUPERSYMMETRIC TODA THEORIES

SILVIA PENATI
Dipartimento di Fisica dell'Università di Milano and INFN, Sezione di Milano, Via Celoria 16, I–20133 Milano, Italy

and

DANIELA ZANON
Dipartimento di Fisica dell'Università di Milano and INFN, Sezione di Milano, Via Celoria 16, I–20133 Milano, Italy

ABSTRACT

We consider two-dimensional supersymmetric Toda theories based on the Lie superalgebras $A(n,n)$, $D(n+1,n)$ and $B(n,n)$ which admit a fermionic set of simple roots and a fermionic untwisted affine extension. In particular, we concentrate on two simple examples, the $B(1,1)$ and $A(1,1)$ theories. Both in the conformal and massive case we address the issue of quantum integrability by constructing the first non trivial conserved currents and proving their conservation to all-loop orders. While the $D(n+1,n)$ and $B(n,n)$ systems are genuine $N=1$ supersymmetric theories, the $A(n,n)$ models possess a global $N=2$ supersymmetry. In the conformal case, we show that the $A(n,n)$ stress-energy tensor, uniquely determined by the holomorphicity condition, has vanishing central charge and it corresponds to the stress-energy tensor of the associated topological theory.

1. Introduction

In the recent past Toda field theories have been the subject of extensive investigations in connection with the study of W-algebra representations, perturbed conformal field theories, 2d gravity and its W-generalizations. The bosonic Toda systems are constructed from a Lie algebra $\mathcal{G}$ and depending on whether the algebra is affine or not the resulting theories are massive or conformally invariant, respectively. They possess an infinite number of conservation laws which are not spoiled by quantum anomalies [1, 2, 3]. In the massive case this implies the existence of factorized, elastic S-matrices [4], which have been determined exactly for all unitary theories based on affine simply-laced [5] as well as nonsimply-laced algebras [6].

In the region of imaginary coupling constant affine Toda theories represent suitable perturbations of $\mathcal{G}$-minimal models along a direction which maintains the integrability properties. The spectrum contains soliton configurations [7] and the corresponding S-matrix has been proposed for some of the relevant theories [8].

It is interesting to study the fermionic generalization of these models. Fermions can be added [9] consistently while maintaining the basic property of classical integrability: in this case the field theory is based on a Lie superalgebra [10]. In general the introduction of fermions does not lead to systems invariant under supersymmetry transformations; indeed all unitary fermionic Toda theories exhibit explicitly broken supersymmetry [9]. In order to obtain a supersymmetric model one has to consider those superalgebras which admit a purely fermionic system of simple roots. By selecting the ones for which the untwisted affine perturbation can be implemented in a supersymmetric way one is led to consider Toda theories based on the $A(n,n)$, $D(n+1,n)$ and $B(n,n)$ superalgebras in addition to the exceptional superalgebras $D(2,1;\alpha)$. The $A(n,n)$, $D(n+1,n)$ and $B(n,n)$ models have been recently studied in Ref. [11, 12]. It has been shown that both in the conformal and in the massive cases there exist quantum higher-spin conserved currents. Therefore the S-matrix of the massive theories is purely elastic and factorizes into two-body processes. At the lagrangian level these theories are nonunitary, nonetheless the particle and soliton mass spectrum is real and the explicit construction of exact S-matrices could be attempted. The $A(n,n)$, $D(n+1,n)$ and $B(n,n)$ models possess a *local* $N=1$ supersymmetry at the conformal point, so they are good candidates to describe $N=1$ superconformal minimal models. This supersymmetry becomes *rigid* when the systems are perturbed by the affine interaction. In addition the $A(n,n)$ Toda theory possesses a second *rigid* supersymmetry [13]. Its improved, holomorphic stress–energy tensor has zero central charge and it is the one associated to the topological version of the theory [14]. Then it is suggestive to interpret the $A(n,n)$ Toda theories as perturbations of the $N=2$ superconformal $A(n,n-1)$ models along integrable, "topological" trajectories where the $N=2$ superconformal invariance is broken down to an $N=1$ superconformal invariance.

In this paper we focus our attention on two explicit examples, i.e. the $B(1,1)$ and $A(1,1)$ Toda field theories. We describe the construction of quantum higher–spin conserved currents both in the conformal and massive cases. In the $A(1,1)$ case we discuss in detail the role of the $N=2$ supersymmetry in connection with topological aspects.

2. Review of the bosonic Toda theories

The general expression of the Minkowski space lagrangian for a Toda theory based on a finite Lie algebra $\mathcal{G}$ is

$$S=\frac{1}{\beta^2}\int d^2x\left[\frac{1}{2}\partial_\nu\vec{\phi}\cdot\partial^\nu\vec{\phi}-\sum_{i=1}^{r}e^{\vec{\alpha}_i\cdot\vec{\phi}}\right] \tag{1}$$

where $\vec{\alpha}_i$, $i=1,\cdots,r$ ($r=$ rank of $\mathcal{G}$) are r–dimensional euclidean vectors which realize the positive simple roots of the algebra, ϕ_a, $a=1,\cdots,r$ are real scalar fields and β is the coupling constant. The classical equations of motion can be written as

$$\Box\phi_j=-\sum_{i=1}^{r}C_{ji}e^{\phi_i} \tag{2}$$

where we have defined $\phi_j \equiv \vec{\alpha}_j \cdot \vec{\phi}$ and $C_{ij} \equiv \vec{\alpha}_i \cdot \vec{\alpha}_j$. Being the symmetric matrix C_{ij} invertible the potential in eq.(1) does not possess minima at finite values of the fields, so that the theory has no classical stable point. The theory is conformally invariant since its stress–energy tensor, suitably improved, is traceless. In order to show this explicitly it is convenient to use light–cone coordinates:

$$z \equiv x^+ = \frac{1}{\sqrt{2}}(x^0 + x^1) \qquad \bar{z} \equiv x^- = \frac{1}{\sqrt{2}}(x^0 - x^1)$$
$$\partial \equiv \partial_z = \frac{1}{\sqrt{2}}(\partial_0 + \partial_1) \qquad \bar{\partial} \equiv \partial_{\bar{z}} = \frac{1}{\sqrt{2}}(\partial_0 - \partial_1) \qquad \Box = 2\partial\bar{\partial} \tag{3}$$

and define light–cone components of the stress–energy tensor

$$\begin{aligned} T \equiv T_{++} &= T_{00} + T_{11} + 2T_{01} \\ \bar{T} \equiv T_{--} &= T_{00} + T_{11} - 2T_{01} \\ \Theta \equiv T_{+-} &= T_{00} - T_{11} \end{aligned} \tag{4}$$

Using the equations of motion from eq.(1) it follows that

$$T = -\partial\vec{\phi} \cdot \partial\vec{\phi} + \frac{2}{\beta}\vec{\rho} \cdot \partial^2\vec{\phi} \tag{5}$$

is conserved and traceless, $\bar{\partial}T = 0$, whenever $\vec{\rho}$ is the Weyl vector of the algebra:

$$\vec{\rho} = \sum_{i=1}^{r} \vec{\omega}_i \qquad \vec{\omega}_i \cdot \vec{\alpha}_j = \delta_{ij} \tag{6}$$

To stabilize the vacuum of the theory one can add new exponential terms to the potential in eq.(1). In general this will introduce a mass scale and destroy the conformal invariance of the theory. However certain classes of perturbations do maintain the integrability property which the theory possesses at its conformal point. This is achieved by adding the perturbing term

$$\frac{1}{\beta^2} e^{\vec{\alpha}_0 \cdot \vec{\phi}} \tag{7}$$

where $\vec{\alpha}_0$ is an extra euclidean r–dimensional vector such that the set $(\vec{\alpha}_0, \vec{\alpha}_1, \cdots, \vec{\alpha}_r)$ forms an extended, but admissible root system described by an affine Dynkin diagram of the algebra. The simplest possibility is to consider the set which realizes the untwisted affine extension of the algebra, for which $\vec{\alpha}_0$ is the negative of the *highest* root of $\mathcal{G}$. In the affine version the Toda potential has a minimum at finite values of the fields. One can shift the ϕ_i's so that the minimum is at $\phi_i = 0$ and the potential can be written as

$$V(\vec{\phi}) = \sum_{i=0}^{r} q_i e^{\vec{\alpha}_i \cdot \vec{\phi}} \tag{8}$$

where q_i are the Kac labels of $\mathcal{G}$ defined by the condition $\sum_{i=0}^{r} q_i\vec{\alpha}_i = 0$, $q_0 = 1$.

The classical integrability of such theories, i.e. the existence of an infinite number of conserved currents, follows from the identification of the Toda field equations with the integrability condition of a linear problem. Given the r generators $\vec{h}$ of the Cartan subalgebra of $\mathcal{G}$ and the basis vectors $e_j^{\pm}$ associated to the positive and negative roots $(\vec{\alpha}_0, \cdots, \vec{\alpha}_r)$ satisfying the commutation relations

$$[\vec{h}, \vec{h}] = 0 \qquad [\vec{h}, e_j^{\pm}] = \pm\vec{\alpha}_j e_j^{\pm} \qquad [e_i^+, e_j^-] = \delta_{ij}\vec{h}\cdot\vec{\alpha}_i \tag{9}$$

one defines the Lie algebra–valued gauge connection

$$A = \partial\vec{\Phi}\cdot\vec{h} + \lambda\sum_{j=0}^{r} e_j^+ \qquad \bar{A} = \frac{1}{2\lambda}\sum_{j=0}^{r} q_j e^{\vec{\alpha}_j\cdot\vec{\Phi}} e_j^- \tag{10}$$

where λ is the spectral parameter. The integrability condition of the linear Lax system

$$(\partial + A)\chi = 0 \qquad (\bar{\partial} - \bar{A})\chi = 0 \tag{11}$$

is then

$$\bar{\partial}A + \partial\bar{A} + [A, \bar{A}] = 0 \tag{12}$$

Using the commutation relations in eq.(9), it is straightforward to check that the zero curvature condition in eq.(12) is equivalent to the Toda equations of motion with the perturbed potential in eq.(8). The system (11) guarantees the existence of an infinite set of classical conserved currents [15]. (The quantum version of these currents has been studied in Refs. [1, 2, 3].)

3. N=1 supersymmetric Toda theories

Toda theories that contain fermionic as well as bosonic fields are constructed from a Lie superalgebra $\mathcal{G}$. Lie superalgebras have been classified by Kac [10] on the basis of the associated Cartan matrix. A distinguishing feature is given by the fact that the system of simple roots for a given Lie superalgebra is not unique (up to a Weyl transformation). In general it is possible to define several unequivalent sets which have a different content of fermionic roots: they are related by generalized Weyl transformations associated to fermionic roots [16].

As for the bosonic case, the classical field equations can be viewed as the zero curvature condition on a Lax connection in $N = 1$ superspace $(z, \bar{z}, \theta, \bar{\theta})$, where z, $\bar{z}$ are the Minkowski space light–cone coordinates defined in eq.(3) and θ, $\bar{\theta}$ are the (1,1) spinor coordinates. The corresponding covariant derivatives

$$D = \partial_\theta + i\theta\partial \qquad \bar{D} = \partial_{\bar{\theta}} - i\bar{\theta}\bar{\partial} \tag{13}$$

satisfy the commutation relations $\{D, \bar{D}\} = 0$, $D^2 = i\partial$, $\bar{D}^2 = -i\bar{\partial}$. We denote by $\vec{\Phi}$ the superfields

$$\Phi^a = \phi^a + \frac{1}{\sqrt{2}}\theta\psi^a + \frac{1}{\sqrt{2}}\bar{\theta}\bar{\psi}^a + \theta\bar{\theta}F^a \qquad a = 1, \ldots, r \tag{14}$$

where r is the rank of $\mathcal{G}$. Introducing the r generators $\vec{H}$ of the Cartan subalgebra of $\mathcal{G}$, and the *even* (*odd*) generators $E_i^\pm$ associated to the positive and negative *bosonic* (*fermionic*) simple roots, satisfying the graded commutation relations

$$[\vec{H},\vec{H}]=0 \qquad [\vec{H},E_j^\pm]=\pm\vec{\alpha}_j E_j^\pm \qquad [E_i^+,E_j^-\}=\delta_{ij}\vec{H}\cdot\vec{\alpha}_i \tag{15}$$

we define the two Lie superalgebra–valued gauge superfields [9, 17]

$$U = D\vec{\Phi}\cdot\vec{H} - \lambda\left(\sum_{j\in\mathcal{F}} E_j^+ + \theta\sum_{j\in\mathcal{B}} E_j^+\right)$$
$$\bar{U} = -\frac{1}{2\lambda}\left(\sum_{j\in\mathcal{F}} q_j e^{\vec{\alpha}_j\cdot\vec{\Phi}} E_j^- + \bar{\theta}\sum_{j\in\mathcal{B}} q_j e^{\vec{\alpha}_j\cdot\vec{\Phi}} E_j^-\right) \tag{16}$$

where $\mathcal{B}$ and $\mathcal{F}$ are the two subsets of bosonic and fermionic simple roots respectively and λ is the spectral parameter. The integrability of the Lax pair

$$(D+U)\chi = 0 \qquad (\bar{D}+\bar{U})\chi = 0 \tag{17}$$

corresponds to the flatness condition on the Lax connection

$$\bar{D}U + D\bar{U} + \{U,\bar{U}\} = 0 \tag{18}$$

Once again, using the commutation relations in eq.(15), one can show that equation (18) coincides with the equations of motion derivable from the Toda action

$$S = \frac{1}{\beta^2}\int d^2z d^2\theta \left[D\vec{\Phi}\cdot\bar{D}\vec{\Phi} + \sum_{i\in\mathcal{F}} q_i e^{\vec{\alpha}_i\cdot\vec{\Phi}} + \theta\bar{\theta}\sum_{i\in\mathcal{B}} q_i e^{\vec{\alpha}_i\cdot\vec{\Phi}}\right] \tag{19}$$

where $d^2z = dzd\bar{z}$, $d^2\theta = \bar{D}D$ and β is the coupling constant. If bosonic roots are present in the Dynkin diagram of $\mathcal{G}$ supersymmetry is explicitly broken by the last term in the action.

At the classical level these theories are integrable as a consequence of the identification of the Toda field equations with the linear problem in eq.(17).

In order to obtain Toda theories with $N = 1$ unbroken supersymmetry one has to consider Toda models associated to Lie superalgebras which admit a purely fermionic root system. In the following we will restrict ourselves to these cases with the additional requirement that the untwisted affine extension be realized by $\vec{\alpha}_0 = -\sum_{i=1}^{r} q_i\vec{\alpha}_i$ with $\vec{\alpha}_0$ fermionic. This selects the following superalgebras:
1) The unitary series $A(n,n) \equiv sl(n+1,n+1;C)$, $n \geq 1$. The superalgebra $A(n,n)$ is not simple; it possesses a one–dimensional graded invariant subalgebra generated by the basis element $\mathbf{I}_{2n+2}$. $A(n,n)$ is then defined as $sl(n+1,n+1;C)$ factored by this subspace.
2) The orthosymplectic series $D(n+1,n) \equiv Osp(2n+2,2n;C)$, $n \geq 1$ and $B(n,n) = Osp(2n+1,2n;C)$, $n \geq 1$. These superalgebras are simple.

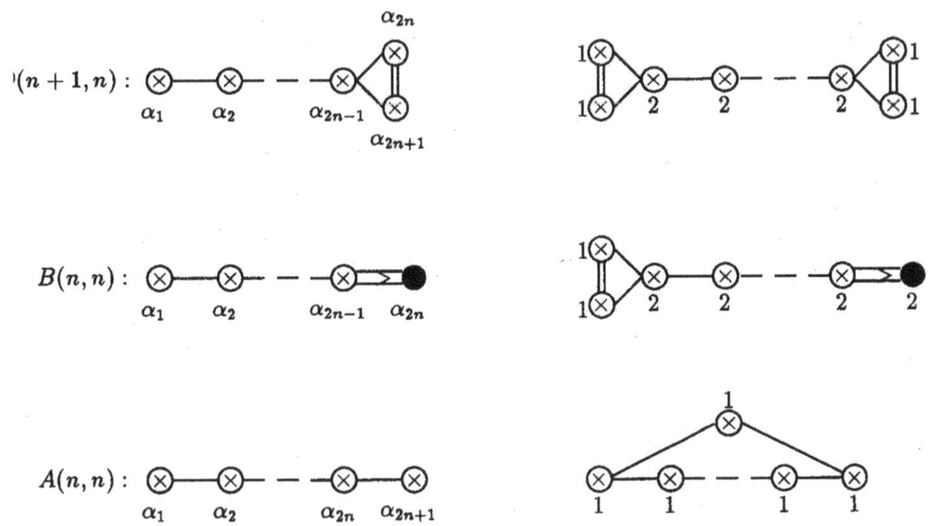

Figure 1: *Fermionic Dynkin diagrams for Lie superalgebras and their affine extensions. The crossed circles denote fermionic roots with vanishing norm. In the affine diagrams the Kac labels are explicitly shown.*

3) Among the exceptional superalgebras only $D(2,1;\alpha)$, $\alpha \neq 0,-1$, have a fermionic Dynkin diagram with the affine extension also fermionic. They are deformations of the $D(2,1)$ superalgebra.

The corresponding purely fermionic Dynkin diagrams and their untwisted extensions are given in Fig. 1.

The quantum conservation laws for the $A(n,n)$, $D(n+1,n)$ and $B(n,n)$ supersymmetric Toda theories have been studied extensively in Refs. [11, 12]. In what follows we will give the results for the two particularly simple cases of the $B(1,1)$ and $A(1,1)$ models.

4. The B(1,1) Toda theory

We consider the $N=1$ supersymmetric Toda theory associated to the Lie superalgebra $B(1,1) \equiv Osp(3,2)$. The fermionic Dynkin diagram contains two simple roots (see Fig.1). They can be realized in terms of two–dimensional euclidean vectors:

$$\vec{\alpha}_1 = (1,-i) \qquad \vec{\alpha}_2 = (0,i) \tag{20}$$

Therefore the corresponding Toda theory contains two superfields whose dynamic is described by the following action:

$$S=\frac{1}{\beta^2}\int d^2zd^2\theta\left[D\Phi_1\bar{D}\Phi_1+D\Phi_2\bar{D}\Phi_2+e^{\Phi_1-i\Phi_2}+2e^{i\Phi_2}\right] \tag{21}$$

The equations of motion for Φ_1 and Φ_2 are

$$\begin{aligned} D\bar{D}\Phi_1 &= \frac{1}{2}e^{\Phi_1-i\Phi_2} \\ D\bar{D}\Phi_2 &= \frac{1}{2}\left(-ie^{\Phi_1-i\Phi_2}+2ie^{i\Phi_2}\right) \end{aligned} \tag{22}$$

As described in the previous section, this theory is classically integrable. The most practical way to obtain explicit expressions for the conserved supercurrents is via the Miura operator [18, 12]

$$\Delta(\vec{\Phi})=(D+D\vec{\Phi}\cdot\vec{\lambda}_5)(D+D\vec{\Phi}\cdot\vec{\lambda}_4)\cdots(D+D\vec{\Phi}\cdot\vec{\lambda}_1)=D^5+\sum_{i=0}^{4}W^{(2-\frac{i}{2})}D^i \tag{23}$$

where $\vec{\lambda}_j$, $j=1,\cdots,5$ are the weights of the fundamental representation, $\vec{\lambda}_1=-\vec{\lambda}_5=\vec{\alpha}_1+\vec{\alpha}_2$, $\vec{\lambda}_2=-\vec{\lambda}_4=\vec{\alpha}_2$ and $\vec{\lambda}_3=0$. Therefore eq.(23) can be written as

$$(D-D\Phi_1)(D-iD\Phi_2)D(D+iD\Phi_2)(D+D\Phi_1)=W^{(2)}+W^{(\frac{3}{2})}D+W^{(1)}D^2+D^5 \tag{24}$$

where

$$\begin{aligned} W^{(1)} &= -iD\Phi_1\partial\Phi_1-iD\Phi_2\partial\Phi_2+D\partial\Phi_2+2iD\partial\Phi_1 \\ W^{(\frac{3}{2})} &= -i\partial^2\Phi_2+iD\Phi_1D\partial\Phi_1-(\partial\Phi_2)^2-iD\Phi_2D\partial\Phi_2 \\ &\quad +2D\Phi_1D\partial\Phi_2-2iD\Phi_1D\Phi_2\partial\Phi_2 \\ W^{(2)} &= -D\partial^2\Phi_1+D\Phi_1\partial^2\Phi_1-iD\Phi_1\partial^2\Phi_2+iD\partial\Phi_2\partial\Phi_1-(\partial\Phi_2)^2D\Phi_1 \\ &\quad +\partial\Phi_2D\Phi_2\partial\Phi_1-iD\Phi_2D\partial\Phi_2D\Phi_1 \\ &= \frac{1}{2}D(W^{(\frac{3}{2})}+DW^{(1)}) \end{aligned} \tag{25}$$

Using the equations of motion (22) it is easy to see that the Miura operator commutes with the spinor derivative $\bar{D}$ so that it generates a set of superholomorphic conserved currents:

$$\bar{D}W^{(s)}=0 \qquad s=1,\frac{3}{2},2 \tag{26}$$

We note that $W^{(1)}$ can be written as

$$W^{(1)}=-iD\vec{\Phi}\cdot\partial\vec{\Phi}+2i\vec{\rho}\cdot D\partial\vec{\Phi} \tag{27}$$

where $\vec{\rho} = (1, -\frac{i}{2})$ is the Weyl vector. It is proportional to the improved super stress–energy tensor of the theory. Due to the conservation law (26) the theory has a manifest $N = 1$ superconformal invariance. The components of $W^{(1)}$

$$\begin{aligned} W^{(1)}\Big|_{\theta=0} &= -\frac{i}{\sqrt{2}}\psi_1\partial\phi_1 - \frac{i}{\sqrt{2}}\psi_2\partial\phi_2 + i\sqrt{2}\partial\psi_1 + \frac{1}{\sqrt{2}}\partial\psi_2 \\ DW^{(1)}\Big|_{\theta=0} &= \partial\phi_1\partial\phi_1 + \partial\phi_2\partial\phi_2 + \frac{i}{2}\psi_1\partial\psi_1 + \frac{i}{2}\psi_2\partial\psi_2 - 2\partial^2\phi_1 + i\partial^2\phi_2 \end{aligned} \quad (28)$$

are respectively the generators of supersymmetry and conformal transformations.

The quantum integrability of the theory has been addressed in Refs. [11, 12] by studying the renormalization and conservation of the supercurrents. There it has been shown that the holomorphic currents in eq.(25) maintain their form at the quantum level, albeit the coefficients of the various terms acquire a coupling constant dependence. We report here a brief summary of the procedure and refer for details of the calculation to Ref.[11].

In order to determine the quantum corrections to the classical currents it is convenient to use massless perturbation theory which is best suited for an all–loop analysis [3]. The massless superspace propagators are given by

$$\left\langle \Phi_i(Z,\bar{Z})\Phi_j(0,0)\right\rangle = -\delta_{ij}\frac{\beta^2}{4\pi}\bar{D}D[log(2z\bar{z})\delta^{(2)}(\theta)] \quad (29)$$

with $(Z,\bar{Z}) \equiv (z,\theta,\bar{z},\bar{\theta})$. The quantum lagrangian is defined by normal ordering the exponentials in eq.(21) (no ultraviolet divergences), which are then treated as interaction terms. The quantum conservation laws defined as

$$\bar{D}_Z\left\langle W^{(s)}(Z,\bar{Z})\right\rangle \equiv \bar{D}_Z\left\langle W^{(s)}(Z,\bar{Z})\exp\left(\frac{i}{\beta^2}\int d^2w d^2\theta' \mathcal{L}_{int}\right)\right\rangle_0 = 0 \quad (30)$$

could be spoiled by local anomalies from Wick contractions of the current $W^{(s)}$ with the first order expansion in $\mathcal{L}_{int}$. In fact, once the D–algebra has been performed we are left with terms of the form

$$\bar{D}_Z\int d^2w\mathcal{A}(z,\theta,\bar{z},\bar{\theta})\bar{D}_Z\frac{1}{(z-w)^n}\mathcal{B}(w,\theta,\bar{w},\bar{\theta}) \quad (31)$$

where $\mathcal{A}$ and $\mathcal{B}$ are products of superfields and their derivatives. Now, using

$$\bar{D}_Z\bar{D}_Z\frac{1}{(z-w)^n} = -i\bar{\partial}_z\frac{1}{(z-w)^n} = \frac{2\pi}{(n-1)!}\partial_w^{n-1}\delta^{(2)}(z-w) \quad (32)$$

we produce local contributions which affect the last equality in eq.(30). In Ref. [11] it has been proved that these potential anomalies can be cancelled by modifying the classical $W^{(s)}$-currents, i.e. by adding coupling constant dependent terms. Since the calculation can be done to all-loop orders, the exact quantum version of the conserved supercurrents is thus obtained. For the $B(1,1)$ case one has

$$W^{(1)} = -iD\Phi_1\partial\Phi_1 - iD\Phi_2\partial\Phi_2 + (1-\frac{\beta^2}{4\pi})D\partial\Phi_2 + 2i(1-\frac{\beta^2}{8\pi})D\partial\Phi_1 \tag{33}$$

$$\begin{aligned} W^{(\frac{3}{2})} &= -i(1-\frac{\beta^2}{4\pi})\partial^2\Phi_2 + i(1-\frac{\beta^2}{2\pi})D\Phi_1 D\partial\Phi_1 - (\partial\Phi_2)^2 - iD\Phi_2 D\partial\Phi_2 \\ &+2(1-\frac{\beta^2}{4\pi})D\Phi_1 D\partial\Phi_2 - 2iD\Phi_1 D\Phi_2\partial\Phi_2 \end{aligned} \tag{34}$$

The quantum spin-2 current can be obtained from eqs.(33) and (34) since $W^{(2)}$ is linearly dependent on $W^{(1)}$ and $W^{(\frac{3}{2})}$. Eq.(33) gives the quantum expression of the super stress-energy tensor. The corresponding central charge is

$$c = 3 + 24\vec{\rho}_q^{\,2} \tag{35}$$

where the "quantum" Weyl vector is $\vec{\rho}_q = \left(1 - \frac{\beta^2}{8\pi}, -\frac{i}{2} + i\frac{\beta^2}{8\pi}\right)$.

A massive perturbation of the $B(1,1)$ Toda theory is realized by adding to the action (21) an exponential term associated to the lowest fermionic root

$$\vec{\alpha}_0 = -(q_1\vec{\alpha}_1 + q_2\vec{\alpha}_2) = (-1, -i) \tag{36}$$

The action for the $B^{(1)}(1,1)$ Toda theory is

$$S = \frac{1}{\beta^2}\int d^2z d^2\theta \left[D\vec{\Phi}\cdot\bar{D}\vec{\Phi} + e^{\Phi_1 - i\Phi_2} + 2e^{i\Phi_2} + e^{-\Phi_1 - i\Phi_2}\right] \tag{37}$$

The classical mass spectrum can be obtained from the second order expansion of the lagrangian:

$$\mathcal{L}^{(2)} \equiv -M_1\Phi_1^2 - M_2\Phi_2^2 = \Phi_1^2 - 2\Phi_2^2 \tag{38}$$

It follows that the bosonic masses are

$$m_1^2 = 2M_1^2 = 2 \quad ; \quad m_2^2 = 2M_2^2 = 8 \tag{39}$$

We note that the classical mass spectrum is real despite the manifest non-unitarity of the Lagrangian.

Due to the precence of a mass scale the perturbed theory is no longer conformally invariant. The super stress-energy tensor acquires a non-vanishing trace and its conservation law takes the form

$$\bar{D}J^{(s)} + D\bar{J}^{(s)} = 0 \tag{40}$$

with $s = 1$. The study of higher-spin ($s > 1$)conserved currents reveals that the affine theory does not have conserved quantities at levels $s = \frac{3}{2}$ and $s = 2$. At the classical level the first non-trivial conserved current has spin 3 and it can be expressed in terms of the W-currents as

$$J^{(3)} = W^{(1)}DW^{(1)} + 2W^{(1)}W^{(\frac{3}{2})} \tag{41}$$

While in the conformal case the two terms are separately holomorphic currents, in the affine case only the linear combination above satisfies the classical conservation law.

At the quantum level a lengthy calculation leads to a quantum spin-3 conserved current [11] and to a corresponding charge

$$Q^{(3)} = \int dz d\theta J^{(3)}(z, \bar{z}, \theta, \bar{\theta}) \tag{42}$$

which satisfies $\bar{D}Q^{(3)} = 0$.

The existence of higher-spin conserved currents insures the quantum integrability of the system and the exact, factorizable S-matrix could be determined. The actual construction of an S-matrix is complicated by the presence of solitons in the spectrum, due to the periodicity of the potential under the shift

$$\Phi_2 \rightarrow \Phi_2 + 2\pi k \tag{43}$$

k being any integer. In particular, soliton solutions are given by $\Phi_1 = 0$ and Φ_2 assuming the field configurations of the super sine-Gordon solitons [19]. The solitonic masses are *real.*

The previous analysis can be extended to the $B(n,n)$ theories with $n > 1$. These are $N = 1$ superconformal models with central charge $c = \frac{3}{2}r + 24\vec{\rho}^{\,2}$, where r is the rank of the algebra and $\vec{\rho}$ its Weyl vector. Massive perturbations associated to the lowest fermionic root give rise to nonunitary integrable systems (the action is not hermitian). However, the bosonic mass spectrum is real

$$\begin{aligned} m_k^2 &= 8\sin^2\frac{2\pi k}{h} \qquad k = 1, \cdots, 2n-1 \\ m_{2n}^2 &= 2 \end{aligned} \tag{44}$$

For theories with $n > 1$ the classical mass ratios are affected by quantum corrections [12].

These theories contain solitons. Indeed, setting to zero the fields which appear in the potential with real couplings the generic $B^{(1)}(n,n)$ theory reduces to n decoupled sine-Gordon systems. Solitons have a real mass spectrum. The reality of the elementary particle and soliton masses leads to the expectation that, even if unitarity is not manifest at the lagrangian level, these theories admit a unitary restriction.

5. The A(1,1) Toda theory

5.1. Conservation laws

We consider now the supersymmetric Toda theory associated to the $A(1,1)$ superalgebra. This algebra is defined as $sl(2,2;C)$ factored by the one-dimensional invariant subspace generated by $\mathbf{I}_4$. The purely fermionic Dynkin diagram and its untwisted fermionic extension are given in Fig.1. Since this superalgebra is not simple, the number of simple roots exceeds the rank by one. Their explicit expressions can be given in terms of two-dimensional vectors as

$$\begin{aligned}\vec{\alpha}_1 &= = (1,-i)\\ \vec{\alpha}_2 &= = (1,i)\\ \vec{\alpha}_3 &= -\vec{\alpha}_1\end{aligned}\tag{45}$$

The corresponding supersymmetric action is

$$S=\frac{1}{\beta^2}\int d^2zd^2\theta\left[D\Phi_1\bar{D}\Phi_1+D\Phi_2\bar{D}\Phi_2+e^{\Phi_1-i\Phi_2}+e^{\Phi_1+i\Phi_2}+e^{-\Phi_1+i\Phi_2}\right]\tag{46}$$

and the equations of motion for the two superfields are

$$\begin{aligned}D\bar{D}\Phi_1 &= \frac{1}{2}\left(e^{\Phi_1-i\Phi_2}+e^{\Phi_1+i\Phi_2}-e^{-\Phi_1+i\Phi_2}\right)\\ D\bar{D}\Phi_2 &= \frac{1}{2}\left(-ie^{\Phi_1-i\Phi_2}+ie^{\Phi_1+i\Phi_2}+ie^{-\Phi_1+i\Phi_2}\right)\end{aligned}\tag{47}$$

To establish the existence of classical conserved currents one can construct the Miura operator (Ref. [12])

$$D(D-D\Phi_1+iD\Phi_2)(D+2iD\Phi_2)(D+D\Phi_1+iD\Phi_2)=D^4+\sum_{i=0}^{2}W^{(\frac{3}{2}-\frac{i}{2})}\,D^i\tag{48}$$

which gives the classical currents

$$\begin{aligned}W^{(\frac{1}{2})} &= -2\partial\Phi_2+2iD\Phi_2D\Phi_1\\ W^{(1)} &= iD\Phi_1\partial\Phi_1+iD\Phi_2\partial\Phi_2-iD\partial\Phi_1-D\partial\Phi_2+iD(D\Phi_2D\Phi_1)\\ W^{(\frac{3}{2})} &= -D(W^{(1)}-DW^{(\frac{1}{2})})\end{aligned}\tag{49}$$

The Miura operator in eq.(48) does *not* commute with the spinor derivative $\bar{D}$, so in general the W–currents are not holomorphic. However, using the equations of motion (47) one obtains

$$\bar{D}W^{(1)}=0\tag{50}$$

and

$$\bar{D}W^{(\frac{1}{2})}=-2De^{-\Phi_1+i\Phi_2}\qquad\qquad \bar{D}W^{(\frac{3}{2})}=-2iD\partial e^{-\Phi_1+i\Phi_2}\tag{51}$$

The spin–1 current is holomorphic and gives the classical traceless super stress–energy tensor of the theory, whereas $W^{(\frac{1}{2})}$ and $W^{(\frac{3}{2})}$ are not holomorphic but still conserved. The existence of a spin–1 holomorphic current is a general property of all the $A(n,n)$ theories which describe then $N=1$ superconformal models.

The expressions in eq.(49) give also the correct currents for the quantum system since in this case no corrections appear in the study of the conservation laws (50),(51).

Writing $W^{(1)}$ in components we obtain the supersymmetry current

$$W^{(1)}\Big|_{\theta=0} = \frac{i}{\sqrt{2}}\psi_1\partial\phi_1 + \frac{i}{\sqrt{2}}\psi_2\partial\phi_2 - \frac{i}{\sqrt{2}}\partial\psi_1 - \frac{1}{\sqrt{2}}\partial\psi_2 - \frac{1}{\sqrt{2}}\psi_1\partial\phi_2 + \frac{1}{\sqrt{2}}\psi_2\partial\phi_1 \quad (52)$$

and the stress–energy tensor which generates conformal tranformations

$$DW^{(1)}\Big|_{\theta=0} = -\partial\vec{\phi}\cdot\partial\vec{\phi} - \frac{i}{2}\vec{\psi}\cdot\partial\vec{\psi} + \partial^2\phi_1 - i\partial^2\phi_2 - \frac{1}{2}\partial(\psi_2\psi_1) \quad (53)$$

We note that in this case the traceless condition requires the addition of the fermionic improvement term $\frac{1}{2}\partial(\psi_1\psi_2)$. This term gives a negative contribution to the central charge which turns out to be identically zero. In the next subsection we will discuss some of the implications of these results.

We turn now to the massive perturbation of the theory realized by adding to the potential the exponential associated to the fermionic lowest root

$$\vec{\alpha}_0 = -(\vec{\alpha}_1 + \vec{\alpha}_2 + \vec{\alpha}_3) = (-1, -i) \quad (54)$$

The affine Toda action

$$S = \frac{1}{\beta^2}\int d^2z d^2\theta \left[D\vec{\Phi}\cdot\bar{D}\vec{\Phi} + e^{\Phi_1 - i\Phi_2} + e^{\Phi_1 + i\Phi_2} + e^{-\Phi_1 + i\Phi_2} + e^{-\Phi_1 - i\Phi_2}\right] \quad (55)$$

is manifestly hermitian and symmetric under $\Phi_1 \to -\Phi_1$ and $\Phi_2 \to -\Phi_2$. The bosonic classical masses are

$$m_1^2 = m_2^2 = 8 \quad (56)$$

The classical mass ratios are stable under quantum corrections [12].

The system is quantum integrable. In addition to the $W^{(\frac{1}{2})}$ current and the super stress–energy tensor $W^{(1)}$ which are conserved also in the affine theory, the first nontrivial conserved current appears again at $s = 3$. Its exact quantum expression is (for details of the calculation see Ref. [12])

$$\begin{aligned} J^{(3)} &= -i\left(1 - \frac{11\beta^4}{16\pi^2}\right)\left[D\partial\Phi_1\partial^2\Phi_1 + D\partial\Phi_2\partial^2\Phi_2\right] - i\left(1 - \frac{3\beta^2}{4\pi}\right)D\Phi_1(\partial\Phi_1)^3 \\ &\quad + i\left(1 + \frac{3\beta^2}{4\pi}\right)D\Phi_2(\partial\Phi_2)^3 - i\left(3 - \frac{9\beta^2}{4\pi}\right)(\partial\Phi_1)^2 D\Phi_2\partial\Phi_2 \\ &\quad + i\left(3 - \frac{3\beta^2}{4\pi}\right)(\partial\Phi_2)^2 D\Phi_1\partial\Phi_1 - \frac{3\beta^2}{\pi}D\Phi_1 D\partial\Phi_1 D\Phi_2\partial\Phi_2 \end{aligned} \quad (57)$$

The spectrum of the theory contains soliton solutions which are the same as in the super sine–Gordon model.

The $A(n,n)$ theories with $n > 1$ describe $N = 1$ superconformal models with central charge $c = 0$. Their untwisted fermionic affinization is described by an action which is not hermitian, nonetheless the bosonic particle mass spectrum is real

$$m_k^2 = m_{2n+1-k}^2 = 8\sin^2\frac{\pi k}{n+1} \qquad k = 1, \cdots, n \quad (58)$$

and stable under quantum corrections. Solitonic solutions with real masses have also been determined [13, 20].

5.1. N=2 supersymmetry

In addition to the explicit $N=1$ supersymmetry the $A(n,n)$ models possess a second supersymmetry [13]. We discuss here the $n=1$ case. The actions in eqs.(46), (55) for the conformal and massive $A(1,1)$ theories respectively, are invariant under $N=2$ supersymmetry transformations

$$\begin{aligned} \delta\Phi_1 &= \varepsilon Q_1 \Phi_1 \qquad\qquad \delta\Phi_1 = \xi Q_2 \Phi_2 \\ \delta\Phi_2 &= \varepsilon Q_1 \Phi_2 \qquad\qquad \delta\Phi_2 = -\xi Q_2 \Phi_1 \end{aligned} \tag{59}$$

generated by the two anticommuting operators

$$Q_1 = \partial_\theta - i\theta\partial \qquad\qquad Q_2 \equiv D = \partial_\theta + i\theta\partial \tag{60}$$

In terms of the superfields

$$\Phi_+ = \Phi_1 + i\Phi_2 \equiv \phi_+ + \tfrac{1}{\sqrt{2}}\theta\psi_+ + \tfrac{1}{\sqrt{2}}\bar\theta\bar\psi_+ + \theta\bar\theta F_+ \tag{61}$$

$$\Phi_- = \Phi_1 - i\Phi_2 \equiv \phi_- + \tfrac{1}{\sqrt{2}}\theta\psi_- + \tfrac{1}{\sqrt{2}}\bar\theta\bar\psi_- + \theta\bar\theta F_- \tag{62}$$

we can write the general action as

$$S = \frac{1}{\beta^2}\int d^2z d^2\theta \left[D\Phi_+\bar D\Phi_- + e^{\Phi_-} + e^{\Phi_+} + g_1 e^{-\Phi_-} + g_2 e^{-\Phi_+}\right] \tag{63}$$

Setting $g_1 = 1$, $g_2 = 0$ it gives the $A(1,1)$ $N=1$ superconformal invariant action in eq.(46); for $g_1 = g_2 = 1$ it gives the affine $A^{(1)}(1,1)$ action (55), whereas for $g_1 = g_2 = 0$ it reduces to the $N=2$ superconformal $A(1,0)$ theory [17]. In components we have

$$\begin{aligned} S = \int d^2z \Big[&\partial\phi_+\bar\partial\phi_- + \frac{i}{2}\psi_+\bar\partial\psi_- - \frac{i}{2}\bar\psi_+\partial\bar\psi_- \\ &-e^{\phi_+ + \phi_-} + g_1 e^{\phi_+ - \phi_-} + g_2 e^{-\phi_+ + \phi_-} - g_1 g_2 e^{-\phi_+ - \phi_-} \\ &+\frac{1}{2}\bar\psi_+\psi_+\left(e^{\phi_+} + g_2 e^{-\phi_+}\right) + \frac{1}{2}\bar\psi_-\psi_-\left(e^{\phi_-} + g_1 e^{-\phi_-}\right)\Big] \end{aligned} \tag{64}$$

Correspondingly we can reexpress the W–currents in eq.(49) as

$$\begin{aligned} W^{(\frac{1}{2})} &= D\Phi_+ D\Phi_- + i\partial\Phi_+ - i\partial\Phi_- \\ W^{(1)} &= iD\Phi_-\partial\Phi_+ - iD\partial\Phi_- \end{aligned} \tag{65}$$

They satisfy the conservation equations

$$\begin{aligned} \bar D W^{(\frac{1}{2})} &= -2D\left(g_1 e^{-\Phi_-} - g_2 e^{-\Phi_+}\right) \\ \bar D W^{(1)} &= -2i\partial\left(g_2 e^{-\Phi_+}\right) \end{aligned} \tag{66}$$

In particular, when $g_1 = g_2 = 0$ they are holomorphic.

We discuss now the various symmetries generated by these currents. At the component level we define

$$
\begin{aligned}
J &\equiv W^{(\frac{1}{2})}\Big|_{\theta=0} = \frac{1}{2}\psi_+\psi_- + i\partial\phi_+ - i\partial\phi_- \\
Q_+ &\equiv \sqrt{2}i\left[\, W^{(1)}\Big|_{\theta=0} - DW^{(\frac{1}{2})}\Big|_{\theta=0}\,\right] = -\psi_+\partial\phi_- + \partial\psi_+ \\
Q_- &\equiv \sqrt{2}i\, W^{(1)}\Big|_{\theta=0} = -\psi_-\partial\phi_+ + \partial\psi_- \\
T &\equiv DW^{(1)}\Big|_{\theta=0} = -\partial\phi_-\partial\phi_+ - \frac{i}{2}\psi_-\partial\psi_+ + \partial^2\phi_-
\end{aligned} \tag{67}
$$

The $Q_\pm$ currents generate the two supersymmetry transformations, J is a $U(1)$ generator and T is the stress–energy tensor of the theory. As previously emphasized, the central charge vanishes. Using the expressions of the currents in eq.(67) we can construct the corresponding charges from which the infinitesimal transformations of the fields can be easily computed. In general, from the conservation equation

$$
\bar{\partial} J^{(s)} = \partial \bar{J}^{(s)} \tag{68}
$$

it follows that the corresponding time independent charge, $\frac{dQ^{(s)}}{dt} = 0$ is

$$
Q^{(s)} = \int dz J^{(s)} + \int d\bar{z} \bar{J}^{(s)} \tag{69}
$$

Using the equations of motion derivable from the action in eq.(64) we have

$$
\begin{aligned}
\bar{\partial} J &= \partial\left[\frac{1}{2}\bar{\psi}_-\bar{\psi}_+ + i\bar{\partial}\phi_+ - i\bar{\partial}\phi_-\right] \\
\bar{\partial} Q_+ &= \partial\left(-2ig_1\bar{\psi}_- e^{-\phi_-}\right) + \partial\left[\bar{\partial}\psi_+ + i\bar{\psi}_-\left(e^{\phi_-} + g_1 e^{-\phi_-}\right)\right] \\
\bar{\partial} Q_- &= \partial\left(-2ig_2\bar{\psi}_+ e^{-\phi_+}\right) + \partial\left[\bar{\partial}\psi_- + i\bar{\psi}_+\left(e^{\phi_+} + g_2 e^{-\phi_+}\right)\right] \\
\bar{\partial} T &= \partial\left[-2g_2\left(e^{-\phi_+ + \phi_-} - g_1 e^{-\phi_+ - \phi_-}\right) + g_2\psi_+\bar{\psi}_+ e^{-\phi_+}\right] \\
&\quad + i\partial\left[\psi_+\bar{\partial}\psi_- + i\psi_+\bar{\psi}_+\left(e^{\phi_+} + g_2 e^{-\phi_+}\right)\right] \\
&\quad + \frac{i}{2}\partial\left[\psi_-\bar{\partial}\psi_+ + i\psi_-\bar{\psi}_-\left(e^{\phi_-} + g_1 e^{-\phi_-}\right)\right] \\
&\quad + \partial\left[\partial\bar{\partial}\phi_- + \left(e^{\phi_+} + g_2 e^{-\phi_+}\right)\left(e^{\phi_-} - g_1 e^{-\phi_-}\right)\right. \\
&\qquad \left. + \frac{1}{2}\psi_+\bar{\psi}_+\left(e^{\phi_+} - g_2 e^{-\phi_+}\right)\right]
\end{aligned} \tag{70}
$$

In the r.h.s. we have included terms which vanish on–shell in such a way that the charges constructed according to eq.(69) do generate invariances of the action. We obtain

1) $U(1)$–transformations:

$$\begin{aligned} \delta_J \phi_\pm &= 0 \\ \delta_J \psi_\pm &= \mp\zeta\psi_\pm \\ \delta_J \bar{\psi}_\pm &= \pm\zeta\bar{\psi}_\pm \end{aligned} \tag{71}$$

2) Q_+–supersymmetry transformations:

$$\begin{aligned} \delta_{Q_+}\phi_+ &= \xi\psi_+ \\ \delta_{Q_+}\phi_- &= 0 \\ \delta_{Q_+}\psi_+ &= 0 \\ \delta_{Q_+}\psi_- &= 2i\xi\partial\phi_- + 2i\partial\xi \\ \delta_{Q_+}\bar{\psi}_+ &= 2\xi\left(e^{\phi_-} - g_1 e^{-\phi_-}\right) \\ \delta_{Q_+}\bar{\psi}_- &= 0 \end{aligned} \tag{72}$$

3) Q_-–supersymmetry transformations:

$$\begin{aligned} \delta_{Q_-}\phi_+ &= 0 \\ \delta_{Q_-}\phi_- &= \varepsilon\psi_- \\ \delta_{Q_-}\psi_+ &= 2i\varepsilon\partial\phi_+ + 2i\partial\varepsilon \\ \delta_{Q_-}\psi_- &= 0 \\ \delta_{Q_-}\bar{\psi}_+ &= 0 \\ \delta_{Q_-}\bar{\psi}_- &= 2\varepsilon\left(e^{\phi_+} - g_2 e^{-\phi_+}\right) \end{aligned} \tag{73}$$

4) Conformal transformations generated by T:

$$\begin{aligned} \delta_T\phi_+ &= \lambda\partial\phi_+ + \partial\lambda \\ \delta_T\phi_- &= \lambda\partial\phi_- \\ \delta_T\psi_+ &= \lambda\partial\psi_+ \\ \delta_T\psi_- &= \lambda\partial\psi_- + \partial\lambda\psi_- \\ \delta_T\bar{\psi}_\pm &= \lambda\partial\bar{\psi}_\pm \end{aligned} \tag{74}$$

In the $A(1,0)$ theory both $W^{(\frac{1}{2})}$ and $W^{(1)}$ are superholomorphic currents and the action is invariant under the *local* transformations given above. The holomorphic currents J, $Q_\pm$ and the "untwisted" stress–energy tensor $T' \equiv T - \frac{1}{2}\partial J$ generate a $N=2$ superconformal algebra. The topological version [21] of this $N=2$ superconformal theory is described by the $c=0$ stress-energy tensor T. The role of the BRST charge is played by the supersymmetry charge Q_+ which satisfies the nihilpotency condition $Q_+^2 = 0$. It is easy to check that the stress-energy tensor can be written as

$$T = \{Q_{\rm BRST}, Q_-\} \tag{75}$$

The $A(1,1)$ Toda system is obtained by adding the g_1 perturbation. As explicitly shown in eq.(66) the $W^{(\frac{1}{2})}$ current is no longer holomorphic, so that the $U(1)$ and Q_+ symmetries become *rigid*. However, the stress-energy tensor T is still holomorphic and the theory preserves a $N = 1$ superconformal invariance. We note that being T the *only* holomorphic spin–1 current, the $A(1,1)$ Toda theory is naturally in its topological, conformal phase.

Finally we can switch the g_2–perturbation on and reach the $A^{(1)}(1,1)$ massive theory. In this case neither current is holomorphic and all the symmetries become *rigid*. The current T is the stress–energy tensor associated to the topological version of the theory.

6. Conclusions

We have analyzed the quantum integrability of the $A(1,1)$ and $B(1,1)$ supersymmteric Toda theories and of their affine untwisted extensions. We have chosen these two specific examples since they are good representatives of the more general supersymmetric theories based on the $A(n,n)$, $D(n+1,n)$ and $B(n,n)$ superalgebras. Most of the relevant issues could be discussed in a simple setting and easily generalized: in particular we have given an explicit lagrangian realization in $N = 1$ superspace and constructed the quantum higher–spin conserved currents both in the conformal and massive case.

The $B(1,1)$ Toda theory is characterized by two holomorphic conserved supercurrents of spin 1 and $\frac{3}{2}$. The components of $W^{(1)}$ are the generators of supersymmetry and conformal transformations, while $W^{(\frac{3}{2})}$ contains the generators of higher–spin transformations.

The supersymmetric theory based on the $A(1,1)$ superalgebra possesses a second supersymmetry in addition to the one manifest in the $N = 1$ superspace formalism. In fact the theory admits two conserved supercurrents of spin $\frac{1}{2}$ and 1. The spin–1 current is holomorphic and its components are one supersymmetry current and the stress-energy tensor. The $W^{(\frac{1}{2})}$–current is not holomorphic, but still conserved: at the component level it contains the generator of $U(1)$–transformations and the second supersymmetry current. As emphasized in the previous section, the stress–energy tensor that we have obtained via the Miura operator construction has zero central charge and it can be written as a BRST–commutator.

We have briefly discussed the implications of the existence of these quantum symmetries for the $A(1,0)$, $A(1,1)$ and $A^{(1)}(1,1)$ theories. A more detailed analysis will be presented in a future publication [20].

7. Acknowledgements

We wish to thank Camillo Imbimbo and Mario Pernici for valuable discussions and suggestions.

8. References

1. P.P. Kulish and E.R. Nisimov, *Theor. Math. Phys.* **29** (1976) 922;
   R. Flume, *Phys. Lett.* **62B** (1976) 93; *Phys. Lett.* **68B** (1977) 487;
   J.H. Lowenstein and E.R. Speer, *Comm. Math. Phys.* **63** (1978) 97;
   R. Sasaki and I. Yamanaka, *Adv. Stud. Pure Math.* **16** (1988) 271; *Progr. Theor. Phys.* **79** (1988) 1167;
   T. Marinucci and S. Sciuto, *Nucl. Phys.* **B156** (1979) 144;
   E.R. Nisimov, *Nucl. Phys.* **B163** (1980) 374;
   A.B. Zamolodchikov, *Int. J. Mod. Phys.* **A3** (1988) 743; Rutherford preprint RAL–89–001;
   A. Bilal and J.L. Gervais, *Phys. Lett.* **206B** (1988) 412; *Nucl. Phys.* **B314** (1989) 646.
2. L. Palla, *Nucl. Phys.* **B341** (1990) 714.
3. G.W. Delius, M.T. Grisaru and D. Zanon, "Quantum conserved currents in affine Toda theories", IFUM 412/FT (1992), to be published in Nucl. Phys. B.
4. A.B. Zamolodchikov and Al.B. Zamolodchikov, *Ann. Phys. (NY)* **120** (1979) 253.
5. A.E. Arinshtein, V.A. Fateev and A.B. Zamolodchikov, *Phys. Lett.* **87B** (1979) 389;
   J.L. Cardy and G. Mussardo, *Phys. Lett.* **225B** (1989) 275;
   P.G.O. Freund, T.R. Klassen and E. Melzer, *Phys. Lett.* **229B** (1989) 243;
   C. Destri and H.J. de Vega, *Phys. Lett.* **233B** (1989) 336;
   P. Christe and G. Mussardo, *Nucl. Phys.* B330 (1990) 465; *Int. J. Mod. Phys.* **A5** (1990) 4581;
   T.R. Klassen and E. Melzer, *Nucl. Phys.* B338 (1990) 485;
   H.W. Braden, E. Corrigan, P.E. Dorey and R. Sasaki, *Nucl. Phys.* B338 (1990) 689;
   G.W. Delius, M.T. Grisaru, S. Penati and D. Zanon, *Phys. Lett.* **256B** (1991) 164; *Nucl. Phys.* **B359** (1991) 125;
   C. Destri, H.J. de Vega and V.A. Fateev, *Phys. Lett.* **256B** (1991) 173.
6. G.W. Delius, M.T. Grisaru and D. Zanon, *Phys. Lett.* **277B** (1992) 414; *Nucl. Phys.* **B382** (1992) 365.
7. T. Hollowood, "Solitons in affine Toda field theories", OUTP–92–04P (1992);
   H.C. Liao, D. Olive and N. Turok, "Topological solitons in A_r affine Toda theories", Imperial/TP/91–92/34 (1992);
   C.P. Constantinidis, L.A. Ferreira, J.F. Gomes and A.H. Zimerman, "Connection between the affine and conformal affine Toda models and their Hirota's solution", IFT–P.26/92 (1992);
   N.J. MacKay and W.A. McGhee, "Affine Toda solitons and automorphisms of Dynkin diagrams", DTP–92–45/RIMS–890 (1992).

8. T. Hollowood, "Quantizing SL(N) solitons and the Hecke algebra", OUTP–92–03P (1992).
9. M. A. Olshanetsky, *Comm. Math. Phys.* **88** (1983) 63.
10. V.G. Kac, *Adv. Math.* **26** (1977) 8; *Adv. Math.* **30** (1978) 85.
11. S. Penati and D. Zanon, *Phys. Lett.* **288B** (1992) 297.
12. A. Gualzetti, S. Penati and D. Zanon, "Quantum conserved currents in supersymmetric Toda theories", IFUM 426/FT (1992).
13. J. Evans and T. Hollowood, *Phys. Lett.* **B293** (1992) 100.
14. E. Witten, *Comm. Math. Phys.* **118** (1988) 411; *Nucl. Phys.* **B340** (1990) 281.
15. A.V. Mikhailov, M.A. Olshanetsky and A.M. Perelomov, *Comm. Math. Phys.* **79** (1981) 473;
    D. Olive and N. Turok, *Nucl. Phys.* **B257[FS14]** (1985) 277.
16. D.A. Leites, M.V. Saviliev and V.V. Serganova, in *"Group theoretical methods in physics"*, VNU Science Press (1986);
    L. Frappat, A. Sciarrino and P. Sorba, *Comm. Math. Phys.* **121** (1989) 457.
17. J. Evans and T. Hollowood, *Nucl. Phys.* **B352** (1991) 723.
18. S. Komata, K. Mohri and H. Nohara, *Nucl. Phys.* **B359** (1991) 168;
    P. Mansfield and B. Spence, *Phys. Lett.* **265B** (1991) 79.
19. P. di Vecchia and S. Ferrara, *Nucl. Phys.* **B130** (1977) 93;
    C. Imbimbo and S. Mukhi, *Nucl. Phys.* **B247** (1984) 471.
20. S. Penati, M. Pernici and D. Zanon, in preparation.
21. T. Eguchi and S.K. Yang, *Mod. Phys. Lett.* **A5** (1990) 1693;
    C. Vafa, *Mod. Phys. Lett.* **A6** (1991) 337;
    R. Dijkgraaf, E. Verlinde and H. Verlinde, *Nucl. Phys.* **B352** (1991) 59.

IS THE GREEN-SCHWARZ SUPERSTRING ANOMALOUS?

Massimo Porrati[1]
Physics Department, NYU, 4 Washington Pl.
New York, NY 10003, USA

and

Peter van Nieuwenhuizen
ITP, SUNY, Stony Brook, NY 11794-3800, USA

ABSTRACT

We show how to cancel the anomalies of the heterotic string in the semi light-cone gauge, by adding suitable $\mathcal{O}(\hbar)$ terms to the local Weyl and target-space (rigid) Lorentz transformation rules. The modified transformations still satisfy the standard target-space Lorentz algebra.

In this contribution to the workshop We report on a calculation published in Phys. Lett. B273 (1991) 47 [1].

Strings with space-time supersymmetry[2] admit two equivalent formulations. In the first one (Neveu-Schwarz formulation [3]) one keeps manifest the *world-sheet* supersymmetry, whereas the space time supersymmetry is recovered in an indirect way by projecting away unwanted states in both the Ramond and Neveu-Schwarz sectors of the free-superstring Hilbert space [4]. This Neveu-Schwarz formulation can be quantized in a covariant way, that is in a way that respects the full 10-D space time symmetries (Lorentz and SUSY) [5]. The second formulation (Green-Schwarz formulation [6]) instead, does away with the two-dimensional supersymmetry, and makes manifest the ten-dimensional space-time supersymmetry. The world-sheet action reads, in the heterotic case, which is the one we shall consider from now on

$$S = \int d^2x \left[-\frac{1}{2}eg^{\alpha\beta}\Pi^\mu_\alpha\Pi_{\beta\mu} + \epsilon^{\alpha\beta}\partial_\alpha X^\mu(\bar{\Theta}\Gamma_\mu\partial_\beta\Theta) + \frac{ie}{\sqrt{2}}\lambda^I e^\alpha_-\partial_\alpha\lambda^I\right]. \tag{1}$$

Here $\Pi^\mu_\alpha = \partial_\alpha X^\mu - \bar{\Theta}\Gamma^\mu\partial_\alpha\Theta$; Θ is the 16-component (left-moving) Majorana-Weyl space-time fermion, λ^I are the 32 world-sheet right-moving (heterotic) fermions.

The Γ^μ are the ten-dimensional gamma-matrices obeying $\{\Gamma^\mu, \Gamma^\nu\} = 2\eta^{\mu\nu}$, $\eta^{\mu\nu} = diag(-1, +1, .., +1)$. This action, which is generally covariant, does not possess a local world-sheet supersymmetry, rather, it is invariant under the *local* fermionic κ-symmetry [7]

$$\delta\Theta = \Gamma_\mu\Pi^\mu_\alpha e^\alpha_-\kappa_+, \quad \delta X^\mu = \bar{\Theta}\Gamma^\mu\delta\Theta, \quad \delta e^-_\alpha = 4e^+_\alpha\bar{\kappa}_+\nabla_+\Theta. \tag{2}$$

[1] Reporting, on leave of absence from INFN, Pisa, Italy.

The covariant quantization of the Green-Schwarz string is still an unsolved problem. The positive results obtained for the superparticle cannot be carried over to the superstring in an obvious manner since the σ-dependence of the string leads to constraints which do not commute [8].

The light-cone quantization [6] of the Green-Schwarz superstring gives rise to no anomalies, in critical dimension. The problem with this gauge-fixing is that it cannot be extended naively beyond one loop. Indeed, starting from genus-two surfaces, the light-cone gauge gives rise to singular world-sheet metrics, whose curvature diverges at isolated points.

As a step towards a fully covariant quantization of the superstring, there exists a partially covariant quantization scheme [9, 10], called semi light-cone quantization, in which the troublesome local fermionic κ symmetry is fixed at the classical level by the gauge choice $\Gamma^+\Theta = 0$, while all other local symmetries are covariantly quantized. Here $\Gamma^\pm = (\Gamma^9 \pm \Gamma^0)/\sqrt{2}$. This gauge-fixing does not give rise to singularities in the two-dimensional curvature at higher genus, since the world-sheet metric is fixed in a covariant way.

By integrating the world-sheet anomalies (corresponding to Einstein, Weyl and local Lorentz gauge invariance), one obtains the effective action for the Green-Schwarz string. The world-sheet anomalies can be cancelled by adding a local counter term depending on the background values u^μ of the space-time coordinate fields X^μ. This effective action vanishes identically in the chiral sector with the 32 heterotic fermions, while in the other chiral sector with the 8 Θ fields one finds [11, 12]

$$\Gamma^{eff} = -\frac{d - 26 + 8/2}{96\pi}\hbar \int d^2x eR \left\{ \frac{1}{e\nabla^2} eR^+ + \log \nabla_- u^+ \right\}, \tag{3}$$

with, of course, $d = 10$. Here $R = R^+ + R^-$, where $R^+ = -2(d_-\omega_+ + \omega_+\omega_-)$ is a local Lorentz and Einstein scalar, and $d_- = e^\alpha_-\partial_\alpha$ while $\nabla_- u^+ = d_- u^+$. Since the expression between curly brackets in (3) is local Lorentz and Weyl invariant, the effective action is Einstein and local Lorentz invariant, but under Weyl transformations one has $\delta(eR) = e\nabla^2 2\sigma$, and the ensuing Weyl anomaly only vanishes on the u-shell, where $\nabla_+\nabla_- u^+ = (d_+ - \omega_+)d_- u^+ = 0$. This fact will be important for what follows below.

The spacetime transformation rules acquire extra Θ-dependent terms in order to maintain the gauge $\Gamma^+\Theta = 0$. It turns out that the only contribution to spacetime anomalies comes from the local counterterm in (3), which shifts the anomalies from the world-sheet to the spacetime sector. Under rigid Lorentz transformations one has $\delta_\Lambda \nabla_- u^+ = \Lambda_{-j}\nabla_- u^j + \Lambda_{-+}\nabla_- u^+$ and

$$\delta\Gamma^{eff} = \frac{\hbar}{8\pi}\int d^2x eR \left(\Lambda_{-+} + \Lambda_{-j}\frac{\nabla_- u^j}{\nabla_- u^+} \right). \tag{4}$$

In ref. [1] this problem was reanalyzed, and it was found that one can write $\delta\Gamma^{eff}$ in a covariant manner proportional to the u field equations. Following [1] we shall show that one can add order $\hbar$ terms to the Weyl and rigid (spacetime) Lorentz transformations rules such that (i) *all* quantum anomalies are cancelled *off-shell*, (ii) the quantum Lorentz transformation rules contain compensating Weyl transformations and close on the *classical* rigid Lorentz algebra without *any* order $\hbar$ leftovers.

The expression in (4) can be written as follows

$$\delta\Gamma^{eff} = \frac{\hbar}{4\pi}\int d^2x \frac{e}{\nabla_- u^+}(\nabla_-\nabla_+ - \nabla_+\nabla_-)\Lambda_{-\mu}\nabla_- u^\mu, \tag{5}$$

since $Rv_- = 2(\nabla_-\nabla_+ - \nabla_+\nabla_-)v_-$ for a vector v_-[2]. By partially integrating the ∇_+ in $\nabla_+\nabla_-$, and the ∇_- in $\nabla_-\nabla_+$, all terms become proportional to the u field equations. Notice that here partial integration is legitimate, since in eq. (5) $\nabla_- u^\mu$ and $1/\nabla_- u^+$ are bona fide local Lorentz vectors[3]. Boundary terms, present when the world-sheet is open, are harmless. A further discussion of this point is presented in the concluding remarks. Normalizing the classical action for X^μ to $eX^\mu\nabla_+\nabla_- X_\mu$, one finds that the anomalies in (5) can be removed by adding the following order $\hbar$ terms to the rigid Lorentz transformation rules of X^μ

$$\hat{\delta}_\Lambda X^+ = 0, \quad \hat{\delta}_\Lambda X^- = \frac{\hbar}{8\pi}\Lambda_{-j}\frac{\nabla_-^2 X^j}{(\nabla_- X^+)^2}, \quad \hat{\delta}_\Lambda X^j = -\frac{\hbar}{8\pi}\Lambda_{-j}\frac{\nabla_-^2 X^+}{(\nabla_- X^+)^2}. \tag{6}$$

Similarly the above mentioned off-shell Weyl anomaly can be removed by adding the following order $\hbar$ corrections to the Weyl transformation law for X^-

$$\hat{\delta}_\sigma X^- = \frac{\hbar}{4\pi}\frac{\nabla_-\sigma}{\nabla_- X^+}. \tag{7}$$

After having cancelled all world-sheet as well as spacetime anomalies, even off-shell, one must chech whether the new transformation laws satisfy the standard rigid Lorentz algebra at order $\hbar$.

Let us define new Lorentz transformation rules $\delta'_\Lambda = \delta_\Lambda + \hat{\delta}_\Lambda + \hat{\delta}_{\sigma(\Lambda)}$, where the last term leaves open the possibility that We shall need a compensating Weyl transformation at order $\hbar$. It results that, that for

$$\sigma(\Lambda) = -\Lambda_{-j}\frac{\nabla_- X^j}{\nabla_- X^+} \tag{8}$$

these δ'_Λ satisfy the spacetime Lorentz algebra at order $\hbar$. Although this can be verified straightforwardly, it may be helpful to give a few details. We want to show that on X^+, X^- and X^j the following identity holds

$$\left[\delta_\Lambda, \hat{\delta}_\lambda + \hat{\delta}_{\sigma(\lambda)}\right] + \left[\hat{\delta}_\Lambda + \hat{\delta}_{\sigma(\Lambda)}, \delta_\lambda\right] = \hat{\delta}_{[\lambda,\Lambda]} + \hat{\delta}_{\sigma([\lambda,\Lambda])}, \tag{9}$$

[2]Use $\nabla_\pm v_- = d_\pm v_- - \omega_\pm v_-$ and $(d_+ - \omega_+)d_- = (d_- + \omega_-)d_+$.

[3]They are Einstein scalars, since we used flat world-sheet indices.

where $[\lambda,\Lambda]_\mu^{\ \nu}=\lambda_\mu^{\ \rho}\Lambda_\rho^{\ \nu}-\Lambda_\mu^{\ \rho}\lambda_\rho^{\ \nu}$. On X^+, for example, $\hat{\delta}_\sigma$ and $\hat{\delta}_\Lambda$ both vanish, hence the left hand side of (9) should vanish on X^+. One indeed finds $-\Lambda^+_{\ \mu}\hat{\delta}_\lambda X^\mu-\lambda\leftrightarrow\Lambda$. Since this is both antisymmetric in λ, Λ and proportional to $\Lambda_{-j}\lambda_{-j}$ it indeed vanishes.

On X^j the terms without $\hat{\delta}_\sigma$ in (9) leave a term

$$-\frac{\hbar}{4\pi}\lambda_{-j}\Lambda_{-k}\frac{1}{\nabla_-X^+}\nabla_-\left(\frac{\nabla_-X^k}{\nabla_-X^+}\right)-\lambda\leftrightarrow\Lambda, \tag{10}$$

which is cancelled by the $\hat{\delta}_\sigma$ contributions, provided σ is given by (8).

Finally, on X^-, one finds that omitting all $\hat{\delta}_\sigma$ in (9), on the left-hand side there is a contribution

$$-\frac{\hbar}{4\pi}\lambda_{-j}\Lambda_{-k}\frac{\nabla_-^2X^j}{(\nabla_-X^+)^2}\nabla_-X^k-\lambda\leftrightarrow\Lambda. \tag{11}$$

The terms with $\hat{\delta}_\sigma$ in (9) cancel this contribution, provided $\sigma(\Lambda)$ is again given by (8), and reconstruct the $\hat{\delta}_{\sigma([\lambda,\Lambda])}$ term on the left-hand side of (9).

Let us conclude with some further remarks.

(i) Usually one tries to remove anomalies by adding a local counter term of order $\hbar$ to the classical action. The modified transformation rules correspond also to a local counter term. Indeed, the order $\hbar$ anomaly Δ can be obtained as

$$\left(\frac{\delta\Gamma^0}{\delta\phi}\frac{\delta}{\delta K}+\frac{\delta\Gamma^0}{\delta K}\frac{\delta}{\delta\phi}\right)\Gamma^1=\Delta \tag{12}$$

where the effective action is expanded as $\Gamma=\sum_n\hbar^n\Gamma^n$and the K are the sources for the symmetry variations of the fields ϕ [13]. The classical variation of ϕ is $\delta\Gamma^0/\delta K$. The first term in (12) is sometimes written as $\delta\Gamma^0/\delta\phi(\delta\phi\cdot\Gamma)^1$ and contains the one-loop corrections to $\delta\phi$. Usually one tries to add a K-independent local counterterm of order $\hbar$ to Γ, which cancels $\hbar\Delta$. However, one can also try to eliminate $\hbar\Delta$ by adding a local K-dependent counterterm. This counterterm corresponds to the order $\hbar$ modifications of the classical transformation laws, given in eq. (6) and (7), and reads $K\hat{\delta}\phi$. It is straightforward to see that our analysis, done at $K=0$, is valid for arbitrary K [1].

(ii) Notice that the cancellation of anomalies, at first sight, seems to work for all space-time dimensions in which the heterotic superstring exists, namely $d=3,4,6$, and 10. However, it is very likely that one would find that $d=10$ is the only dimension in which no anomalies (SUSY anomalies, in particular) are present on a non-trivial *fermionic* background.

(iii) Previously, the conformal (Weyl) anomaly of the Green-Schwarz superstring was shown to cancel in a path-integral approach by Carlip [10], and Kallosh and Morozov [9]. These authors did not need a counterterm of the form (3) since they used a regularization scheme different from the one of ref. [12], and made a field redefinition on the Θ variables. Their analysis was not complete, though, since they did not consider spacetime anomalies.

(iv) For the Green-Schwarz open or type II superstring [2, 6], the same cancellation of anomalies holds, but now in both chiral sectors.
(v) The boundary terms mentioned below eq. (5) are proportional to

$$\int d^2x \partial_\alpha \left[\epsilon^{\alpha\beta} \Lambda_{-\mu} \frac{1}{\nabla_- u^+} \nabla_- \nabla_\beta u^\mu \right], \tag{13}$$

where we used that ∇_- and ∇_β commute when acting on the world-sheet scalar u^μ. For closed strings one has boundary terms at $\tau = \pm\infty$, which vanish. For open strings, one has another term in eq. (13), with all ∇_- replaced by ∇_+, coming from the other chiral sector, and also boundaries at $\sigma = 0, 2\pi$ for all τ. In this case, the boundary terms vanish due to the open string boundary conditions $\partial_\sigma u^\mu = 0$.

1. M. Porrati and P. van Nieuwenhuizen, *Phys. Lett.* **B273** (1991) 47.
2. M.B. Green and J.H. Schwarz, *Phys. Lett.* **B136** (1984) 367;
   D.J. Gross, J. Harvey, E. Martinec and R. Rohm, *Phys. Rev. Lett.* **54** (1985) 502; *Nucl. Phys.* **B256** (1985) 253; **B267** (1986) 75.
3. A. Neveu and J.H. Schwarz, *Nucl. Phys.* **B31** (1971) 86; *Phys. Rev.* **D4** (1971) 1109.
4. F. Gliozzi, J. Scherk and D.I. Olive, *Phys. Lett.* **B65** (1976) 282; *Nucl. Phys.* **B122** (1977) 253.
5. D. Friedan, E. Martinec and S. Shenker, *Nucl. Phys.* **B271** (1986) 93.
6. M.B. Green and J.H. Schwarz, *Nucl. Phys.* **B243** (1984) 285.
7. W. Siegel, *Phys. Lett.* **B128** (1983) 397.
8. See e.g. the review talk by W. Siegel at the conference *Strings and Symmetries*, (Stony Brook, NY (1991)).
9. R. Kallosh and A. Morozov, *Int. Jou. of Mod. Phys.* **A3** (1988) 1943.
10. S. Carlip, *Nucl. Phys.* **B284** (1987) 365.
11. U. Kraemmer and A. Rebhan, *Phys. Lett.* **B236** (1990) 255.
12. F. Bastianelli, P. van Nieuwenhuizen and A. Van Proeyen, *Phys. Lett.* **B253** (1991) 67.
13. C. Becchi, in *Les Houches 1983*, (North-Holland, Amsterdam, 1984).

PECULIARITY OF STRING THEORY ON ORBIFOLDS IN THE PRESENCE OF AN ANTISYMMETRIC BACKGROUND FIELD

MAKOTO SAKAMOTO‡
The Niels Bohr Institute, University of Copenhagen
Blegdamsvej 17, DK-2100 Copenhagen Ø, Denmark

ABSTRACT

We study string theory on orbifolds in the presence of an antisymmetric constant background field and discuss some of new aspects of the theory. It is shown that the term with the antisymmetric field has a topological nature like a Chern-Simons term or a WZW term. Due to this property, the theory exhibits various anomalous behavior: Zero mode variables obey nontrivial quantization conditions. Coordinate transformations which define orbifolds are modified at the quantum level. Wavefunctions of twisted strings acquire phase factors when they move around non-contractible loops on orbifolds. Zero mode eigenvalues are shifted from naively expected values, in favor of modular invariance.

1. Introduction

Geometrical notion of string theory is far from obvious. For instance in toroidal compactification we can obtain enhanced gauge symmetries by tuning the moduli of the torus[1,2] and there exists a striking isomorphism between the large radius and small radius tori[3]. Another surprising example is that strings on some orbifolds[4] can equivalently be described as strings on tori[5−8] although orbifolds are geometrically quite different from tori. An antisymmetric background field has first been introduced by Narain, Sarmadi and Witten[9] to explain Narain torus compactification[10] in the conventional canonical approach. Orbifold models based on Narain torus compactification have been discussed by many authors and various "realistic" models have been proposed[11−14]. Nevertheless, one interesting topological nature of the antisymmetric field has not been discussed before. Our aim of this paper is to discuss some of new aspects of string theory on orbifolds in the presence of the antisymmetric background field. As we will explain below, orbifold models considered before are *trivial* in a topological point of view. Our concern of this paper is topologically *nontrivial* orbifolds. Since fermionic degrees of freedom play no important role in our discussion, we will restrict our considerations to bosonic strings.

‡ On leave from Department of Physics, Kobe University, Nada, Kobe 657, Japan

The action from which we shall start is[9]

$$S[X] = \int d\tau \int_0^\pi d\sigma \frac{1}{2\pi} \{\eta^{\alpha\beta}\partial_\alpha X^I \partial_\beta X^I + B^{IJ}\varepsilon^{\alpha\beta}\partial_\alpha X^I \partial_\beta X^J\}, \tag{1.1}$$

where B^{IJ} $(I, J = 1, \cdots, D)$ is the antisymmetric constant background field and $X^I(\tau, \sigma)$ $(I, \cdots, D)$ describes a string coordinate on a D-dimensional torus T^D which is defined by identifying a point X^I to $X^I + \pi w^I$ for all $w^I \in \Lambda$, where Λ is a D-dimensional lattice. An orbifold [4] is obtained by dividing a torus by the action of a discrete symmetry group G of the torus. Any element g of G can be represented (for symmetric orbifolds) by[4,15]

$$g = (U, v), \tag{1.2}$$

where U denotes a rotation and v is a translation. To simplify the discussion we will set the shift vector $v = 0$ throughout this paper. The shift vector will play no essential role in our discussion. In section 2 we clarify a topological nature of the second term of the action (1.1): Consider a transformation,

$$g: \quad X^I \longrightarrow U^{IJ}X^J. \tag{1.3}$$

It is shown in the next section that a Euclidean action $S_E[X]$ in the path integral formalism[16] is not invariant under the transformation (1.3) unless the rotation matrix U^{IJ} commutes with B^{IJ} but in general

$$S_E[X] - S_E[UX] = i2\pi n \quad \text{with} \quad n \in \mathbf{Z}. \tag{1.4}$$

This means that the transformation (1.3) is not a symmetry of the Euclidean action $S_E[X]$ unless $[U, B] = 0$ but is a symmetry of the theory at the quantum level. Thus, th second term of the action (1.1) can be regarded as a topological term like a Chern-Simons term or a WZW term. So far, much attention has been paid to the case

$$[U, B] = 0, \tag{1.5}$$

or simply $B^{IJ} = 0$, i.e., to topologically "trivial" orbifold models. The generalization to the case

$$[U, B] \neq 0, \tag{1.6}$$

is not, however, trivial, as expected from the result (1.4). A naive construction of orbifold models with Eq.(1.6) would give modular non-invariant partition functions and also destroy the duality of amplitudes[15,17,18].

Due to the property (1.4), the orbifold models with Eq.(1.6) exhibit various anomalous behavior. In section 3 we discuss the duality of amplitudes and clarify cocycle properties of vertex operators. Then we find that zero modes of strings should obey nontrivial quantization conditions, which are the origin of the anomalous behavior of the theory in an operator formalism point of view.

In section 4 we see that the (classical) transformation should be modified at the quantum level and that a vertex operator $V(k_L, k_R; z)$ transforms as

$$g: \quad V(k_L, k_R; z) \quad \longrightarrow \quad \rho\, V(U^T k_L, U^T k_R; z). \tag{1.7}$$

The phase factor ρ can be regarded as a kind of a quantum effect and plays an important role in extracting physical states. It is pointed out that in an algebraic point of view the phase ρ has a connection with automorphisms of algebras.

In section 5 we find an Aharonov-Bohm[19] like effect in our system. Strings on orbifolds in the presence of the antisymmetric background field are very similar to electrons in the presence of an infinitely long solenoid: The space is not simply-connected. The antisymmetric background field plays the same role as an external gauge field. If an electron moves around the solenoid, a wavefunction of the electron in general acquires a phase. The same thing happens to twisted strings. A wavefunction of a twisted string is not, in general, periodic with respect to a torus shift but

$$\Psi(x^I + \pi w^I) = e^{i\pi w \cdot v}\Psi(x^I), \tag{1.8}$$

for $w^I \in \Lambda$ such that $w^I = U^{IJ} w^J$. The constant shift vector v^I depends on the commutator $[U, B]$. We also discuss a physical implication of Eq.(1.8) and see how a naively expected spectrum is modified.

Section 6 is devoted to conclusions.

2. Topological Nature of a $\mathbf{B^{IJ}}$-Term

In this section, we shall clarify a topological nature of the second term of the action (1.1) from a path integral point of view.

In the Euclidean path integral formalism, the one-loop vacuum amplitude of the closed bosonic string theory on a torus is given by the functional integral[16],

$$\int \frac{[dg_{\alpha\beta}][dX^I]}{\mathcal{V}_{ol}} \;\exp\{-S[X,g]\}, \tag{2.1}$$

where $g_{\alpha\beta}$ is a Euclidean metric of the two dimensional would sheet with the topology of a torus and $\mathcal{V}_{ol}$ is a volume of local gauge symmetry groups. The Euclidean action is given by

$$S[X,g] = \int_0^1 d^2\sigma \frac{1}{2\pi}\{\sqrt{g}g^{\alpha\beta}\partial_\alpha X^I \partial_\beta X^I - iB^{IJ}\varepsilon^{\alpha\beta}\partial_\alpha X^I \partial_\beta X^J\}. \tag{2.2}$$

It should be emphasized that the imaginary number i appears in the second term of the Euclidean action (2.2) due to the antisymmetric property of $\varepsilon^{\alpha\beta}$. Since strings propagate on a torus defined by identifying a point X^I with $X^I + \pi w^I$ for all

$w^I \in \Lambda$, where Λ is a D-dimensional lattice, the string coordinate $X^I(\sigma^1, \sigma^2)$ obeys the following boundary condition:

$$\begin{aligned} X^I(\sigma^1+1, \sigma^2) &= X^I(\sigma^1, \sigma^2) + \pi w^I, \\ X^I(\sigma^1, \sigma^2+1) &= X^I(\sigma^1, \sigma^2) + \pi w'^I, \end{aligned} \tag{2.3}$$

for some $w^I, w'^I \in \Lambda$.

Let us consider a transformation,

$$g: \quad X^I \longrightarrow U^{IJ} X^J \quad \text{with} \quad U^T U = 1, \tag{2.4}$$

where U^{IJ} is an orthogonal matrix. The first term of the action (2.2) is trivially invariant under the transformation (2.4) but the second term is not invariant if U^{IJ} does not commute with B^{IJ}. Since B^{IJ} is a constant antisymmetric field, the second term of the action (2.2) can be written as a total divergence, i.e.,

$$B^{IJ}\varepsilon^{\alpha\beta}\partial_\alpha X^I \partial_\beta X^J = \partial_\alpha (B^{IJ}\varepsilon^{\alpha\beta} X^I \partial_\beta X^J). \tag{2.5}$$

It turns out that the difference between $S[X,g]$ and $S[UX,g]$ is given by

$$S[X,g] - S[UX,g] = i\pi w'^I (B - U^T B U)^{IJ} w^J. \tag{2.6}$$

From this relation, we come to an important conclusion that the transformation (2.4) is a *quantum* symmetry of the theory in a path integral point of view if

$$(B - U^T B U)^{IJ} w^J \in 2\Lambda^* \quad \text{for all} \quad w^I \in \Lambda, \tag{2.7}$$

where Λ^* is the dual lattice of Λ, although the Euclidean action (2.2) itself is not invariant under the transformation (2.4) if $[U, B] \neq 0$. Then, the second term of the action (2.2) can be regarded as a topological term like a Chern-Simons term or a WZW term. In the following, we will show that the condition (2.7) is nothing but a consistency condition for constructing string theory on orbifolds.

An orbifold [4] is obtained by dividing a torus by the action of a suitable discrete group G. In the construction of an orbifold model, we start with a D-dimensional toroidally compactified closed bosonic string theory which is specified by a $(D+D)$-dimensional Lorentzian even self-dual lattice $\Gamma^{D,D}$[9,10], on which the left- and right-moving momentum (p_L^I, p_R^I) $(I = 1, \cdots, D)$ lies. In general any group element g of G (for symmetric orbifolds) can be represented by

$$g = (U, v), \tag{2.8}$$

where U is a rotation and v is a translation. As mentioned in the introduction, we will set the shift vector $v = 0$ for simplicity. To define an orbifold consistently, G

must be a discrete symmetry group of the torus. This means that any element g of G must be an automorphism of the lattice $\Gamma^{D,D}$,i.e.,

$$g: \quad (p_L^I, p_R^I) \longrightarrow (U^{IJ}p_L^J, U^{IJ}p_R^J) \in \Gamma^{D,D}. \tag{2.9}$$

for all $(p_L^I, p_R^I) \in \Gamma^{D,D}$. The left- and the right- moving momenta, p_L^I and p_R^I, are related to the center of mass momentum p^I and the winding number w^I of a string as follows[9]:

$$\begin{aligned} p_L^I &= \frac{1}{2}p^I + \frac{1}{2}(1-B)^{IJ}w^J, \\ p_R^I &= \frac{1}{2}p^I - \frac{1}{2}(1+B)^{IJ}w^J. \end{aligned} \tag{2.10}$$

The winding number w^I, by definition, lies on the lattice Λ, i.e.,

$$w^I \in \Lambda. \tag{2.11}$$

Since a wavefunction $\Psi(x^I)$ of a string on a torus must be periodic, i.e., $\Psi(x^I + \pi w^I) = \Psi(x^I)$ for all $w^I \in \Lambda$, the allowed momentum is

$$p^I \in 2\Lambda^*. \tag{2.12}$$

It follows from Eqs. (2.9) and (2.10) that g acts on w^I and p^I as follows:

$$\begin{aligned} g: \quad w^I &\longrightarrow U^{IJ}w^J, \\ p^I &\longrightarrow U^{IJ}p^J - [U,B]^{IJ}w^J. \end{aligned} \tag{2.13}$$

For the action of g on w^I to be well-defined, it must be an automorphism of Λ,i.e.,

$$U^{IJ}w^J \in \Lambda \quad \text{for all} \quad w^I \in \Lambda. \tag{2.14}$$

For the action of g on p^I to be well-defined, we further require the following condition:

$$[U,B]^{IJ}w^J \in 2\Lambda^*,$$

or equivalently

$$(B - U^TBU)^{IJ}w^J \in 2\Lambda^* \quad \text{for all} \quad w^I \in \Lambda. \tag{2.15}$$

This is just the condition (2.7), as announced before.

In the following sections, we shall construct the operator formalism of string theory on orbifolds in the presence of the antisymmetric constant background field and reveal various anomalous behavior of the theory. We will see that the necessity of nontrivial cocycle operators in vertex operators is the origin of the anomalous behavior in the operator formalism.

Before closing this section, it may be instructive to present some examples of B^{IJ} and U^{IJ}, which satisfy the condition (2.15). Let us consider the following

$(D+D)$-dimensional Lorentzian even self-dual lattice, which has been introduced by Englert and Neveu[20],

$$\Gamma^{D,D} = \{ \ (p_L^I, p_R^I) \mid p_L^I, p_R^I \in \Lambda_W(\mathcal{G}), \ p_L^I - p_R^I \in \Lambda_R(\mathcal{G}) \ \}, \tag{2.16}$$

where $\Lambda_R(\mathcal{G})$ ($\Lambda_W(\mathcal{G})$) is the root (weight) lattice of a simply-laced Lie algebra $\mathcal{G}$ with rank D and the squared length of the root vectors is normalized to two. In this normalization, the weight lattice $\Lambda_W(\mathcal{G})$ is just the dual lattice of $\Lambda_R(\mathcal{G})$. It should be emphasized that the lattice (2.16) plays a crucial role of gauge symmetries in closed string theory[1,2]. The lattice (2.16) can be obtained through the relations (2.10) by choosing Λ and B^{IJ} as follows:

$$\Lambda = \Lambda_R(\mathcal{G}), \tag{2.17}$$

and

$$\alpha_i^I B^{IJ} \alpha_j^J = \alpha_i^I \alpha_j^I \mod 2, \tag{2.18}$$

where α_i is a simple root of $\mathcal{G}$ which is normalized to $(\alpha_i)^2 = 2$. If we choose the rotation matrix U to be an automorphism of the root lattice $\Lambda_R(\mathcal{G})$, i.e.,

$$U^{IJ} w^J \in \Lambda_R(\mathcal{G}) \quad \text{for all} \quad w^I \in \Lambda_R(\mathcal{G}), \tag{2.19}$$

then the matrix U always satisfies the condition (2.15), i.e.,

$$(B - U^T B U)^{IJ} w^J \in 2\Lambda_W(\mathcal{G}) \quad \text{for all} \quad w^I \in \Lambda_R(\mathcal{G}). \tag{2.20}$$

3. Cocycle Properties of Vertex Operators

In this section we shall investigate cocycle properties of vertex operators and show that zero modes of strings should obey nontrivial quantization conditions.

Let us consider a vertex operator $V(k_L, k_R, z)$ which describes the emission of a state with the momentum $(k_L^I, k_R^I) \in \Gamma^{D,D}$. The vertex operator will be of the form,

$$V(k_L, k_R; z) = f_{k_L,k_R}(z) : e^{ik_L \cdot X_L(z) + ik_R \cdot X_R(\bar{z})} C_{k_L,k_R} :, \tag{3.1}$$

where $X_L^I(z)$ ($X_R^I(\bar{z})$) are the left- (right-) moving string coordinate and $f_{k_L,k_R}(z)$ is a normalization factor, which will in general depend on k_L^I, k_R^I, z and $\bar{z}$ in twisted sectors. The C_{k_L,k_R} is, if necessary, some extra (cocycle) operator. The product of two vertex operators

$$V(k_L, k_R; z) V(k'_L, k'_R; z'), \tag{3.2}$$

is well-defined if $|z| > |z'|$. The different ordering of the two vertex operators corresponds to the different "time"-ordering. To obtain scattering amplitudes, we must sum over all possible "time"-ordering for the emission of states. We must then establish that each contribution is independent of the order of the vertex operators to enlarge the regions of integrations over z variables[21]. Thus the product (3.2), with respect to z and z', has to be analytically continued to the region $|z'| > |z|$ and to be identical to

$$V(k'_L, k'_R; z')V(k_L, k_R; z), \tag{3.3}$$

for $|z'| > |z|$. In terms of zero modes, the above statement can be translated into the following condition:

$$V_0(k_L, k_R)V_0(k'_L, k'_R) = \eta\ V_0(k'_L, k'_R)V_0(k_L, k_R), \tag{3.4}$$

where

$$V_0(k_L, k_R) = e^{ik_L\cdot\hat{x}_L + ik_R\cdot\hat{x}_R} C_{k_L,k_R}. \tag{3.5}$$

The $\hat{x}^I_L$ ($\hat{x}^I_R$) denotes the left-(right-) moving "center of mass" coordinate. The phase factor η is required to compensate a phase appearing in reversing the order of the oscillator modes of the vertex operators. In the untwisted sector, the phase η is given by[1,2]

$$\eta = (-1)^{k_L\cdot k'_L - k_R\cdot k'_R}. \tag{3.6}$$

In the U-twisted sector with $U^N = \mathbf{1}$, the phase η is given by[22]

$$\eta = \exp\{i\pi k^I_L(1 + \sum_{\ell=1}^{N}\frac{\ell}{N}(U^{-\ell} - U^{\ell}))^{IJ}k'^J_L - i\pi k^I_R(1 + \sum_{\ell=1}^{N}\frac{\ell}{N}(U^{-\ell} - U^{\ell}))^{IJ}k'^J_R\}. \tag{3.7}$$

In the untwisted sector, Frenkel and Kač[1,2] have shown that the cocycle operator can be constructed without introducing any more degrees of freedom. However, instead of introducing the cocycle operator C_{k_L,k_R} it is possible to achieve the relation (3.4). In ref.23, it has been shown that multiplying vertex operators by cocycle operators is equivalent to modifying commutation relations for zero modes and that the following commutation relations ‡ with $C_{k_L,k_R} = 1$:

$$\begin{aligned} [\hat{x}^I_L, \hat{x}^J_L] &= i\pi B^{IJ}, \\ [\hat{x}^I_R, \hat{x}^J_R] &= i\pi B^{IJ}, \\ [\hat{x}^I_L, \hat{x}^J_R] &= i\pi(1 - B)^{IJ}, \end{aligned} \tag{3.8}$$

lead to the relation (3.4) with the correct phase (3.6). A geometrical meaning of the above commutation relations has also been discussed in ref.23.

In twisted sectors, the relation (3.4) with the phase (3.7) may require new degrees of freedom, which are called fixed points (fixed lines, fixed surfaces,$\cdots$) by physicists. The explicit realization has been obtained in ref.24 in the case of $[U, B] =$

‡ The normalization of $\hat{x}^I_L$ and $\hat{x}^I_R$ is different from that in ref.23 by factor two.

0 and $det(1-U) \neq 0$. In ref.24, the quantization of zero modes of twisted strings has been clarified from a geometrical point of view and the nontrivial commutation relations of the zero modes with $C_{k_L,k_R}=1$ have been shown to naturally satisfy the relation (3.4) with the correct phase (3.7). For $[U,B] \neq 0$ and/or $det(1-U)=0$ we need to modify the results given in ref.24. It turns out that the following commutation relations with $C_{k_L,k_R}=1$ satisfy the relation (3.4) with the phase (3.7):

$$
\begin{aligned}
[\hat{x}_L^I, \hat{x}_L^J] &= i\pi(B - \sum_{\ell=1}^{N} \frac{\ell}{N}(U^{-\ell} - U^{\ell}))^{IJ}, \\
[\hat{x}_R^I, \hat{x}_R^J] &= i\pi(B + \sum_{\ell=1}^{N} \frac{\ell}{N}(U^{-\ell} - U^{\ell}))^{IJ}, \\
[\hat{x}_L^I, \hat{x}_R^J] &= i\pi(1-B)^{IJ},
\end{aligned}
\tag{3.9}
$$

These commutation relations reduce to those given in ref. 24 for $[U,B]=0$ and $det(1-U) \neq 0$ and also reduce to the relations (3.8) for $U=\mathbf{1}$. We will see in the following sections that the above nontrivial commutation relations are the origin of anomalous features of the theory.

4. Anomalous Transformations

In this section, we shall show that the action of g on the string coordinate becomes anomalous at the quantum level if $[U,B] \neq 0$.

The relevant operators in string theory are the momentum operators, $P_L^I(z) \equiv i\partial_z X_L^I(z)$ and $P_R^I(\bar{z}) \equiv i\partial_{\bar{z}} X_R^I(\bar{z})$, and the vertex operator $V(k_L,k_R;z)$. Other operators can be obtained from the operator products of these operators. For example the energy-momentum tensors of the left- and the right- movers are given by

$$
\begin{aligned}
T(z) &= \lim_{w\to z} \frac{1}{2}\Big(P_L^I(w)P_L^I(z) - \frac{D}{(w-z)^2}\Big), \\
\bar{T}(\bar{z}) &= \lim_{\bar{w}\to \bar{z}} \frac{1}{2}\Big(P_R^I(\bar{w})P_R^I(\bar{z}) - \frac{D}{(\bar{w}-\bar{z})^2}\Big).
\end{aligned}
\tag{4.1}
$$

Under the action of g, the momentum operators transform as

$$
g: \quad (P_L^I(z), P_R^I(\bar{z})) \longrightarrow (U^{IJ}P_L^I(z), U^{IJ}P_R^I(\bar{z})), \tag{4.2}
$$

which leaves the energy-momentum tensors invariant, as it should be. One might expect that under the action of g the left- and the right- moving string coordinates, $X_L^I(z)$ and $X_R^I(\bar{z})$, transform in a similar way to $P_L^I(z)$ and $P_R^I(\bar{z})$. However, it is not the case if U^{IJ} does not commute with B^{IJ}. Suppose that under the action of g, $\hat{x}_L^I$ and $\hat{x}_R^I$ would transform as

$$
\begin{aligned}
g: \quad \hat{x}_L^I &\longrightarrow U^{IJ}\hat{x}_L^J, \\
\hat{x}_R^I &\longrightarrow U^{IJ}\hat{x}_R^J.
\end{aligned}
\tag{4.3}
$$

It is easy to see that the transformations (4.3) is inconsistent with the commutation relations (3.8) or (3.9) unless U^{IJ} commutes with B^{IJ}. In ref.15, it has been shown that in the untwisted sector the correct action of g on $\hat{x}_L^I$ and $\hat{x}_R^I$ is given by ‡

$$\begin{aligned} g:\quad \hat{x}_L^I &\longrightarrow U^{IJ}\hat{x}_L^J + \pi U^{IJ}(\frac{1}{2}(B - U^TBU) - C_U)^{JK}\hat{w}^K, \\ \hat{x}_R^I &\longrightarrow U^{IJ}\hat{x}_R^J - \pi U^{IJ}(\frac{1}{2}(B - U^TBU) - C_U)^{JK}\hat{w}^K, \end{aligned} \tag{4.4}$$

where $\hat{w}^k$ is the winding number operator, which satisfies the following commutation relations[24]:

$$\begin{aligned} [\hat{x}_L^I, \hat{w}^J] &= i\delta^{IJ}, \\ [\hat{x}_R^I, \hat{w}^J] &= -i\delta^{IJ}. \end{aligned} \tag{4.5}$$

The C_U^{IJ} is a symmetric matrix defined through the relation,

$$w^I C_U^{IJ} w'^J = \frac{1}{2} w^I (B - U^TBU)^{IJ} w'^J \quad \text{mod } 2, \tag{4.6}$$

for all $w^I, w'^I \in \Lambda$. The existence of a symmetric matrix C_U is guaranteed by the fact that

$$\frac{1}{2} w^I (B - U^TBU)^{IJ} w'^J \in \mathbf{Z} \quad \text{for all } \ w^I, w'^I \in \Lambda. \tag{4.7}$$

In ref.15, the transformation (4.4) has been derived from the requirement that the action of g on vertex operators has to preserve the duality property of amplitudes. It is important to verify that the transformation (4.4) is consistent with the commutation relations (3.8). We can further show that the transformation (4.4) is still correct in the U-twisted sector and is consistent with the commutation relations (3.9).

It seems that the transformation (4.4) loses its geometrical meaning unless U^{IJ} commutes with B^{IJ}. However, the transformation (4.4) is geometrically still well-defined because the anomalous terms in Eq.(4.4) can be regarded as a torus shift, i.e.,

$$g:\quad (\hat{x}_L^I, \hat{x}_R^I) \longrightarrow (U^{IJ}\hat{x}_L^J, U^{IJ}\hat{x}_R^J) + \text{torus shift}. \tag{4.8}$$

To see this, let us consider the action of g on $k_L \cdot \hat{x}_L + k_R \cdot \hat{x}_R$ for $(k_L^I, k_R^I) \in \Gamma^{D,D}$.

$$g:\quad k_L\cdot\hat{x}_L + k_R\cdot\hat{x}_R \longrightarrow k_L\cdot U\hat{x}_L + k_R\cdot U\hat{x}_R + \pi(k_L - k_R)\cdot U(\frac{1}{2}(B - U^TBU) - C_U)\hat{w}. \tag{4.9}$$

The last term of the right hand side may be regarded as a trivial shift because

$$(k_L - k_R)\cdot U(\frac{1}{2}(B - U^TBU) - C_U)\hat{w} = 0 \quad \text{mod } 2, \tag{4.10}$$

‡ Set $U_L = U_R \equiv U$ and assume the form (6-13) in the first reference of ref.15.

for all $(k_L^I, k_R^I) \in \Gamma^{D,D}$ and $\hat{w}^I \in \Lambda$. (Recall that $k_L^I - k_R^I \in \Lambda$.)

What role do the anomalous terms in Eq.(4.4) play? To answer this question, let us consider the action of g on a vertex operator. The result is

$$g: \quad V(k_L, k_R; z) \quad \longrightarrow \quad e^{-i\frac{\pi}{2}(k_L-k_R)\cdot UC_UU^T(k_L-k_R)}\, V(U^Tk_L, U^Tk_R; z). \quad (4.11)$$

It should be noted that to derive the relation (4.11) we cannot use the relation (4.10) directly in the exponent because $\hat{w}^I$ is not a c-number but a q-number. We can use the relation (4.10) only after separating the terms depending on $\hat{w}^I$ from $V(U^Tk_L, U^Tk_R; z)$ by use of the Hausdorff formula. The phase factor appearing in the transformation (4.11) plays an important role in extracting physical states, which must be g-invariant. The following example may be helpful to understand a role of the phase in Eq.(4.11): Let us consider the (left-moving) momentum operator $H^I(z) \equiv P_L^I(z)$ $(I = 1,2)$ and the vertex operators $E^\alpha(z) \equiv V(\alpha, 0; z)$, where α's are root vectors of $SU(3)$ with $\alpha^2 = 2$. They will form level one $SU(3)$ Kač-Moody algebra[1,2]. Let α_i $(i = 1,2)$ be a simple root of $SU(3)$. Consider the following transformation:

$$g: \quad \alpha_1 \leftrightarrow \alpha_2. \quad (4.12)$$

This is clearly an automorphism of the root lattice of $SU(3)$. Then one might expect that under the transformation (4.12) the generators of $SU(3)$ Kač-Moody algebra would transform as follows:

$$\begin{aligned} \alpha_1 \cdot H(z) &\longleftrightarrow \alpha_2 \cdot H(z), \\ E^{\pm\alpha_1}(z) &\longleftrightarrow E^{\pm\alpha_2}(z), \end{aligned} \quad (4.13)$$

and $E^{\pm(\alpha_1+\alpha_2)}(z)$ would be left invariant. If it were ture, we would have a strange result: The invariant generators, $(\alpha_1+\alpha_2)\cdot H(z)$, $E^{\pm\alpha_1}(z)+E^{\pm\alpha_2}(z)$ and $E^{\pm(\alpha_1+\alpha_2)}(z)$ could form a subalgebra of $SU(3)$. This is not, however, acceptable because five generators must form an algebra with rank more than 3 while rank of $SU(3)$ is 2. The key to resolve this inconsistency is the phase in Eq.(4.11): We may take an Englert-Neveu lattice (2.16) with $\Lambda_R(SU(3))$ in order for $E^\alpha(z)$ to be well-defined[1,2]. This means that we take a nontrivial antisymmetric field B^{IJ} through the relation (2.18). It turns out that B^{IJ} does not commute with the transformation (4.12). Then, our formalism tells us that the transformations (4.13) are still true but $E^{\pm(\alpha_1+\alpha_2)}(z)$ must transform as

$$g: \quad E^{\pm(\alpha_1+\alpha_2)}(z) \longrightarrow -E^{\pm(\alpha_1+\alpha_2)}(z). \quad (4.14)$$

Therefore, $E^{\pm(\alpha_1+\alpha_2)}(z)$ are not invariant under the action of g although $\alpha_1 + \alpha_2$ is. In an algebraic point of view the phase in Eq.(4.11) has a connection with automorphisms of algebras rather than automorphisms of lattices.

5. Aharonov-Bohm like Effect

Strings on orbifolds in the presence of the antisymmetric background field may be in a similar situation to electrons in the presence of an infinitely long solenoid. Both underlying spaces are not simply-connected and possess non contractible loops. The antisymmetric background field B^{IJ} may play a similar role to an external gauge field A_μ. The gauge field A_μ is not gauge invariant and B^{IJ} is not invariant under the transformation $B^{IJ} \to (U^T B U)^{IJ}$ if U^{IJ} does not commute with B^{IJ}. The relation (2.6) means that B^{IJ} cannot be defined as a single-valued "function" for twisted strings with $[U, B] \neq 0$. This fact suggests that if a twisted string moves around a non-contractible loop on the orbifold the wavefunction may acquire a nontrivial phase. In fact we can show that

$$\Psi(x^I + \pi w^I) = e^{-i\frac{\pi}{2} w^I C_U^{IJ} w^J} \Psi(x^I), \tag{5.1}$$

where

$$w^I \in \Lambda_U = \{w^I \in \Lambda \mid w^I = U^{IJ} w^J\}. \tag{5.2}$$

What is a physical implication of the relation (5.1) ? To see this, we first note that the left hand side of Eq.(5.1) can be expressed as

$$\Psi(x^I + \pi w^I) = e^{-i\pi w \cdot \hat{p}_{/\!/}} \Psi(x^I), \tag{5.3}$$

where $\hat{p}^I_{/\!/}$ is the canonical conjugate momentum restricted to the U-invariant subspace, i.e., $\hat{p}^I_{/\!/} = U^{IJ} \hat{p}^J_{/\!/}$. Introduce a vector v^I with $v^I = U^{IJ} v^J$ through the relation [15]

$$w^I v^I = \frac{1}{2} w^I C_U^{IJ} w^J \mod 2, \tag{5.4}$$

for all $w^I \in \Lambda_U$. Comparing Eq.(5.1) with (5.3), we conclude that

$$\hat{p}^I_{/\!/} \in v^I + \Lambda_U{}^*, \tag{5.5}$$

where $\Lambda_U{}^*$ is the dual lattice of Λ_U. Thus, allowed eigenvalues of $\hat{p}^I_{/\!/}$ are different from naively expected values[6] by v^I. This result exactly agrees with the one expected from the argument of modular invariance in ref.15. Further we can show that Eq.(5.5) is consistent with the single-valuedness of g-invariant operators with respect to z.

Before deriving Eq.(5.1) or (5.5), it may be instructive to recall quantum mechanics on a circle with a radius L, where a point x is identified with $x + 2\pi n L$ for all $n \in \mathbf{Z}$. It is important to understand that the coordinate operator $\hat{x}$ itself is not well-defined on the circle while the canonical momentum operator $\hat{p}$ is well-defined. However, the following operator

$$e^{ik\hat{x}} \quad \text{with } k = \frac{m}{L} \quad (m \in \mathbf{Z}), \tag{5.6}$$

is well-defined on the circle because it is consistent with the torus identification. Physically, the operator (5.6) plays a role of a momentum shift by $k = \frac{m}{L}$. It turns out that an operator $e^{i2\pi L\hat{p}}$ commutes with all operators, i.e., $\hat{p}$ and $e^{ik\hat{x}}$, and hence it must be a c-number. Indeed we have

$$e^{i2\pi L\hat{p}} = 1. \tag{5.7}$$

This leads to the well-known result,

$$\hat{p} = \frac{m}{L} \quad \text{with} \quad m \in \mathbf{Z}. \tag{5.8}$$

In string theory on orbifolds, $\hat{p}$ and $e^{ik\hat{x}}$ will be replaced by the momentum operators $P_L^I(z)$,$P_R^I(\bar{z})$ and the vertex operator $V(k_L, k_R; z)$, respectively, and an analog of the identity (5.7) in the U-twisted sector is

$$\xi_k \, e^{-ik_L\cdot U\hat{x}_L - ik_R\cdot U\hat{x}_R} \, e^{ik_L\cdot\hat{x}_L + ik_R\cdot\hat{x}_R} \, e^{i2\pi(k_L\cdot\mathcal{P}_U\hat{p}_L - k_R\cdot\mathcal{P}_U\hat{p}_R)} = 1, \tag{5.9}$$

where

$$\xi_k = \exp\{i\pi(k_L\cdot\mathcal{P}_U k_L - k_R\cdot\mathcal{P}_U k_R) + i\frac{\pi}{2}(k_L - k_R)\cdot UC_U U^T(k_L - k_R)\}. \tag{5.10}$$

The $\mathcal{P}_U$ is a projection matrix defined by ($U^N = \mathbf{1}$)

$$\mathcal{P}_U = \frac{1}{N}\sum_{\ell=1}^{N} U^\ell. \tag{5.11}$$

It is not difficult to show that the left hand side of Eq.(5.9) commutes with all the operators by use of the commutation relations (3.9). A physical meaning of the identity (5.9) is obvious for $[U, B] = 0$. The identity (5.9) then reduces to

$$e^{i\pi k\cdot\hat{w}} e^{i\pi w\cdot U\hat{p}} = 1, \tag{5.12}$$

where

$$\begin{aligned} w^I &= k_L^I - k_R^I \in \Lambda, \\ k^I &= (1+B)^{IJ}k_L^J + (1-B)^{IJ}k_R^J \in 2\Lambda^*. \end{aligned} \tag{5.13}$$

The $\hat{p}^I$ is the canonical momentum conjugate to $\hat{x}^I$ in the U-twisted sector ‡. The identity (5.12) implies that

$$\begin{aligned} \hat{w}^I &\in \Lambda, \\ \hat{p}^I &\in 2\Lambda^*. \end{aligned} \tag{5.14}$$

‡ For the precise definition of $\hat{p}^I$ in the U-twisted sector, see Eq.(3.15) in ref.17.

These results are consistent with the identification $x^I \sim x^I + \pi w^I$ for all $w^I \in \Lambda$. For $[U, B] \neq 0$, the first relation of (5.14) still holds while the second one does not. In fact, putting $k^I = 0$ and $w^I \in \Lambda_U$ in Eq.(5.9), we have

$$\exp\{i\pi w \cdot \hat{p} - i\frac{\pi}{2} w \cdot C_U w\} = 1. \tag{5.15}$$

This is equivalent to Eq. (5.1) or (5.5).

6. Conclusions

In this paper, we have discussed some of new aspects of string theory on orbifolds in the presence of the antisymmetric constant background field B^{IJ} both in a path integral point of view and in an operator formalism point of view, although we have omitted some of the details. This work is, however, far from complete. We have shown that the transformation (2.4) is a quantum symmetry of string theory on *tori*. To clarify the full topological nature of the B^{IJ}-term, it is necessary to construct the path integral formalism of string theory on *orbifolds* in the presence of the antisymmetric background field and also beyond one-loop. For topologically nontrivial cases, i.e., $[U, B] \neq 0$, we have not yet found a Euclidean action of string theory on orbifolds, which respects the duality and modular invariance of amplitudes [17,25]. In the operator formalism, we also have to construct Hilbert spaces of twisted sectors and to verify modular invariance of amplitudes. Subtle part for constructing twisted Hilbert spaces is the zero modes, which obey the nontrivial algebras (3.9). A necessary condition for modular invariance is left-right level matching conditions [26,27]. In ref.15, the level matching conditions have been proved to be satisfied. The level matching conditions may be necessary and sufficient for modular invariance as discussed in ref. 26 and 27 although the discussions may not directly be applicable to our case. The details of this paper and the subjects addressed above will be reported elsewhere.

Acknowledgements

I would like to thank K. Ito, H. Ooguri, J. Petersen, F. Ruiz and M. Tabuse for useful discussions and also would like to acknowledge the hospitality for the Niels Bohr Institute where this work was done.

References

1. I. Frenkel and V. Kač, *Invent. Math.* **62** (1980) 23;
   G. Segal, Commun. *Math. Phys.* **80** (1981) 301;
2. P. Goddard and D. Olive, *Intern. J. Mod. Phys.* **A1** (1986) 303.
3. K. Kikkawa and M. Yamasaki, *Phys. Lett.* **B149** (1984) 357;

N. Sakai and I. Senda, *Prog. Theor. Phys.* **75** (1986) 692.
4. L. Dixon, J.A. Harvey, C. Vafa and E. Witten, *Nucl. Phys.* **B261** (1985) 678; **B274** (1986) 285.
5. I. Frenkel, J. Lepowsky and A. Meurman, Proc. Conf. on *Vertex operators in mathematics and physics*, ed. J. Lepowsky et al. (Springer 1984).
6. K.S. Narain, M.H. Sarmadi and C. Vafa, *Nucl. Phys.* **B288** (1987) 551.
7. K. Kobayashi, *Phys. Rev. Lett.* **58** (1987) 2507;
K. Kobayashi and M. Sakamoto, *Z. Phys.* **C41** (1988) 55;
M. Sakamoto, *Mod. Phys. Lett.* **A5** (1990) 1131; *Prog. Theor. Phys.* **84** (1990) 351.
8. R. Dijkgraaf, E. Verlinde and H. Verlinde, *Commun. Math. Phys.* **115** (1988) 649;
P. Ginsparg, *Nucl. Phys.* **B295** (1988) 153.
9. K.S. Narain, M.H. Sarmadi and E. Witten, *Nucl. Phys.* **B279** (1987) 369.
10. K.S. Narain, *Phys. Lett.* **B169** (1986) 41.
11. A. Font, L.E. Ibáñez, F. Quevedo and A. Sierra, *Nucl. Phys.* **B331** (1990) 421.
12. J.A. Casas and C. Muñoz, *Nucl. Phys.* **B332** (1990) 189.
13. Y. Katsuki, Y. Kawamura, T. Kobayashi, N. Ohtsubo, Y. Ono and K. Tanioka, *Nucl. Phys.* **B341** (1990) 611.
14. A. Fujitsu, T. Kitazoe, M. Tabuse and H. Nishimura, *Intern. J. Mod. Phys.* **A5** (1990) 1529.
15. M. Sakamoto and M. Tabuse, Kobe preprint KOBE-92-02 (1992);
T. Horiguchi, M. Sakamoto and M. Tabuse, Kobe preprint KOBE-92-03 (1992).
16. A.M. Polyakov, *Phys .Lett.* **B103** (1981) 207;
J. Polchinski, *Commun. Math. Phys.* **104** (1986) 37.
17. K. Inoue, S. Nima and H. Takano, *Prog. Theor. Phys.* **80** (1988) 881.
18. J. Erler, D. Jungnickel, J. Lauer and J. Mas, preprint SLAC-PUB-5602 (1991).
19. Y. Aharonov and D. Bohm, *Phys. Rev.* **115** (1959) 485.
20. F. Englert and A. Neveu, *Phys. Lett.* **B163** (1985) 349.
21. J.H. Schwarz, *Phys. Rep.* **8C** (1973) 269; **89** (1982) 223;
J. Scherk, *Rev. Mod. Phys.* **47** (1975) 123.
22. V. Kač and D.Peterson, in *Anomalies, geometry and topology,* ed. A. White (World Scientific, Singapore, 1985);
J. Lepowsky, *Proc. Nat. Acad. Sci.* (USA) **82** (1985) 8295.
23. M. Sakamoto, *Phys. Lett.* **B231** (1989) 258.
24. K. Itoh, M. Kato, H. Kunitomo and M. Sakamoto, *Nucl. Phys.* **B306** (1988) 362.
25. K. Inoue and S. Nima, *Prog. Theor. Phys.* **84** (1990) 702.
26. C. Vafa, *Nucl. Phys.* **B273** (1986) 592.
27. K.S. Narain, M.H. Sarmadi and C. Vafa, *Nucl. Phys.* **B356** (1991) 163.

Noncritical Dimensions for Critical String Theory: Life beyond the Calabi–Yau Frontier

Rolf Schimmrigk

Institut für Theoretische Physik, Universität Heidelberg
Philosophenweg 16, 6900 Heidelberg, FRG

ABSTRACT

A recently introduced framework for the compactification of supersymmetric string theory involving noncritical manifolds of complex dimension $2k + D_{crit}$, $k \geq 1$, is reviewed. These higher dimensional manifolds are spaces whose nonzero Ricci curvature is quantized in a particular way and therefore do not, a priori, describe consistent string vacua. It is nevertheless possible to derive from these manifolds the massless spectra of critical string groundstates. For a subclass of these noncritical theories it is also possible to explicitly construct Calabi–Yau manifolds from the higher dimensional spaces. Thus the new class of theories makes contact with the standard framework of string compactification. This class of manifolds is more general than that of Calabi–Yau manifolds because it contains spaces which correspond to critical string vacua with no Kähler deformations, i.e. no antigenerations, thus providing mirrors of rigid Calabi–Yau manifolds. The constructions reviewed here lead to new insight into the relation between exactly solvable models and their mean field theories on the one hand and Calabi–Yau manifolds on the other, leading, for instance, to a modification of Gepner's conjecture. They also raise fundamental questions about the Kaluza–Klein concept of string compactification, in particular regarding the rôle played by the dimension of the internal theories.

1. Introduction

String theory remains the only viable candidate for a unified theory of quantum gravity. One of the attractions of this theory is the fact that it describes a rather tight framework. A consequence is that there are severe restrictions on the internal part of the theory which to a large extent determines the observable low energy physics in four dimensions. Based on the conventional framework formulated in [1] it is believed that in left–right symmetric compactifications without torsion the internal space of the heterotic string is described by a space which has to be a compact manifold which is

- complex,
- Kähler, and admits a
- covariantly constant spinor,

i.e. has vanishing first Chern class, so-called Calabi–Yau manifolds.

Such manifolds are particularly simple, a fact that is encoded concisely in the spectrum of the theory described, in part, by the cohomology of the space. Because the space is complex the real cohomology can be decomposed, via the Hodge decomposition, into complex cohomology groups. Thus the Betti numbers $b_i = \dim_{\mathbb{R}} H^i(\mathcal{M}, \mathbb{R})$ can be expressed in terms of the Hodge numbers $h^{(p,q)} = \dim_{\mathbb{C}} H^{p,q}(\mathcal{M}, \mathbb{C})$:

$$b_i = \sum_{p+q=i} h^{(p,q)}. \tag{1}$$

Because the manifold is Kähler the Hodge numbers are symmetric, $h^{(p,q)} = h^{(q,p)}$, and because the first Chern class vanishes it follows that $h^{(p,0)} = 0 = h^{(0,p)}$ for $p = 1, 2$ and $h^{(3,0)} = 1 = h^{(0,3)}$. Hence the cohomology of the internal space, summarized in the Hodge diamond

$$\begin{array}{ccccccc}
 & & & 1 & & & \\
 & & 0 & & 0 & & \\
 & 0 & & h^{(1,1)} & & 0 & \\
1 & & h^{(2,1)} & & h^{(2,1)} & & 1 \\
 & 0 & & h^{(1,1)} & & 0 & \\
 & & 0 & & 0 & & \\
 & & & 1 & & &
\end{array}$$

contains only two independent elements $h^{(1,1)} = h^{(2,2)}$ and $h^{(2,1)} = h^{(1,2)}$ which parametrize the number of antigenerations and generations, respectively, that are observed in low energy physics.

It is also believed that this class of string vacua features an unexpected symmetry, mirror symmetry, which has been discovered in the context of Landau–Ginzburg vacua in [2]. Independent evidence for this symmetry has been found in the context of orbifolds of exactly solvable tensor models by Greene and Plesser [3]. The effect of this symmetry is that for each string vacuum with some number $h^{(1,1)}$ of antigenerations and some number $h^{(2,1)}$ of generations there exists a mirror vacuum for which these number are exchanged: the spectrum of the mirror vacuum consists of $h^{(2,1)}$ antigenerations and $h^{(1,1)}$ generations. Mirror symmetry thus flips the Hodge diamond along the offdiagonal.

Mirror symmetry is by now well established: beyond the class of exactly solvable models discussed in [3], in which mirror symmetry is understood best, lies the much larger class of Landau–Ginzburg theories constructed explicitly in [2] [1] which clearly indicates that mirror symmetry is a property of string theory. That this symmetry is not accidental in this wider context has been proven in [4] where it was shown that by a combination of orbifolding and fractional transformations a mirror construction can be established between a priori independent pairs of Landau–Ginzburg theories with opposite spectrum. Mirror symmetry is at present being used as a hypothesis to obtain

[1] The construction of all quasihomogeneous $N = 2$ Landau–Ginzburg with an isolated singularity was recently completed in [5,6].

results in algebraic geometry and has been shown to be correct in all computations that have been performed sofar [10,11,12].

This symmetry creates a puzzle. There exist well-known Calabi–Yau vacua which are rigid, i.e. they do not have string modes corresponding to complex deformations of the manifold. Since mirror symmetry exchanges complex deformations and Kähler deformations of a manifold it would seem that the mirror of a rigid Calabi–Yau manifold cannot be Kähler and hence does not describe a consistent string vacuum. In fact, it appears, using Zumino's result [13] that $N = 2$ supersymmetry of a σ–model requires that the target manifold is Kähler, that the mirror vacuum cannot even be $N = 1$ spacetime supersymmetric. It follows that the class of Calabi–Yau manifolds is not the appropriate setting by a long shot in which to discuss mirror symmetry and the question arises what the proper framework might be.

In this review I discuss recent work [14] which shows the existence of a new class of manifolds which generalizes the class of Calabi–Yau spaces of complex dimension D_{crit} in a natural way. The manifolds involved are of complex dimension $(2k + D_{crit})$ and have a positive first Chern class which is quantized in multiples of the degree of the manifold. Thus they do not describe, a priori, consistent string groundstates. Surprisingly however, it is possible to derive from these higher dimensional manifolds the spectrum of critical string vacua. This can be done not only for the generations but also for the antigenerations. For particular types of these new manifolds it is also possible to construct the corresponding D_{crit}–dimensional Calabi–Yau manifold directly from the $(2k + D_{crit})$–dimensional space.

This new class of manifolds is, however, not in one to one correspondence with the class of Calabi–Yau manifolds as it also contains manifolds which describe string vacua that do not contain massless modes corresponding to antigenerations. It is precisely this new type of manifold that is needed in order to construct mirrors of rigid Calabi–Yau manifolds without generations.

2. Higher Dimensional Manifolds with Quantized Positive First Chern Class

Consider the class of manifolds of complex dimension N embedded in a weighted projective space $\mathbb{P}_{(k_1,\ldots,k_{N+2})}$ as hypersurfaces

$$M_{N,d} \equiv \{p(z_1,\ldots,z_{N+2}) = 0\} \cap \mathbb{P}_{(k_1,\ldots,k_{N+2})} \tag{2}$$

defined as the zero set of some transverse polynomial p of degree d. Here the k_i are the weights of the ambient weighted projective space. The set of hypersurfaces determined by such polynomials will be denoted by

$$\mathbb{P}_{(k_1,k_2,\ldots\ldots,k_{N+2})}[d] \tag{3}$$

and called a configuration. Assume that for the hypersurfaces (3) the weights k_i and the degree d are related via the constraint

$$\sum_{i=1}^{N+2} k_i = Qd, \tag{4}$$

where Q is a positive integer. Relation (4) is the defining property of the class of spaces to be considered below. It is a rather restrictive condition in that it excludes many types of varieties which are transverse and even smooth but are not of physical relevance [2]. A simple example is the Fermat hypersurface

$$\mathbb{C}_{(420,280,210,168,140,120,105)}[840] \ni \{p = \sum_{i=1}^{7} z_i^{i+1} = 0\} \tag{5}$$

which is a rather nice, transverse, i.e. quasismooth manifold. It is also interesting from a different point of view. A curious aspect of Calabi–Yau hypersurfaces is that they are automatically what is called *well formed*, i.e. they do not contain orbifold singularities that are surfaces (in the case of threefolds). More generally this fact translates into the statement that the only resolutions that have to be performed are so–called small resolutions, i.e. the singular set are of codimension larger than one. The same is true for the higher dimensional manifolds defined above whereas the manifold (5) contains the singular 4–fold $S = \mathbb{C}_{(210,140,105,84,70,60)}[420]$.

Alternatively, manifolds of the type above may be characterized via a curvature constraint. Because of (4) the first Chern class is given by

$$c_1(M_{N,d}) = (Q-1)\ c_1(\mathcal{N}) \tag{6}$$

where $c_1(\mathcal{N}) = d\ h$ is the first Chern class of the normal bundle of the hypersurface $M_{N,d}$ and h is the pullback of the Kähler form $\mathrm{H} \in \mathrm{H}^{(1,1)}\left(\mathbb{P}_{(k_1,\ldots,k_{N+2})}\right)$ of the ambient space. Hence the first Chern class is quantized in multiples of the degree of the hypersurface $M_{N,d}$. For $Q = 1$ the first Chern class vanishes and the manifolds for which (4) holds are Calabi–Yau manifolds, defining consistent groundstates of the supersymmetric closed string. For $Q > 1$ the first Chern class is nonvanishing and therefore these manifolds cannot possibly describe vacua of the critical string, or so it seems.

It turns out that these spaces are closely related to string vacua of complex critical dimension

$$D_{crit} = N - 2(Q-1) \tag{7}$$

i.e. the critical dimension is offset by twice the coefficient of the first Chern class of the normal bundle of the hypersurface. The evidence for this is twofold. First it is possible to derive from these higher dimensional manifolds the massless spectrum of critical vacua. Furthermore it is shown that for certain subclasses of hypersurfaces of type (4) it is possible to construct Calabi–Yau manifolds M_{CY} of dimension D_{crit} and complex codimension

$$codim_{\mathbb{C}}(M_{CY}) = Q \tag{8}$$

directly from these manifolds. In terms of the critical dimension and the codimension the class of manifolds to be investigated below can be described as the projective

[2]It will become clear below that this definition is rather natural in the context of the theory of Landau–Ginzburg string vacua with an arbitrary number of scaling fields. A particular simple manifold in this class, the cubic sevenfold $\mathbb{P}_8[3]$, has been the subject of recent investigations [7][8][9].

configurations

$$\mathbb{P}_{(k_1,\ldots,k_{(D_{crit}+2Q)})}\left[\frac{1}{Q}\sum_{i=1}^{D_{crit}+2Q} k_i\right]. \tag{9}$$

As mentioned already in the introduction the class of spaces defined by (9) contains manifolds with no antigenerations and hence it is necessary to have some way other than Calabi–Yau manifolds to represent string groundstates if one wants to compare them with the higher dimensional manifolds. One possible way to do this is to relate them to Landau-Ginzburg theories: any manifold of type (3) can be viewed as a projectivization via a weighted equivalence defined on an affine noncompact hypersurface defined by the same polynomial

$$\mathbb{C}_{(k_1,\ldots,k_{N+2})}[d] \ni \{p(z_1,\ldots,z_{N+2})=0\}. \tag{10}$$

Because the polynomial p is assumed to be transverse in the projective ambient space the affine variety has a very mild singularity: it has an isolated singularity at the origin defining what is called a catastrophe in the mathematics literature.

The complex variables z_i parametrizing the ambient space are to be viewed as the field theoretic limit $\varphi_i(z,\bar{z}) = z_i$ of the lowest components of the order parameters $\Phi_i(z_i,\bar{z}_i,\theta_i^{\pm},\bar{\theta}_i^{\pm})$, described by chiral $N=2$ superfields of a 2–dimensional Landau–Ginzburg theory defined by the action

$$\int d^2z d^2\theta d^2\bar{\theta}\; K(\Phi_i,\bar{\Phi}_i) + \int d^2z d^2\theta\; W(\Phi_i) + c.c. \tag{11}$$

where K is the Kähler potential and W is the superpotential. It was the important insight of Martinec [15] and Vafa and Warner [16] that such Landau–Ginzburg theories are useful for the understanding of string vacua and also that much information about such groundstates is already encoded in the associated catastrophe (10). A crucial piece of information about a vacuum, e.g., is its central charge. Using a result from singularity theory, it is easy to derive that the central charge of the conformal fixed point of the LG theory is

$$c = 3\sum_{i=1}^{N+2}(1-2q_i), \tag{12}$$

where $q_i = k_i/d$ are the U(1) charges of the superfields. It is furthermore possible to derive the massless spectrum of the GSO projected fixed of the LG theory, defining the string vacuum, directly from the catastrophe (10) via a procedure described by Vafa [17].

The manifolds (9) therefore correspond to LG theories of central charge

$$c = 3(N-2(Q-1)) = 3D_{crit} \tag{13}$$

where the relation (7) has been used.

In certain benign situations the subring of monomials of charge 1 in the chiral ring describes the generations of the vacuum [18]. For this to hold at all it is important that the GSO projection is the canonical one with respect to the cyclic group $\mathbb{Z}_d$, the order of which is the degree d of the superpotential [3]. Thus the generations are easily derived for this subclass of theories in (9) because the polynomial ring is identical to the chiral ring of the corresponding Landau–Ginzburg theory. In general a more sophisticated analysis, involving the resolution of higher dimensional singularities, will have to be done [20].

It remains to extract the second cohomology. In a Calabi–Yau manifold there are no holomorphic 2–forms and hence all of the second cohomology is in $H^{(1,1)}$. Because of Kodaira's vanishing theorem the same is true for manifolds with positive first Chern class and therefore for the manifolds under discussion. At first sight it might appear hopeless to find a construction corresponding to the analysis of (2,1)–forms because of the following example which involves the orbifold of a 3–torus.

Consider the orbifold $T_1^3/\mathbb{Z}_3^2$ where the two actions are defined as $(z_1, z_4) \longrightarrow (\alpha z_1, \alpha^2 z_4)$, all other coordinates invariant and $(z_1, z_7) \longrightarrow (\alpha z_1, \alpha^2 z_7)$, all other invariant. Here α is the third root of unity. The resolution of the singular orbifold leads to a Calabi–Yau manifold with 84 antigenerations and no generations [21]. This is precisely the mirror flipped spectrum of the exactly solvable tensor model 1^9 of 9 copies of $N = 2$ superconformal minimal models at level $k = 1$ [22] which can be described in terms of the Landau–Ginzburg potential $W = \sum z_i^3$ which belongs to the configuration $\mathbb{C}_{(1,1,1,1,1,1,1,1,1)}[3]$. After imposing the GSO projection by modding out a $\mathbb{Z}_3$ symmetry this Landau–Ginzburg theory leads to the same spectrum as the 1^9 theory.

This Landau–Ginzburg theory clearly is a mirror candidate for the resolved torus orbifold just mentioned [7][8][9] and the question arises whether a manifold corresponding to this LG potential can be found. Since the theory does not contain modes corresponding to (1,1)–forms it appears that the manifold cannot be Kähler and hence not projective. Thus it appears that the 7–dimensional manifold $\mathbb{P}_8[3]$ whose polynomial ring is identical to the chiral ring of the LG theory is merely useful as an auxiliary device in order to describe one sector of the critical LG string vacuum. Even though there exists a precise identity between the Hodge numbers in the middle cohomology group of the higher dimensional manifold and the middle dimension of the cohomology of the Calabi–Yau manifold this is not the case for the second cohomology group.

3. Noncritical Manifolds and Critical Vacua

It turns out however, that by looking at the manifolds (9) in a slightly different way it is nevertheless possible to extract the second cohomology in a canonical manner (even if there is *none*). The way this works is as follows: the manifolds of type (9) will,

[3] It does not hold for projections that involve orbifolds with respect to different groups such as those discussed in [19]. This is to be expected as these modified projections can be understood as orbifolds of canonically constructed vacua. The additional moddings generate singularities the resolution of which introduces, in general, additonal modes in both sectors, generations and antigenerations.

in general, not be described by smooth spaces but will have singularities which arise from the projective identification. The basic idea now is to associate the existence of antigenerations in a *critical* string vacuum with the existence of singularities in these higher dimensional *noncritical* spaces.

Because the structure of these geometrical singularities depends on the precise form of the polynomial constraint it is difficult to prove the correctness of this idea in full generality. Instead I will, in the following, make this idea more precise and illustrate how it works with a few particular simple classes of theories, leaving a more detailed investigation of other types of manifolds for a more extensive discussion [20]. As an unexpected bonus this derivation will provide new insight into the Landau–Ginzburg/Calabi–Yau connection.

Consider again the simple example related to the tensor model 1^9. Its LG theory is described by $\mathbb{C}_9^*[3]$ the naive compactification of which leads to

$$\mathbb{P}_8[3] \ni \{\sum_{i=1}^{9} z_i^3 = 0\}. \tag{14}$$

Counting monomials leads to the spectrum of 84 generations found previously for the corresponding string vacuum and because this manifold is *smooth* there *no* antigenerations are expected in this model! Hence there does not exist a Calabi–Yau manifold that describes the groundstate [4] . A second theory in the space of all LG vacua with no antigenerations is

$$(2^6)_{A^6}^{(0,90)} \equiv \mathbb{C}^*_{(1,1,1,1,1,1,2)}[4] \ni \{\sum_{i=1}^{6} z_i^4 + z_7^2 = 0\} \tag{15}$$

with an obviously smooth manifold $\mathbb{P}_{(1,1,1,1,1,1,2)}[4]$.

Vacua without antigenerations are rather exceptional however; the generic groundstate will have both sectors, generations and antigenerations. The idea described above to derive the antigenerations works for higher dimensional manifolds corresponding to different types of critical vacua [5] but in the following we will illustrate it with two types of such manifolds.

To be concrete consider the exactly solvable tensor theory $(1 \cdot 16^3)_{A\otimes E^3}$ with 35 generations and 8 antigenerations which corresponds to a Landau–Ginzburg theory belonging to the configuration

$$\mathbb{C}^*_{(2,3,2,3,2,3,3)}[9]^{(8,35)} \tag{16}$$

and which induces, via projectivization, a 5–dimensional weighted hypersurface

$$\mathbb{P}_{(2,2,2,3,3,3,3)}[9] \ni \{p = \sum_{i=1}^{3}(y_i^3 x_i + x_i^3) + x_4^3 = 0\}. \tag{17}$$

[4]It would seem that a generalization of this 7–dimensional smooth manifold is the infinite class of models $\mathbb{C}_{(1,1,1,1,1,1,1,1,1+3q)}[3+q]$, but since the manifolds (9) are required to be transverse the only possibility is $q = 0$.

[5]A more detailed analysis can be found in [20]

with orbifold singularities

$$
\begin{aligned}
\mathbb{Z}_3 \ &: \ \mathbb{P}_3[3] \ni \{p_1 = \sum_{i=1}^{4} x_i^3 = 0\} \\
\mathbb{Z}_2 \ &: \ \mathbb{P}_2.
\end{aligned} \tag{18}
$$

The $\mathbb{Z}_3$-singular set is a smooth cubic surface which supports seven (1,1)-form as can be easily shown. The $\mathbb{Z}_2$ singular set is just the projective plane and therefore adds one further (1,1)-form. Hence the singularities induced on the hypersurface by the singularities of the ambient weighted projective space give rise to a total of eight (1,1)-forms. A simple count leads to the result that the subring of monomials of charge 1 is of dimension 35. Thus we have derived the spectrum of the critical theory from the noncritical manifold (17).

It is presumably possible to derive this result via a surgery process on the singular space (17) but more important is, at this point, that the idea introduced above of relating the spectrum of the string vacuum to the singularity structure of the noncritical manifold also makes it possible to derive from these higher dimensional manifolds the Calabi–Yau manifold of critical dimension! Thus a canonical prescription is obtained which also allows to pass from the Landau–Ginzburg theory to its geometrical counterpart.

This works as follows: Recall that the structure of the singularities of the weighted hypersurface just involved part of the superpotential, namely the cubic polynomial p_1 which determined the $\mathbb{Z}_3$ singular set described by a surface. The superpotential thus splits naturally into the two parts

$$
p = p_1 + p_2 \tag{19}
$$

where p_2 is the remaining part of the polynomial. The idea now is to consider the product $\mathbb{P}_3[3] \times \mathbb{P}_2$ where the factors are determined by the singular sets of the higher dimensional space and to impose on this 4-dimensional space a constraint described by the remaining part of the polynomial which did not take part in constraining the singularities of this ambient space. In the case at hand this leaves a polynomial of bidegree $(3,1)$ and hence we are lead to a manifold embedded in

$$
\begin{matrix} \mathbb{P}_2 \\ \mathbb{P}_3 \end{matrix} \begin{bmatrix} 3 & 0 \\ 1 & 3 \end{bmatrix} \tag{20}
$$

defined by polynomials

$$
\begin{aligned}
p_1 &= y_1^3 x_1 + y_2^3 x_2 + y_3^3 x_3 \\
p_2 &= \sum_{i=1}^{4} x_i^3
\end{aligned} \tag{21}
$$

which is precisely the manifold constructed in [23], the exactly solvable model of which was later found in [24]. Thus we have found how to construct from the noncritical manifold (17) the critical Calabi–Yau manifold.

A subclass of manifolds of a different type which can be discussed in this framework rather naturally is defined by

$$\mathbb{P}_{(2k,K-k,2k,K-k,2k_3,2k_4,2k_5)}[2K] \tag{22}$$

where $K = k + k_3 + k_4 + k_5$ and it is assumed, for simplicity, that K/k and K/k_i are integers. The potentials are

$$W = \sum_{i=1}^{2}(x_i^{K/k} + x_i y_i^2) + x_3^{K/k_3} + x_4^{K/k_4} + x_5^{K/k_5}. \tag{23}$$

The singularities in these manifolds are of two types,

$$\begin{aligned} \mathbb{Z}_2 \quad &: \quad \mathbb{P}_{(k,k,k_3,k_4,k_5)}[K] \\ \mathbb{Z}_{K-k} \quad &: \quad \mathbb{P}_1. \end{aligned} \tag{24}$$

where the constraint of degree K is given by

$$p_1 = \sum_{i=1}^{5} x^{K/k_i}. \tag{25}$$

The $\mathbb{Z}_2$-singular set is 3–fold with positive first Chern class embedded in weighted $\mathbb{P}_4$ whereas the $\mathbb{Z}_{K-k}$ singular set is just the sphere $S^2 \sim \mathbb{P}_1$.

To construct the corresponding critical manifolds note that the structure of the singularities of the weighted hypersurface just involved part of the superpotential, namely the quartic polynomial p_1 which determined the $\mathbb{Z}_2$ singularity set described by a 3–fold. The superpotential thus splits naturally into the two parts $p = p_1 + p_2$ where p_2 is the remaining part of the polynomial. The idea now is to consider again the product $\mathbb{P}_{(k,k,k_3,k_4,k_5)}[K] \times \mathbb{P}_1$ of singular sets of the higher dimensional space and to impose on this 4–dimensional space a constraint described, as before, by the remaining part p_2 of the polynomial which did not take part in constraining the singularities of this ambient space. In the case at hand this leaves a polynomial of bidegree $(k, 2)$ and hence we are lead to a manifold embedded in

$$\begin{matrix} \mathbb{P}_1 \\ \mathbb{P}_{(k,k,k_3,k_4,k_5)} \end{matrix} \begin{bmatrix} 2 & 0 \\ k & K \end{bmatrix} \tag{26}$$

defined by polynomials

$$\begin{aligned} p_1 &= y_1^2 x_1 + y_2^2 x_2 \\ p_2 &= x_1^{K/k} + x_2^{K/k} + x_3^{K/k_3} + x_4^{K/k_4} + x_5^{K/k_5}. \end{aligned} \tag{27}$$

That this correspondence is in fact correct can be inferred from the work of [25] where it was shown that these codimension–2 weighted CICYs correspond to $N = 2$ minimal exactly solvable tensor models of the type

$$\left[2\left(\frac{K}{k} - 1\right)\right]_D^2 \cdot \prod_{i=3}^{5} \left(\frac{K}{k_i} - 2\right)_A. \tag{28}$$

where the subscripts indicate the affine invariants chosen for the individual levels.

The general picture that emerges from these construction then is the following: embedded in the higher dimensional manifold is a submanifold which is fibered where the base and the fibres are determined by the singular sets of the ambient manifold. The Calabi–Yau manifold itself is a hypersurface embedded in this fibered submanifold. A heuristic sketch of the geometry is shown in the Figure 1.

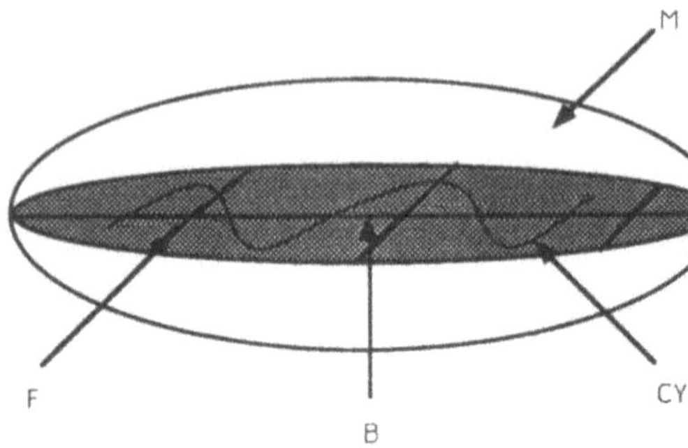

Figure 1: *The Calabi–Yau manifold CY is embedded in the fibered submanifold $B \times F$ of the higher dimensional manifold M. Both the base B and the fibres F of the bundle (indicated by the grey surface) are determined by the singular sets of the manifold M.*

The examples above illustrate the simplest situation that can appear. In more complicated manifolds the singularity structure will consist of hypersurfaces whose fibers and/or base themself are fibered, leading to an iterative procedure. The submanifold to be considered will, in those cases, be of codimension larger than one and the Calabi–Yau manifold will be described by a submanifold with codimension larger than one as well. To illustrate this point consider the 7–fold

$$\mathbb{P}_{(1,1,6,6,2,2,2,2,2)}[8] \ni \left\{ \sum_{i=1}^{2} \left(x_i^2 y_i + y_i z_i + z_i^4 \right) + z_3^4 + z_4^4 + z_5^4 = 0 \right\} \tag{29}$$

which leads to the $\mathbb{Z}_2$ fibering $\mathbb{P}_1 \times \mathbb{P}_{(3,3,1,1,1,1,1)}[4]$ which in turn leads to the $\mathbb{Z}_3$ fibering $\mathbb{P}_1 \times \mathbb{P}_1 \times \mathbb{P}_4[4]$. Following the splits of the potential thus leads to the Calabi–Yau configuration

$$\begin{matrix} \mathbb{P}_1 \\ \mathbb{P}_1 \\ \mathbb{P}_4 \end{matrix} \begin{bmatrix} 2 & 0 & 0 \\ 1 & 1 & 0 \\ 0 & 1 & 4 \end{bmatrix} \ni \left\{ \begin{array}{rl} p_1 = & \sum_{i=1}^2 x_i^2 y_i = 0 \\ p_2 = & \sum_{i=1}^2 y_i z_i = 0 \\ p_3 = & \sum_{j=1}^5 z_i^4 = 0 \end{array} \right\} \tag{30}$$

which is of codimension 3. This example also shows that there are nontrivial relations between these higher dimensional manifolds. The way to see this is via the process of splitting and contraction of Calabi–Yau manifolds introduced in ref. [26]. It can be shown in fact that the Calabi–Yau manifold (30) is an ineffective split of a Calabi–Yau manifold in the class (22). Thus there also exists a corresponding relation between the higher dimensional manifolds.

4. Generalization to Arbitrary Critical Dimensions

Even though the examples discussed in the previous section were all concerned with 6–dimensional Calabi–Yau manifolds and the way they are embedded, in a sense, in the new class of spaces, it should be clear that the ideas presented are not specific to this dimension. Instead of considering compactifications down to the physical dimension, 4, one might contemplate compactifying down to 2, 6 or 8 dimensions, or else, discuss the class of manifolds above not in the context of compactification at all.

To illustrate this point consider the infinite class of manifolds

$$\mathbb{P}_{(2,n-1,2,n-1,2,\ldots,2)}[2n] \tag{31}$$

of complex dimension $n+1$, defined by polynomials

$$p = \sum_{i=1}^{2}(x_i^n + x_i y_i^2) + x_3^n + \cdots + x_{n+1}^n. \tag{32}$$

According to the considerations above these spaces are related to Calabi–Yau manifolds embedded in

$$\begin{matrix}\mathbb{P}_1 \\ \mathbb{P}_n\end{matrix}\begin{bmatrix}2 & 0\\ 1 & n\end{bmatrix} \tag{33}$$

via the equations

$$\begin{aligned} p_1 &= y_1^2 x_1 + y_2^2 x_2 \\ p_2 &= \sum_{i=1}^{n+1} x_i^n. \end{aligned} \tag{34}$$

The simplest example is, of course, the case $n=2$ where the higher dimensional manifold is a 3–fold described by

$$\mathbb{P}_{(2,1,2,1,2)}[4] \ \ni \ \{\sum_{i=1}^{2}(z_i^2 + z_i y_i^2) + z_3^2 = 0\} \tag{35}$$

with a $\mathbb{Z}_2$ singular set isomorphic to the sphere $\mathbb{P}_2[2] \sim \mathbb{P}_1$ which contributes one (1,1)–form, the remaining one being provided by the $\mathbb{P}_1$ defined by the remaining coordinates. The singularity structure of the 3–fold then relates this space to the complex torus described by the algebraic curve

$$\begin{matrix}\mathbb{P}_1 \\ \mathbb{P}_2\end{matrix}\begin{bmatrix}2 & 0\\ 1 & 2\end{bmatrix}. \tag{36}$$

It should be remarked that the Landau–Ginzburg theory corresponding to this theory derives from an exactly solvable tensor model $(2^2)_{D^2}$ described by the product of two $N = 2$ superconformal minimal theories at level $k = 2$ equipped with the affine D–invariant.

It is of interest to consider the cohomology groups of the 3–fold itself. With the third Chern class $c_3 = 2h^3$ the Euler number of the singular space is

$$\chi_s = \int c_3 = 1 \tag{37}$$

and hence the Euler number of the resolved manifold is

$$\tilde{\chi} = 1 - (2/2) + 2 \cdot 2 = 4. \tag{38}$$

Since the singular set is a sphere its resolution contributes just one (1,1)–form and hence the second Betti number of the becomes $b_2 = 2$. With $\tilde{\chi} = 2(1 + h^{(1,1)}) - 2h^{(2,1)}$ it follows that

$$h^{(2,1)} = 1. \tag{39}$$

The case $n = 3$ is particularly illuminating because it involves a higher dimensional manifold that is smooth

$$\mathbb{P}_5[3] \tag{40}$$

and hence it is easy to determine the cohomology groups of this space and to compare it with the spectrum of K3

$$K3 = \begin{matrix} \mathbb{P}_1 \\ \mathbb{P}_3 \end{matrix} \begin{bmatrix} 2 & 0 \\ 1 & 3 \end{bmatrix}, \tag{41}$$

which consists of $h^{(0,0)} = h^{(2,2)} = h^{(2,0)} = h^{(0,2)} = 1$ and $h^{(1,1)} = 20$, all other Hodge numbers are zero. Hence the Euler number becomes $\chi(K3) = 24$.

The Euler number for the noncritical manifold is easily computed to be $\chi = 27$. Since the manifold is Kähler $h^{(p,q)} = h^{(q,p)}$ and because of Poincaré duality $b_p = b_{8-p}$. Because the manifold has positive first Chern class it follows from Kodaira's vanishing theorem that $h^{(p,0)} = 0$ for $p \neq 0$ and via Lefshetz' hyperplane theorem it is known that below the middle dimension all the cohomology is inherited from the ambient space and therefore the only nonvanishing cohomology groups lead to $h^{(0,0)} = h^{(1,1)} = 1$. It can be shown that $h^{(3,1)} = h^{(1,3)} = 1$ [6] and therefore the only remaining cohomology is in $H^{(2,2)}$. Since

$$\chi = 2(b_0 + b_2) + b_4 = 6 + h^{(2,2)} \tag{42}$$

it follows that $h^{(2,2)} = 21 = 20 + 1$. Thus we have obtained the spectrum of K3 plus one additional mode which always appears in this construction.

This example is also useful because it indicates a generalization of the considerations of the previous section. The surprising new feature of this manifold is that even though the higher dimensional manifold did not have any orbifold singularities it was nevertheless possible to split it in such a way as to construct a Calabi–Yau manifold from it. This was possible because the defining equation was not of Fermat type but

[6] I'm grateful to P.Candelas and T.Hübsch for explanations regarding this computation.

involved couplings between the fields. Because of this the manifold featured a new $\mathbb{Z}_2$ symmetry not present in the Fermat hypersurface and it is this new symmetry that dictated how to perform the split. This indicates that even for smooth higher dimensional manifolds it is possible to relate them to Calabi–Yau manifolds once one moves away from the symmetric point. Table 1 containes a few other examples of how to relate different singular 4–folds to K3 representations.

ECFT	Projective Manifold	Singularities	CY
$(1^6)_{D^2\otimes A^4}$	$\mathbb{P}_5[3]$ $\sum(z_i^3 + z_i y_i^2) + z_3^3 + z_4^3$	$\mathbb{Z}_2 : \mathbb{P}_1$	$\begin{matrix} \mathbb{P}_1 \\ \mathbb{P}_3 \end{matrix} \begin{bmatrix} 2 & 0 \\ 1 & 3 \end{bmatrix}$
$(6^2 \cdot 2)_{D^2\otimes A}$	$\mathbb{P}_{(2,3,2,3,2,4)}[8]$ $\sum(z_i^4 + z_i y_i^2) + z_3^4 + z_4^2$	$\mathbb{Z}_2 : \mathbb{P}_{(1,1,1,2)}[4]$ $\mathbb{Z}_3 : \mathbb{P}_1$	$\begin{matrix} \mathbb{P}_1 \\ \mathbb{P}_{(1,1,1,2)} \end{matrix} \begin{bmatrix} 2 & 0 \\ 1 & 4 \end{bmatrix}$
$(10^2 \cdot 1)_{D^2\otimes A}$	$\mathbb{P}_{(2,5,2,5,4,6)}[12]$ $\sum(z_i^6 + z_i y_i^2) + z_3^3 + z_4^2$	$\mathbb{Z}_2 : \mathbb{P}_{(1,1,2,3)}[6]$ $\mathbb{Z}_5 : \mathbb{P}_1$	$\begin{matrix} \mathbb{P}_1 \\ \mathbb{P}_{(1,1,2,3)} \end{matrix} \begin{bmatrix} 2 & 0 \\ 1 & 6 \end{bmatrix}$

Table 1: *Examples of dimension* $D_{\mathbb{C}} = 2$

An example involving a 4–dimensional critical manifold of a different type is defined by the polynomial

$$p = \sum_{i=1}^{3} \left(x_i^3 + x_i y_i^3\right) + \sum_{j=4}^{5} x_i^6 \tag{43}$$

which corresponds to the tensor model $(16^3 \cdot 4)_{E^3\otimes D^2}$ with central charge $c = 12$ and belongs to the configuration

$$\mathbb{P}_{(6,4,6,4,6,4,3,3)}[18]. \tag{44}$$

The critical manifold derived from this 6–fold belongs to the configuration class

$$\begin{matrix} \mathbb{P}_2 \\ \mathbb{P}_{(2,2,2,1,1)} \end{matrix} \begin{bmatrix} 3 & 0 \\ 2 & 6 \end{bmatrix} \tag{45}$$

which is indeed a Calabi–Yau deformation class.

Again it should be emphasized that the construction is not restricted to the infinite series defined in (32) as the final example illustrates. A five–dimensional crtical vacuum is obtained by considering the Landau–Ginzburg potential

$$W = \sum_{j=1}^{2} \left(u_i^3 + u_i v_i^2\right) + \sum_{i=3}^{5} \left(u_i^3 + u_i w_i^3\right) \tag{46}$$

which corresponds to the exactly solvable model $(16^3 \cdot 4^2)_{E_7^3\otimes D^2}$. The nine–dimensional noncritical manifold

$$\mathbb{P}_{(3,2,3,2,3,2,3,3,3,3)}[9] \tag{47}$$

leads, via its singularity structure, to the five-dimensional critical manifold

$$\begin{matrix} \mathbb{P}_1 \\ \mathbb{P}_2 \\ \mathbb{P}_4 \end{matrix} \begin{bmatrix} 2 & 0 & 0 \\ 0 & 3 & 0 \\ 1 & 1 & 3 \end{bmatrix}. \tag{48}$$

It is again crucial that a non-Fermat polynomial was chosen for the last four coordinates in the noncritical manifold.

5. Conclusion

Mirror symmetry cannot be understood in the framework of Calabi–Yau manifolds. Thus, beyond the class of such spaces, there must exist a space of a new type of noncritical manifolds which contain information about critical vacua, such as the mirrors of rigid Calabi–Yau manifolds. Mirrors of spaces with both sectors, antigeneration and generations, are again of Calabi–Yau type and hence the noncritical manifolds which correspond to such groundstates should make contact with Calabi–Yau manifolds in some manner.

What has been shown in [14] is that the class (9) of higher dimensional Kähler manifolds with positive first Chern class, quantized in a particular way, generalizes the framework of Calabi–Yau vacua in the desired way: For particular types of such noncritical manifolds Calabi–Yau manifolds of critical dimension are embedded algebraically in a fibered submanifold. For string vacua which cannot be described by Kähler manifolds and which are mirror candidates of rigid Calabi–Yau manifolds the higher dimensional manifolds still lead to the spectrum of the critical vacuum and a rationale emerges that explains why a Calabi–Yau representation is not possible in such theories. Thus these manifolds of dimension $c/3 + 2k$ define an appropriate framework in which to discuss mirror symmetry.

There are a number of important consequences that follow from the results of the previous sections. First it should be realized that the relevance of noncritical manifolds suggests the generalization of a conjecture regarding the relation between (2,2) superconformal field theories of central charge $c = 3D$, $D \in \mathbb{N}$, with N=1 spacetime supersymmetry on the one hand and Kähler manifolds of complex dimension D with vanishing first Chern class on the other. It was suggested by Gepner that this relation is 1–1. It follows from the results above that instead superconformal theories of the above type are in correspondence with Kähler manifolds of dimension $c/3 + 2k$ with a first Chern class quantized in multiples of the degree.

A second consequence is that the ideas of section 3 lead, for a large class of Landau–Ginzburg theories, to a new canonical prescription for the construction of the critical manifold, if it exists, directly from the 2D field theory.

Recently Batyrev [27] introduced a new construction of mirrors of Calabi–Yau manifolds based on dual polyhedra. His method appears to apply only to manifolds defined by one polynomial in a weighted projective space or products thereof. The method of toric geometry that is used in [27] is however not restricted to Calabi–Yau manifolds [28] and therefore the constructions described in sections 3 and 4 lead to the exciting possibility of extending Batyrev's results to Calabi–Yau manifolds of codimension larger than one by proceeding via noncritical manifolds.

A final remark is that in this framework the role played by the dimension of the manifolds becomes of secondary importance. This is as it should be, at least for an effective theory, which tests only matter content and couplings. It is, however, somewhat mysterious that via ineffective splittings manifolds of different dimension describe one and the same critical vacuum.

It is clear that the emergence in string theory of manifolds with quantized first Chern class should be understood better. The results described here are a first step in this direction. They indicate that these manifolds are not just auxiliary devices but may be as physical as Calabi–Yau manifolds of critical dimension. In order to probe the structure of these models in more depth it is important to get further insight into the complete spectrum of these theories and to compute the Yukawa couplings of the fields. It is clear from the results presented here that the spectra of the higher dimensional manifolds contain additional modes beyond those that are related to the generations and antigenerations of the critical vacuum and the question arises what physical interpretation these fields have.

A better grasp on the complete spectrum of these spaces should also give insight into a different, if not completely independent, approach toward a deeper understanding of these higher dimensional manifold, which is to attempt the construction of consistent σ–models defined via these spaces. Control of the complete spectrum will shed light on the precise relation between the σ–models associated to the noncritical manifolds and critical σ–models.

5. Acknowledgements

I'm grateful to CERN for hospitality and the theoretical theorists there for lively discussions, in particular Per Berglund, Philip Candelas, Wolfgang Lerche and Jan Louis. It is a pleasure to thank Herbert Clemens, Tristan Hübsch, Cumrun Vafa and Nick Warner for conversations and R.Hartshorne for asking the right question. I'm also grateful to the Aspen Center of Physics for hospitality.

6. References

1. P.Candelas, G.Horowitz, A.Strominger and E.Witten, Nucl.Phys. **B258**(1985)49
2. P. Candelas, M.Lynker and R.Schimmrigk, Nucl.Phys. **B341**(1990)383
3. B.R.Greene and R.Plesser, Nucl.Phys. **B338**(1990)15
4. M.Lynker and R.Schimmrigk, Phys.Lett. **249B**(1990)237
5. A.Klemm and R.Schimmrigk, *Landau–Ginzburg String Vacua*, CERN preprint CERN–TH–6459/92 and Heidelberg preprint HD–THEP–92–13
6. M.Kreuzer and H.Skarke, *No Mirror Symmetry among Landau–Ginzburg Vacua*, CERN preprint CERN–TH–6461/92

7. P.Candelas, Talk at the Workshop on Geometry and Quantum Field Theory, Baltimore, March 1992
8. P.Candelas, E.Derrick and L.Parkes, in preparation
9. C.Vafa, *Topological Mirrors and Quantum Rings*, Harvard preprint HUTP-91/A059
10. P.Candelas, X.de la Ossa, P.Green and L.Parkes, Phys.Lett **258B**(1991)118; Nucl.Phys. **B359**(1991)21
11. G.Ellingsrud and S.A.Stromme, *The Number of Twisted Cubic Curves on the General Quintic 3-Fold*
12. S.Katz, private communication
13. B.Zumino, Phys.Lett. **87B**(1979)203
14. R.Schimmrigk, *Critical String Vacua from Noncritical Manifolds: A Novel Framework for String Compactification*, Heidelberg preprint HD-THEP-92-29
15. E.Martinec, Phys.Lett. **217**(1989)431
16. C.Vafa and N.Warner, Phys.Lett. **218**(1989)51
17. C.Vafa, Mod.Phys.Lett. **A4**(1989)1169
18. P.Candelas, Nucl.Phys. **B298**(1988)458
19. P.Berglund, B.R.Greene and T.Hübsch, *Classical vs. Landau-Ginzburg Geometry of Compactification*, CERN preprint, CERN-TH-6381/92
20. R.Schimmrigk, work in progress
21. B.R.Greene, C.Vafa and N.Warner, Nucl.Phys. **B324**(1989)371
22. D.Gepner, Nucl.Phys. **B296**(1988)757
23. R.Schimmrigk, Phys.Lett. **193B**(1987)175
24. D.Gepner, *String Theory on Calabi-Yau Manifolds: The Three Generation Case*, Princeton University preprint, December 1987
25. R.Schimmrigk, Phys.Lett. **229B**(1989)227
26. P.Candelas, A.Dale, C.A.Lütken and R.Schimmrigk, Nucl.Phys. **B298**(1988)493
27. V.V.Batyrev, *Dual Polyhedra and Mirror Symmetry for Calabi-Yau Hypersurfaces in Toric Varieties*, University of Essen preprint
28. V.V.Batyrev, *Variations of the Mixed Hodge Structure of Affine Hypersurfaces in Algebraic Tori*, University of Essen preprint

DILATON–AXION SYMMETRY

JOHN H. SCHWARZ
California Institute of Technology
Physics Department, Pasadena, California 91125, U.S.A.

ABSTRACT

The heterotic string compactified on a six-torus is described by a low-energy effective action consisting of N=4 supergravity coupled to N=4 super Yang-Mills, a theory that was studied in detail many years ago. By explicitly carrying out the dimensional reduction of the massless fields, we obtain the bosonic sector of this theory. In the Abelian case the action is written with manifest global O(6,6+n) symmetry. A duality transformation that replaces the antisymmetric tensor field by an axion brings it to a form in which the axion and dilaton parametrize an SL(2,R)/SO(2) coset, and the equations of motion have SL(2,R) symmetry. This symmetry, which combines Peccei–Quinn translations with Montonen–Olive duality transformations, has been exploited in several recent papers to construct black hole solutions carrying both electric and magnetic charge. Our purpose is to explore whether, as various authors have conjectured, an SL(2,Z) subgroup could be an exact symmetry of the full quantum string theory. If true, this would be of fundamental importance, since this group transforms the dilaton nonlinearly and can relate weak and strong coupling.

1. Introduction

The unexpected appearance of noncompact global symmetries was one of the most intriguing discoveries to emerge from the study of supergravity theories in the 1970's. The first appearance of a noncompact symmetry was the discovery of a global $SU(1,1)$ invariance in an appropriate formulation of $N = 4$, $D = 4$ supergravity.[1] The qualification "appropriate formulation" refers to the fact that duality transformations allow n-forms to be recast as $(D-n-2)$-forms in D dimensions ($d\tilde{A} = *dA$), interchanging the role of Bianchi identities and equations of motion. Only after appropriate transformations is the full noncompact symmetry exhibited. In the $SU(1,1)$ theory there are two scalar fields, nowadays called the "dilaton" and the "axion," which parametrize the coset space $SU(1,1)/U(1)$.

We shall also focus on theories with half the maximum possible supersymmetry ($N = 1$ in $D = 10$ or $N = 4$ in $D = 4$), as these are most relevant to heterotic string theory. Ref. 2 showed that 10-dimensional $N = 1$ supergravity, dimensionally reduced to $D \geq 4$ dimensions (by dropping the dependence of the fields on $d = 10-D$ dimensions), has global $O(d,d)$ symmetry. Moreover, when the original $N = 1$, $D = 10$ theory has n Abelian vector supermultiplets, in addition to the supergravity multiplet, the global symmetry of the dimensionally reduced theory becomes extended to $O(d, d+n)$. The coupling of $N = 1$, $D = 10$ supergravity to vector supermultiplets require the inclusion of a Chern–Simons term ($H = dB - \omega_3$) in order to achieve supersymmetry. This was shown in the Abelian case by Bergshoeff *et al.*,[3] and in the non-Abelian case by Chapline and Manton.[4]

In this paper we will focus on the bosonic sector, which can be formulated in any dimension. In section 2 we show that dimensional reduction from $D+d$ dimensions to D dimensions gives rise to a theory with global $O(d, d)$ symmetry when there are no vector fields in $D+d$ dimensions. Adding n Abelian vector fields in $D+d$ dimensions gives rise to a dimensionally reduced theory with $O(d, d+n)$ symmetry, provided that the Chern–Simons term (described above) is included. This construction has been discussed in detail elsewhere[5] and is summarized here to present the theory that is studied in the subsequent sections.

In section 3 we specialize to the $D = 4$ case and review the duality transformation that replaces the antisymmetric two-form field by the axion and gives rise to global $SL(2, R)$ (or $SU(1, 1)$) symmetry of the equations of motion. The symmetry is not present in the action for those terms involving the vector gauge fields. In section 4 we explore whether this symmetry could be generally valid in the heterotic string theory compactified to four dimensions, or whether it is a special feature of the low-energy effective action. The question is both subtle and profound, because the symmetry gives a nonlinear transformation of the the dilaton, whose expectation value gives the coupling constant (loop expansion parameter). Thus, even if the symmetry is exact, one should not expect to find it order-by-order in perturbation theory. By the same token, the question is certainly of fundamental importance, since such a symmetry is potentially a powerful tool for obtaining non-perturbative information about the theory. We examine the classical string equations of motion in the presence of appropriate background fields and demonstrate that the linearly realized subgroup of $SL(2, R)$ is a symmetry, but the full group is not. However, this is all the symmetry that should appear at this order, so the question remains open.

2. Noncompact Global Symmetry from Dimensional Reduction

In the 1970's it was noted that noncompact global symmetries are a generic feature of supergravity theories containing scalar fields. One of the useful techniques that was exploited in these studies was the method of "dimensional reduction." In its simplest form, this consists of considering a theory in a spacetime M×K, where M has D dimensions and K has d dimensions, and supposing that the fields are independent of the coordinates y^α of K. For this to be a consistent procedure it is necessary that K-independent solutions be able to solve the classical field equations. Then one speaks of "spontaneous compactification" (at least when K is compact). In a gravity theory this implies that K is flat, a torus for example. Of course, in recent times more interesting possibilities, such as Calabi–Yau spaces, have received a great deal of attention. In such a case, the analog of dropping y dependence is to truncate all fields to their zero modes on K.

Explicit formulas for dimensional reduction were given in a 1979 paper by Joël Scherk and me[6] and subsequently explored by Cremmer.[7] The main purpose of ref. 6 was to introduce a "generalized" method of dimensional reduction, but here we will stick to the simplest case in which the fields are taken to be independent of the K coordinates. The notation is as follows: Local coordinates of M are x^μ ($\mu = 0, 1, \ldots, D-1$) and local coordinates of K are y^α ($\alpha = 1, \ldots, d$). The tangent space Lorentz metric has signature $(-+\ldots+)$. All fields in $D+d$ dimensions are written with hats on the fields and the indices ($\hat{\phi}$, $\hat{g}_{\hat{\mu}\hat{\nu}}$, etc.). Quantities without hats are reserved for D dimensions. Thus, for example, the Einstein action on M×K (with a dilaton field $\hat{\phi}$) is

$$S_{\hat{g}} = \int\limits_M dx \int\limits_K dy \, \sqrt{-\hat{g}} \, e^{-\hat{\phi}} \left[\hat{R}(\hat{g}) + \hat{g}^{\hat{\mu}\hat{\nu}} \partial_{\hat{\mu}} \hat{\phi} \partial_{\hat{\nu}} \hat{\phi} \right] \ . \tag{1}$$

If K is assumed to be a torus we can choose the coordinates y^α to be periodic with unit periods, so that $\int_K dy = 1$. The radii and angles that characterize the torus are then encoded in the metric tensor. In terms of a $(D+d)$-dimensional vielbein, we can use local Lorentz invariance to choose a triangular parametrization

$$\hat{e}_{\hat{\mu}}^{\hat{r}} = \begin{pmatrix} e_\mu^r & A_\mu^{(1)\beta} E_\beta^a \\ 0 & E_\alpha^a \end{pmatrix} . \tag{2}$$

The "internal" metric is $G_{\alpha\beta} = E_\alpha^a \delta_{ab} E_\beta^b$ and the "spacetime" metric is $g_{\mu\nu} = e_\mu^r \eta_{rs} e_\nu^s$. As usual, $G^{\alpha\beta}$ and $g^{\mu\nu}$ represent inverses. In terms of these quantities the complete $(D+d)$-dimensional metric is

$$\hat{g}_{\hat{\mu}\hat{\nu}} = \begin{pmatrix} g_{\mu\nu} + A_{\mu\gamma}^{(1)} G^{\gamma\delta} A_{\nu\delta}^{(1)} & A_{\mu\beta}^{(1)} \\ A_{\nu\alpha}^{(1)} & G_{\alpha\beta} \end{pmatrix} . \tag{3}$$

A convenient property of this parametrization is that $\sqrt{-\hat{g}} = \sqrt{-g}\sqrt{\det G}$. If all fields are assumed to be y independent, one finds the D-dimensional action

$$\begin{aligned} S_{\hat{g}} = \int\limits_M dx \sqrt{-g} \, e^{-\phi} \Big\{ & R + g^{\mu\nu} \partial_\mu \phi \partial_\nu \phi \\ & + \frac{1}{4} g^{\mu\nu} \partial_\mu G_{\alpha\beta} \partial_\nu G^{\alpha\beta} - \frac{1}{4} g^{\mu\rho} g^{\nu\lambda} G_{\alpha\beta} F_{\mu\nu}^{(1)\alpha} F_{\rho\lambda}^{(1)\beta} \Big\} , \end{aligned} \tag{4}$$

where we have introduced a shifted dilaton field[8,9] $\phi = \hat{\phi} - \frac{1}{2}\log \det G_{\alpha\beta}$ and $F_{\mu\nu}^{(1)\alpha} = \partial_\mu A_\nu^{(1)\alpha} - \partial_\nu A_\mu^{(1)\alpha}$.

Another field that is of interest in string theory is a second-rank antisymmetric tensor $\hat{B}_{\hat{\mu}\hat{\nu}}$ with field strength $\hat{H}_{\hat{\mu}\hat{\nu}\hat{\rho}} = \partial_{\hat{\mu}}\hat{B}_{\hat{\nu}\hat{\rho}} + \text{cyc. perms}$. The Chern–Simons terms that appear in superstring theory are not present here, since we are not yet including $(D+d)$-dimensional vector fields. The Lorentz Chern–Simons term[10] is higher order in derivatives than we are considering. The action for the $\hat{B}$ term is

$$S_{\hat{B}} = -\frac{1}{12}\int\limits_M dx \int\limits_K dy\, \sqrt{-\hat{g}}\, e^{-\hat{\phi}}\, \hat{g}^{\hat{\mu}\hat{\mu}'}\, \hat{g}^{\hat{\nu}\hat{\nu}'}\, \hat{g}^{\hat{\rho}\hat{\rho}'}\, \hat{H}_{\hat{\mu}\hat{\nu}\hat{\rho}}\, \hat{H}_{\hat{\mu}'\hat{\nu}'\hat{\rho}'}\, . \tag{5}$$

Again dropping y dependence, one finds that

$$S_{\hat{B}} = -\int\limits_M dx\sqrt{-g}\, e^{-\phi}\left\{\frac{1}{4}H_{\mu\alpha\beta}H^{\mu\alpha\beta} + \frac{1}{4}H_{\mu\nu\alpha}H^{\mu\nu\alpha} + \frac{1}{12}H_{\mu\nu\rho}H^{\mu\nu\rho}\right\}, \tag{6}$$

where $H_{\mu\alpha\beta} = \partial_\mu B_{\alpha\beta}$ and $H_{\mu\nu\alpha} = F^{(2)}_{\mu\nu\alpha} - B_{\alpha\beta}F^{(1)\beta}_{\mu\nu}$. Also, $\hat{B}_{\alpha\beta} = B_{\alpha\beta}$, $F^{(2)}_{\mu\nu\alpha} = \partial_\mu A^{(2)}_{\nu\alpha} - \partial_\nu A^{(2)}_{\mu\alpha}$, and $A^{(2)}_{\mu\alpha} = \hat{B}_{\mu\alpha} + B_{\alpha\beta}A^{(1)\beta}_\mu$. The gauge transformations of the vector fields are simply $\delta A^{(1)\alpha}_\mu = \partial_\mu\Lambda^{(1)\alpha}$ and $\delta A^{(2)}_{\mu\alpha} = \partial_\mu\Lambda^{(2)}_\alpha$, under which $H_{\mu\nu\alpha}$ is invariant. Also,

$$H_{\mu\nu\rho} = \partial_\mu B_{\nu\rho} - \frac{1}{2}(A^{(1)\alpha}_\mu\, F^{(2)}_{\nu\rho\alpha} + A^{(2)}_{\mu\alpha}F^{(1)\alpha}_{\nu\rho}) + \text{cyc. perms.}, \tag{7}$$

where

$$B_{\mu\nu} = \hat{B}_{\mu\nu} + \frac{1}{2}A^{(1)\alpha}_\mu A^{(2)}_{\nu\alpha} - \frac{1}{2}A^{(1)\alpha}_\nu A^{(2)}_{\mu\alpha} - A^{(1)\alpha}_\mu B_{\alpha\beta}A^{(1)\beta}_\nu\, . \tag{8}$$

In this case gauge invariance of Eq. (7) requires that under the $\Lambda^{(1)}$ and $\Lambda^{(2)}$ transformations $\delta B_{\mu\nu} = \frac{1}{2}(\Lambda^{(1)\alpha}F^{(2)}_{\mu\nu\alpha} + \Lambda^{(2)}_\alpha F^{(1)\alpha}_{\mu\nu})$. The extra terms in $H_{\mu\nu\rho}$, which have arisen as a consequence of the dimensional reduction, are Abelian Chern–Simons terms.

To recapitulate the results so far, the dimensionally reduced form of $S = S_{\hat{g}} + S_{\hat{B}}$ has been written in the form

$$S = \int\limits_M dx\sqrt{-g}\, e^{-\phi}(\mathcal{L}_1 + \mathcal{L}_2 + \mathcal{L}_3 + \mathcal{L}_4)\, , \tag{9}$$

where

$$
\begin{aligned}
\mathcal{L}_1 &= R + g^{\mu\nu}\partial_\mu\phi\partial_\nu\phi \\
\mathcal{L}_2 &= \frac{1}{4}g^{\mu\nu}(\partial_\mu G_{\alpha\beta}\partial_\nu G^{\alpha\beta} - G^{\alpha\beta}G^{\gamma\delta}\partial_\mu B_{\alpha\gamma}\partial_\nu B_{\beta\delta}) \\
\mathcal{L}_3 &= -\frac{1}{4}g^{\mu\rho}g^{\nu\lambda}(G_{\alpha\beta}F^{(1)\alpha}_{\mu\nu}F^{(1)\beta}_{\rho\lambda} + G^{\alpha\beta}H_{\mu\nu\alpha}H_{\rho\lambda\beta}) \\
\mathcal{L}_4 &= -\frac{1}{12}H_{\mu\nu\rho}H^{\mu\nu\rho}.
\end{aligned}
\tag{10}
$$

We now claim that there is an $O(d,d)$ global symmetry that leaves each of these four terms separately invariant. The first term ($\mathcal{L}_1$) is trivially invariant since $g_{\mu\nu}$ and ϕ are. It should be noted, however, that the individual terms in $\phi = \hat{\phi} - \frac{1}{2}\log\det G_{\alpha\beta}$ are not invariant. To investigate the invariance of $\mathcal{L}_2$ we first rewrite it, using matrix notation, as

$$
\mathcal{L}_2 = \frac{1}{4}\mathrm{tr}(\partial_\mu G^{-1}\partial^\mu G + G^{-1}\partial_\mu B G^{-1}\partial^\mu B) . \tag{11}
$$

Then we introduce two $2d \times 2d$ matrices, written in $d \times d$ blocks, as follows:[12]

$$
\eta = \begin{pmatrix} 0 & 1 \\ 1 & 0 \end{pmatrix} \quad \text{and} \quad M = \begin{pmatrix} G^{-1} & -G^{-1}B \\ BG^{-1} & G - BG^{-1}B \end{pmatrix} \tag{12}
$$

Since η has d eigenvalues $+1$ and d eigenvalues -1, it is a metric for the group $O(d,d)$ in a basis rotated from the one with a diagonal metric. Next we note that $M \in O(d,d)$, since $M^T\eta M = \eta$. In fact, M is a *symmetric* $O(d,d)$ matrix, which implies that $M^{-1} = \eta M \eta$. It is now a simple exercise to verify that

$$
\mathcal{L}_2 = \frac{1}{8}\mathrm{tr}(\partial_\mu M^{-1}\partial^\mu M) . \tag{13}
$$

Thus $\mathcal{L}_2$ is invariant under a global $O(d,d)$ transformation $M \to \Omega M \Omega^T$, where $\Omega^T\eta\Omega = \eta$. This transformation acts nonlinearly on $G_{\alpha\beta}$ and $B_{\alpha\beta}$, which parametrize the coset space $O(d,d)/O(d) \times O(d)$.

Next we consider the $\mathcal{L}_3$ term:

$$
\begin{aligned}
\mathcal{L}_3 &= -\frac{1}{4}[F^{(1)\alpha}_{\mu\nu}G_{\alpha\beta}F^{(1)\mu\nu\beta} + (F^{(2)}_{\mu\nu\alpha} - B_{\alpha\gamma}F^{(1)\gamma}_{\mu\nu})G^{\alpha\beta}(F^{(2)\mu\nu}_{\beta} - B_{\beta\delta}F^{(1)\mu\nu\delta})] \\
&= -\frac{1}{4}\mathcal{F}^i_{\mu\nu}(M^{-1})_{ij}\mathcal{F}^{\mu\nu j} ,
\end{aligned}
\tag{14}
$$

where $\mathcal{F}^i_{\mu\nu}$ is the $2d$-component vector of field strengths

$$\mathcal{F}^i_{\mu\nu} = \begin{pmatrix} F^{(1)\alpha}_{\mu\nu} \\ F^{(2)}_{\mu\nu\alpha} \end{pmatrix} = \partial_\mu \mathcal{A}^i_\nu - \partial_\nu \mathcal{A}^i_\mu \, . \tag{15}$$

Thus $\mathcal{L}_3$ is manifestly $O(d,d)$ invariant provided that the vector fields transform according to the vector representation of $O(d,d)$, *i.e.*, $\mathcal{A}^i_\mu \to \Omega^i{}_j \mathcal{A}^j_\mu$. The demonstration of $O(d,d)$ symmetry is completed by noting that $\mathcal{L}_4$ is invariant (if we require that $B_{\nu\rho}$ is invariant), since $H_{\mu\nu\rho}$ can be written in the manifestly invariant form

$$H_{\mu\nu\rho} = \partial_\mu B_{\nu\rho} - \frac{1}{2}\mathcal{A}^i_\mu \eta_{ij} \mathcal{F}^j_{\nu\rho} + (\text{cyc. perms.}) \, . \tag{16}$$

Previous work in supergravity[13,14] and superstring theory[15] suggests that if we add n Abelian $U(1)$ gauge fields to the original $(D+d)$-dimensional theory, that $O(d,d+n)$ symmetry should result from dimensional reduction to D dimensions. The additional term to be added to the action is

$$S_{\hat{A}} = -\frac{1}{4}\int\limits_M dx \int\limits_K dy \, \sqrt{-\hat{g}}\, e^{-\hat{\phi}}\, \hat{g}^{\hat{\mu}\hat{\rho}} \hat{g}^{\hat{\nu}\hat{\lambda}} \delta_{IJ} \hat{F}^I_{\hat{\mu}\hat{\nu}}\, \hat{F}^J_{\hat{\rho}\hat{\lambda}} \, , \tag{17}$$

where $\hat{F}^I_{\hat{\mu}\hat{\nu}} = \partial_{\hat{\mu}} \hat{A}^I_{\hat{\nu}} - \partial_{\hat{\nu}} \hat{A}^I_{\hat{\mu}}$ and the index I takes the values $I = 1, 2, \cdots, n$. The most important point to note is that the original $(D+d)$-dimensional theory should have $O(n)$ symmetry with $M_{IJ} = \eta_{IJ} = \delta_{IJ}$. Looking at the various pieces of the Lagrangian, we see that $\mathcal{L}_1$ has the usual form, $\mathcal{L}_2 = 0$, and $\mathcal{L}_3$ gives $S_{\hat{A}}$. The crucial observation concerns $\mathcal{L}_4$, which is built from the square of $\hat{H}_{\hat{\mu}\hat{\nu}\hat{\rho}}$. This contains the Chern–Simons term (for the $U(1)$ gauge fields), a feature that is clearly crucial for the symmetries we wish to implement.

The dimensional reduction of $S_{\hat{g}}$ is unchanged from before. For the vectors we obtain

$$S_{\hat{A}} = -\frac{1}{4}\int dx \sqrt{-g}\, e^{-\phi}\Big\{ F^I_{\mu\nu} F^{I\mu\nu} + 2F^I_{\mu\alpha} F^{I\mu\alpha} \Big\} \, , \tag{18}$$

where we define

$$\begin{aligned} A^{(3)I}_\mu &= \hat{A}^I_\mu - a^I_\alpha A^{(1)\alpha}_\mu \\ F^{(3)I}_{\mu\nu} &= \partial_\mu A^{(3)I}_\nu - \partial_\nu A^{(3)I}_\mu \\ a^I_\alpha &= \hat{A}^I_\alpha \\ F^I_{\mu\nu} &= F^{(3)I}_{\mu\nu} + F^{(1)\alpha}_{\mu\nu} a^I_\alpha \\ F^I_{\mu\alpha} &= \partial_\mu a^I_\alpha \, . \end{aligned} \tag{19}$$

The reduction of $S_{\hat{B}}$ is still given by Eq. (6), but including the Chern–Simons term gives

$$H_{\mu\alpha\beta} = \partial_\mu B_{\alpha\beta} + \frac{1}{2}(a^I_\alpha \partial_\mu a^I_\beta - a^I_\beta \partial_\mu a^I_\alpha)$$
$$H_{\mu\nu\alpha} = -C_{\alpha\beta} F^{(1)\beta}_{\mu\nu} + F^{(2)}_{\mu\nu\alpha} - a^I_\alpha F^{(3)I}_{\mu\nu} \tag{20}$$
$$H_{\mu\nu\rho} = \partial_\mu B_{\nu\rho} - \frac{1}{2}\mathcal{A}^i_\mu \eta_{ij} \mathcal{F}^j_{\nu\rho} + \text{cyc. perms.}\ ,$$

where we have used the definitions

$$A^{(2)}_{\mu\alpha} = \hat{B}_{\mu\alpha} + B_{\alpha\beta} A^{(1)\beta}_\mu + \frac{1}{2} a^I_\alpha A^{(3)I}_\mu \tag{21}$$

$$C_{\alpha\beta} = \frac{1}{2} a^I_\alpha a^I_\beta + B_{\alpha\beta}\ . \tag{22}$$

and $A^{(3)I}_\mu A^{(1)\alpha}_\nu a^I_\alpha - A^{(3)I}_\nu A^{(1)\alpha}_\mu a^I_\alpha$ should be added to the definition of $B_{\mu\nu}$ in Eq. (8). We have introduced a $(2d+n)$-component vectors $\mathcal{A}^i_\mu = (A^{(1)\alpha}_\mu,\ A^{(2)}_{\mu\alpha},\ A^{(3)I}_\mu)$ and $\mathcal{F}^i_{\mu\nu} = \partial_\mu \mathcal{A}^i_\nu - \partial_\nu \mathcal{A}^i_\mu$. The $O(d, d+n)$ metric η, written in blocks, now takes the form

$$\eta = \begin{pmatrix} 0 & 1 & 0 \\ 1 & 0 & 0 \\ 0 & 0 & 1 \end{pmatrix}\ . \tag{23}$$

With these definitions, $H_{\mu\nu\rho}$ has manifest $O(d, d+n)$ symmetry.

Now we can combine all terms that are quadratic in field strengths in the form

$$\mathcal{L}_3 = -\frac{1}{4}\ \mathcal{F}^i_{\mu\nu} (M^{-1})_{ij}\ \mathcal{F}^{j\mu\nu}\ . \tag{24}$$

Contributions come from $S_{\hat{g}}$ (as before), from $\frac{1}{4} F^I_{\mu\nu} F^{I\mu\nu}$, and from $\frac{1}{4} H_{\mu\nu\alpha} H^{\mu\nu\alpha}$. Altogether, we read off the result

$$M^{-1} = \begin{pmatrix} G + C^T G^{-1} C + a^T a & -C^T G^{-1} & C^T G^{-1} a^T + a^T \\ -G^{-1} C & G^{-1} & -G^{-1} a^T \\ a G^{-1} C + a & -a G^{-1} & 1 + a G^{-1} a^T \end{pmatrix} \tag{25}$$

Since $M^{-1}\eta M^{-1}\eta = 1$, M^{-1} and $M = \eta M^{-1} \eta$ are symmetric $O(d, d+n)$ matrices.

The last remaining check of $O(d,d+n)$ symmetry is to verify that we recover $\mathcal{L}_2 = \frac{1}{8}\mathrm{tr}(\partial_\mu M^{-1}\partial^\mu M)$, with the matrix M given above. Relevant contributions come from $S_{\hat{g}}$, $-\frac{1}{2}(F^I_{\mu\alpha})^2$, and $-\frac{1}{4}(H_{\mu\alpha\beta})^2$. The calculation is a bit tedious, but the desired result is obtained.

3. $SL(2,R)$ Symmetry in Four Dimensions

In four-dimensional supersymmetric models arising from string theory, it is well known that the dilaton and axion belong to the same chiral supermultiplet and can be described by a coset construction based on $SL(2,R)$. The same coset construction turns out to be true for the class of models under consideration here, even though no supersymmetry is assumed (just as with the Chern–Simons terms). As we will see, the symmetry is realized on the equations of motion, but not on the action. This symmetry first appeared in the "$SU(4)$ formulation" of N=4 supergravity,[1], and it was extended to include the coupling to n Abelian vector supermultiplets by de Roo.[14] He showed that the global $O(6,6+n)$ symmetry is present off-shell (*i.e.*, as a symmetry of the action), whereas the $SU(1,1)$ symmetry is only present on-shell (*i.e.*, as a symmetry of the equations of motion). The extension to non-Abelian Yang–Mills supermultiplets has also been investigated.[16] These authors found that the non-Abelian gauge interactions cause the $SU(1,1)$ symmetry to be broken, even on shell. The present analysis confirms these results for the bosonic portions of the theories. From the supergravity studies, we know that there are no surprises when the fermions are added. In the next section we will discuss whether higher-order string effects destroy the symmetry in the Abelian case, or (more optimistically) whether they could restore it in the non-Abelian case.

We begin by setting $D=4$ and performing a Weyl rescaling that brings the Einstein term to canonical form (let $g_{\mu\nu} = e^\phi g'_{\mu\nu}$ and drop the prime). Then the action becomes

$$S^{(4)} = \int_M dx\, \sqrt{-g}\Big\{R - \frac{1}{2}g^{\mu\nu}\partial_\mu\phi\partial_\nu\phi + \mathcal{L}_2 + e^{-\phi}\mathcal{L}_3 + e^{-2\phi}\mathcal{L}_4\Big\}\,, \tag{26}$$

with $\mathcal{L}_2$, $\mathcal{L}_3$, and $\mathcal{L}_4$ as defined in the preceding section.

The next step is to perform a duality transformation, which replaces the field $B_{\mu\nu}$ by a scalar field χ. This is achieved by first forming the $B_{\mu\nu}$ equation of motion

$$\partial_\mu\big(\sqrt{-g}\, e^{-2\phi}H^{\mu\nu\rho}\big) = 0\,, \tag{27}$$

and solving it by setting

$$\sqrt{-g}\, e^{-2\phi}H^{\mu\nu\rho} = \gamma\epsilon^{\mu\nu\rho\lambda}\partial_\lambda\chi\,, \tag{28}$$

where χ is the "axion" and γ is a constant to be fixed later. In the language of differential forms,

$$H = \gamma e^{2\phi} * d\chi \tag{29}$$

or, using $H = dB - \frac{1}{2}\eta_{ij}\mathcal{A}^i_\wedge\mathcal{F}^j$,

$$dB = \frac{1}{2}\eta_{ij}\mathcal{A}^i_\wedge\mathcal{F}^j + \gamma e^{2\phi} * d\chi \ . \tag{30}$$

The Bianchi identity ($d^2B = 0$) now turns into the χ field equation

$$\frac{1}{2}\eta_{ij}\mathcal{F}^i_\wedge\mathcal{F}^j + \gamma d(e^{2\phi} * d\chi) = 0 \ , \tag{31}$$

or, in terms of components, (choosing a convenient value for γ)

$$\partial_\mu(e^{2\phi}\sqrt{-g}\ g^{\mu\nu}\partial_\nu\chi) - \frac{1}{8}\eta_{ij}\epsilon^{\mu\nu\rho\lambda}\mathcal{F}^i_{\mu\nu}\mathcal{F}^j_{\rho\lambda} = 0 \ . \tag{32}$$

This is an equation of motion if we replace the $\mathcal{L}_4$ term in $S^{(4)}$ by

$$S_\chi = -\int dx\sqrt{-g}\left(\frac{1}{2}e^{2\phi}g^{\mu\nu}\partial_\mu\chi\partial_\nu\chi + \frac{1}{4}\chi\mathcal{F}\cdot\tilde{\mathcal{F}}\right) \ , \tag{33}$$

where

$$\mathcal{F}\cdot\tilde{\mathcal{F}} \equiv \frac{1}{2\sqrt{-g}}\epsilon^{\mu\nu\rho\lambda}\mathcal{F}^i_{\mu\nu}\eta_{ij}\mathcal{F}^j_{\rho\lambda} \ . \tag{34}$$

Let us now regroup the terms in the dual action in the following way:

$$\tilde{S}^{(4)} = \int_M dx\sqrt{-g}(R + \mathcal{L}_2) + S_D + S_F \ , \tag{35}$$

where

$$\begin{aligned} S_D &= -\frac{1}{2}\int_M dx\sqrt{-g}g^{\mu\nu}\left(\partial_\mu\phi\partial_\nu\phi + e^{2\phi}\partial_\mu\chi\partial_\nu\chi\right) \\ S_F &= -\frac{1}{4}\int_M dx\sqrt{-g}\ \left(e^{-\phi}\mathcal{F}^2 + \chi\mathcal{F}\cdot\tilde{\mathcal{F}}\right) \\ \mathcal{F}^2 &\equiv g^{\mu\rho}g^{\nu\lambda}\mathcal{F}^i_{\mu\nu}(M^{-1})_{ij}\mathcal{F}^j_{\rho\lambda} \end{aligned} \tag{36}$$

The claim now is that S_D is given by a $SL(2,R)/SO(2)$ coset construction.

Starting with the $SL(2,R)$ matrix

$$T = \begin{pmatrix} e^{-\phi/2} & 0 \\ e^{\phi/2}\chi & e^{\phi/2} \end{pmatrix} , \tag{37}$$

the idea is that under a global $SL(2,R)$ transformation M and a local $SO(2)$ transformation A, $T \rightarrow ATM^T$. For a given M, one can always choose A to preserve the triangular form of T. We compute the symmetric $SL(2,R)$ "metric"

$$S = T^T T = \begin{pmatrix} e^{-\phi} + e^{\phi}\chi^2 & e^{\phi}\chi \\ e^{\phi}\chi & e^{\phi} \end{pmatrix} . \tag{38}$$

Then one finds that

$$\mathrm{tr}(\partial_\mu S \partial^\mu S^{-1}) = -2(\partial_\mu \phi \partial^\mu \phi + e^{2\phi}\partial_\mu \chi \partial^\mu \chi) . \tag{39}$$

Therefore we learn that

$$S_D = \frac{1}{4}\int_M dx\sqrt{-g}\, g^{\mu\nu} tr(\partial_\mu S \partial_\nu S^{-1}) , \tag{40}$$

showing that ϕ and χ parametrize the coset $SL(2,R)/SO(2)$.

Another way of describing the $SL(2,R)$ symmetry of the dilaton and axion kinetic terms is to introduce a complex modular parameter

$$\tau = \chi + ie^{-\phi} , \tag{41}$$

which has the nice property that under a linear fractional transformation

$$\tau \rightarrow \frac{a\tau + b}{c\tau + d} \tag{42}$$

the combination

$$\frac{g^{\mu\nu}\partial_\mu \tau \partial_\nu \bar{\tau}}{(\mathrm{Im}\ \tau)^2} = g^{\mu\nu}(\partial_\mu \phi \partial_\nu \phi + e^{2\phi}\partial_\mu \chi \partial_\nu \chi) \tag{43}$$

is invariant. (In $N = 1$ supersymmetry models one often introduces a chiral superfield

S, whose bosonic part is $i\tau$.) It follows that*

$$S_D = -\frac{1}{2}\int_M dx\sqrt{-g}\,\frac{g^{\mu\nu}\partial_\mu\tau\partial_\nu\bar{\tau}}{(\mathrm{Im}\,\tau)^2}\,. \tag{44}$$

Let us now turn to the last remaining piece of the theory, namely S_F, the terms that depend on the gauge fields. This part of the story is of particular interest, because the $SL(2,R)$ transformations give rise to an electric-magnetic duality rotation. This fact has been exploited in a number of recent works, which construct black hole solutions with both electric and magnetic charge by applying $SL(2,R)$ transformations to known solutions with electric or magnetic charge only.[17,18,19] (Note that there are no fields in the theory that carry electric or magnetic charge. Indeed, it is not known how to maintain $SL(2,R)$ symmetry when such fields are added. Still, one can construct charged black hole solutions.)

To see how the $SL(2,R)$ symmetry works for S_F, we define

$$\mathcal{F}^{\pm}_{\mu\nu} = M\eta\mathcal{F}_{\mu\nu} \pm i\tilde{\mathcal{F}}_{\mu\nu}\,. \tag{45}$$

Then, using the identity $\mathcal{F}^{+\mu\nu}M^{-1}\mathcal{F}^{-}_{\mu\nu} = 0$, we can rewrite S_F in the form

$$S_F = -\frac{1}{16i}\int_M dx\sqrt{-g}\Big(\tau\mathcal{F}^{+\mu\nu}M^{-1}\mathcal{F}^{+}_{\mu\nu} - \bar{\tau}\mathcal{F}^{-\mu\nu}M^{-1}\mathcal{F}^{-}_{\mu\nu}\Big)\,. \tag{46}$$

The $\mathcal{A}_\mu$ equation of motion is

$$\nabla^\mu(\tau\mathcal{F}^{+}_{\mu\nu} - \bar{\tau}\mathcal{F}^{-}_{\mu\nu}) = 0 \tag{47}$$

and the Bianchi identity is

$$\nabla^\mu(\mathcal{F}^{+}_{\mu\nu} - \mathcal{F}^{-}_{\mu\nu}) = 0\,. \tag{48}$$

To exhibit $SL(2,R)$ symmetry it is necessary to have $\mathcal{A}_\mu$ transform at the same time as τ. The appropriate choice is to require that $\mathcal{F}^{\pm}_{\mu\nu}$ transform as modular

---

⋆ The $SU(1,1)$ formulation of refs. [14, 16] is obtained by the change of variables $Z = \frac{\tau-i}{\tau+i}$, which maps the upper half plane to the unit disk. In this formulation

$$S_D = -2\int_M dx\sqrt{-g}\,\frac{g^{\mu\nu}\partial_\mu Z\partial_\nu\bar{Z}}{(1-|Z|^2)^2}.$$

forms as follows

$$\mathcal{F}^{+}_{\mu\nu} \to (c\tau + d)\mathcal{F}^{+}_{\mu\nu}\,, \quad \mathcal{F}^{-}_{\mu\nu} \to (c\bar{\tau} + d)\mathcal{F}^{-}_{\mu\nu}\,. \tag{49}$$

This implies that

$$\tau\mathcal{F}^{+}_{\mu\nu} \to (a\tau + b)\mathcal{F}^{+}_{\mu\nu}\,, \quad \bar{\tau}\mathcal{F}^{-}_{\mu\nu} \to (a\bar{\tau} + b)\mathcal{F}^{-}_{\mu\nu}\,. \tag{50}$$

Thus the equation of motion (47) and the Bianchi identity (48) transform into linear combinations of one another and are preserved. In particular, the negative of the unit matrix sends $\mathcal{F}^{\pm}_{\mu\nu} \to -\mathcal{F}^{\pm}_{\mu\nu}$. This result is acceptable if we identify the symmetry as $SL(2,R)$, not just $PSL(2,R) = SL(2,R)/Z_2$. Note that $SL(2,R)$ is not a symmetry of the action. The transformation in (49) is a nonlocal transformation of $\mathcal{A}_\mu$, and such transformations can do strange things to the action. For example, the total derivative $\mathcal{F} \cdot \tilde{\mathcal{F}}$ transforms into an expression that is not a total derivative.

To complete the demonstration of $SL(2,R)$ symmetry one should also examine the equations of motion of the other fields in the theory. Each of them works nicely, as has been amply discussed by previous authors. For example, in forming the Einstein equation one needs to show that the contribution of S_F to the energy–momentum tensor is $SL(2,R)$ invariant. After a short calculation one finds that only terms of the structure $e^{-\phi}\mathcal{F}^{+}\mathcal{F}^{-}$ survive, and these are invariant since $e^{-\phi} \to |c\tau + d|^{-2}e^{-\phi}$.

At this point we can note the problem that arises when one attempts to generalize the discussion to allow non-Abelian gauge field interactions. In this case the divergences in Eqs. (47) and (48) become covariant derivatives involving the vector potentials. Since they undergo horrible non-linear transformations, implied by Eq. (49) it is quite clear that the Bianchi identity and the equation of motion can no longer be preserved.

To recapitulate, the $SL(2,R)$ symmetry described in this section differs from the noncompact symmetries in section 2 in several respects: 1) It is special to four dimensions. More precisely, it is compatible with Lorentz invariance in four dimensions. (It might be present in higher dimensions.) 2) It is realized in terms of nonlocal field transformations. 3) The coupling strength $< e^{\phi} >$ is involved nonlinearly in the transformations. 4) It is destroyed by non-Abelian gauge field interactions.

4. Could SL(2,R) Be a Symmetry of Heterotic String Theory?

It is an important question whether $SL(2,R)$, or at least an $SL(2,Z)$ subgroup, is a symmetry of string theory, as has been conjectured by Font *et al.*,[20] and emphasized once again in the recent work of Sen.[19] Sen has argued that since the effective field theory admits string-like solutions, whose zero modes correspond to the dynamical degrees of freedom of four-dimensional heterotic strings,[21] the symmetry of the effective field theory might carry over to the full heterotic string theory. This reasoning is analogous to that introduced long ago by Montonen and Olive.[22] Indeed the $SL(2,Z)$ transformation $\tau \to -1/\tau$, evaluated at $\chi = 0$, corresponds to $e^{\phi} \to e^{-\phi}$ and hence $\kappa \to 1/\kappa$, which is Montonen–Olive duality. Indeed, their reasoning is most compelling in the context of N=4 super Yang–Mills theories.[23] The Montonen–Olive duality transformation exchanges elementary fields with monopoles. The $SL(2,Z)$ symmetry under consideration is precisely that duality combined with Peccei–Quinn symmetry and generalized to the supergravity and string contexts.

In order to investigate this question further, let us examine the equations for strings propagating in the presence of background fields satisfying the equations of the previous sections. These equations were derived in ref. [5], and so we simply sketch the derivation here. The $D+d$ string coordinates $X^{\hat{\mu}}(\sigma,\tau)$ decompose into two sets $\{X^{\mu}\}$ and $\{Y^{\alpha}\}$ where $\mu = 0,1,\ldots,D-1$ and $\alpha = 1,2,\ldots,d$. In order to make contact with the low-energy theory of the preceding sections, we consider $(D+d)$-dimensional massless background fields $\hat{g}_{\hat{\mu}\hat{\nu}}$ and $\hat{B}_{\hat{\mu}\hat{\nu}}$ that depend only on the X^{μ} coordinates.[24–28] The world sheet action is

$$S = \frac{1}{2}\int d^2\sigma(\hat{g}_{\hat{\mu}\hat{\nu}}\eta^{ab} + \hat{B}_{\hat{\mu}\hat{\nu}}\epsilon^{ab})\partial_a X^{\hat{\mu}}\partial_b X^{\hat{\nu}} \; . \tag{51}$$

Even though this is a bosonic string action, it is also the relevant part of the heterotic string action, as well. Varying S with respect to $X^{\hat{\mu}}(\sigma,\tau)$ gives the classical equation of motion for the string

$$\begin{aligned}\frac{\delta S}{\delta X^{\hat{\mu}}} = & -\hat{\Gamma}_{\hat{\mu}\hat{\nu}\hat{\rho}}\partial^a X^{\hat{\nu}}\partial_a X^{\hat{\rho}} - \hat{g}_{\hat{\mu}\hat{\nu}}\partial^a\partial_a X^{\hat{\nu}} \\ & + \frac{1}{2}\epsilon^{ab}(\partial_{\hat{\mu}}\hat{B}_{\hat{\nu}\hat{\rho}} + \partial_{\hat{\nu}}\hat{B}_{\hat{\rho}\hat{\mu}} + \partial_{\hat{\rho}}\hat{B}_{\hat{\mu}\hat{\nu}})\partial_a X^{\hat{\nu}}\partial_b X^{\hat{\rho}} = 0 \; ,\end{aligned} \tag{52}$$

where

$$\hat{\Gamma}_{\hat{\mu}\hat{\nu}\hat{\rho}} = \frac{1}{2}(\partial_{\hat{\nu}}\hat{g}_{\hat{\mu}\hat{\rho}} + \partial_{\hat{\rho}}\hat{g}_{\hat{\mu}\hat{\nu}} - \partial_{\hat{\mu}}\hat{g}_{\hat{\nu}\hat{\rho}}) \; . \tag{53}$$

To analyze these equations it is convenient to consider X^{μ} and Y^{α} separately. Since the Y^{α} equation is somewhat simpler we begin with that. Indeed for that case, let us back up and focus on those terms in S that are Y dependent. These are

$$S_Y = \int d^2\sigma\left\{\frac{1}{2}(\eta^{ab}G_{\alpha\beta}(X)\partial_a Y^{\alpha}\partial_b Y^{\beta} + \epsilon^{ab}B_{\alpha\beta}(X)\partial_a Y^{\alpha}\partial_b Y^{\beta}) + \Gamma^a_{\alpha}(X)\partial_a Y^{\alpha}\right\}, \tag{54}$$

where

$$\begin{aligned}\Gamma^a_{\alpha} &\equiv \eta^{ab}\hat{g}_{\mu\alpha}\partial_b X^{\mu} - \epsilon^{ab}\hat{B}_{\mu\alpha}\partial_b X^{\mu} \\ &= \eta^{ab}G_{\alpha\beta}A^{(1)\beta}_{\mu}\partial_b X^{\mu} - \epsilon^{ab}(A^{(2)}_{\mu\alpha} - B_{\alpha\beta}A^{(1)\beta}_{\mu})\partial_b X^{\mu} \; .\end{aligned} \tag{55}$$

Since the backgrounds are independent of Y^{α}, the Euler–Lagrange equations for the

Y coordinates take the form

$$\partial_a \left(\frac{\delta S}{\delta \partial_a Y^\alpha} \right) = 0 . \tag{56}$$

Therefore, locally, we can write

$$\frac{\delta S}{\delta \partial_a Y^\alpha} = \eta^{ab} \partial_b Y^\beta G_{\alpha\beta} + \epsilon^{ab} \partial_b Y^\beta B_{\alpha\beta} + \Gamma^a_\alpha = \epsilon^{ab} \partial_b \tilde{Y}_\alpha , \tag{57}$$

where $\tilde{Y}_\alpha$ are the dual coordinates. If we define an enlarged manifold combining the coordinates Y^α and $\tilde{Y}_\alpha$ such that $\{Z^i\} = \{Y^\alpha, \tilde{Y}_\alpha\}$, $i = 1, 2, \ldots, 2d$, then we obtain[5]

$$M \eta D_a Z = \epsilon_a{}^b D_b Z , \tag{58}$$

where

$$(D_a Z)^i = \partial_a Z^i + \mathcal{A}^i_\mu \partial_a X^\mu , \tag{59}$$

and $\mathcal{A}^i_\mu$ is comprised of $A^{(1)\alpha}_\mu$ and $A^{(2)}_{\mu\alpha}$, as in section 2. This equation (which appears in ref. [27] for the special case $\mathcal{A}_\mu = 0$) has manifest $O(d,d)$ invariance provided the transformation rules $M \to \Omega M \Omega^T$ and $\mathcal{A}_\mu \to \Omega \mathcal{A}_\mu$, obtained in section 2, are supplemented with $Z \to \Omega Z$.

Even though these equations have continuous $O(d,d)$ invariance, the symmetry is broken to the discrete subgroup $O(d,d,Z)$ by the boundary conditions $Y^\alpha \simeq Y^\alpha + 2\pi$ and $\tilde{Y}_\alpha \simeq \tilde{Y}_\alpha + 2\pi$. The fundamental point is that all geometries related by $O(d,d,Z)$ transformations correspond to the same conformal field theory and are physically equivalent. The moduli space of conformally inequivalent (and hence physically inequivalent) classical solutions is given by the coset space $O(d,d)/O(d) \times O(d) \times O(d,d,Z)$ and is parametrized locally by the scalar fields $G_{\alpha\beta}$ and $B_{\alpha\beta}$.

The X^μ equation of motion is obtained by considering Eq. (52) for the case of $\hat{\mu} = \mu$ and substituting the various definitions given in section 2. After a certain amount of algebra, one finds that the X^μ equation of motion can be written in the manifestly $O(d,d)$ invariant form[5]

$$\begin{aligned} &\frac{1}{2} D_+ Z (\partial_\mu M^{-1}) D_- Z + \epsilon^{ab} \partial_a X^\nu \mathcal{F}_{\mu\nu} \, \eta D_b Z \\ &- \Gamma_{\mu\nu\rho} \partial^a X^\nu \partial_a X^\rho - g_{\mu\nu} \partial^a \partial_a X^\nu + \frac{1}{2} \epsilon^{ab} H_{\mu\nu\rho} \partial_a X^\nu \partial_b X^\rho = 0 . \end{aligned} \tag{60}$$

Together with Eq. (58), this gives the classical dynamics of strings moving in an arbitrary X-dependent background.

Eqs. (58) and (60) continue to hold for the $O(d, d+n)$ generalization, provided that M, η, and $\mathcal{A}_\mu^i$ are defined as in section 2. Also, Z^i now becomes a $(2d+n)$-component vector made by combining $Y^\alpha, \tilde{Y}_\alpha$, and Y^I, where Y^I are n additional internal coordinates. One must require that

$$\partial_- Y^I + A_\mu^{(3)I} \partial_- X^\mu = 0 \ , \tag{61}$$

as a "gauge invariant" generalization of what we know to be true for the heterotic string with vanishing $A_\mu^{(3)I}$ background fields, *viz.* that the Y^I are left-moving.

Let us now consider how eqs. (58) and (60) transform under the $SL(2,R)$ transformations introduced in section 3. For this purpose we should specialize to $D = 4$ and eliminate $H_{\mu\nu\rho}$ in favor of the axion field χ using Eq. (28). Also, the space-time metric that appears in Eq. (60) is the "string metric" of section 2, and it needs to be Weyl rescaled, as in section 3 ($g_{\mu\nu}^S = e^\phi g_{\mu\nu}$) in order to make contact with the equations of that section. In this way the dilaton field enters the equation. (The usual coupling of the dilaton to the world-sheet curvature is a higher-order effect than is being considered here.)

For symmetry of Eq. (58), we need a covariant interpretation of Eq. (59). We learned in section 3 that $\mathcal{F}_{\mu\nu}^+ \to (c\tau + d)\mathcal{F}_{\mu\nu}^+$ under an $SL(2,R)$ transformation. This has no simple solution for $\mathcal{A}_\mu$ unless $c = 0$ in which case we have $\mathcal{A}_\mu \to d\mathcal{A}_\mu$. Therefore Eqs. (58) and (59) are covariant for this subgroup ($\tau \to (a\tau+b)/d$) provided that we simultaneously transform $Z \to dZ$. It is straightforward to verify that Eq. (60), with the dilaton and axion introduced as described above, is also invariant under the same linear subgroup of $SL(2,R)$.

What should we conclude about the status of $SL(2,R)$ (or $SL(2,Z)$ when we restrict to discrete translations of the axion field) in string theory? Infinitesimal $SL(2,R)$ transformations can be written in the form $\delta\tau = \alpha + \beta\tau + \gamma\tau^2$, and we have found that the α and β transformations are okay, but the γ one is not. However, the γ transformation mixes up different powers of the string loop expansion parameter e^ϕ, and the equations we are studying are only lowest-order equations. So, it is still possible that the symmetry could be restored when higher-order corrections are taken into account. I am not very sanguine about this, but the evidence so far is not sufficient to exclude this possibility.

Even if this works, we would still need to understand what happens when non-Abelian gauge fields are present. In heterotic string theory, the Yang–Mills coupling constant $g^2 \sim \kappa^2/\alpha' \sim < e^\phi >$. So, the extra gauge field terms are also of higher order, and should not be included in the leading order analysis. Therefore, is is conceivable that they could also be reconciled with the symmetry.

One piece of evidence in support of the conjecture that the $SL(2,Z)$ symmetry is present in the quantum theory was presented recently by Sen.[29] He showed that, when all normalizations are carefully accounted for, the quantization conditions of electric and magnetic charge for dyons[30] are preserved under by $SL(2,Z)$ transformations. Electric and magnetic charges play much the same role for the $SL(2,Z)$ group as momentum and winding modes do for the $O(d,d+n,Z)$ groups.

5. Conclusion

This work has reviewed the noncompact $O(d,d)$ group that appears in toroidal compactification of oriented closed bosonic strings, as well as the $O(d,d+n)$ generalization that is required for the heterotic string. Using methods of dimensional reduction, we showed that these noncompact groups are exact symmetries of the (classical) low-energy effective field theory that is obtained when one truncates the dependence on the internal coordinates y^α keeping zero modes only. Starting from the two-dimensional sigma model describing string world sheet dynamics in the presence of background fields, we found that the classical string equations of motion also have the full noncompact symmetry, but that in string theory it is broken to the discrete subgroup $O(d,d+n,Z)$ by the boundary conditions that describe the toroidal topology of the compactified dimensions. These subgroups are, in fact, "discrete gauge symmetries,"[31] which means that they should be quite robust, surviving the plethora of phenomena that typically lead to explicit breaking of global symmetries. (However, they are broken *spontaneously* in general.)

A much deeper question is the status of the axion–dilaton $SL(2,Z)$ symmetry, which combines Peccei–Quinn symmetry with Montonen–Olive duality. If it is an exact symmetry of heterotic string theory (compactified to four dimensions), that is very profound. Unlike the $O(d,d+n)$-type symmetries, it transforms the dilaton field nonlinearly, and therefore has non-perturbative implications, rather like those that have been suggested for five-branes.[32] In particular, an understanding of how it works could give insight into how the dilaton acquires mass and supersymmetry is broken.[20] I think it could even shed light on the question of why the cosmological constant remains zero when supersymmetry is broken by non-perturbative effects. Unlike $O(d,d+n)$-type symmetries, this symmetry is presumably not a discrete gauge symmetry of the heterotic string. If it were, this would make its fundamental status much more convincing, and it could also have phenomenological benefits.[33] Lacking that, it seems more likely that it is broken (at least) by small non-perturbative effects.[34]

Let me conclude with a more optimistic remark. Toroidal compactification and N=4 supersymmetry certainly simplify the analysis, but if the $SL(2,Z)$ symmetry is really fundamental, it should also apply in more realistic situations, such as Calabi–Yau compactification.

6. Acknowledgments

Much of this work was carried out in collaboration with J. Maharana. I also wish to acknowledge the hospitality of the Aspen Center for Physics, where I had helpful discussions with R. Kallosh and A. Strominger.

7. References

1. E. Cremmer, J. Scherk, and S. Ferrara, *Phys. Lett.* **B74** (1978) 61.
2. A. Chamseddine, *Nucl. Phys.* **B185** (1981) 403.
3. E. Bergshoeff, M. de Roo, B. de Wit, and P. Van Nieuwenhuizen, *Nucl. Phys.* **B195** (1982) 97.
4. G. Chapline and N. Manton, *Phys. Lett.* **B120** (1983) 105.
5. J. Maharana and J. H. Schwarz, Caltech preprint CALT-68-1790 (1992), to be published in *Nucl. Phys.* **B**.
6. J. Scherk and J. H. Schwarz, *Nucl. Phys.* **B153** (1979) 61.
7. E. Cremmer, in *Supergravity '81*, ed. S. Ferrara and J.G. Taylor (Cambridge Univ. Press, Cambridge, 1982).
8. T. Buscher, *Phys. Lett.* **B194** (1987) 59; *Phys. Lett.* **B201** (1988) 466.
9. E. Smith and J. Polchinski, *Phys. Lett.* **B263** (1991) 59; A. A. Tseytlin, *Mod. Phys. Lett.* **A6** (1991) 1721; G. Veneziano, *Phys. Lett.* **B265** (1991) 287.
10. M. B. Green and J. H. Schwarz, *Phys. Lett.* **B149** (1984) 117.
11. E. Witten, *Phys. Lett.* **B149** (1984) 351.
12. A. Shapere and F. Wilczek, *Nucl. Phys.* **B320** (1989) 669; A. Giveon, E. Rabinovici, and G. Veneziano, *Nucl. Phys.* **B322** (1989) 167; A. Giveon, N. Malkin, and E. Rabinovici, *Phys. Lett.* **B220** (1989) 551; W. Lerche, D. Lüst, and N. P. Warner, *Phys. Lett.* **B231** (1989) 417.
13. N. Marcus and J. H. Schwarz, *Nucl. Phys.* **B228** (1983) 145.
14. M. de Roo, *Nucl. Phys.* **B255** (1985) 515.
15. K. S. Narain, *Phys. Lett.* **B169** (1986) 41.
16. M. de Roo, *Phys. Lett.* **B156** (1985) 331; E. Bergshoeff, I. G. Koh, and E. Sezgin, *Phys. Lett.* **B155** (1985) 71; M. de Roo and F. Wagemans, *Nucl. Phys.* **B262** (1985) 644.
17. A. Shapere, S. Trivedi, and F. Wilczek, *Mod. Phys. Lett.* **A6** (1991) 2677.
18. R. Kallosh, A. Linde, T. Ortín, A. Peet, and A. Van Proeyen, Stanford preprint SU-ITP-92-13; T. Ortín, Stanford preprint SU-ITP-92-24.
19. A. Sen, preprint TIFR-TH-92-41 (hepth@xxx/9207053).
20. A. Font, L. E. Ibáñez, D. Lüst, and F. Quevedo, *Phys. Lett.* **B249** (1990) 35.
21. A. Dabholkar and J. Harvey, *Phys. Rev. Lett.* **63** (1989) 719; A. Dabholkar, G. Gibbons, J. Harvey, and F. R. Ruiz, *Nucl. Phys.* **B340** (1990) 33; A. Sen preprint TIFR-TH-92-39 (hepth@xxx/9206016).

22. C. Montonen and D. Olive, *Phys. Lett.* **B72** (1977) 117.

23. D. Olive and E. Witten, *Phys. Lett.* **B78** (1978) 97; H. Osborn, *Phys. Lett.* **B83** (1979) 321.

24. K. S. Narain, M. H. Sarmadi, and E. Witten, *Nucl. Phys.* **B279** (1987) 369.

25. S. Cecotti, S. Ferrara, and L. Girardello, *Nucl. Phys.* **B308** (1988) 436.

26. J. Molera and B. Ovrut, *Phys. Rev.* **D40** (1989) 1146.

27. M. Duff, *Nucl. Phys.* **B335** (1990) 610.

28. A. Giveon and M. Roček, *Nucl. Phys.* **380** (1992) 128.

29. A. Sen, preprint TIFR-TH-92-46 (hepth@xxx/9209016).

30. E. Witten, *Phys. Lett.* **B86** (1979) 283.

31. M. Dine, P. Huet, and N. Seiberg, *Nucl. Phys.* **B322** (1989) 301.

32. M. Duff, *Class. Quant. Grav.* **5** (1988) 189; A. Strominger *Nucl. Phys.* **B343** (1990) 167; M. J. Duff and J. X. Lu, *Nucl. Phys.* **B354** (1991) 129, 141; *Nucl. Phys.* **B357** (1991) 534; *Class. Quant. Grav.* **9** (1992) 1.

33. L. E. Ibáñez and G. G. Ross, *Phys. Lett.* **B260** (1991) 291 and *Nucl. Phys.* **B368** (1992) 95.

34. J. Preskill, S. Trivedi, F. Wilczek, and M. B. Wise, *Nucl. Phys.* **B363** (1991) 207; T. Banks and M. Dine, *Phys. Rev.* **D45** (1992) 1424.

LIST OF PARTICIPANTS

— E. Abdalla, *Univ. Sao Paulo, CP 20516, Sao Paulo, BRAZIL*

— M. Ademollo, *Univ. di Firenze, Largo E. Fermi 2, 50125 Firenze, ITALY*

— J. Alfaro, *CERN, CH-1211 Geneva 23, SWITZERLAND*

— D. Amati, *SISSA, V. Beirut 2-4, 34014 Trieste, ITALY*

— D. Anselmi, *SISSA, V. Beirut 2-4, 34014 Trieste, ITALY*

— I. Antoniadis, *École Polytechnique, 91128 Palaiseau, FRANCE*

— C. Bachas, *École Polytechnique, 91128 Palaiseau, FRANCE*

— L. Baulieu, *Univ. Paris VI, 4 place Jussieu, 75252 Paris Cedex 05, FRANCE*

— M. Bianchi, *Univ. di Roma "Tor Vergata", V. Ric. Sci. 1, 00133 Roma ITALY*

— M. Bochicchio, *CERN, CH-1211 Geneva 23, SWITZERLAND*

— R. Brooks, *M.I.T., Cambridge, MA 02139, U.S.A.*

— N. Cabibbo, *Univ. di Roma "Tor Vergata", V. Ric. Sci. 1, 00133 Roma ITALY*

— L. Castellani, *Univ. di Torino, V. P. Giuria 1, 10125 Torino, ITALY*

— A. Cattaneo, *Univ. di Milano, V. Celoria 16, 20133 Milano, ITALY*

— A. Ceresole, *Univ. di Torino, V. P. Giuria 1, 10125 Torino, ITALY*

— G. Cleaver, *California Institute of Technology, Pasadena, CA 91125, U.S.A.*

— A. D'Adda, *Univ. di Torino, V. P. Giuria 1, 10125 Torino, ITALY*

— R. D'Auria, *Politecnico di Torino, C. Duca degli Abruzzi 24, 10129 Torino, ITALY*

— M. De Groot, *Univ. di Roma "Tor Vergata", V. Ric. Sci. 1, 00133 Roma ITALY*

— K. Dienes, *McGill Univ., 3600 University St., Montreal, Quebec H3A-2T8, CANADA*

— R. Dijkgraaf, *Univ. of Amsterdam, Plantage Muidergracht 24, NL-1018 Amsterdam, THE NETHERLANDS*

— P. Di Vecchia, *NORDITA, Blegdamsvej 17, DK-2100 Copenhagen, DENMARK*

— V. Dotsenko, *Univ. Paris VI 4, place Jussieu, F-75252 Paris Cedex 05, FRANCE*

— F. Englert, *Univ. Libre de Bruxelles, Bd. du Triomphe, B-1050 Bruxelles, BELGIUM*

— N. Fabiano, *Univ. di Roma "Tor Vergata", V. Ric. Sci. 1, 00133 Roma ITALY*

— J. Feinberg, *Technion, Haifa 32000, ISRAEL*

— S. Ferrara, *CERN, CH-1211 Geneve 23, SWITZERLAND*

— V. Ferrari, *Univ. di Roma "La Sapienza", P. Aldo Moro 2, 00185 Roma, ITALY*

— P. Fré, *SISSA, V. Beirut 2/4, 34014 Trieste, ITALY*

— F. Fucito, *Univ. di Roma "Tor Vergata", V. Ric. Sci. 1, 00133 Roma ITALY*

— J.L. Gervais, *E.N.S., 24 Rue Lhomond, 75231 Paris Cedex 05, FRANCE*

— L. Girardello, *Univ. di Milano, V. Celoria 16, 20133 Milano, ITALY*

— D.J. Gross, *Princeton University, Princeton, NJ 08544, U.S.A.*

— J.A. Harvey, *Univ. of Chicago, 5640 S. Eliis Av., Chicago, IL 60637, U.S.A.*

— L. Houart, *Univ. Libre de Bruxelles, Bd du Triomphe, B-1050 Brexelles, BELGIUM*

— K. Huitu, *Univ. of Helsinki, Siltavuorenpenger 20C, SF-00170 Helsinki, FINLAND*

— S. Hwang, *Chalmers Inst. Tech., 41296 Göteborg, SWEDEN*

— K. Ito, *Niels Bohr Inst., Blegdamsvej 17, DK-2100 Copenhagen, DENMARK*

— D. Johnston, *Heriot-Watt Univ., Edimburgh EH14 4AS, Scotland, U.K.*

— G. Jona-Lasinio, *Univ. di Roma "La Sapienza", P. Aldo Moro 2, 00185 Roma, ITALY*

— D. Kallifatides, *California Institute of Technology, Pasadena, CA 91125 U.S.A.*

— E. Kiritsis, *E.N.S., 24, rue Lhomond 75231 Paris Cedex 05, FRANCE*

— G. Korchemsky, *Univ. di Parma, V. delle Scienze, 43100 Parma, ITALY*

— K. Kounnas, *E.N.S., 24, rue Lhomond, 75231 Paris Cedex 05, FRANCE*

— L. Lusanna, *Univ. di Firenze, Largo E. Fermi 2, 50125 Firenze, ITALY*

— J. Madsen, *Niels Bohr Institute, Blegdamsvej 17, DK-2100 Copenhagen, DENMARK*

— N. Marcus, *Univ. of Tel-Aviv, 69778 Tel-Aviv, ISRAEL*

— E. Marinari, *Univ. di Roma "Tor Vergata", V. Ric. Sci. 1, 00133 Roma ITALY*

— A. Marshakov, *Lebedev Inst. of Physics, Leninsky pr.53, 1117924 Moscow, RUSSIA*

— M. Martellini, *Univ. di Milano, V. Celoria 16, 20133 Milano, ITALY*

— P. Menotti, *Univ. di Pisa, P. Torricelli 2, 56100 Pisa, ITALY*

— J.A. Minahan, *Univ. of Virginia, McCormick Rd., Charlottesville, VA 22901, U.S.A.*

— R. Myers, *McGill Univ., 3600 University St., Montreal, Quebec H3A-2T8, CANADA*

— D. Olive, *Newton Institute, 20 Clarkson Road, Cambridge CB3 0EH, U.K.*

— G. Parisi, *Univ. di Roma "Tor Vergata", V. Ric. Sci. 1, 00133 Roma ITALY*

— S. Penati, *Univ. di Milano, V. Celoria 16, 20133 Milano, ITALY*

— V. Periwal, *Inst. for Adv. Study, Olden Lane, Princeton, NJ 08540-4920, U.S.A.*

— I. Pesando, *Univ. di Torino, V. P. Giuria 1, 10125 Torino, ITALY*

— J.L. Petersen, *Niels Bohr Inst., Blegdamsvej 17, DK-Copenhagen,DENMARK*

— R. Petronzio, *Univ. di Roma "Tor Vergata", V. Ric. Sci. 1, 00133 Roma ITALY*

— M. Petropoulus, *École Polytechnique, 91128 Palaiseau, FRANCE*

— R. Pettorino, *Univ. di Napoli, Pad. 19 Mostra d'Oltremare, 80125 Napoli, ITALY*

— M. Porrati, *CERN, CH-1211 Geneve 23, SWITZERLAND*

— G. Pradisi, *Univ. di Roma "Tor Vergata", V. Ric. Sci. 1, 00133 Roma ITALY*

— R. Ricci, *Univ. di Roma "Tor Vergata", V. Ric. Sci. 1, 00133 Roma ITALY*

— J. Rizos, *École Polytechnique, 91128 Palaiseau, FRANCE*

— P. Roche, *École Polytechnique, 91128 Palaiseau, FRANCE*

— G.C. Rossi, *Univ. di Roma "Tor Vergata", V. Ric. Sci. 1, 00133 Roma ITALY*

— A. Sagnotti, *Univ. di Roma "Tor Vergata", V. Ric. Sci. 1, 00133 Roma ITALY*

— M. Sakamoto, *Niels Bohr Inst., Blegdamsvej 17, DK-2100 Copenhagen, DENMARK*

— R. Schimmrigk, *Univ. Heidelberg, Philisophenweg 16, 6900 Heidelberg, GERMANY*

— M.G. Schmidt, *Univ. Heidelberg Philosophenweg 16, 6900 Heidelberg, GERMANY*

— J.H. Schwarz, *California Institute of Technology, Pasadena, CA 91125, U.S.A.*

— R. Seneor, *École Polytechnique, 91128 Palaiseau, FRANCE*

— A. Tamburrino, *V. Picenza 101, 84100 Salerno, ITALY*

— G. Weigt, *Desy-IfH Zeuthen, Platanenallee 6, 1615 Zeuthen, GERMANY*

— P. Windey, *Univ. Paris VI, 4, place Jussieu, 75252 Paris Cedex 05, FRANCE*

— K. Yoshida, *Univ. di Roma "La Sapienza", P. Aldo Moro 2, 00185 Roma, ITALY*

— D. Zanon, *Univ. di Milano, V. Celoria 16, 20133 Milano, ITALY*

www.ingramcontent.com/pod-product-compliance
Lightning Source LLC
LaVergne TN
LVHW051054210926
848486LV00002B/104
* 9 7 8 9 8 1 0 2 1 1 6 1 5 *